KEY FORMULAS

Definition | For $a > 0$, $a \neq 1$, and $x > 0$ the **logarithmic function** of x with base a is

$$\log_a x = y \text{ and is equivalent to } a^y = x$$

Logarithmic Properties

For $a > 0$, $a \neq 1$, and $x > 0$,

(a) $a^{\log_a x} = x$

(b) $\log_a a^x = x$

(c) $\log_a 1 = 0$

(d) $\log_a (n_1 \cdot n_2) = \log_a n_1 + \log_a n_2$

(e) $\log_a \left(\dfrac{n_1}{n_2} \right) = \log_a n_1 - \log_a n_2$

(f) $\log_a (n_1)^r = r \log_a n_1$

In an **arithmetic sequence** $a_1, a_2, a_3, \cdots, a_n, \cdots$,

$$a_k - a_{k-1} = d \qquad \text{(The common difference)}$$

$$a_n = a_1 + (n-1)d \qquad \text{(The nth term)}$$

$$S_n = \sum_{i=1}^{n} a_i = \frac{n}{2}[2a_1 + (n-1)d] \quad \text{and}$$

$$S_n = \sum_{i=1}^{n} a_i = \frac{n}{2}(a_1 + a_n) \qquad \text{(The nth partial sum)}$$

In a **geometric sequence** $a_1, a_2, a_3, \cdots, a_n, \cdots$,

$$\frac{a_k}{a_{k-1}} = r \qquad \text{(The common ratio)}$$

$$a_n = a_1 r^{n-1} \qquad \text{(The nth term)}$$

$$S_n = \sum_{i=1}^{n} a_i = \frac{a_1(1 - r^n)}{1 - r} \qquad \text{(The nth partial sum)}$$

$$S_\infty = \sum_{i=1}^{\infty} a_i = \frac{a_1}{1 - r} \qquad \text{when } |r| < 1 \quad \text{(The infinite sum)}$$

The Binomial Theorem | For $a \in R$, $b \in R$ and $n \in N$,

$$(a + b)^n = \sum_{i=0}^{n} \frac{n!}{(n-i)! \, i!} a^{n-i} b^i$$

PRECALCULUS
Functions and Graphs

PRECALCULUS
Functions and Graphs

A. Robert Marshall • Cuesta College

Addison-Wesley Publishing Company

READING, MASSACHUSETTS • MENLO PARK, CALIFORNIA
NEW YORK • DON MILLS, ONTARIO • WOKINGHAM, ENGLAND
AMSTERDAM • BONN • SYDNEY • SINGAPORE • TOKYO • MADRID • SAN JUAN

Executive Editor: *David F. Pallai*
Production Administrator: *Catherine Felgar*
Text Designer: *Nancy Blodget*
Copyeditor: *Barbara Willette*
Composition: *Jonathan Peck Typographers*
Art Consultant: *Loretta Bailey*
Illustrator: *Hardlines*
Manufacturing Supervisor: *Roy Logan*
Cover Designer: *Marshall Henrichs*

Library of Congress Cataloging-in-Publication Data

Marshall, A. Robert
    Precalculus: functions and graphs / by A. Robert Marshall.
       p.   cm.
    ISBN 0-201-19095-8
    1. Functions   2. Algebra—Graphic Methods.   3. Trigonometry.
  I. Title.
  QA331.3.M37   1990
  512′.1—dc20   89-17675
                                       CIP

ABCDEFGHIJ–DO–943210

To my former students: *Thank you for your good questions. Your questions have made you better students and encouraged me to write this book.*

To all students using this book: *Study well so you will have the skills you need to be successful in calculus.*

PREFACE

Precalculus has become a necessary course to prepare many students for calculus. Some students need to review; others need to relearn; the rest need to discover the details of algebra and analytic geometry essential for success in calculus. As a refresher course, precalculus helps students understand the mathematics they have completed by rote in the past and introduces them to the more interesting, finer points of analytic geometry. For all students pre-calculus can provide proficiency through practice.

Almost everything in this precalculus textbook will be used again in calculus. This book emphasizes functions and their graphs. By understanding and analyzing functions, students learn to sketch their graphs. In addition to the standard precalculus topics, this book contains the following special features:

Highlight Features

Chapter Introductions Each chapter introduction provides a "real world" use of some of the mathematics in that chapter. These introductions are included to provide students a brief glimpse of the many uses of mathematics in today's world.

Algebra Review Chapter 1 contains the standard review topics as well as sets, set notation, the properties of real numbers, complex numbers, graphing lines, variations, and proportionality.

Functions Chapter 2 introduces and explains the function concept using different functions (including linear, quadratic, cubic, absolute value, and greatest integer functions). Then translation, multiples, powers, and reciprocals of functions are applied to these functions. This conceptual approach is applied to exponential, logarithmic, and trigonometric functions in later chapters.

Graphing Chapters 2 and 3 introduce and develop techniques for sketching the graph of a function. This method starts with point-plotting and moves quickly to conceptual graphing using the basic shape of functions. Then the basic shape is modified using the methods for translation, multiples, powers, and the reciprocals of functions. These conceptual techniques are extended in later chapters to graphing functions in polar coordinates. Many problems begin with an equation and then develop its graph, while others start with a graph and then determine the equation for the function. This helps students understand the properties of functions.

Examples and Guidelines Examples have been carefully chosen and explained to teach both the general principles as well as the finer points of new material. Special guidelines and strategies are provided to clarify more complicated procedures.

Exercise Sets and Applications All exercise sets are of graded difficulty and include drill problems to help establish competency in each new skill. (Seventy-five percent of the students should always be able to complete at least 75% of the exercises.) Applications and proofs are spread throughout the text and appear in most exercise sets. Application exercises include problems trditional to calculus as well as many newer, interesting problems. As mentioned earlier many exercises start with the graph of a function and then generate its equation.

Trigonometry Trigonometry is introduced with the right triangle approach but moves quickly to the circle definition of trigonometric functions. The three trigonometry chapters include angular speed, vectors, vector projections, the law of sines, the law of cosines, inverse trigonometric functions, DeMoivre's theorem, and many trigonometric identities and equations.

Three-Dimensional Graphing The graphing of lines and planes is an extension of the solution to dependent systems and linear programming. Similarly graphing quadratic surfaces is an extension of two-dimensional conic sections.

Sequences and Mathematical Induction The last chapter starts with sequences in general including the summation notation for their partial sums. Thus the summation notation can be used with mathematical induction,

arithmetic sequences, geometric sequences, and the binomial theorem that follow. Mathematical induction is used to prove earlier theorems as well as theorems with arithmetic sequence, geometric sequences, and the binomial theorem.

Limits The informal concept of the limit is introduced with the discussion of the asymptotes of rational functions and hyperbolas. The limit is also used again with sequences.

Calculator This book is written with the assumption that scientific calculators are readily available to students. Calculator problems are sprinkled throughout the book. Whenever calculation becomes cumbersome, the calculator should be used to facilitate computation. However, many problems in the trigonometry chapter require exact values rather than calculator approximations. Because of the many differences in calculators, this book does not provide step by step calculator instructions. This allows the freedom to use a variety of calculators with this book. Calculator instruction is left to the student, calculator instruction book, and sometimes a helpful instructor.

Features—Quick Reference

Chapter Introduction:
  A mathematics occupation and work ethic: p. 1–2
  Biology: p. 243–244, 283–284, 401–402, and 503–504
  Research and development: p. 85–86, 173–174, 453–454, and 503–504
  Light waves: p. 355–356
  Fractals: p. 575–576
Guidelines and Strategies: p. 181, 206, 228, 328, 338, 344, 464, 522, 558, 563, and 568
Sets and set notation: p. 3, 4, 37, 40, 44, 155, 216, 367, and 443
Complex numbers as a review topic: p. 63-67
Variation and proportion: p. 77–80, 88 Example 2, 102 Example 6
Multifunction approach in Chapter 2:
  quadratic functions: p. 92 Example 7, 108–110, 129–130 Example 11, 136 Example 1b, 137 Example 3a,

139 Examples 4, 5, and 6, and exercise sets on 118–120, 131–133, 145–147, 156–158, and 168–171
  cubic function: p. 113, 128, 137 Example 2b, 140 Example 7, 152 Example 5 and exercise sets on 118–120, 131–133, 145–147, 156–158, and 168–171
  absolute value function: p. 114 and exercise sets on 118–120, 131–133, 145–147, 156–158, and 168–171
  greatest integer function: p. 115, 116 and exercise sets on 118–120, 131–133, 145–147, 156–158, and 168–171
Function Composition: Introduced p. 150–157 and used with
  Inverse functions: p. 158–167
  Exponential functions: p. 253, 254
  Logarithmic functions: p. 278
  Trigonomometric functions: p. 310

Supplements

Instructor's Solutions Manual
Complete worked-out solutions for every exercise in the text.

Student's Solutions Manual
Worked-out solutions to selected odd problems

Printed Test Bank
Includes multiple choice, open-ended essay problems, and True/False questions

Answer Book
Answers to even questions.

Computerized Testing—Test Edit
Contains over 1000 problems easily accessed by computer in either multiple-choice or open-ended format. Features include editing of problems, option to leave space for working problems on the text, scrambling problem and answer orders, and printed answer keys (for IBM-PC®).

Transparency Masters
One hundred twenty printed transparency masters covering key definitions, theorems, and graphs from the text.

Precalculus Acetate Package
Twenty acetates to be used with overhead projectors. Includes key figures to illustrate precalculus concepts, four-color transparencies with overlays. Available to adoptions of 500 copies or more.

Master Grapher/3D Grapher
A powerful, interactive graphing utility for functions, polar equations, parametric equations, conic equations, and functions of two variables. Available for IBM-PC®, Apple®, and Macintosh®.

Acknowledgments

Initially I wish to thank the many students with whom I have taught and learned during the past 22 years. Their questions, difficulties, and solutions have provided important insights.

Throughout its development, mathematics professors reviewed this manuscript and provided much appreciated help. I sincerely thank all reviewers.

Anthony Barkauskas
University of Wisconsin-La Crosse

Steven Blasberg
West Valley College

Sr. Gabriel Mary Donohue
College of Saint Elizabeth

Elayn Gay
University of New Orleans

Dorothy Goldberg
Kean College of New Jersey

Kevin Hastings
Knox College

Peter Horn
Northern Arizona State University

William Keils
San Antonio College

Ted Moore
Mohawk Valley Community College

Richard Plagg
Highline Community College

Marcia Siderow
California State University–Northridge

John Soptick

I cannot express the depth of my appreciation to my editor David Pallai. His encouragement, support, and guidance were invaluable in developing this book. I also sincerely thank Catherine Felgar and the Addison-Wesley staff for their diligent work.

Last but most importantly I thank my wife June and our boys Aaron and Andy. Their understanding, encouragement, and support were essential and have been sincerely appreciated.

A. Robert Marshall

Contents

6

Analytic Trigonometry   355

7

Other Applications of Trigonometry   401

8

Solving Systems of Equations   453

9

Analytic Geometry   503

10

Sequences and Sums   575

PRECALCULUS
Functions and Graphs

Algebraic Foundations

EARTHBOUND ANALYST

Photo by: Chad Slattery

In the movie *Top Gun* a Navy fighter pilot played by Tom Cruise falls in love with a beautiful astrophysicist and instructor, Charlie Blackwood. Although this heroine's role may be perceived as a fantasy, it was actually inspired by the professional career of Christine Fox.

Christine Fox is a mathematician and tactical analyst working for the Center for Naval Analysis (CNA), an independent civilian organization that conducts research for the U.S. Navy and Marine Corps. Her work brings the scientific approach of computers and applied mathematics to the complexities of battle exercises and tactics. Christine is not a pilot instructor as depicted in *Top Gun*, but her tactical suggestions were often incorporated into lectures by *Top Gun* instructors.

As a woman advising a male-dominated military, she observes, "I must be sure my information is right and I can justify it." Even when people doubt her ability because of her gender, she feels that, "If you do your homework and your job, eventually they won't care who you are."

She likes to interact with the squadron as much as possible. "The pilots come back and tell me 'Hey, you're crazy if you think this will work,' or 'It's true, it did work.' It's a lot of fun. If you're going to try to help someone improve his tactics, you don't have a chance of getting them to listen to anything you say if you don't understand what they do." Though the Navy hesitates to allow women to participate in combat exercises, Christine Fox was the first woman from CNA to be placed aboard an aircraft carrier during mock combat. When asked how she feels about working for the military, she replies, "I think what they do is important and I'm proud if I can contribute to it. I don't like war but I believe in a strong defense to prevent war. It's hard to explain that to someone who doesn't share the same philosophy."

Christine Fox has an undergraduate degree in mathematics and a graduate degree in applied mathematics from George Mason University in Fairfax, Virginia. She credits her interest in mathematics to her father, a retired nuclear engineer. "My father felt that math was the key to the sciences. He always said that if he had acquired a better understanding of math, nuclear engineering would have come easier to him." Christine also has a desire for other women to experience a career as exciting as the one she has. "After the movie came out I thought about trying to let other women know what my job was really like. I want to tell them it's really fun and they might think about doing it."

SOURCE: "Role Model," by J. E. Ferrell, *Air & Space*, Smithsonian, June/July 1987: pp. 59–65. Reprinted by permission.

1.1 Sets, Numbers, and Their Properties

Mathematics, especially algebra, contains special notation, logic, and patterns. In this section we will quickly review the basic ideas of sets, the properties of real numbers, and the properties of equality. The concepts of sets and set notation are unifying building blocks for all of mathematics.

Sets and Their Properties

Definition | A **set** is a collection or list of well-defined things, objects, or numbers.

Definitions | *Roster Notation.* $A = \{1, 2, 3, 4, 5\}$ is read "A is the set that contains the numbers (or elements) 1, 2, 3, 4, 5." This is the **roster notation** for a set.
Set Builder Notation. $A = \{x : x$ is a positive integer smaller than 6$\}$ is read "A is the set of all numbers x such that each x is a positive integer less than 6." This is the **rule or set builder notation** of the set A listed above with the roster method. When the roster method is impractical, the rule notation can be more concise.

Elements of a Set

$2 \in A$ is read "2 is an element of the set *A*" or "2 belongs to set *A*."

$7 \notin A$ is read "7 is not an element of set *A*" or "7 does not belong to set *A*."

$a \in A$ is read "*a* is an element of set *A*." For $A = \{1, 2, 3, 4, 5\}$, $a \in A$ means that *a* equals 1, 2, 3, 4, or 5.

The Empty Set. $\emptyset = \{\ \}$ is read "the empty set equals the set containing no elements." ($\emptyset$ is not the Greek letter phi but comes from the Scandinavian alphabet, probably through usage by Niels Henrik Abel, 1802–1829.)

We usually use lowercase letters to denote an element in a set, whereas we usually use capital letters to denote sets. When we list the elements of a set in roster notation, we list each element only once. The following are ways in which we can combine and compare sets.

SET OPERATIONS AND COMPARISONS

For $A = \{2, 4, 6\}$ and $B = \{1, 2, 3, 4, 5\}$,

$A \cup B$ is read "*A* **union** *B*." $A \cup B$ is a new set that contains all elements that are in *A* together with all elements that are in *B*:

$$A \cup B = \{1, 2, 3, 4, 5, 6\}$$

$A \cap B$ is read "*A* **intersect** *B*." $A \cap B$ is a new set that contains all elements that are in both *A* *and* *B* (elements common to both sets).

$$A \cap B = \{2, 4\}.$$

$C \subseteq A$ is read "*C* is a subset of *A*." *C* is a subset of *A* when every element of *C* is also an element of *A*. For example, $\{2, 4\} \subseteq A$, $\{4\} \subseteq A$, $\{2\} \subseteq A$, $\{2, 4, 6\} \subseteq A$, and $\emptyset \subseteq A$, since all of the elements in the subset on the left are also in *A*. Thus $\{2, 4\}$, $\{4\}$, $\{2\}$, $\{2, 4, 6\}$, and $\emptyset$ are subsets of *A*. (Note that $\emptyset$ is a subset of every set.)

$A \nsubseteq B$ is read "*A* is not a subset of *B*." *A* is not a subset of *B*, since $6 \in A$ but $6 \notin B$.

There are two ways to describe the size of a set. A set can be classified as a **finite set** or an **infinite set**. The sets $\{1, 2, 3, 4\}$ and $\{1, 2, 3, 4, \ldots, 100\}$ are finite sets. The first set has four elements, and the second set has 100 elements. When we count the elements in a finite set, the counting process always stops at some integer. Since these two sets have four and 100 elements, respectively, they are finite. The set $\{2, 4, 6, 8, 10, 12, \ldots\}$ is an infinite set. The three dots mean "continue the indicated pattern." When we count the elements of a set in an infinite set, the counting process never stops. As we see in $\{2, 4, 6, 8, 10, 12, \ldots\}$, we can never stop counting the elements in this set, and thus the set is infinite.

We will find it useful to have names for specific sets of numbers. The following sets of numbers are useful to know by name and are all infinite.

Numbers and Their Properties

<table>
<tr><td rowspan="14">SETS OF NUMBERS</td><td>Natural numbers (N) or counting numbers</td><td>$N = \{1, 2, 3, 4, 5, 6, \ldots\}$</td></tr>
<tr><td>Whole numbers (W)</td><td>$W = \{0, 1, 2, 3, 4, 5, \ldots\}$</td></tr>
<tr><td>Integers (Z)</td><td>$Z = \{\ldots, -4, -3, -2, -1, 0, 1, 2, 3, 4, \ldots)$</td></tr>
<tr><td>Positive integers</td><td>$\{1, 2, 3, 4, 5, \ldots\}$</td></tr>
<tr><td>Negative integers</td><td>$\{-1, -2, -3, -4, -5, \ldots\}$</td></tr>
<tr><td>Nonnegative integers</td><td>$\{0, 1, 2, 3, 4, 5, \ldots\}$</td></tr>
<tr><td>Rational numbers (Q)</td><td>A rational number is a real number that can be written as the ratio (fraction) of two integers $\frac{a}{b}$, where $a \in Z$, $b \in Z$, and $b \neq 0$.

Rational numbers include $\frac{7}{8}, \frac{17}{3}, \frac{1}{2}, \frac{3}{4}, \frac{-7}{8}$, and $\frac{-1}{4}$, as well as $5 = \frac{5}{1}$, $0.3 = \frac{3}{10}$, and $0 = \frac{0}{1}$. $Q = \left\{\frac{a}{b} : a \in Z, b \in Z, b \neq 0\right\}$. The decimal representation of a rational number is described below.</td></tr>
<tr><td>Irrational numbers ($\overline{Q}$)</td><td>Irrational numbers are real numbers that are not rational. Numbers like $\sqrt{2}$, $\sqrt{5}$, and π cannot be written as the ratio of two integers and are irrational numbers. The decimal representation of an irrational number is described below.</td></tr>
<tr><td>Real numbers (R)</td><td>The set of real numbers is the union of set of the rational and the set of irrational numbers, $R = Q \cup \overline{Q}$.</td></tr>
<tr><td>Prime numbers</td><td>Integers greater than one whose only positive integer factors are one and itself. $\{2, 3, 5, 7, 11, 13, 17, 19, \ldots\}$</td></tr>
<tr><td>Composite numbers</td><td>Integers that are not prime.</td></tr>
</table>

We notice the subset relationship between these sets of numbers. Specifically, we have $N \subseteq Z \subseteq Q \subseteq R$ and $\overline{Q} \subseteq R$. Upon further investigation we can also identify rational numbers by their decimal representation.

Rational numbers have a decimal representation that either (a) terminates (or stops) or (b) repeats a pattern of digits forever. The repeated pattern of digits must be exactly the same digits each time with no other digits included.

$\frac{1}{2} = 0.5$ (terminates) $\frac{1}{3} = 0.33333\ldots$ (repeats a pattern)

$\frac{1}{4} = 0.25$ (terminates) $\frac{53}{99} = 0.535353\ldots$ (repeats a pattern)

These rational numbers have decimal representations that either terminate or repeat a pattern. We can check these and other examples with a calculator. A calculator can identify a repeating pattern only in the finite number of digits on the calculator display. Thus calculator results must be used cautiously.

Irrational numbers, on the other hand, have decimal representations that (a) never terminate *and* (b) never repeat a pattern forever. The following examples are irrational numbers. It is possible to check some of these values with your calculator.

$5.78123421876\ldots$ $\sqrt{2} = 1.414213562\ldots$

$\pi = 3.14159265\ldots$ $\sqrt{5} = 2.236067978\ldots$

$e = 2.71828182845904523536\ldots$ $\sqrt{12} = 3.464101615\ldots$

Note that the infinite decimal representation of these numbers does not repeat a pattern; thus they are irrational numbers.

EXAMPLE 1   Is $1.010010001\ldots$ a rational or an irrational number?

Solution   This is an interesting example, since $1.010010001\ldots$ appears to have a repeating pattern. But because an extra zero is added to the pattern each time it is repeated, the pattern changes and the number is not rational. Thus $1.010010001 \ldots$ is irrational.   ■

EXAMPLE 2   Verify that the rational number $\frac{2}{7}$ has a decimal representation that repeats a finite pattern.

Solution   This is another interesting problem. Using a calculator, we divide 2 by 7 with the result $\frac{2}{7} = 0.285714285$. This answer does not conclusively show a repeating pattern generated by the calculator. But $\frac{2}{7}$, as the ratio of two integers, is rational by definition. Using paper, pencil, and calculator, we have

$$
\begin{array}{r}
0.285714285714285714\ldots \\
7\overline{)2.000000000000000000} \\
\underline{1\,999998} \\
2000000 \\
\underline{1999998} \\
2000000 \\
\underline{1999998} \\
2
\end{array}
$$

and $\frac{2}{7} = 0.285714285714285714\ldots$ with a repeating pattern of 285714 continued forever.   ■

Care must be exercised in using a calculator. It is also important to note that the calculator can only approximate irrational numbers and even some rational numbers because the decimal representations on calculators have only a finite number of digits.

Each number in the set of real numbers, the rational and irrational numbers, can be given a unique point on a line. To do this, we construct a **real number line** by picking a point on a line, naming it zero, and then marking off equal unit increments to the left and right of zero. Each successive unit marked to the right of zero is given an integer name starting with $+1$ and continuing with $+2$, $+3$, $+4$, $\cdots$. Each successive unit marked to the left of zero is given an integer name starting with -1 and continuing with -2, -3, -4, $\cdots$. Thus we construct the following real number line:

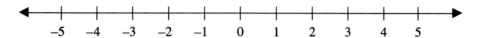

Any real number names a unique point on the real number line. Each real number is called the coordinate of the point it names. Each point on the real number line has a real number coordinate, and every real number is the coordinate for one point on the real number line. The following real number line shows that even irrational numbers are the coordinate of points on the real number line:

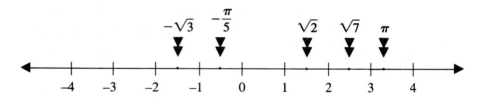

The real number line will be most helpful when we graph solutions, especially solutions of inequalities.

The following properties of real numbers are also important building blocks for our system of algebra.

The Field Properties of the Real Numbers

	For Addition	**For Multiplication**
Closure Property	Every pair of real numbers a and b has a unique sum $a + b$ that is also in the set R. For $-2 \in R$ and $3 \in R$, $-2 + 3 = 1 \in R$.	Every pair of real numbers a and b has a unique product $a \cdot b$ that is also in the set R. For $-2 \in R$ and $3 \in R$, $(-2)(3) = -6 \in R$.

Commutative Property

For real numbers a and b:

$$a + b = b + a \qquad a \cdot b = b \cdot a$$

In adding or multiplying two or more numbers, either number can be used first:

$$\sqrt{2} + 3 = 3 + \sqrt{2} \qquad 2 \cdot 3 = 3 \cdot 2$$
$$3 + x = x + 3 \qquad x \cdot 2 = 2 \cdot x$$

Associative Property

For real numbers a, b, and c:

$$(a + b) + c = a + (b + c) \qquad (a \cdot b)c = a(b \cdot c)$$

When adding or multiplying three real numbers, either the first two or the last two can be combined first by changing the parentheses:

$$(2 + 3) + 5 = 2 + (3 + 5) \qquad (2 \cdot 3) \cdot 5 = 2 \cdot (3 \cdot 5)$$
$$5 + 5 = 2 + 8 \qquad\qquad 6 \cdot 5 = 2 \cdot 15$$
$$10 = 10 \qquad\qquad\qquad 30 = 30$$

Identity Property

For every real number a, there is a unique number 0, the additive identity, such that $a + 0 = a$.

For every real number a, there is a unique number 1, the multiplicative identity, such that $a \cdot 1 = a$.

Inverse Property

For every real number a, there is a unique number $-a$, the additive inverse of a, such that $a + (-a) = 0$.

For every real number a, $a \neq 0$, there is a unique number $\dfrac{1}{a}$, the multiplicative inverse of a, such that $a\left(\dfrac{1}{a}\right) = 1$.

Distributive Property

For real numbers a, b, and c,

$$a(b + c) = ab + ac$$

The distributive property will be used to change a product to a sum of products as in $2(x + 3) = 2x + 6$ or to change a sum to a product as in $20x + 35 = 5(4x + 7)$.

We use the following properties for solving equations.

Properties of Equality

For each of the following, a, b, and c are real numbers.

Reflexive Property of Equality	$a = a$
Symmetric Property of Equality	If $a = b$, then $b = a$.
Transitive Property of Equality	If $a = b$ and $b = c$, then $a = c$.
Substitution Property of Equality	If $a = b$, then a may replace b and b may replace a in any statement.
Addition Property of Equality	If $a = b$, then $a + c = b + c$, or If $a = b$, then $a + (-c) = b + (-c)$.
Multiplication Property of Equality	If $a = b$, then $ac = bc$, or If $a = b$, then $\dfrac{a}{c} = \dfrac{b}{c}$ for $c \neq 0$.

The following miscellaneous properties are helpful in working with negatives.

For each of the following, a, b, and c are real numbers.

Properties of Negatives	$-(-a) = a$
	$(-a)b = a(-b) = -ab$
	$(-a)(-b) = ab$
Multiplication	$(-1)(a) = -a$
Definition of Subtraction	$a - b = a + (-b)$

We will now state the properties for rational expressions.

Rational Expression Properties

For each of the following, a, b, c, and d are real numbers with $b \neq 0$ and $d \neq 0$:

Changing denominators or reducing

$$\frac{ad}{bd} = \frac{a}{b}$$

Addition

$$\frac{a}{b} + \frac{c}{b} = \frac{a+c}{b} \quad \text{or} \quad \frac{a}{b} + \frac{c}{d} = \frac{ad}{bd} + \frac{cb}{bd} = \frac{ad+cb}{bd}$$

Subtraction

$$\frac{a}{b} - \frac{c}{b} = \frac{a-c}{b} \quad \text{or} \quad \frac{a}{b} - \frac{c}{d} = \frac{ad}{bd} - \frac{cb}{bd} = \frac{ad-cb}{bd}$$

Multiplication

$$\frac{a}{b} \cdot \frac{c}{d} = \frac{ac}{bd}$$

Division

$$\frac{a}{b} \div \frac{c}{d} = \frac{a}{b} \cdot \frac{d}{c} = \frac{ad}{bc} \quad \text{for } c \neq 0.$$

Properties of Negatives

$$-\frac{a}{b} = \frac{-a}{b} = \frac{a}{-b}$$

Remember that the letters b, d, and sometimes c above do not equal zero, since division by zero is undefined.

We will now use these properties to solve some examples.

EXAMPLE 3   Solve the equation

$$\frac{5}{7x} - \frac{1}{7} = \frac{3}{7x}.$$

Solution   Using the properties previously listed, we have

$$\frac{5}{7x} - \frac{1}{7} = \frac{3}{7x}$$

and $x \neq 0$, since division by zero is undefined.

$$7x\left(\frac{5}{7x} - \frac{1}{7}\right) = \left(\frac{3}{7x}\right)7x \qquad \textit{(Multiply both sides by 7x—the multiplication property of equality)}$$

$$\frac{7x \cdot 5}{7x} - \frac{1 \cdot 7x}{7} = \frac{3 \cdot 7x}{7x} \qquad \textit{(Distributive property)}$$

$$(1)\frac{5}{1} - (1)\frac{x}{1} = \frac{3}{1}(1) \qquad \textit{(Multiplicative inverse)}$$

$$5 - x = 3 \qquad \textit{(Multiplicative identity)}$$

$$5 + (-5) - x = 3 + (-5) \qquad \textit{(Addition property of equality)}$$

$$-x = -2$$

$$(-1)(-x) = -2(-1) \qquad \textit{(Multiplication property of equality)}$$

$$x = 2$$

EXAMPLE 4   Solve $P = 2l + 2w$ for w, where P is the perimeter, l is the length, and w is the width of a rectangle. Find the width when the perimeter is 34 inches and the length is 7 inches.

Solution   Solving $P = 2l + 2w$ for w, we have

$$P - 2l = 2w \qquad \textit{(Addition property of equality)}$$

$$\frac{1}{2}(P - 2l) = \frac{1}{2}(2w) \qquad \textit{(Multiplication property of equality)}$$

$$\frac{P}{2} - \frac{2l}{2} = \frac{2w}{2} \qquad \textit{(Distributive property)}$$

$$\frac{P}{2} - l = w \qquad \textit{(Multiplicative inverse)}$$

Substituting $P = 34$ and $l = 7$, we have

$$w = \frac{34}{2} - 7 = 17 - 7 = 10$$

Therefore the width of the rectangle is 10 inches.

By completing Examples 5 and 6 we will prove an "if and only if" theorem.

EXAMPLE 5   Prove that if $a = -a$, then $a = 0$.

Solution   If $a = -a$, then by the addition property of equality we have

$$a + a = a + (-a)$$

$$(1 + 1)a = a + (-a) \qquad \textit{(By the distributive property, } (a + a) = (1 + 1)a)$$

$$(1 + 1)a = 0 \qquad \textit{(By the additive inverse property, } a + (-a) = 0)$$

$$2a = 0$$

$$\frac{2a}{2} = \frac{0}{2} \qquad \textit{(By the multiplication property of equality)}$$

$$a = 0 \qquad \textit{(By the multiplicative inverse and } 0 \div a = 0)$$

Therefore if $a = -a$, then $a = 0$.

EXAMPLE 6   Prove that if $a = 0$, then $a = -a$.

Solution   This proof is a little easier than Example 5. Since $a = 0$, we have

$$(-1)a = (-1)0 \quad \textit{(By the multiplication property of equality)}$$
$$-a = 0$$
$$a = -a \quad \textit{(By substitution, since } a = 0 \text{ and } -a = 0\text{)}$$

Therefore if $a = 0$, then $a = -a$.   ■

When we put the results of Examples 5 and 6 together, we have
(a) if $a = -a$, then $a = 0$, and **(b)** if $a = 0$, then $a = -a$. With both of these implications true, we can say that

$$a = -a \quad \text{if and only if} \quad a = 0$$

Thus the phrase "if and only if" means that there are two true statements:
(a) if $a = -a$, then $a = 0$, and **(b)** if $a = 0$, then $a = -a$. We will work with "if and only if" statements throughout this book.

Absolute Value

The **absolute value** of a number provides a result that is a nonnegative number (positive or zero). Although this value can be thought of as the distance a number is from zero, we will define the absolute value algebraically.

Definition │   The **absolute value of x**, denoted $|x|$, is defined as

$$|x| = \begin{cases} x \text{ when } x \geq 0 \\ -x \text{ when } x < 0 \end{cases}$$

This simple definition guarantees that $|x| \geq 0$ (is always positive or zero), since

$$|x| = \begin{cases} x \text{ when } x \geq 0 \text{ means that } x \text{ is a positive number or zero} \\ -x \text{ when } x < 0 \text{ means that } -x \text{ is } -(\text{negative}) \text{ or a positive number} \end{cases}$$

Examples like $|3| = 3$, $|-3| = -(-3) = 3$, $|5| = 5$, $|-5| = -(-5) = 5$, and $|0| = 0$ are all positive numbers or zero. We will find that the definition is necessary when we work with problems that have variables inside the absolute value symbols. The following examples show ways in which we can use the absolute value.

EXAMPLE 7   Evaluate **(a)** $|7| + |-7|$ and   **(b)** $|8| - |-8|$.

Solution   Using the definition of the absolute, we have
(a) Since $|7| = 7$ and $|-7| = -(-7) = 7$,

$$|7| + |-7| = 7 + 7 = 14$$

(b) Since $|8| = 8$ and $|-8| = -(-8) = 8$,

$$|8| - |-8| = 8 - (8) = 0 \qquad \blacksquare$$

EXAMPLE 8   Evaluate $|\pi - 4|$.

Solution   For $|\pi - 4|$ we notice that $\pi \approx 3.14$ and $\pi - 4 \approx -0.86 < 0$. (We use $\approx$ to indicate a round-off answer; it is read "approximately equal to.") Therefore by the definition,

$$|\pi - 4| = -(\pi - 4) = -\pi - (-4) = -\pi + 4$$
$$|\pi - 4| = 4 - \pi \qquad \blacksquare$$

We will do much more with absolute value in later sections.

Exercises 1.1

Complete Exercises 1–10 when $A = \{1, 3, 5, 7\}$, $B = \{2, 4, 6, 8\}$, and $C = \{1, 2, 3, 4, 5\}$.

1. $A \cup B$

2. $A \cup C$

3. $A \cap B$

4. $A \cap C$

5. $B \cup C$

6. $B \cap C$

7. $A \cap (B \cap C)$

8. $A \cup (B \cup C)$

9. Is $(A \cap B) \subseteq (A \cap C)$? Why or why not?

10. Is $(A \cap C) \cup (B \cap C) \subseteq C$? Is $C \subseteq (A \cap C) \cup (B \cap C)$? Why or why not?

For Exercises 11–14, write the set in roster notation.

11. $\{x : x = 2n, n \in N\}$

12. $\{x : x = 3w, w \in Z\}$

13. $\{x : x = 2n + 5, n \in N\}$

14. $\{x : x = 2n + 1, n \in N\}$

In Exercises 15–24, label each number rational or irrational. Demonstrate your answer by showing the decimal representation. (Use a calculator approximation if necessary.)

15. $\frac{71}{90}$

16. $\frac{16}{18}$

17. $\frac{\pi}{2}$

18. $\sqrt{3}$

19. $\frac{\sqrt{7}}{2}$

20. $\frac{1}{\sqrt{7}}$

21. $\frac{4}{11}$

22. $\frac{3}{16}$

23. $\sqrt{15}$

24. $\sqrt{7}$

The calculator is limited to a finite number of digits in the display. Find the following quotients, using paper and pencil with the help of a calculator. Each is a rational number by definition and thus should be a decimal number with a repeating pattern.

25. $\frac{1}{7}$

26. $\frac{3}{7}$

27. $\frac{71}{81}$

28. $\frac{51}{81}$

Complete and simplify Exercises 29–42. Always reduce your answers.

29. $\frac{17}{34}$

30. $\frac{13}{91}$

31. $\frac{5}{8} + \frac{2}{8}$

32. $\frac{6}{17} + \frac{9}{17}$

33. $\frac{3}{4}\left(\frac{8}{15}\right)$

34. $\frac{7}{8}\left(\frac{4}{21}\right)$

35. $5 - 4(7 + 4)$

36. $4 - 5(7 - 3)$

37. $2 - 3(4 - 7)$

38. $3 - 2(4 - 6)$

39. $\frac{1}{2}\left(\frac{3}{4} - \frac{1}{4}\right)$

40. $\frac{2}{3}\left(\frac{2}{8} + \frac{1}{8}\right)$

41. $\frac{5}{4} \div \left(\frac{5}{4} - \frac{1}{2}\right)$

42. $\frac{7}{6} \div \left(\frac{5}{8} + \frac{1}{4}\right)$

In Exercises 43–54, solve each equation using the properties of equality.

43. $5x - 3 = 2x + 9$

44. $17y - 5 = 3y + 5$

45. $\frac{2}{3}x = \frac{1}{3}x + \frac{1}{6}$

46. $\frac{3}{4}x - 4 = \frac{1}{2}x$

47. $\frac{1}{3} = \frac{5}{y} - \frac{2}{3}$

48. $\frac{4}{x} + \frac{3}{4} = \frac{1}{4}$

49. $\frac{2}{x} - 3 = \frac{11}{x}$

50. $3 - \frac{4}{y} = \frac{5}{y}$

51. $\frac{5}{x} + 3 = \frac{7}{x}$

52. $\frac{1}{x+1} + 2 = \frac{3}{x+1}$

53. $\frac{3}{y-1} + 1 = \frac{2}{y-1}$

54. $1 - \frac{2x}{x-2} = \frac{3}{x-2}$

55. Solve $a = \left(\frac{1}{2}\right)gt$ for t.

56. Solve $i = prt$ for t.

57. Solve $E = mc^2$ for m.

58. Solve $A = \frac{1}{2}bh$ for h.

59. Solve $P = 2l + 2w$ for l.

60. Solve $C = 2\pi r$ for r.

61. Solve $S = \frac{a}{1-r}$ for r.

62. Solve $P = \frac{n-r}{r}$ for n.

Evaluate the expressions in Exercise 63–76.

63. $|3| - |4|$

64. $|4| - |3|$

65. $-|5| + |4|$

66. $-|4| + |5|$

67. $|-5| + |4|$

68. $|-4| + |5|$

69. $16 - |-7 - 2|$

70. $-16 + |-7 - 2|$

71. $|-5|(-5)$

72. $\frac{-5}{|-5|}$

73. $|\sqrt{3} - \pi|$

74. $|\pi - \sqrt{3}|$

75. $|\pi - 3|$

76. $|3 - \pi|$

77. The circumference C of a circle is 48 square units and $C = 2\pi r$. Solve for the radius r (Fig. 1.1).

Figure 1.1

78. If the area of a parallelogram is 54 square units with a base of 7 units, solve for the height h of this parallelogram (Fig. 1.2).

Figure 1.2

79. A trapezoid has a height of 6 units, an area of 51 square units, and one base of 7 units. Find the length of the other base (Fig. 1.3).

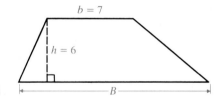

Figure 1.3

80. The area of a triangle is 34 square units. Find the height if the base is 5 (Fig. 1.4).

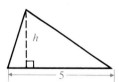

Figure 1.4

81. Prove that $a \cdot 0 = 0$.

82. Prove that if $a + a = a$, then $a = 0$.

83. Prove the zero factor theorem: $a \cdot b = 0$ if and only if $a = 0$ or $b = 0$.

84. Prove that $a(b - c) = ab - ac$.

85. Is subtraction of the real numbers closed? Why or why not?

86. Is division of the real numbers closed? Why or why not?

87. Are the numbers π, $\frac{22}{7}$, and 3.14 rational or irrational? Which is the largest and which is the smallest of these three numbers?

1.2 Exponents, Radicals, and Their Properties

We use exponents extensively in algebra. Exponents can simplify various mathematical operations. In this section we will review the use of integers and rational numbers as exponents.

Exponential Expressions

Definition | *Natural Number Exponents.* For $n \in N$, the set of natural numbers, a^n (read "the nth power of a") is the number with a as a factor n times (that is, $a^n = a \cdot a \cdot a \cdot a \cdots a \cdot a$, n factors each equal to a).

When we work with the nth power of a, written a^n, a is the **base** and n is the **exponent**. Thus for x^3, x is the base for the exponent of 3 with $x^3 = x \cdot x \cdot x$. The expression x^3 is read "the third power of x" or "the cube of x." Likewise, for $5x^2$, x is the base for the exponent 2 and $5x^2 = 5 \cdot x \cdot x$. The expression $5x^2$ can be read "5 times the second power of x" or "the product of 5 and the square of x."

We can extend the definition of natural number exponents to include all exponents that are nonzero integers, using the following definition.

Definition | *Negative Integer Exponents.* For $a \neq 0$ and $n \in N$, the set of natural numbers, a^{-n} is defined as $a^{-n} = \dfrac{1}{a^n}$.

Now that we can use nonzero integers as exponents, we will consider other properties of exponents. These properties will make it easier to manipulate exponents in multiplication and division.

Properties of Exponents

For real numbers a and b and nonzero integers m and n ($n \neq 0$, $m \neq 0$),

(a) $a^n \cdot a^m = a^{n+m}$

(b) $(a^n)^m = a^{nm}$

(c) $a^n \div a^m = \dfrac{a^n}{a^m} = a^{n-m}$ for $a \neq 0$

(d) $(ab)^n = a^n b^n$

(e) $\left(\dfrac{a}{b}\right)^n = \dfrac{a^n}{b^n}$ for $b \neq 0$

EXAMPLE 1   Simplify each of the following, using the properties of exponents:
(a) $a^3 \cdot a^4$ **(b)** $(b^2)^3$ **(c)** $x^7 \div x^3$ **(d)** $(2y)^4$
(e) $\left(\frac{2}{3}\right)^3$ **(f)** -2^2 **(g)** $(-2)^{-2}$

Solution
(a) Using property (a), $a^3 \cdot a^4 = a^{3+4} = a^7$
(b) Using property (b), $(b^2)^3 = b^{2(3)} = b^6$
(c) Using property (c), $x^7 \div x^3 = \dfrac{x^7}{x^3} = x^{7-3} = x^4$
(d) Using property (d), $(2y)^4 = 2^4 y^4 = 2 \cdot 2 \cdot 2 \cdot 2y^4 = 16y^4$
(e) Using property (e),

$$\left(\frac{2}{3}\right)^3 = \frac{2^3}{3^3} = \frac{2 \cdot 2 \cdot 2}{3 \cdot 3 \cdot 3} = \frac{8}{27}$$

(f) Using the first definition, $-2^2 = -2 \cdot 2 = -4$
(g) Using both definitions,

$$(-2)^{-2} = \frac{1}{(-2)^2} = \frac{1}{(-2)(-2)} = \frac{1}{4}$$ ■

Notice in Examples 1(f) and 1(g) that the negative sign is not included with the exponent unless it is included inside the parentheses.
We can now evaluate the value of a^0.

Theorem   *Zero as an Exponent.* For any nonzero real number a,

$$a^0 = 1$$

We note that we must have $a \ne 0$ to write a^0. We cannot evaluate 0^0 here, since the proof of $a^0 = 1$ is based on dividing by zero, and dividing by zero is undefined.

PROOF: When $a \ne 0$, using property (c) of exponents we have

$$a^n \div a^n = \frac{a^n}{a^n} = a^{n-n} = a^0$$

and also

$$\frac{a^n}{a^n} = a^n\left(\frac{1}{a^n}\right) = 1 \quad \textit{(By the multiplicative inverse property)}$$

Therefore by substitution we have $a^0 = 1$ for $a \ne 0$.   □

We now have another property of exponents:

property (f) $a^0 = 1$ for $a \ne 0$

We will now use all definitions and properties of exponents to simplify expressions whenever the exponents are integers.

EXAMPLE 2   Simplify each expression:
(a) $(2xy^3)(3x^3y^2)$ **(b)** $(5a^2b^3)^2$

Solution   Using the definitions and properties of exponents, we have

(a) $(2xy^3)(3x^3y^2) = (2 \cdot 3)(x \cdot x^3)(y^3 \cdot y^2)$ *(Associative and commutative properties of multiplication)*

$$= 6x^{1+3}y^{3+2}$$ *(Property (a) of exponents)*

$$= 6x^4y^5$$

(b) $(5a^2b^3)^2 = (5)^2(a^2)^2(b^3)^2$ *(Property (d) of exponents)*

$$= 5^2a^4b^6$$ *(Property (b) of exponents)*

$$= 5 \cdot 5a^4b^6$$

$$= 25a^4b^6$$

■

EXAMPLE 3   Simplify and write the result with positive exponents:

(a) $(x^{-3}y^2)^{-4}$ **(b)** $(a^2b^{-3})(a^{-2}b^5)$ **(c)** $\dfrac{(2xy^{-2})^2}{(x^2y^{-3})^{-1}}$

Solution   Using the definitions and properties of exponents, we have

(a) $(x^{-3}y^2)^{-4} = (x^{-3})^{-4}(y^2)^{-4}$ **(b)** $(a^2b^{-3})(a^{-2}b^5) = (a^2 \cdot a^{-2})(b^{-3} \cdot b^5)$

$$= x^{-3(-4)}y^{2(-4)} \qquad\qquad\qquad = a^{2-2}b^{-3+5}$$

$$= x^{12}y^{-8} \qquad\qquad\qquad\qquad = a^0b^2$$

$$= x^{12} \cdot \frac{1}{y^8} \qquad\qquad\qquad\quad = 1 \cdot b^2$$

$$= \frac{x^{12}}{y^8} \qquad\qquad\qquad\qquad = b^2$$

(c) There are several ways to complete this problem:

(i) $\dfrac{(2xy^{-2})^2}{(x^2y^{-3})^{-1}} = \dfrac{2^2x^2(y^{-2})^2}{(x^2)^{-1}(y^{-3})^{-1}}$ or   (ii) $\dfrac{(2xy^{-2})^2}{(x^2y^{-3})^{-1}} = \dfrac{2^2x^2(y^{-2})^2}{(x^2)^{-1}(y^{-3})^{-1}}$

$$= \frac{4x^2y^{-4}}{x^{-2}y^3} \qquad\qquad\qquad\qquad = \frac{4x^2y^{-4}}{x^{-2}y^3}$$

$$= 4x^{2-(-2)}y^{-4-3} \qquad\qquad\qquad = \frac{4x^2x^2}{y^3y^4}$$

$$= 4x^4y^{-7} \qquad\qquad\qquad\qquad = \frac{4x^4}{y^7}$$

$$= \frac{4x^4}{y^7}$$

or (iii) $\dfrac{(2xy^{-2})^2}{(x^2y^{-3})^{-1}} = (2xy^{-2})^2 \cdot (x^2y^{-3})^1$

$$= 2^2x^2(y^{-2})^2 \cdot x^2y^{-3}$$

$$= 4x^2y^{-4} \cdot x^2y^{-3} = 4x^2 \cdot x^2y^{-4}y^{-3}$$

$$= 4x^4y^{-7}$$

$$= \dfrac{4x^4}{y^7}$$

∎

Radical Expressions

We will now review the meaning and simplification of roots and radical expressions.

Definition

For $x = \sqrt[n]{a}$, x is **the principal nth root** of the number a and can be written as $x^n = a$. The **index** is n, the **radical** is the symbol $\sqrt{}$, and a is the **radicand**. The index n meets the condition that $n \in N$, $n \geq 2$, and when n is even, $a \geq 0$.

We can also express radical expressions using exponents. The properties of exponents are valid for both integer exponents and rational number exponents. Radical expressions will have rational number exponents, as we can see in the following examples.

We remember that

$\sqrt{9} = 3$
since $3^2 = 9$

$\sqrt[3]{8} = 2$
since $2^3 = 8$

$\sqrt[4]{81} = 3$
since $3^4 = 81$

We notice that

simplifying $9^{1/2}$ yields
$(9)^{1/2} = (3^2)^{1/2} = 3^{2/2}$
$= 3^1 = 3$

simplifying $8^{1/3}$ yields
$(8)^{1/3} = (2^3)^{1/3} = 2^{3/3}$
$= 2^1 = 2$

simplifying $81^{1/4}$ yields
$(81)^{1/4} = (3^4)^{1/4} = 3^{4/4}$
$= 3^1 = 3$

We conclude that

for $\sqrt{9} = 3$
and $9^{1/2} = 3$,
we have $\sqrt{9} = 9^{1/2}$

for $\sqrt[3]{8} = 2$,
and $8^{1/3} = 2$,
we have $\sqrt[3]{8} = 8^{1/3}$

for $\sqrt[4]{81} = 3$
and $(81)^{1/4} = 3$,
we have $\sqrt[4]{81} = (81)^{1/4}$

Our conclusions from these examples correctly imply that

$$x^{1/2} = \sqrt{x}, \qquad x^{1/3} = \sqrt[3]{x}, \qquad x^{1/4} = \sqrt[4]{x}$$

and in general

$$x^{1/n} = \sqrt[n]{x} \quad (\text{for } n \in N)$$

Definition

Rational Numbers as Exponents. If $n \in N$ and $n \geq 2$, then

(a) for $a \geq 0$, $a^{1/n} = \sqrt[n]{a}$, and
(b) for $a < 0$, $a^{1/n} = \sqrt[n]{a}$ only when the index n is an odd integer.

(When $a < 0$ and n is an even integer, the expression $a^{1/n} = \sqrt[n]{a}$ is not a real number.) Also if $n \in N$ and $n \geq 2$, then

(a) for $a \geq 0$, $\sqrt[n]{a^n} = a^{n/n} = a$, and
(b) for $a < 0$, if n is odd, then $\sqrt[n]{a^n} = a^{n/n} = a$, and
    if n is even, then $\sqrt[n]{a^n} = a^{n/n} = |a|$.

In the expression $\sqrt[n]{x}$, x is nonnegative when n is even (until we discuss complex numbers later). It is also important to remember that whenever n is even, $\sqrt[n]{x^n}$ is a nonnegative number, and

$$\sqrt[n]{x^n} = |x| \quad \text{for } n \text{ even}$$

Using the definitions and properties of exponents, we have

$$x^{m/n} = (x^m)^{1/n} = (x^{1/n})^m$$
$$x^{m/n} = \sqrt[n]{x^m} = (\sqrt[n]{x})^m$$

when $m \in Z$, $n \in N$, $\sqrt[n]{x}$ is defined, and $\sqrt[n]{x^m}$ is defined.

We can now work with radical expressions using rational numbers as exponents. All the properties of exponents listed earlier also apply to rational number exponents. **With rational numbers as exponents, the numerator of the exponent indicates the power of the base and the denominator of the exponent indicates the index of the root (or the index of the radical). Also it does not matter which is calculated first, the radical or the power.** We will now use the definitions and properties of exponents to simplify expressions even with rational exponents.

EXAMPLE 4

Simplify each expression:

(a) $(4x^4y^6)^{1/2}$ **(b)** $(81x^8)^{3/4}$ **(c)** $\dfrac{(a^{1/2}b^{1/3})^6}{(a^{1/2}b)^2}$

Solution   Using the definitions and properties of exponents, we have

(a) $(4x^4y^6)^{1/2} = 4^{1/2}(x^4)^{1/2}(y^6)^{1/2}$
$$= \sqrt{4}\,x^{4/2}y^{6/2}$$
$$= 2x^2y^3$$

(b) $(81x^8)^{3/4} = 81^{3/4}(x^8)^{3/4}$
$$= (3^4)^{3/4}(x^8)^{3/4}$$
$$= 3^{(3\cdot4)/4}x^{(8\cdot3)/4}$$
$$= 3^3x^6 = 27x^6$$

(c) $\dfrac{(a^{1/2}b^{1/3})^6}{(a^{1/2}b)^2} = \dfrac{(a^{1/2})^6(b^{1/3})^6}{(a^{1/2})^2 b^2}$

$\qquad\qquad = \dfrac{a^{6/2}b^{6/3}}{a^{2/2}b^2} = \dfrac{a^3 b^2}{ab^2}$

$\qquad\qquad = a^{3-1}b^{2-2}$

$\qquad\qquad = a^2 b^0 = a^2 \cdot 1 = a^2$ ∎

We can use the properties of exponents to establish the properties of radicals as follows.

For nonzero integers m and n and real numbers a and b, with $a > 0$ and $b > 0$ when n and/or m are even, we have

Properties of Exponents	**Properties of Radicals**
$(ab)^{1/n} = a^{1/n}b^{1/n}$	$\sqrt[n]{ab} = \sqrt[n]{a}\,\sqrt[n]{b}$
$\left(\dfrac{a}{b}\right)^{1/n} = \dfrac{a^{1/n}}{b^{1/n}}$	$\sqrt[n]{\dfrac{a}{b}} = \dfrac{\sqrt[n]{a}}{\sqrt[n]{b}}$
$(a^{1/n})^{1/m} = a^{1/mn}$	$\sqrt[m]{\sqrt[n]{a}} = \sqrt[nm]{a}$

EXAMPLE 5   Simplify the following radical expressions:

(a) $\sqrt{75}$ **(b)** $\sqrt{6}\sqrt{6}$ **(c)** $\sqrt[3]{-24}$ **(d)** $\sqrt[3]{4}\sqrt[3]{6}$ **(e)** $\sqrt[3]{8}\sqrt{8}$

Solution   Using the properties of radicals, we have

(a) $\sqrt{75} = \sqrt{25 \cdot 3}$ **(b)** $\sqrt{6}\sqrt{6} = \sqrt{6 \cdot 6}$

$\qquad\quad = \sqrt{25}\sqrt{3} = 5\sqrt{3}$ $\qquad\qquad\quad = \sqrt{36}$

$\qquad\qquad\qquad\qquad\qquad\qquad\qquad\qquad\quad = 6$

(c) $\sqrt[3]{-24} = \sqrt[3]{-8 \cdot 3}$ **(d)** $\sqrt[3]{4}\,\sqrt[3]{6} = \sqrt[3]{24}$

$\qquad\qquad = \sqrt[3]{-8}\,\sqrt[3]{3}$ $\qquad\qquad\qquad = \sqrt[3]{8 \cdot 3}$

$\qquad\qquad = \sqrt[3]{(-2)^3}\,\sqrt[3]{3}$ $\qquad\qquad\qquad = \sqrt[3]{8}\,\sqrt[3]{3}$

$\qquad\qquad = -2\sqrt[3]{3}$ $\qquad\qquad\qquad = 2\sqrt[3]{3}$

(e) $\sqrt[3]{8}\,\sqrt{8} = \sqrt[3]{8}\,\sqrt{4 \cdot 2}$

$\qquad\qquad = \sqrt[3]{2^3}\,\sqrt{4 \cdot 2}$ *(Notice that the indices are different)*

$\qquad\qquad = (2^3)^{1/3} \cdot \sqrt{4}\,\sqrt{2}$

$\qquad\qquad = 2 \cdot 2\sqrt{2} = 4\sqrt{2}$ ∎

Sometimes it is easier to simplify a radical expression by changing the radical to a rational exponent as in Example 6.

EXAMPLE 6

Simplify the following radical expressions:

(a) $\sqrt{4^6 x^2}$ (b) $\sqrt[3]{\sqrt{64x^{12}}}$ (c) $\dfrac{\sqrt[3]{x^2}}{\sqrt{x^3}}$

Solution   Changing radicals to exponents, we have

(a) $\sqrt{4^6 x^2} = (4^6 x^2)^{1/2}$

$\qquad\qquad = (4^6)^{1/2}(x^2)^{1/2}$

$\qquad\qquad = 4^3 |x|$ *(Remember that $\sqrt{x^2} = |x|$)*

$\qquad\qquad = 64|x|$

(b) $\sqrt[3]{\sqrt{64x^{12}}} = [(64x^{12})^{1/2}]^{1/3}$

$\qquad\qquad = (64^{1/2} x^{12/2})^{1/3}$

$\qquad\qquad = (8x^6)^{1/3}$

$\qquad\qquad = 8^{1/3} x^{6/3}$

$\qquad\qquad = 2x^2$ *(No absolute values are needed, since $x^2 \geq 0$)*

(c) $\dfrac{\sqrt[3]{x^2}}{\sqrt{x^3}} = \dfrac{x^{2/3}}{x^{3/2}}$

$\qquad\qquad = x^{2/3 - 3/2}$

$\qquad\qquad = x^{4/6 - 9/6}$

$\qquad\qquad = x^{-5/6}$

$\qquad\qquad = \dfrac{1}{x^{5/6}}$ for $x \geq 0$, since the index of the radical is even

The variable x must be greater than zero for this solution to **(c)** since the index of the radical is 6, an even integer.   ■

Simplified form for a radical expression includes a rational denominator (no radicals in the denominator). We call the process of making the denominator a rational number **rationalizing the denominator**. When the irrational denominator is a single term, we eliminate the radical from the denominator by multiplying the numerator and denominator by the radical expression that will eliminate the radical from the denominator. The following example demonstrates this process.

EXAMPLE 7

Rationalize the denominators in each of the following:

(a) $\dfrac{7}{\sqrt{3}}$ (b) $\dfrac{\sqrt{2}}{\sqrt{5}}$ (c) $\dfrac{\sqrt{3}}{\sqrt{12}}$ (d) $\dfrac{5}{\sqrt[3]{4}}$ (e) $\sqrt{\dfrac{x}{2y}}$

Solution For a and b we multiply the numerator and denominator by the term that makes the denominator a perfect square.

(a) $\dfrac{7}{\sqrt{3}} = \dfrac{7}{\sqrt{3}} \cdot \dfrac{\sqrt{3}}{\sqrt{3}} = \dfrac{7\sqrt{3}}{\sqrt{9}} = \dfrac{7\sqrt{3}}{3}$ or $\dfrac{7}{3}\sqrt{3}$

(b) $\dfrac{\sqrt{2}}{\sqrt{5}} = \dfrac{\sqrt{2}}{\sqrt{5}} \cdot \dfrac{\sqrt{5}}{\sqrt{5}} = \dfrac{\sqrt{10}}{\sqrt{25}} = \dfrac{\sqrt{10}}{5}$ or $\dfrac{1}{5}\sqrt{10}$

(c) $\dfrac{\sqrt{3}}{\sqrt{12}} = \dfrac{\sqrt{3}}{\sqrt{4 \cdot 3}} = \dfrac{\sqrt{3}}{\sqrt{4}\sqrt{3}} = \dfrac{\sqrt{3}}{2\sqrt{3}} = \dfrac{1}{2}$

(d) Since the denominator is a cube root, the radicand, 4, must become a perfect cube. Thus we multiply the numerator and denominator by $\sqrt[3]{2}$:

$$\frac{5}{\sqrt[3]{4}} = \frac{5}{\sqrt[3]{4}} \frac{\sqrt[3]{2}}{\sqrt[3]{2}} = \frac{5\sqrt[3]{2}}{\sqrt[3]{4 \cdot 2}} = \frac{5\sqrt[3]{2}}{\sqrt[3]{8}} = \frac{5\sqrt[3]{2}}{2} \quad \text{or} \quad \frac{5}{2}\sqrt[3]{2}$$

(e) $\sqrt{\dfrac{x}{2y}} = \sqrt{\dfrac{x}{2y} \dfrac{2y}{2y}} = \sqrt{\dfrac{2xy}{4y^2}} = \dfrac{\sqrt{2xy}}{\sqrt{4y^2}} = \dfrac{\sqrt{2xy}}{2|y|}$ or $\dfrac{1}{2|y|}\sqrt{2xy}$ ∎

Scientific Notation

One application of exponents is scientific notation. Scientific notation provides a way to write very large and very small numbers in a concise, condensed, and standardized manner. For example, we can write $750{,}000{,}000$ and 0.0000006 as

$$750{,}000{,}000 = 7.5 \times (100{,}000{,}000) = 7.5 \times 10^8$$

$$0.0000006 = \frac{6}{1{,}000{,}000} = \frac{6}{10^7} = 6 \times 10^{-7}$$

SCIENTIFIC NOTATION | To write a number in **scientific notation**, rewrite the number as the product of

(a) a number between 1 and 10 (including 1 but not including 10) and
(b) 10^n, where n is the correct integer.

As we saw in the examples above, multiplying with positive powers of 10 (multiplying by 10) make a number large, while multiplying with negative powers of 10 (dividing by 10) make a number smaller.

EXAMPLE 8   Write each number in scientific notation:

(a) $753{,}000{,}000$ **(b)** 0.0000000751 **(c)** $\dfrac{180{,}000}{0.00006}$

Solution

(a) $753{,}000{,}000 = 7.53 \times 100{,}000{,}000$

$= 7.53 \times 10^8$ *(Multiplying by 10^8 makes 7.53 larger by 8 decimal places)*

(b) $0.0000000751 = \dfrac{7.51}{100{,}000{,}000}$

$= 7.51 \times 10^{-8}$ *(Multiplying by 10^{-8} makes 7.51 smaller by 8 decimal places)*

(c) $\dfrac{180{,}000}{0.00006} = \dfrac{18 \times 10^4}{6 \times 10^{-5}}$

$= 3 \times 10^{4-(-5)}$

$= 3 \times 10^{4+5}$

$= 3 \times 10^9$ ∎

EXAMPLE 9

The diameter of the Milky Way, our galaxy, is about 110,000 light years. If one light year is 5,880,000,000,000 miles, find the diameter of our galaxy in miles.

Solution   The galaxy's diameter $= 110{,}000 = 1.1 \times 10^5$ light years. One light year $= 5{,}880{,}000{,}000{,}000 = 5.88 \times 10^{12}$ miles. Then the galaxy's diameter is

$$1.1 \times 10^5 \text{ light years} = (1.1 \times 10^5 \text{ light years})\left(5.88 \times 10^{12}\frac{\text{miles}}{\text{light year}}\right)$$

$$= 6.468 \times 10^{17} \text{ miles}$$

$$\approx 6.5 \times 10^{17} \text{ miles}$$ ∎

Exercises 1.2

Simplify each of the following expressions. Write answers with positive exponents. All variables are positive real numbers.

1. $x^2(xy^3)$

2. $x^2(xyz)$

3. $x^3(-x^2y^3)$

4. $-x^2(x^3y^2)$

5. $(2x)^3(-3xy)^2$

6. $(-3x)^2(4xy)^2$

7. $\left(\dfrac{2xy}{3x}\right)^3$

8. $\left(\dfrac{3xy}{4y}\right)^2$

9. $(2a^2b)^3$

10. $(3xy^3)^2$

11. $(-2)^2 - 3^2$

12. $(-3)^2 - 2^2$

13. $\left(\dfrac{2x}{y}\right)^3\left(\dfrac{y}{x^2}\right)^2$

14. $\left(\dfrac{3a}{b^2}\right)^3\left(\dfrac{a^2}{b}\right)^2$

15. $\left(\dfrac{3a}{b^2}\right)^3 \div \left(\dfrac{a^2}{b}\right)^2$

16. $\left(\dfrac{2x}{y}\right)^3 \div \left(\dfrac{y}{x^2}\right)^2$

17. $(-3x^2y^{-3})^{-3}$

18. $(-3x^{-2}y^3)^{-2}$

19. $(2x^3y^{-2})^{-5}$

20. $(2a^2b^{-3})^{-3}$

21. $\dfrac{(-4x^2y^{-3})^2}{(-2x^3y^{-2})^{-3}}$

22. $\dfrac{(-9x^{-2}y^3)^{-2}}{(-3x^{-3}y^2)^{-3}}$

23. $\dfrac{(-27a^4b^{-2})^{-1}}{(-3a^3b^2)^{-2}}$

24. $\dfrac{(-2a^3b^2)^{-2}}{(-2a^4b^{-2})^{-3}}$

25. $(x^{1/2}y^{1/3})^6$

26. $(x^{1/3}y^{1/4})^{12}$

27. $(x^{-1/3}y^{1/4})^{-12}$

28. $(x^{1/2}y^{-1/3})^{-6}$

29. $(9x^6y^2)^{1/2}$

30. $(8x^6y^9)^{1/3}$

31. $(16x^8y^{-4})^{1/4}$

32. $(27x^3y^{-9})^{1/3}$

33. $(x^4y^{1/3})^{3/4}$

34. $(x^3y^{1/2})^{2/3}$

35. $(x^4y)^{3/4}$

36. $(xy^3)^{2/3}$

37. $\left(\dfrac{-8x^3}{y^{-6}}\right)^{2/3}$

38. $\left(\dfrac{-125x^{-3}}{y^9}\right)^{2/3}$

39. $\dfrac{(x^{1/3}y^{1/4})^6}{(x^{1/2}y^{1/3})^2}$

40. $\dfrac{(x^{1/2}y^{3/8})^4}{(x^{1/3}y^{1/2})^3}$

41. $\sqrt{50}$

42. $\sqrt{27}$

43. $\sqrt[3]{54}$

44. $\sqrt[3]{24}$

45. $\sqrt[3]{-24}$

46. $\sqrt[3]{-54}$

47. $\sqrt{6}\sqrt{30}$

48. $\sqrt{10}\sqrt{30}$

49. $\sqrt[3]{16}\sqrt{4}$

50. $\sqrt[3]{27}\sqrt{27}$

51. $\sqrt{8}\sqrt[3]{27}$

52. $\sqrt{27}\sqrt[3]{8}$

53. $\sqrt{7^6x^2y^4}$

54. $\sqrt{3^6x^4y^2}$

55. $\sqrt[3]{\sqrt{16x^6y^2}}$

56. $\sqrt[3]{\sqrt{64x^6y^{12}}}$

57. $\dfrac{\sqrt{x^3}}{\sqrt[4]{x^2}}$

58. $\dfrac{\sqrt[4]{x^5}}{\sqrt{x}}$

Rationalize the denominator in problems 59–74.

59. $\dfrac{5}{\sqrt{7}}$

60. $\dfrac{7}{\sqrt{5}}$

61. $\dfrac{\sqrt{2}}{\sqrt{3}}$

62. $\dfrac{\sqrt{3}}{\sqrt{2}}$

63. $\dfrac{\sqrt{3}}{\sqrt{8}}$

64. $\dfrac{\sqrt{2}}{\sqrt{27}}$

65. $\sqrt{\dfrac{1}{2}}$

66. $\sqrt{\dfrac{1}{3}}$

67. $\dfrac{3}{\sqrt[3]{9}}$

68. $\dfrac{3}{\sqrt[4]{8}}$

69. $\dfrac{5}{\sqrt[5]{16}}$

70. $\dfrac{2}{\sqrt[3]{4}}$

71. $\dfrac{7}{\sqrt[4]{27}}$

72. $\dfrac{5}{\sqrt[3]{25}}$

73. $\sqrt{\dfrac{2x}{y}}$

74. $\sqrt{\dfrac{y}{3x}}$

Write each of the following in scientific notation.

75. 76,100,000

76. 0.00000643

77. 0.0000323

78. 5,010,000

79. $\dfrac{2,100,000}{0.0007}$

80. $\dfrac{16,000,000}{0.0008}$

81. One light year is about 9,460,000,000,000,000 meters. Write the distance of a light year in scientific notation.

82. The diameter of the Milky Way, our galaxy, is approximately 110,000 light years. Find the diameter and radius of our galaxy in meters, using scientific notation.

83. The sea otter is a small mammal that spends most of its life in cold ocean waters. Its dense fur averages 300,000 hairs per square inch. If a sea otter has 700 square inches of hair, approximately how many hairs does it have? (Use scientific notation.)

84. A red blood cell is 0.000007 meter in size. Write this in scientific notation.

85. With white light the best microscopes have a resolution of 0.00000025 meter. An electronic microscope is theoretically capable of a resolution of about 0.000000000005 meter. Write both distances in scientific notation.

86. The mass of a hydrogen atom is approximately 0.00000000000000000000001673 gram. Write this in scientific notation.

87. The maximum length for the tail of a comet is 160,000,000 kilometers. Write this length in meters, using scientific notation. What part of a light year would this distance be?

88. The space shuttle makes an orbit in 90 minutes with a speed of 93,000,000 feet per hour. Using 21,000,000 feet as the radius of the earth, find the shuttle's altitude in feet.

89. The space shuttle makes an orbit in 90 minutes with a speed of 28,000,000 meters per hour (Fig. 1.5). Using 6,380,000 meters as the radius of the earth, find the shuttle's altitude in meters.

Figure 1.5

1.3　Polynomials and Rational Expressions

As well as using letters to represent sets and their elements, we also use letters to represent numbers. Letters used to represent numbers are called **variables**. In the following example, both the variables and the value of the resulting expression must be real numbers.

EXAMPLE 1

Find the values for x that provide a real number result in these expressions:

(a) $\sqrt{x}$　　**(b)** $\dfrac{1}{x-2}$

Solution

(a) Since $\sqrt{x}$ must be a real number, we have $x \geq 0$. (Even integer roots have nonnegative radicands; for example, $\sqrt[4]{-16}$ is not a real number.)

(b) Since division by zero is not possible (undefined by the closure property), we have $x - 2 \neq 0$ and $x \neq 2$ or $x \in \{x : x \in R \text{ and } x \neq 2\}$.　　∎

A polynomial **term** is either a constant, a variable with a nonnegative integer as an exponent, or the product of these (this includes products of constants with variables as well as the products of variables with other variables). The following are polynomial terms:

$$5, \quad x^7, \quad -5x^3, \quad x^3y^2z, \quad \tfrac{1}{2}xy^2, \quad \text{and} \quad \sqrt{2}x$$

The term $\sqrt{2}x$ is a polynomial term even with the $\sqrt{2}$, since x is not part of the radicand. The following expressions are *not* polynomial terms:

$$x^{-1}, \quad 5\sqrt{x}, \quad \text{and} \quad \frac{1}{x-1}$$

They are not polynomial terms because the exponent in each is not a nonnegative integer:

$$x^{-1}, \quad 5\sqrt{x} = 5x^{1/2}, \quad \text{and} \quad \frac{1}{x-1} = (x-1)^{-1}$$

Polynomial terms *do not* contain fractional exponents of variables, *do not* contain variables in the radicand, and *do not* have variables in the denominators.

Polynomials

The sum of polynomial terms is called a **polynomial**. We will investigate polynomials as well as some interesting polynomial applications in this section and later sections. In this section we will review polynomials and their properties.

Definition

A **polynomial** in x is an expression that can be written in the form

$$a_n x^n + a_{n-1} x^{n-1} + \cdots + a_2 x^2 + a_1 x + a_0$$

where n is a nonnegative integer, each coefficient a_i is a real number, and i is a nonnegative integer.

The i in a_i is called a **subscript** and acts only as an identification number, like a license plate, indicating the power of x for which a_i is the coefficient. (Thus a_i is the coefficient of x^i in the term $a_i x^i$.) The **leading coefficient** of a polynomial is a_n, when n is the largest integer such that $a_n \neq 0$. The **degree** of a polynomial in x is the exponent n of $a_n x^n$ when a_n is the leading coefficient. A polynomial is in **standard form** when the exponents of the variable decrease in value term by term reading from left to right.

Thus $2x^5 + 5x^4 - 2x^3 + x^2 - x + 1$ is a fifth-degree polynomial in standard form with a leading coefficient of 2. The polynomial $3x^2 + x - 1$ is also in standard form and has degree 2. The polynomial consisting of the one term a_0, where a_0 is a nonzero constant, has degree 0, since the exponent for x in the term $a_0 = a_0 x^0$ is zero. The constant term 5 is a term of degree zero, since $5 = 5 \cdot 1 = 5x^0$. We say that the polynomial term $a_0 = 0$ has no degree. Although the variable x is used here, the same statements are also true for other variables.

We will now add and subtract polynomials using the distributive property. In polynomials we will add and subtract **like terms**, terms with exactly the same powers of the same variables.

EXAMPLE 2

Subtract $2x^4 - 2x^3 + 3x^2 + 2x - 2$ from $5x^4 + 3x^3 - x + 4$.

Solution Use the distributive property to complete this subtraction:

$$(5x^4 + 3x^3 - x + 4) - (2x^4 - 2x^3 + 3x^2 + 2x - 2)$$
$$= (5x^4 + 3x^3 - x + 4) - (2x^4) - (-2x^3) - (3x^2) - (2x) - (-2)$$
$$= 5x^4 + 3x^3 - x + 4 - 2x^4 + 2x^3 - 3x^2 - 2x + 2$$
$$= 5x^4 - 2x^4 + 3x^3 + 2x^3 - 3x^2 - x - 2x + 4 + 2$$
$$= (5 - 2)x^4 + (3 + 2)x^3 - 3x^2 + (-1 - 2)x + 6$$
$$= 3x^4 + 5x^3 - 3x^2 - 3x + 6$$

Now we will use the distributive property to multiply polynomials.

EXAMPLE 3

Find the product of $x^3 - x + 1$ and $x^2 + 2x + 1$.

Solution Distribute the first factor to each term in the second factor:

$$(x^3 - x + 1)(x^2 + 2x + 1)$$
$$= x^2(x^3 - x + 1) + 2x(x^3 - x + 1) + 1(x^3 - x + 1)$$
$$= x^2(x^3) + x^2(-x) + x^2(1) + 2x(x^3) + 2x(-x) + 2x(1) + x^3 - x + 1$$

$$= x^5 - x^3 + x^2 + 2x^4 - 2x^2 + 2x + x^3 - x + 1$$
$$= x^5 + 2x^4 + (-x^3 + x^3) + (x^2 - 2x^2) + (2x - x) + 1$$
$$= x^5 + 2x^4 - x^2 + x + 1$$

We can multiply two polynomials using the column method as in Example 4.

EXAMPLE 4   Find the product of $x^2 - xy + y^2$ and $x + y$.

Solution   Using the column method, we have

$$x^2 - xy + y^2$$

Multiplication: $x + y$

$x(x^2 - xy + y^2) = x^3 - x^2y + xy^2$

$y(x^2 - xy + y^2) = \quad\quad + x^2y - xy^2 + y^3$

Addition: $x^3 \quad\quad\quad\quad + y^3$

Thus $(x^2 - xy + y^2)(x + y) = x^3 + y^3$.

We can use the column method above to find the product of two binomials, but it is faster to use the **FOIL method**. In the FOIL method, each letter stands for the following product:

F $\longrightarrow$ **first**   term times **first**   term

O $\longrightarrow$ **o**utside term times **o**utside term

I $\longrightarrow$ **i**nside   term times **i**nside   term

L $\longrightarrow$ **last**   term times **last**   term

$$(x - 2)(3x + 4) = 3x^2 - 2x - 8$$

This last streamlined form of the FOIL method is excellent for multiplying two binomials and will also be helpful in factoring trinomials later in this section.

EXAMPLE 5   Find the product of $(2x - 5)(3x - 4)$ using the FOIL method.

Solution   Using the FOIL method, we have

$$(2x - 5)(3x - 4) = 6x^2 - 23x + 20$$

We will now move on to polynomial division. Division of a polynomial by a single term also utilizes the distributive property.

EXAMPLE 6   Divide $5x^3y^3z + 15x^3y^2z^2 - 20xy^2z^3$ by $5xy^2z$.

Solution   Using the distributive property and simplifying, we have

$$(5x^3y^3z + 15x^3y^2z^2 - 20xy^2z^3) \div (5xy^2z)$$

$$= \frac{(5x^3y^3z + 15x^3y^2z^2 - 20xy^2z^3)}{5xy^2z}$$

$$= \frac{5x^3y^3z}{5xy^2z} + \frac{15x^3y^2z^2}{5xy^2z} - \frac{20xy^2z^3}{5xy^2z}$$

$$= x^2y + 3x^2z - 4z^2$$ ■

Using these techniques, we can verify the following special patterns. These patterns should be memorized so that we can use them in future work.

SPECIAL PRODUCTS   For variables x and y,

    (a) $(x + y)(x - y) = x^2 - y^2$
    (b) $(x + y)^2 = x^2 + 2xy + y^2$
    (c) $(x - y)^2 = x^2 - 2xy + y^2$
    (d) $(x + y)^3 = x^3 + 3x^2y + 3xy^2 + y^3$
    (e) $(x - y)^3 = x^3 - 3x^2y + 3xy^2 - y^3$

These special products are helpful in multiplying polynomials as in the following example.

EXAMPLE 7   Multiply the resulting polynomials with results in standard form:
(a) $(x - 2)(x^2 + 4)(x + 2)$ **(b)** $(2x + 3)^2$ **(c)** $[(x - 1)(x + 1)]^3$

Solution
(a) Using the commutative property and special product (a) twice, we have

$$(x - 2)(x^2 + 4)(x + 2) = (x - 2)(x + 2)(x^2 + 4) \quad \textit{(Commutative property)}$$

$$= [(x - 2)(x + 2)](x^2 + 4) \quad \textit{(Associative property)}$$

$$= (x^2 - 4)(x^2 + 4) \quad \textit{(Special product (a))}$$

$$= (x^2)^2 - (4)^2 \quad \textit{(Special product (a))}$$

$$= x^4 - 16$$

(b) Using special product (b), we have

$$(2x + 3)^2 = (2x)^2 + 2(2x)(3) + 3^2$$
$$= 4x^2 + 12x + 9$$

(c) Using special products (a) and then (e), we have

$$[(x - 1)(x + 1)]^3 = (x^2 - 1)^3$$
$$= (x^2)^3 - 3(x^2)^2(1) + 3(x^2)(1)^2 - (1)^3$$
$$= x^6 - 3x^4 + 3x^2 - 1$$ ∎

As in Example 7, using these special products will speed up the multiplication process.

Polynomial Factoring

We multiply when we find the product of two factors. When we change an expression into the product of its factors, we call the process **factoring**.

$$\overrightarrow{\textit{multiplying}}$$
$$(x - 2y)(x + 2y) = x^2 - 4y^2$$
$$\overleftarrow{\textit{factoring}}$$

It is always possible to factor any nonzero monomial from a polynomial. Notice that

$$5x^3 + 3x^2 - 2x + 1 = c\left(\frac{5}{c}x^3 + \frac{3}{c}x^2 - \frac{2}{c}x + \frac{1}{c}\right)$$

where c is any nonzero number. We can check this answer with the distributive property.

EXAMPLE 8   Factor the following problems as indicated:
(a) Factor 5 from $(5x^3 + 15x^2 - 20x + 10)$.
(b) Factor $\frac{3}{4}x^3$ from $\left(\frac{3}{4}x^5 - \frac{1}{4}x^4 + x^3\right)$.
(c) Factor $2x^{-2}$ from $(2x^2 + 4 + 6x^{-2})$.

Solution

(a) $5x^3 + 15x^2 - 20x + 10 = 5\left(\dfrac{5x^3}{5} + \dfrac{15x^2}{5} - \dfrac{20x}{5} + \dfrac{10}{5}\right)$

$$= 5(x^3 + 3x^2 - 4x + 2)$$

(b) $\dfrac{3}{4}x^5 - \dfrac{1}{4}x^4 + x^3 = \dfrac{3}{4}x^3\left(\dfrac{\frac{3}{4}x^5}{\frac{3}{4}x^3} - \dfrac{\frac{1}{4}x^4}{\frac{3}{4}x^3} + \dfrac{x^3}{\frac{3}{4}x^3}\right)$

$$= \frac{3}{4}x^3\left[x^2 - \frac{1}{4}\left(\frac{4}{3}\right)x + 1\left(\frac{4}{3}\right)\right]$$

$$= \frac{3}{4}x^3\left(x^2 - \frac{1}{3}x + \frac{4}{3}\right).$$

(c) Although $2x^{-2}$ is not a monomial, we use the same process:

$$2x^2 + 4 + 6x^{-2} = 2x^{-2}\left(\frac{2x^2}{2x^{-2}} + \frac{4}{2x^{-2}} + \frac{6x^{-2}}{2x^{-2}}\right)$$

$$= 2x^{-2}(x^4 + 2x^2 + 3)$$

$$= \frac{2}{x^2}(x^4 + 2x^2 + 3) \qquad ■$$

Looking back at the special products listed in this section and the methods of multiplication, we have the following patterns for factoring. These should also be committed to memory.

SPECIAL FACTORS | For variables x and y,

(a) $x^2 - y^2 = (x + y)(x - y)$
(b) $x^2 + 2xy + y^2 = (x + y)^2$
(c) $x^2 - 2xy + y^2 = (x - y)^2$
(d) $x^3 + y^3 = (x + y)(x^2 - xy + y^2)$
(e) $x^3 - y^3 = (x - y)(x^2 + xy + y^2)$

Notice that special factors (d) and (e) are very different from the previously listed special products (d) and (e).

EXAMPLE 9   Factor each completely:
(a) $8x^3 - y^3$ **(b)** $4x^2 - y^2$ **(c)** $3x^2 - 6xy + 3y^2$
(d) $8x^5 - 8x^3 + x^2 - 1$

Solution   Using the special factoring patterns, we have
(a) $8x^3 - y^3 = (2x)^3 - (y)^3$

$$= (2x - y)[(2x)^2 + (2x)(y) + y^2] \quad \textit{(Special factor (e))}$$

$$= (2x - y)(4x^2 + 2xy + y^2)$$

(b) $4x^2 - y^2 = (2x)^2 - y^2$

$$= (2x + y)(2x - y) \qquad\qquad \textit{(Special factor (a))}$$

(c) $3x^2 - 6xy + 3y^2 = 3(x^2 - 2xy + y^2) \qquad \textit{(Factor the common factor of 3)}$

$$= 3(x - y)^2 \qquad\qquad \textit{(Special factor (c))}$$

(d) With four terms it is sometimes possible to group by pairs and factor (described as factoring by grouping):

$$8x^5 - 8x^3 + x^2 - 1 = (8x^5 - 8x^3) + (x^2 - 1) \quad \textit{(Group by twos and factor common monomial terms)}$$

$$= 8x^3(x^2 - 1) + 1(x^2 - 1) \quad \textit{(Factor the common factor } x^2 - 1\textit{)}$$

$$= (8x^3 + 1)(x^2 - 1) \quad \textit{(Special factors (d) and (a))}$$

$$= (2x + 1)(4x^2 - 2x + 1)(x - 1)(x + 1) \quad \blacksquare$$

Factoring a trinomial that is not a special pattern is also an important skill. To factor a trinomial, we use the FOIL method in reverse order (that is, we "undo" the FOIL).

STEPS FOR FACTORING TRINOMIALS	Complete the following steps to factor a trinomial after factoring the common factors: STEP 1: Use the first term in the trinomial to find the possible first terms of the two binomial factors. STEP 2: Use the sign of the last term in the trinomial to find the correct signs for the two last terms of the two binomial factors. STEP 3: Use the last term of the trinomial to find the possible last terms of the two binomial factors. STEP 4: Use these possibilities to find combinations of first and last terms in the binomial factors that yield the correct middle term (outside and inside of FOIL).

The following example demonstrates this process.

EXAMPLE 10   Factor $4x^2 - 21x + 5$.

Solution   Factoring $4x^2 - 21x + 5$, we have

STEP 1: $4x^2 = (4x)(x)$ or $(2x)(2x)$.

STEP 2: The sign of 5, the third term in the trinomial, is positive. Then the last terms in the binomial factors are both positive or both negative. But since the middle term of the trinomial is $-21x$, the last terms of the binomials must also be negative to add up to $-21x$ (outside + inside = $-21x$). Thus $(4x -)(x -)$ or $(2x -)(2x -)$.

STEP 3: The factors of 5, the third term in the trinomial, are 5 and 1. Thus the possible binomial factors are $(4x - 5)(x - 1)$, $(4x - 1)(x - 5)$, and $(2x - 5)(2x - 1)$.

STEP 4: The only pair of factors with a middle term of $-21x$ is $(4x - 1)(x - 5)$. Thus $4x^2 - 21x + 5 = (4x - 1)(x - 5)$. $\blacksquare$

Rational Expressions

The quotient of two polynomials is called a **rational expression**. The following are examples of rational expressions:

$$\frac{x^3 + 2x - 1}{x^2 + 3}, \quad \frac{xy}{x^2 + 5x - 2}, \quad \frac{1}{4x^2y^3}, \quad \text{and} \quad \frac{x^3 - 1}{x^2 + 1}$$

When adding, subtracting, multiplying, or dividing rational expressions, we will use the properties of rational expression from Section 1.1. Remember that two rational expressions cannot be added or subtracted unless they both have the same denominator. To find the best common denominator, we will find the least common denominator (LCD).

Definition

The **least common denominator (LCD)** is an expression that is

(a) a multiple of each denominator and
(b) contains the least number of factors to accomplish part (a).

Thus to find the LCD, we will factor each denominator and then create the LCD. The LCD will contain all factors of each denominator with as few factors as possible. The following examples shows this process.

EXAMPLE 11

Find the LCD for

(a) $\dfrac{1}{6x} + \dfrac{1}{16x^2}$ (b) $\dfrac{2}{x + 3} - \dfrac{x}{x + 2}$ (c) $\dfrac{1}{(x + 2)^2} = \dfrac{1}{x^2 - 4}$

Solution We factor each denominator and then construct the LCD:
(a) For

$$\frac{1}{6x} + \frac{1}{16x^2}$$

the factored denominators are

$$6x = 2 \cdot 3 \cdot x$$
$$16x^2 = 2 \cdot 2 \cdot 2 \cdot 2 \cdot x \cdot x$$

Thus the LCD has factors of $2 \cdot 3 \cdot x$ from $6x$ as well as the factors of $2 \cdot 2 \cdot 2 \cdot x$, the factors of $16x^2$ not yet represented in the LCD. Therefore we have LCD $= (2 \cdot 3 \cdot x)(2 \cdot 2 \cdot 2 \cdot x) = 48x^2$. We notice that LCD $= 48x^2 = (6x)(8x)$, a multiple of $6x$, and LCD $= 48x^2 = 3(16x^2)$, a multiple of $16x^2$.
(b) For

$$\frac{2}{x + 3} - \frac{x}{x + 2}$$

the denominators do not factor and their product is the LCD.

$$\text{LCD} = (x + 3)(x + 2)$$

(c) For

$$\frac{1}{(x + 2)^2} = \frac{1}{x^2 - 4}$$

we have

$$(x + 2)^2 = (x + 2)(x + 2)$$
$$(x^2 - 2) = (x - 2)(x + 2)$$
$$\text{LCD} = (x + 2)(x + 2)(x - 2)$$

Thus this LCD is a multiple of $(x + 2)^2$ and $(x^2 - 4)$. ∎

We will use the LCD to simplify problems with rational expressions. When we simplify a rational expression, we will always factor the numerator and denominator to reduce to lowest terms.

EXAMPLE 12   Complete the following operations. Always reduce to lowest terms and include in your answers any restrictions on the variables.

(a) $\dfrac{x^3 - 1}{x + 1} \cdot \dfrac{x^2 - 1}{(x - 1)^2}$ **(b)** $\dfrac{x^3 - y^3}{x + 1} \div \left(\dfrac{x^2}{x + 1} - \dfrac{y^2}{x + 1} \right)$

(c) $\dfrac{2(x^2 + 2x - 1)}{x^2 + 5x + 6} + \dfrac{2}{x + 3} - \dfrac{x}{x + 2}$

Solution   Factor and use the properties of rational expressions for each.

(a) $\dfrac{x^3 - 1}{x + 1} \cdot \dfrac{x^2 - 1}{(x - 1)^2} = \dfrac{(x^3 - 1)(x^2 - 1)}{(x + 1)(x - 1)^2}$ *(Now factor and reduce)*

$$= \frac{(x - 1)(x^2 + x + 1)(x - 1)(x + 1)}{(x + 1)(x - 1)(x - 1)}$$

$$= \frac{x^2 + x + 1}{1}$$

$$= x^2 + x + 1 \qquad \text{with } x \neq \pm 1$$

since these values make the denominator zero in the original expression.

(b) $\dfrac{x^3 - y^3}{x + 1} \div \left(\dfrac{x^2}{x + 1} - \dfrac{y^2}{x + 1} \right) = \dfrac{x^3 - y^3}{x + 1} \div \dfrac{x^2 - y^2}{x + 1}$ *(Subtracting)*

$$= \frac{x^3 - y^3}{x + 1} \cdot \frac{x + 1}{x^2 - y^2} \qquad \text{\textit{(Dividing)}}$$

$$= \frac{(x - y)(x^2 + xy + y^2)}{x + 1} \cdot \frac{x + 1}{(x - y)(x + y)} \qquad \text{\textit{(Factor and reduce)}}$$

$$= \frac{x^2 + xy + y^2}{x + y} \qquad \text{with } x \neq -1, \ x \neq y, \text{ and } x \neq -y$$

since these values make a denominator zero for this expression.

(c) $\dfrac{2(x^2 + 2x - 1)}{x^2 + 5x + 6} + \dfrac{2}{x + 3} - \dfrac{x}{x + 2} = \dfrac{2(x^2 + 2x - 1)}{(x + 3)(x + 2)} + \dfrac{2}{x + 3} - \dfrac{x}{x + 2}$

(Now use the LCD of (x + 3)(x + 2) to add and subtract)

$$= \frac{2(x^2 + 2x - 1)}{(x + 3)(x + 2)} + \frac{2(x + 2)}{(x + 3)(x + 2)} - \frac{x(x + 3)}{(x + 2)(x + 3)}$$

$$= \frac{2(x^2 + 2x - 1) + 2(x + 2) - x(x + 3)}{(x + 2)(x + 3)}$$

$$= \frac{2x^2 + 4x - 2 + 2x + 4 - x^2 - 3x}{(x + 2)(x + 3)}$$

$$= \frac{x^2 + 3x + 2}{(x + 2)(x + 3)}$$

$$= \frac{(x + 2)(x + 1)}{(x + 2)(x + 3)}$$

$$= \frac{x + 1}{x + 3} \qquad \text{for } x \neq -2 \text{ and } x \neq -3$$

since these values make the denominator zero in the original expression. ∎

Complex Fractions

A **complex fraction** is a quotient (fraction) containing rational expressions in the numerator and denominator. One method of simplification is to simplify the numerator and denominator and then divide. A more effective method is to find the least common denominator (LCD) for the numerator and the denominator. Then multiply the numerator and denominator by the LCD. We will find this second method most useful in future work as in the following examples.

EXAMPLE 13   Simplify the complex fraction

$$\frac{\dfrac{5}{x + 1}}{x^{-1} + 4(x + 1)^{-1}}$$

Solution   First change the expressions with negative exponents to fractional form:

$$\frac{\dfrac{5}{x + 1}}{x^{-1} + 4(x + 1)^{-1}} = \frac{\dfrac{5}{x + 1}}{\dfrac{1}{x} + \dfrac{4}{x + 1}} \qquad$$ *(The LCD of both the numerator and denominator is x(x + 1). Multiply the numerator and denominator by x(x + 1))*

$$= \frac{\dfrac{5}{x + 1}}{\dfrac{1}{x} + \dfrac{4}{x + 1}} \cdot \frac{x(x + 1)}{x(x + 1)}$$

$$= \frac{\dfrac{5x(x + 1)}{x + 1}}{\dfrac{x(x + 1)}{x} + \dfrac{4x(x + 1)}{x + 1}}$$

$$= \frac{5x}{(x + 1) + 4x}$$

$$= \frac{5x}{5x + 1} \qquad \text{with } x \neq 0, \ x \neq -1, \text{ and } x \neq -\tfrac{1}{5}$$

since these values make a denominator zero in the original expression.   ■

We can also use the LCD to solve equations with rational expressions.

EXAMPLE 14   A small airplane flies 560 miles to college and back again. The airplane travels at 75 miles per hour in still air, and the round trip takes a total of 15 hours. If the wind blows in the same direction as the airplane is flying when it is traveling to the college and at the same rate all day, what is the rate of the wind?

Solution   Since the rate of the wind and the time of each trip are unknowns, we will let,

$$x = \text{the rate of the wind}$$

$$t_1 = \text{the time of the flight into the wind}$$

$$t_2 = \text{the time of the flight with the wind}$$

Then we fill in this chart utilizing the formula rate · time = distance.

	R	· T	= D
Into the wind	$75 - x$	t_1	560
With the wind	$75 + x$	t_2	560
Total		15	

Thus

$$(75 - x)t_1 = 560 \qquad \text{and} \qquad (75 + x)t_2 = 560 \qquad \text{or}$$

$$t_1 = \frac{560}{75 - x} \qquad \text{and} \qquad t_2 = \frac{560}{75 + x}$$

Since $t_1 + t_2 = 15$, we have

$$\frac{560}{75 - x} + \frac{560}{75 + x} = 15 \quad \text{\textit{(Now multiply both sides by the LCD} = (75 - x)(75 + x))}$$

$$560(75 + x) + 560(75 - x) = 15(75 - x)(75 + x)$$

$$42{,}000 + 560x + 42{,}000 - 560x = 15(5625 - x^2)$$

$$84{,}000 = 84{,}375 - 15x^2$$

$$15x^2 - 375 = 0$$

$$15(x^2 - 25) = 0$$

$$15(x - 5)(x + 5) = 0$$

Thus $x = \pm 5$, and the rate of the wind is 5 miles per hour.   ■

As we saw in this section, we can use the LCD to add and subtract fractions, simplify complex fractions, and solve equations with rational expressions.

Exercises 1.3

Complete the following operations and simplify each answer as much as possible.

1. $(3x^3 - 2x^2 + x - 5) + (x^4 - 3x^2 + 3x + 4)$

2. $(x^5 - x^3 + x - 1) + (2x^4 + x^3 + 3x + 2)$

3. $(3x^3 - 2x^2 + x - 5) - (x^4 - 3x^2 + 3x + 4)$

4. $(x^5 - x^3 + x - 1) - (2x^4 + x^3 + 3x + 2)$

5. $(y^2 - 3)(2y^3 + y^2 - 1)$ **6.** $(3y^2 - 2)(y^3 + y^2 - 1)$

7. $(x^3 - y^3)(x^3 + y^3)$ **8.** $(x^2 - y)(x^2 + y)$

9. $(x^3 - x^2 + x - 1)(x + 1)$

10. $(x^4 + x^2 - 1)(x^2 + 2x - 1)$

11. $(3y - 2)^2$ **12.** $(2y + 1)^2$

13. $(3t + 2)^3$ **14.** $(2t + 3)^3$

15. $(x^2 - y^2)^3$ **16.** $(x^2 + y^2)^3$

17. $(x + y^2)^2$ **18.** $(x - y^2)^2$

19. $(x + y^2)(x - y^2)$ **20.** $(2x + 3)(2x - 3)$

21. $(x^2 + xy + y^2)(x + y)$ **22.** $(x^2 - xy + y^2)(x - y)$

23. $(15x^3y^2 - 3x^2y^3 + 9x^3y) \div (3x^2y)$

24. $(4x^3y^3 + 12x^2y^4 - 6xy^3) \div (2xy^3)$

25. Factor $8xy$ from $(8x^2y - 6xy^2 + xy)$.

26. Factor $25x^2$ from $(25x^4 - 15x^3 - x^2)$.

27. Factor $3x^{-2}$ from $(3x^4 - 5x^2 - 9 + 12x^{-2})$.

28. Factor $5y^{-2}$ from $(25y^3 - 15y + 5y^{-2})$.

29. Factor $7y^{-3}$ from $(7y^3 - y + 14y^{-1} - 21y^{-3})$.

30. Factor $4x^{-3}$ from $(4x^3 - x - x^{-1} + 12x^{-3})$.

31. Factor $(x + 1)^{-1}$ from $3(x + 1)^3 - 2(x + 1) + 5(x + 1)^{-1}$.

32. Factor $(x - 1)^{-2}$ from $5(x + 1)^2 - 6 + 2(x - 1)^{-2}$.

Completely factor each of the following.

33. $8x^3 - y^3$ **34.** $x^3 + 27y^3$

35. $x^6 - y^2$ **36.** $x^2 - y^8$

37. $x^3 + 27y^3$ **38.** $8x^3 + y^3$

39. $x^2 + 6x + 9$ **40.** $y^2 - 4y + 4$

41. $5x^2 - 20x + 20$ **42.** $12x^2 - 24x + 12$

43. $x^3 + 3x^2 + 2x + 6$ **44.** $5x^3 - 5x^2 + 2x - 2$

45. $x^4 - x^2 - x^2y^2 + y^2$ **46.** $y^4 - 4y^2 - 9y^2 + 36$

Complete the following operations and simplify each. Always reduce to lowest terms and note restrictions on the variables.

47. $\dfrac{x^2 - 9}{x^2 - x - 6} \cdot \dfrac{x^2 + 5x + 6}{x^2 + x - 6}$

48. $\dfrac{x^2 - 1}{x^2 - 4} \cdot \dfrac{x^2 - 5x + 6}{x^2 - 2x - 3}$

49. $\dfrac{y^3 - 1}{(y - 1)(y + 2)} \div \dfrac{y^2 + y + 1}{y + 2}$

50. $\dfrac{y^2 - 1}{y^2 - y + 1} \div \dfrac{y^2 + 2y + 1}{y^3 + 1}$

51. $\dfrac{x^2 - 4}{x - 1} \cdot \left(\dfrac{5x}{x - 2} - \dfrac{5}{x - 2} \right)$

52. $\dfrac{x^2 - 1}{x - 2} \cdot \left(\dfrac{3x}{x - 1} - \dfrac{6}{x - 1} \right)$

53. $\dfrac{4y^2 + y - 6}{y^2 + 3y + 2} - \dfrac{3y}{y + 1} + \dfrac{5}{y + 2}$

54. $\dfrac{6y^2 + 17y - 40}{y^2 + y - 20} + \dfrac{3}{-4} - \dfrac{5y}{5}$

55. $\dfrac{3x^2 + 3x + 3}{x^3 - 1} - \dfrac{2}{x - 1}$

56. $\dfrac{4x^2 - 8x - 16}{x^3 + 8} - \dfrac{3}{x + 2}$

Simplify the following complex fractions.

57. $\dfrac{\dfrac{5}{8} - 2}{1 - \dfrac{5}{6}}$

58. $\dfrac{\dfrac{3}{4} - 3}{1 + \dfrac{5}{18}}$

59. $\dfrac{\dfrac{1}{x} + \dfrac{1}{y}}{\dfrac{1}{x + y}}$

60. $\dfrac{\dfrac{1}{3} + \dfrac{3}{x}}{\dfrac{3}{x} - \dfrac{1}{3}}$

61. $\dfrac{\dfrac{1}{x} + \dfrac{1}{y}}{\dfrac{1}{y} - \dfrac{1}{x}}$

62. $\dfrac{\dfrac{1}{3} + \dfrac{3}{x}}{\dfrac{1}{x + 3}}$

63. $\dfrac{x^{-1} + y^{-1}}{x^{-2} - y^{-2}}$

64. $\dfrac{x^{-1} - y^{-1}}{x^{-2} - y^{-2}}$

65. $\dfrac{\dfrac{1}{(x + h)^2} - \dfrac{1}{x^2}}{h}$

66. $\dfrac{\dfrac{1}{(x + h)^3} - \dfrac{1}{x^3}}{h}$

67. $\dfrac{\dfrac{1}{2x + h} - \dfrac{1}{2x}}{h}$

68. $\dfrac{\dfrac{5}{x + h} - \dfrac{5}{x}}{h}$

69. One train leaves SLO Town at a speed of 55 miles per hour heading toward BIG City. At the same time, another train leaves BIG City at a speed of 75 miles per hour heading toward SLO Town. If SLO Town and BIG City are 2000 miles apart, when will the trains pass each other?

70. To exercise, an athlete rides a bike the same amount of time she runs. If her riding rate is 25 miles per hour and her running rate is 6 miles per hour, how long does it take her to travel a total of 20 miles?

71. A bus leaves for a football game 100 miles away at 10 o'clock and will arrive at the stadium at noon. A student's auto travels 30 miles per hour faster than the bus. When should the student leave to arrive at the stadium at the same time as the bus?

72. A Model A leaves on a 180-mile tour at 8 o'clock in the morning. A trailer truck travels 15 miles per hour faster than the Model A. When should the truck leave to meet the Model A at the end of its $4\frac{1}{2}$-hour tour?

73. A small fishing boat travels upstream for 5 miles and then returns. If the total trip takes 6 hours and the current is 2 miles per hour, what is the rate of the boat in still water?

74. A canoeist paddles upstream 6 miles and then returns. If he can paddle at a rate of 5 miles per hour in still water and can make the total trip in 5 hours, what is the rate of the current?

1.4 Solving Inequalities with First-Degree Polynomials

Inequality Notation

The equality of two expressions has been our prime consideration up to this point. Now we will work with two numbers or expressions that are not equal. We will use the following notation with inequalities:

$a > b$ is read "a is greater than b,"
$a \geq b$ is read "a is greater than or equal to b,"
$a < b$ is read "a is less than b," and
$a \leq b$ is read "a is less than or equal to b."

Notice that $a \geq b$ is equivalent to $b \leq a$. We interpret the inequalities $\geq$, greater than or equal to, and $\leq$, less than or equal to, as in this example:

$$\{x : x \in Z \text{ and } x \geq 3\} = \{3, 4, 5, 6, 7, \cdots\}.$$

We note that this set includes both values of x equal to 3 and values of x greater than 3. Thus the inequalities $\geq$ and $\leq$ are interpreted as the union of two sets, in this case, $\{x : x \in Z \text{ and } x \geq 3\} = \{x : x = 3\} \cup \{x : x \in Z \text{ and } x > 3\}$.

Inequality Properties

As with equality, we also have properties for inequalities. We will use the following properties to solve linear inequalities. (We will use $>$ here, but these properties also apply to $\geq$, $<$, and $\leq$.)

Properties of Inequalities

For any real numbers a, b, and c:

Transitive Property of Inequality	If $a > b$ and $b > c$, then $a > c$.
Addition Property of Inequality	If $a > b$, then $a + c > b + c$, or If $a > b$, then $a - c > b - c$.
Multiplication Property of Inequality	With $c \neq 0$:

(a) If $c > 0$ and $a > b$,
　　　then $ac > bc$.
OR If $c > 0$ and $a > b$,
　　　then $\dfrac{a}{c} > \dfrac{b}{c}$.

(b) If $c < 0$ and $a > b$,
　　　then $ac < bc$.
OR If $c < 0$ and $a > b$,
　　　then $\dfrac{a}{c} < \dfrac{b}{c}$.

It is important to notice that **multiplying or dividing both sides of an inequality by a negative number, $c < 0$, will change the direction of the inequality**. We see this result when we multiply both sides of the inequalities $-1 < 1$ or $2 < 3$ by $c = -2$. For $c = -2 < 0$ we have

$$-1 < 1 \qquad \text{and} \qquad 2 < 3$$
$$(-2)(-1) > (-2)1 \qquad (-2)(2) > (-2)3$$
$$+2 > -2 \qquad -4 > -6$$

Thus multiplying both sides of an inequality changes the direction of the inequality. Using the properties of inequalities, we can easily define a positive expression.

Definition | *Positive Expression.* The expression $x - y$ is a positive number if and only if $x - y > 0$ and $x > y$.

We can use these properties to solve inequalities.

EXAMPLE 1 Solve $(x + 2)(x - 5) < (x - 1)^2$.

Solution Expand both sides and then solve for x using the properties of inequalities:

$$(x + 2)(x - 5) < (x - 1)^2$$
$$x^2 - 3x - 10 < x^2 - 2x + 1 \quad \text{\textit{(Subtract x^2 from both sides and add $2x$ to both sides of the inequality)}}$$
$$x^2 - x^2 - 3x + 2x - 10 < x^2 - x^2 - 2x + 2x + 1$$
$$-x - 10 < 1 \quad \text{\textit{(Add 10 to both sides)}}$$
$$-x - 10 + 10 < 1 + 10$$
$$-x < 11 \quad \text{\textit{(Multiply both sides by (-1), which}}$$
$$(-1)(-x) > (-1)(11) \quad \text{\textit{is less than zero and changes the}}$$
$$x > -11 \quad \text{\textit{direction of the inequality)}} \quad \blacksquare$$

We will also find it useful to combine two inequalities into one statement called a **compound inequality**. An example of this occurs when $a < x$ and $x < b$ are combined into the compound inequality $a < x < b$. The solution to the inequality $a < x < b$ consists of all values of x greater than a and at the same time less than b. By the transitive property of inequalities it is essential that $a < b$. Thus for $a < x < 2$, a must be a number smaller than x and x must be smaller than 2.

With the properties of inequalities, especially the transitive property, we can extend the properties of inequalities to compound inequalities.

Properties of Compound Inequalities

For algebraic expressions a, b, c, and d we have the following inequalities:

If $a < b < c$, then $a + d < b + d < c + d$.
If $a < b < c$, then $a - d < b - d < c - d$.
If $a < b < c$ and $d > 0$, then $ad < bd < cd$.
If $a < b < c$ and $d < 0$, then $ad > bd > cd$.

(Notice that the direction of both inequalities change when multiplying by a negative number d, $d < 0$.)

There are many ways in which we can use inequalities to solve practical problems. The next examples demonstrate these properties.

EXAMPLE 2   A candy store makes a mixture of chocolates. Some chocolates are priced at $4 per pound, and others are priced at $7 per pound. How much of the $4 per pound chocolate must be used in a 20-pound mixture if the mixture is to sell between $5.50 and $6.50 per pound?

Solution   We let x = the amount of $4 per pound chocolate, and then
$20 - x$ = the amount of $7 per pound chocolate.

	Cost	Amount =	Value
$4 chocolate	4	x	$4x$
$7 chocolate	7	$20 - x$	$7(20 - x)$
Total	4.50 to 5.50	20	$90 to $110

Then the value of the candy must be between $90 and $110 and can be written as

$$90 < 4x + 7(20 - x) < 110 \quad \textit{(Distributive property)}$$
$$90 < 4x + 140 - 7x < 110$$
$$90 < 140 - 3x < 110 \qquad \textit{(Addition property)}$$
$$90 - 140 < -3x < 110 - 140$$
$$-50 < -3x < -30 \qquad \textit{(Dividing by -3 changes the direction}$$
$$\frac{-50}{-3} > \frac{-3x}{-3} > \frac{-30}{-3} \qquad \textit{of the inequalities)}$$
$$\frac{50}{3} > x > 10$$
$$16\frac{2}{3} > x > 10$$

Thus the amount of $4 per pound chocolate in the 20-pound mixture must be between 10 pounds and $16\frac{2}{3}$ pounds.   ∎

Graphing Inequalities

An inequality can be represented graphically as a segment or ray on the real number line. We call this solution an interval.

Definition

An **interval** is a set of points on the real number line that can be represented by one of the inequalities of the form listed in these sets:

$$\{x : a < x < b\}, \quad \{x : a \le x \le b\}, \quad \{x : a < x \le b\}, \quad \{x : a \le x < b\},$$
$$\{x : x \ge a\}, \quad \{x : x \le a\}, \quad \{x : x < a\}, \quad \text{and} \quad \{x : x > a\}$$

We can represent these sets, inequalities, and intervals with a graph on the real number line and also with interval notation. On the real number line we use a closed dot, •, to include a point at a segment's end and an open dot, ∘, to exclude a point at a segment's end. With interval notation we use square brackets, [on the left end and] on the right end of the interval to include an end point, while we use parentheses, (on the left end and) on the right end of the interval to exclude an end point. Study Table 1.1 carefully to better understand the real number line graph and interval notation.

TABLE 1.1

Set Notation	The Graph	Interval Notation	Description
$\{x : 2 < x < 5\}$		$(2, 5)$	An open interval
$\{x : 2 \le x \le 5\}$		$[2, 5]$	A closed interval
$\{x : 2 < x \le 5\}$		$(2, 5]$	A half-open, half-closed interval
$\{x : 2 \le x < 5\}$		$[2, 5)$	A half-open, half-closed interval
$\{x : x \ge 1\}$		$[1, \infty)$	A closed ray (extending infinitely to the right)
$\{x : x \le 1\}$		$(-\infty, 1]$	A closed ray (extending infinitely to the left)
$\{x : x < 1\}$		$(-\infty, 1)$	An open ray (extending infinitely to the left)
$\{x : x > 1\}$		$(1, \infty)$	An open ray (extending infinitely to the right)

We read the symbol ∞ as "infinity." This symbol, ∞, does not stand for a number but for an idea. Infinity is the concept of continuing on forever, getting larger and larger without bound. Thus we use parentheses, (on the left end and) on the right end of the interval, with infinity in interval notation. The symbol ∞ *is not a number* but rather the idea of always getting larger without bound in a positive direction (to the right) on the real number line. Similarly, the symbol $-\infty$ *is not a number* but rather the idea of getting smaller without bound in a negative direction (to the left) on the real number line.

EXAMPLE 3   Solve each inequality, graph the solution set, and write the solution set in interval notation:

(a) $5x + 14 > 6 - 3x$ **(b)** $14 - 5x \geq -6$

Solution   Using the properties of inequalities and interval notation, we have

(a)
$$5x + 14 > 6 - 3x$$
$$5x + 14 - 14 > 6 - 3x - 14 \quad \text{(Addition property of inequality)}$$
$$5x > -3x - 8$$
$$5x + 3x > -3x - 8 + 3x \quad \text{(Addition property of inequality)}$$
$$8x > -8$$
$$\frac{8x}{8} > \frac{-8}{8} \quad \text{(Multiplication property of inequality)}$$
$$x > -1$$

Therefore the graph of the solution set is
and the solution is all x in the interval $(-1, \infty)$.

(b)
$$14 - 5x \geq -6$$
$$14 - 14 - 5x \geq -6 - 14 \quad \text{(Addition property of inequality)}$$
$$-5x \geq -20$$
$$\frac{-5x}{-5} \leq \frac{-20}{-5} \quad \text{(Multiplication property of inequalities. Notice that division by a negative number changes the direction of the inequality)}$$
$$x \leq 4$$

Therefore the graph of the solution set is
and the solution set is all x in the interval $(-\infty, 4]$.   ■

EXAMPLE 4   Solve each inequality, graph the solution set, and write the solution set in interval notation:

(a) $-13 < 2x - 5 < 21$ **(b)** $-1 \leq 2 - 3x \leq 5$

Solution   We now use the properties of compound inequalities.

(a) $-13 < 2x - 5 < 21$

$-13 + 5 < 2x - 5 + 5 < 21 + 5$ *(Addition property of inequality)*

$-8 < 2x < 26$

$\dfrac{-8}{2} < \dfrac{2x}{2} < \dfrac{26}{2}$ *(Multiplication property of inequality)*

$-4 < x < 13$

Therefore the graph of the solution set is

and x is in the interval $(-4, 13)$.

(b) $-1 \leq 2 - 3x \leq 5$

$-1 -2 \leq - 3x \leq -2 + 5$ *(Addition property of inequality)*

$-3 \leq -3x \leq 3$

$\dfrac{-3}{-3} \geq \dfrac{-3x}{-3} \geq \dfrac{3}{-3}$ *(Multiplication property of inequality)*

$1 \geq x \geq -1$ or $-1 \leq x \leq 1$

Therefore the graph of the solution set is

and x is in the interval $[-1, 1]$.   ■

Absolute Value

As we noticed before, the absolute value provides a way to keep an expression positive. In this section we will use the absolute value with equality and inequality.

Definition   The **absolute value** of x, denoted $|x|$, is defined as follows:

$$|x| = \begin{cases} x \text{ when } x \geq 0 \\ -x \text{ when } x < 0 \end{cases}$$

This simple definition guarantees that $|x| \geq 0$ (is always positive or zero), since

$$|x| = \begin{cases} x \text{ when } x \geq 0 \text{ means that } x \text{ is a positive number or zero} \\ -x \text{ when } x < 0 \text{ means that } -x \text{ is } -(\text{negative}) \text{ or a positive number} \end{cases}$$

Thus examples like $|3| = 3$, $|-3| = -(-3) = 3$, $|5| = 5$, $|-5| = -(-5) = 5$, and $|0| = 0$ are all positive numbers or zero. We will now use the definition of the absolute value to evaluate some examples.

EXAMPLE 5   Evaluate **(a)** $|\pi - \sqrt{10}|$ and   **(b)** $|\sqrt{10} - \pi|$

Solution The approximate values of $\sqrt{10}$ and π are $\sqrt{10} \approx 3.1623$ and $\pi \approx 3.1416$. Thus we have $\sqrt{10} > \pi$.
(a) Then $\pi - \sqrt{10} < 0$, and we have

$$\begin{aligned}
|\pi - \sqrt{10}| &= -(\pi - \sqrt{10}) \\
&= -\pi + \sqrt{10} \\
&= \sqrt{10} - \pi
\end{aligned}$$

(b) Similarly, $\sqrt{10} - \pi > 0$ and

$$|\sqrt{10} - \pi| = \sqrt{10} - \pi \qquad \blacksquare$$

We can use the absolute value with inequalities. The following theorem shows how to evaluate the absolute value in equations and inequalities. The proof of this theorem is a direct result of the absolute value definition.

Theorem | *Absolute Value Theorem.* For $d > 0$,

(a) $|x| = d$ if and only if $x = \pm d$,

(b) $|x| < d$ if and only if $-d < x < d$, and

(c) $|x| > d$ if and only if $x < -d$ or $x > d$.

It is not difficult to prove this theorem by using the definition of the absolute value, but we will not do it here. We will now use this new theorem to solve some examples.

EXAMPLE 6   Solve the following equalities:
(a) $|x| = 7$ **(b)** $|x + 5| = -3$ **(c)** $|x - a| = 5$

Solution Use part (a) of the absolute value theorem to solve these equations.
(a) If $|x| = 7$, then $x = \pm 7$. Thus $x \in \{7, -7\}$.
(b) For $|x + 5| = -3$ we note that $-3 < 0$. Thus there is no value for x that will make $|x + 5| < 0$. Therefore $x \in \varnothing$.
(c) If $|x - a| = 5$, then $x - a = \pm 5$.
Thus we have

$$\begin{aligned}
x - a &= 5 & \text{or} && x - a &= -5 \\
x &= a + 5 & && x &= a - 5
\end{aligned}$$

Therefore $x \in \{a + 5, a - 5\}$. $\qquad \blacksquare$

We can write the results of the absolute value theorem in the following three ways: with a graph on the real number line, with set notation, and with interval notation. When $d = 5$, in the absolute value theorem, we have the results shown in Table 1.2.

TABLE 1.2

The Statement	The Graph of the Solution Set	The Solution in Set Notation	The Solution in Interval Notation
$\lvert x \rvert = 5$		$\{-5, 5\}$	
$\lvert x \rvert < 5$		$\{x : -5 < x < 5\}$	$(-5, 5)$
$\lvert x \rvert > 5$		$\{x : x < -5 \text{ or } 5 < x\}$	$(-\infty, -5) \cup (5, \infty)$

We will now use this notation in some examples.

EXAMPLE 7   Solve the following inequalities, write the solution in interval notation, and graph the solution set:

(a) $\lvert x - 3 \rvert < 5$ **(b)** $\lvert 2x - 4 \rvert \geq 2$

Solution   We use the absolute value theorem to solve these inequalities.
(a) For $\lvert x - 3 \rvert < 5$ we use part (b) of the absolute value theorem. Then

$$-5 < x - 3 < 5$$
$$-5 + 3 < x - 3 + 3 < 5 + 3$$
$$-2 < x < 8$$

Thus x is in the interval $(-2, 8)$ and is graphed as
(b) For $\lvert 2x - 4 \rvert \geq 2$ we use part (c) of the absolute value theorem. Then

$$
\begin{array}{lcl}
2x - 4 \leq -2 & \text{or} & 2x - 4 \geq 2 \\
2x \leq -2 + 4 & & 2x \geq 2 + 4 \\
2x \leq 2 & & 2x \geq 6 \\
x \leq 1 & & x \geq 3
\end{array}
$$

Thus x is in the interval $(-\infty, 1] \cup [3, \infty)$ and is graphed as ■

EXAMPLE 8   Solve the following inequalities and write the solution in interval notation:
(a) $\lvert x + 1 \rvert \leq -4$ **(b)** $\lvert 3x - 1 \rvert \leq \varepsilon$ for $\varepsilon > 0$

Solution   We use the absolute value theorem to solve these inequalities.
(a) For $\lvert x + 1 \rvert \leq -4$ we notice that $-4 < 0$. Then $x \, \varepsilon \, \emptyset$, since the absolute value must be greater than zero.

(b) For $|3x - 1| \leq \varepsilon$ with $\varepsilon > 0$, we use part (b) of the absolute value theorem. Then we have

$$-\varepsilon \leq 3x - 1 \leq \varepsilon$$

$$1 - \varepsilon \leq 3x \leq 1 + \varepsilon$$

$$\frac{1 - \varepsilon}{3} \leq x \leq \frac{1 + \varepsilon}{3}$$

$$\frac{1}{3} - \frac{\varepsilon}{3} \leq x \leq \frac{1}{3} + \frac{\varepsilon}{3}$$

and x is in interval

$$\left(\frac{1}{3} - \frac{\varepsilon}{3}, \frac{1}{3} + \frac{\varepsilon}{3} \right)$$ ∎

We can also use the absolute value to describe the distance between two points on the real number line. The distance between the point with coordinate a and some point with coordinate x is either the positive real number $x - a$, when $x - a \geq 0$, or $a - x$, when $x - a > 0$. Thus we will define this distance as follows.

Definition | *Distance on the Real Number Line.* The distance c between the two points with coordinates x and a can be written as $|x - a| = c$.

This distance description using the absolute value will be useful in calculus. The following example shows how to use this distance description of absolute value.

EXAMPLE 9   Solve the following inequalities using the distance description of the absolute value:
(a) $|x - 3| < 5$ **(b)** $|x - 4| \geq 2$

Solution
(a) $|x - 3| < 5$ represents all of the points with coordinate x less than five units from the point 3. The solution can be drawn on the real number line as shown:

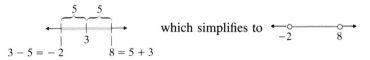

or x is in the interval $(-2, 8)$.

(b) $|x - 4| \geq 2$ represents all of the points two or more units from 4. The solution can be drawn on the real number line as shown:

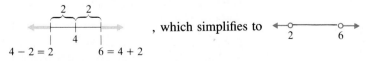

and in interval notation, x is in $(-\infty, 2] \cup [6, \infty)$.   ∎

In calculus we will use the description of distance, $|x - a| = c$, to describe the intervals where a is the center of the interval and c is the radius of the interval.

EXAMPLE 10   Write the following intervals using the absolute value:
(a) $(-7, 2)$ (b) $[4 - \varepsilon, 4 + \varepsilon]$, where $\varepsilon > 0$.

Solution
(a) For $(-7, 2)$ the center of the interval is $a = \frac{(-7 + 2)}{2} = -\frac{5}{2}$, and the radius of the interval is $d = |2 - \frac{5}{2}| = |2 - 2\frac{1}{2}| = |-\frac{1}{2}| = \frac{1}{2}$. Therefore the interval $(-7, 2)$ is equivalent to

$$|x - (-\tfrac{5}{2})| < \frac{1}{2} \qquad \text{or} \qquad |x + \tfrac{5}{2}| < \frac{1}{2}$$

There is no equality, since the interval is open and does not include the endpoints.
(b) For $[4 - \varepsilon, 4 + \varepsilon]$, where $\varepsilon > 0$, the center of the interval is

$$a = \frac{(4 - \varepsilon) + (4 + \varepsilon)}{2} = \frac{8}{2} = 4$$

and the radius is $d = |(4 + \varepsilon) - 4| = |\varepsilon| = \varepsilon$. Therefore $[4 - \varepsilon, 4 + \varepsilon]$ is equivalent to $|x - 4| \leq \varepsilon$. The equality is included, since the interval is closed.

■

Sometimes we can solve rational expression inequalities by simply considering when they are positive and when they are negative.

EXAMPLE 11   Solve these inequalities:
(a) $\dfrac{3}{x + 2} > 0$ (b) $\dfrac{2}{1 - x} > 0$

Solution   Since these inequalities are either greater than zero or less than zero, we will find out when the quotient is positive or negative.
(a) $\dfrac{3}{x + 2} > 0$ means that $\dfrac{3}{x + 2}$ is a positive number. A quotient that is positive must be either

$$a \; \frac{\text{positive number}}{\text{positive number}} \qquad \text{or} \qquad a \; \frac{\text{negative number}}{\text{negative number}}$$

Since 3 is positive, $x + 2$ must be positive also. Thus we conclude that $x + 2 > 0$, $x > -2$, or x is in the interval $(-2, \infty)$.
(b) $\dfrac{2}{1 - x} < 0$ means that $\dfrac{2}{1 - x}$ is negative. Since 2 is positive, $1 - x$ must be negative to make the quotient negative. Therefore

$$1 - x < 0$$
$$1 - x + x < 0 + x$$
$$1 < x, \qquad \text{or} \qquad x \text{ is in the interval } (1, \infty)$$

■

Exercises 1.4

Simplify the following.

1. $|\sqrt{3} - 2|$

2. $|2 - \sqrt{3}|$

3. $|\pi - \sqrt{11}|$

4. $|\sqrt{11} - \pi|$

5. Find $|3 - x|$ for $x > 3$.

6. Find $|x - 5|$ for $x < 5$.

Solve the following absolute value equalities.

7. $|x| = 5$

8. $|x| = 1$

9. $|x + 3| = 7$

10. $|x + 5| = 1$

11. $|2x - 3| = 6$

12. $|2x - 5| = 2$

13. $|x - 3| = -2$

14. $|x - 5| = -6$

15. $|2x + 1| = |3x - 2|$

16. $|x - 5| = |5x - 1|$

17. For (a) $x = 5$, $y = -3$ and (b) $x = -5$, $y = -4$, show that

    (a) $|xy| = |x||y|$,

    (b) $\left|\dfrac{x}{y}\right| = \dfrac{|x|}{|y|}$, and

    (c) $|x - y| = |y - x|$.

18. For (a) $x = -5$, $y = 3$ and (b) $x = 5$, $y = 3$ show that

    (a) $|xy| = |x||y|$,

    (b) $\left|\dfrac{x}{y}\right| = \dfrac{|x|}{|y|}$, and

    (c) $|x - y| = |y - x|$.

Solve the following inequalities. Graph the solution and also write each solution in interval notation.

19. $6x - 14 > 5x + 4$

20. $7x - 10 > 6x - 9$

21. $(y - 2)(y - 3) > (y - 1)^2$

22. $(y - 5)(y + 2) < (y + 2)^2$

23. $7 \leq 3x + 5 \leq 20$

24. $10 \leq 4x + 2 \leq 26$

25. $-3 < 2 - 5t < 17$

26. $-5 < 1 - 2t < 15$

27. $-5 < 1 - 3x < 19$

28. $-4 < 1 - 5x < 21$

29. $|x| \leq 4$

30. $|x| \leq 7$

31. $|y| \geq 17$

32. $|y| \leq -2$

33. $|x| \leq -1$

34. $|x| \geq 1$

35. $|x| \geq -2$

36. $|x| \geq -4$

37. $|t - 2| \leq 12$

38. $|t - 3| \leq 15$

39. $|3x + 1| \leq 2$

40. $|2x + 1| \leq 4$

41. $|2x| > 5$

42. $|3x| > 2$

43. $|x - 1| > 3$

44. $|x - 1| > 4$

45. $|y + 1| > 3$

46. $|y + 1| > 4$

47. $|5x - 1| < \dfrac{\varepsilon}{2}$

48. $|7x + 1| < \dfrac{\varepsilon}{3}$

49. $\dfrac{5}{x + 1} > 0$

50. $\dfrac{1}{x + 2} < 0$

51. $\dfrac{1}{x + 3} < 0$

52. $\dfrac{-5}{x + 1} < 0$

53. $\dfrac{-3}{y - 5} < 0$

54. $\dfrac{-4}{y - 2} < 0$

55. $\dfrac{-5}{x + 1} > 0$

56. $\dfrac{-4}{x + 2} > 0$

57. $\dfrac{7}{x + 7} > 0$

58. $\dfrac{5}{x + 5} > 0$

Write the following intervals in absolute value notation.

59. $[1, 9]$

60. $[2, 6]$

61. $(-2, 4)$

62. $(1, 5)$

63. $[-5, 2]$

64. $[-4, 1]$

65. $(-2 - \varepsilon, -2 + \varepsilon)$

66. $(3 - \varepsilon, 3 + \varepsilon)$

67. $(5 - \delta, 5 + \delta)$

68. $(-1 - \delta, -1 + \delta)$

69. A retired couple wishes to invest \$12,000 in a no-risk account yielding 8% interest and in a higher-risk stock yielding 10% interest. How much must they invest in the 8% account to receive a combined annual interest not lower than \$1000 per year?

70. A couple wants to invest a \$15,000 inheritance in a 9% savings account and in a higher-risk 12% loan. How much should they invest in the 9% account so that the combined annual interest is never less than \$1500?

71. A candy store has a special mix of two candies selling separately for \$1 and \$5 per pound. In making 50 pounds of this mixture, how much \$5 candy must be used so that the value of the mixture is to be at least \$3.40 per pound?

72. A nut store has a special mix of two types of nuts selling separately for \$2 and \$6 a pound. In making 30 pounds of this mixture, how much of the \$6 nuts must be used if the value of the mixture is to be at least \$4 per pound?

73. A nut store has a special mix of two types of nuts selling separately for \$2 and \$6 per pound. In making 20 pounds of this mixture, how many pounds of the \$6 nuts are needed if the value of the mixture is to be between \$4 and \$5 per pound?

74. A candy store has a special mix of two candies selling separately for \$1 and \$5 per pound. In making 30 pounds of this mixture, how many pounds of the \$5 candy are needed if the value of the mixture is to be between \$3.50 and \$4.50 per pound?

75. What is the area of a circle if the radius is between 4.7 and 5.1 meters?

76. What is the volume of a spherical balloon if the radius varies between 15.5 and 16.5 feet?

77. What is the volume of a spherical balloon if the radius is between 2.8 and 3.2 meters?

78. What is the surface area of a cylindrical storage tank of height h with a base radius between 3.9 and 4.1 inches (Fig. 1.6)?

79. What is the surface area of a cylindrical storage tank of radius h with a base radius between 2.4 and 2.6 inches (Fig. 1.6)?

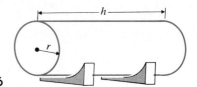

Figure 1.6

1.5 Solving Equalities and Inequalities with Polynomials and Rational Expressions

In this section we will solve inequalities with polynomials and rational expressions. We will first review solving second-degree polynomial equations. Remember the zero factor theorem, which states that for real numbers a and b

$$a \cdot b = 0 \quad \text{if and only if} \quad a = 0 \quad \text{or} \quad b = 0$$

We will use this property to solve equalities that factor.

EXAMPLE 1   Solve these equalities: **(a)** $x^2 - x - 6 = 0$ **(b)** $y^3 + y^2 - 4y - 4 = 0$

Solution   Factor first and then set each factor equal to zero.

(a) For $x^2 - x - 6 = 0$ we have

$$(x + 2)(x - 3) = 0$$
$$(x + 2) = 0 \quad \text{or} \quad (x - 3) = 0$$
$$x = -2 \quad \text{or} \quad x = 3$$

Therefore $x \in \{-2, 3\}$.

(b) Factor by grouping the terms two at a time:

$$y^3 + y^2 - 4y - 4 = 0$$
$$(y^3 + y^2) - (4y + 4) = 0 \quad \textit{(Factor the common factors from each of these two groups)}$$
$$y^2(y + 1) - 4(y + 1) = 0 \quad \textit{(Factor the common factor y + 1)}$$
$$(y + 1)(y^2 - 4) = 0 \quad \textit{(Factor the difference of two squares)}$$
$$(y + 1)(y - 2)(y + 2) = 0$$

Then $y + 1 = 0$, $y - 2 = 0$, or $y + 2 = 0$, and

$$y = -1, \quad y = 2 \quad \text{or} \quad y = -2$$

Therefore $y \in \{-2, -1, 2\}$. ∎

The square root method is another way of solving a second-degree equation. This method is most helpful when one side of the equation is a perfect square. When this occurs, we take the square root of both sides of the equation. It is important to recall from Section 1.2 that $\sqrt{x^2} = |x|$. Thus when $\sqrt{x^2} = |x| = d$, from Section 1.4 we know that $x = \pm d$. Therefore when $(x + 2)^2 = 3$, we take the square root of both sides (the square root method), and we have

$$(x + 2)^2 = 3$$
$$\sqrt{(x + 2)^2} = \sqrt{3} \qquad \textit{(Take the square root of both sides)}$$
$$|x + 2| = \sqrt{3} \qquad \textit{(Remember that $\sqrt{x^2} = |x|$)}$$
$$x + 2 = \pm\sqrt{3}$$
$$x = -2 \pm \sqrt{3} \qquad \textit{(Subtract 2 from both sides)}$$

Thus for $(x + 2)^2 = 3$ we have $x \in \{-2 - \sqrt{3}, -2 + \sqrt{3}\}$.

This process is the square root method defined below.

THE SQUARE ROOT METHOD	If $d > 0$ and $x^2 = d$, then $$\sqrt{x^2} = |x| = \sqrt{d}$$ and $$x = \pm\sqrt{d}$$ (Take the square root of both sides with $a \pm$ on one side of the equation.)

PROOF: If $d > 0$ and $x^2 = d$, then by the definition of the square root we have

$$\sqrt{x^2} = |x| = \sqrt{d}$$

Next, using the definition of the absolute value, we have

$$x = \pm\sqrt{d} \qquad \square$$

We will now use the new method to solve some equations.

EXAMPLE 2   Solve $(x - 1)^2 = 5$.

Solution   Using the square root method, we have

$$(x - 1)^2 = 5 \qquad \textit{(Take the square root of both sides)}$$
$$\sqrt{(x - 1)^2} = \sqrt{5} \qquad \textit{(Remember that $\sqrt{x^2} = |x|$)}$$
$$|x - 1| = \sqrt{5} \qquad \textit{(Remember that if $|x| = d > 0$, then $x = \pm d$)}$$
$$x - 1 = \pm\sqrt{5} \qquad \textit{(Add 1 to both sides)}$$
$$x = 1 \pm \sqrt{5}$$

Therefore for $(x - 1)^2 = 5$, $x \in \{1 - \sqrt{5}, 1 + \sqrt{5}\}$.   ■

Completing the Square

We can use completing the square and the square root method to solve any type of second-degree equation.

COMPLETING THE SQUARE	To complete the square for a second degree equation in one variable, use the following steps: STEP 1: The coefficient of the squared term must be 1. If it is not 1, divide both sides of the equation by the coefficient of the squared term. STEP 2: Add or subtract the constant to both sides of the equation so that only the squared and first-degree terms are on one side together. STEP 3: Take half of the coefficient (including sign) of the first-degree term, square it, and add it to both sides of the equation. STEP 4: Use the square root method to solve the resulting equation.

We will now complete the square to solve the equations in the following example.

EXAMPLE 3

Solve each of the following by completing the square:
(a) $x^2 - 6x + 4 = 0$ **(b)** $2x^2 - 3x - 8 = 0$

Solution   Using the steps to complete the square, we have:
(a) STEP 1: The coefficient of x^2 is already 1.

STEP 2: Subtract 4 from both sides of the equation.

$$x^2 - 6x + 4 = 0$$
$$x^2 - 6x \quad = -4$$

STEPS 3 AND 4: Take half of (-6), square $\left(\frac{1}{2}\right)(-6)$, and add it to both sides of the equation. Then use the square root method.

$$x^2 - 6x + \mathbf{9} = -4 + \mathbf{9} \quad \textit{(Complete the square)}$$

$$(x - 3)^2 = 5$$
$$\sqrt{(x - 3)^2} = \sqrt{5} \quad \textit{(The square root method)}$$
$$|x - 3| = \sqrt{5}$$
$$x - 3 = \pm\sqrt{5}$$
$$x = 3 \pm \sqrt{5}$$

(b) For $2x^2 - 3x - 8 = 0$ the coefficient of x^2 is 2.

$$\frac{2x^2 - 3x - 8}{2} = \frac{0}{2} \qquad \textit{(Divide both sides by 2)}$$

$$\frac{2x^2}{2} - \frac{3x}{2} - \frac{8}{2} = \frac{0}{2}$$

$$x^2 - \frac{3}{2}x - 4 = 0$$

$$x^2 - \frac{3}{2}x = 4 \qquad \textit{(Add 4 to both sides)}$$

$$x^2 - \frac{3}{2}x + \frac{9}{16} = 4 + \frac{9}{16} \qquad \begin{array}{l}\textit{(Take half of } -\frac{3}{2}\textit{, the coefficient of } x.\\ \textit{Square } \left(\frac{1}{2}\right)\left(-\frac{3}{2}\right) \textit{ and add it to both}\\ \textit{sides of the equation)}\end{array}$$

$$\cdot \tfrac{1}{2} \qquad \textit{square} \qquad \textit{(Completing the square)}$$

$$\left(x - \frac{3}{4}\right)^2 = \frac{64}{16} + \frac{9}{16}$$

$$\left(x - \frac{3}{4}\right)^2 = \frac{73}{16}$$

$$\sqrt{\left(x - \frac{3}{4}\right)^2} = \sqrt{\frac{73}{16}}$$

$$\left|x - \frac{3}{4}\right| = \frac{\sqrt{73}}{\sqrt{16}}$$

$$x - \frac{3}{4} = \pm\frac{\sqrt{73}}{4}$$

$$x = \frac{3}{4} \pm \frac{\sqrt{73}}{4}$$

$$x = \frac{3 \pm \sqrt{73}}{4}$$

Thus for $2x^2 - 3x - 8 = 0$,

$$x \in \left\{\frac{3 - \sqrt{73}}{4}, \frac{3 + \sqrt{73}}{4}\right\}$$

The Quadratic Formula

A quadratic equation is a second-degree equation in one variable. Quadratic equations can always be solved by completing the square, but we will now develop a formula for the solution of any quadratic equation. The quadratic formula is a useful solution to any quadratic equations.

> THE QUADRATIC
> FORMULA
>
> If a, b, and c are real numbers with $a \neq 0$ and $ax^2 + bx + c = 0$, then
>
> $$x = \frac{-b \pm \sqrt{b^2 - 4ac}}{2a}$$

To prove this theorem, we start with the general quadratic equation, $ax^2 + bx + c = 0$, and complete the square. This proof will be a problem in the exercise set. We will now use the quadratic formula to solve the following problems.

EXAMPLE 4

Solve $x^2 = 2x + 4$.

Solution For $x^2 = 2x + 4$ we first change this equation to standard form, $x^2 - 2x - 4 = 0$ with $a = 1$, $b = -2$, $c = -4$. Then using the quadratic formula,

$$x = \frac{-(-2) \pm \sqrt{(-2)^2 - 4(1)(-4)}}{(2)(1)} = \frac{2 \pm \sqrt{4 + 16}}{2}$$

$$x = \frac{2 \pm \sqrt{20}}{2} = \frac{2 \pm \sqrt{4}\sqrt{5}}{2}$$

$$= \frac{2 \pm 2\sqrt{5}}{2}$$

$$x = \frac{2}{2} \pm \frac{2\sqrt{5}}{2} = 1 \pm \sqrt{5}$$

EXAMPLE 5

Solve **(a)** $x^2 + 2x + 4 = 0$ and   **(b)** $(x - 3)(x + 2) = 4$.

Solution Apply the quadratic formula:
(a) For $x^2 + 2x + 4 = 0$ we have $a = 1$, $b = 2$, $c = 4$, and

$$x = \frac{-2 \pm \sqrt{2^2 - 4(1)4}}{(2)(1)} = \frac{-2 \pm \sqrt{4 - 16}}{2}$$

$$x = \frac{-2 \pm \sqrt{-12}}{2}$$

Since $\sqrt{-12} \notin R$, there is no real number solution for x. We will solve this equation in the next section after we have discussed complex numbers.
(b) For $(x - 3)(x + 2) = 4$,

$$x^2 - x - 6 = 4$$

$$x^2 - x - 10 = 0$$

so $a = 1$, $b = -1$, and $c = -10$. Thus

$$x = \frac{-(-1) \pm \sqrt{(-1)^2 - 4(1)(-10)}}{2(1)}$$

$$x = \frac{1 \pm \sqrt{1 + 40}}{2} = \frac{1 \pm \sqrt{41}}{2}$$ ∎

It is interesting to note that in Example 5(a) the answer is not a real number, since the radicand is negative. The radicand is the value of $b^2 - 4ac$. This expression $b^2 - 4ac$ is called the **discriminant** of the quadratic expression. When the discriminant is negative, there is no real number solution for the variable.

Quadratic or second-degree polynomials with negative discriminants will always be either positive for all values of the variable or negative for all values of the variable on the basis of the sign of the leading coefficient (the coefficient of the squared term). As an example we consider $x^2 + 2x + 4$, which has a negative discriminant. Any value we substitute for x produces a value of $x^2 + 2x + 4 > 0$, since the leading coefficient is positive. This conclusion is demonstrated for a few arbitrary values of x in Table 1.3.

TABLE 1.3

x	$x^2 +$	$2x + 4$
1	$1^2 +$	$2(1) + 4 = 1 + 2 + 4 = 7$
-3	$(-3)^2 +$	$2(-3) + 4 = 9 - 6 + 4 = 7$
-1	$(-1)^2 +$	$2(-1) + 4 = 1 - 2 + 4 = 3$
$-\dfrac{1}{2}$	$\left(-\dfrac{1}{2}\right)^2 +$	$2\left(-\dfrac{1}{2}\right) + 4 = \dfrac{1}{4} - 1 + 4 = 3\dfrac{1}{4}$

Whenever a quadratic factor has a negative discriminant ($b^2 - 4ac < 0$), the quadratic expression is either always positive or always negative for all real number values of x. When the leading coefficient is positive, the quadratic factor is positive; when the leading coefficient is negative, the quadratic factor is negative.

Inequalities

Solving second- and higher-degree polynomial inequalities is not a difficult task when we work with polynomials that are positive or negative. To solve polynomial inequalities, we will use the new technique of the sign chart. This sign chart uses the positive and negative intervals of the polynomial's factors to find the positive and negative intervals of the polynomial.

SOLVING
POLYNOMIAL
INEQUALITIES

Complete the following steps to solve second- and higher-degree polynomial inequalities:

STEP 1: Rearrange the terms with zero on one side of the inequality.

STEP 2: Factor the polynomial.

STEP 3: Make a sign chart as follows:

(a) construct a real number line;

(b) under this real number line, use one line for each factor of the polynomial;

(c) on the line of each factor, mark where the factor is zero, with a 0;

(d) then indicate with + signs the side of the zero where the factor is positive and indicate with − signs the sides of the zero where the factor is negative (do this for each factor);

(e) the product of the signs of the factors in each interval provides the sign of the polynomial in that interval.

We will now solve $x^2 > x + 6$ using these steps.

STEP 1: Rearrange terms so that zero is on one side of the inequality:

$$x^2 > x + 6$$
$$x^2 - x - 6 > x + 6 - x - 6$$
$$x^2 - x - 6 > 0$$

STEP 2: Factor: $(x - 3)(x + 2) > 0$.

STEP 3: (a) Construct a real number line.

(b) The factor $x - 3$ is zero when $x = 3$, positive for $x > 3$, and negative for $x < 3$.

(c) For $x - 3$, on the line under the real number line, place a 0 under 3, + signs to the right, and − signs to the left. Thus the sign chart is now

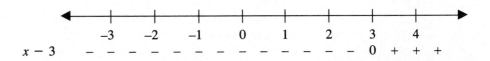

Now repeat Steps 3(b) and 3(c) for the factor $x + 2$:

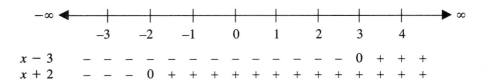

(We add $-\infty$ to the left end of the real number line and ∞ to the right end of the real number line to help identify the intervals in the results.)

(d) We now consider the signs for the factors in the column under the interval $(-\infty, -2)$. These signs are $-$ and $-$. Thus in the interval $(-\infty, -2)$, the product of the factors $x - 3$ and $x + 2$ is $(-)(-)$, which is a positive $(+)$ number. Next, in the interval $(-2, 3)$ the product of the factors $x - 3$ and $x + 2$ is $(-)(+)$, which is a negative $(-)$ number. Likewise, in the interval $(3, \infty)$ the product of the factors $x - 3$ and $x + 2$ is $(+)(+)$, which is a positive $(+)$ number. These results are displayed in the following sign chart:

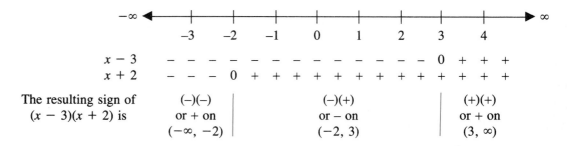

Thus the solution to $x^2 > x + 6$ is any x in the interval $(-\infty, -2) \cup (3, \infty)$. Note that if the problem were different and $(x - 3)(x + 2) \geq 0$, then the solution would include the points -2 and 3. Thus if $(x - 3)(x + 2) \geq 0$, the solution would be any x in the interval $(-\infty, -2] \cup [3, \infty)$.

The sign chart provides a quick and easy way to visualize all combinations of positive and negative factors. This will be even more helpful when we have more than two factors. We will use the technique of the sign chart in future work.

EXAMPLE 6   Use the steps above to solve the following inequalities:

(a) $x^2 - 6x + 8 > 0$ (b) $\dfrac{x^2 - 1}{x + 3} > 0$

Solution   We will factor and then use the sign chart.
(a) Thus $x^2 - 6x + 8 > 0$ is $(x - 2)(x - 4) > 0$. Now constructing the sign chart, we have

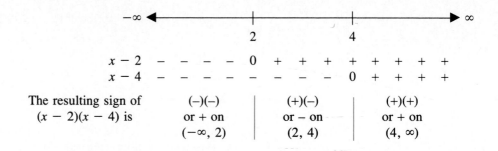

The resulting sign of
$(x - 2)(x - 4)$ is

| | $(-)(-)$ or $+$ on $(-\infty, 2)$ | $(+)(-)$ or $-$ on $(2, 4)$ | $(+)(+)$ or $+$ on $(4, \infty)$ |

The solution consists of the intervals where $(x - 2)(x - 4)$ is positive, since $x^2 - 6x + 8 > 0$. There is no equality involved, so we have $x^2 - 6x + 8 > 0$ for all x in the interval $(-\infty, 2) \cup (4, \infty)$.

(b) Solve

$$\frac{x^2 - 1}{x + 3} > 0$$

Even though this expression is a quotient, we can solve by factoring and using the sign chart.

$$\frac{x^2 - 1}{x + 3} = \frac{(x - 1)(x + 1)}{(x + 3)} > 0$$

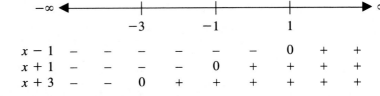

The resulting sign of
$\dfrac{x^2 - 1}{x + 3}$ is

| | $-$ on $(-\infty, -3)$ | $+$ on $(-3, -1)$ | $-$ on $(-1, 1)$ | $+$ on $(1, \infty)$ |

Therefore $\dfrac{x^2 - 1}{x + 3} > 0$ for x in the interval $(-3, -1) \cup (1, \infty)$.

EXAMPLE 7   Use the steps above to solve the following inequalities:
(a) $x^3 - 1 \geq 0$ **(b)** $(x - 1)(2 - x) \leq 0$

Solution   We will factor and then use the sign chart.
(a) Factoring $x^3 - 1 \geq 0$ yields $(x - 1)(x^2 + x + 1) \geq 0$. Before using the sign chart we notice that $x^2 + x + 1$ has a negative discriminant ($b^2 - 4ac < 0$) and is always positive, since its leading coefficient is positive.

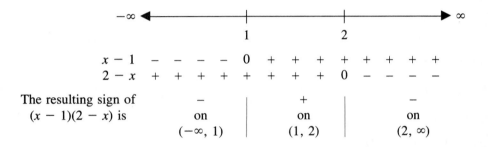

Thus $x^3 - 1 \geq 0$ for x in the interval $[1, \infty)$.

(b) When we make the sign chart for $(x - 1)(2 - x) \leq 0$, we must be careful with the signs on $2 - x$. Because of the $-x$, we must reverse the signs for $2 - x$ in the sign chart.

The solutions are all values of x in the intervals where $(x - 1)(2 - x)$ is negative or zero. The values 1 and 2 are included, since the product is zero at 1 and 2. Therefore $(x - 1)(2 - x) \leq 0$ for x in the interval $(-\infty, 1] \cup [2, \infty)$.   ■

When solving equalities containing rational expressions, we usually multiply both sides by the LCD. With an inequality the LCD may be positive or negative; we will multiply both sides by the square of the LCD (or more generally a common denominator with all factors squared) as in the next example.

EXAMPLE 8   Solve

$$\frac{1}{x - 1} \geq \frac{1}{x}$$

Solution   We can solve both of these inequalities by adding the same term to both sides of the inequality, forcing one side to be zero. Then we add or subtract using the LCD, factor, and use the sign chart. Although this yields the correct answer, it is often quicker and less confusing to multiply both sides of the *square* of the LCD. This second method ensures that both sides are multiplied by the same positive number, since an expression to the second power is always greater than zero. Then we have

$$\frac{1}{x-1} \geq \frac{1}{x}$$ *(Multiply both sides by the positive number $x^2(x-1)^2$)*

$$x^2(x-1)^2\frac{1}{(x-1)} \geq \frac{1}{x}x^2(x-1)^2$$ *(Reduce)*

$$x^2(x-1) \geq x(x-1)^2$$ *(Rearrange)*

$$x^2(x-1) - x(x-1)^2 \geq 0$$ *(Factoring the common factor $x(x-1)$)*

$$x(x-1)[x-(x-1)] \geq 0$$ *(Simplify)*

$$x(x-1)(x-x+1) \geq 0$$

$$x(x-1) \cdot 1 \geq 0$$

$$x(x-1) \geq 0$$

Using the sign chart, we have

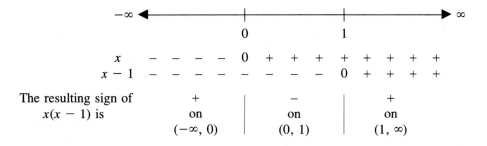

	$(-\infty, 0)$	$(0, 1)$	$(1, \infty)$
The resulting sign of $x(x-1)$ is	$+$ on	$-$ on	$+$ on

Since x cannot equal 0 or 1 (both numbers make the original denominator zero), the solution is all x in the interval $(-\infty, 0) \cup (1, \infty)$. ■

EXAMPLE 9   Solve

$$\frac{16}{|x+4|} \geq 8$$

Solution   Since $|x+4|$ is positive, we multiply both sides by $|x+4|$:

$$|x+4|\frac{16}{|x+4|} \geq 8|x+4|$$

$$16 \geq 8|x+4|$$

$$\frac{16}{8} \geq \frac{8}{8}|x+4|$$

$$2 \geq |x+4|$$

Using the absolute value theorem, we have

$$-2 \le x + 4 \le 2$$
$$-2 - 4 \le x \le 2 - 4$$
$$-6 \le x \le -2$$

Since $x = -4$ makes the denominator zero, it must not be included in the solution. Thus the solution is $\{x: -6 \le x \le -2, x \neq -4\}$, or all x in the interval $[-6, -4)$ $\cup$ $(-4, -2]$, which can be graphed as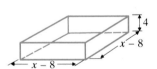

EXAMPLE 10

A square sheet of cardboard is folded into a candy box with an open top by cutting 4-inch by 4-inch squares out of each corner (Fig. 1.7). Find the size of the square sheet if the volume of the box is not less than 16 cubic inches.

Solution   Let x = the length of the side of the square sheet of cardboard. Then for $h = 4$ the volume is

$$V = lwh$$
$$= (x - 8)(x - 8)4$$
$$= 4(x^2 - 16x + 64)$$
$$= 4x^2 - 64x + 256$$

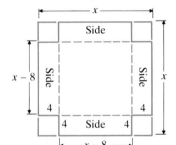

If the volume is greater than 16 cubic inches,

$$V = 4x^2 - 64x + 256 \ge 16$$
$$4x^2 - 64x + 240 \ge 0$$
$$4(x^2 - 16x + 60) \ge 0$$
$$4(x - 10)(x - 6) \ge 0$$
$$(x - 10)(x - 6) \ge 0$$

Figure 1.7   Now find the values for x using the sign chart.

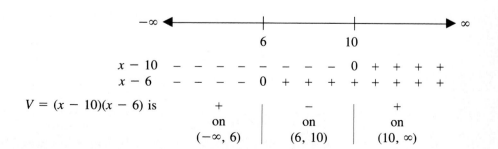

Thus when the volume is not less than 16 cubic inches, x is in the interval $(-\infty, 6] \cup [10, \infty)$. Since x must be positive and greater than 8 (the length of the cutouts is $2h = 2 \cdot 4 = 8$), the square sheet must have a side greater than or equal to 10 inches ($x \geq 10$). ∎

Exercises 1.5

Solve the following equalities.

1. $x^2 + 4x - 5 = 0$

2. $x^2 - 5x + 6 = 0$

3. $x^2 + 5x - 14 = 0$

4. $x^2 + x - 20 = 0$

5. $y^2 + y = 7$

6. $y^2 = 9y + 14$

7. $x^2 = 9x - 20$

8. $x^2 = 6x - 5$

9. $(y + 1)(y^2 - 4) = 0$

10. $(y - 2)(y^2 - 1) = 0$

11. $(x + 1)(x - 2)^2 = 0$

12. $(x - 2)(x - 4)^2 = 0$

13. $r^3 + 3r^2 - r - 3 = 0$

14. $r^3 + 5r^2 - 4r - 20 = 0$

15. $x^3 - x^2 - 5x + 5 = 0$

16. $x^3 - 3x^2 - 3x + 9 = 0$

Solve the following nonlinear inequalities. Use the sign chart for Exercises 17–38. Write all answers with interval notation.

17. $x^2 + 5x - 14 \leq 0$

18. $x^2 + x - 20 \leq 0$

19. $x^2 \geq 9x - 20$

20. $x^2 \geq 6x - 5$

21. $(y + 1)(y - 2)^2 < 0$

22. $(y - 2)(y - 4)^2 \leq 0$

23. $(3 - y)(y + 1) \geq 0$

24. $(5 - y)(y - 1) \geq 0$

25. $\dfrac{x - 5}{x + 1} < 0$

26. $\dfrac{x - 3}{x + 1} < 0$

27. $\dfrac{x^2 - 1}{x - 3} \geq 0$

28. $\dfrac{x - 5}{x^2 - 4} \geq 0$

29. $\dfrac{5}{x - 1} > 2$

30. $\dfrac{1}{x + 5} > 3$

31. $\dfrac{3}{y + 2} \leq 5$

32. $\dfrac{5}{y + 5} \leq 1$

33. $\dfrac{1}{y + 5} \leq 5$

34. $\dfrac{1}{y + 6} \geq 4$

35. $\dfrac{1}{x + 1} \geq \dfrac{1}{x}$

36. $\dfrac{5}{x} \geq \dfrac{3}{x - 1}$

37. $\dfrac{4}{t + 2} \leq \dfrac{4}{t + 1}$

38. $\dfrac{1}{t + 3} \leq \dfrac{1}{t - 1}$

39. $\dfrac{1}{|x - 2|} \geq 5$

40. $\dfrac{1}{|x + 4|} \geq 1$

41. $\dfrac{5}{|y - 5|} \geq 1$

42. $\dfrac{6}{|y + 2|} \geq 3$

43. A square sheet of cardboard is folded into a box with an open top by cutting 4-inch by 4-inch squares out of each corner as shown in Fig. 1.8. Find the size of the square sheet if the volume of the box is not less than 36 cubic inches.

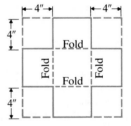

Figure 1.8

44. A square sheet of cardboard is folded into a box with an open top by cutting 4-inch by 4-inch squares out of each corner. Find the size of the square sheet if the volume of the box is not less than 64 cubic inches.

45. A square sheet of cardboard is folded into a box with an open top by cutting 3-inch by 3-inch squares out of each corner. Find the size of the square sheet if the volume of the box is not less than 48 cubic inches.

46. The height of a missile (Fig. 1.9) is measured by the equation $s = 4800t - 16t^2$ feet. During what time interval is the missile above 89,600 feet?

47. The height of a missile is measured by the equation $s = 4800t - 16t^2$ feet. During what time interval is the missile above 166,400 feet?

48. The height of a missile is measured by the equation $s = 4800t - 16t^2$ feet. During what time interval is the missile above 46,400 feet?

49. Prove that if a, b, and c are real numbers with $a \neq 0$ and $ax^2 + bx + c = 0$, then

$$x = \frac{-b \pm \sqrt{b^2 - 4ac}}{2a}$$

50. Why must $a \neq 0$ for the quadratic formula (Exercise 49)? How would you solve the equation if $a = 0$?

Figure 1.9

1.6   Irrational Sums and Complex Numbers

Earlier, we simplified, multiplied, and divided irrational expressions. In completing these operations we will again use the following properties.

Properties of Irrational Expressions

For any integer n and real numbers a and b, with $a > 0$ and $b > 0$ when n is even,

$$\sqrt[n]{a}\sqrt[n]{b} = \sqrt[n]{ab} \qquad \text{and} \qquad \frac{\sqrt[n]{a}}{\sqrt[n]{b}} = \sqrt[n]{\frac{a}{b}}, \qquad b \neq 0$$

We will also add and subtract irrational expressions. We will use the distributive property to add and subtract irrational expressions much as we add and subtract polynomial terms. We add and subtract polynomial terms when they are like terms: terms of exactly the same variables with the same exponents. Likewise, we will add and subtract irrational terms when they contain the same radicands with the same index for the radicals. These types of irrational terms can be added and subtracted easily with the use of the distributive property as in the following example.

EXAMPLE 1   Simplify each of the following:

(a) $\sqrt{2} - 3\sqrt{2}$ (b) $2\sqrt[5]{x} - \sqrt[3]{x} + 5\sqrt[5]{x}$ (c) $\sqrt[3]{2} - 2 - \sqrt[3]{16}$

Solution

(a) We use the distributive with the common factor of $\sqrt{2}$:

$$\sqrt{2} - 3\sqrt{2} = (1 - 3)\sqrt{2} \quad \textit{(Distributive property)}$$
$$= -2\sqrt{2}$$

(b) We use the commutative property and then the distributive property to add the two terms containing a factor of $\sqrt[5]{x}$:

$$\begin{aligned}
2\sqrt[5]{x} - \sqrt[3]{x} + 5\sqrt[5]{x} &= 2\sqrt[5]{x} + (-\sqrt[3]{x}) + 5\sqrt[5]{x} \\
&= 2\sqrt[5]{x} + 5\sqrt[5]{x} + (-\sqrt[3]{x}) \quad \textit{(Commutative property)} \\
&= (2 + 5)\sqrt[5]{x} - \sqrt[3]{x} \quad\quad \textit{(Distributive property)} \\
&= 7\sqrt[5]{x} - \sqrt[3]{x}
\end{aligned}$$

(c) We first simplify and then add and subtract like terms:

$$\begin{aligned}
\sqrt[3]{2} - 2 - \sqrt[3]{16} &= \sqrt[3]{2} - 2 - \sqrt[3]{8 \cdot 2} \\
&= \sqrt[3]{2} - 2 - \sqrt[3]{8}\sqrt[3]{2} \\
&= \sqrt[3]{2} - 2 - 2\sqrt[3]{2} \\
&= -2 + \sqrt[3]{2} - 2\sqrt[3]{2} \\
&= -2 + (1 - 2)\sqrt[3]{2} \\
&= -2 - \sqrt[3]{2}
\end{aligned}$$

  ■

Now that we can find irrational sums, we can work with irrational products as in the following example.

EXAMPLE 2   Multiply and simplify each of the following:
(a) $(2 - \sqrt{3})(2 + 3\sqrt{3})$ **(b)** $(3 - \sqrt{2})(3 + \sqrt{2})$

Solution We use the FOIL method to multiply. Then we simplify and combine terms where possible.

(a) $(2 - \sqrt{3})(2 + 3\sqrt{3}) = 2(2) + 4\sqrt{3} - 3\sqrt{3}(\sqrt{3})$

$$\begin{aligned}
&= 4 + 4\sqrt{3} - 3\sqrt{9} \\
&= 4 + 4\sqrt{3} - 3 \cdot 3 \\
&= 4 - 9 + 4\sqrt{3} \\
&= -5 + 4\sqrt{3}
\end{aligned}$$

(b) $(3 - \sqrt{2})(3 + \sqrt{2}) = 3(3) + 0 - \sqrt{2}(\sqrt{2})$

$$\begin{aligned}
&= 9 - \sqrt{4} \\
&= 9 - 2 \\
&= 7
\end{aligned}$$

  ■

Complex Numbers

In earlier work we avoided the square root of a negative number, but equations like $x^2 = -1$ do occur. There is no real number solution for $x^2 = -1$. To solve this and similar equations, we need a new number.

Definition

The number i is defined as

$$i = \sqrt{-1} \quad \text{and} \quad i^2 = -1$$

Since $i = \sqrt{-1}$ involves a square root, the properties of radicals are still valid. For example, we still have $\sqrt{ab} = \sqrt{a}\sqrt{b}$. But when we take the square root of a negative number, we first factor the square root of negative -1:

$$\sqrt{-9} = \sqrt{(-1)(9)} = \sqrt{-1}\sqrt{9} = i \cdot 3 = 3i$$

Always take $\sqrt{-1}$ first when working with $\sqrt{-a}$ with $a > 0$.

EXAMPLE 3

Simplify the following radicals:
(a) $\sqrt{-5}$ (b) $\sqrt{-4}$ (c) $\sqrt{-20}$

Solution   Using the definition of i, we have

(a) $\sqrt{-5} = \sqrt{(-1)5} = \sqrt{-1}\sqrt{5} = i\sqrt{5}$

(b) $\sqrt{-4} = \sqrt{(-1)4} = \sqrt{-1}\sqrt{4} = i \cdot 2 = 2i$

(c) $\sqrt{-20} = \sqrt{(-1)4 \cdot 5} = \sqrt{-1}\sqrt{4}\sqrt{5} = i \cdot 2\sqrt{5} = 2i\sqrt{5}$ ∎

When we introduce this new number i, we still keep all the properties of our number system. We now consider the closure property of addition and multiplication with the new number i. Multiplication, addition, and subtraction with i create other new numbers such as

$$5 + i, \quad 3i, \quad \frac{1}{2} + i, \quad \text{and} \quad 2 - 3i$$

All of these possible numbers make up the set of complex numbers.

Definition

A **complex number** is any number that can be written as $z = a + bi$, where a and b are real numbers and $i = \sqrt{-1}$. The set of all complex numbers is denoted by C, and

$$C = \{z : z = a + bi, a \in R, b \in R, i^2 = -1\}$$

The set of complex numbers obeys all the previously stated field properties and the properties of equality. This makes it possible to work algebraically with the complex numbers. We also note that $\{x : x = a + bi \text{ when } b = 0\} = R$, the set of real numbers. Thus $R \subseteq C$.

It can be interesting to work with powers of i. When we simplify powers of i, we use the highest possible power of i^2 as a factor as in the following example.

EXAMPLE 4   Simplify **(a)** i^{71},   **(b)** i^{104}, and   **(c)** $3i^{78} - 2i^7$.

Solution

(a) $i^{71} = (i^2)^{35}i$

$\qquad = (-1)^{35}i$

$\qquad = (-1)i$

$\qquad = -i$

(b) $i^{104} = (i^2)^{52}$

$\qquad = (-1)^{52}$

$\qquad = 1$

(c) $3i^{78} - 2i^7 = 3(i^2)^{39} - 2(i^2)^3 i$

$\qquad = 3(-1)^{39} - 2(-1)^3 i$

$\qquad = 3(-1) - 2(-1)i$

$\qquad = -3 + 2i$ ■

The following theorem shows when complex numbers are equal as well as how to add, subtract, and multiply complex numbers.

Theorem | *Properties of Complex Numbers.* For $i = \sqrt{-1}$ and real numbers a, b, c, and d,

(a) $a + bi = c + di$ if and only if $a = c$ and $b = d$.
(b) $(a + bi) + (c + di) = (a + c) + (b + d)i$,
(c) $(a + bi) - (c + di) = (a - c) + (b - d)i$, and
(d) $(a + bi)(c + di) = (ac - bd) + (ad + bc)i$.

The proofs of the first three parts of this theorem are part of the exercises at the end of this section. We will now prove part (d).

PROOF: To prove part (d), we start with $(a + bi)(c + di)$, multiply using the FOIL method, simplify, and combine like terms:

$$(a + bi)(c + di) = ac + adi + bci + bdi^2$$
$$= ac + (ad + bc)i + bd(-1)$$
$$= ac + (ad + bc)i - bd$$
$$= ac - bd + (ad + bc)i$$
$$= (ac - bd) + (ad + bc)i$$

Do not memorize the equation in part (d) of this theorem. The result can always be found quickly by using the FOIL method, simplifying, and combining like terms.

□

The following examples demonstrate the operations of complex numbers.

EXAMPLE 5

Simplify the following complex expressions:
(a) $(3 + 2i) + (5 - i)$ **(b)** $(3 + 2i) - (5 - i)$ **(c)** $-5i(2 - 3i)$
(d) $(3 - 2i)(2 + 3i)$

Solution Using the properties of complex numbers, we have

(a)
$$(3 + 2i) + (5 - i) = (3 + 5) + (2i - i) \quad \textit{(Property (b))}$$
$$= 8 + (2 - 1)i \qquad \textit{(Distributive property)}$$
$$= 8 + 1i$$
$$= 8 + i$$

(b)
$$(3 + 2i) - (5 - i) = (3 - 5) + [2i - (-i)] \quad \textit{(Property (c))}$$
$$= -2 + (2i + i)$$
$$= -2 + (2 + 1)i \qquad \textit{(Distributive property)}$$
$$= -2 + 3i$$

(c)
$$-5i(2 - 3i) = (-5i)2 - (-5i)(3i) \quad \textit{(Distributive property)}$$
$$= -10i - (-15i^2)$$
$$= -10i + 15i^2 \qquad \textit{(Remember that } i^2 = -1\textit{)}$$
$$= -10i + 15(-1)$$
$$= -15 - 10i$$

(d)
$$(3 - 2i)(2 + 3i) = 3 \cdot 2 + 3(3i) + (-2i)2 + (-2i)(3i) \quad \textit{(FOIL)}$$
$$= 6 + 9i - 4i - 6i^2$$
$$= 6 + 5i - 6(-1) \quad \textit{(Remember that } i^2 = -1\textit{)}$$
$$= 6 + 5i + 6$$
$$= 12 + 5i$$

EXAMPLE 6

Find x and y such that $x^2 + 2x - 9i = -1 + (y^2 - 6y)i$.

Solution From part (a) of the previous theorem, when $a + bi = c + di$, the real parts are equal, $a = c$, and the coefficients of i are equal, $b = d$. Thus for $x^2 + 2x - 9i = -1 + (y^2 - 6y)i$ we have

$$
\begin{array}{cc}
a = c & b = d \\
x^2 + 2x = -1 & -9 = y^2 - 6y \\
x^2 + 2x + 1 = 0 & 0 = y^2 - 6y + 9 \\
(x + 1)(x + 1) = 0 & 0 = (y - 3)(y - 3) \\
x + 1 = 0 & 0 = y - 3 \\
x = -1 \quad \text{and} \quad & y = 3
\end{array}
$$

This new number i is also useful in solving some quadratic equations.

EXAMPLE 7   Solve $x^2 - 3x + 5 = 0$.

Solution   When $x^2 - 3x + 5 = 0$, we have $a = 1$, $b = -3$, and $c = 5$:

$$x = \frac{-(-3) \pm \sqrt{(-3)^2 - 4(1)(5)}}{2} = \frac{3 \pm \sqrt{9 - 20}}{2}$$

$$x = \frac{3 \pm \sqrt{-11}}{2} = \frac{3 \pm \sqrt{(-1)(11)}}{2}$$

$$x = \frac{3 \pm i\sqrt{11}}{2} \qquad \text{or} \qquad \frac{3}{2} \pm \frac{\sqrt{11}}{2} i$$

Rationalizing the Denominator

To divide complex numbers, we will rationalize the denominator using the complex conjugate.

Definition ▍ The **complex conjugate** of $z = a + bi$ is $\bar{z} = a - bi$.

The complex conjugate is useful because the product of a complex number $a + bi$ and its complex conjugate $a - bi$ is a real number. To verify this, we use the FOIL method and

$$
\begin{aligned}
(a + bi)(a - bi) &= a \cdot a + abi - abi - b \cdot b \cdot i^2 \\
&= a^2 + (ab - ab)i - b^2 i^2 \\
&= a^2 + (0)i - b^2(-1) \\
&= a^2 + b^2
\end{aligned}
$$

Dividing by $a + bi$ is the same as having $a + bi$ in the denominator. We do not usually allow $a + bi$ in the denominator, since we would then have $\sqrt{-1}$ in the denominator. Thus we would need to rationalize the denominator. When $a + bi$ is in the denominator, we multiply the numerator and denominator by its complex conjugate $a - bi$. This process is called rationalizing the denominator.

EXAMPLE 8   Simplify the following (rationalize the denominator):

(a) $\dfrac{1}{2 + i}$ **(b)** $\dfrac{2 - i}{2 - 3i}$ **(c)** $\dfrac{2}{3i}$

Solution Multiplying the numerator and denominator by the complex conjugate of the denominator, we have

(a) $\dfrac{1}{2+i} = \dfrac{1}{(2+i)} \cdot \dfrac{(2-i)}{(2-i)}$

$\qquad = \dfrac{2-i}{2^2 - i^2}$

$\qquad = \dfrac{2-i}{4 - i^2} = \dfrac{2-i}{4 - (-1)} = \dfrac{2-i}{4+1}$

$\qquad = \dfrac{2-i}{5} \quad \text{or} \quad \dfrac{2}{5} - \dfrac{1}{5}i$

(b) $\dfrac{2-i}{2-3i} = \dfrac{(2-i)}{(2-3i)} \cdot \dfrac{(2+3i)}{(2+3i)}$

$\qquad = \dfrac{2(2) + 2(3i) - i(2) - i(3i)}{2^2 - (3i)^2}$

$\qquad = \dfrac{4 + 6i - 2i - 3i^2}{4 - 9i^2}$

$\qquad = \dfrac{4 + 4i - 3(-1)}{4 + (9)}$

$\qquad = \dfrac{7 + 4i}{13} \quad \text{or} \quad \dfrac{7}{13} + \dfrac{4}{13}i$

(c) When the denominator is a complex number $a + bi$ with $a = 0$, we multiply the numerator and denominator by i and simplify. (It is not necessary to use the complex conjugate).

$$\dfrac{2}{3i} = \dfrac{2}{3i} \cdot \dfrac{i}{i}$$

$$= \dfrac{2i}{3i^2}$$

$$= \dfrac{2i}{3(-1)}$$

$$= \dfrac{2i}{-3} \quad \text{or} \quad -\dfrac{2}{3}i$$

Rationalizing the denominator in Example 8 is similar to rationalizing the denominator when the denominator is an irrational number. Multiply the numerator and denominator by the conjugate of the denominator.

EXAMPLE 9   Rationalize the denominators in these expressions:

(a) $\dfrac{1}{5 - \sqrt{3}}$ **(b)** $\dfrac{5 - \sqrt{2}}{\sqrt{2} - \sqrt{3}}$

Solution

(a) $\dfrac{1}{5 - \sqrt{3}} = \dfrac{1}{(5 - \sqrt{3})} \cdot \dfrac{(5 + \sqrt{3})}{(5 + \sqrt{3})}$

$\qquad = \dfrac{5 + \sqrt{3}}{5^2 - (\sqrt{3})^2} = \dfrac{5 + \sqrt{3}}{25 - 3}$

$\qquad = \dfrac{5 + \sqrt{3}}{22}$ or $\dfrac{5}{22} + \dfrac{\sqrt{3}}{22}$ or $\dfrac{5}{22} + \dfrac{1}{22}\sqrt{3}$

(b) $\dfrac{5 - \sqrt{2}}{\sqrt{2} - \sqrt{3}} = \dfrac{(5 - \sqrt{2})}{(\sqrt{2} - \sqrt{3})} \cdot \dfrac{(\sqrt{2} + \sqrt{3})}{(\sqrt{2} + \sqrt{3})}$

$\qquad = \dfrac{5\sqrt{2} + 5\sqrt{3} - \sqrt{2}\sqrt{2} - \sqrt{2}\sqrt{3}}{(\sqrt{2})^2 - (\sqrt{3})^2}$

$\qquad = \dfrac{5\sqrt{2} + 5\sqrt{3} - \sqrt{4} - \sqrt{6}}{2 - 3}$

$\qquad = \dfrac{5\sqrt{2} + 5\sqrt{3} - 2 - \sqrt{6}}{-1}$

$\qquad = -5\sqrt{2} - 5\sqrt{3} + 2 + \sqrt{6}$ or $2 + \sqrt{6} - 5\sqrt{2} - 5\sqrt{3}$ ■

Exercises 1.6

Simplify the following radicals.

1. $\sqrt{5} - 7\sqrt{5}$ **2.** $5\sqrt{7} - 2\sqrt{7}$

3. $\sqrt{45} - \sqrt{20}$ **4.** $\sqrt{27} - \sqrt{12}$

5. $\sqrt{2} - 2 - \sqrt{18}$ **6.** $\sqrt{27} + 3 - \sqrt{3}$

7. $\sqrt[3]{x} - \sqrt[3]{2x^3} + \sqrt[3]{27x}$ **8.** $\sqrt[3]{3x^3} - x - \sqrt[3]{24x^3}$

9. $\sqrt{-7}$ **10.** $\sqrt{-6}$

11. $\sqrt{-9}$ **12.** $\sqrt{-16}$

13. $\sqrt{-81}$ **14.** $\sqrt{-121}$

15. $\sqrt{-18}$ **16.** $\sqrt{-24}$

17. $\sqrt{-32}$ **18.** $\sqrt{-48}$

19. $\sqrt{-45}$ **20.** $\sqrt{-80}$

21. i^{17} **22.** i^{21}

23. i^{56} **24.** i^{38}

25. i^{233} **26.** i^{302}

Combine the following complex numbers as indicated.

27. $(5 + \sqrt{2})(2 - \sqrt{5})$ **28.** $(5 + \sqrt{2})(3 - \sqrt{7})$

29. $(7 - \sqrt{2})(5 - \sqrt{2})$ **30.** $(4 + \sqrt{2})(3 - \sqrt{2})$

31. $(\sqrt{2} - \sqrt{7})(\sqrt{2} - \sqrt{7})$ **32.** $(\sqrt{3} + \sqrt{5})(\sqrt{3} + \sqrt{5})$

33. $(\sqrt{3} - \sqrt{5})(\sqrt{3} + \sqrt{5})$ **34.** $(\sqrt{2} + \sqrt{7})(\sqrt{2} - \sqrt{7})$

35. $(5 - 2i) + (-3 - 7i)$ **36.** $(5 + 2i) + (2 - 5i)$

37. $(7 - 2i) - (5 - 2i)$ **38.** $(4 + 2i) - (-1 + 7i)$

39. $(15 + 2i) - (6 - 7i)$ **40.** $(-5 + 12i) - (-5 + 2i)$

41. $5i(7 - 3i)$ **42.** $6i(5 - 2i)$

43. $-3i(-2 + 5i)$ **44.** $-2i(3 - 5i)$

45. $(2 + i)(3 - 2i)$ **46.** $(6 - i)(6 + i)$

47. $(3 - 2i)(3 + 2i)$ **48.** $(5 - 6i)(5 - i)$

Solve the following quadratic equations.

49. $x^2 - 3x + 4 = 0$ **50.** $x^2 + 2x + 5 = 0$

51. $x^2 - 2x + 4 = 0$ **52.** $x^2 - 2x + 2 = 0$

53. $2x^2 - x + 2 = 0$ **54.** $3x^2 - 2x + 3 = 0$

Rationalize the denominator in the following problems.

55. $\dfrac{1}{1 + i}$ **56.** $\dfrac{1}{2 - i}$

57. $\dfrac{1}{3 - i}$ **58.** $\dfrac{1}{4 + i}$

59. $\dfrac{1}{2 + 3i}$

60. $\dfrac{1}{1 - 2i}$

61. $\dfrac{2 + i}{2 - i}$

62. $\dfrac{3 + 2i}{3 - 2i}$

63. $\dfrac{1}{1 - \sqrt{3}}$

64. $\dfrac{4}{1 + \sqrt{2}}$

65. $\dfrac{1}{\sqrt{2} - \sqrt{3}}$

66. $\dfrac{4}{\sqrt{3} + \sqrt{2}}$

67. $\dfrac{1 - \sqrt{3}}{\sqrt{3} + \sqrt{5}}$

68. $\dfrac{\sqrt{5} - 1}{\sqrt{3} - \sqrt{5}}$

69. For real numbers a, b, c, and d, prove that

$$a + bi = c + di$$

if and only if $a = c$ and $b = d$.

70. Prove for complex numbers $a + bi$ and $c + di$, where a, b, c, and d are real numbers, that

$$(a + bi) + (c + di) = (a + c) + (b + d)i.$$

71. Prove for complex numbers $a + bi$ and $c + di$, where a, b, c, and d are real numbers, that

$$(a + bi) - (c + di) = (a - c) + (b - d)i.$$

1.7   Graphing, Lines, and Variation

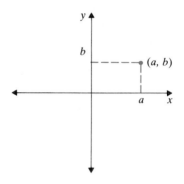

Figure 1.10

During the European Renaissance (ca. 1300–1600 A.D.), European scholars began to combine both geometry and algebra. It was René Descartes who popularized a method for naming every geometric point in a plane with a unique algebraic ordered pair. He accomplished this by using two perpendicular lines to form the coordinate system described below. This coordinate system is called the Cartesian (or rectangular) coordinate system.

The Coordinate System

The **Cartesian** (or **rectangular**) **coordinate system** is made up of two perpendicular real number lines that intersect at their origins. Usually, one real number line is horizontal and is called the x**-axis**, while the other real number line is vertical and is called the y**-axis**. The x-axis has positive coordinates to the right of zero and negative coordinates to the left of zero. The y-axis has positive coordinates up from zero and negative coordinates down from zero. The two axes intersect at the point zero on both number lines. The ordered pair (a, b) is a unique point in the plane created by the Cartesian coordinate system. For the point (a, b), a is the x**-coordinate** (or **abscissa**) and b is the y**-coordinate** (or **ordinate**). Locating the point (a, b) on the coordinate system is called plotting the point (a, b). We plot an ordered pair (a, b) by placing a dot at the point that is the intersection of a line perpendicular to the x-axis at the coordinate a and another line perpendicular to the y-axis at the coordinate b (Fig. 1.10).

We find the ordered pair for a point in a similar manner. We start at the given point and sketch a line from that point perpendicular to the x-axis to find the x-coordinate of the ordered pair and another line perpendicular to the y-axis to find the y-coordinate of the point. In this way we can find a unique ordered pair for every point in the plane, and every point in the plane corresponds to a unique ordered pair. The intersection of the x- and y-axes is the point $(0, 0)$, called the **origin**.

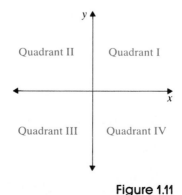

Figure 1.11

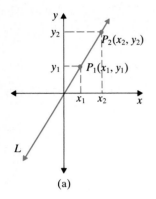

(a)

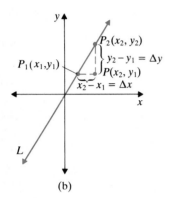

(b)

Figure 1.12

The axes in the Cartesian coordinate system divide the plane into four quadrants as noted in Fig. 1.11: Quadrant I, Quadrant II, Quadrant III, and Quadrant IV.

Plotting all the points whose ordered pairs (x, y) make an equation true on this coordinate system is called **graphing the equation**. Graphing an equation is simply a matter of first determining all ordered pairs that are solutions to the equation and then plotting these ordered pairs on the Cartesian coordinate system. If we know enough ordered pairs, we can graph the equation.

Lines

The equations and graphs of lines are a familiar part of your past work in algebra and geometry. In this section we will work with lines (their equations, their graphs, and some relationships between lines). Although these ideas might seem like review, it will be important for you, in this course and in calculus, to be fluent in working with lines, their equations, and their graphs.

A line (always a straight line) is uniquely determined by two points. In Fig. 1.12(a) the two points P_1 and P_2 with ordered pairs (x_1, y_1) and (x_2, y_2), respectively, determine the line L. Notice that point P forms a right triangle with P_1 and P_2 and P is the ordered pair (x_1, y_2) (Fig. 1.12b). The distance from P_1 to P is $x_2 - x_1$, and the distance from P to P_2 is $y_2 - y_1$. The ratio of $y_2 - y_1$ and $x_2 - x_1$ is called the **slope** of a line and is denoted by m. Thus the slope is

$$m = \frac{y_2 - y_1}{x_2 - x_1} = \frac{\text{the change in } y \text{ values}}{\text{the change in } x \text{ values}} = \frac{\Delta y}{\Delta x}$$

where Δy is the change in the ordinate (y-values) and Δx is the change in the abscissa (x-values). These changes in values may be positive or negative, and Δy may be zero, but $\Delta x \neq 0$. The changes in values may be found by moving from P_1 to P_2 or from P_2 to P_1. Thus the slope of the line from P_1 to P_2 is $m_{P_1P_2}$ and is found by

$$m_{P_1P_2} = \frac{\Delta y}{\Delta x} = \frac{y_2 - y_1}{x_2 - x_1} \quad \text{or} \quad \frac{y_1 - y_2}{x_1 - x_2}$$

Now we will find the slope of line L through the points $P_3(x_3, y_3)$ and $P_4(x_4, y_4)$ as shown in Fig. 1.13.

$$m_{P_3P_4} = \frac{\Delta y}{\Delta x} = \frac{y_4 - y_3}{x_4 - x_3} \quad \text{or} \quad \frac{y_3 - y_4}{x_3 - x_4}$$

We also see in Fig. 1.13 that triangle P_1PP_2 is similar to triangle P_3MP_4, and since corresponding sides of similar triangles are proportional, we have

$$\frac{y_2 - y_1}{x_2 - x_1} = \frac{y_4 - y_3}{x_4 - x_3}$$

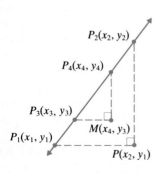

Figure 1.13

Therefore

$$m_{P_1 P_2} = m_{P_3 P_4}$$

Since P_1, P_2, P_3, and P_4 are four different and arbitrary points on line L, we conclude that a nonvertical line has a unique value for its slope. (We say "a nonvertical line" because the slope of any vertical line has $\Delta x = 0$ and the slope is undefined.)

Definition

A nonvertical line through the distinct points (x_1, y_1) and (x_2, y_2) has a unique **slope**, denoted m, calculated as follows:

$$m = \frac{\text{the change in } y}{\text{the change in } x} = \frac{\Delta y}{\Delta x} = \frac{y_2 - y_1}{x_2 - x_1}$$

It is important to notice that any two points on the same line generate the same value for the slope.

For a horizontal line there is no change in y-values from point to point, and $\Delta y = 0$. Thus the slope of a horizontal line is

$$m = \frac{\Delta y}{\Delta x} = \frac{0}{\Delta x} = 0$$

Similarly, a vertical line has no change in x-values from point to point, and $\Delta x = 0$. Thus the slope of a vertical line is

$$m = \frac{\Delta y}{\Delta x} = \frac{\Delta y}{0} \qquad \text{which is undefined}$$

Therefore we know that **horizontal lines have $m = 0$ and vertical lines have no slope number** (or the slope is undefined).

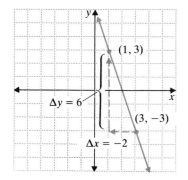

Figure 1.14

EXAMPLE 1

Sketch the graph and find the slope of the lines through the following pairs of points:
(a) $(1, 3)$ and $(3, -3)$ (b) $\left(5\frac{1}{2}, \frac{3}{4}\right)$ and $\left(-1\frac{1}{2}, \frac{1}{4}\right)$ (c) $(5, 0)$ and $(5, 5)$

Solution
(a) For $(1, 3)$ and $(3, -3)$,

$$m = \frac{\Delta y}{\Delta x} = \frac{-3 - 3}{3 - 1} = \frac{-6}{2} = -3 \qquad \text{or}$$

$$m = \frac{\Delta y}{\Delta x} = \frac{3 - (-3)}{1 - 3} = \frac{6}{-2} = -3$$

Therefore $m = -3$. (See Fig. 1.14.)

(b) For $\left(5\frac{1}{2}, \frac{3}{4}\right)$ and $\left(-1\frac{1}{2}, \frac{1}{4}\right)$ (Fig. 1.15),

$$m = \frac{\Delta y}{\Delta x} = \frac{\frac{3}{4} - \frac{1}{4}}{5\frac{1}{2} - \left(-1\frac{1}{2}\right)} = \frac{\frac{2}{4}}{7}$$

$$m = \frac{\frac{1}{2}}{7} = \frac{1}{14}$$

(c) For $(5, 0)$ and $(5, 5)$ (Fig. 1.16),

$$m = \frac{0 - 5}{5 - 5} = \frac{-5}{0}, \qquad \text{undefined}$$

The slope is undefined. This vertical line has no slope number.

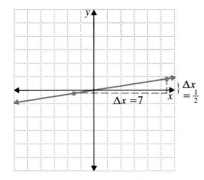

Figure 1.15

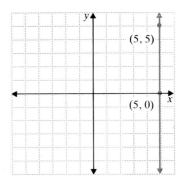

Figure 1.16

EXAMPLE 2   Find the slope for each of the lines containing the following pairs of points:
(a) $(3, 5)$ and $(-2, 3)$ **(b)** $(a + 1, a + 5)$ and $(a - 2, a + 3)$.

Solution   Since the slope m is the change in y divided by the change in x, we calculate the slope as follows:
(a) For $(3, 5)$ and $(-2, 3)$,

$$m = \frac{\Delta y}{\Delta x} = \frac{5 - 3}{3 - (-2)} = \frac{2}{5}$$

(b) For the ordered pairs $(a + 1, a + 5)$ and $(a - 2, a + 3)$,

$$m = \frac{\Delta y}{\Delta x} = \frac{(a + 5) - (a + 3)}{(a + 1) - (a - 2)}$$

$$= \frac{a + 5 - a - 3}{a + 1 - a + 2} = \frac{2}{3}$$

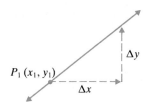

Figure 1.17

We can also use the slope when graphing lines. We start at a point on the line such as $P_1(x_1, y_1)$. Then we use the slope $m = \dfrac{\Delta y}{\Delta x}$ to move Δx units (+ to the right and − to the left) parallel to the x-axis and Δy units (+ up and − down) parallel to the y-axis (Fig. 1.17).

EXAMPLE 3

Sketch the graphs of **(a)** the line through $(1, 1)$ with slope $m = 3$ and **(b)** the line through $(-3, 3)$ with slope $m = -\dfrac{2}{5}$.

Solution   We will use the point and slope to graph these lines as described above.
(a) First plot the point $(1, 1)$. For

$$m = 3 = \frac{3}{1} = \frac{\Delta y}{\Delta x}$$

$\Delta y = 3$ and $\Delta x = 1$ is one of many possibilities for Δy and Δx. From $(1, 1)$, first move $\Delta x = 1$ units to the right (parallel to the x-axis) and then move $\Delta y = 3$ units up (parallel to the y-axis). The result is the point $(2, 4)$. Connecting $(1, 1)$ and $(2, 4)$, we have the graph of the desired line, as shown in Fig. 1.18.
(b) For

$$m = \frac{\Delta y}{\Delta x} = -\frac{2}{5} = \frac{-2}{5}$$

we let $\Delta y = -2$ and $\Delta x = 5$. Starting at the given point $(-3, 3)$, we move $\Delta x = 5$ units parallel to the x-axis (to the right) and then move $\Delta y = -2$ units parallel to the y-axis (down). The result is the point $(2, 1)$. Connecting $(-3, 3)$ and $(2, 1)$, we have the graph of the desired line, as shown in Fig. 1.19.

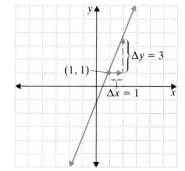

Figure 1.18

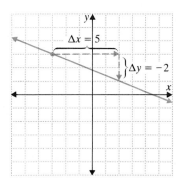

Figure 1.19

With the slope of a line we can write the general equation for any nonvertical line. We will write the equation of a line L that has a slope m with $P_1(x_1, y_1)$ as a

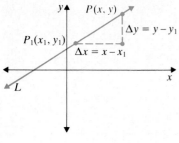

Figure 1.20

point on L. To do this, we let $P(x, y)$ be any point on L different from P_1 and write the slope of line L (Fig. 1.20). Then

$$\frac{(y - y_1)}{(x - x_1)} = m$$

$$(y - y_1) = m(x - x_1)$$

This equation $(y - y_1) = m(x - x_1)$ is the **point-slope form** for the equation of a line through point (x_1, y_1) with slope m. This equation is very useful in writing the equations of lines.

THE POINT-SLOPE FORM FOR THE EQUATION OF A LINE	The equation of the line through the point (x_1, y_1) with slope m is $$(y - y_1) = m(x - x_1)$$

EXAMPLE 4

Write the equation of a line when **(a)** $m = 3$ and one point on the line is $(2, 5)$ and **(b)** two points on the line are $(1, 1)$ and $(3, -1)$.

Solution   We will use the point-slope form to write the equation of these lines.
(a) For $m = 3$ and $(2, 5) = (x_1, y_1)$ we write

$$(y - y_1) = m(x - x_1)$$
$$(y - 5) = 3(x - 2)$$
$$y - 5 = 3x - 6$$
$$y = 3x - 1 \quad \text{or} \quad 0 = 3x - y - 1$$

(b) We first use the points $(1, 1)$ and $(3, -1)$ to find the slope of the line:

$$m = \frac{-1 - 1}{3 - 1} = \frac{-2}{2} = -1$$

To write the equation of the line, we use the slope and either point in the point-slope form. Using $(1, 1)$, we have

$$(y - y_1) = m(x - x)$$
$$(y - 1) = -1(x - 1)$$
$$y - 1 = -x + 1$$
$$y = -x + 2 \quad \text{or} \quad x + y - 2 = 0$$

We would also obtain the same result if we used the point $(3, -1)$ instead of the point $(1, 1)$.

To write the equation of a horizontal line through (a, b), we remember that the slope of a horizontal line is $m = 0$. Using point-slope form, we have

$$y - b = m(x - a)$$
$$y - b = 0(x - a)$$
$$y - b = 0 \quad \text{or} \quad y = b$$

On the other hand, the equation of the vertical line through (a, b) has undefined slope, but it has an equation of $x = a$.

HORIZONTAL AND VERTICAL LINES	The equation of a horizontal line through (a, b) is $y = b$. The equation of a vertical line through (a, b) is $x = a$. 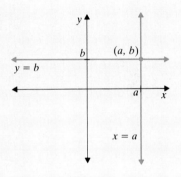

We can easily identify two special points for most lines. They are the points where the line crosses the x- and y-axes. The x-coordinate of the point where the graph intersects (touches) the x-axis is called the **x-intercept**, and the y-coordinate of the point where the graph intersects the y-axis is called the **y-intercept**. Thus the line through $(a, 0)$ and $(0, b)$ in Fig. 1.21 has an x-intercept of a and a y-intercept of b. We can use these two intercepts to write two other interesting forms for the equations of a line.

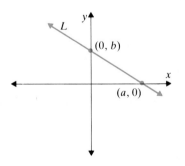

Figure 1.21

THE SLOPE-INTERCEPT FORM FOR THE EQUATIONS OF A LINE	The equation of the line with slope m and y-intercept b is $$y = mx + b$$

THE INTERCEPT FORM FOR THE EQUATION OF A LINE	The equation of the line with an x-intercept of a and a y-intercept of b, when $a \neq 0$ and $b \neq 0$, is $$\frac{x}{a} + \frac{y}{b} = 1$$

Although these equations are fairly intuitive, we will prove them as a part of the exercises at the end of this section. We will also use standard form for writing the equation of a line.

Definition	The **standard form for the equation of a line** is $$ax + by + c = 0$$ where a, b, and c are real numbers with a and b not both equal to zero.

Sometimes we have pairs of lines that are parallel or perpendicular. The slopes of these pairs of lines have special relationships.

PARALLEL AND PERPENDICULAR LINES	For two nonvertical lines L_1 and L_2 with slopes m_1 and m_2, (a) the two lines are parallel if and only if $m_1 = m_2$ (equal slopes), and (b) the two lines are perpendicular if and only if $m_1 \cdot m_2 = -1$. $\left(\text{One slope is the negative reciprocal of the other, } m_1 = \dfrac{-1}{m_2}\right).$

The fact that parallel lines have the same slope is fairly intuitive, while the conclusion for perpendicular lines is not very intuitive. The proof of part (b) will be done in a later exercise set after we have defined the distance formula.

EXAMPLE 5   Find the equation of the line through $(4, 5)$
(a) parallel to the line $6x - 3y = 3$,
(b) perpendicular to the line $6x - 3y = 3$.

Solution For both parts of this problem we need the slope of the given line $6x - 3y = 3$. We write this equation in slope-intercept form as follows:

$$6x - 3y = 3$$
$$-3y = -6x + 3$$
$$\frac{-3y}{-3} = \frac{-6x + 3}{-3}$$
$$y = 2x - 1$$

Thus the slope of the given line is $m = 2$.

(a) The line through (4, 5) parallel to the given line must also have slope $m = 2$. Therefore its equation is

$$y - 5 = 2(x - 4)$$
$$y - 5 = 2x - 8$$
$$y = 2x - 3 \quad \text{or} \quad 2x - y - 3 = 0$$

(b) The line through (4, 5) perpendicular to the given line must have a slope of $m = -\frac{1}{2}$, the negative reciprocal of 2. Thus the equation of the perpendicular line is

$$y - 5 = \left(-\frac{1}{2}\right)(x - 4)$$
$$y - 5 = \left(-\frac{1}{2}\right)x + 2$$
$$y = -\frac{1}{2}x + 7 \quad \text{or} \quad \frac{1}{2}x + y - 7 = 0$$

Variations

A variation is an equation with many practical uses. The word *variation*, or *varies*, describes a specific change in one variable with respect to the change in another variable. These variations have specific meanings associated with specific words as in the following definition.

Definition

For variables x and y, a real number constant k, and a positive real number n we define the following terms.

(a) **"varies directly"** or **"is directly proportional"**:

y varies directly with x^n is defined as $y = kx^n$

(b) **"varies inversely"** or **"is inversely proportional"**:

y varies inversely with x^n is defined as $y = \dfrac{k}{x^n}$

(c) **"varies jointly"** or **"is jointly proportional"**:

y varies jointly with x and y is defined as $y = kxy$

The constant k is often called the constant of proportionality.

As we see in this definition, these words (*varies directly*, *varies inversely*, and *varies jointly*) all have special meaning. Remembering the special meaning for each phrase is essential in working variation problems.

To solve variation problems, we will usually be given a set of data that make it possible to find the value of the constant k for that problem. Once we have the value of k, we can then solve for the desired unknown. The following examples demonstrate this process.

EXAMPLE 6

The value of y varies directly with the square of x. If $x = 3$ when $y = 12$, find y when $x = 15$.

Solution   Since y varies directly with x^2, we have

$$y = kx^2$$

Since $x = 3$ when $y = 12$, we can find k for this problem:

$$12 = k(3)^2$$
$$12 = 9k$$
$$k = \frac{12}{9} = \frac{4}{3}$$

With this value for k we can write this direct variation as

$$y = \frac{4}{3}x^2$$

We now use this equation to find y when $x = 15$ as follows:

$$y = \frac{4}{3}(15)^2$$

$$y = \frac{4}{3}(15)(15)$$

$$y = 4(5)15$$

$$y = 300$$

Therefore when $y = \frac{4}{3}x^2$ and $x = 15$, $y = 300$. ■

EXAMPLE 7

In a photographic enlarger the intensity of light I is inversely proportional to the square of the distance d from the light. If intensity I is 2 lumens when the distance d is 18 inches, find I when **(a)** $d = 2$ feet and **(b)** $d = 6$ inches.

Solution   Since I is inversely proportional to d^2, we have

$$I = \frac{k}{d^2}$$

We first find k for $I = 2$ when $d = 18$ inches $= \frac{3}{2}$ feet:

$$I = \frac{k}{\left(\frac{3}{2}\right)^2}$$

$$\left(\frac{3}{2}\right)^2 2 = k$$

$$\left(\frac{9}{4}\right) 2 = k$$

$$k = \frac{9}{2}$$

Thus

$$I = \frac{9}{2d^2}$$

(a) For $d = 2$ feet we have

$$I = \frac{9}{2(2)^2}$$

$$= \frac{9}{8} = 1\frac{1}{8} \text{ lumens}$$

(b) For $d = \frac{1}{2}$ foot we have

$$I = \frac{9}{2\left(\frac{1}{2}\right)^2}$$

$$= \frac{9}{\frac{1}{2}}$$

$$= 18 \text{ lumens} \qquad \blacksquare$$

EXAMPLE 8   Write the following in words using the new idea of variations:

(a) $xy^3 = 3$ **(b)** $\dfrac{x^2}{y} = 5$

Solution
(a) Rewriting $xy^3 = 3$, we have

$$x = \frac{3}{y^3}$$

This is the equation for an inverse variation, so we can conclude that "x varies inversely with the cube of y, and the constant of proportionality is 3."

(b) Rewriting

$$\frac{x^2}{y} = 5$$

we have

$$y = \frac{1}{5}x^2$$

Thus we can say that "y varies directly with x^2, and the constant of proportionality is $\frac{1}{5}$."

We will work with variations again in Chapter 2.

Exercises 1.7

Find the slope in Exercises 1–4 and sketch the graph of each line described in Exercises 1–8.

1. The points $(1, 3)$ and $(-2, 4)$ are on the line.

2. The points $(5, -3)$ and $(-7, 5)$ are on the line.

3. The points $(2.4, -4.2)$ and $(-3.1, 5.8)$ are on the line.

4. The points $(2.7, 4.3)$ and $(.3, -3.7)$ are on the line.

5. The line contains $(3, -1)$ with slope $m = -3$.

6. The line contains $(-2, -3)$ with slope $m = 2$.

7. The line contains $(-2, -3)$ with slope $m = \frac{3}{4}$.

8. The line contains $(-4, -2)$ with slope $m = \frac{2}{3}$.

Find the equations for the lines in slope-intercept form in Exercises 9–16.

9.

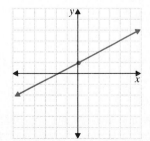

10.

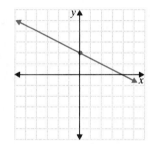

11.

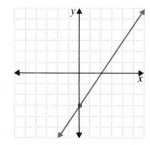

12.

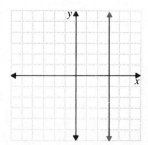

13.

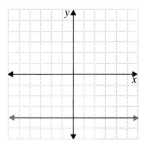

14.

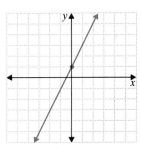

15.

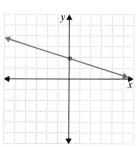

16.

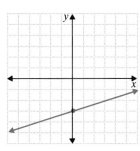

17. Find the equations of the line through the point $(1, -3)$
  (a) parallel and
  (b) perpendicular to the line $y = -2x - 3$.

18. Find the equation of the line through the point $(-2, 5)$
  (a) parallel and
  (b) perpendicular to the line $y = 4x + 1$.

19. Show that the diagonals of the square $ABCD$ are perpendicular when A is $(-1, 1)$, B is $(2, 4)$, C is $(5, 1)$, and D is $(2, -2)$.

20. Show that the diagonals of a square $PQRS$ are perpendicular when P is $(1, -1)$, Q is $(0, 2)$, R is $(3, 3)$, and S is $(4, 0)$.

21. Verify that the line with a as the nonzero x-intercept and b as the nonzero y-intercept has the equation

$$\frac{x}{a} + \frac{y}{b} = 1$$

22. Verify that the line with a y-intercept b and slope m has the equation $y = mx + b$.

23. If x is directly proportional to y and $x = 15$ when $y = \frac{3}{5}$,
  (a) find x when $y = \frac{1}{5}$ and
  (b) find y when $x = 50$.

24. If x is directly proportional to y and $x = 4$ when $y = 12$,
  (a) find x when $y = \frac{1}{3}$ and
  (b) find y when $x = \frac{1}{3}$.

25. If y varies inversely with the square root of x and $x = 4$ when $y = 6$,
  (a) find y when $x = 9$ and
  (b) find x when $y = 3$.

26. If y varies inversely with the cube of x and $y = 16$ when $x = \frac{1}{8}$,
  (a) find y when $x = 9$ and
  (b) find x when $y = 6$.

27. If w varies jointly with x and the square of y and $w = 12$ when $x = 2$ and $y = 3$,
  (a) find w when $x = 3$ and $y = 2$,
  (b) find y when $w = 8$ and $x = 3$, and
  (c) find x when $w = 18$ and $y = 3$.

28. If w is directly proportional to x and inversely proportional to y^2 and $w = 15$ when $x = 12$ and $y = 2$,
  (a) find w when $x = 5$ and $y = 5$,
  (b) find x when $w = 2$ and $x = 5$, and
  (c) find y when $w = 30$ and $x = 3$.

Write each of the following equations in words using the ideas of variation. Also identify the constant of proportionality for each equation.

29. $xy = 5$

30. $xy^2 = 5$

31. $\dfrac{x}{y^2} = 3$

32. $\dfrac{x^2}{y} = 3$

33. $7x = y^2w^3$

34. $3y = x^2w^3$

35. $wy^2 = 3x$

36. $7x^2y = w^2$

37. The current I in an electrical circuit varies directly with the voltage V and inversely with the resistance R. Find this equation involving I, V, and R.

38. Write the equation for the slope of a line in words using the idea of variation. What is the value of the constant of proportionality?

39. The conductance G of a resistor varies inversely with the resistance R.
   (a) Find the equation for G and R if $G = 0.02$ siemen when $R = 50$ ohms.
   (b) Find G when $R = 75$ ohms.

40. The conductance G of a resistor varies inversely with the resistance R.
   (a) Find the equation for G and R if $G = 0.05$ siemen when $R = 20$ ohms.
   (b) Find G when $R = 50$ ohms.

41. The pitch, or frequency, p of a vibrating string varies directly with the square root of the tension t of the string. If the tension triples, what happens to the value of the pitch?

42. In a photographic enlarger the intensity of light I is inversely proportional to the square of the distance d from the light. What must be the change in the distance d to double the intensity I?

43. In a photographic enlarger the intensity of light I is inversely proportional to the square of the distance d from the light. What is the change in intensity I if distance d is doubled?

44. The force F required to stretch a spring the distance d from its at-rest length, $d = 0$, is directly proportional to d. If

$F = 6$ newtons when $d = 0.3$ meter, find the additional force required to stretch the spring 0.2 meter more (a total distance of 0.5 meter) (Fig. 1.22).

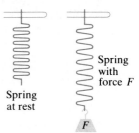

Figure 1.22

Spring at rest

Spring with force F

F

45. For n identical energy cells in series, Ohm's law states that the current I varies jointly with the number of cells n and the electromotive force E of each cell and varies inversely with the sum of the external resistance R_0 and the product of the number of cells and the internal resistance b of each cell (Fig. 1.23). Find this equation if the constant of proportionality is $k = 1$.

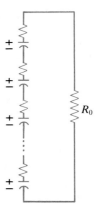

Figure 1.23

R_0

CHAPTER SUMMARY AND REVIEW

KEY TERMS

Set notation

Set operations: $A \cup B$, $A \cap B$, $A \subseteq B$.

Special sets of numbers: N, Z, Q, $\overline{Q}$, R, C

Scientific notation

Properties of inequalities

Compound inequalities

Field properties of real numbers:
   Closure, commutative, associative, identities,
      inverses, distributive.
Properties of equality:
   Addition, multiplication, substitution, symmetric,
      reflexive.
The absolute value of x
Integral exponents
Properties of exponents
Principal root
Rational numbers exponents
Special products
Special factors

The absolute value theorem
The square root method
Completing the square
The quadratic formula
Solving polynomial inequalities
The number i
Complex numbers
Properties of complex numbers
The complex conjugate
Lines and slope
The equations of lines
Parallel and perpendicular lines
Variation or proportion

CHAPTER EXERCISES

When $A = \{5, 6, 7\}$ and $B = \{3, 4, 5\}$:

1. Find $A \cup B$

2. Find $A \cap B$

3. Is $A \subseteq B$? Why or why not?

4. Is $(A \cap B) \subseteq (A \cup B)$? Why or why not?

5. Write this set in roster notation: $\{x : x \in Z, 2 < x \leq 7\}$.

6. Solve $i = prt$ for p.

7. Solve $P = 2L + 2W$ for W.

Simplify the following, using the properties of exponents and radicals.

8. $(2x)^2(-3xy)^3$

9. $(2x^{-1})^3(-3xy)^{-3}$

10. $(x^2y^{-4})^{-2}$

11. $(x^{-3}y^2)^{-3}$

12. $(x^2y^4)^{1/2}$

13. $(x^{-3}y^{-6})^{-1/3}$

14. $\sqrt[3]{-27}$

15. $\sqrt{36x^2y^6}$

Rationalize the denominators.

16. $\dfrac{\sqrt{2}}{\sqrt{3}}$

17. $\dfrac{\sqrt[3]{2}}{\sqrt[3]{4}}$

Change to scientific notation.

18. 56,000,000

19. 0.000000075

Complete the following operations.

20. $(y^2 + y - 1) - (y^2 - y + 3)$

21. $(y^2 + y - 1)(y^2 - 1)$

22. $(3x^3 - 9x^2 + 27x) \div 3x$

23. $(5x - 7)(5x + 7)$

Factor each.

24. $4x^2 - 9$

25. $x^3 - 8y^3$

26. Factor $2x^{-2}$ from $(16x^2 - 18 + 6x^{-2})$.

Simplify.

27. $\dfrac{\dfrac{1}{x} + \dfrac{1}{x-1}}{\dfrac{1}{x-1} - \dfrac{1}{x}}$

28. $\dfrac{\dfrac{1}{x-1} + \dfrac{1}{x+1}}{\dfrac{1}{x+1} - \dfrac{1}{x-1}}$

Solve the following, graph the solution, and write the solution in interval notation.

29. $5 < 7 - 2x < 17$

30. $|y| \leq 7$

31. $|x - 3| < 5$

32. $|y - 1| \geq 4$

33. $\dfrac{5}{x + 3} > 0$

34. $\dfrac{1}{y - 2} < 0$

Solve the following.

35. $y^2 - 5y - 2 = 0$ **36.** $r^2 - 4r + 4 = 0$

37. $y^3 + 6y^2 - 9y - 54 = 0$

38. $\dfrac{3}{x-2} > 1$ **39.** $\dfrac{5}{t+5} < 3$

40. $\dfrac{6}{|x-1|} \geq 3$

Simplify.

41. $\sqrt{-9}$ **42.** $\sqrt{-20}$

43. Solve $x^2 - 4x + 5 = 0$

Simplify.

44. $(3 - i) - (3 + 2i)$ **45.** $(3 - i)(3 + 2i)$

Rationalize these denominators.

46. $\dfrac{1}{5-i}$ **47.** $\dfrac{1}{5-\sqrt{2}}$

48. Find the perimeter of a square if the area of the square is 72 square inches.

49. In a race the contestants swim, ride a bike, and run on three different parts of the race. One contestant took the same amount of time to complete each part of the race on a 55-mile course. If the contestant's average rate is 4 miles per hour when swimming, 12 miles per hour when bicycling, and 6 miles per hour when running, how many hours did it take to finish this race?

50. A boat's speed in still water is 15 miles per hour. A boat travels for 11 hours up the Mississippi River against a constant current and then makes the return trip in 4 hours with the same constant current. What is the rate of the current?

51. A square sheet of cardboard is folded into a box with an open top by cutting 4-inch by 4-inch squares out of each corner (Fig. 1.24). Find the size of the square sheet if the volume of the box is not less than 100 cubic inches.

52. A square sheet of cardboard is folded into a box with an open top by cutting 4-inch by 4-inch squares out of each corner (Fig. 1.24). Find the size of the square sheet if the volume of the box is not less than 12 cubic inches.

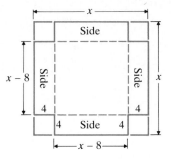

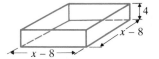

Figure 1.24

53. Find the equation of the line through the points $(2, 3)$, $(5, 9)$.

54. Find the equation of the line through $(2, 5)$ perpendicular to the line $2x - 4y = 8$.

55. Find the equation of the line through $(2, 5)$ parallel to the line $2x - 4y = 8$.

56. The force F required to stretch a spring the distance d from its at-rest length, $d = 0$, is directly proportional to d. If $F = 2$ newtons when $d = 0.5$ meter, find the additional force required to stretch the spring 0.5 meter more (a total distance of 1 meter) (Fig. 1.25).

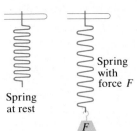

Figure 1.25

2

Functions and Graphs

EQUATIONS FOR EFFICIENCY

Without air turbulence in the cylinder of a gasoline engine, combustion would take place too slowly for the engine to function. The ability to design engines that provide the necessary fuel efficiency depends on an understanding of the relationship between the turbulence of gas motion and the burning rate. Quantifying this relationship is a complex task that is made more difficult by the need to measure events occurring in a few milliseconds within a small, confined space.

As senior engineers in the Engine Research Department at the General Motors Research Laboratories, Drs. Frederic Matekunas and Edward Groff have incorporated experimental observations into a model that successfully predicts the effects of engine design and operating conditions on power and fuel economy. In the first phase they determined flame speeds over a broad range of operating conditions. More than 400,000 pieces of data were processed for each 10-second measurement period. In the second phase, the combustion phase, over 100 operating conditions with varied spark timing,

spark plug location, engine speed, and intake valve geometry were considered.

The researchers used the measured flame speeds, turbulence intensities, and conditions under which they occurred to formulate a burning law for engine flames. They divided the combustion part into the four stages, which are shown in graph form in the Flame Speed Behavior graph and the accompanying illustrations. The initiation stage, Stage I, begins with ignition and ends as the flame grows to consume 1% of the fuel mass. In Stage II the flame accelerates, and it thickens in response to the turbulence field. Stage III exhibits peak flame speed. In the final stage, Stage IV, the thick flame interacts with the chamber walls and decelerates.

The turbulence velocity, S_T, during Stage III can be described as

$$S_T = 2.0S_L + 1.2u'P_R^{0.82}\beta$$

where S_L is the flame speed without turbulence, u' is the turbulence intensity, P_R is a pressure ratio

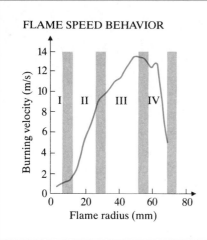

FLAME SPEED BEHAVIOR

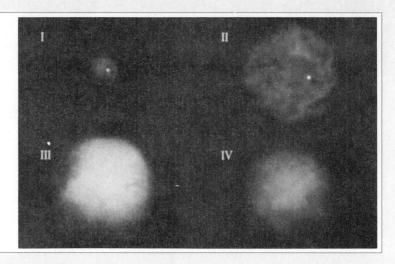

Burning velocity plotted as a function of flame radius. Combustion stages are indicated by roman numerals.

High-speed photographs showing flame evolution (lasting 6 milliseconds) through four stages: initiation (I); flame growth (II); full development (III); termination (IV).

for the combustion-induced compression of the unburned mixture, and the dimensionless β accounts for the effect of spark timing and the geometric distortion of the flame during Stage I. The researchers also observed that the burning velocity in Stage II increases proportionally to the flame radius and that it is necessary to know the finite thickness of the flame front to predict the energy release rate from the burning velocity.

"The form of our burning equation," says Dr. Matekunas, "shows a satisfying resemblance to the expressions for nonengine flames. This helps link complex engine combustion phenomena to the existing body of knowledge on turbulent flames."

"We see this extension," adds Dr. Groff, "as a significant step toward optimizing fuel economy in automotive engines."

SOURCE: "The Turbulence Parameter," General Motors Research and Development advertising, Detroit, Michigan. Reprinted by permission.

In the preceding example you can see how people in research and industry use experimental data to solve a problem. The experimental data were displayed in the Flame Speed Behavior graph that shows the relationship of the flame radius on the horizontal axis to the burning velocity on the vertical axis. In this chapter we will learn a lot about graphing. The example also shows us how experimental data can be used to generate equations that describe the data. One such equation describes the turbulence velocity S_T as

$$S_T = 2.0S_L + 1.2u'P_R^{0.82}\beta$$

This equation represents the value of S_T as the result, or function, of the other variables and constants in this equation that determine the turbulence velocity. This concept of a *function* is a major topic that we will explore in this chapter.

2.1 Functions

As we saw in the example at the beginning of the chapter, the power of mathematics to solve problems can be the direct result of the ability to identify patterns in experimental data. Much of your previous work in mathematics involved special relationships between variables, such as x and y. (For example, $y = mx + b$, $x^2 + y^2 = r^2$, $y = x^3$, and $x = y^3$.) In general, equations that define a relationship between variables are called relations. Some special relations are called functions. One such relation is the relationship of the volume of gas in a spherical weather balloon to the radius of the balloon. Since the volume V of a sphere with radius r is $V = \left(\frac{4}{3}\right)\pi r^3$, Table 2.1 shows that each radius corresponds to a specific balloon volume.

TABLE 2.1

Radius (feet)	Approximate Volume (cubic feet)
r	$V = \dfrac{4}{3}\pi r^3$
1	4.189
3	113.097
5	523.599
7	1436.755

In this example we can pick any desired length for the radius and use this length to calculate the corresponding volume of the balloon. We call the radius r the **independent variable**, since its value can be freely (independently) chosen. Similarly, the balloon's volume V is called the **dependent variable**, since the value of V is determined by (depends on) each specific value chosen for r.

Relations

We will first consider the most general type of *relation* as defined here.

Definition

A **relation** from A to B is a set of ordered pairs (x, y) with $x \in A$ and $y \in B$. The set of all values of x, the set A, is called the **domain** of the relation. The set of all values of y, B or a subset of B, is called the **range** of the relation.

We have some special words that describe the relationship between the x and y in the ordered pair (x, y). For a relation we say that each value of the domain is **mapped** (matched or corresponds) to at least one value of the range. The element x mapped to y can be denoted as $x \rightarrow y$. The range is sometimes called the *image* of the domain. The variable x is the **independent variable**, and the variable y is the **dependent variable**. Thus the set of all values for the independent variable x is the domain, and the set of all values for the dependent variable y (which depend on x) is the range. Consider the following example.

EXAMPLE 1

For $A = \{2, 4, 6\}$ and $B = \{-1, 3, 7\}$, consider the relation $F = \{(x, y): x \in A, y \in B, \text{ and } x < y\}$. Show all ordered pairs in the relation F. Find the range and domain of F.

Solution   We must find all points (x, y) for which $x < y$, $x \in A$, and $y \in B$. Since $2 < 3$, $2 < 7$, $4 < 7$, and $6 < 7$, the relation F is $F = \{(2, 3), (2, 7), (4, 7), (6, 7)\}$. The domain is $\{2, 4, 6\}$, and the range is $\{3, 7\}$.   ∎

We can also make a diagram of the mapping of a relation. The mapping of A to B by F from the last example can be diagrammed as follows:

$$\text{Domain Elements} \xrightarrow{\;F\;} \text{Range Elements}$$

Diagrams such as this make it easier to visualize and understand a relation. Notice in this example that the domain of the relation F is A but the range of F is $\{3, 7\} \subseteq B = \{-1, 3, 7\}$. The range does not include -1, since $x \leq -1$ is not true for any $x \in A$. The range contains only elements in B that are used in the mapping of the elements from A. When the relation F does not use all elements of B as the range, it is said that F maps A *into* B as in Example 1. A relation will either map the domain *into* or *onto* a given set as described in the following definition.

Definition
  | *Into and Onto Mappings.* Let F be a relation that maps A to B. When the relation F does not use all elements of B as the range, we say that **F maps A into B**. When the relation F uses all elements of B as the range, we say that **F maps A onto B**.

Now we will work some examples to better understand the ideas of into and onto.

EXAMPLE 2

The conductance G, the measure of the ability of a resistor to conduct current, varies inversely with the resistance R, the measure of the opposition to current. For a particular resistor, $G = 0.02$ siemen when $R = 50$ ohms. Is this a relation? If so, find the range such that G maps the domain $\{25, 100, 200\}$ onto the range.

Solution   Since G varies inversely with R, we remember from Chapter 1 that $G = \dfrac{k}{R}$ where k is a constant. Since we are given that $G = 0.02$ siemen when $R = 50$ ohms, we can use these values to find k. Thus

$$G = \frac{k}{R}$$

$$0.02 = \frac{k}{50} \quad \text{and}$$

$$k = (0.02)(50) = 1$$

Therefore $G = \dfrac{1}{R}$. Now use the domain $\{25, 100, 150\}$ to find the range. When $R = 25$ ohms, $G = \frac{1}{25} = 0.04$ siemen; when $R = 100$ ohms, $G = \frac{1}{100} = 0.01$ siemen; and when $R = 200$ ohms, $G = \frac{1}{200} = 0.005$ siemen. Thus the range is $\{0.04, 0.01, 0.005\}$, and G maps $\{25, 100, 200\}$ onto $\{0.04, 0.01, 0.005\}$, since each element in the range is matched with an element in the domain. ■

EXAMPLE 3

For $A = \{0, 1, 4\}$ and $B = \{-1, 0, 1, 2\}$, find the domain, range, and mapping diagram for the relation $H = \{(x, y): x \in A, y \in B, \text{ and } y^2 = x\}$. Also find the ordered pairs of H.

Solution   For $x \in \{0, 1, 4\} = A$ and $y \in \{-1, 0, 1, 2\} = B$ and $y^2 = x$, or $y = \pm\sqrt{x}$, we know that

$$x \to y = \pm\sqrt{x} \qquad \text{or} \qquad x \in A \xrightarrow{H} y \in B$$

$$x = 0 \to y = \pm\sqrt{0} = 0 \qquad 0 \;.{\swarrow}. \; 0$$

$$x = 1 \to y = \pm\sqrt{1} = \pm1 \qquad 1 \;.{\diagup}. \; 1$$

$$x = 4 \to y = \pm\sqrt{4} = \pm2 \qquad 4 \;.\to. \; 2$$

Thus $H = \{(0, 0), (1, 1), (1, -1), (4, 2)\}$. Note that $(4, -2) \notin H$, since $-2 \notin B$. The domain of H is $\{0, 1, 4\} = A$, and the range of H is $\{-1, 0, 1, 2\} = B$. Thus H maps A *onto* B. (Note that H maps A *into* the set of integers, but H maps A *onto* B.) ■

Functions

Now that we have done some work with relations we can explore that special and most useful type of relation, a *function*. The mapping that is a function is very similar to the receipt you are given when you purchase groceries (Fig. 2.1). The grocery store receipt maps each item purchased to a specific price. The fact that each item purchased maps to only one price is a distinctive property of a function, as we can see in the following definition.

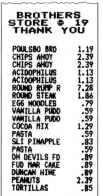

Figure 2.1

Definition | A **function** is a relation in which every domain value is mapped to one and only one range value.

Sometimes it is easier to work with functions as a special set of ordered pairs. An alternative definition provides this additional perspective for functions in terms of ordered pairs.

Alternative Definition | The relation f with $(x, w) \in f$ and $(x, z) \in f$ is a **function** if and only if $w = z$.

The alternative definition provides us with a more concrete method for determining whether a relation is or is not a function, as seen in these examples.

EXAMPLE 4   Are the relations F, G, and H in Examples 1–3 functions?

Solution   We look back at Examples 1–3 to identify the ordered pairs in each relation.
(a) The relation $F = \{(2, 3), (2, 7), (4, 7), (6, 7)\}$ is not a function, since 2 maps to both 3 and 7 [or $(2, 3) \in F$ and $(2, 7) \in F$ but $3 \neq 7$]. In a function, each domain value must map to only one range value.
(b) The relation $G = \{(25, 0.04), (100, 0.01), (200, 0.005)\}$ is a function, since each domain value maps to only one range value.
(c) The relation $H = \{(0, 0), (1, 1), (1, -1), (4, 2)\}$ is not a function, since 1 maps to both 1 and -1. Each domain element of a function must map to only one range element.   ∎

EXAMPLE 5   Is the relation $K = \{(x, y): x \in Z, y \in Z, \text{ and } 2x = |y|\}$ a function? Find the domain and range for K.

Solution   If K is a function, then each value for x will provide only one value for y. Since $x \in Z$ is the independent variable, choose any integer for x. If $x = 0$ then $2(0) = |y|$ and $y = 0$. Thus $(0, 0) \in K$. This result could seem to imply that K is a function, but K must be a function for *all* values of x. To check further, let $x = 1$. If $x = 1$, then $2(1) = |y|$ and $y = \pm 2$. Thus $(1, 2) \in K$ and $(1, -2) \in K$, implying that K is not a function, since $2 \neq -2$. *Therefore K is a relation but not a function.* When considering the domain of K, we notice that $2x = |y| \geq 0$. Hence the domain of K is the set of all integers greater than or equal to zero, which is $D = \{0, 1, 2, 3, 4, \ldots\}$. The range elements are positive and negative even integers, since $|y| = 2x$ and $x \in D$. Thus the range of K is $\{\ldots, -6, -4, -2, 0, 2, 4, 6, \ldots\}$.   ∎

We will also write one variable as a function of another variable as seen in this next example.

EXAMPLE 6   Use the function for the surface area of a sphere to write the radius of a weather balloon as a function of its surface area.

Solution   Since the surface area of a balloon $S = 4\pi r^2$, then

$$4\pi r^2 = S$$

$$r^2 = \frac{S}{4\pi}$$

Since r is a positive distance, we have

$$r = \frac{\sqrt{S}}{\sqrt{4\pi}}$$

$$r = \frac{\sqrt{S}\sqrt{\pi}}{2\sqrt{\pi}\sqrt{\pi}} \quad \textit{(rationalizing the denominator)}$$

$$r = \frac{\sqrt{S\pi}}{2\pi}$$

$$r = \frac{1}{2\pi}\sqrt{S\pi}$$

Thus r is a function of S.   ∎

Function Notation

Notation is important and will help you to manipulate and understand functions and relations. Here are some different ways to visualize a function (or relation) and some special notation for writing a function (or relation). The relation $G = \{(1, 2), (3, 5), (4, 3)\}$ is a function, since every domain element maps to only one range element. Often functions are named with lowercase letters, so this function can also be written as $g = \{(1, 2), (3, 5), (4, 3)\}$. The function $g = \{(1, 2), (3, 5), (4, 3)\}$ has a domain $A = \{1, 3, 4\}$ and a range $B = \{2, 3, 5\}$ and can be described as follows:

(a)

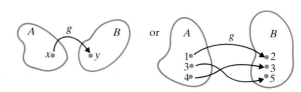

(b) $A \xrightarrow{g} B$ or $g : A \to B$

(c) $g(x) = y$ or $g(1) = 2$ or $g : x \to g(x)$
    with $x \in A$ $g(3) = 5$
    and $y \in B$ $g(4) = 3$

In (a), (b), and (c) the function g maps A to B. The first representation, (a), gives us a pictorial overview of this mapping. In (b) this overview is written more symbolically and less pictorially. Representation (b) is a good way to write the mapping showing both the domain A and the set B. The range is B or a subset of B. The third representation, (c), is the most commonly used notation. In (c) the independent variable x is written in parentheses after the letter for the function as $g(x)$, and $g(x)$ equals the dependent variable y. We read this notation $g(x) = y$ as "g of x equals y" or "g of x is y." Also $g(1) = 2$ is read "g of 1 equals 2" or "g of 1 is 2." It is important for you to notice that the notation $g(x) = y$ represents the ordered pair $(x, y) \in g$ as well as $(x, g(x))$. Likewise, $g(1) = 2$ indicates that $(1, 2) \in g$.

We will now use this new notation to find ordered pairs of a function.

EXAMPLE 7   Find the values of $f(-5)$ and $f(a + 1)$, where $f(x) = x^2 + x + 3$. Write each answer with function notation and as an ordered pair.

Solution   For $f(x) = x^2 + x + 3$:
(a) To find $f(-5)$, substitute (-5) for x in the equation

$$f(x) = x^2 + x + 3$$

Then

$$f(-5) = (-5)^2 + (-5) + 3$$

$$f(-5) = 25 - 5 + 3 = 23$$

$$f(-5) = 23 \quad \text{and} \quad (-5, 23) \in f$$

(b) To find $f(a + 1)$, substitute $(a + 1)$ for x in the equation

$$f(x) = x^2 + x + 3$$

Then

$$f(a + 1) = (a + 1)^2 + (a + 1) + 3$$

$$f(a + 1) = (a^2 + 2a + 1) + (a + 1) + 3$$

$$f(a + 1) = a^2 + 3a + 5 \quad \text{and} \quad (a + 1, a^2 + 3a + 5) \in f \quad \blacksquare$$

We will now use function notation to evaluate a special value L in the next example.

EXAMPLE 8   When $f(x) = 3x - 5$ evaluate

$$L = \frac{f(x + h) - f(x)}{h}$$

Solution   To find L, first evaluate $f(x + h)$ and then substitute $f(x)$ and $f(x + h)$ into the expression for L:

$$f(x + h) = 3(x + h) - 5 = 3x + 3h - 5$$
$$f(x) = 3x - 5$$

Then

$$L = \frac{f(x + h) - f(x)}{h}$$

$$L = \frac{(3x + 3h - 5) - (3x - 5)}{h}$$

$$L = \frac{3x + 3h - 5 - 3x + 5}{h} = \frac{3h}{h} = 3$$ ■

One-to-One Functions

When we worked with relations and functions, we classified them as either into or onto mappings. We will now define another classification for relations and functions. We will call this new type of function a *one-to-one* function as defined here.

Definition   *One-to-One Mapping.* Let f be a function that maps A onto B. The function f is **one-to-one** if every element in B corresponds to only one element in the domain A.

We notice that the concept of a one-to-one mapping is similar to the idea of a function. In a function, each domain element maps to only one range element. In a one-to-one function, every range element is the image of (is mapped from) only one domain element. Again, we have an alternative definition that uses ordered pairs to classify a function as one-to-one or not one-to-one.

Alternative Definition   *One-to-One Mapping.* The function f is **one-to-one** if whenever $(u, y) \in f$ and $(v, y) \in f$. Then $u = v$.

We will now use these definitions to classify the relations and functions in the following examples:

EXAMPLE 9   For $A = B = \{-1, 0, 1\}$, is $g = \{(x, y): x \in A, y \in B, \text{ and } y = x^3\}$ a one-to-one, onto function?

Solution   For $A = B = \{-1, 0, 1\}$,

$$g(-1) = (-1)^3 = -1$$
$$g(0) \quad = (0)^3 \quad = \quad 0$$
$$g(1) \quad = (1)^3 \quad = \quad 1$$

and

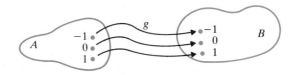

(a) The mapping g is a function, since every value in the domain maps (or corresponds) to only one value in the range.
(b) The function g is one-to-one, since every value in the range is the image of (corresponds to) a unique value in the domain.
(c) The function g is onto B, since all values of B are in the range.   ■

EXAMPLE 10   When $h = \{(x, y) : x \in R, y \in R,$ and $y = |x|\}$, is h a one-to-one, onto function?

Solution   For $h = \{(x, y) : x \in R, y \in R,$ and $y = |x|\}$:
(a) The mapping h is a function, since every domain element corresponds to exactly one range element. For each a, $(a, |a|) \in h$.
(b) The h is not one-to-one, since at least one range element is the image of the two domain elements (or $(1, 1) \in h$ and $(-1, 1) \in h$ but $-1 \neq 1$).
(c) The h maps into R, since not all real numbers are in the range. The relation h maps onto the nonnegative real numbers, since the set of nonnegative real numbers is the range of h.   ■

One-to-one functions are essential to our work with inverse functions later in this chapter.

Domain and Range Restrictions

In future work it will also be important for you to be able to find the domains and ranges of relations and functions. Values that cannot be in the domain or range are eliminated or restricted from the domain or range, respectively. Restrictions will include values for x from the domain and values for the range that

(a) make the denominator zero or
(b) make the expression under a radical with an even index negative.

In this chapter we will limit the domain and range to subsets of the real numbers.
    Some restrictions on the range can be observed from function notation such as $f(x) = |x|$. For $f(x) = |x| = y$, the range is $\{y: y \geq 0\}$, since $f(x) = |x| \geq 0$. The

function $k(x) = \dfrac{x + 3}{x - 2}$ has the domain restriction that $x \neq 2$, since the denominator is zero and k is undefined when $x = 2$. Other range restrictions for $k(x) = \dfrac{x + 3}{x - 2} = y$ are found by solving for x as a function of y as seen in Example 11(d).

EXAMPLE 11

Find the domain and range of the following functions (the domain and range are subsets of the real numbers):

(a) $f(x) = x + 3$,　　**(b)** $g(x) = x^2 + 1$,　　**(c)** $h(x) = \sqrt{x + 3}$, and

(d) $k(x) = \dfrac{x - 3}{x + 2}$. Find the range so that the mapping is onto.

Solution

(a) For $f(x) = x + 3$, x can be any real number, since any real number can be substituted for x. Thus the domain is the set of real numbers, R. If x can be any real number, then $y = x + 3$ can also be any real number. Thus the range, the set of all $y = f(x) = x + 3$, is also R.

(b) For $g(x) = x^2 + 1$, x could be any real number, and the domain is the set of real numbers, R. The range is the set of all $y = g(x) = x^2 + 1$. Since $x^2 \geq 0$, we have $y = g(x) = x^2 + 1 \geq 1$. Thus the range is $\{y: y \geq 1\}$.

(c) For $h(x) = \sqrt{x + 3}$, x can be any value such that $x + 3 \geq 0$. (Other values would make the expression under the radical negative.) Thus $x \geq -3$, and the domain is $\{x: x \geq -3\}$. The elements of the range must be nonnegative (positive or zero), since this is the nonnegative square root. Zero is in the range, since $h(-3) = 0$. Thus the range of h is $\{y: y \geq 0\}$.

(d) For $k(x) = \dfrac{x + 3}{x - 2}$, x can be any real number except 2 (2 would make the denominator zero). The range has no "obvious" limitations. To check in more detail, we must write x as a function of y to observe the range limitations, if any, in $k(x) = \dfrac{x - 3}{x + 2} = y$.

$$\frac{x + 3}{x - 2} = y$$

$$x + 3 = y(x - 2)$$

$$x + 3 = xy - 2y$$

$$x - xy = -2y - 3$$

$$x(1 - y) = -(2y + 3)$$

$$x = \frac{-(2y + 3)}{1 - y} = \frac{2y + 3}{y - 1}$$

Thus the range cannot include $y = 1$. The domain of k is $\{x: x \in R$ and $x \neq 2\}$, and the range of k is $\{y: y \in R$ and $y \neq 1\}$. ■

As we saw in Example 11, function notation and equations are a compact way to write functions. Functions and equations can also be written with several variables. Sometimes it is necessary to solve for one specific variable so that the variable can be seen as a function of the other variables. This is done in part (d) of Example 11.

Now we will be able to apply the concepts of relations, functions, and their notation to the more complicated and interesting ideas in the rest of this chapter and book.

Exercises 2.1

For each of the following sets, find (a) the domain and (b) the smallest set that is the range. State whether the mapping is (c) a function and (d) a one-to-one function. Then draw a mapping diagram for each relation.

1. $f = \{(1, 2), (2, 4), (3, 6), (4, 8)\}$

2. $g = \{(2, 3), (4, 5), (6, 7), (8, 9)\}$

3. $h = \{(1, 0), (1, 1), (2, 3), (4, 5)\}$

4. $k = \{(1, 0), (2, 3), (4, 1), (4, 5)\}$

5. $f = \{(1, 1), (2, 1), (3, 2), (4, 2)\}$

6. $g = \{(3, 0), (4, 1), (5, 0), (1, 1)\}$

List the set of ordered pairs that make up the following relations.

7. $A = \{2, 4, 6\}$, $B = \{1, 5\}$, and $f = \{(x, y): x \in A, y \in B,$ and $x \le y\}$

8. $A = \{2, 4, 6\}$, $B = \{1, 5\}$, and $g = \{(x, y): x \in A, y \in B,$ and $y \le x\}$

9. $h = \{(x, y): x \in N, y \in N,$ and $x = 2y\}$

10. $k = \{(x, y): x \in N, y \in N,$ and $x = 3y\}$

11. $r = \{(x, y): x \in Z, y \in Z,$ and $y = |x|\}$

12. $s = \{(x, y): x \in Z, y \in Z,$ and $x = |y|\}$

13. In Problems 7, 9, and 11, which mappings are (a) functions and (b) one-to-one?

14. In Problems 8, 10, and 12, which mappings are (a) functions and (b) one-to-one?

15. For $f = \{(x, y): y = x^2, x \in R\}$, calculate and simplify each of the following:

(a) $f(x)$ (b) $f(1)$

(c) $f(-2)$ (d) $f\left(\frac{1}{2}\right)$

(e) $f(x + 3)$ (f) $\dfrac{f(x + h) - f(x)}{h}$

16. For $g = \{(x, y): y = 3x - 2, x \in R\}$, calculate and simplify each of the following:

(a) $g(x)$ (b) $g(-1)$

(c) $g(3)$ (d) $g\left(\frac{1}{3}\right)$

(e) $g(x + 3)$ (f) $\dfrac{g(x + h) - g(x)}{h}$

For the functions f in Problems 17–22, evaluate each of the following:

(a) $f(1)$ (b) $f(-3)$

(c) $f\left(\frac{1}{2}\right)$ (d) $f(h)$

(e) $-f(h)$ (f) $f(-h)$

(g) $f\left(\dfrac{1}{x}\right)$ (h) $f(x + h)$

(i) $\dfrac{f(x + h) - f(x)}{h}$

17. $f(x) = 5x - 2$ 18. $f(x) = x^2$

19. $f(x) = \dfrac{1}{x}$ 20. $f(x) = \sqrt{x + 5}$

21. $f(x) = 2x^2$ 22. $f(x) = x^3$

23. A circle has radius r, circumference C, and area A. Find
(a) the radius as a function of the circumference C and
(b) the radius as a function of the area A.

24. A triangle has a base b, a height h to this base, and an area A. Find
(a) the base as a function of the triangle's area and height and
(b) the height as a function of the triangle's area and base.

25. A right circular cone has a volume V, a height h, and a base radius r. Find
    (a) the radius of the base as a function of the volume and height and
    (b) the height as a function of the volume and the radius of the base.

26. The intensity I of a light varies inversely with the square of the distance d from a light. Find the intensity of the light as a function of the distance from the light. What are the restrictions on the domain and range?

27. For a constant temperature the volume V of a gas varies inversely with its pressure P. Find the pressure of the gas as a function of its volume. What are the restrictions on the domain and range?

28. The current I varies directly with the voltage V and inversely with the resistance R in an electrical circuit. Find
    (a) the current as a function of the voltage and resistance and
    (b) the resistance as a function of the voltage and current.

29. According to Kirchhoff's law, the equivalent electrical capacity, C_{eq} (in farads), of two capacitors, with capacity of C_1 and C_2, combined in series as in Fig. 2.2 is described by

$$\frac{1}{C_{eq}} = \frac{1}{C_1} + \frac{1}{C_2}$$

Find C_{eq} as a function of C_1 and C_2. Simplify the answer.

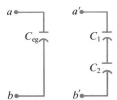

Figure 2.2

30. Find C_1 as a function of C_{eq} and C_2 for the formula in Problem 29. Simplify the answer.

31. For energy cells in series (Fig. 2.3), Ohm's law defines the relationship

$$I = \frac{nE}{R_0 + nb}$$

where I amperes is the current and n is the number of cells, each having an electromotive force of E volts, an internal resistance of b ohms, and an external resistance of R_0 ohms. Find R_0 as a function of the other variables.

32. Find b as a function of the other variables for Ohm's law in Problem 31.

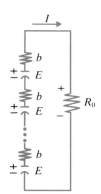

Figure 2.3

33. The equation $H_T = \left(\dfrac{H_S}{A}\right) + \left(\dfrac{H_M}{A}\right)$ is used in solar design to calculate the total heat gain H_T per square foot of floor area, where H_S is the heat gain through unshaded skylights and windows; H_M is the heat gain through thermal storage walls, roof ponds, or walls adjacent to an attached greenhouse; and A is the floor area of the total space. Find H_S as a function of H_T, H_M, and A. Simplify the answer.

34. For the solar heat gain equation in Problem 33, find H_M as a function of H_T, H_S, and A. Simplify your answer.

For the following functions, find (a) the real number interval that is the domain of f and (b) the smallest real number interval that is the range of f.

35. $f(x) = \sqrt{4 - 5x}$

36. $f(x) = \sqrt{2 - 7x}$

37. $f(x) = \dfrac{x - 1}{x + 1}$

38. $f(x) = \dfrac{x + 2}{x - 1}$

39. $f(x) = \dfrac{x - 3}{2x - 4}$

40. $f(x) = \dfrac{x - 2}{3x - 9}$

41. $f(x) = \dfrac{5x + 3}{x - 1}$

42. $f(x) = \dfrac{2x + 5}{x + 1}$

43. $f(x) = \dfrac{1}{|x| + 1}$

44. $f(x) = \dfrac{1}{|x + 1|}$

45. $f(x) = \dfrac{2}{\sqrt{x^2 - 9}}$

46. $f(x) = \dfrac{1}{\sqrt{x^2 - 5x + 6}}$

47. A rectangle is inscribed in a circle (Fig. 2.4) with a diameter of 12 inches. Find the area of the rectangle as a function of one side of the rectangle.

48. A rectangle with a 4-inch side is inscribed in a circle with a diameter of d inches (Fig. 2.4). Find the area of the rectangle as a function of the diameter of the circle.

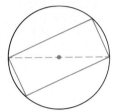

Figure 2.4

49. A Norman window is a rectangle surmounted by a semi-circle as shown in Fig 2.5. If the rectangle has a width of x and a height of $2x$, find the area of the window as a function of x.

Figure 2.5

50. A Norman window is a rectangle surmounted by a semi-circle. If the rectangle has a width of $2x$ and a height of $3x$, find the area of the window as a function of x.

51. A Norman window has a perimeter of 200 inches. Let r be the radius of the semicircle and find the area of the window as a function of r.

2.2   Graphing Functions

In the example at the beginning of this chapter we saw how experimental data can be displayed on a graph. The graph in that example was drawn on the **Cartesian** (or **rectangular**) **coordinate system**, with which we worked in Chapter 1. The Cartesian coordinate system helps us to see the interrelationship between the domain and range values of a relation or function. All of the relations and functions that we worked with in Section 2.1 are sets of ordered pairs that we can easily display on the Cartesian (or rectangular) coordinate system.

Another example of this relationship between range and domain values is shown in Fig. 2.6. Here we see the relationship of the acid in rain to different times of a rainstorm during three different rainstorms at Gainesville, Florida. The pH level is tested every hour for a 5-minute interval. Since 7 is neutral on the pH scale and 0 is a strongly acidic solution, as the pH moves from 7 to 0, the rain becomes a more acidic solution. Figure 2.6 shows the relationship between the length of the storm and the acidity of the rain. One conclusion that we can reach from looking at this graph is that in all three storms the least acidic value was measured at the beginning of each storm.

As we can see here, graphing functions and relations on the Cartesian coordinate system can be a useful tool.

The Coordinate System

As we noted earlier, the Cartesian (or rectangular) coordinate system is made up of two perpendicular real number lines that intersect at their origins. Usually, one real number line is horizontal and is called the **x-axis**, while the other real number

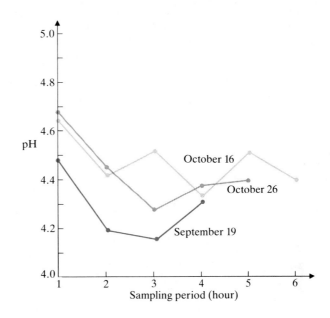

Figure 2.6

Source: "Acid Rain Information Book," by GCA Corporation for U.S. Dept. of Energy, Dec. 1980.

line is vertical and is called the **y-axis**. The *x*-axis is the *domain* axis, and the *y*-axis is the *range* axis. The ordered pair (a, b) has a unique point in the plane created by the Cartesian coordinate system. For the point (a, b) the *x*-coordinate (or **abscissa**) a is the *domain* element and the *y*-coordinate (or **ordinate**) b is the *range* element.

Graphing Functions

Graphing a function is a matter of determining and plotting all ordered pairs on the Cartesian coordinate system. If we know the ordered pairs, it is easy to graph a function (or relation).

EXAMPLE 1   Graph the relations **(a)** $g = \{(1, 2), (3, 5), (4, 3)\}$ and **(b)** $h = \{(-1, 1), (0, 0), (1, 1)\}$.

Solution

(a)

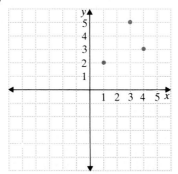

(b)

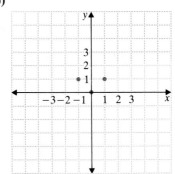

In Chapter 1 we worked with the different forms for the equation of a line as well as the graph of a line. Using the slope-intercept form for the equation of a line, $y = mx + b$, we can define a *linear function*.

Definition | The function $f(x) = ax + b$ is called a **linear function** when a and b are real numbers and $a \neq 0$. If $a = 0$ in $f(x) = ax + b$, then $f(x) = b$ is called a **constant function**.

EXAMPLE 2   Is the line graphed in Fig. 2.7 a linear function of x? If it is, find the linear function.

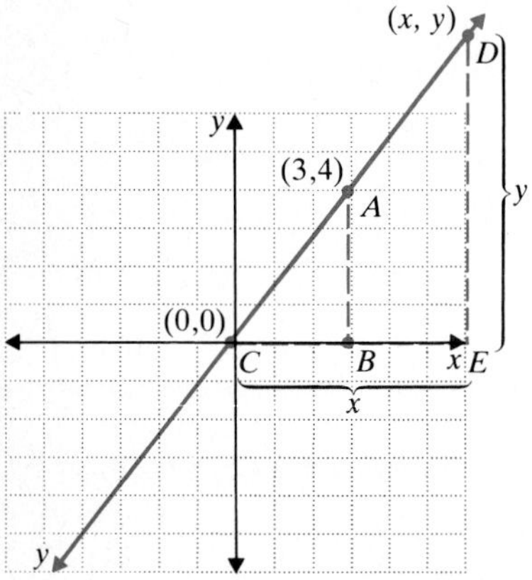

Solution   Since triangle ABC is similar to triangle DEC and corresponding sides of similar triangles are proportional, we have

$$\frac{DE}{AB} = \frac{EC}{BC}$$

$$\frac{y - 0}{4 - 0} = \frac{x - 0}{3 - 0}$$

$$y = \left(\frac{4}{3}\right)x \qquad \text{or}$$

$$f(x) = \left(\frac{4}{3}\right)x$$

Thus y is a linear function, since it is in the form $f(x) = ax + b$, where $a = \frac{4}{3}$ and $b = 0$. ∎

Figure 2.7

EXAMPLE 3   Find the linear function containing the point $(-2, 5)$
(a) parallel and   **(b)** perpendicular to the line $6x + 2y + 3 = 0$.

Solution   To find the slope of the given line $6x + 2y + 3 = 0$, we write the equation in the slope-intercept form ($y = mx + b$).

$$6x + 2y + 3 = 0$$
$$2y = -6x - 3$$
$$y = -3x - \frac{3}{2}$$

Thus the line $6x + 2y + 3 = 0$ has a slope of -3.
(a) The line containing $(-2, 5)$ and parallel to the given line also has a slope of $m = -3$. Its equation is

$$(y - y_1) = m(x - x_1)$$
$$(y - 5) = -3[x - (-2)]$$
$$y = -3(x + 2) + 5$$
$$y = -3x - 6 + 5$$
$$y = -3x - 1$$

Therefore the linear function is $f(x) = -3x - 1$.

(b) The perpendicular line through $(-2, 5)$ has a slope of $m = -\frac{1}{-3} = \frac{1}{3}$, since the slope is the negative reciprocal of -3. Now we can write the equation of the perpendicular line as

$$(y - y_1) = m(x - x_1)$$

$$y - 5 = \frac{1}{3}[x - (-2)]$$

$$y = \frac{1}{3}(x + 2) + 5$$

$$y = \frac{1}{3}x + \frac{2}{3} + 5 = \frac{1}{3}x + 5\frac{2}{3}$$

Therefore

$$f(x) = \frac{1}{3}x + 5\frac{2}{3}$$ ■

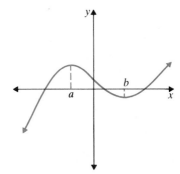

Figure 2.8

Increasing and Decreasing Functions

We can classify functions in many ways, but we will now define them as increasing or decreasing functions. Intuitively, you might think that an increasing function would move uphill when moving from left to right (as x increases), and that is correct for an increasing function. Likewise, a decreasing function moves downhill from left to right (as x increases). These intuitive ideas provide the basis for the definition that follows.

Definition

Let f be a function where x_1 and x_2 are any two elements from the domain of f. The function f is

(a) an **increasing function** if $f(x_1) \leq f(x_2)$ for every $x_1 < x_2$;
(b) a **decreasing function** if $f(x_1) \geq f(x_2)$ for every $x_1 < x_2$;
(c) a **strictly increasing function** if $f(x_1) < f(x_2)$ for every $x_1 < x_2$;
(d) a **strictly decreasing function** if $f(x_1) > f(x_2)$ for every $x_1 < x_2$.

You will be able to easily identify domain intervals on which the function is increasing or decreasing by examining its graph, as we will see in the next example.

EXAMPLE 4

Identify the intervals for the function in Fig. 2.8 in which the function is increasing or decreasing.

Solution   From the graph this function is increasing (and strictly increasing) in the interval $(-\infty, a]$ and also in the interval $[b, \infty)$. Likewise, this function is decreasing (and strictly decreasing) in the interval $[a, b]$. ■

As you can see from the definition and Example 4, as x increases, the range value $f(x)$ for an increasing function stays the same or increases; in a strictly increasing function, as x increases, the range value $f(x)$ always increases. Likewise, for decreasing and strictly decreasing functions, as x increases, the range value of the decreasing function stays the same or decreases while the range value of the strictly decreasing function only decreases.

We will now work with some functions that are increasing or decreasing.

EXAMPLE 5   Find the distance d a car can travel as a function of the time t it travels with a speed of 35 miles per hour. Is this function increasing, decreasing, strictly increasing, strictly decreasing, or none of these?

Solution   This is the function $d = rt$ or in this case $d = 35t$, a linear function. Graphing this function (Fig. 2.9), we see that $d = 35t$ is strictly increasing.   ■

EXAMPLE 6   The price of gasoline is inversely proportional to the percentage of refinery production capacity. If refineries are at 85% of production capacity when the pump price of gasoline is $1 per gallon, find the price of gasoline as a function of refinery production capacity. Is the function increasing, decreasing, strictly increasing, strictly decreasing, or none of these?

Solution   Let C be the price, or cost, of a gallon of gasoline and let r be the percentage of refinery production capacity in decimal or fractional form. With C inversely proportional to r and $C = 1$ when $r = .85$, we have

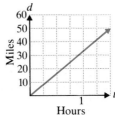

$$C = \frac{k}{r}$$

$$1 = \frac{k}{.85}$$

$$k = .85$$

Figure 2.9

Therefore

$$C(r) = \frac{.85}{r}$$

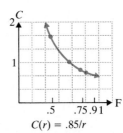

$C(r) = .85/r$

Figure 2.10

We will now find some other points on the graph of $C(r)$:

$$C(.50) = \frac{.85}{.50} = 1.70$$

$$C(.90) = \frac{.85}{.90} = .94$$

$$C(.75) = \frac{.85}{.75} = 1.13$$

The domain of C must be all $r \geq 0$. From the graph of $C(r) = \dfrac{.85}{r}$ shown in Fig. 2.10 it appears that C is strictly decreasing.   ■

Graphical Tests

Often, you can identify the type of function you are working with by evaluating the properties of its graph. As we saw with increasing and decreasing functions, it is possible to identify the characteristics of a function from its graph. We will now find a graphical test for identifying a function and also identifying a one-to-one relation (or function).

THE VERTICAL LINE TEST (FUNCTION TEST)	If every vertical line intersects (or crosses) the graph of a relation at most once, then the graph of a relation is a *function*. This test guarantees that no *x*-value in the domain corresponds to more than one *y*-value in the range.

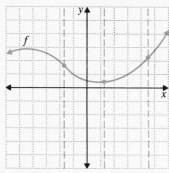

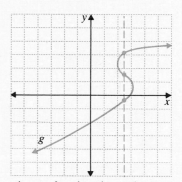

f is a function, since each vertical line intersects the graph only once

g is not a function, since at least one vertical line crosses the graph more than once

THE HORIZONTAL LINE TEST (ONE-TO-ONE TEST)	If every horizontal line intersects the graph at most once, then the graph is a *one-to-one relation*. This test guarantees that no *y*-value corresponds to more than one *x*-value.

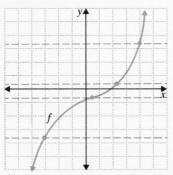

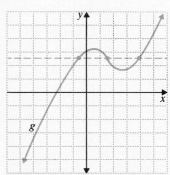

f is one-to-one, since each horizontal line intersects the graph only once

g is not one-to-one, since at least one horizontal line crosses the graph more than once

We can now look at a graph and quickly determine whether we have a function, a one-to-one relation, or a one-to-one function. We will apply these graphical tests in the following example.

EXAMPLE 7   Identify each of the following as a function or a relation and one-to-one or not one-to-one, using the vertical and horizontal line tests:

(a)

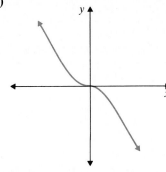

(b)

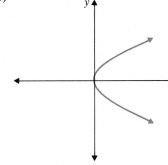

(c) $x + 3y = 5$

(d) $f(x) = -x + 3$

Solution

(a) This is a function by the vertical line test, since every vertical line crosses the graph at most once. It is also one-to-one by the horizontal line test, since each horizontal line crosses the graph only once. Therefore (a) is a one-to-one function.

(b) This is not a function, since it fails the vertical line test. It is one-to-one, since every horizontal line intersects the graph at most once. Therefore (b) is a one-to-one relation.

(c) First we graph the line $x + 3y = 5$, using the points $(5, 0)$, $(2, 1)$, and $(-1, 2)$ on $x + 3y = 5$ as in Fig. 2.11. The graph passes both the horizontal and vertical line tests; therefore this line is a one-to-one function.

(d) First graph the linear function $f(x) = -x + 3$. This is the line $y = -x + 3$, which has a slope of $m = -1$ and y-intercept of 3. As we see in Fig. 2.12, the graph passes the horizontal and vertical line tests and is therefore a one-to-one function.   ■

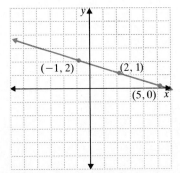

Figure 2.11

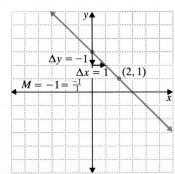

Figure 2.12

Exercises 2.2

Find the domain and the smallest range of each of the following relations.

For each of the following relations find the real number interval that is (a) the domain and (b) the smallest range.

1.

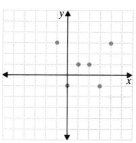

The relation F

5.

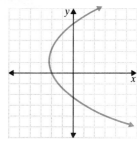

The relation F

2.

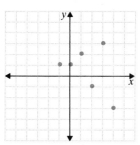

The relation G

6.

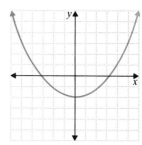

The relation G

3.

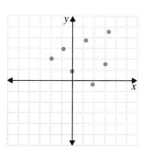

The relation H

7.

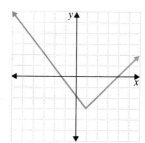

The relation H

4.

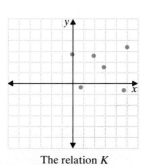

The relation K

8.

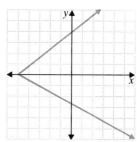

The relation K

9.

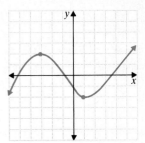

The relation *F*

10.

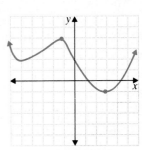

The relation *G*

11.

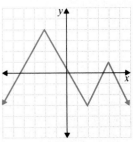

The relation *H*

12.

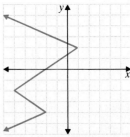

The relation *K*

13.

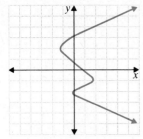

The relation *F*

14.

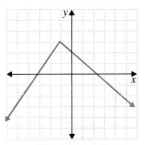

The relation *G*

15.

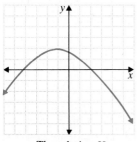

The relation *H*

16.

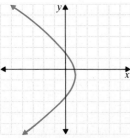

The relation *K*

17. Use the appropriate graphical tests and list the relations in Problems 1, 3, 5, 7, 9, 11, 13, and 15 that are (a) functions and (b) one-to-one function.

18. Use the appropriate graphical tests and list the relations in Problems 2, 4, 6, 8, 10, 12, 14, and 16 that are (a) functions and (b) one-to-one functions.

19. For the functions in Problems 7, 9, and 15, indicate the domain interval in which the function is (a) increasing and (b) decreasing.

20. For the functions in Problems 6, 10, and 14, indicate the domain interval in which the function is (a) increasing and (b) decreasing.

21. Find the linear function f with $(2, 5) \in f$ and slope $m = 2$.

22. Find the linear function g with $(3, 1) \in g$ and slope $m = 5$.

23. Find the linear function h with $(2, 5) \in h$ and $(3, -2) \in h$.

24. Find the linear function k with $(7, -4) \in k$ and $(-1, 6) \in k$.

25. Find the linear function f with $(2.7, -1.5) \in f$ and $(-5.1, 3.4) \in f$.

26. Find the linear function g with $(-4.6, 4.3) \in g$ and $(1.7, 2.1) \in g$.

27. Find the linear function f when $f(2) = -7$ and $f(-3) = 8$.

28. Find the linear function f when $f(-1) = 2$ and $f(2) = 11$.

29. Find the linear function g when $g(a + 1) = a - 3$ and $g(a + 2) = a - 2$.

30. Find the linear function g when $g(a + 5) = a + 1$ and $g(a + 1) = a - 3$.

31. Find the linear function containing the point $(1, -3)$ (a) parallel and (b) perpendicular to the line $y = -2x - 3$.

32. Find the linear function containing the point $(-2, 5)$ (a) parallel and (b) perpendicular to the line $y = 4x + 1$.

33. Find the linear function containing the point $(-3, 2)$ (a) parallel and (b) perpendicular to the line $2x - 3y = 5$.

34. Find the linear function containing the point $(1, -7)$ (a) parallel and (b) perpendicular to the line $5x - 2y = 7$.

35. Two trains start toward each other from two stations with 300 miles of track between the stations. One train averages 72 miles per hour, and the other train averages 49 miles per hour. Find the distance between the two trains as a function of time.

36. Two trains start from two stations with 300 miles of track between the stations. One train averages 72 miles per hour, and the other train averages 49 miles per hour. Find the miles of track between the two trains as a function of time when the trains are headed in opposite directions.

37. In the two right triangles shown in Fig. 2.13, $a = 4$, $b = x$, and $d = y$. Find y as a function of x. Is this a linear function? Is this function increasing, decreasing, or neither?

38. In the two right triangles shown in Fig. 2.13, $a = x$, $b = y$, and $d = 4$. Find y as a function of x. Is this a linear function? Is this function increasing, decreasing, or neither?

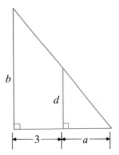

Figure 2.13

39. The sum of the radii of the two circles in Fig. 2.14 is 10 inches. Find the sum of the areas of the two circles as a function of the radius r of the smaller circle.

Figure 2.14

40. A rectangular sheet of metal is 30 feet long by x feet wide. A rain gutter is formed by folding a 30-foot by 6-inch rectangle on each 30-foot side as shown in Fig. 2.15. Find the volume of the rain gutter as a function of x. Is this a linear function?

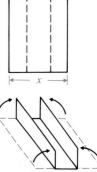

Figure 2.15

41. A rectangular sheet of metal is 20 feet long by 1 foot wide. A rain gutter is formed by folding a 20-foot by x-foot rectangle on each 20 foot side as shown in Fig. 2.15. Find the volume of the rain gutter as a function of x. Is this a linear function?

42. Business assets can often be depreciated over a specific period of time to account for their replacement cost. One method of depreciation is called linear depreciation. Find the linear depreciation function that describes the value y of an asset after x years when the asset is depreciated for a total of n years if the original cost is C and the final cost after n years is 0.

43. Since antiquity, many philosophers, artists, and mathematicians have been intrigued by the special ratio of two sides of a rectangle called the Golden Ratio (or the Golden Sector). When the sides of the rectangle are a and b, the Golden Ratio is $\dfrac{a}{b} = \dfrac{b}{(a+b)}$. Find a as a function of b for a rectangle with sides that provide the Golden Ratio

(Fig. 2.16). Use this function to find the value of the Golden Ratio of $\dfrac{a}{b}$.

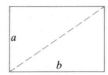

Figure 2.16

44. Using Problem 43, find the diagonal of this rectangle as a function of a for the rectangle with sides a and b whose ratio $\dfrac{a}{b}$ is the Golden Ratio.

45. The price of gasoline is inversely proportional to the percentage of refinery production capacity. If refineries are at 80% of production capacity when the pump price of gasoline is $1.10 per gallon, find the price of gasoline as a function of refinery production capacity. Is this function increasing, decreasing, strictly increasing, strictly decreasing, or none of these?

2.3   Graphing Some Special Functions

In Section 2.2 we concentrated on graphing linear function (lines) on the Cartesian coordinate system. In this section we will learn how to graph some other functions (the quadratic function, a cubic function, the absolute value function, and the greatest integer function) as well as multiples of these functions. These new functions are common functions in future mathematics courses and are also important in many real-world applications.

Each new function in this section has a stereotyped shape. Knowing the general shape of each function will provide you with a mental picture that you can use to graph this type of function in the future. Study the examples and explanations in this section to recognize these basic shapes. Recognizing a function's shape will shorten the time you spend graphing a similar function.

The Quadratic Function

Quadratic functions have many applications, including the description of projectile paths, falling objects, and the shape of reflecting surfaces such as solar reflectors and satellite-receiving dishes.

Definition | The **quadratic function** is $f(x) = ax^2 + bx + c$ where a, b, and c are real numbers and $a \neq 0$.

Earlier, we identified the quadratic function $f(x) = ax^2 + bx + c$ as a second-degree polynomial in one variable. This observation will help to identify a quadratic function.

EXAMPLE 1   The altitude s of a missile is measured as a function of time by $s(t) = 4800t - 16t^2$, where altitude $s(t)$ is measured in feet and time t is measured in seconds. Is the function s a quadratic function? What is the altitude of the missile after **(a)** 100 seconds, **(b)** 150 seconds, **(c)** 200 seconds, and **(d)** 5 minutes?

Solution   The function s is a quadratic function, since $s(t) = 4800t - 16t^2$ is a second-degree polynomial,
(a) If $t = 100$ seconds, then $s(100) = 4800(100) - 16(100)^2 = 320{,}000$ feet;
(b) If $t = 150$ seconds, then $s(150) = 4800(150) - 16(150)^2 = 360{,}000$ feet;
(c) If $t = 200$ seconds, then $s(200) = 4800(200) - 16(200)^2 = 320{,}000$ feet;
(d) If $t = 5$ minutes $= 300$ seconds, then $s(300) = 4800(300) - 16(300)^2 = 0$ feet, and the missile is back on the ground again.   ■

We can find the basic shape of the quadratic function by examining the graph of $f(x) = ax^2 + bx + c$ when $b = 0$ and $c = 0$. This graph is developed in the following example.

EXAMPLE 2   Graph the quadratic function $h(x) = x^2$. Is it a one-to-one function? On what domain intervals is h increasing and decreasing?

Solution   Careful consideration of the equation $h(x) = x^2$ tells us that the domain element x can be any real number while the range elements are the nonnegative real numbers. Thus all $h(x) = y \geq 0$. It is now necessary to plot some points to have the correct shape for this graph. Since x is the independent variable, we pick several values for x, compute the value for y, and then plot the resulting ordered pair.

x	$h(x) = x^2$	(x, y)
0	$f(0) = 0^2 = 0$	$(0, 0)$
1	$f(1) = 1^2 = 1$	$(1, 1)$
-1	$f(-1) = (-1)^2 = 1$	$(-1, 1)$
$\sqrt{2} \approx 1.414$	$f(\sqrt{2}) = (\sqrt{2})^2 = 2$	$(1.414, 2)$
$-\sqrt{2} \approx 1.414$	$f(-\sqrt{2}) = (-\sqrt{2})^2 = 2$	$(-1.414, 2)$
2	$f(2) = 2^2 = 4$	$(2, 4)$
-2	$f(-2) = (-2)^2 = 4$	$(-2, 4)$
3	$f(3) = 3^2 = 9$	$(3, 9)$

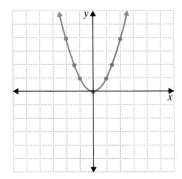

Figure 2.17

Connecting the plotted points from this chart provides a sketch of the graph of $h(x) = x^2$ for all domain elements (Fig. 2.17). By the vertical line test, $h(x) = x^2$

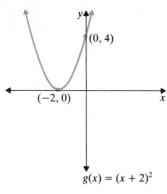

$$g(x) = (x + 2)^2$$

Figure 2.18

appears to be a function, but by the horizontal line test it is not one-to-one. From this graph we see that h is decreasing on the interval $(-\infty, 0]$ and increasing on the interval $[0, \infty)$. ■

The shape in Fig. 2.17 is the correct shape for the quadratic function $h(x) = x^2$. This shape is called a **parabola**. The point where the quadratic function changes from increasing to decreasing or decreasing to increasing is called the **vertex** of the parabola. In Fig. 2.17 the vertex of $h(x) = x^2$ is $(0, 0)$.

In general, this basic shape (the parabola in Fig. 2.17) is the shape of any quadratic function $f(x) = ax^2 + bx + c$. When you recognize the equation of a quadratic function, recall the shape of the parabola. Then combine the shape of the parabola with the limits on the range and make a quick sketch of the quadratic function. A few points on the graph may be calculated to provide more accuracy, but it is important not to rely on the calculation of points to provide the basic shape of the graph. This process is shown in the next example.

EXAMPLE 3

Sketch the graph $g(x) = (x + 2)^2$.

Solution   First, recognize that the equation $g(x) = (x + 2)^2$ is a quadratic function with all $g(x) \geq 0$. In fact, $g(x) = 0$ when $x = -2$, so $(-2, 0) \in g$ and all other $g(x) > 0$. Then use the shape of the parabola, $(-2, 0)$, and $g(x) > 0$ for $x \neq -2$ to sketch the graph (Fig. 2.18). Accuracy is improved a little by noting the y-intercept $g(0) = (0 + 2)^2 = 4$. ■

EXAMPLE 4

Sketch the graph of $k(x) = -x^2 - 4x - 4$.

Solution   First factor this form of the quadratic function $k(x)$:

$$k(x) = -x^2 - 4x - 4$$
$$k(x) = -(x^2 + 4x + 4)$$
$$k(x) = -(x + 2)^2$$

This parabola's range elements are $k(x) \leq 0$ and $k(x) = 0$ when $x = -2$. The y-intercept is $k(0) = -4$. Also notice that $k(x)$ is the negative of $g(x)$ in Example 2. Thus $k(x) = -g(x)$, and the graph of k could be sketched quickly by sketching the negative of all range values from g for each appropriate domain element, as shown in Fig. 2.19. ■

Now that we can recognize the quadratic function as the parabolic shape, we will also consider multiples of the basic parabolic shape.

Multiples of Functions

The functions $\frac{1}{4}f$, $2f$, and $5f$ are all multiples of f. Understanding the relationships between f, $\frac{1}{4}f$, $2f$, and $5f$ will make it easier to graph these multiples of f. In general,

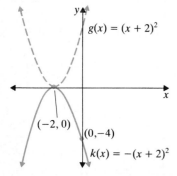

$$g(x) = (x + 2)^2$$
$$(-2, 0)$$
$$(0, -4)$$
$$k(x) = -(x + 2)^2$$

Figure 2.19

it will be helpful to understand how the constant a times a function f changes the shape of the function. To see this change in shape, we will use the basic parabolic shape $f(x) = x^2$ as the standard for comparison. The following example illustrates the relationship between different multiples of a function.

EXAMPLE 5

Graph the following functions on one coordinate system:
(a) $f(x) = x^2$,
(b) $g(x) = \frac{1}{4}x^2$,
(c) $h(x) = 3x^2$, and
(d) $k(x) = -x^2$.

Solution

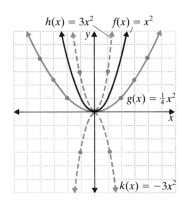

	(a)	**(b)**	**(c)**	**(D)**
x	$f(x) = x^2$	$g(x) = \dfrac{1}{4}x^2$	$h(x) = 3x^2$	$k(x) = -3x^2$
0	0	0	0	0
± 1	1	$\dfrac{1}{4}$	3	-3
± 2	4	1	12	-12
± 3	9	$\dfrac{9}{4} = 2\dfrac{1}{4}$	27	-27
± 4	16	4	48	-48

Figure 2.20

Plotting these points provides the graphs in Fig. 2.20.   ■

Using the chart for f, g, h, and k, we can make the following conclusions. For each specific domain value of g and f, notice that $g(x) = \frac{1}{4}x^2$ is smaller than $f(x) = x^2$. Also notice that for each specific domain value, $h(x) = 3x^2$ is three times larger than that of $f(x) = x^2$. Since $k(x) = -h(x)$, the range values of k are the negative of all the range values of h.

Example 5 demonstrates that for the constant a,

(a) if $a > 1$, then $a \cdot f(x)$ gets larger more quickly than $f(x)$
    [$a \cdot f(x)$ is "skinnier" than $f(x)$];
(b) if $0 < a < 1$, then $a \cdot f(x)$ gets larger more slowly than $f(x)$
    [$a \cdot f(x)$ is "fatter" than $f(x)$]; and
(c) if $a < 0$, then the graph of $a \cdot f(x)$ is the negative of
    each value of $|a| \cdot f(x)$.

Using the absolute value, we can write these results more generally as follows.

MULTIPLES OF A
FUNCTION

For functions g and f where $g(x) = a \cdot f(x)$ and constant a,

(a) when $a > 1$, then $|g(x)| > |f(x)|$;
(b) when $0 < a < 1$, then $|g(x)| < |f(x)|$; and
(c) when $a < 0$, then $g(x) = -|a|f(x)$.

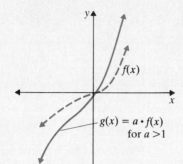
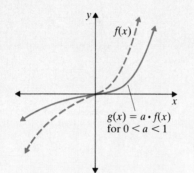

Now we can graph the multiple of a function by making the range values larger or smaller than the range values of the basic function.

EXAMPLE 6

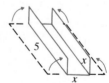

A rectangular piece of sheet metal is used to make a rain gutter. The width is divided into three equal lengths and bent to form a three-sided rain gutter (a bottom and two sides) as seen in Fig. 2.21. If the rectangular piece of sheet metal is 5 feet long, write the volume of the rain gutter as a function of its depth. Then sketch a graph of the volume function.

Solution   As seen in Fig. 2.21, the rain gutter has a length of 5 feet, a depth of x, and a width of x. Thus the volume is $V = (5)(x)(x) = 5x^2$ or, in function notation, $V(x) = 5x^2$. We can now sketch this function as a multiple of the function $f(x) = x^2$ (Fig. 2.22). Notice that $V(x)$ is skinnier than $f(x)$, and the domain must be limited to $x > 0$.   ■

Figure 2.21

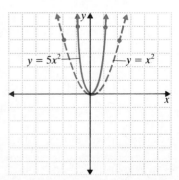

Figure 2.22

These techniques for graphing the multiples of a function can also be used on other functions. We will now find the stereotyped shape of another basic function.

A Cubic Function

A cubic function is common to algebra and calculus. Although there are many cubic functions, this section will deal with the basic cubic function.

Definition ▌ A **cubic function** is $f(x) = x^3$.

We must first find the basic shape of $f(x) = x^3$.

EXAMPLE 7   Sketch the graph $f(x) = x^3$. Is it a one-to-one function? Find the intervals on which this function is increasing and decreasing.

Solution   The domain and range elements for $f(x) = x^3$ can be any real number. To better see this, we will compute and plot some points on the graph of $f(x) = x^3$.

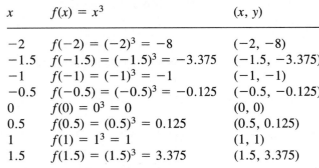

x	$f(x) = x^3$	(x, y)
-2	$f(-2) = (-2)^3 = -8$	$(-2, -8)$
-1.5	$f(-1.5) = (-1.5)^3 = -3.375$	$(-1.5, -3.375)$
-1	$f(-1) = (-1)^3 = -1$	$(-1, -1)$
-0.5	$f(-0.5) = (-0.5)^3 = -0.125$	$(-0.5, -0.125)$
0	$f(0) = 0^3 = 0$	$(0, 0)$
0.5	$f(0.5) = (0.5)^3 = 0.125$	$(0.5, 0.125)$
1	$f(1) = 1^3 = 1$	$(1, 1)$
1.5	$f(1.5) = (1.5)^3 = 3.375$	$(1.5, 3.375)$

We now sketch the graph by connecting these points as in Fig. 2.23. By the horizontal and vertical line tests, $f(x) = x^3$ is a one-to-one function. As we see in the graph, the cubic function $f(x) = x^3$ is an increasing function.   ■

Figure 2.23

Although we needed to plot points the first time we sketched the graph $f(x) = x^3$, we can use this cubic shape to graph other cubic functions of this type.

EXAMPLE 8   Sketch the graph of $g(x) = \frac{1}{4}x^3$.

Solution   We first notice that $g(x)$ is a multiple of the basic cubic shape of $f(x) = x^3$. Thus all the range values for $g(x)$ will be $\frac{1}{4}$ the range values of $f(x) = x^3$ for respective domain elements (Fig. 2.24).   ■

Now we can recognize two basic functions, the quadratic function and the cubic function, as well as their respective shapes.

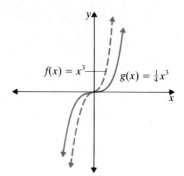

$f(x) = x^3$

$g(x) = \frac{1}{4}x^3$

Figure 2.24

The Absolute Value Function

Another common and useful function is the absolute value function.

Definition ▌ The **absolute value function** is $f(x) = |x|$.

We have two ways to graph the absolute value function. We could sketch the graph by calculating and plotting individual points, but it is quicker and more accurate to use the definition of the absolute value, as seen in the following example.

EXAMPLE 9   Sketch the graph of $f(x) = |x|$. Is $f(x) = |x|$ a one-to-one function? On what intervals is this function increasing and decreasing?

Solution   In evaluating this function, notice that the domain is the set of all real numbers and that $f(x) = |x| \geq 0$ with $f(0) = 0$ (thus $(0, 0) \in f$). By definition of the absolute value,

$$f(x) = \begin{cases} x & \text{for } x \geq 0 \\ -x & \text{for } x < 0 \end{cases}$$

Thus the absolute value function is a combination of two linear functions, one for each part of the domain. When $x \geq 0$, $f(x) = x$ (Fig. 2.25a); and when $x < 0$, $f(x) = -x$ (Fig. 2.25b). Putting both parts together, we have the graph of the absolute value function $f(x) = |x|$ (Fig. 2.25c). As we see in the graph, the absolute value function is strictly decreasing in the interval $(-\infty, 0]$ and strictly increasing in the interval $[0, \infty)$. ∎

As we see in Fig. 2.25(c), the basic shape of the absolute value function is a V-shape, created by two linear functions on their limited domain intervals. We could obtain the same results by calculating and plotting points; but in the long run, plotting points as a primary means of graphing is not the best method. We can now recognize the graph of the absolute value as a V-shape. We will apply this result in the next example.

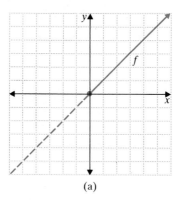

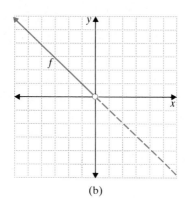

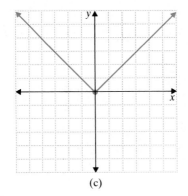

(a)   (b)   (c)

Figure 2.25

EXAMPLE 10   Sketch the graph of $g(x) = \frac{1}{2}|x + 1|$.

Solution   The domain of $g(x)$ is the set of all real numbers, and the range is the set of nonnegative real numbers, since $g(x) \geq 0$ with $g(-1) = 0$ (thus $(-1, 0) \in g$). Notice that this is a multiple of the absolute value function. Since the multiplier is $\frac{1}{2}$, the range values of the sides of the V-shape will be $\frac{1}{2}$ those of $f(x) = |x + 1|$ in Example 9. We now graph the two linear functions that make up this absolute value function. According to the definition,

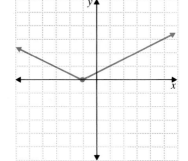

Figure 2.26

$$g(x) = \begin{cases} \frac{1}{2}(x + 1) = \frac{1}{2}x + \frac{1}{2} & \text{for} \quad (x + 1) \geq 0 \quad \text{or} \quad x \geq -1 \\[2mm] -\frac{1}{2}(x + 1) = -\frac{1}{2}x - \frac{1}{2} & \text{for} \quad (x + 1) < 0 \quad \text{or} \quad x < -1 \end{cases}$$

Then graph $g(-1) = 0$, $g(x) = \frac{1}{2}x + \frac{1}{2}$ when $x \geq -1$, and $g(x) = -\frac{1}{2}x - \frac{1}{2}$ when $x < -1$. Notice that the slopes of both sides of the V-shape for $g(x)$ (Fig. 2.26) are $\frac{1}{2}$ the slope of the sides of the V-shape for $f(x)$ (Fig. 2.25c). ∎

Now we can recognize a third function, the absolute value function, and its V-shaped graph.

The Greatest Integer Function

Various forms of this new function, the greatest integer function, are very practical functions in problems that require rounding off of range values. The greatest integer function is also known as the step function or the postage function. An application of the greatest integer function would be a postage system that charged 25¢ for a delivery less than 1 ounce and 25¢ for each ounce or fraction of an ounce thereafter. Thus all charges are multiples of 25¢. We will be able to write this cost of postage as a function by using the greatest integer function.

Definition | The **greatest integer function**, denoted by $[\![x]\!]$, is defined as $[\![x]\!] = n$, where n is an integer and $n \leq x < n + 1$ (n is the largest integer such that $n \leq x$). In words, $[\![x]\!] = n$ means that n is the largest integer less than or equal to x.

It is important for us to note that the domain of $[\![x]\!]$ is the set of real numbers and the range of $[\![x]\!]$ is the set of integers. This function maps all rational and irrational numbers onto the set of integers. Of course, multiples of the greatest integer function do not necessarily have range elements that are limited to integers, but the range of $[\![x]\!]$ is the set of integers. To graph the $[\![x]\!]$, it is important for us to consider individual domain intervals as seen in the following example.

EXAMPLE 11   Sketch the graph of $f(x) = [\![x]\!]$.

Solution   Whenever n is an integer, $[\![n]\!] = n$ is also an integer. But when x is not an integer, the value of $[\![x]\!]$ is the largest integer less than x. Thus for any

$$x_1 \text{ in } [-3, -2), \text{ such as } x_1 = -2.3, [\![x_1]\!] = -3 \text{ (since } -3 \leq x_1);$$
$$x_2 \text{ in } [-2, -1), \text{ such as } x_2 = -1.6, [\![x_2]\!] = -2 \text{ (since } -2 \leq x_2);$$
$$x_3 \text{ in } [-1, 0), \text{ such as } x_3 = -0.2, [\![x_3]\!] = -1 \text{ (since } -1 \leq x_3);$$
$$x_4 \text{ in } [0, 1), \text{ such as } x_4 = 0.5, [\![x_4]\!] = 0 \text{ (since } 0 \leq x_4);$$
$$x_5 \text{ in } [1, 2), \text{ such as } x_5 = 1.7, [\![x_5]\!] = 1 \text{ (since } 1 \leq x_5);$$
$$x_6 \text{ in } [2, 3), \text{ such as } x_6 = 2.4, [\![x_6]\!] = 2 \text{ (since } 2 \leq x_6);$$

and so forth for each such interval of the real numbers. Graphing these values and continuing this pattern provides the graph of $[\![x]\!]$. ∎

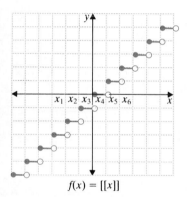

$f(x) = [[x]]$

Figure 2.27

As you can see in Fig. 2.27, the greatest integer function is appropriately called a step function because of its stair-step shape. The next example is the postage cost function application that we saw earlier in this section.

EXAMPLE 12   Sketch the graph of the postage function $f(x) = 25[\![x + 1]\!]$ for $x \geq 0$ where x is in ounces and $f(x)$ is in cents.

Solution   We first recognize that this is a multiple of the greatest integer function with a domain limited to the interval $[0, \infty)$ function. Thus it will maintain the stair-step shape on appropriate intervals of $[\![x]\!]$. Thus for any

$$x_1 \text{ in } [0, 1), f(x_1) = 25[\![x_1 + 1]\!] = 25 \cdot 1 = 25¢;$$
$$x_2 \text{ in } [1, 2), f(x_2) = 25[\![x_2 + 1]\!] = 25 \cdot 2 = 50¢;$$
$$x_3 \text{ in } [2, 3), f(x_3) = 25[\![x_3 + 1]\!] = 25 \cdot 3 = 75¢.$$

Graph these conclusions and then continue the pattern for the graph of $f(x)$, as shown in Fig. 2.28. Notice that the range units on this coordinate system are in multiples of 25 to accommodate this function. ∎

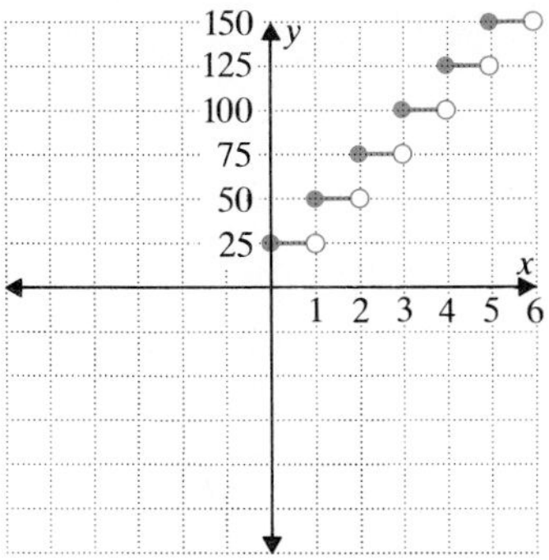

Figure 2.28

If you remember that the shape of the greatest integer function, $[\![x]\!]$, is a stair-step graph, you can graph functions involving the greatest integer function fairly quickly.

Piecewise-Defined Functions

Sometimes a function is a different function on various intervals of its domain. We saw an example of this kind of function earlier in the absolute value function; it was made up of two linear functions for two different domain intervals. We will call this type of function a *piecewise-defined function*.

Definition | A **piecewise-defined function** is the union of more than one function on nonintersecting domain intervals. To graph a piecewise-defined function, you will

   (i) graph each of its function parts in the prescribed domain interval as a separate function and
   (ii) then combine the results in one graph.

In the next example you will see how to graph a piecewise-defined function.

EXAMPLE 13   Graph the piecewise-defined function

$$g(x) = \begin{cases} x^2 & \text{for } x \le 0 \\ |x - 2| & \text{for } 0 < x \le 3 \\ 1 & \text{for } x > 3 \end{cases}$$

Solution   To graph

$$g(x) = \begin{cases} x^2 & \text{for } x \le 0 \\ |x - 2| & \text{for } 0 < x \le 3 \\ 1 & \text{for } x > 3 \end{cases}$$

examine and graph the function in the three domain intervals, $(-\infty, 0]$, $(0, 3]$, and $(3, \infty)$. Then combine these three interval graphs together on one coordinate system for the graph of $g(x)$. For x_1 in $(-\infty, 0]$, $g(x_1) = x^2$, a parabola (Fig. 2.29a). For x_2 in $(0, 3]$, $g(x_2) = |x - 2|$, the V-shaped graph with the two linear functions meeting at $f(2) = 0$ (Fig. 2.29b). Then for x_3 in $(3, \infty)$, $g(x_3) = 1$ the horizontal line $y = 1$ (Fig. 2.29c). Combining these three graphs provides the graph of $g(x)$ in Fig. 2.29(d). ∎

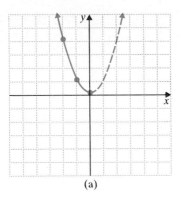

(a)

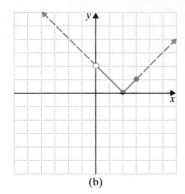

(b)

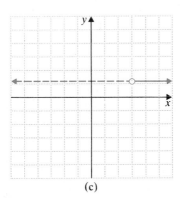

(c)

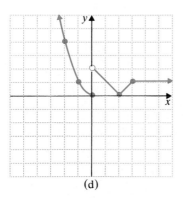

(d)

Figure 2.29

As you can see from this example, graphing piecewise-defined functions requires you to be able to graph many different functions quickly. Now is a good time to study the different types of functions in this section and review those covered earlier in this chapter. When you can match each function with its basic shape, you will be able to graph these functions efficiently.

Exercises 2.3

For each of the following, find (a) the real number interval that is the domain and (b) the smallest real number interval that is the range.

1. $f(x) = 2x^2$

2. $f(x) = 3x^2$

3. $g(x) = \frac{1}{3}x^3$

4. $g(x) = \frac{1}{2}x^3$

5. $k(x) = 2[\![x]\!]$

6. $k(x) = 3[\![x]\!]$

Sketch the graph or graphs for each problem on a single coordinate system.

7. $f(x) = 3x^2$ and $h(x) = \frac{1}{3}x^2$

8. $f(x) = 2x^2$ and $h(x) = \frac{1}{2}x^2$

9. $g(x) = \frac{1}{2}x^3$ and $f(x) = -x^3$

10. $g(x) = \frac{1}{3}x^3$ and $f(x) = -\frac{1}{3}x^3$

11. $k(x) = -2x^2$ **12.** $k(x) = -3x^2$

13. $f(x) = 2|x|$ and $h(x) = -\frac{1}{2}|x|$

14. $f(x) = \frac{1}{2}|x|$ and $h(x) = -2|x|$

15. $f(x) = 2[\![x]\!]$ **16.** $f(x) = 3[\![x]\!]$

17. $g(x) = [\![x + 1]\!]$ **18.** $g(x) = [\![x - 1]\!]$

19. $h(x) = -[\![x]\!]$ **20.** $h(x) = -2[\![x]\!]$

21. $k(x) = [\![\frac{1}{2}x]\!]$ **22.** $k(x) = [\![\frac{1}{3}x]\!]$

23. $f(x) = \begin{cases} x & \text{for } x \leq -1 \\ x^2 & \text{for } -1 < x < 2 \\ 4 & \text{for } 2 \leq x \end{cases}$

24. $f(x) = \begin{cases} 1 & \text{for } x \leq -1 \\ x^2 & \text{for } -1 < x < 2 \\ 12 - 4x & \text{for } x \geq 2 \end{cases}$

25. $g(x) = \begin{cases} -3x - 4 & \text{for } x \leq -1 \\ x^3 & \text{for } -1 < x < 1 \\ 4 - 3x & \text{for } x \geq 1 \end{cases}$

26. $g(x) = \begin{cases} x^2 & \text{for } x \leq -1 \\ |x| & \text{for } -1 < x < 2 \\ 4 - x & \text{for } x \geq 2 \end{cases}$

27. $k(x) = \begin{cases} x^2 & \text{for } x < -1 \\ 1 & \text{for } -1 \leq x \leq 1 \\ x^3 & \text{for } x > 1 \end{cases}$

28. $h(x) = \begin{cases} |x| & \text{for } x < 1 \\ 1 & \text{for } 1 \leq x \leq 3 \\ 16 - 5x & \text{for } x > 3 \end{cases}$

29. $f(x) = \begin{cases} \dfrac{|x|}{x} & \text{for } x \neq 0 \\ 0 & \text{for } x = 0 \end{cases}$

30. $f(x) = \begin{cases} \dfrac{|x - 1|}{|1 - x|} & \text{for } x \neq 1 \\ 2 & \text{for } x = 1 \end{cases}$

31. A Navy intelligence aircraft drops sonar buoys to track an enemy submarine (Fig. 2.30). The aircraft pilot reports

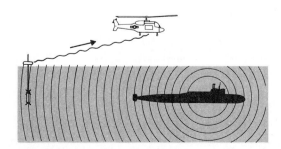

Figure 2.30

dropping, in order, buoys at the following points: $(-3, 1)$, $(-1, 1)$, $(0, 0)$, $(1, 1)$, $(2, 2)$, and $(3, 3)$. Graph the position of these buoys and find the equation for the flight path of the aircraft.

32. Complete Problem 31 with the following ordered pairs, in order: $(-3, 9)$, $(-2, 4)$, $(-1, 1)$, $(0, 0)$, $(1, 1)$, and $(2, 4)$.

33. Complete Problem 31 with the following ordered pairs, in order: $(3, 27)$, $(2, 8)$, $(1, 1)$, $(0, 0)$, $(-1, -1)$, and $(-2, -8)$.

34. Complete Problem 31 with the following ordered pairs, in order: $(-4, 10)$, $(-2, 6)$, $(-1, 4)$, $(0, 2)$, $(1, 0)$, and $(3, 4)$.

35. Complete Problem 31 with the following ordered pairs, in order: $(-3, 9)$, $(-2, 4)$, $(-1, 1)$, $(0, 0)$, $(1, 0)$, $(2, 0)$, and $(4, 0)$.

36. Complete Problem 31 with the following ordered pairs, in order: $(-3, 1)$, $(-2, 1)$, $(-1, 1)$, $(0, 1)$, $(1, 1)$, $(2, 2)$, $(3, 3)$, $(4, 4)$, and $(5, 5)$.

37. A toy rocket is launched upward, from ground level, with an initial velocity of 48 feet per second. The distance of the rocket above level ground is given by the function $s(t) = -16t^2 + 48t$, where t is the time in seconds. Find the domain and range for $s(t) \geq 0$. Sketch a graph of s.

38. A baseball is thrown upward with a velocity of 32 feet per second when released by a throwing machine at ground level. The distance of the baseball above level ground is given by the function $s(t) = -16t^2 + 32t$, where t is the time in seconds. Find the domain and range for $s(t) \geq 0$. Sketch the graph of s.

39. A postal service charges 25¢ for a delivery of less than 3 ounces and 20¢ for each ounce or fraction of an ounce greater than or equal to 3 ounces. Find the piecewise-defined function for the cost of postage as a function of the number of ounces mailed. Sketch a graph of the result.

40. A taxicab company charges $6.00 for riding less than 1 mile and $2.00 for each mile or part of a mile greater than or equal to one mile. Find the piecewise-defined function for the price of a taxi ride as a function of the miles traveled. Sketch a graph of the result.

41. A taxicab company charges $2.00 for each time interval less than 2 minutes that you are in the cab. Find the price of a ride as a function of the number of minutes in the cab. Sketch a graph of this result.

42. For a bargain ad a newspaper charges a $2.00 base rate plus another $2.00 for every two lines or fraction thereof.

Find the price of a bargain ad as a function of the number of lines in the ad. Sketch a graph of this result.

43. A rectangular sheet of metal is 20 feet long by 1 foot wide. A rain gutter is formed by turning up a 20-foot by x-foot rectangle on each 20-foot side as shown in Fig. 2.31. Find the volume as a function of x. What type of function is this? Find the domain interval for $V(x) > 0$.

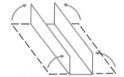

Figure 2.31

44. An illustrator wants to compare the area of a rectangular picture to the area of the mat around the picture, as shown in Fig. 2.32. The outside dimensions of the mat are 8 inches by 10 inches, and the mat is x inches wide on all four sides. Find f where f is the area of the mat as a function of x. Find g where g is the area of the picture as a function of x. What type of functions are f and g?

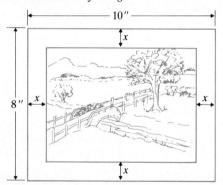

Figure 2.32

45. A credit card company charges interest on an account's unpaid balance from the previous month. Interest on the

first \$5000 is 1.825% per month, and the interest for any amount over \$5000 is 1.4%. Find the interest paid as a function of the unpaid balance.

46. A salesman is paid a commission of 10% for the first \$10,000 in orders each month and 15% commission on each dollar ordered over \$10,000. Find the function for the commission as a function of dollars in orders.

47. A kite is launched with 20 feet of line out and rises straight up at 1 foot per second in a strong breeze (Fig. 2.33). Find the amount of line out as a function of time disregarding any curvature in the line to the kite. Is this a quadratic function?

Figure 2.33

48. An open-topped candy box is made from a 16-inch by 20-inch sheet of cardboard by cutting squares with sides of x inches out of each corner and turning up the sides as shown in Fig. 2.34. Find the volume as a function of x. What kind of function is this?

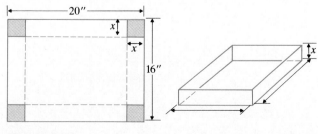

Figure 2.34

2.4   Graphing Some Relations and Symmetry

In this section we will learn how to graph relations that are not functions and develop some techniques to make graphing easier. There are several interesting and useful relations that are not functions to consider.

One example is the relationship between the number of items produced and the price of those items to be considered when manufacturing a certain product. The number of items produced, n, is often related to the wholesale cost of the product, c. If the number of items produced must be between 2000 and 4500 each month and the wholesale cost of the product ranges between \$2.50 and \$4.00, then the relation between the number produced and the wholesale cost would be

$$K = \{(n, c): 2000 \leq n \leq 4500 \quad \text{and} \quad \$2.50 \leq c \leq \$4.00\}$$

The relation K is not a function, and its graph is called a *region*.

Regions

A **region** is the set of all interior points inside certain boundaries; the region may also include some or all of its boundaries. For example, one region would be the set of all points inside the sides of the triangle, the interior of the triangle. We will now graph the region K described above.

EXAMPLE 1   Sketch the graph of

$$K = \{(n, c): 2000 \leq n \leq 4500 \quad \text{and} \quad \$2.50 \leq c \leq \$4.00\}$$

Solution   First, we graph all points such that $2000 \leq n \leq 4500$. This is the vertical region in Fig. 2.35(a). Second, graph all points such that $\$2.50 \leq c \leq \$4.00$. This is the horizontal region in Fig. 2.35(b). Third, graph the intersection of the two regions in Fig. 2.35(a) and 2.35(b). The result is the rectangular region K (Fig. 2.35c), which includes the boundary lines. This graph of K shows the region of production.

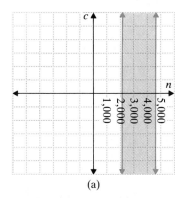

(a)

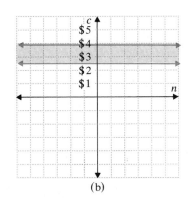

(b)

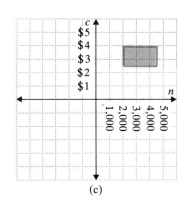

(c)

Figure 2.35

We can also describe regions using other relations, as we will see in the next example.

EXAMPLE 2   Graph the relation $H = \{(x, y): |x| \leq 1 \quad \text{and} \quad |y| \geq 3\}$.

Solution   To graph $H = \{(x, y): |x| \leq 1 \quad \text{and} \quad |y| \geq 3\}$, we complete the following steps:

(a) Graph all points where $|x| \leq 1$, which is $-1 \leq x \leq 1$. This is the vertical region in Fig. 2.36(a).

(b) Graph all points where $|y| \geq 3$, which is $y \leq -3$ or $y \geq 3$. These are the two horizontal regions in Fig. 2.36(b).

(c) Then graph the region H as the intersection of the points in Figs. 2.36(a) and 2.36(b). The result is Fig. 2.36(c).   ■

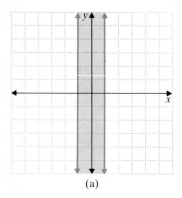
(a)

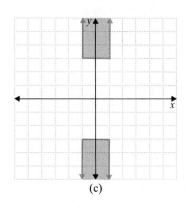
(b)

(c)

Figure 2.36

Distance Between Points and the Midpoint of a Segment

In this book we will study several classical mathematical shapes that are related to the distances between points (circles, parabolas, ellipses, and hyperbolas). The equation of a circle is a good example of a relation based on the distance between points. Before we discuss the circle, we must first develop the formula for the distance between two points.

We will start with two arbitrary points P_1 and P_2. For $P_1 = (x_1, y_1)$ and $P_2 = (x_2, y_2)$ there is a unique distance between P_1 and P_2. By using the Pythagorean Theorem and Fig. 2.37 the distance d between P_1 and P_2 is

$$d^2 = |x_2 - x_1|^2 + |y_2 - y_1|^2$$

(Since the absolute values and the squares both make the expression positive, the absolute value symbols may be omitted.) Thus $d^2 = (x_2 - x_1)^2 + (y_2 - y_1)^2$, and

$$d = \sqrt{(x_2 - x_1)^2 + (y_2 - y_1)^2}$$

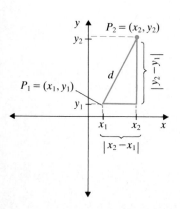

Figure 2.37   is the distance between P_1 and P_2.

THE DISTANCE FORMULA	The distance d between the points $P_1(x_1, y_1)$ and $P_2(x_2, y_2)$ is

$$d = \sqrt{(x_2 - x_1)^2 + (y_2 - y_1)^2} \quad \text{or} \quad \sqrt{(\Delta x)^2 + (\Delta y)^2}$$

where Δx is the change in x and Δy is the change in y.

Sometimes we will need to calculate the midpoint of a line segment. The midpoint is halfway between the endpoints of the segment. The formula for calculating the midpoint is stated here, and its proof will be a problem in the exercises.

MIDPOINT FORMULA	The midpoint M between $P_1(x_1, y_1)$ and $P_2(x_2, y_2)$ is

$$\left(\frac{x_1 + x_2}{2}, \frac{y_1 + y_2}{2} \right)$$

EXAMPLE 3

Find the midpoint between **(a)** $(1, 5)$ and $(3, 8)$, **(b)** $(2, 3)$ and $(-1, 4)$, and **(c)** $(5.79, 3.25)$ and $(-2.15, 7.41)$.

Solution   We use the midpoint formula above to calculate the following midpoints:

(a) For points $(1, 5)$ and $(3, 8)$ the midpoint M is

$$\left(\frac{1 + 3}{2}, \frac{5 + 8}{2} \right) = \left(2, \frac{13}{2} \right) \quad \text{or} \quad \left(2, 6\frac{1}{2} \right)$$

(b) For points $(2, 3)$ and $(-1, 4)$ the midpoint M is

$$\left(\frac{2 - 1}{2}, \frac{3 + 4}{2} \right) = \left(\frac{1}{2}, \frac{7}{2} \right) \quad \text{or} \quad \left(\frac{1}{2}, 3\frac{1}{2} \right)$$

(c) For points $(5.79, 3.25)$ and $(-2.15, 7.41)$ the midpoint M is

$$\left(\frac{5.79 - 2.15}{2}, \frac{3.25 + 7.41}{2} \right) = \left(\frac{3.64}{2}, \frac{10.66}{2} \right) = (1.82, 5.33) \quad \blacksquare$$

The Circle

The circle is a special geometric shape that is defined by the distance between points. Specifically, a **circle** is defined as the set of all points in a plane that are a given distance from a certain point. Thus we can use the distance formula to generate the equation of a circle.

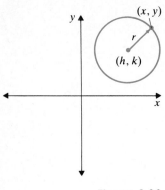

Figure 2.38

In general, to find the equation of the circle with its center at (h, k) and a radius of r, we let (x, y) be any point on the desired circle. Using Fig. 2.38 and the distance formula, we write the equation of this circle as

$$r = \sqrt{(x - h)^2 + (y - k)^2}$$

or

$$r^2 = (x - h)^2 + (y - k)^2$$

Therefore $(x - h)^2 + (y - k)^2 = r^2$ is an equation of a circle with center at (h, k) and radius r.

EQUATION OF A CIRCLE

The equation of the circle with center at (h, k) and radius r is

$$(x - h)^2 + (y - k)^2 = r^2.$$

Given the center and radius of a circle, we can easily find the equation of that circle. The following examples illustrate this process.

EXAMPLE 4

Find the equation of a space shuttle's circular orbit if its altitude is 203 miles and the radius of the earth is about 3950 miles. Pick a convenient point for the origin of its axis system.

Solution   Let the center of the earth be the origin of our coordinate axis system. Then the radius of the shuttle's circular orbit is $r = 3950 + 203 = 4153$ miles. Thus the equation of the shuttle's orbit is $x^2 + y^2 = 4153^2$.   ■

EXAMPLE 5

Find the equations of the following two circles:
(a) the circle with center at $(2, -3)$ and radius of 4 and
(b) the circle with the points $(2, 5)$ and $(-4, 3)$ as the endpoints of a diameter.

Solution   We first identify the center and radius of the circle and then use the equation for each circle.
(a) For $r = 4$, $(h, k) = (2, -3)$, and $(x - h)^2 + (y - k)^2 = r^2$,

$$(x - 2)^2 + [y - (-3)]^2 = 4^2$$

or

$$(x - 2)^2 + (y + 3)^2 = 16$$

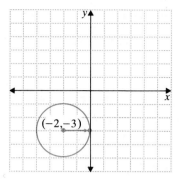

Figure 2.39

(b) Use the endpoints of the diameter, (2, 5) and (−4, 3), and the midpoint formula to find the center of the circle. Thus the center is

$$\left(\frac{2-4}{2}, \frac{5+3}{2}\right) = (-1, 4)$$

Using one endpoint of the diameter (2, 5), the center of the circle (−1, 4), and the distance formula, we calculate the radius of the circle,

$$r = \sqrt{[2-(-1)]^2 + (5-4)^2}$$
$$= \sqrt{3^2 + 1^2} = \sqrt{10}$$
$$r^2 = 10$$

Therefore the equation of the circle is

$$[x-(-1)]^2 + (y-4)^2 = r^2$$
$$(x+1)^2 + (y-4)^2 = 10 \qquad ■$$

EXAMPLE 6   Find the equation of the circle with center at (−2, −3) tangent to the y-axis.

Solution   Sketch a graph of the circle with center at (−2, −3) that is tangent to the y-axis (Fig. 2.39). We see the distance from the center (−2, −3) to the point of tangency is the radius, $r = 2$. Thus the equation of the circle is

$$[x-(-2)]^2 + [y-(-3)]^2 = 2^2$$

or

$$(x+2)^2 + (y+3)^2 = 4 \qquad ■$$

EXAMPLE 7   Find the center and radius of the circle $x^2 - 4x + y^2 = 12$. Graph the result. Is it a function or a relation?

Solution   To change the equation to the form $(x-h)^2 + (y-k)^2 = r^2$, we will *complete the square* as in Section 1.5 (since the coefficient of x^2 is 1, take $\frac{1}{2}$ the coefficient of x, then add the square of this value to both sides of the equation):

$$x^2 - 4x \qquad + y^2 = 12$$
$$(x^2 - 4x + 4) + y^2 = 12 + 4$$
$$\underset{\frac{1}{2}\quad \text{squared}}{\longrightarrow}$$
$$(x-2)^2 + y^2 = 16$$

and

$$(x-2)^2 + (y-0)^2 = 4^2$$

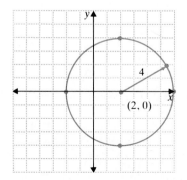

Figure 2.40

This is a circle with its center at (2, 0) and a radius of 4 (Fig. 2.40). Using the vertical line test, we note that this relation is not a function, and by the horizontal line test it is not a one-to-one relation. ■

In calculus we will work mainly with functions. Since the circle is not a function, we sometimes need to divide the equation of a circle into the equations of two functions. The following example shows how we can do this.

EXAMPLE 8   Divide the equation of the circle $(x - 2)^2 + y^2 = 16$ into two functions, each of which describes half a circle, and graph each. State the domain and range of each.

Solution   First solve the equation for y in terms of x:

$$(x - 2)^2 + y^2 = 16$$
$$y^2 = 16 - (x - 2)^2$$
$$y = \pm\sqrt{16 - (x - 2)^2}$$

or

$$y = \pm\sqrt{12 - x^2 + 4x}$$

This can be divided into the two functions f and g as follows:
(a) $f(x) = +\sqrt{12 - x^2 + 4x}$ with a domain of $[-2, 6]$ and a range of $[0, 4]$ (Fig. 2.41a), and
(b) $g(x) = -\sqrt{12 - x^2 + 4x}$ with a domain of $[-2, 6]$ and a range of $[-4, 0]$ (Fig. 2.41b).   ∎

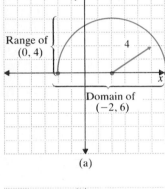

Range of $(0, 4)$

4

Domain of $(-2, 6)$

(a)

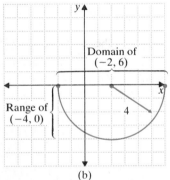

Domain of $(-2, 6)$

Range of $(-4, 0)$

4

(b)

Figure 2.41

Symmetry

Symmetry can be a useful tool in generating points when graphing. A graph has symmetry when one part of the graph is the mirror image of another part of the same graph. A butterfly is a good example of symmetry in the natural world (Fig. 2.42). With symmetry, calculating ordered pairs for one part of the graph will then make it much easier to graph points on another part of the graph.

With this intuitive view of symmetry we can now be more precise with the following two definitions.

SYMMETRY OF POINTS   Two points P and P' are said to be **symmetric with respect to a line**, L, if and only if the line L is the perpendicular bisector of the line segment between P and P'.

Two points P and P' are said to be **symmetric with respect to a given point**, Q, if and only if the point Q is the midpoint of the segment between P and P'.

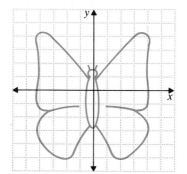

Figure 2.42

We will use symmetry with respect to a line most often when the line is the x-axis, the y-axis, or the line $y = x$. In Fig. 2.43 the points (3, 4) and (−3, 4) are symmetric with respect to the y-axis, since the y-axis is the perpendicular bisector of the line segment between points (3, 4) and (−3, 4). Likewise, in Fig. 2.43 the points (3, 4) and (3, −4) are symmetric with respect to the x-axis. In Fig. 2.44 the points (2, 4) and (4, 2) are symmetric with respect to the line $y = x$ because the line $y = x$ is the perpendicular bisector of the line segment from (2, 4) to (4, 2).

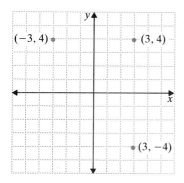

Figure 2.43

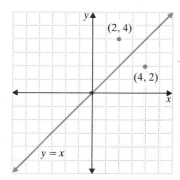

Figure 2.44

Symmetry with respect to a point is easiest to work with when the point is (0, 0), the origin. In Fig. 2.43 the points (−3, 4) and (3, −4) are symmetric with respect to the origin, since the origin is the midpoint of the line segment between points (−3, 4) and (3, −4). We will find that the symmetry of pairs of points will be most helpful in graphing all points of an equation.

SYMMETRY OF A GRAPH	The graph of an equation is **symmetric with respect to a line** L if and only if for every point P on the graph there is a point P' on the graph such that P and P' are symmetric with respect to L.
	The graph of an equation is **symmetric with respect to a point** Q if and only if for every point P on the graph there is a point P' such that P and P' are symmetric with respect to Q.

We can see the symmetry of a graph with respect to a line by using the graph of the butterfly (see top of page 128). Figure 2.45(a) shows symmetry with respect to the y axis, Fig. 2.45(b) shows symmetry with respect to the x-axis, and Fig. 2.45(c) shows symmetry with respect to $y = x$.

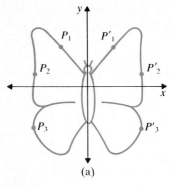

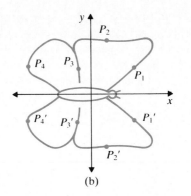

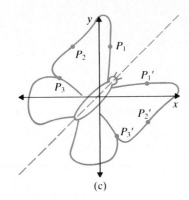

(a)   (b)   (c)

Figure 2.45

The graph of $y = x^3$ in Fig. 2.46 shows symmetry with respect to the origin.

We can also identify symmetry with respect to a line or a point using specific algebraic tests. With these tests we can establish symmetry as a step in sketching a graph. Establishing symmetry will make it easier for you to sketch the graphs of many relations and functions.

TEST FOR SYMMETRY

The graph of an equation in x and y is

(a) symmetric with respect to the x-axis if and only if replacing y with $(-y)$ in the equation produces an equivalent equation,

(b) symmetric with respect to the y-axis if and only if replacing x with $(-x)$ in the equation produces an equivalent equation,

(c) symmetric with respect to the origin if and only if replacing x with $(-x)$ and y with $(-y)$ in the equation produces an equivalent equation, and

(d) symmetric with respect to the line $y = x$ if and only if replacing x with y and y with x in the equation produces an equivalent equation.

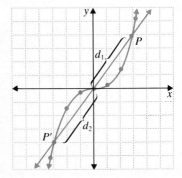

Figure 2.46

We can easily prove these tests using the midpoint formula. The proof of part (a) is presented here; the proofs of parts (b), (c), and (d) are a part of the exercises at the end of this section.

PROOF: Proof of the test for x-axis symmetry: If replacing y with $(-y)$ results in an equivalent equation, then for all points (x, y) on the graph, the point $(x, -y)$ is also on the graph. On the other hand, symmetry with respect to the x-axis means that the midpoint of the line segment from (x, y) to $(x, -y)$ must be on the x-axis. The midpoint is

$$\left(\frac{x + x}{2}, \frac{y - y}{2} \right) = (x, 0)$$

a point on the x-axis. Also, the line through (x, y) and $(x, -y)$ is vertical and perpendicular to the horizontal x-axis. Therefore the x-axis is the perpendicular bisector of the line segment from (x, y) to $(x, -y)$.   ☐

EXAMPLE 9

Determine whether the graph of $y = x^4$ has symmetry with respect to any or all of the following: the x-axis, the y-axis, the origin, or the line $y = x$.

Solution   We need to apply the four tests for symmetry with each equation.
(a) Replacing y with $(-y)$ yields $(-y) = x^4$, which is not equivalent to $y = x^4$. Therefore, the graph of $y = x^4$ is not symmetric with respect to the x-axis.
(b) Replacing x with $(-x)$ yields $y = (-x)^4$, or $y = x^4$, which is equivalent to $y = x^4$. Therefore, the graph of $y = x^4$ is symmetric to the y-axis.
(c) Replacing x with $(-x)$ and y with $(-y)$ yields $-y = (-x)^4$, or $-y = x^4$, which is not equivalent to $y = x^4$. Therefore, the graph of $y = x^4$ is not symmetric with respect to the origin.
(d) Replacing x with y and y with x yields $x = y^4$, which is not equivalent to $y = x^4$. Therefore, the graph of $y = x^4$ is not symmetric with respect to the line $y = x$.   ■

EXAMPLE 10

Determine whether the graph of $xy = 1$ has symmetry with respect to any or all of the following: the x-axis, the y-axis, the origin, or the line $y = x$.

Solution   We need to apply the four tests for symmetry with each equation.
(a) Replacing y with $(-y)$ yields $x(-y) = 1$, or $-xy = 1$, which is not equivalent to $xy = 1$. Therefore, the graph of $xy = 1$ is not symmetric with respect to the x-axis.
(b) Replacing x with $(-x)$ yields $(-x)y = 1$, or $-xy = 1$, which is not equivalent to $xy = 1$. Therefore, the graph of $xy = 1$ is not symmetric to the y-axis.
(c) Replacing x with $(-x)$ and y with $(-y)$ yields $(-x)(-y) = 1$, or $xy = 1$, which is equivalent to $xy = 1$. Therefore, the graph of $xy = 1$ is symmetric with respect to the origin.
(d) Replacing x with y and y with x yields $yx = 1$ which is equivalent to $xy = 1$. Therefore, the graph of $xy = 1$ has symmetry with respect to the line $y = x$.   ■

EXAMPLE 11

A yacht uses a symmetrical spinnaker sail when racing with the wind. The particular spinnaker sail shown in Fig. 2.47 is defined in quadrant I by $f(x) = \left(-\frac{4}{9}\right)$ $(x^2 + 10x - 200)$ for $0 \le x \le 10$ and is defined in quadrant IV by $g(x) = \left(\frac{1}{20}\right)$

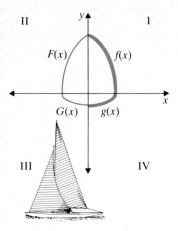

II   y   I

F(x)   f(x)

x

G(x)   g(x)

III   IV

Figure 2.47

$(x^2 - 100)$ for $0 \leq x \leq 10$. Since the spinnaker is symmetrical with respect to the y-axis, write the $F(x)$ and $G(x)$ that describe the spinnaker in quadrants II and III, respectively.

Solution   From Fig. 2.47 we note that the spinnaker has symmetry with respect to the y-axis. Thus we substitute $-x$ for x to obtain

$$F(x) = f(-x) = \left(-\frac{4}{9}\right)[(-x)^2 + 10(-x) - 200]$$

$$= \left(-\frac{4}{9}\right)(x^2 - 10x - 200) \quad \text{for} \quad -10 \leq x \leq 0$$

and

$$G(x) = g(-x) = \left(\frac{1}{20}\right)[(-x)^2 - 100]$$

$$= \left(\frac{1}{20}\right)(x^2 - 100) \quad \text{for} \quad -10 \leq x \leq 0 \quad \blacksquare$$

Even and Odd Functions

Functions that have symmetry with respect to the x-axis are called *even functions*, and functions that have symmetry with respect to the origin are called *odd functions*. Some functions are even functions, some are odd functions, and some are neither odd nor even functions. We define even and odd functions as follows.

Definition | If $f(-x) = f(x)$ for every x in the domain of f, then f is an **even function**.
If $f(-x) = -f(x)$ for every x in the domain of f, then f is an **odd function**.

EXAMPLE 12   Are the following functions even, odd, or neither?
(a) $f(x) = x^3 - 3x$,
(b) $g(x) = x^4 - x^2$, and
(c) $h(x) = x^3 + x + 1$

Solution
(a) Find $f(-x)$ and compare it with $f(x)$ and $-f(x)$. For $f(x) = x^3 - 3x$ we have

$$f(-x) = (-x)^3 - 3(-x) = -x^3 + 3x$$
$$-f(x) = -(x^3 - 3x) = -x^3 + 3x$$

From this we see that $f(-x) \neq f(x)$ but $f(-x) = -f(x)$. Therefore $f(x)$ is an odd function.

(b) Find $g(-x)$ and compare it with $g(x)$ and $-g(x)$. For $g(x) = x^4 - x^2$ we have

$$g(-x) = (-x)^4 - (-x)^2 = x^4 - x^2$$
$$-g(x) = -(x^4 - x^2) = -x^4 + x^2$$

From this we see that $g(-x) = g(x)$ but $g(-x) \neq -g(x)$. Therefore g is an even function.

(c) Find $h(-x)$ and compare it with $h(x)$ and $-h(x)$. For $h(x) = x^3 + x + 1$ we have

$$h(-x) = (-x)^3 + (-x) + 1 = -x^3 - x + 1$$
$$-h(x) = -(x^3 + x + 1) = -x^3 - x - 1$$

We see that h is neither an odd nor an even function, since $h(-x) \neq h(x)$ and $h(-x) \neq -h(x)$.   ∎

Functions with only even-power variables, as in Example 11(b), are even functions. Functions with only odd-power variables, as in Example 11(a), can be odd functions. The function in Example 12(c) is a combination of odd and even powers, since $1 = 1 \cdot x^0$, and in this case the function is neither even nor odd. Other functions without an even exponent can also be even functions if they satisfy the definition.

Graphs of even functions are symmetric with respect to the y-axis, while graphs of odd functions are symmetric with respect to the origin. You will prove these statements in the exercises for this section.

Exercises 2.4

Graph the following regions. State the domain and the smallest range for each problem.

1. $\{(x, y): x \leq 1, y > 1\}$ **2.** $\{(x, y): x \geq 2, y < 3\}$

3. $\{(x, y): -2 < x < 3, y \leq 2\}$

4. $\{(x, y): -3 < x < 1, y > -1\}$

5. $\{(x, y): |x| \leq 3, |y| > 2\}$ **6.** $\{(x, y): |x| > 2, |y| \geq 3\}$

7. $\{(x, y): |x| \geq 3, |y| > 1\}$ **8.** $\{(x, y): |x| \geq 1, |y| < 3\}$

9. $\{(x, y): x^2 + y^2 < 4\}$ **10.** $\{(x, y): x^2 + y^2 \leq 1\}$

11. $\{(x, y): x^2 + y^2 \geq 9\}$ **12.** $\{(x, y): x^2 + y^2 > 16\}$

Graph each of the following circles. State the domain and smallest range intervals of each. Complete the square when necessary.

13. $x^2 + y^2 = 16$ **14.** $x^2 + y^2 = 25$

15. $(x - 2)^2 + (y - 1)^2 = 9$ **16.** $(x - 1)^2 + (y - 3)^2 = 4$

17. $x^2 + 6x + y^2 = -5$ **18.** $x^2 + 4x + y^2 = 9$

19. $x^2 + y^2 - 2y = 8$ **20.** $x^2 + y^2 + 4y = 21$

21. $x^2 + 2x + y^2 - 4y = 4$ **22.** $x^2 + 4x + y^2 + 6y = 3$

23. $x^2 + 2x + y^2 - 2y = 7$ **24.** $x^2 - 4x + y^2 + 4y = 8$

25. The points $(2, 5)$ and $(-4, 3)$ are the endpoints of the circle's diameter. Find the equation of a circle.

26. The points $(-1, 5)$ and $(-5, 1)$ are the endpoints of the circle's diameter. Find the equation of the circle.

27. Find the equation of the circle with center $(2, -3)$ that is tangent to the x-axis (intersects the x-axis at only one point).

28. Find the equation of the circle with center $(-2, 4)$ that is tangent to the y-axis (intersects the y-axis at only one point).

29. Find all of the points that are on the y-axis a distance of 7 from $(3, -1)$. [*Hint:* Solve for the points $(0, y)$.]

30. Find all of the points that are on the y-axis a distance of 10 from $(-3, 5)$. [*Hint:* Solve for the points $(x, 0)$.]

State the type of symmetry (x-axis symmetry, y-axis symmetry, origin symmetry, or symmetry to the line $y = x$) of each of the following equations:

31. $4x^2 + 3y^2 = 12$ **32.** $4x^2 - 3y^2 = 12$

33. $x^2 = 3y$ **34.** $x^3 = 2xy$

35. $x^2 = 2xy$ **36.** $xy = y^3$

37. $x^2 + y^2 = 10$ **38.** $x^2 - y^2 = 10$

39. $y = |x| + 1$ **40.** $y = |x - 1|$

41. $y = |x + 2|$ **42.** $y = |x| - 3$

43. A yacht uses a symmetrical spinnaker sail when racing with the wind (Fig. 2.48). A particular spinnaker sail is defined in quadrant I by $f(x) = -\frac{9}{40}(x^2 + 10x - 375)$ for $0 \le x \le 15$ and is defined in quadrant IV by $g(x) = \left(\frac{1}{45}\right)$ $(x^2 - 225)$ for $0 \le x \le 15$. Since the spinnaker is symmetrical with respect to the y-axis, find $F(x)$ and $G(x)$ that describe the spinnaker in quadrants II and III, respectively.

44. A yacht uses a symmetrical spinnaker sail when racing with the wind (Fig. 2.48). A particular spinnaker sail is defined in quadrant I by $f(x) = -\frac{1}{10}(9x^2 + 18x - 891)$ for $0 \le x \le 9$ and is defined in quadrant IV by $g(x) = \left(\frac{1}{81}\right)$ $(x^2 - 81)$ for $0 \le x \le 9$. Since the spinnaker is symmetric with respect to the y-axis, find $F(x)$ and $G(x)$ that describe the spinnaker in quadrants II and III, respectively.

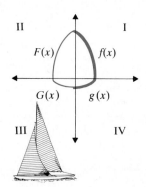

Figure 2.48

45. The profile of one side of a rocket is $g(x) = -\frac{1}{512}x^2 + 8$ for $0 \le x \le 64$ (Fig. 2.49). Use symmetry to find a function $G(x)$ for the rocket's profile in quadrant IV.

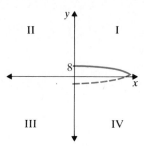

Figure 2.49

46. The profile of one side of a bullet is $g(x) = -\frac{1}{8}(x^2 - 1)$ for $0 \le x \le 1$ (Fig. 2.50). Use symmetry to find a function $G(x)$ for the bullet's profile in quadrant IV.

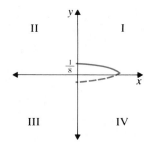

Figure 2.50

47. Show that the midpoint of a segment

$$\left(\frac{x_1 + x_2}{2}, \frac{y_1 + y_2}{2}\right)$$

is the same distance from each endpoint, (x_1, y_1) and (x_2, y_2), of the segment.

48. The circle $x^2 + y^2 = a^2$ is a relation but not a function. Find the equation of the top half of this circle, which is a function. What are the domain and smallest range for this function?

49. The circle $x^2 + y^2 = a^2$ is a relation but not a function. Find the equation of the bottom half of this circle, which is a function. What are the domain and smallest range for this function?

50. Find two functions that together make up the relation

$$x^2 - y^2 = 1.$$

51. Use the distance formula and the Pythagorean Theorem to show that the lines $y = m_1 x$ and $y = m_2 x$ are perpendicular only when $m_1 \cdot m_2 = -1$ (Fig. 2.51).

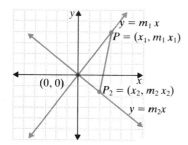

Figure 2.51

52. Prove that the midpoint M between the points $P_1(x_1, y_1)$ and $P_2(x_2, y_2)$ (Fig. 2.52) is

$$\left(\frac{x_1 + x_2}{2}, \frac{y_1 + y_2}{2} \right)$$

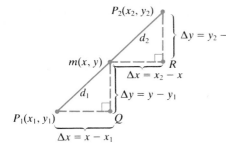

Figure 2.52

53. Prove that replacing x with $(-x)$ in the original equation will produce an equivalent equation if the graph has symmetry with respect to the y-axis.

54. Prove that replacing x with $(-x)$ and y with $(-y)$ in the original equation will produce an equivalent equation if the graph is symmetric with respect to the origin.

Determine whether the following are even functions, odd functions, or neither.

55. $f(x) = x^2 + 1$

56. $f(x) = x^3 - 3x$

57. $f(x) = 3x^2$

58. $f(x) = -\frac{1}{2}x^3$

59. $f(x) = 3x^3 - 5x$

60. $f(x) = 2x^4 - x^2$

61. $f(x) = |2x|$

62. $f(x) = [\![x]\!]$

63. $f(x) = [\![x + 1]\!]$

64. $f(x) = |x + 1|$

65. Prove that an even function has symmetry with respect to the y-axis.

66. Prove that an odd function has symmetry with respect to the origin.

2.5 Translation and Other Graphing Techniques

Earlier in this chapter we found the basic shapes of some common functions. Recognition of these basic shapes makes graphing much easier. Another technique that simplifies graphing of more complicated versions of these functions is *translation*. This is the moving of a graph to the left or right and/or up or down. We will first graph the basic shape of a relation (or function) and then translate this shape to the left, right, up, and down. In this section we will not only develop the graphing techniques for translation but also develop techniques for graphing the power of functions and the techniques for graphing the reciprocal of functions.

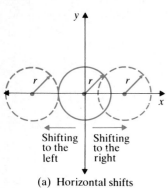

(a) Horizontal shifts

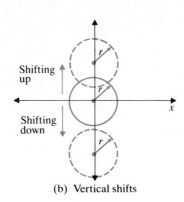

(b) Vertical shifts

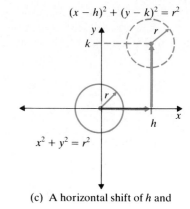

(c) A horizontal shift of h and

Figure 2.53

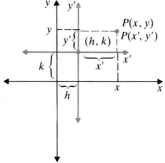

Figure 2.54

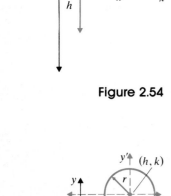

Figure 2.55

Translation

Translation of relations and functions refers to moving or shifting the graph of a relation or function. Translation to the left or right is called a **horizontal shift**. Translation up or down is called a **vertical shift**. Figure 2.53 shows the results of horizontal and vertical shifts on the basic graph of a circle. Once you understand the mechanics of translation, you will also be able to write the equations of a translated relation or function.

To understand translation in the x-y coordinate systems, we construct an x'-y' coordinate system with (h, k) as the origin of the new x'-y' coordinate system (Fig. 2.54). The point P can be named by using either coordinate system. Thus the same point P can be written as $P(x, y)$ or $P(x', y')$. To find the relationship between x and x' as well as that between y and y', we notice in Fig. 2.54 that

$$x = h + x' \quad \text{and} \quad y = k + y'$$

or

$$x' = x - h \quad \text{and} \quad y' = y - k$$

These equivalent expressions with substitution are used to write an equation in terms of the x-y axes or the x'-y' axes. For example, the circle

$$(x - h)^2 + (y - k)^2 = r^2$$

can be written as

$$(x')^2 + (y')^2 = r^2$$

when (h, k) is the origin for the x'-y' axes system (Fig. 2.55).

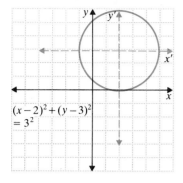

$(x-2)^2+(y-3)^2$
$=3^2$

Figure 2.56

With this translation notation we will graph an equation from its basic shape and its new translated position. To graph the circle $(x - 2)^2 + (y - 3)^2 = 9$, we let $x' = x - 2$ and $y' = y - 3$. We then graph the circle $(x')^2 + (y')^2 = 9$, of radius 3 on the x'-y' coordinate axes with origin at $(2, 3)$, as shown in Figure 2.56. As we see in this example, when $x' = x - 2$, $h = 2$ (since $x' = x - h$), and the vertical axis (the y-axis) is translated 2 units to the right becoming the new y'-axis. Also for $y' = y - 3$, $k = 3$ (since $y' = y - k$), and the horizontal axis is moved 3 units up and becomes the new x'-axis. Once we have translated the axes, we can easily graph $(x')^2 + (y')^2 = 9$ on the translated x'-y' axes.

GUIDELINES FOR TRANSLATION OF AXES

To graph a translated equation in x and y, complete the following steps:

STEP 1: Substitute $x' = x - h$ and $y' = y - k$ into the x-y equation to get an equation in x' and y' (h and k are real numbers).

STEP 2.: Now sketch the x'-y' axes with origin at (h, k). To do this, sketch the translated x'-y' axes by

(a) locating the new y'-axis and moving the y-axis h units to the left or right (the translation is to the right when $h > 0$ and to the left when $h < 0$) and

(b) locating the x'-axis and moving the x-axis k units up or down (the translation is up when $k > 0$ and down when $k < 0$).

STEP 3: Sketch the basic shape seen in the x'-y' equation on the x'-y' axes that has its origin at the point (h, k) of the x-y axes.

We must be careful to find the correct values for h and k when translating axes. For the circle with equation $(x - 2)^2 + (y - 3)^2 = 9$ we let $h = 2$ and $k = 3$, which produces the equation $(x')^2 + (y')^2 = 3^2$. We then recognize this as the basic shape of a circle with a radius of 3 and center at $(h, k) = (2, 3)$, the origin of the x'-y' axes system. The following example shows how to sketch a graph by using the translation of axes.

EXAMPLE 1

Using translation of axes, sketch the graph of
(a) $y - 2 = \frac{1}{2}(x - 5)^3$
(b) $y + 2 = \frac{1}{4}(x - 3)^2$

Solution   Use the steps for translation of axes.
(a) For $y - 2 = \frac{1}{2}(x - 5)^3$:
   STEP 1: For $h = 5$ and $k = 2$ we have

$$x' = x - h \quad \text{is} \quad x' = x - 5$$

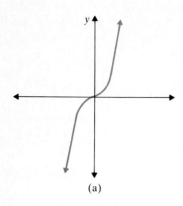

(a)

and

$$y' = y - k \quad \text{is} \quad y' = y - 2$$

Substituting into $(y - 2) = \frac{1}{2}(x - 5)^3$ yields

$$y' = \frac{1}{2}(x')^3$$

STEP 2: Sketch the x'-y' axes with origin at $(5, 2)$.
STEP 3: Next we sketch the basic shape of the parabola $y = \frac{1}{2}x^3$ (Fig. 2.57a) on the translated axes with origin at $(h, k) = (5, 2)$ (Fig. 2.57b).

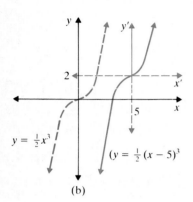

(b)

Figure 2.57

(b) For $y + 2 = \frac{1}{4}(x - 3)^2$:

STEP 1: For $h = 3$ and $k = -2$ we have

$$x' = x - h \quad \text{is} \quad x' = x - (3) \quad \text{or} \quad x' = x - 3$$

and

$$y' = y - k \quad \text{is} \quad y' = y - (-2) \quad \text{or} \quad y' = y + 2$$

Substituting into $y + 2 = \frac{1}{4}(x - 3)^2$ yields

$$y' = \frac{1}{4}(x')^2$$

STEP 2: Sketch the x'-y' axes with origin at $(3, -2)$.
STEP 3: We now sketch the basic shape of the graph $y = \frac{1}{4}x^2$ on the translated axes with origin at $(h, k) = (3, -2)$ (Fig. 2.58).

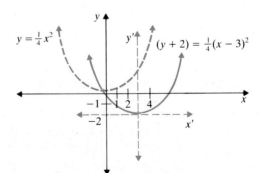

Figure 2.58

EXAMPLE 2   In Section 2.4 we wrote the equation of the space shuttle orbiting at an altitude of 203 miles as $x^2 + y^2 = 4153^2$. Find the equation of this orbit with the new origin at Houston, Texas, which has coordinates of $(-3950, 0)$ on the x-y axis system. Disregard the rotation of the earth.

Solution   With Houston, Texas, as the origin of the coordinate system we notice that the center of the circle is shifted 3950 miles to the left, so $x' = x - (-3450)$

and $y' = y$. The equation

$$(x')^2 + (y')^2 = 4153^2$$

becomes

$$(x + 3450)^2 + y^2 = 4153^2$$

This would be the equation of the orbit with Houston as the origin. ■

EXAMPLE 3   In the following coordinate systems the solid graphs are translations of the basic dashed graph. Use the appropriate values for h and k to write the equation of each solid graph in terms of the x-y coordinate system.

Solution

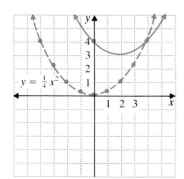

(a)

(a) The basic graph of $y = \frac{1}{4}x^2$ has been translated so that the origin of the x'-y' axes is $(h, k) = (2, 3)$. Therefore

$$y' = \frac{1}{4}(x')^2$$

becomes

$$(y - 3) = \frac{1}{4}(x - 2)^2$$

the equation of the solid graph in (a).

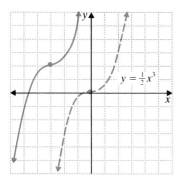

(b)

(b) The basic graph of $y = \frac{1}{2}x^3$ has been translated so that the center of the x'-y' axes is $(h, k) = (-3, 2)$. Therefore

$$y' = \frac{1}{2}(x')^3$$

becomes

$$(y - 2) = \frac{1}{2}[x - (-3)]^3$$

or

$$(y - 2) = \frac{1}{2}(x + 3)^3$$

the equation of the solid graph in (b).

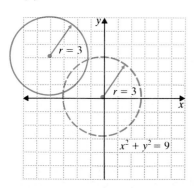

(c)

(c) The basic graph of $x^2 + y^2 = 9$ has been translated so that $(h, k) = (-4, 3)$ is the origin of the x'-y' axes and the center of the circle. Therefore

$$(x')^2 + (y')^2 = 3^2$$

is

$$[x - (-4)]^2 + (y - 3)^2 = 3^2$$

or

$$(x + 4)^2 + (y - 3)^2 = 9$$

the equation of the solid graph in (c). ■

Now we will put these translation ideas into function notation.

TRANSLATION WITH FUNCTION NOTATION

The graph of the function $f(x - h) + k$ is the graph of $f(x')$ with its origin at (h, k) and $x' = x - h$ and $y' = y - k$. Thus to graph $f(x - h) + k$, sketch the translated axes (translated h units horizontally and k units vertically) and then sketch $f(x')$ on the translated x'-y' axes.

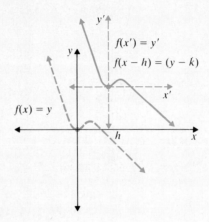

We have the function notation above because

$$f(x') = y' \quad \text{with} \quad x' = x - h \quad \text{and} \quad y' = y - k$$

is

$$f(x - h) = y - k$$

or

$$f(x - h) + k = y$$

Now we will use this notation to solve the following examples.

EXAMPLE 4

A fountain in a hotel lobby is made with jets that spray water. The water from a jet placed at the center of the fountain at ground level traces out a path of $w(x) = 160x - 16x^2$, where x is the horizontal distance in feet from the center of the fountain in feet and $w(x)$ is the height of the water for each value of x. Write a function that describes the height of the water as a function of x if the jet is placed 10 feet above the center of the fountain.

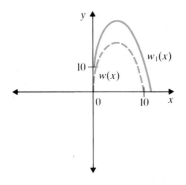

Figure 2.59

Solution In this problem the height is written as $w(x) = 160x - 16x^2$. The change of the height at $x = 0$ from 0 feet to 10 feet represents a translation of the origin to $(0, 10)$ (Fig. 2.59). Thus the translated function $w_1(x)$ is

$$w_1(x) = w(x) + 10$$

$$w_1(t) = 160x - 16x^2 + 10$$

or

$$w_1(t) = 10 + 160x - 16x^2$$ ■

EXAMPLE 5   What does $w(x - 7)$ indicate when $w(x)$ is the height of the water from the water jet described in the hotel fountain of Example 4?

Solution The expression $w(x - 7)$ represents a horizontal change in x, the distance from the center of the fountain. Thus $w(x - 7)$ indicates that the water jet has been moved 7 feet from the center of the fountain. The origin has been translated to $(7, 0)$ (Fig. 2.60). ■

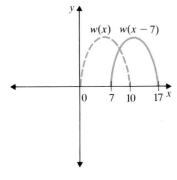

Figure 2.60

Powers of Functions

It will also be helpful in graphing if we understand the relationships between $f(x)$, $[f(x)]^3$, and $[f(x)]^6$. In this notation the function $[f(x)]^n$ is the nth power of the function $f(x)$. We will now consider some examples of this type of function.

EXAMPLE 6   Graph $f(x) = x^2$ and $g(x) = x^4$ on the same coordinate systems.

Solution We first calculate some values and then sketch the graphs of these functions (Fig. 2.61).

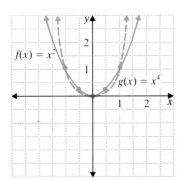

Figure 2.61

x	$f(x) = x^2$	$g(x) = x^4$
0	0	0
$\pm\frac{1}{2}$	$\frac{1}{4}$	$\frac{1}{16}$
± 1	1	1
± 2	4	16
± 3	9	81

■

In Example 6 we see the relationship between a function and an even power of that function, $g(x) = [f(x)]^2$. Figure 2.62 demonstrates the following conditions and conclusions for $f(x) > 0$.

Conditions	*Conclusions*
when $0 < f(x) < 1$,	$f(x) > [f(x)]^n$
when $f(x) \geq 1$,	$f(x) \leq [f(x)]^n$

If we take the absolute value of the functions, we can state this more generally for $f(x) > 0$ as well as $f(x) < 0$.

THE POWER OF A FUNCTION

For a function $f(x)$ and $n > 1$,

(a) when $|f(x)| < 1$, $|f(x)| > |[f(x)]^n|$ for $n > 1$ or $|[f(x)]^n|$ is closer to the x-axis than $|f(x)|$, and

(b) when $|f(x)| \geq 1$, $|f(x)| \leq |[f(x)]^n|$ for $n > 1$ or $|f(x)|$ is closer to the x-axis than $|[f(x)]^n|$.

EXAMPLE 7   Apply the conclusions above to sketch the graph $f(x) = x^3$ and $g(x) = x^5$ on the same coordinate system.

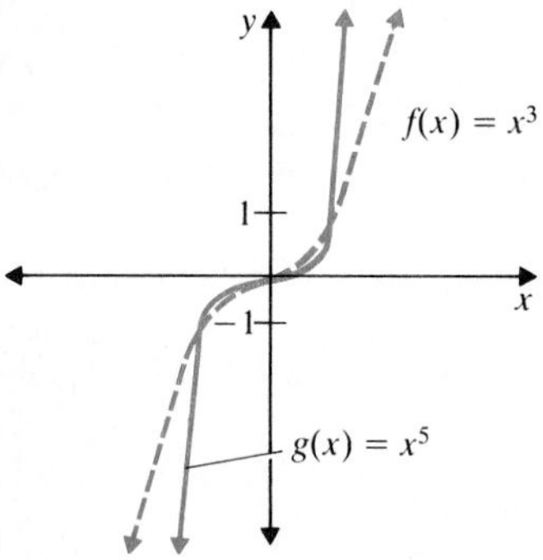

Solution   First graph $f(x) = x^3$ from its basic shape, which includes $(-1, -1)$, $(0, 0)$, and $(1, 1)$. To graph $g(x) = x^5$, compare it to $f(x) = x^3$. For $|x| < 1$, $|f(x)| = |x^3| < 1$ and $|x^3| > |x^5|$; therefore $g(x) = x^5$ is closer to the x-axis than $f(x) = x^3$ in the interval $(-1, 1)$. Similarly, for $|x| > 1$, $|f(x)| = |x^3| > 1$ and $|x^3| < |x^5|$; therefore $f(x) = x^3$ is closer to the x-axis than $g(x) = x^5$ in the intervals $(-\infty, -1)$ and $(1, \infty)$ (Fig. 2.62). ∎

Figure 2.62

Reciprocal Function Graphing

We can graph some functions very quickly because we know how to graph their reciprocal easily. To graph reciprocal functions, we must remember the following properties.

PROPERTIES OF FUNCTION RECIPROCALS

For a function $f(x)$ where x is an element of the domain such that $f(x) \neq 0$,

(a) if $f(x) = 1$, the reciprocal $\dfrac{1}{f(x)} = 1$;

(b) if $|f(x)| > 1$, the reciprocal $\dfrac{1}{|f(x)|} < 1$;

(c) if $|f(x)| < 1$, the reciprocal $\dfrac{1}{|f(x)|} > 1$; and

(d) if $|f(x)| > 1$ as $|f(x)|$ becomes larger and larger, the reciprocal $\dfrac{1}{|f(x)|}$ gets smaller and smaller. Also, if $|f(x)| < 1$ as $f(x)$ becomes smaller and smaller, $\dfrac{1}{|f(x)|}$ gets larger and larger.

We can see how these properties work by setting up two charts that provide a numerical example.

$f(x)$	$\dfrac{1}{f(x)}$		$f(x)$	$\dfrac{1}{f(x)}$
1	$\dfrac{1}{1} = 1$	and	$\dfrac{1}{2}$	2
2	$\dfrac{1}{2}$		$\dfrac{1}{3}$	3
3	$\dfrac{1}{3}$		$\dfrac{1}{4}$	4
100	$\dfrac{1}{100}$		$\dfrac{1}{100}$	100
-1	$-\dfrac{1}{-1} = -1$		$-\dfrac{1}{2}$	-2
-2	$-\dfrac{1}{-2} = -\dfrac{1}{2}$		$-\dfrac{1}{3}$	-3
-3	$\dfrac{1}{-3} = -\dfrac{1}{3}$		$-\dfrac{1}{4}$	-4
-100	$\dfrac{1}{-100} = -\dfrac{1}{100}$		$-\dfrac{1}{100}$	-100

We will apply these properties in the next examples.

EXAMPLE 8   Sketch the graph of $g(x) = \dfrac{1}{(x^2 + 1)}$ using its reciprocal function $f(x) = x^2 + 1$.

Solution   Since $g(x) = \dfrac{1}{f(x)}$, $g(x)$ is the reciprocal of $f(x)$. Sketch the graph of $f(x) = x^2 + 1$ with a dashed curve and notice that $f(x) \geq 1$ for every x in the domain (Fig. 2.63a). Since $f(0) = 1$, we have $g(0) = \dfrac{1}{1} = 1$ and $(0, 1) \in g$. Also note that $0 < g(x) < 1$, since $f(x) > 1$ for all $x \neq 0$. Thus the range of g is limited to values between 0 and 1 (including 1, since $g(0) = 1$). When sketching these values, remember that as $f(x)$ gets larger and larger, $g(x) = \dfrac{1}{f(x)}$ gets smaller and smaller. The resulting sketch of $g(x) = \dfrac{1}{(x^2 + 1)}$ is shown in Fig. 2.63(b). For a more accurate graph of $g(x) = \dfrac{1}{f(x)}$ we can graph points. For $(\pm 1, 2) \in f$ we have $\left(\pm 1, \frac{1}{2}\right) \in g$. ∎

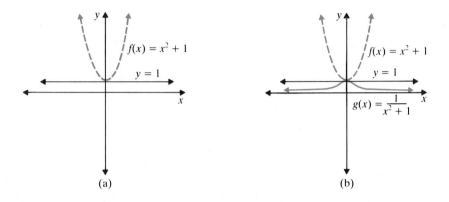

Figure 2.63   (a)   (b)

EXAMPLE 9

Sketch the graph of $h(x) = \dfrac{1}{x}$.

Solution   When we graph $h(x)$, we use the reciprocal of the function $k(x) = x$, since $h(x) = \dfrac{1}{k(x)} = \dfrac{1}{x}$. Also notice that $h(x)$ is undefined when $k(x) = 0$. Thus the domain of h is the interval $(-\infty, 0) \cup (0, \infty)$, since the denominator of $h(x)$ is zero when $x = 0$. As the $|x|$ gets larger and larger, the value of $|h(x)| = \dfrac{1}{|x|}$ gets closer and closer to zero (Fig. 2.64a). Also as $|x|$ gets closer and closer to zero, $|h(x)| = \dfrac{1}{|x|}$ gets larger and larger, as in the chart and Fig. 2.64(b).

for x	we have $|x|$	and $|h(x)| = \dfrac{1}{|x|}$
$\pm\dfrac{1}{10}$	$\dfrac{1}{10}$	10
$\pm\dfrac{1}{100}$	$\dfrac{1}{100}$	100
$\pm\dfrac{1}{1000}$	$\dfrac{1}{1000}$	1000

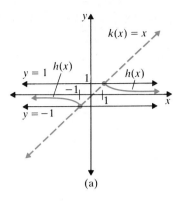

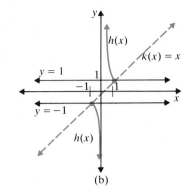

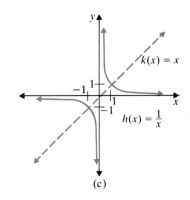

(a)   (b)   (c)

Figure 2.64

We combine this information in one graph for the graph $h(x) = \dfrac{1}{x}$ in Fig. 2.64(c). ∎

In Example 9 the line $x = 0$ is called a **vertical asymptote** of $h(x) = \dfrac{1}{x}$. We can have vertical asymptotes when domain values cause the denominator of the range to equal zero. Since the vertical asymptotes occur for x values that cannot be in the domain, the function will never intersect a vertical asymptote.

Also in Example 9 the line $y = 0$ is called a **horizontal asymptote** of $h(x) = \dfrac{1}{x}$. A horizontal asymptote occurs when $|x|$ gets very large and the value of $h(x)$ gets closer and closer to some constant value K. In the case of $h(x) = \dfrac{1}{x}$, as $|x|$ gets larger and larger, the value of $h(x)$ gets closer and closer to $K = 0$. Thus the horizontal asymptote of $h(x) = \dfrac{1}{x}$ is $y = 0$. (It is possible for the function to intersect its horizontal asymptote, and we will discuss this in the next chapter.) In general, asymptotes approximate the behavior of the function for large values of x and y in the function $y = f(x)$.

These new ideas associated with asymptotes can be more succinctly described by using the notation of limits as defined here. (The limit concept will be defined in more detail and used to develop important ideas in calculus.)

LIMIT NOTATION

For a function $f(x)$,

(a) as x approaches c from the left, $x < c$, and $f(x)$ approaches L_1, we write

$$\lim_{x \to c^-} f(x) = L_1$$

(b) as x approaches c from the right, $x > c$, and $f(x)$ approaches L_2, we write

$$\lim_{x \to c^+} f(x) = L_2$$

If L_1 or L_2 is ∞, the value of $f(x)$ gets larger and larger (approaches ∞); and if L_1 or L_2 is $-\infty$, the value of $f(x)$ gets smaller and smaller (approaches $-\infty$).

Using the limit notation for the vertical asymptote in Example 9, we have

$$\lim_{x \to 0^-} h(x) = -\infty \qquad \text{and} \qquad \lim_{x \to 0^+} h(x) = \infty$$

as seen in Fig. 2.64(b). The horizontal asymptote in Example 9 is described by

$$\lim_{x \to -\infty} h(x) = 0^-$$

where 0^- means that $h(x)$ approaches 0 as a negative number, and

$$\lim_{x \to \infty} h(x) = 0^+$$

where 0^+ means that $h(x)$ approaches 0 as a positive number, as seen in Fig. 2.64(a).

In practice, the value c used in the limit does not need to be an element of the domain as we saw in the vertical asymptote in Example 9. As in Example 9 it is possible for c, L_1, and L_2 to be real numbers or $\pm\infty$. We will now use the limit concept to complete the next example.

EXAMPLE 10

Sketch the graph of $f(x) = \dfrac{1}{|x - 2|}$.

Solution

Since $f(x) = \dfrac{1}{|x - 2|} = \dfrac{1}{g(x)}$, we first graph $g(x) = |x - 2|$. Notice that $g(x) \geq 0$ and $g(1) = g(3) = 1$. Thus $f(x) = \dfrac{1}{g(x)} > 0$ and $f(1) = f(3) = 1$. Now sketch $f(x) < 1$ for $x < 1$ and $x > 3$; as $|x|$ gets larger and larger, $f(x)$ decreases to zero. Thus

$$\lim_{x \to -\infty} f(x) = 0 \qquad \text{and} \qquad \lim_{x \to \infty} f(x) = 0$$

Thus the horizontal asymptote is $y = 0$. Next sketch $f(x) > 1$ when $1 < x < 3$ and $x \neq 2$; as x gets closer and closer to 2, the value of $f(x) = \dfrac{1}{|x - 2|}$ gets larger and larger (Fig. 2.65). Using limit notation, we have

$$\lim_{x \to 2^-} f(x) = \infty \qquad \text{and} \qquad \lim_{x \to 2^+} f(x) = \infty$$

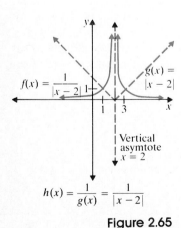

$$f(x) = \frac{1}{|x - 2|}$$

$$g(x) = |x - 2|$$

Vertical asymtote $x = 2$

$$h(x) = \frac{1}{g(x)} = \frac{1}{|x - 2|}$$

Figure 2.65

Thus the line $x = 2$ is a vertical asymptote and is not in the domain of f because this value makes the denominator of $f(x)$ zero. ■

As we have seen in these examples, graphing the reciprocal function provides us with another method for graphing the shape of a function without calculating and graphing a large number of points.

Exercises 2.5

For each of the following graphs the equation of the dashed graph is given. The dashed graph is translated to the position of the solid graph. Find the horizontal shift h, the vertical shift k, and the equation of the translated equation in x and y.

1.

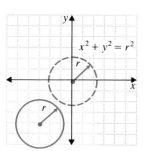

2.

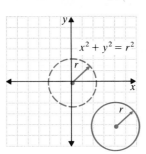

3.

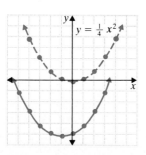

4.

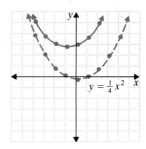

5.

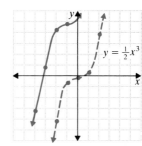

6.

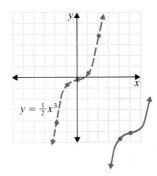

7.

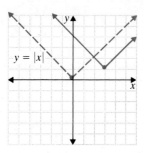

8.

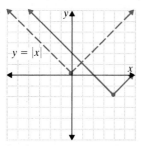

9.

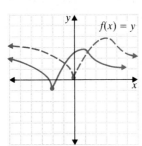

10.

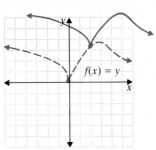

11.

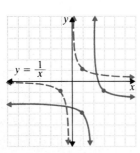

12.

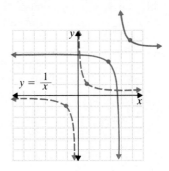

Use translation of axes to graph the following relations and functions.

13. $(x - 2)^2 + (y + 3)^2 = 25$

14. $(x - 1)^2 + (y + 2)^2 = 36$

15. $(y - 1) = x^2$ **16.** $(y + 3) = x^2$

17. $y = (x + 3)^2$ **18.** $y = (x - 1)^2$

19. $(y + 3) = (x - 1)^2$ **20.** $(y - 1) = (x + 3)^2$

21. $y = (x - 1)^3$ **22.** $y = (x + 3)^3$

23. $(y - 3) = (x + 2)^3$ **24.** $(y - 2) = (x - 3)^3$

25. $(y + 1) = |x - 2|$ **26.** $(y + 3) = |x + 3|$

27. $(y - 3) = |x + 1|$ **28.** $(y - 1) = |x + 2|$

Use the properties of the power of a function to sketch the graph of the following.

29. $y = x^4$ **30.** $y = x^6$

31. $y = \frac{1}{4}x^6$ **32.** $y = \frac{1}{8}x^4$

33. $y = |x|^3$ **34.** $y = |x|^5$

35. $y = [\![2x]\!]^2$ **36.** $y = [\![2x]\!]^3$

37. A Navy intelligence aircraft drops sonar buoys to track an enemy submarine (Fig. 2.66). The aircraft pilot reports dropping, in order, buoys at the following points: $(-3, 1)$, $(-1, -1)$, $(0, -2)$, $(1, -1)$, $(2, 0)$, and $(3, 1)$. Graph the position of these buoys and find the equation for the flight path of the aircraft.

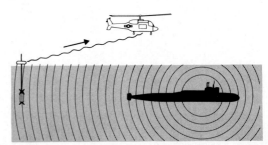

38. As in Problem 37 a Navy aircraft drops sonar buoys, in order, at $(-3, 4)$, $(-2, 3)$, $(-1, 2)$, $(0, 1)$, $(1, 2)$, and $(2, 3)$. Graph the position of these buoys and find the equation for the flight path of the aircraft.

39. A fountain at City Hall Plaza is constructed with jets that spray water. The water from a jet placed at the center of the fountain at ground level traces out a path of $w(x) = 240x - 16x^2$, where x is the horizontal distance in feet from the center of the fountain in feet and $w(x)$ is the height of the water for each value of x. Write a function that describes the height of the water as a function of x if the jet is placed 8 feet above the center of the fountain.

40. What does $w(x - 10)$ indicate when $w(x)$ is the height of the water from the water jet described in the hotel fountain of Problem 39?

41. A Calaveras County jumping frog can jump on a vertical height of $f_1(x) = 4x - x^2$ for a horizontal distance x in feet with $0 \le x \le 4$. For the next three jumps, find the height of the frog as a function of its horizontal distance from the starting point of the first jump. Let the height of second, third, and fourth jumps be the functions f_2, f_3, and f_4, respectively.

42. A missile is launched from sea level. The missile's altitude in feet as a function of time in seconds is given by the function $s(t) = 8000t - 16t^2$. A similar missile is launched 30 seconds later at an altitude of 3500 feet. Find the altitude of the second missile as a function of time, using the function S.

43. A fighter bomber flies a course described by $y = \left(\frac{1}{100}\right)x^2$ and always launches its bombs at the point $(0, 0)$ in this flight curve. Find the equation for the translated flight curve so that the fighter bomber can release its bombs at $\left(1, \frac{1}{2}\right)$.

44. A fighter bomber flies a course described by $y = \left(\frac{1}{100}\right)x^2$ and always launches its bombs at the point $(0, 0)$ in this flight pattern. Find the equation for the translated flight curve so that the fighter bomber can release its bombs at $\left(1, \frac{1}{4}\right)$.

45. The current I of an electrical circuit varies directly with the voltage V and inversely with the resistance R. Using a constant of proportionality of 1, graph I as a function of R when the voltage is 9 volts.

46. Using the distance formula, $D = r \cdot t$, sketch the graph of time as a function of the rate of speed on a 225-mile trip.

Use the properties of the power of a function and the reciprocal of a function to sketch the graphs of the following functions.

47. $f(x) = \dfrac{1}{x^2}$

48. $f(x) = \dfrac{1}{x^3}$

49. $f(x) = \dfrac{1}{|x|}$

50. $f(x) = \dfrac{1}{x + 1}$

51. $f(x) = \dfrac{1}{x + 2}$

52. $f(x) = \dfrac{1}{|x + 2|}$

53. $f(x) = \dfrac{1}{|x - 1|}$

54. $f(x) = \dfrac{1}{|x + 3|}$

55. $f(x) = \dfrac{1}{(x - 2)^2}$

56. $f(x) = \dfrac{1}{(x + 2)^2}$

57. $f(x) = \dfrac{1}{x^2 + 2}$

58. $f(x) = \dfrac{1}{x^2 + 3}$

2.6　Operations with Functions

Using operations is not new. In previous work we used operations to combine two entities and produce a single result many times:

(a) When working with real numbers, we combined two real numbers to create a third real number, using the operations of addition, subtraction, multiplication, and division.

(b) When working with sets, we combined two sets A and B to create a third set using the set operations of union ($A \cup B = C$) and intersection ($A \cap B = D$).

(c) When working with polynomials, we combined two polynomials to create a third polynomial, using the operations of addition, subtraction, and multiplication. (Division is also an operation for polynomials, but the result is not always a polynomial.)

Now we will extend the use of operations to functions. In business, for example, we need to combine the revenue function $R(x)$, which defines the income, and the cost function $C(x)$, which defines the cost of doing business, to find the profit function $P(x)$, which defines the business's profit. We will use the operation of subtraction to find the profit function $P(x)$, where

$$P(x) = R(x) - C(x)$$

Thus we can combine the revenue function R and the cost function C using the operation of subtraction to obtain the profit function P. We will use this to work with revenue, cost, and profit functions in the future. We will now define some operations for functions.

Definition

Operation of Functions. If f and g are functions where $f : A \to B$ and $g : C \to D$, then their **sum** $f + g$, **difference** $f - g$, **product** $f \cdot g$, and **quotient** $f \div g$, for $g(x) \neq 0$, exist for all $x \in A \cap C$. These function operations are defined as follows for every $x \in A \cap C$:

$$(f + g)(x) = f(x) + g(x)$$
$$(f - g)(x) = f(x) - g(x)$$
$$(f \cdot g)(x) = f(x) \cdot g(x)$$

and for $g(x) \neq 0$

$$(f \div g)(x) = \frac{f(x)}{g(x)}$$

We can now use these operations with functions to complete the examples that follow.

EXAMPLE 1

Kennedy's Computer Products can produce up to 4000 disks a month. Under current conditions the monthly revenue and cost functions for this item are $R(x) = -0.01x^2 + 40.1x$ and $C(x) = 0.1x + 100$. What is the profit function $P(x)$?

Solution   Remembering that profit is the difference of the revenue and the cost, we have $P(x) = R(x) - C(x)$. The production interval [0, 4000] is common to $R(x)$ and $C(x)$, and thus $P(x)$ has the same domain interval. Then

$$
\begin{aligned}
P(x) &= R(x) - C(x) \\
&= (-0.01x^2 + 40.1x) - (0.1x + 100) \\
&= -0.01x^2 + 40x - 100
\end{aligned}
$$

Thus the profit is now a function of the number of disks produced. The profit function is a quadratic function, and from the basic shape of the parabola the profit function will have a maximum or minimum profit for some value of $x \in R$. ■

EXAMPLE 2   For $f(x) = x^2 - 1$ and $g(x) = x + 1$, find $(f + g)(x)$, $(f - g)(x)$, $(f \cdot g)(x)$, and $(f \div g)(x)$ and the domain of these functions.

Solution   First we will find the correct domain. The domain for $f(x) = x^2 - 1$ is the set of real numbers, and the domain for $g(x) = x + 1$ is also the set of real numbers. Therefore the following can be calculated for an $x \in R$:

$$(f + g)(x) = f(x) + g(x)$$
$$= (x^2 - 1) + (x + 1)$$
$$= x^2 + x$$

$$(f - g)(x) = f(x) - g(x)$$
$$= (x^2 - 1) - (x + 1)$$
$$= x^2 - x - 2$$

$$(f \cdot g)(x) = f(x) \cdot g(x)$$
$$= (x^2 - 1)(x + 1)$$
$$= x^3 + x^2 - x - 1$$

$$(f \div g)(x) = f(x)/g(x)$$
$$= \frac{x^2 - 1}{x + 1}$$
$$= \frac{(x - 1)(x + 1)}{(x + 1)} = x - 1 \qquad \begin{array}{l} \text{for } x \neq -1, \\ \text{since } g(x) \neq 0. \end{array}$$ ■

EXAMPLE 3   For $f(x) = \sqrt{x^2 - 9}$ and $g(x) = \sqrt{x - 3}$, find $(f + g)(x)$, $(f - g)(x)$, $(f \cdot g)(x)$, and $(f \div g)(x)$ and the domain of these functions.

Solution   First we will find the correct domain. The domain for $f(x) = \sqrt{x^2 - 9}$ is the set of all x-values such that $x^2 - 9 \geq 0$. By using the sign chart from Chapter 1, $x^2 - 9 \geq 0$ for x in $(-\infty, -3] \cup [3, \infty)$. Thus the domain of f is $(-\infty, -3] \cup [3, \infty)$.

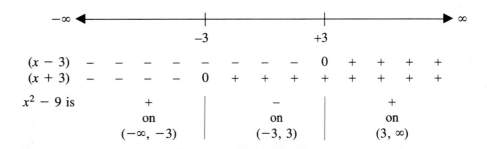

The domain for $g(x) = \sqrt{x - 3}$ is the set of all x-values such that $x - 3 \geq 0$ or x is in $[3, \infty)$. Therefore the domain for addition, subtraction, and multiplication is

$$\{(-\infty, -3] \cup [3, \infty)\} \cap [3, \infty) = [3, \infty).$$

For division the domain is $(3, \infty)$ so that $g(x) \neq 0$ for x in the domain. Now combining $f(x) = \sqrt{x^2 - 9}$ and $g(x) = \sqrt{x - 3}$, we have

$$(f + g)(x) = f(x) + g(x)$$
$$= \sqrt{x^2 - 9} + \sqrt{x - 3} \quad \text{for } x \text{ in } [3, 0)$$
$$(f - g)(x) = f(x) - g(x)$$
$$= \sqrt{x^2 - 9} - \sqrt{x - 3} \quad \text{for } x \text{ in } [3, 0)$$
$$(f \cdot g)(x) = f(x) \cdot g(x)$$
$$= \sqrt{x^2 - 9}\sqrt{x - 3}$$
$$= \sqrt{(x^2 - 9)(x - 3)}$$
$$= \sqrt{(x + 3)(x - 3)^2}$$
$$= (x - 3)\sqrt{x + 3} \quad \text{for } x \text{ in } [3, \infty)$$
$$(f \div g)(x) = f(x) \div g(x)$$
$$= \frac{\sqrt{x^2 - 9}}{\sqrt{x - 3}}$$
$$= \sqrt{\frac{(x - 3)(x + 3)}{(x - 3)}}$$
$$= \sqrt{x + 3} \quad \text{for } x \text{ in } (3, \infty) \quad \blacksquare$$

Composition of Functions

The operations that we worked with above (addition, subtraction, multiplication, and division) are not the only operations we can use to combine two functions. A new function operation will combine two functions by mapping with one function first and then mapping with another function. Figure 2.67 shows how $f: A \to B$ and then $g: B \to C$. Mapping with f first and then with g matches an element of A with an element of C or $A \to C$.

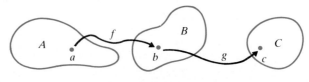

Figure 2.67 $f: A \longrightarrow B$ then $g: B \longrightarrow C$

This operation is called **function composition** and is denoted $g \circ f$. Specifically, in this case, $g \circ f : A \to C$ (Fig. 2.68). Note that in this notation the function that is applied first is written second. We have this order because $(g \circ f)(x) = g(f(x))$.

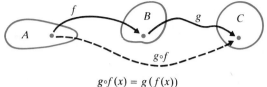

Figure 2.68 $g \circ f(x) = g(f(x))$

We will now explore how function composition can be used to compute the cost of filling a swimming pool as a function of the depth of the pool.

During a water shortage a pool owner wants to calculate the cost of water needed to fill a swimming pool as a function of the depth of the water in the pool. The city charges are based on the volume of water used. The city's price function p maps volume to dollars:

$$p : \text{volume} \to \$$$

We can find the volume of water in the pool as a function of the depth of the water. This volume function V maps depth of the water to volume:

$$V : \text{depth} \to \text{volume}$$

Thus the pool owner can calculate the cost as a function of the depth of the water by calculating the cost function $C = p \circ V$, since

$$\text{depth} \xrightarrow{V} \text{volume} \xrightarrow{p} \$ \quad \text{or} \quad \text{depth} \xrightarrow{p \circ V} \$$$

We will complete a problem like this in the exercises at the end of this section.

Definition

> *Function Composition.* If the function $f : A \to B$, the function $g : C \to D$, and $C \subseteq B$, then the **composite function** $g \circ f$ is defined by
>
> $$(g \circ f)(x) = g(f(x))$$
>
> (read g composed with f of x equals g of f of x). The domain of $g \circ f$ is all x in the domain of f such that $f(x)$ is in the domain of g, $(C \subseteq B)$.

Care must be taken in finding the domain of a composite function $g \circ f$, since the domain of $g \circ f$ contains only those elements x from the domain of f whose images $f(x)$ are in the domain of g.

Sometimes we will find it helpful to consider the mapping of composite functions through its ordered pairs. The alternative definition provides this perspective.

Alternative Definition

> *Function Composition.* For functions $f : A \to B$ and $g : B \to C$, g composed with f, $g \circ f$, is defined as follows: $(a, c) \in g \circ f$ if and only if there exist some b such that $(a, b) \in f$ and $(b, c) \in g$.

The alternative definition provides a way to work with function composition on a point-by-point basis, as we will see in the next example, as well as to prove the last theorem in this section.

EXAMPLE 4

When $f = \{(2, 4), (3, 6), (4, 8)\}$ and $g = \{(4, 9), (6, 13), (8, 17)\}$, draw a mapping diagram for $g \circ f$ and write $g \circ f$ as a set of ordered pairs.

Solution   Notice that for each x in the domain of f, $f(x)$ is in the domain of g. Therefore the domain of $g \circ f$ is $\{2, 3, 4\}$.

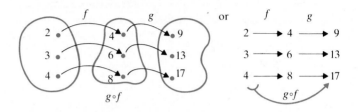

Then $g \circ f = \{(2, 9), (3, 13), (4, 17)\}$.   ■

The following examples show several applications and details of function composition:

EXAMPLE 5

For $f(x) = x^3$ and $g(x) = x + 1$, find **(a)** $f \circ g$ and **(b)** $g \circ f$. In this case, does $g \circ f = f \circ g$?

Solution   Since $f: R \to R$ and $g: R \to R$, the domain of both $f \circ g$ and $g \circ f$ is the set of real numbers R.

(a) $(f \circ g)(x) = f(g(x)) = f(x + 1)$
$$= (x + 1)^3$$
$$= x^3 + 3x^2 + 3x + 1$$

(b) $(g \circ f)(x) = g(f(x)) = g(x^3)$
$$= (x^3) + 1$$
$$= x^3 + 1$$

Therefore

$$(f \circ g)(x) = x^3 + 3x^2 + 3x + 1$$
$$(g \circ f)(x) = x^3 + 1 \quad \text{and} \quad g \circ f \neq f \circ g \quad ■$$

EXAMPLE 6

Rough-cut lumber as full 2-inch by 4-inch (2 by 4) beams costs 50¢ per board foot, and 1 board foot is the volume of a 1-foot by 1-foot by 1-inch block of wood. Find the following functions:

L: (linear length of a 2 by 4) → (board feet in the 2 by 4)

p: (board feet) → cost

$C = p \circ L$: length → cost

Solution   Since rough-cut 2-inch by 4-inch beams cost 50¢ per board foot, the price function p is

$$p(x) = \frac{1}{2}x \quad \text{dollars}$$

where x is in board feet. From the definition of a board foot we know that

$$1 \text{ board foot} = 1 \cdot 1\left(\frac{1}{12}\right) \text{ cubic foot}$$
$$= \frac{1}{12} \text{ cubic foot}$$

In finding the function L we first find the volume of the rough-cut 2 by 4 in cubic feet, using 2 inches $= \frac{1}{6}$ foot, 4 inches $= \frac{1}{3}$ foot, as a function of x linear feet.

$$L(x) = \left(\frac{1'}{6}\right)\left(\frac{1'}{3}\right)x' = \frac{x}{18} \text{ cubic feet}$$

Now changing cubic feet to board feet, we have

$$L(x) = \frac{x}{18} \text{ cubic feet} \left(\frac{1 \text{ board foot}}{\frac{1}{12} \text{ cubic feet}}\right)$$
$$= \frac{12x}{18} \text{ board feet}$$
$$= \frac{2x}{3} \text{ board feet}$$

Thus x feet of a 2 by 4 is $\frac{2}{3}x$ board feet and $L(x) = \frac{2}{3}x$, where x is the number of linear feet of the 2 by 4. Therefore

$$C(x) = (p \circ L)(x) = p(L(x))$$
$$= p\left(\frac{2x}{3}\right) = \frac{1}{2}\left(\frac{2x}{3}\right) = \frac{x}{3} \quad \text{dollars}$$

and $C(x) = \frac{1}{3}x$ is the cost of x linear feet of 2 by 4 lumber when the cost is 50¢ a board foot. ■

It will also be helpful if you can recognize a function as the composition of two simpler functions. The following is an example of that process.

EXAMPLE 7   Find functions f and g such that **(a)** $(f \circ g)(x) = (x + 1)^2$ and **(b)** $(f \circ g)(x) = |2x - 3|$.

Solution

(a) For $(f \circ g)(x) = (x + 1)^2$ there are two operations (adding 1 and squaring). If we let $g(x) = x + 1$ and $f(x) = x^2$, then

$$(f \circ g)(x) = f(g(x))$$
$$= f(x + 1)$$
$$= (x + 1)^2$$

Therefore when $g(x) = x + 1$ and $f(x) = x^2$, we have $(f \circ g)(x) = (x + 1)^2$.

(b) For $(f \circ g)(x) = |2x - 3|$ there are two basic functions (the absolute value and twice x less 3). Thus for

$$g(x) = 2x - 3 \quad \text{and} \quad f(x) = |x|$$

we have

$$(f \circ g)(x) = f(g(x))$$
$$= f(2x - 3)$$
$$= |2x - 3|$$

Therefore when $g(x) = 2x - 3$ and $f(x) = |x|$, $(f \circ g)(x) = |2x - 3|$.   ∎

In Example 7 the functions f and g are not the only possible functions that would work, but they are convenient choices.

The next example shows some of the details of function composition with more complicated functions. Notice the coordination between the range and domain in composition.

EXAMPLE 8   For $f(x) = \sqrt{x - 1}$ and $g(x) = x^2 - 1$, find **(a)** $g \circ f$ and **(b)** $f \circ g$.

Solution   We notice that the domain of $f(x) = \sqrt{x - 1}$ is $\{x: x \geq 1\}$ and the range is $\{y: y \geq 0\}$. The domain of $g(x) = x^2 - 1$ is $\{x: x \in R\}$, and the range is $\{y: y \geq -1\}$. Thus

$$f: \{x: x \geq 1\} \rightarrow \{y: y \geq 0\} \quad \text{and}$$
$$g: \{x: x \in R\} \rightarrow \{y: y \geq -1\}$$

(a) $(g \circ f)(x)$ is defined for all x such that $f(x)$ is in the domain of g. Since the range of f is a subset of the domain of g, $\{y: y \geq 0\} \subseteq \{x: x \in R\}$, the domain of $g \circ f$ is the same as the domain of f, $\{x: x \geq 1\}$. Then for $x \geq 1$,

$$(g \circ f)(x) = g(f(x))$$
$$= g(\sqrt{x - 1})$$
$$= (\sqrt{x - 1})^2 - 1$$
$$= (x - 1) - 1$$
$$= x - 2$$

Therefore $(g \circ f)(x) = x - 2$ for x in $[1, \infty)$.

(b) $(f \circ g)(x)$ is defined for all x such that $g(x)$ is in the domain of f. Since the range of g, $\{y\colon y \geq -1\}$, is not a subset of the domain of f, $\{x\colon x \geq 1\}$, the domain of $f \circ g$ is limited to all x such that $g(x) \geq 1$. Thus the domain of f limits the domain of $f \circ g$ to all x such that

$$x^2 - 1 \geq 1$$
$$x^2 - 2 \geq 0$$

or

$$(x - \sqrt{2})(x + \sqrt{2}) \geq 0$$

Using the sign chart, we have $x^2 - 2 \geq 0$ for x in $(-\infty, -\sqrt{2}] \cup [\sqrt{2}, \infty)$.

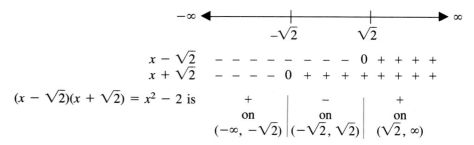

Thus the domain of $f \circ g$ can be $\{x\colon x \text{ is in } (-\infty, -\sqrt{2}] \cup [\sqrt{2}, \infty)\}$. Therefore

$$(f \circ g)(x) = f(x^2 - 1)$$
$$= \sqrt{x^2 - 1 - 1}$$
$$= \sqrt{x^2 - 2} \qquad \text{for } x \text{ in } (-\infty, -\sqrt{2}] \cup [\sqrt{2}, \infty)$$

It is also interesting to note that $f \circ g \neq g \circ f$ even when their domains are the same.   ■

The next theorem states a concept that we have assumed to be true up to this point.

Theorem | *Function composition.* If $f\colon A \to B$ and $g\colon B \to C$ are functions, then $g \circ f\colon A \to C$ is also a function.

This theorem can be proved by using the alternative definition of function composition, but we will not prove it here.

Now we can work with functions and their operations of addition, subtraction, multiplication, division, and composition.

Exercises 2.6

Find $f \circ g$ for the following problems.

1. $f = \{(2, 4), (3, 6), (4, 8), (5, 7)\}$
   $g = \{(1, 2), (2, 5), (3, 3), (5, 4)\}$

2. $f = \{(1, 2), (3, 4), (5, 6), (7, 8)\}$
   $g = \{(1, 3), (2, 1), (3, 7), (4, 5)\}$

3. $f = \{(1, 5), (2, 4), (3, 1), (4, 0)\}$
   $g = \{(0, 1), (1, 4), (2, 3), (3, 2)\}$

4. $f = \{(0, 1), (3, 5), (4, 8), (5, 1)\}$
   $g = \{(0, 4), (2, 0), (3, 5), (5, 3)\}$

Find (a) $f + g$, (b) $f - g$, (c) $f \cdot g$, (d) $f \div g$, (e) $f \circ g$, and (f) $g \circ f$ for each of the following.

5. $f(x) = 2x + 1$
   $g(x) = 3x - 2$

6. $f(x) = 2x - 1$
   $g(x) = 4x - 4$

7. $f(x) = 5x + 6$
   $g(x) = x^2 - 1$

8. $f(x) = 3x - 1$
   $g(x) = x^2 + 1$

9. $f(x) = x^2 - 1$
   $g(x) = (x + 1)^2$

10. $f(x) = x^3 - 1$
   $g(x) = (x - 1)^3$

Find (a) $f \circ g$ and (b) $g \circ f$ for each of the following.

11. $f(x) = |x + 2|$
   $g(x) = 3x - 2$

12. $f(x) = x - 1$
   $g(x) = |x + 1|$

13. $f(x) = x^2 + 1$
   $g(x) = \sqrt{x - 1}$ for $x \geq 1$

14. $f(x) = \sqrt{x + 1}$ for $x \geq -1$
   $g(x) = x^2 - 1$

15. $f(x) = \dfrac{1}{x}$ for $x \neq 0$
   $g(x) = x + 2$

16. $f(x) = x^2$
   $g(x) = \dfrac{1}{x + 1}$ for $x \neq -1$

17. $f(x) = \dfrac{1}{x - 1}$ for $x \neq 1$
   $g(x) = \dfrac{1}{x}$ for $x \neq 0$

18. $f(x) = \dfrac{1}{x + 1}$ for $x \neq -1$
   $g(x) = \dfrac{1}{x^2}$ for $x \neq 0$

Find one pair of functions f and g such that:

19. $(f \circ g)(x) = (x + 1)^3$

20. $(f \circ g)(x) = (x - 3)^2$

21. $(f \circ g)(x) = |x + 5|$

22. $(f \circ g)(x) = |3x|$

23. $(f \circ g)(x) = [\![3x]\!]$

24. $(f \circ g)(x) = [\![x - 1]\!]$

25. $(f \circ g)(x) = x^2 - 5$

26. $(f \circ g)(x) = (x + 2)^3$

27. $(f \circ g)(x) = (x - 5)^2$

28. $(f \circ g)(x) = x^3 + 2$

29. $(f \circ g)(x) = \dfrac{1}{(x + 1)^2}$ for $x \neq -1$

30. $(f \circ g)(x) = \dfrac{1}{(x + 2)^3}$ for $x \neq -2$

31. $(f \circ g)(x) = (x - 5)^{5/3}$

32. $(f \circ g)(x) = (x + 2)^{4/3}$

33. A yacht uses a symmetrical spinnaker sail when racing with the wind. A particular spinnaker sail is defined as shown in quadrant I (Fig. 2.69) by $f(x) = -\frac{4}{9}(x^2 + 10x - 200)$ for $0 \leq x \leq 10$ and is defined in quadrant IV by $g(x) = \frac{1}{20}(x^2 - 100)$ for $0 \leq x \leq 10$. Find $h(x)$ that measures the vertical height of the sail as a function of x for any x in the interval $[0, 10]$.

34. A yacht uses a symmetrical spinnaker sail when racing with the wind. A particular spinnaker sail is defined as shown in quadrant I (Fig. 2.69) by $f(x) = -\frac{1}{10}(9x^2 + 18x - 891)$ for $0 \leq x \leq 9$ and is defined in quadrant IV by $g(x) = \frac{1}{81}(x^2 - 81)$ for $0 \leq x \leq 9$. Find $h(x)$ that measures the vertical height of the sail as a function of x for any x in the interval $[0, 9]$.

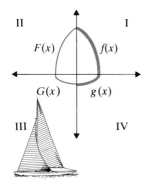

Figure 2.69

35. Bowen's Boats produces a small recreational sailboat. Bowen's can produce up to 150 boats a month with a cost function of $C(x) = 400 - 0.001x^2$ and a revenue function of $R(x) = 0.1x^3 - 21.001x^2 + 1100x + 400$, where x is the number of boats produced. Find the profit function $P(x)$.

36. Kennedy's Computer Products can produce up to 4000 computer disks a month. Under current conditions the monthly cost function is $C(x) = 0.1x + 100$, and the monthly revenue function is $R(x) = -0.01x^2 + 40.1x$. Find the monthly profit function $P(x)$.

37. Martinez Management Consultants operates with a daily revenue function of $R(x) = \dfrac{(2000x + 1000)}{x}$ and a daily cost function of $C(x) = \dfrac{(1000x + 9000)}{x + 5}$ for x days of consultation. Find the daily profit function $P(x)$ for x days of consultation.

38. A rough-cut 2-inch by 6-inch beam costs 75¢ per board foot, and 1 board foot is the volume of a 1-foot by 1-foot by 1-inch block of wood. Find the following three functions:

L: length of 2 by 10 $\rightarrow$ board feet in the 2 by 10
P: board feet $\rightarrow$ cost
$C = P \circ L$: length $\rightarrow$ cost

39. A rough-cut 2-inch by 10-inch beam costs $1.25 per board foot, and 1 board foot is the volume of a 1-foot by 1-foot by 1-inch block of wood. Find the following functions:

L: length of 2 by 10 $\rightarrow$ board feet in the 2 by 10
P: board feet $\rightarrow$ cost
$C = P \circ L$: length $\rightarrow$ cost

40. In the production of an item the cost of product $c(x)$ decreases as the number of items produced, x, increases. For this item, $c(x) = 20 + \left(\dfrac{100}{x}\right)$, where c: (number of items produced) $\rightarrow$ (cost per item). At the same time the profit for each item is a function of the cost such that $p(x) = 50 - x$ where p: (cost per item x) $\rightarrow$ (profit per item).
(a) Find $(p \circ c)(x)$.
(b) Evaluate and explain $c(x)$ and $(p \circ c)(x)$ when $x = 10$, 50, 100, and 200.
(c) For what x value does $(p \circ c)(x) = 0$? What is the significance of this domain value?

41. In the production of an item the cost of product $c(x)$ decreases as the number of items produced, x, increases. For this item, $c(x) = 7.50 + \left(\dfrac{100}{x}\right)$ where c: (number of items produced) $\rightarrow$ (cost per item). At the same time the profit for each item is a function of the item's cost such that $p(x) = 10 - x$ where p: (cost per item x) $\rightarrow$ (profit per item).
(a) Find $(p \circ c)(x)$.
(b) Evaluate and explain $c(x)$ and $(p \circ c)(x)$ when $x = 10$, 50, 100, and 200.
(c) For what x value does $(p \circ c)(x) = 0$? What is the significance of this domain value?

42. The oil slick created by a leaking ocean tanker has a circular area A. If the length of the radius r is a function of time t such that $r = \sqrt{t}$ feet, find the area of the oil slick as a function of time.

43. The oil slick created by a leaking ocean tanker has a circular area A. If the length of the radius r is a function of time t such that $r = \sqrt[3]{t}$ feet, find the area of the oil slick as a function of time.

44. Find one-to-one functions f and g such that $f + g$ is a function but not one-to-one.

45. Find one-to-one functions f and g such that $f - g$ is a function but not one-to-one.

46. Find two functions f and g such that $(f \div g)(x) = x$.

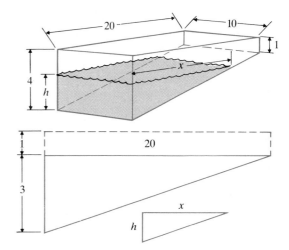

Figure 2.70

47. Find two functions f and g such that $(f \circ g)(x) = x$.

48. A 15- by 20-meter swimming pool is 1 meter deep at one end and 3 meters deep at the other end (Fig. 2.70). As the pool is being filled with water, find the volume of the pool as a function of the depth h of the water from the bottom of the pool. When water costs $1.10 per 100 cubic meters, find the cost of the water as a function of the depth of the pool.

49. A 15- by 20-meter swimming pool is 1 meter deep at one end and 4 meters deep at the other end (Fig. 2.70). As the pool is being filled with water, find the volume of the pool as a function of the depth h of the water from the bottom of the pool. When the cost of water is $1.40 per 100 cubic meters, find the cost of water as a function of the depth of the pool.

2.7 Inverse Functions

Earlier, we worked with the volume of a spherical balloon as a function of the balloon's radius, $V = \frac{4}{3}\pi r^3$ (or $V(x) = \frac{4}{3}\pi x^3$, where x is the radius). We can also rewrite this so that the radius is written as a function of the volume. Then $V = \frac{4}{3}\pi r^3$ becomes

$$\frac{3V}{4\pi} = r^3$$

and

$$r = \sqrt[3]{\frac{3V}{4\pi}}$$

or

$$r(y) = \sqrt[3]{\frac{3y}{4\pi}} \qquad \text{where } y \text{ is a volume}$$

Thus we can write the volume as a function of the radius and we can also write the radius as a function of the volume. Notice how V and r map between the two sets in Fig. 2.71.

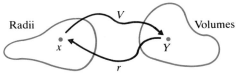

Figure 2.71 V: radius ⟶ volume and r : volume ⟶ radius

Both of these functions provide important information about the balloon, each from a different perspective. For any given radius x the function $V(x)$ provides the corresponding volume, and for any desired volume y the function $r(y)$ provides the corresponding radius. This pair of functions, $V(x)$ and $r(y)$, are called **inverse functions**. We will find that inverse functions are necessary and useful tools.

The composition of a function and its inverse function produces a special result. Although this result is illustrated by the mappings of V and r in Fig. 2.71, we can see it more clearly by calculating $r \circ V$ and $V \circ r$:

$$(r \circ V)(x) = r(V(x)) = r\left(\frac{4}{3}\pi x^3\right)$$

$$= \sqrt[3]{\frac{3\left(\frac{4}{3}\pi x^3\right)}{4\pi}} = \sqrt[3]{\frac{4\pi x^3}{4\pi}} = \sqrt[3]{x^3} = x$$

and

$$(V \circ r)(y) = V(r(y)) = V\left(\sqrt[3]{\frac{3y}{4\pi}}\right)$$

$$= \frac{4}{3}\pi\left(\sqrt[3]{\frac{3y}{4\pi}}\right)^3 = \frac{4}{3}\pi\left(\frac{3y}{4\pi}\right) = y$$

These compositions illustrate that when we start with a radius x, map with V, and then map with r, the result is x, the starting domain element. Thus the function r seems to "undo" the mapping of the function V as $r \circ V : x \rightarrow x$. Likewise, when we start with a volume y, map with r, and then map with V, the result is y, the starting domain element. Here also the function V seems to "undo" the mapping of the function r as $V \circ r : y \rightarrow y$. This behavior of the composition of two functions is one way to identify a pair of inverse functions.

In this section we use the following three ways to work with inverse functions: the definition, ordered pairs, and graphing.

Definition

For a one-to-one onto function $f : X \rightarrow Y$, there is a one-to-one onto function $g : Y \rightarrow X$, called the **inverse function** of f, such that

$$g(f(x)) = x \qquad \text{for every } x \in X, \text{ the domain of } f$$
$$f(g(x)) = x \qquad \text{for every } x \in Y, \text{ the domain of } g$$

The inverse function of $f(x)$ is usually written $g(x) = f^{-1}(x)$. Thus

$$f(f^{-1}(x)) = x \qquad \text{and} \qquad f^{-1}(f(x)) = x$$

The domain of f is the range of f^{-1}, and the range of f is the domain of f^{-1}.

It is important for you to note that f^{-1} is the inverse function of f, not the reciprocal function of $f(x)$. Remember that

$$f^{-1}(x) \neq \frac{1}{f(x)}, \text{ but } f^{-1}(f(x)) = x \text{ and } f(f^{-1}(x)) = x$$

are both necessary conditions for a function and its inverse function (Fig. 2.72).

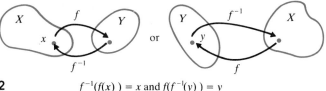

Figure 2.72 $f^{-1}(f(x)) = x$ and $f(f^{-1}(y)) = y$

Example 1 shows how we can use the definition to identify a pair of inverse functions.

EXAMPLE 1

For $f(x) = x^2$ for $x \geq 0$ and $g(x) = \sqrt{x}$ for $x \geq 0$, show that f and g are inverse functions by evaluating $(f \circ g)(x)$ and $(g \circ f)(x)$.

Solution   First notice that for $x \geq 0$, f and g both map the nonnegative real numbers onto the nonnegative real numbers. (The domains and ranges of f and g

are each the nonnegative real numbers.) It is also important to note that f and g are both one-to-one functions. Then

$$(f \circ g)(x) = f(g(x)) = f(\sqrt{x}) = (\sqrt{x})^2 = |x| = x$$
$$(g \circ f)(x) = g(f(x)) = g(x^2) = \sqrt{(x^2)} = |x| = x, \qquad \text{since } x \geq 0.$$

Therefore f and g are inverse functions. ■

We should also note from the definition that being one-to-one is a necessary condition for an inverse function. The function $h = \{(1, 4), (2, 0), (3, 4)\}$ is not a one-to-one function; both 1 and 3 map to 4 (Fig. 2.73). But if there exists an inverse function h^{-1} for h, then $h^{-1}(4) = 1$ or 3. Thus $(4, 1) \in h^{-1}$ and $(4, 3) \in h^{-1}$, and the inverse is not a function (Fig. 2.73).

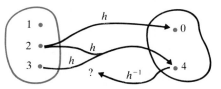

Figure 2.73   When h is not one-to-one, h^{-1} does not exist.

This mapping of h reinforces the necessity for a function to be one-to-one if it has an inverse function.

In the next example you will see one way to use ordered pairs to work with inverse functions.

EXAMPLE 2   Find f^{-1} when $f = \{(1, 5), (2, 3), (3, 1), (4, 0)\}$, where $A = \{1, 2, 3, 4\}$ and $B = \{0, 1, 3, 5\}$.

Solution   To see the relationships $f(f^{-1}(x)) = x$ and $f^{-1}(f(x)) = x$ better, we draw a diagram of the mapping for $f^{-1}(f(x))$. Notice that f is a one-to-one function.

$$
\begin{array}{cl}
f \quad f^{-1} & \text{since } f^{-1}(f(x)) = x, \\
1 \rightarrow 5 \rightarrow a_1 & f^{-1}(f(1)) = f^{-1}(5) = a_1 = 1 \\
2 \rightarrow 3 \rightarrow a_2 & f^{-1}(f(2)) = f^{-1}(3) = a_2 = 2 \\
3 \rightarrow 1 \rightarrow a_3 & f^{-1}(f(3)) = f^{-1}(1) = a_3 = 3 \\
4 \rightarrow 0 \rightarrow a_4 & f^{-1}(f(4)) = f^{-1}(0) = a_4 = 4
\end{array}
$$

Thus from this diagram, $f^{-1} = \{(5, a_1), (3, a_2), (1, a_3), (0, a_4)\}$, which is $f^{-1} = \{(5, 1), (3, 2), (1, 3), (0, 4)\}$. ■

We notice in Example 2 that if (x, y) is an element of f, then (y, x) is an element of f^{-1} (that is, the same values in reverse order). This observation provides an important second way to view and evaluate inverse functions.

Theorem

Inverse Function. The one-to-one onto function f has an inverse function f^{-1} if and only if for every $(a, b) \in f$, $(b, a) \in f^{-1}$.

PROOF: The following two parts are required to prove this "if and only if" statement.

(a) Show that if the one-to-one onto function f has an inverse function f^{-1}, then for every $(a, b) \in f$ we have $(b, a) \in f^{-1}$. If $(a, b) \in f$, then $f(a) = b$. Since f is one-to-one, we can use function composition on both sides of $f(a) = b$ with f^{-1}. We then have

$$f^{-1}f(a) = f^{-1}(b),$$

and

$$a = f^{-1}(b),$$

since f and f^{-1} are inverse functions. Thus for every $(a, b) \in f$ we have $(b, a) \in f^{-1}$.

(b) Show that for the one-to-one onto function f if every $(a, b) \in f$ implies $(b, a) \in f^{-1}$, then f^{-1} is the inverse function of f. Since $(a, b) \in f$ and $(b, a) \in f^{-1}$, we have $f(a) = b$ and $f^{-1}(b) = a$. Using function composition,

$$f^{-1}f(a) = f^{-1}(b) = a$$

so that

$$(f^{-1} \circ f)(a) = a$$

Likewise, we use function composition with f on $f^{-1}(b) = a$ to obtain

$$f(f^{-1}(b)) = f(a) = b$$

Thus

$$(f \circ f^{-1})(b) = b$$

We can then conclude by the definition of an inverse function that f^{-1} is the inverse function of f.

Therefore by parts (a) and (b) a one-to-one onto function f has an inverse function f^{-1} if and only if for every $(a, b) \in f$, $(b, a) \in f^{-1}$.   □

As we saw in the spherical balloon functions at the beginning of this section and as we will see in the next example, we can get new and important information from a function by looking at its inverse.

EXAMPLE 3

Rough-cut lumber as full 2-inch by 4-inch beams costs 50¢ per board foot, and 1 board foot is the volume of a 1-foot by 1-foot by 1-inch block of wood. Find the number of linear feet you can purchase as a function of the cost you can afford to pay. How many linear feet of a 2 by 4 can you purchase for $17?

Solution   In the last section we found the functions

$$L(x) = \frac{2}{3}x \text{ board feet}$$

$$p(x) = \frac{1}{2}x \text{ dollars}$$

$$C(x) = p \circ L(x) = \frac{1}{3}x \text{ dollars}$$

where

L: (linear length of a 2 × 4) → (board feet in the 2 by 4)
p: (board feet) → cost
$C = p \circ L$: length → cost.

To map cost to board feet, we could find C^{-1} or $L^{-1} \circ p^{-1}$. It will be much easier to find C^{-1}: cost → length. To do this, take C and solve for x in terms of C. Thus $C(x) = \left(\frac{1}{3}\right)x$ is

$$C = \frac{1}{3}x$$

$$3C = 3\left(\frac{1}{3}\right)x$$

$$3C = x$$

Then $x = 3C$, where C is a domain value and x is a range value of C^{-1}. Therefore $C^{-1}(x) = 3x$ linear feet, and \$17 will buy $C^{-1}(17) = 3(17) = 51$ linear feet of a 2 by 4.   ■

In Example 4 we will use the inverse function theorem to find the inverse function by reversing the order of the ordered pairs of the original function (reversing the roles of x and y). In Example 5 we will also find the inverse functions using function composition, $(f \circ f^{-1})(x) = x = (f^{-1} \circ f)(x)$. Either method can be used to find the inverse function.

EXAMPLE 4

Find the inverse functions and check the result for $g(x) = x^2 + 1$ for x ≥ 0.

Solution   We note that $g(x) = x^2 + 1$ for $x \geq 0$ is a one-to-one function on this interval and thus has an inverse function. We rewrite $g(x) = x^2 + 1$ as $y = x^2 + 1$. Changing the order of the ordered pairs, we switch the roles of x and y and then solve for y. Then

$g(x) = y$ where $y = x^2 + 1$ for $x \geq 0$ changes into
$g^{-1}(x) = y$ where $x = y^2 + 1$ for $y \geq 0$.

Thus for $g^{-1}(x) = y$ we have

$$y^2 = x - 1 \qquad \text{for } y \geq 0$$

or

$$y = \sqrt{x - 1} \qquad \text{for } x \geq 1$$

Therefore $g^{-1}(x) = \sqrt{x - 1}$ for $x \geq 1$. Only the positive square root is considered, since the restriction is $y \geq 0$. The $x - 1$ under the square root must also be nonnegative so that $x \geq 1$. As we see in this example, switching the roles of x and y even helps us to preserve the bounds on the range and domain. Thus $g(x) = x^2 + 1$ for $x \geq 0$ and $g^{-1}(x) = \sqrt{x - 1}$ for $x \geq 1$.

CHECK: Does $g(g^{-1}(x)) = x$ for $x \geq 1$?

$$\begin{aligned}
g(g^{-1}(x)) &= g(\sqrt{x - 1}) \\
&= (\sqrt{x - 1})^2 + 1 \\
&= x - 1 + 1 \\
&= x \qquad \text{Yes, } g(g^{-1}(x)) = x.
\end{aligned}$$

Does $g^{-1}(g(x)) = x$ for $x \geq 1$?

$$\begin{aligned}
g^{-1}(g(x)) &= g^{-1}(x^2 + 1) \\
&= \sqrt{(x^2 + 1) - 1} \\
&= \sqrt{x^2} \\
&= |x| = x \quad \text{since } x \geq 0 \qquad \text{Yes, } g^{-1}(g(x)) = x.
\end{aligned}$$

Therefore when $g(x) = x^2 + 1$ for $x \geq 0$, we have

$$g^{-1}(x) = \sqrt{x - 1} \text{ for } x \geq 1 \qquad \blacksquare$$

EXAMPLE 5   Find the inverse functions and check the result for $f(x) = 2x - 3$.

Solution   First notice that $f(x) = 2x - 3$ is a one-to-one function. Using function composition, we find the inverse function by allowing $f^{-1}(x)$ to be the domain value for f. With $f^{-1}(x)$ as the domain element for $f(x) = 2x - 3$ we have

$$\begin{aligned}
f(f^{-1}(x)) &= 2(f^{-1}(x)) - 3 \qquad (f(f^{-1}(x)) = x \text{ by the definition of the inverse function}) \\
x &= 2f^{-1}(x) - 3 \\
x + 3 &= 2f^{-1}(x)
\end{aligned}$$

and

$$f^{-1}(x) = \frac{1}{2}(x + 3)$$

Therefore when $f(x) = 2x - 3$, we have $f^{-1}(x) = \frac{1}{2}(x + 3)$.

CHECK: Does $f(f^{-1}(x)) = x$?

$$f(f^{-1}(x)) = f\left[\frac{1}{2}(x + 3)\right]$$

$$= 2\left[\frac{1}{2}(x + 3)\right] - 3$$

$$= (x + 3) - 3 = x \qquad \text{Yes}, f(f^{-1}(x)) = x.$$

Does $f^{-1}(f(x)) = x$?

$$f^{-1}(f(x)) = f^{-1}(2x - 3)$$

$$= \frac{1}{2}[(2x - 3) + 3]$$

$$= \frac{1}{2}(2x) = x \qquad \text{Yes}, f^{-1}(f(x)) = x.$$

Therefore $f(x) = 2x - 3$ and $f^{-1}(x) = \frac{1}{2}(x + 3)$ are inverse functions. ∎

The fact that $(a, b) \in f$ and $(b, a) \in f^{-1}$ means that the graphs of the functions $f(x)$ and $f^{-1}(x)$ have symmetry with respect to the line $y = x$ (Fig. 2.74). As discussed in Section 2.4, when we switch the order of the ordered pair of point P on f, there is a point P' on f^{-1} such that $y = x$ is the perpendicular bisection of the segment from P to P', as seen in Fig. 2.75. In general, the graphs of f and f^{-1} have symmetry with respect to the line $y = x$.

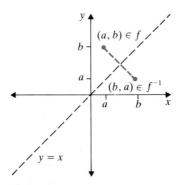

Figure 2.74

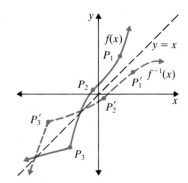

Figure 2.75

Using symmetry with respect to the line $y = x$ will make it easier to sketch the graphs of inverse functions.

EXAMPLE 6   Graph these functions and their inverse functions from Example 4 and 5.

(a) $f(x) = 2x - 3$
$f^{-1}(x) = \frac{1}{2}(x + 3)$

(b) $g(x) = x^2 + 1$ for $x \geq 0$
$g^{-1}(x) = \sqrt{x - 1}$ for $x \geq 1$

Solution   We will use symmetry to the line $y = x$ to sketch these graphs.

(a) For $f(x) = 2x - 3$ and $f^{-1}(x) = \frac{1}{2}(x + 3)$, sketch the graph of f (Fig. 2.76a). Then switch the order of the ordered pairs to graph f^{-1}. This makes f^{-1} the mirror image of f on the other side of the line $y = x$.

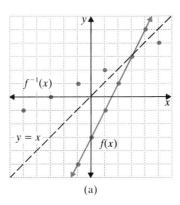

Figure 2.76   (a)   (b)

(b) For $g(x) = x^2 + 1$ for $x \geq 0$ and $g^{-1}(x) = \sqrt{x - 1}$ for $x \geq 1$, sketch as in part (a). (Figs. 2.77a, 2.77b).

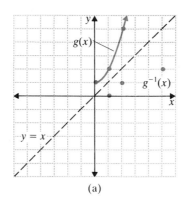

Figure 2.77   (a)   (b)

INVERSE FUNCTIONS

We can work with inverse functions in three ways:

(a) If f and f^{-1} are inverse functions, then
$$f(f^{-1}(x)) = x \qquad \text{and} \qquad f^{-1}(f(x)) = x$$

(b) If f and f^{-1} are inverse functions, then for every a in the domain of f there exists b in the range of f such that
$$(a, b) \in f \qquad \text{and} \qquad (b, a) \in f^{-1}.$$

(c) The graphs of f and f^{-1} have symmetry with respect to the line $y = x$.

The next application is more difficult but shows an interesting use of the inverse relation.

EXAMPLE 7

The altitude s of a missile is measured as a function of time by $s(t) = 4800t - 16t^2$, where the altitude s is measured in feet and the time t is measured in seconds. Find an equation for the time into the flight for any given altitude.

Solution   We have $s(t) = 4800t - 16t^2$, which maps the time to an altitude, and we need to find an equation that maps the altitude to a time. Since we must find t as a function of s, the roles of s and t will be reversed; s will become the domain, and t will become the range. Thus we must find the inverse of s that is t where function $t = s^{-1}$. To find the inverse function s^{-1}, we solve the altitude equation for t. Then

$$s(t) = 4800t - 16t^2$$

becomes

$$16t^2 - 4800t + s(t) = 0$$

or

$$16t^2 - 4800t + s = 0$$

We now use the Quadratic Theorem to solve for t.

$$t = \frac{4800 \pm \sqrt{(4800)^2 - 4(16)s}}{2(16)}$$

$$t = \frac{4800 \pm \sqrt{64[(600)^2 - s]}}{32}$$

$$t = \frac{8(600) \pm 8\sqrt{(600)^2 - s}}{4(8)}$$

$$t = \frac{600 \pm \sqrt{(600)^2 - s}}{4}$$

or

$$t(s) = \frac{600 \pm \sqrt{(600)^2 - s}}{4}$$

It is interesting to note that for each altitude s there are two time values as a result of the plus-or-minus sign. One time occurs when the missile is going up, and the other when the missile is descending. Thus the inverse t of the function s is not a function. ∎

As we see in this example, the inverse of a function is not a function unless the original function is one-to-one.

Exercises 2.7

Find the inverse function for each of the following one-to-one functions.

1. $f = \{(2, 3), (3, 1), (1, 5), (5, 2)\}$
2. $f = \{(0, 3), (1, 2), (3, 1), (2, 0)\}$
3. $g = \{(2, 5), (3, 4), (4, 2), (5, 3)\}$
4. $g = \{(-1, 0), (0, 5), (6, -1), (5, 4)\}$
5. $h = \{(1, 7), (3, 5), (7, 3), (5, 1)\}$
6. $h = \{(6, 2), (2, 7), (7, 8), (8, 2)\}$

Find the inverse for each of the following one-to-one functions. Check your answers by showing that $f(f^{-1}(x)) = x$ and $f^{-1}(f(x)) = x$.

7. $f(x) = \frac{1}{2}(x + 5)$ 8. $f(x) = \frac{1}{3}(x - 4)$

9. $g(x) = 2x - 1$ 10. $g(x) = 5 - 2x$

11. $h(x) = 3 - 4x$ 12. $h(x) = 3x - 2$

13. $f(x) = \dfrac{1}{x^3}$ for $x \neq 0$ 14. $f(x) = \dfrac{1}{x^2}$ for $x > 0$

15. $g(x) = \dfrac{1}{x + 1}$ for $x \neq -1$ 16. $g(x) = \dfrac{1}{x - 1}$ for $x \neq 1$

17. $h(x) = x^3 - 1$ 18. $h(x) = x^3 + 1$
19. $f(x) = \sqrt{x + 3}$ for $x \geq -3$
20. $f(x) = (x - 3)^2$ for $x \geq 3$
21. $g(x) = \sqrt[3]{x - 3}$ 22. $g(x) = \sqrt[3]{x + 3}$

23. $h(x) = -x$ 24. $h(x) = \dfrac{1}{x}$ for $x \neq 0$

25. Graph f and f^{-1} from Exercise 7.

26. Graph f and f^{-1} from Exercise 8.

27. Graph g and g^{-1} from Exercise 9.

28. Graph g and g^{-1} from Exercise 10.

29. Graph h and h^{-1} from Exercise 11.

30. Graph h and h^{-1} from Exercise 12.

31. Graph f and f^{-1} from Exercise 13.

32. Graph f and f^{-1} from Exercise 14.

33. Graph h and h^{-1} from Exercise 17.

34. Graph h and h^{-1} from Exercise 18.

35. Prove that the linear function $f(x) = ax + b$, for $a \neq 0$, has an inverse function. Then find $f^{-1}(x)$.

36. For $h(x) = ax + b$ and $g(x) = cx + d$, when are h and g inverse functions? (Find the ratio of b to a.)

37. Prove that if f is a one-to-one function, $f : X \rightarrow Y$, $g : Y \rightarrow X$, and $(f \circ g)(x) = x$, then $(g \circ f)(x) = x$. Is g a one-to-one function?

38. Find the area of a square as a function of one of its sides. What are the limitations on its domain? Find the inverse of this function. What are the limitations on the range of this inverse function?

39. Find the perimeter of a square as a function of one of its sides. What are the limitations on its domain? Find the inverse of this function. What are the limitations on the range of this inverse function?

40. Find the circumference of a circle as a function of its radius. What are the limitations on its domain? Find the inverse of this function. What are the limitations on the range of this inverse function?

41. Find the area of a circle as a function of its radius. What are the limitations on its domain? Find the inverse of this function. What are the limitations on the range of this inverse function?

42. Find the perimeter of a square as the function of its area, using function composition.

43. Find the circumference of a circle as a function of its area, using function composition.

44. The height of a missile at time t is found by the function $s(t) = -16t^2 + 4800t$. When $0 \leq t \leq 150$ in $s(t)$, find an expression for s^{-1}.

45. The height of a missile at time t is found by the function $s(t) = -16t^2 + 3520t$. When $0 \leq t \leq 110$ in $s(t)$, find an expression for s^{-1}.

46. The oil slick created by a leaking ocean tanker has a circular area A. If the length of the radius r is a function of time t such that $r = \sqrt{t}$ feet, find time as a function of the area of the oil slick.

47. The oil slick created by a leaking ocean tanker has a circular area A. If the length of the radius r is a function of time t such that $r = \sqrt[3]{t}$ feet, find time as a function of the area of the oil slick.

CHAPTER SUMMARY AND REVIEW

KEY TERMS

Relation	The cubic function
Function	The absolute value function
Into and onto mapping	The greatest integer function
One-to-one mapping	Piecewise-defined function
Point-slope form for the equation of a line	The distance formula
Slope-intercept form for the equation of a line	The midpoint formula
Horizontal and vertical lines	The equation of a circle
The slopes of parallel and perpendicular lines	Test for symmetry
Linear function	Even and odd functions
Increasing and decreasing functions	Translation of axes
The vertical line test (function test)	Function composition
The horizontal line test (one-to-one test)	Inverse functions
The quadratic function	

CHAPTER EXERCISES

Find the domain and range of the relations in Exercises 1 and 2, and state whether each relation is a function.

1. $f = \{(1, 3), (2, 3), (3, 1), (4, 0)\}$

2. $g = \{(1, 3), (2, 2), (2, 3), (4, 3)\}$

For the functions in Exercises 3–6, evaluate and simplify the following expressions:

(a) $f(-1)$

(b) $f\left(\frac{1}{2}\right)$

(c) $f\left(\frac{1}{x}\right)$

(d) $f(-a)$

(e) $-f(a)$

(f) $f(x + a)$

(g) $f(x + a) - f(x)$

(h) $\dfrac{f(x + a) - f(x)}{a}$

3. $f(x) = 5x - 4$

4. $f(x) = 3 - 2x$

5. $f(x) = 3x^2$

6. $f(x) = \dfrac{1}{x}$ for $x \neq 0$

Find the domain intervals for each relation.

7. $f(x) = \dfrac{1}{x^2 - 5x + 6}$

8. $f(x) = (x^2 - 4)^{-1}$

9. $f(x) = \sqrt{x^2 - 5x + 6}$

10. $f(x) = \dfrac{1}{\sqrt{x^2 - 1}}$

Find (a) the domain interval and (b) the smallest range interval for each relation in Exercises 11 and 12.

11. $g(x) = |x| + 1$

12. $g(x) = \dfrac{x + 2}{x + 4}$

13. From the similar triangles in Fig. 2.78, find y as a function of x. Is this a linear function?

Figure 2.78

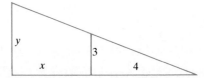

14. A Navy intelligence aircraft drops sonar buoys to track an enemy submarine (Fig. 2.79). Find the equation for the flight path of the aircraft if it drops buoys at $(-1, 5)$, $(0, 4)$, $(1, 3)$, $(2, 2)$, $(3, 3)$, and $(5, 5)$.

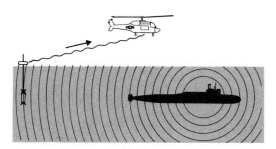

Figure 2.79

15. As in Exercise 14 a Navy intelligence aircraft drops sonar buoys, in order, at the following points: $(-1, 4)$, $(0, 3)$, $(1, 2)$, $(2, 1)$, $(3, 2)$, $(4, 3)$, and $(5, 4)$. Graph the position of these buoys and find the equation for the flight path of the aircraft.

16. As in Exercise 14 a Navy aircraft drops sonar buoys, in order, at $(-4, 1)$, $(-3, 0)$, $(-2, -1)$, $(-1, 0)$, $(0, 1)$, $(1, 2)$, and $(3, 4)$. Graph the position of these buoys and find the equation for the flight path of the aircraft.

Graph the functions in Exercises 17–20 on the Cartesian coordinate system.

17. $f(x) = 5x - 3$

19. $f(x) = |x + 1|$

18. $f(x) = x^2 + 1$

20. $f(x) = [\![x + 2]\!]$

Determine whether the following graphs are functions and/or one-to-one.

21.

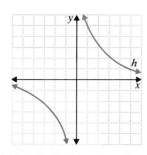

22.

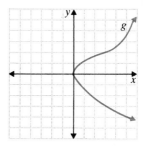

23.

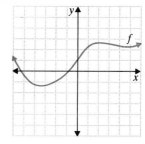

24.

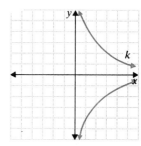

25. Write the equation of the line through $(2, 4)$ and $(-2, 5)$. The final answer should be in slope-intercept form.

26. Write the equation of the line perpendicular to $x + 2y + 3 = 0$ through the point $(1, 1)$. The final answer should be in standard form with integer coefficients.

27. Is $f(x) = |2x|$ an even function, an odd function, or neither? Why?

28. Is $f(x) = x^3 + x$ an even function, an odd function, or neither? Why?

Graph each of the following.

29. $\{(x, y) : x > -1 \text{ and } |y| \leq 3\}$

30. $\{(x, y) : (x - 2)^2 + (y + 1)^2 = 3^2\}$

31. $\{(x, y) : x^2 - 6x + y^2 - 4y = -12\}$

32. $\{(x, y) : x^2 + 2x + y^2 + 2y = 7\}$

What type of symmetry (*x*-axis, *y*-axis, or origin symmetry) does each of the following have, if any?

33. $x^2 + y^2 = 9$

34. $x = y^2 + 5$

35. $y = 5x - 3$

36. $y = [\![x]\!]$

37. $y = x^2 - 2$

38. $x = \dfrac{1}{y}$

Write the equation for the solid graph, given the dashed graph in Exercises 39–42.

39.

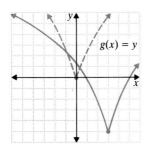

$g(x) = y$

40.

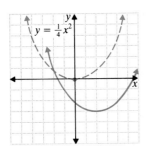

$y = \frac{1}{4}x^2$

41.

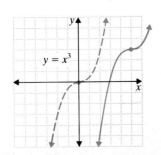

$y = x^3$

42.

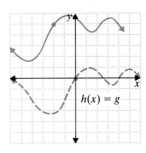

$h(x) = g$

Graph the functions in Exercises 43–48.

43. $f(x) = \frac{1}{5}x^2$ **44.** $f(x) = 5x^3$

45. $f(x) = x^6$ **46.** $f(x) = x^5$

47. $f(x) = \dfrac{1}{x + 1}$ **48.** $f(x) = \dfrac{1}{|x + 2|}$

Find (a) $f \circ g$ and (b) $g \circ f$ for the following.

49. $f(x) = x^2 + 2x$ **50.** $f(x) = x^3 + x^2$
      $g(x) = 5x - 1$ $g(x) = x + 1$

Find function f and g for the following.

51. $(f \circ g)(x) = (x - 1)^3$ **52.** $(f \circ g)(x) = (x + 1)^2 - 1$

53. An oil slick has a circular area A. If the radius r is a function of time t such that $r = \dfrac{3}{t}$ feet, find the area of the oil slick as a function of time. Also find time as a function of the area of the oil slick.

Find the inverse function f^{-1} for f. Verify the answer by computing $f(f^{-1}(x)) = x$ and $f^{-1}(f(x)) = x$.

54. $f(x) = 2x - 3$

55. $f(x) = x^2 - 5$ for $x \geq 0$

56. $f(x) = \dfrac{1}{x - 5}$ for $x \neq 5$

57. The conductance G of a resistor varies inversely with the resistance $R(G = 0.02$ siemen when $R = 50$ ohms and $R > 0$). Find this function. Is this function increasing, decreasing, strictly increasing, strictly decreasing, or none of these?

58. For energy cells in series (Fig. 2.80), Ohm's law defines the relationship $I = \dfrac{nE}{R_0 + nb}$, where I is the current, and n is the number of cells, each having an electromotive force of E and an internal resistance of b, and an external resistance R_0. Find n as a function of the other variables (that is, find an expression for n in terms of I, E, b, and R_0).

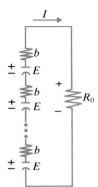

Figure 2.80

the price of water at \$1.25 per 100 cubic meters, find the cost of filling the pool as a function of the depth of the pool.

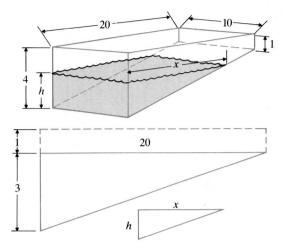

Figure 2.81

59. The altitude s in feet of a missile as a function of time t in seconds is given by $s(t) = 4800t - 16t^2$. This function is for a missile that is launched at $t = 0$ seconds and $s(0) = 0$ feet (sea level). Write a relation for the altitude of the same missile when it is launched from a site with an altitude of 2000 feet.

60. What does $s(t - 7)$ indicate when $s(t)$ is the altitude of the missile in Exercise 59.

61. A 10- by 20-meter swimming pool is 1 meter deep at one end and 4 meters deep at the other end as shown in Fig. 2.81. As the pool is being filled with water, find the volume of the pool as a function of the depth h of the pool. With

62. Prove that replacing x with y and y with x will provide an equivalent equation if the graph has symmetry with respect to the origin.

3

Polynomial and Rational Functions

THE DYNAMICS OF CHEMICAL BOUNDARIES

With the increased use of adhesives in industrial production it has become more important to understand the dynamics of these materials. The use of adhesives provides two advantages in the production of automobiles and aircraft: a reduction in weight and simplified production procedures.

One obstacle to the efficient bonding of adhesives to metals is the ability of adhesives to displace contaminants and to wet the metal surface. Ensuring intimate contact between adhesive and metal requires detailed knowledge of adhesive surface tension, since this controls displacement of contaminants and wetting. The surface tension of an adhesive was formerly assumed to be a constant. Upon study it was found that surface-active components in the adhesive collect preferentially at the interface and react, causing the surface composition to vary with time. This gives rise to dynamic

Experimental measurements of spreading pressure versus time for dialkylaminopropylamines with various Damköhler numbers (λ), and corresponding theoretical calculations of surface concentrations.

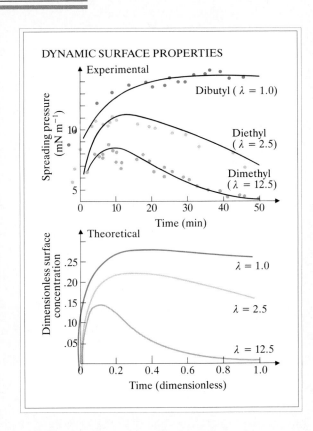

DYNAMIC SURFACE PROPERTIES

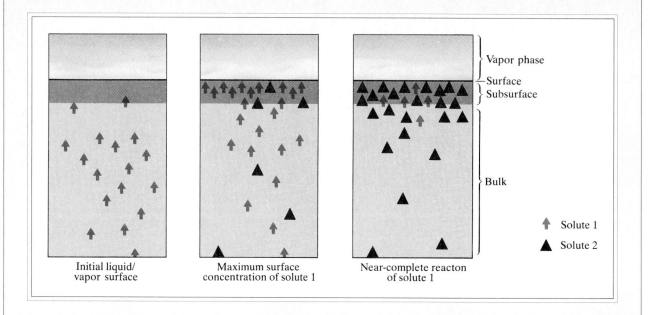

| | Solute 1 |
| | Solute 2 |

Initial liquid/vapor surface

Maximum surface concentration of solute 1

Near-complete reacton of solute 1

Evolution of an adhesive surface: Surface-active Solute 1 (blue arrows) reacts with host resin (gray background) to form surface active Solute 2 (black triangles).

surface tension. Large variations in surface tension can significantly affect adhesive performance. The work of Dr. Robert Foister, a scientist at General Motors Research Laboratories, has provided a general theory of absorption kinetics in binary, chemically reacting surfactant systems.

In a typical adhesive that polymerizes, or "cures" by chemical reaction, a surface-active curing agent (Solute 1) reacts with the host resin from a second surface-active species (Solute 2) that is also reactive. Diffusion to the surface is driven by a potential energy gradient between the surfaces and the bulk, the solute molecules experiencing a lower energy at the surface.

Dr. Foister derived appropriate transport equations to describe diffusion and chemical reaction in the bulk, in subsurface regions, and at the surface itself. Modifying the transport equations to include binary absorption iso-

therms allowed for consideration of competitive absorption of the two reacting and diffusing solutes. By solving the equations numerically and conducting dimension analysis, Dr. Foister identified various dimensionless parameters as predictors of behavior. The most important of these, the dimensionless number λ, of the Damköhler type, is

$$\lambda = \frac{k(\Gamma_m a)^2}{4D}$$

where Solute 1 has a reaction rate k, diffusivity D, "surface capacity" Γ_m (the maximum number of molecules absorbed per unit surface area), and "surface affinity" a (a measure of energy absorption). For an adhesive, lowering λ by reducing k (the reactivity of the curing agent) would prolong the time to maximum and would increase value of the surface concentration at the maximum. As a practical consequence, this would improve wetting by minimizing the surface tension.

"I expect," says Dr. Foister, "that the physical insights gained from this analysis can be applied to other reactive surfactant systems by using specifically tailored isotherms and chemical reaction schemes. Predicting surface behavior can certainly help us design better adhesives for specific applications. . . . Applied to interfaces in biological systems a suitable modified theory may prove valuable in understanding the phenomenon of enzyme activity."

In this chapter we will consider classical methods for solving and graphing similar but less complicated equations.

SOURCE: General Motors Research and Development advertising, Detroit, Michigan, "The Boundary Dynamic." Reprinted by permission.

The example of chemical properties of adhesives shows that functions can be very useful in solving research and development problems in industry. In this example the ability to solve the equation

$$\lambda = \frac{k(\Gamma_m a)^2}{4D}$$

with different values of λ is important in varying the values of k and Γ_m. Thus we can study k as a function of Γ_m or Γ_m as a function of k.

One method of analyzing functions is to find the domain elements x such that $f(x) = 0$ or $g(x) = 0$. These domain values of x are called the **zeros** of functions f or g. In this chapter we will learn how to find the rational zero of polynomials and extend the work of Chapter 2 with linear, quadratic, and cubic functions to polynomial functions in general. We will not only find the zeros of a polynomial but will use these zeros, along with some of the ideas developed in Chapters 1 and 2, to graph polynomial functions. Later in the chapter we will consider functions obtained from the quotient of two polynomials. These functions are called rational functions. Like polynomial functions, rational functions have interesting graphs and can be used to describe several aspects of our world.

3.1 Graphing Polynomial Functions

We recall from Chapter 2 that linear functions are first-degree polynomials in one variable, such as $f(x) = ax + b$, and quadratic functions are second-degree polynomials in one variable, such as $f(x) = ax^2 + bx + c$, ($a \neq 0$ in both cases). For example, $f(x) = 2x + 5$ is a linear function, while $f(x) = x^2 - x - 6$ is a quadratic function. Each of these functions belongs to a larger class of functions known as **polynomial functions**. Some other examples of polynomial functions are

$$f(x) = x^3 - 3x^3 + 3x + 1$$
$$y = 32x^5 - 20x^3 + 5x$$
$$g(x) = x^3(x - 3)(x + 2)$$

Here is the formal definition for a polynomial function.

Definition

The function f is a **polynomial function** of degree n if

$$f(x) = a_n x^n + a_{n-1}x^{n-1} + \ldots + a_2 x^2 + a_1 x + a_0$$

where n is a nonnegative integer and $a_n \neq 0$. Unless specified otherwise, the coefficients $a_n, a_{n-1}, \ldots, a_1, a_0$ are real numbers. Furthermore, the coefficient a_n is called the **leading coefficient**, while a_0 is called the **constant term**.

In sketching the graph of polynomial functions we will restrict the domain to the set of real numbers. That is, if $y = f(x)$ is a polynomial function, then $x \in R$.

As we see in the definition, linear functions are polynomial functions of degree 1, while quadratic functions are polynomial functions of degree 2. Table 3.1 reviews the terminology of polynomial functions, which we discussed in detail in Section 1.3. Review Section 1.3 for the detailed description of these terms if you are not sure what each means.

TABLE 3.1

Polynomial Functions	Degree	Leading Coefficient	Constant Term	Other Names
$f(x) = 2x + 5$	1	2	5	Linear function
$y = \frac{1}{2}x^2 - x - 6$	2	$\frac{1}{2}$	-6	Quadratic function
$g(x) = x^3 - 3x^2 + 3x - 1$	3	1	-1	Cubic function
$P(x) = -0.1x^5 + 40x - 200$	5	-0.1	-200	Fifth-degree polynomial function
$f(x) = 7$	0	7	7	Constant function

The polynomial function $f(x) = 7$ has degree 0 because $f(x) = 7 = 7x^0$. Likewise, all constant polynomial functions $f(x) = k$, where k is a nonzero real number, have degree 0. The function $f(x) = 0$ is called the **zero function**. We do not assign a degree to the polynomial function $f(x) = 0$.

Zeros of a Polynomial Function

It is important to be able to solve for x when $f(x) = 0$ and $g(x) = 0$. To do this, we must find domain values whose corresponding range values are 0. These domain values are called **zeros** of the function f and **solutions** or **roots** of the equation $f(x) = 0$. Here is the formal definition.

Definition | The number d is called a **zero of the function** f if and only if $f(d) = 0$. If d is a real number and $f(d) = 0$, then d is called a **real zero** of f.

If d is a real zero of f, then $(d, 0) \in f$. This means that the graph of f will cross or touch the x-axis at d. Thus each zero d is an x-intercept of f. There are also other applications of zeros.

EXAMPLE 1   Kennedy's Computer Products can produce up to 4000 computer disks a month. From past performance and the present economic conditions the management knows that the monthly profit function is $P(x) = -0.01x^2 + 40.9x - 100$, where x is the number of disks produced. Find the zeros of $P(x)$ and then show that these zeros are the domain values of the break-even points, the points where revenue equals cost, $R(x) = C(x)$.

Solution   To find the zeros of $P(x)$, we solve the equation

$$P(x) = 0 \qquad \text{or}$$
$$-0.01x^2 + 40.9x - 100 = 0$$

One way to solve this equation is to multiply both sides by -1 and use the quadratic formula. Multiplying both sides by -1 gives us

$$0.01x^2 - 40.9x + 100 = 0 \qquad \text{and}$$
$$x = \frac{40.9 \pm \sqrt{(40.9)^2 - 4(0.01)(100)}}{2(0.01)}$$

Simplifying and rounding off, we have $x \approx 2.446$ and $x \approx 4087.55$. Thus the zeros of the profit function are 2.45 and 4087.55 to the nearest tenth; that is, both $P(2.45)$ and $P(4087.55)$ are 0.

The break-even points occur at domain values where $R(x) = C(x)$. Recall from Chapter 2 that $P(x) = R(x) - C(x)$. Now, if d is a zero of P then $P(d) = 0$ and $P(d) = R(d) - C(d) = 0$. Thus $R(d) = C(d)$ means that the zero d is the domain of a break-even point of $P(x)$. Therefore the break-even point occurs when $x = d = 2.45$ and $x = d = 4087.55$.   ∎

Graphing Polynomial Functions

Like most functions, polynomial functions can be graphed quite accurately with the aid of a computer or calculator. However, graphing in this manner can be very time-consuming. In fact, an exact graph of a polynomial function is not always necessary. Many times a sketch showing the general tendencies and characteristics of a graph will do. To graph polynomial functions without extensive use of a computer or calculator, we will need to make some assumptions about the shapes of the graphs. To introduce these assumptions, we will first consider the graph of a fairly simple polynomial function.

EXAMPLE 2

Sketch the graph of $f(x) = x^3 - x^2 - 4x + 4$.

Solution To begin, we find the real zeros of f by solving the equation $f(x) = 0$. Doing so will give us the x-intercepts of the graph of f.

$$x^3 - x^2 - 4x + 4 = 0 \qquad \textit{(Factor by grouping)}$$
$$x^2(x - 1) - 4(x - 1) = 0$$
$$(x^2 - 4)(x - 1) = 0 \qquad \textit{(Factor } x^2 - 4)$$
$$(x + 2)(x - 2)(x - 1) = 0$$

The graph of f will intersect the x-axis at -2, 2, and 1. But how does the graph behave for values of x between the x-intercepts? And what does the graph look like for values of x that are less than -2? Is $f(x)$ large or small when x is greater than 1? To help answer some of these questions, we can construct a sign chart for f as we did in Chapter 2:

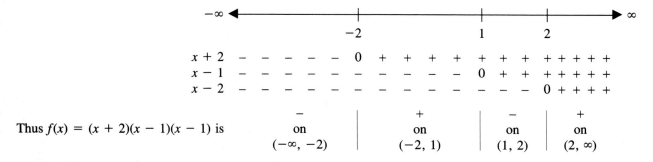

Thus $f(x) = (x + 2)(x - 1)(x - 1)$ is

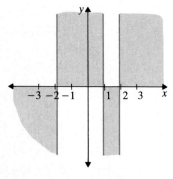

Figure 3.1

As the sign chart indicates, $f(x)$ is positive for all x in the interval $(-2, 1)$ and in the interval $(2, \infty)$. For x in these intervals the graph of f is above the x-axis. Likewise, the sign chart shows that $f(x) < 0$ for x in the interval $(-\infty, -2)$ and in the interval $(1, 2)$. Therefore the graph of f must lie below the x-axis, for values of x from these intervals. Figure 3.1 summarizes these ideas by showing the regions in which we can expect to find the graph of f. Next, we make a table to find a few ordered pairs that are solutions to $y = f(x) = (x + 2)(x - 2)(x - 1)$. In the table we use only convenient values of x, that is, values of x for which $f(x)$ is easy to calculate.

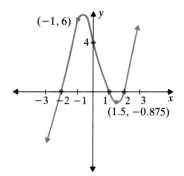

x	$f(x)$
-3	-20
-2	0
-1	6
0	4
1	0
1.5	-0.875
2	0
3	10

Figure 3.2

To make a reasonable sketch of the graph based on the information we have to this point, we must assume that the graph is a smooth, unbroken curve with no sharp turning points. Figure 3.2 gives the graph of f. The graph shown in Fig. 3.2 is the simplest graph we can draw that takes into account the information we obtained from finding the zeros of f, the sign chart for f, and the table of ordered pairs that are solutions to $y = f(x)$. ∎

Before we do any more graphing we need to add more detail to the assumptions we made about the graph in Example 2. We will generalize these assumptions so that they apply to any polynomial function.

ASSUMPTIONS | The graph of a polynomial function is a smooth, continuous curve. It contains smooth turning points with no breaks or sharp points as shown in Fig. 3.3.

To illustrate, a polynomial function can have a graph like the graph shown in Fig. 3.3(a) but can never look like the graphs shown in Figs. 3.3(b) and 3.3(c).

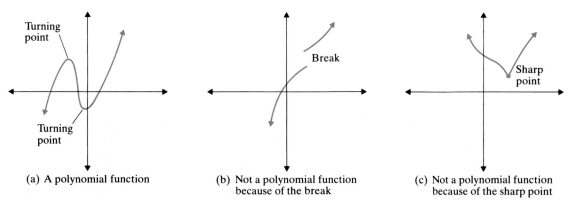

(a) A polynomial function

(b) Not a polynomial function because of the break

(c) Not a polynomial function because of the sharp point

Figure 3.3

Although we will not prove these assumptions, they can be proven with the aid of calculus.

EXAMPLE 3   Sketch the graph of $g(x) = (x^2 - 3x - 4)(x + 1)$.

Solution   We begin by factoring:

$$g(x) = (x^2 - 3x - 4)(x + 1)$$
$$= (x - 4)(x + 1)(x + 1)$$

The zeros are 4 and -1. We now sketch a sign chart for g.

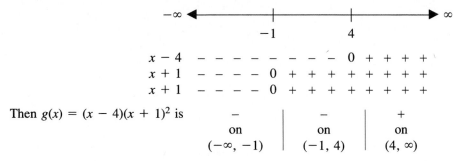

Since $g(x)$ is positive for x in the interval $(4, \infty)$, the graph of g will lie above the x-axis for all x in that interval. When x is in either $(-\infty, -1)$ or $(-1, 4)$, the graph of g will appear below the x-axis.

We obtain a table by substituting some convenient values of x into

$$g(x) = (x - 4)(x + 1)^2.$$

x	$g(x)$
-4	-72
-3	-28
-2	-6
-1	0
0	-4
1	-12
2	-18
3	-16
4	0
5	36

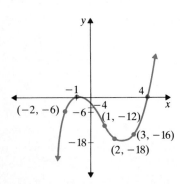

Figure 3.4

Assuming that the graph is a smooth, continuous curve with no breaks or sharp turning points, we have the graph shown in Fig. 3.4. Although we used ten points to sketch the graph of g in Fig. 3.4, we can sketch the same general tendencies of the graph of g using a smaller number of points, the zeros of g, and the sign chart. ∎

We summarize what we have done so far by listing some guidelines that can be used for graphing polynomial functions.

Guidelines for Graphing a Polynomial Function f of Degree n

STEP 1: Find the real zeros of f to obtain the x-intercepts, if there are any.

STEP 2: Use a sign chart to find the domain intervals on which $f(x)$ is positive and the domain intervals on which $f(x)$ is negative.

STEP 3: Make a table of a few ordered pairs that are solutions to f. Include x-values to the left, to the right, and between the zeros of f.

STEP 4: Assuming that the graph is a smooth, unbroken curve, make a reasonable sketch of f from the results of Steps 1 through 3.

These steps might not give the *exact* graph of a polynomial function, but they will give us the *general* shape of the graph. The study of calculus will provide the exact turning points as well as the curvature tendencies for each polynomial function. *In this course we are concerned mainly with the general shape of the graph of a polynomial function (zeros, positive values, and negative values).*

EXAMPLE 4   Sketch the graph of $h(x) = x^3 - 4x^2 + x - 4$.

Solution   We follow the steps for graphing a polynomial listed above.

STEP 1: Find the zeros, in this case by factoring:

$$
\begin{aligned}
h(x) &= x^3 - 4x^2 + x - 4 \\
&= x^2(x - 4) + 1(x - 4) \\
&= (x^2 + 1)(x - 4)
\end{aligned}
$$

The factor $x^2 + 1$ gives the complex zeros $x = i$ and $x = -i$. However, these complex zeros do not appear on the graph of h. The only real zero is 4. It comes from the factor $x - 4$.

STEP 2: Construct a sign chart for h:

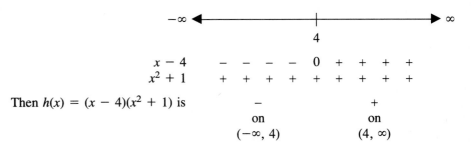

Then $h(x) = (x - 4)(x^2 + 1)$ is

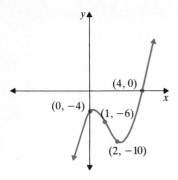

Figure 3.5

STEP 3: Make a table using convenient values of x:

x	$h(x)$
0	-4
1	-6
2	-10
4	0
5	26

STEP 4: We assume that the graph is a smooth, continuous curve and sketch the graph of h as in Fig. 3.5. ■

The graph of $h(x) = x^3 - 4x^2 + x - 4$ might have looked a little different if we had chosen different values for x in our table or connected the dots differently, but in general the graph will still be similar to the graph in Fig. 3.5. As a general rule, we choose values of x for our table that are to the left, to the right, and between the zeros of the function.

EXAMPLE 5   Sketch the graph of $f(x) = x^4 - 7x^2$.

Solution   Factoring first, we have

$$f(x) = x^4 - 7x^2 = x^2(x^2 - 7)$$
$$= x \cdot x(x - \sqrt{7})(x + \sqrt{7})$$

Thus f has zeros of 0, $\sqrt{7}$, and $-\sqrt{7}$, which provide the points $(0, 0)$, $(\sqrt{7}, 0)$ and $(-\sqrt{7}, 0)$ on the graph of f. The sign chart is

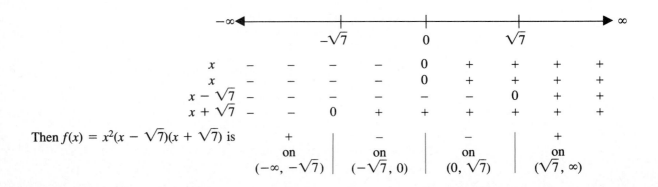

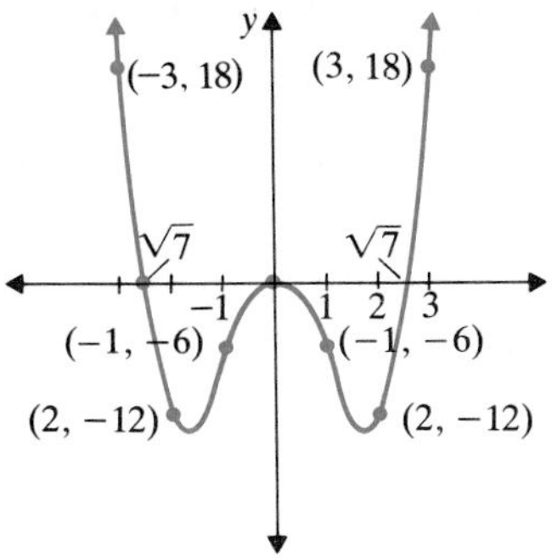

Figure 3.6

A table showing some points on the graph of $f(x) = x^2(x^2 - 7)$ follows.

x	$f(x)$
0	0
± 1	-6
± 2	-12
± 3	18

Note also that since $f(-x) = f(x)$, we can expect the graph of f to have symmetry about the y-axis, as in Fig. 3.6. ■

There are two interesting shortcuts for graphing polynomial functions. If the polynomial has a zero that comes from a single factor, then the graph of f crosses the x-axis. On the other hand, when a zero results from a double factor (the same factor twice), then the graph of f only meets the x-axis and does not cross it. We see both of these tendencies in Fig. 3.6. In Fig. 3.6, 0 is a zero twice, and thus the graph meets but does not cross the line x-axis at $x = 0$. On the other hand, the zeros $\sqrt{7}$ and $-\sqrt{7}$ are each a zero once, and thus the graph crosses the x-axis at $x = \sqrt{7}$ and $x = -\sqrt{7}$. In general, we have the following guidelines for graphing polynomial zeros.

GRAPHING
POLYNOMIAL ZEROS

For the polynomial $f(x)$ with a factor of $(x - d)^n$,

(a) if n is a positive odd integer, then the graph of f crosses the x-axis at $x = d$, and

(b) if n is a positive even integer, then the graph of f intersects but does not cross the x-axis at $x = d$.

Recognizing the number of times a number is a zero will make graphing polynomial functions easier.

Here are some other problems with polynomial zeros.

EXAMPLE 6   For what values of k is 2 a zero of $f(x) = x^2 + kx - (k^2 + 1)$? Find the resulting polynomial(s) for $f(x)$.

Solution   If $f(x) = x^2 + kx - (k^2 + 1)$ has a zero of 2, then $f(2) = 0$. Substitute $x = 2$ for x and 0 for $f(2)$ to find k:

$$f(x) = x^2 + kx - (k^2 + 1)$$
$$f(2) = 2^2 + k \cdot 2 - (k^2 + 1)$$
$$0 = 4 + 2k - k^2 - 1$$
$$0 = k^2 - 2k - 3$$
$$0 = (k - 3)(k + 1)$$
$$k \in \{-1, 3\}$$

Therefore when $k = -1$,

$$f(x) = x^2 + (-1)x - [(-1)^2 + 1] = x^2 - x - 2$$

and when $k = 3$,

$$f(x) = x^2 + (3)x - (3^2 + 1) = x^2 + 3x - 10$$

Each of the polynomials $f(x) = x^2 - x - 2$ and $f(x) = x^2 + 3x - 10$ has a zero of 2.   ■

EXAMPLE 7   A rectangular package is to be constructed with a width of x inches, a length of $2x - 2$ inches, and a height of $3x - 12$ inches (Fig. 3.7). Find x such that the volume of the box is not greater than 120 cubic inches.

Solution   We have the width $= x$ inches,

the length $= 2x - 2$ inches,

the height $= 3x - 12$ inches, and therefore

the volume is

$$V(x) = x(2x - 2)(3x - 12)$$

Thus for a volume not greater than 120 cubic inches we have

$$V(x) \le 120$$
$$x(2x - 2)(3x - 12) \le 120$$
$$6x^3 - 30x^2 + 24x \le 120$$
$$6x^3 - 30x^2 + 24x - 120 \le 0$$
$$6x^2(x - 5) + 24(x - 5) \le 0$$
$$(6x^2 + 24)(x - 5) \le 0$$
$$6(x^2 + 4)(x - 5) \le 0$$

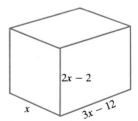

Figure 3.7

Since $6 > 0$ and $x^2 + 4 \geq 0$ we can conclude that $V(x) \leq 120$ when $(x - 5) \leq 0$. Thus $x \leq 5$. We cannot forget that the height $3x - 12$ must be positive. Thus we have $x > 4$ and $x \leq 5$. Therefore the width in inches is $\{x : 4 < x \leq 5\}$. ∎

Exercises 3.1

Find the zeros and sketch the graphs of the following polynomial functions.

1. $f(x) = (x - 2)(x + 3)$ **2.** $f(x) = (x - 5)(x + 1)$

3. $h(x) = (x - 2)(x + 3)(x + 4)$

4. $h(x) = (x + 2)(x + 3)(x - 4)$

5. $f(x) = x^3 - 4x^2$ **6.** $f(x) = x^3 - 9x$

7. $g(x) = x^3 - x^2 - 4x + 4$

8. $g(x) = x^3 - 2x^2 - 9x + 18$

9. $h(x) = x^4 - 9x^2$ **10.** $h(x) = x^4 - 4x^2$

11. $f(x) = x^3 - 8$ **12.** $f(x) = x^3 + 8$

13. $g(x) = (x^2 - x - 6)(x^2 - 1)$

14. $g(x) = (x^2 - 5x + 6)(x^2 - 1)$

15. $h(x) = -(x - 5)(x + 1)$ **16.** $h(x) = -(x - 2)(x + 3)$

17. $f(x) = (x - 5)^3(x + 1)^3$ **18.** $f(x) = (x - 2)^2(x + 1)^2$

19. $g(x) = (x - 5)^2(x + 1)^2$ **20.** $g(x) = (x - 2)^3(x + 1)^3$

Find the zeros of the following functions.

21. $f(x) = 3x^2 + 49x - 34$ **22.** $f(x) = 2x^2 - 9x - 26$

23. $g(x) = x^2 + x - 3$ **24.** $g(x) = x^2 + 3x - 2$

25. $h(x) = x^2 + 2x - 3$ **26.** $h(x) = x^2 + 3x + 2$

Find k such that the following functions have only one zero.

27. $f(x) = 3x^2 + kx + 12$ **28.** $f(x) = 3x^2 + kx + 3$

29. $g(x) = 2x^2 + kx + 3$ **30.** $g(x) = 3x^2 + kx + 5$

31. $h(x) = 2x^2 + 4x + k$ **32.** $h(x) = 3x^2 + 6x + k$

33. $f(x) = 2x^2 + 5x + k$ **34.** $f(x) = 2x^2 + 3x + k$

35. For which value(s) of k is -2 a zero of the polynomial $g(x) = x^2 + kx + (k^2 - 4)$? Find the resulting polynomial(s) for $g(x)$.

36. For which value(s) of k is 3 a zero of the polynomial $g(x) = x^2 + kx + (1 - k^2)$? Find the resulting polynomial(s) for $g(x)$.

Greathouse Construction plans to manufacture prefabricated log cabin homes. Under current economic conditions they can produce a maximum of 365 kits a year with cost and revenue functions of $C(x) = -290x^2 + 106{,}000x + 400{,}000$ and $R(x) = x^3 - 700x^2 - 150{,}000x$, respectively, where x is the number of kits produced.

37. Sketch the graph of the cost function $C(x)$ for Greathouse Construction.

38. For Greathouse Construction find the interval when the revenue function $R(x)$ is greater than zero and find the equation of the profit function $P(x)$.

Bowen's Boats produces a small recreational sailboat. From past experience and current economic conditions they know that they can produce up to 150 boats a month with cost and profit functions of $C(x) = 400 - 0.001x^2$ and $P(x) = 0.1x^3 - 21x^2 + 1100x$, respectively, where x is the number of boats produced.

39. Find the zeros and sketch the graph of the profit function for Bowen's Boats.

40. Sketch the graph of the cost function for Bowen's Boats.

From past experience and current economic conditions, Kennedy's Computer Products knows that they can produce up to 4000 computer disks a month. Under current conditions the monthly revenue function is $R(x) = -0.01x^2 + 40.1x$.

41. Find the zeros and sketch the graph of the revenue function for Kennedy's Computer Products. What value of x provides a maximum monthly revenue?

42. A rectangular package is to be constructed with a width of x inches, a length of $2x - 6$ inches, and a height of $x - 4$ inches. Find x such that the volume of the box is not greater than 168 cubic inches.

43. A rectangular package is to be constructed with a width of x inches, a length of $2x - 6$ inches, and a height of $x - 4$ inches. Find x such that the volume of the box is not less than 168 cubic inches.

3.2 The Division Algorithm, the Remainder Theorem, and the Factor Theorem

In Section 3.1 we used factoring to find the zeros of polynomial functions. In this section we will continue earlier work by using division of polynomials to find polynomial zeros.

Division with polynomials is similar to long division with whole numbers. To review, dividing 5355 by 25 using long division looks like this:

$$
\begin{array}{r}
214 \leftarrow quotient \\
divisor \to 25\overline{)5355} \leftarrow dividend \\
\underline{50} \\
35 \\
\underline{25} \\
105 \\
\underline{100} \\
5 \leftarrow remainder
\end{array}
$$

The result of this division process can be summarized as follows:

$$
\frac{5355}{25} = 214 + \frac{5}{25} \qquad \frac{\text{dividend}}{\text{divisor}} = \text{quotient} + \frac{\text{remainder}}{\text{divisor}}
$$

If we multiply each side of each expression by the divisor, we have

$$
5355 = 25(214) + 5 \qquad \text{dividend} = \text{divisor(quotient)} + \text{remainder}
$$

To divide a polynomial by a polynomial, we use a process similar to the long division process shown above. The steps are as follows.

STEP 1: Estimate.
STEP 2: Multiply.
STEP 3: Subtract.
STEP 4: Bring down the next term.

The process is repeated until the degree of the remainder is less than the degree of the divisor or the remainder is zero. Here is an example that illustrates these four steps.

EXAMPLE 1   Divide $x^3 + 5x^2 + 3x - 10$ by $x + 2$.

Solution   The dividend is $x^3 + 5x^2 + 3x - 10$, and the divisor is $x + 2$. The problem can be rewritten as

$$
x + 2\overline{)x^3 + 5x^2 + 3x - 10}
$$

STEP 1: Our first estimate must be the expression by which we multiply x to obtain the leading term x^3. That expression is x^2:

$$
\begin{array}{r}
x^2 \\
x + 2\overline{)x^3 + 5x^2 + 3x - 10}
\end{array}
$$

STEP 2: Using the distributive property, multiply x^2 and $x + 2$:

$$\begin{array}{r} x^2 \hphantom{{}+5x^2+3x-10} \\ x + 2 \overline{)x^3 + 5x^2 + 3x - 10} \\ \underline{x^3 + 2x^2 \hphantom{{}+3x-10}} \end{array}$$

STEP 3: Subtract $x^3 + 2x^2$ from $x^3 + 5x^2$ by changing the signs of $x^3 + 2x^2$ and adding:

$$\begin{array}{r} x^2 \hphantom{{}+5x^2+3x-10} \\ x + 2 \overline{)x^3 + 5x^2 + 3x - 10} \\ \underline{-x^3 - 2x^2 \hphantom{{}+3x-10}} \\ 3x^2 \hphantom{{}+3x-10} \end{array}$$ *(Since $-(x^3 + 2x^2) = -x^3 - 2x^2$)*

STEP 4: Bring down the next term, which is $+ 3x$:

$$\begin{array}{r} x^2 \hphantom{{}+5x^2+3x-10} \\ x + 2 \overline{)x^3 + 5x^2 + 3x - 10} \\ \underline{x^3 + 2x^2 \hphantom{{}+3x-10}} \\ 3x^2 + 3x \hphantom{{}-10} \end{array}$$

Since the degree of the remainder, $3x^2 + 3x$, is higher than the degree of the divisor, we repeat the process by dividing $3x^2 + 3x$ by $x + 2$.
STEP 1: Our estimate is $3x$, since $3x \cdot x = 3x^2$.
STEP 2: Multiply $3x$ and $x + 2$.
STEP 3: Subtract $3x^2 + 6x$ from $3x^2 + 3x$.
STEP 4: Bring down the $- 10$.

$$\begin{array}{r} x^2 + 3x \hphantom{{}-10} \\ x + 2 \overline{)x^3 + 5x^2 + 3x - 10} \\ \underline{x^3 + 2x^2 \hphantom{{}+3x-10}} \\ 3x^2 + 3x \hphantom{{}-10} \\ \underline{3x^2 + 6x \hphantom{{}-10}} \\ - 3x - 10 \end{array}$$

Again, the degree of our remainder is equal to or higher than the degree of the divisor, so we repeat the process a third time. Without listing each step, this is how the complete problem looks:

$$\begin{array}{r} x^2 + 3x - 3 \\ x + 2 \overline{)x^3 + 5x^2 + 3x - 10} \\ \underline{x^3 + 2x^2 \hphantom{{}+3x-10}} \\ 3x^2 + 3x \hphantom{{}-10} \\ \underline{3x^2 + 6x \hphantom{{}-10}} \\ -3x - 10 \\ \underline{-3x - 6} \\ - 4 \end{array}$$

The quotient is $x^2 + 3x - 3$, and the remainder is -4. We summarize our results as follows:

$$\frac{x^3 + 5x^2 + 3x - 10}{x + 2} = x^2 + 3x - 3 + \frac{-4}{x + 2}$$

Multiplying both sides of this expression by the divisor $x + 2$, we have

$$x^3 + 5x^2 + 3x - 10 = (x^2 + 3x - 3)(x + 2) + (-4) \qquad ■$$

Note that this last line indicates how to check the result of the long division. To check, we add the remainder to the product of the quotient and the divisor. This result should be the original dividend. If not, we have made a mistake.

The following theorem describes the type of division shown in Example 1. We state it here without proof.

THE DIVISION ALGORITHM FOR POLYNOMIALS	For polynomials $f(x)$ and $g(x)$, where $g(x) \neq 0$, there exist polynomials $q(x)$ and $r(x)$ such that

(i) $\dfrac{f(x)}{g(x)} = q(x) + \dfrac{r(x)}{g(x)}$

where the degree of $r(x)$ is less than the degree of $g(x)$ or $r(x)$ is the zero polynomial,

or, equivalently,

(ii) $f(x) = g(x)q(x) + r(x)$

where the degree of $r(x)$ is less than the degree of $g(x)$ or $r(x)$ is the zero polynomial.

As we will see, the equation $f(x) = g(x)q(x) + r(x)$ will be the most useful form of the division algorithm.

EXAMPLE 2   For $f(x) = x^4 + x^3 + 4x^2 + 5x + 4$ and $g(x) = x^2 + 3$, find the quotient $q(x)$ and the remainder $r(x)$ when $f(x)$ is divided by $g(x)$.

Solution   Using long division, we have

$$
\begin{array}{r}
x^2 + x + 1 \\
x^2 + 3 \overline{\smash{\big)}\ x^4 + x^3 + 4x^2 + 5x + 4} \\
\underline{x^4 \qquad\quad + 3x^2} \\
x^3 + x^2 + 5x \\
\underline{x^3 \qquad\quad + 3x} \\
x^2 + 2x + 4 \\
\underline{x^2 \qquad + 3} \\
2x + 1
\end{array}
$$

Using the division algorithm, we write our results as

$$x^4 + x^3 + 4x^2 + 5x + 4 = (x^2 + 3)(x^2 + x + 1) + (2x + 1)$$

The quotient is $q(x) = x^2 + x + 1$, and the remainder is $r(x) = 2x + 1$. Note that the degree of the remainder $2x + 1$ is less than the degree of the divisor $x^2 + 3$.

∎

EXAMPLE 3   Find the quotient and remainder when $x^5 + x^3 + x - 6$ is divided by $x - 2$.

Solution

$$
\begin{array}{r}
x^4 + 2x^3 + 5x^2 + 10x + 21 \\
x - 2 \overline{\smash{\big)}\ x^5 + 0x^4 + x^3 + 0x^2 + x - 6} \\
\underline{x^5 - 2x^4} \\
2x^4 + x^3 \\
\underline{2x^4 - 4x^3} \\
5x^3 + 0x^2 \\
\underline{5x^3 - 10x^2} \\
10x^2 + x \\
\underline{10x^2 - 20x} \\
21x - 6 \\
\underline{21x - 42} \\
36
\end{array}
$$

The result can be written as

$$\frac{x^5 + x^3 + x - 6}{x - 2} = (x^4 + 2x^3 + 5x^2 + 10x + 21) + \frac{36}{x - 2}$$

or as

$$x^5 + x^3 + x - 6 = (x - 2)(x^4 + 2x^3 + 5x^2 + 10x + 21) + 36$$

In either case the quotient is $q(x) = x^4 + 2x^3 + 5x^2 + 10x + 21$, and the remainder is $r(x) = 36$, a constant.

∎

Note that in Example 3 the degree of the remainder is less than the degree of the divisor, which is consistent with the division algorithm. The degree of the remainder is always smaller than the degree of the divisor, or the remainder is the zero polynomial.

We see an interesting and useful result of the division algorithm in the following theorem.

Theorem

The Remainder Theorem. If a polynomial $f(x)$ is divided by $x - d$, then the remainder is $f(d)$.

PROOF: By the division algorithm, dividing $f(x)$ by $x - d$ yields

$$f(x) = (x - d)q(x) + r(x)$$

where $q(x)$ is the quotient and $r(x)$ is the remainder. Since the degree of $r(x)$ is less than the degree of $x - d$ or $r(x)$ is the zero polynomial, $r(x)$ is a constant; call it r. Thus

$$f(x) = (x - d)q(x) + r$$

But if we substitute d for x in this expression, we have

$$f(d) = (d - d)q(d) + r$$
$$f(d) = 0 + r = r$$

Thus $f(d)$ is the remainder when $f(x)$ is divided by $x - d$.   □

One useful result of the remainder theorem is the fact that the point (d, r), where r is the remainder when $f(x)$ is divided by $x - d$, is always a point on the graph of $y = f(x)$. We will use this fact in the next section.

EXAMPLE 4

If $f(x) = x^5 + x^4 + x - 6$, find $f(2)$.

Solution   In Example 3 we divided $f(x) = x^5 + x^3 + x - 6$ by $x - 2$ and obtained a remainder of 36. Therefore by the remainder theorem it must be true that $f(2) = 36$. Furthermore, if we were to graph $y = f(x)$, the point $(2, 36)$ would be on the graph.   ■

Another useful theorem is the factor theorem. It establishes the relationship between the linear factors of a polynomial and the zeros of the associated polynomial function.

Theorem ▮ *The Factor Theorem.* A polynomial $f(x)$ has a factor $x - d$ if and only if $f(d) = 0$.

PROOF: Since the phrase "if and only if" appears, there are two parts to this proof. We must prove that

(a) if $x - d$ is a factor of $f(x)$, then $f(d) = 0$, and
(b) if $f(d) = 0$, then $x - d$ is a factor of $f(x)$.

To prove part (a), if $x - d$ is a factor of $f(x)$, then $r(x) = 0$. Since $x - d$ is a factor of $f(x)$, the remainder is 0, and we write

$$f(x) = (x - d)q(x)$$

By calculating $f(d)$, we have

$$f(d) = (d - d)q(x) = 0$$

finishing this part of the proof.

To prove part (b), if $f(x)$ is divided by $x - d$, then by the division algorithm,

$$f(x) = (x - d)q(x) + r(x)$$

By the remainder theorem, $f(d) = r(x)$, so

$$f(x) = (x - d)q(x) + f(d)$$

But with $f(d) = 0$ we have

$$f(x) = (x - d)q(x) + 0 = (x - d)q(x)$$

Thus $x - d$ is a factor of $f(x)$.   ☐

We can use the factor theorem in a number of ways, as we will see in the following examples.

EXAMPLE 5   For a positive integer n, show that $f(x) = x^n - 7^n$ has a factor of $x - 7$.

Solution   By the factor theorem, if $f(7) = 0$, then $x - 7$ is a factor of $f(x)$. Since $f(7) = 7^n - 7^n = 0$, then $x - 7$ is a factor of $f(x) = x^n - 7^n$.   ■

EXAMPLE 6   Write an equation for a third-degree polynomial, $f(x)$, with zeros of $1 + i$, $1 - i$, and -2.

Solution   By the factor theorem, if d is a zero of the polynomial f, then $x - d$ is a factor of $f(x)$. Thus if the polynomial f has zeros of $1 + i$, $1 - i$, and -2, then it also has the three factors $x - (1 + i)$, $x - (1 - i)$, and $x - (-2)$. We can now find $f(x)$:

$$f(x) = [x - (1 + i)][x - (1 - i)][x - (-2)]$$
$$= (x - 1 - i)(x - 1 + i)(x + 2)$$
$$= [(x - 1) - i][(x - 1) + i](x + 2)$$
$$= [(x - 1)^2 - i^2](x + 2)$$
$$= [x^2 - 2x + 1 - (-1)](x + 2)$$
$$= (x^2 - 2x + 2)(x + 2)$$
$$= x^3 - 2x + 4$$

Therefore $f(x) = x^3 - 2x + 4$ is a third-degree polynomial function with the given roots. Note that any constant multiple of f is also a third-degree polynomial function with the given roots. For example, $g(x) = 2f(x) = 2x^3 - 4x + 8$, or in general,

$$g(x) = af(x) = a(x^3 - 2x + 4),$$

where a is a nonzero real number. ∎

EXAMPLE 7   Find all zeros of $f(x) = x^3 + 3x^2 - 10x - 24$ if $x = -2$ is one of the zeros.

Solution   By the factor theorem, if $x = -2$ is a zero of f, then $x + 2$ is a factor, meaning that $x + 2$ divides f evenly. To factor $f(x)$ completely, we simply divide $f(x)$ by $x + 2$ and then factor the resulting quotient:

$$
\begin{array}{r}
x^2 + x - 12 \\
\hline
x + 2 \overline{)\,x^3 + 3x^2 - 10x - 24} \\
\underline{x^3 + 2x^2} \\
x^2 - 10x \\
\underline{x^2 + 2x} \\
-12x - 24 \\
\underline{-12x - 24} \\
0
\end{array}
$$

We have $f(x) = (x + 2)(x^2 + x - 12) = (x + 2)(x + 4)(x - 3)$. The zeros of f are -2, -4, and 3. ∎

EXAMPLE 8   Greathouse Construction plans to manufacture prefabricated log cabin homes. They can produce a maximum of 365 kits a year. From the graph of the profit function $P(x)$ (Fig. 3.8), determine the start-up costs, the domain value for the break-even points, and the equation of $P(x)$.

Solution   The start-up costs are the initial loss in profit at the beginning of construction when $x = 0$. From the graph, $P(0) = -400,000$, which is a start-up cost of $400,000. The break-even points occur when $P(x) = R(x) - C(x) = 0$, which are the zeros of $P(x)$. From the graph the break-even points are at $x = 10$, 200, and 200 (200 is a zero twice, since the graph meets but does not cross the

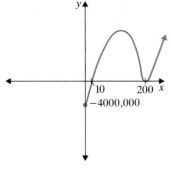

Figure 3.8

x-axis). We use the factor theorem with these zeros and some real number *a* to write the profit function:

$$P(x) = a(x - 10)(x - 200)(x - 200)$$
$$= a(x - 10)(x^2 - 400x + 400,000)$$
$$= a(x^3 - 410x^2 + 44,000x - 400,000)$$

Using the fact that $P(0) = -400,000$ to find the value of *a*, we have

$$P(0) = a(0 - 10)(0 - 200)(0 - 200)$$
$$-400,000 = a(-400,000)$$
$$1 = a$$

Therefore the profit function is

$$P(x) = x^3 - 410x^2 + 44,000x - 400,000 \qquad \blacksquare$$

Now use these ideas and techniques in the following exercises.

Exercises 3.2

For Exercises 1–16, find the quotient $q(x)$ and the remainder $r(x)$ by dividing $f(x)$ by $g(x)$. Write the results in both forms shown in the division algorithm. Use these results to find the coordinates of a point on the graph of $y = f(x)$.

1. $f(x) = x^2 + 3x + 5$
$g(x) = x + 2$

2. $f(x) = x^2 + 5x + 7$
$g(x) = x - 3$

3. $f(x) = 2x^3 + 5x^2 - 4x - 20$
$g(x) = x^2 - 4$

4. $f(x) = 2x^3 + 5x^2 - 4x - 20$
$g(x) = x^2 + 3x - 10$

5. $f(x) = 2x^3 + 5x^2 - 4x - 20$
$g(x) = x - 2$

6. $f(x) = 3x^3 + 5x^2 - 4x - 20$
$g(x) = x + 5$

7. $f(x) = x^3 + 5x^2 - 4x - 20$
$g(x) = 3x - 3$

8. $f(x) = x^3 + 5x^2 - 4x - 20$
$g(x) = 2x + 3$

9. $f(x) = x^3 + 4x^2 + x - 6$
$g(x) = x - \frac{1}{2}$

10. $f(x) = x^3 + 4x^2 + x - 6$
$g(x) = x + \frac{1}{2}$

11. $f(x) = x^5 + x - 1$
$g(x) = x^2 - 1$

12. $f(x) = x^5 + x + 1$
$g(x) = x^2 - 1$

13. $f(x) = x^7 - 1$
$g(x) = x + 1$

14. $f(x) = x^9 + 1$
$g(x) = x - 1$

15. $f(x) = x^8 - 1$
$g(x) = x - 1$

16. $f(x) = x^8 - 1$
$g(x) = x + 1$

Use the factor theorem in Exercises 17–24 to find a third-degree polynomial function with the indicated zeros.

17. 1, −1, and 3

18. 2, −2, and 3

19. $\sqrt{7}$, $-\sqrt{7}$, and 3

20. $\sqrt{2}$, $-\sqrt{2}$, and 1

21. $1 + 2i$, $1 - 2i$, and −1

22. $2 + i$, $2 - i$, and 2

23. $1 + i$, 1, and −1

24. $1 - i$, 2, and −2

25. Find *k* such that $x - 2$ is a factor of $f(x) = x^2 - kx - (k^2 + 1)$. Write the resulting polynomial(s) for $f(x)$.

26. Find *k* such that $x + 3$ is a factor of $f(x) = x^2 + kx - 7 + k^2$. Write the resulting polynomial(s) for $f(x)$.

27. Show that $x - 1$ is a factor of $g(x) = x^{71} - 1$.

28. Show that $x - 2$ is a factor of $g(x) = x^8 - 256$.

29. Show that $x + 5$ is a factor of $h(x) = x^7 + 78,125$.

30. Show that $x + 3$ is a factor of $h(x) = x^{11} + 177,147$.

31. Find the remainder when $f(x) = x^6 - 5x^4 + 3x^2 + 5$ is divided by $x - 2$. Use the remainder theorem.

32. Find the remainder when $f(x) = x^6 - 5x^4 + 3x^2 + 5$ is divided by $x - 3$. Use the remainder theorem.

33. Find the remainder when $f(x) = x^7 - 3x^4 + 5x^3 - x^2 + 5$ is divided by $x + 1$. Use the remainder theorem.

34. Find the remainder when $f(x) = x^7 - 3x^4 + 5x^3 - x^2 + 5$ is divided by $x - 1$. Use the remainder theorem.

35. For any positive integer n, show that $x - a$ is a factor of $f(x) = x^n - a^n$.

36. For any positive even integer n, show that $x + a$ is a factor of $f(x) = x^n - a^n$.

37. For any positive odd integer n, show that $x + a$ is a factor of $f(x) = x^n + a^n$.

38. For any positive integer n, show that $x + a$ is a factor of $f(x) = x^{2n-1} + a^{2n-1}$.

39. For any positive integer n, show that $x + a$ is a factor of $f(x) = x^{2n} - a^{2n}$.

40. Show that $f(x) = 2x^2 + x + 5$ has no factors $x - d$ where d is a real number.

41. Show that $f(x) = 5x^2 + 3x + 2$ has no factors $x - d$ where d is a real number.

42. Find the remainder when $f(x) = x^7 + 2x^6 + 3x^5 + 4x^4 + 4x^3 + 3x^2 + 2x + 1$ is divided by (a) $x - 1$ and (b) $x + 1$.

43. Find the remainder when $f(x) = ax^5 + bx^4 + cx^3 + cx^2 + bx + a$ is divided by (a) $x - 1$ and (b) $x + 1$.

44. Find the third-degree polynomial with zeros of $2 - i$, $2 + i$, and -1, where $f(-1) = 15$.

45. Find the third-degree polynomial with zeros $-1 - i$, $-1 + i$, and -2, where $f(1) = 15$.

46. Kennedy's Computer Products can produce up to 4000 computer disks a month. Use the graph of the profit function $P(x)$ in Fig. 3.9 to find (a) the start-up costs, (b) the domain values of the break-even points, and (c) the polynomial function $P(x)$.

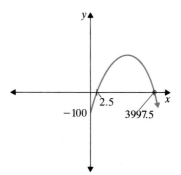

Figure 3.9

47. Bowen's Boats produces a small recreational sailboat. They can produce up to 150 boats a month with a profit of $P(x)$. Use the fact that the profit for 50 boats is $15,000 and the graph of the profit function in Fig. 3.10 to find the profit function $P(x)$.

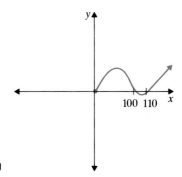

Figure 3.10

3.3 Synthetic Division and the Intermediate Value Theorem

In this section we will develop a shortcut method of doing division with polynomials. To begin, we will review long division as shown in the previous sections. This is how it looks when we divide $f(x) = 2x^4 - 3x^3 - 4x^2 - 11x - 5$ by $g(x) = x - 3$:

$$
\begin{array}{r}
2x^3 + 3x^2 + 5x\ + 4 \\
x - 3\overline{)2x^4 - 3x^3 - 4x^2 - 11x - 5} \\
\underline{2x^4 - 6x^3} \\
3x^3 - 4x^2 \\
\underline{3x^3 - 9x^2} \\
5x^2 - 11x \\
\underline{5x^2 - 15x} \\
4x - \ 5 \\
\underline{4x - 12} \\
7 \quad \textit{the remainder}
\end{array}
$$

This method involves a lot of writing. We will take out all the unnecessary elements in this division process. The result is called synthetic division.

Since the variables are always written in descending powers (**standard form**), they can be eliminated from the problem without affecting the result. Here is the same problem written without the variables:

$$
\begin{array}{r}
2 \quad 3 \quad 5 \quad 4 \quad \textit{the quotient} \\
1 - 3\overline{)2 - 3 - 4 - 11 - 5} \\
\underline{\mathbf{2} - 6} \\
3 - \mathbf{4} \\
\underline{\mathbf{3} - 9} \\
5 - \mathbf{11} \\
\underline{\mathbf{5} - 15} \\
4 - \mathbf{5} \\
\underline{\mathbf{4} - 12} \\
7 \quad \textit{the remainder}
\end{array}
$$

Notice that the bold-printed numbers in this condensed form are repetitions of the numbers above them. This is not coincidental and can be used to shorten this division process even more. Here is the problem again with the bold-printed numbers omitted:

$$
\begin{array}{r}
2 \quad 3 \quad 5 \quad 4 \quad \textit{the quotient} \\
1 - 3\overline{)2 - 3 - 4 - 11 - 5} \\
\underline{- 6} \\
3 \\
\underline{- 9} \\
5 \\
\underline{- 15} \\
4 \\
\underline{- 12} \\
7 \quad \textit{the remainder}
\end{array}
$$

To condense even further, we move each number vertically upward to eliminate the open space. We also bring down the initial 2 from the dividend to the third line. This is how it looks when we have condensed with the quotient on both the first and third line:

$$
\begin{array}{r}
\ 2\ \ \ 3\ \ \ 5\ \ \ 4 \qquad \text{quotient} \\
1-3\overline{)2\ -3\ -4\ -11\ -\ 5} \qquad \text{divisor)dividend} \\
-6\ -9\ -15\ -12 \qquad \text{multiplication} \\
\hline
2\ \ \ 3\ \ \ 5\ \ \ \ 4\ \ \big|\ 7 \qquad \text{quotient and remainder}
\end{array}
$$

Since the coefficient of the variable in the divisor $x - 3$ is 1, the 1 will also be eliminated in our condensed form. Also notice that the last row gives the coefficients of the quotient, along with the remainder. Thus the top row can be eliminated in condensed form. The condensed form can now be written as:

$$
\begin{array}{r}
-3\big|\ 2\ -3\ -4\ -11\ -\ 5 \\
-6\ -9\ -15\ -12 \\
\hline
2\ \ \ 3\ \ \ 5\ \ \ \ 4\ \ \big|\ 7
\end{array}
$$

To obtain the numbers in the bottom row, we multiplied by -3 and then subtracted that product from the number above it. The same result can be accomplished by changing the sign of -3, multiplying, and *adding instead of subtracting*. Here is how the problem looks when this final modification has been made:

$$
\begin{array}{r}
3\big|\ 2\ -3\ -4\ -11\ -\ 5 \\
6\ \ \ 9\ \ \ 15\ \ \ 12 \\
\hline
2\ \ \ 3\ \ \ 5\ \ \ \ 4\ \ \big|\ 7
\end{array}
\qquad \textit{(The remainder is 7)}
$$

Here is the step-by-step procedure that will give the results shown above. The procedure is called synthetic division:

$$
\begin{array}{r}
3\big|\ 2\ -3\ -4\ -11\ -\ 5 \\
\textit{Bring down} \qquad \downarrow \\
\hline
2
\end{array}
$$

$$
\begin{array}{r}
3\big|\ 2\ -3\ -4\ -11\ -\ 5 \\
6 \qquad\qquad \textit{Add} \\
\hline
\textit{Multiply} \qquad 2\ \ \ 3
\end{array}
$$

$$
\begin{array}{r}
3\big|\ 2\ -3\ -4\ -11\ -\ 5 \\
6\ \ \ 9 \qquad\qquad \textit{Add} \\
\hline
\textit{Multiply} \qquad 2\ \ \ 3\ \ \ 5
\end{array}
$$

$$
\begin{array}{r}
3\big|\ 2\ -3\ -4\ -11\ -\ 5 \\
6\ \ \ 9\ \ \ 15 \qquad\qquad \textit{Add} \\
\hline
\textit{Multiply} \qquad 2\ \ \ 3\ \ \ 5\ \ \ 4
\end{array}
$$

$$
\begin{array}{r}
3\big|\ 2\ -3\ -4\ -11\ -\ 5 \\
6\ \ \ 9\ \ \ 15\ \ \ 12 \qquad\qquad \textit{Add} \\
\hline
\textit{Multiply} \qquad 2\ \ \ 3\ \ \ 5\ \ \ 4\ \ \big|\ 7
\end{array}
$$

As you can see, synthetic division gives the same result as long division, but with considerably less writing.

It is important to notice that in dividing by $x - c$, c appears as the divisor in synthetic division. Likewise, in dividing by $x + c$, $-c$ appears as the divisor in the synthetic division table.

When one of the terms in the divisor or dividend is missing, the missing term must be represented by 0. The following example illustrates this.

EXAMPLE 1

Use synthetic division to find the quotient of $f(x) = x^5 - 2x^4 + x^2 - 3x + 5$ and $g(x) = x - 1$.

Solution   Both polynomials are in standard form, but the x^3 term is missing in $f(x)$. Rewrite $f(x)$ as $f(x) = x^5 - 2x^4 + 0x^3 + x^2 - 3x + 5$. Using synthetic division, divide $f(x)$ by $g(x) = x - 1$. Change the constant -1 of the divisor $x - 1$ to 1 and complete the synthetic division by multiplying and adding:

$$\begin{array}{r|rrrrrr} 1 & 1 & -2 & 0 & 1 & -3 & 5 \\ & & 1 & -1 & -1 & 0 & -3 \\ \hline & 1 & -1 & -1 & 0 & -3 & \boxed{2} \end{array}$$

$$f(x) \div g(x) = x^4 - x^3 - x^2 - 3 + \frac{2}{x - 1}$$ ■

We summarize our discussion as follows.

Guidelines for Synthetic Division

(a) Both polynomials must be in standard form.
(b) A missing term must be represented by 0.
(c) The divisor must be a first-degree polynomial with 1 as the leading coefficient.
(d) The sign of the constant in the linear divisor is changed in synthetic division.

The following examples show how the remainder theorem and synthetic division can be used with polynomials.

EXAMPLE 2

Use the remainder theorem to find $f(5)$ and $f(-2)$ when $f(x) = x^3 - 25x^2 + 105x - 10$.

Solution   By the remainder theorem, $f(5)$ is the remainder when $f(x) = x^3 - 25x^2 + 105x - 10$ is divided by $x - 5$. Using synthetic division, we have

$$\underline{5|}\ 1\ -25\quad 105\ -10$$
$$\quad\quad\quad 5\ -100\quad 25$$
$$\overline{\quad\quad 1\ -20\quad\quad 5\ \ \boxed{15}}$$

Then $f(5) = 15$ and $(5, 15) \in f$.

Similarly, $f(-2)$ is the remainder when $f(x) = x^3 - 25x^2 + 105x - 10$ is divided by $x + 2$. Use synthetic division:

$$\underline{-2|}\ 1\ -25\quad 105\quad -10$$
$$\quad\quad\quad\quad -2\ +54\ -318$$
$$\overline{\quad\quad 1\ -27\quad 159\ \boxed{-328}}$$

Then $f(-2) = -328$ and $(-2, -328) \in f$.   ∎

EXAMPLE 3   A rectangular package has a width of x inches, a length of $2x - 2$ inches, and a height of $3x - 12$ inches. Use the remainder theorem, synthetic division, and your calculator to find the volume when x equals (a) 4.5 inches and (b) 4.75 inches.

Solution   As we found earlier, the volume of this package is

$$V(x) = 6x^3 - 30x^2 + 24x + 0$$

To find $V(4.5)$ and $V(4.75)$, we use a calculator and synthetic division:

$$\underline{4.5|}\ 6\ -30\quad 24\quad\quad 0 \qquad\qquad \underline{4.75|}\ 6\ -30\quad\quad 24\quad\quad\quad 0$$
$$\quad\quad\quad 27\ -13.5\quad 47.25 \qquad\qquad\qquad 28.5\ -7.125\quad 80.15625$$
$$\overline{\quad\quad 6\ -3\quad 10.5\ \boxed{47.25}} \qquad\qquad \overline{\quad\quad 6\ -1.5\quad 16.875\ \boxed{80.15625}}$$

Therefore $V(4.5) = 47.25$ cubic inches, and $V(x) = 80.15625$ cubic inches.   ∎

EXAMPLE 4   Use synthetic division to find the quotient of $f(x) = x^3 + 5x^2 + 5x + 5$ and $g(x) = x - i$.

Solution   Divide $f(x) = x^3 + 5x^2 + 5x + 5$ by $g(x) = x - i$, using synthetic division. (Remember, $i^2 = -1$.)

$$\underline{i|}\ 1\ \ 5 \qquad\qquad\quad 5 \qquad\quad 5$$
$$\quad\quad\quad (\ + i)\ (5i + i^2)\ (5i^2 + 4i)$$
$$\overline{\quad\ 1\ (5 + i)\ (5i + 4)\ \boxed{0 + 4i}}$$

Thus

$$\frac{f(x)}{g(x)} = x^2 + (5 + i)x + (4 + 5i) + \frac{4i}{x - 1}$$ ∎

EXAMPLE 5   Greathouse Construction plans to manufacture prefabricated log cabin homes. They can produce a maximum of 365 kits a year. Use the profit function $P(x) = x^3 - 410x^2 + 44,000x - 400,000$ and the cost function $C(x) = -290x^2 + 106,000x + 400,000$ to find the profit $P(x)$, cost $C(x)$, and revenue $R(x)$ when $x = 50$ kits.

Solution   Use synthetic division to find $P(50)$ and $C(50)$:

$$
\begin{array}{r|rrrr}
50 & 1 & -410 & 44,000 & -400,000 \\
 & & 50 & -18,000 & 1,300,000 \\
\hline
 & 1 & -360 & 26,000 & 900,000
\end{array}
$$

Thus P(50) = \$900,000.

$$
\begin{array}{r|rrr}
50 & -290 & 106,000 & 400,000 \\
 & & -14,500 & 4,575,000 \\
\hline
 & -290 & 91,500 & 4,975,000
\end{array}
$$

Thus C(50) = \$4,975,000.

Remembering that $R(x) - C(x) = P(x)$, we have

$$R(50) = P(50) + C(50)$$
$$= \$5,875,000$$

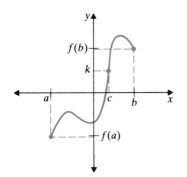

Figure 3.11

We can use the next theorem and synthetic division to approximate zeros of polynomial functions. This method will make it possible to approximate both rational and irrational zeros.

Theorem   *Intermediate Value Theorem for Polynomials.* Let f be a polynomial function on $[a, b]$ with $f(a) \neq f(b)$. For any value k between $f(a)$ and $f(b)$ there exists at least one value c in the interval (a, b) such that $f(c) = k$.

The concept expressed in the intermediate value theorem for polynomials is fairly intuitive. We will not prove this theorem, but we can offer some justification for the theorem by considering the graph shown in Fig. 3.11. The intermediate value theorem states that when $f(a) \neq f(b)$, the polynomial function f takes on each value between $f(a)$ and $f(b)$ at least once.

EXAMPLE 6   For the polynomial function $f(x) = x^4 - 10x^2 + 21$,
(a) show that f has zeros in the intervals $[1, 2]$ and $[-2, -1]$, and
(b) use synthetic division to approximate at least one zero in each interval to the nearest tenth.

Solution   For $f(x) = x^4 - 10x^2 + 21$,
(a) consider the interval $[1, 2]$:

$$f(1) = 1^4 - 10(1)^2 + 21 = 22 - 10 = 12$$
$$f(2) = 2^4 - 10(2)^2 + 21 = 16 - 40 + 21 = -3$$

Since $f(1) \neq f(2)$ and 0 is between $f(1) = 12$ and $f(2) = -3$, by the intermediate value theorem for polynomials there exists at least one zero in the interval $[1, 2]$. Next, consider the interval $[-2, -1]$. Calculating $f(-x)$, we have

$$f(-x) = (-x)^4 - 10(-x)^2 + 21$$
$$= x^4 - 10x^2 + 21$$
$$= f(x)$$

Thus f has symmetry with respect to the y-axis. Because there is at least one zero in the interval $[1, 2]$ and f has symmetry with respect to the y-axis, there is also at least one zero in the interval $[-2, -1]$.

(b) Use synthetic division to find one zero in the interval $[1, 2]$ to the nearest tenth. You can arbitrarily pick a number in the interval $[1, 2]$ with one decimal place. The number 1.5 is a good choice, since it is the midpoint of the interval $[1, 2]$. (By picking the midpoint we divide the interval into two equal intervals. This can save time in approximating zeros.) Divide synthetically with a calculator:

$$
\begin{array}{r|rrrrr}
1.5 & 1 & 0 & -10 & 0 & 21 \\
 & & 1.5 & 2.25 & -11.625 & -17.4375 \\
\hline
 & 1 & 1.5 & -7.75 & -11.625 & 3.5625 \\
\end{array}
$$

Since $f(1) = 12$, $f(1.5) = 3.5625$, and $f(2) = -3$, with $f(d) = 0$ between $f(1.5)$ and $f(2)$, we know that there is a zero in the interval $(1.5, 2)$. Arbitrarily pick a one-place decimal number near the midpoint of the interval $(1.5, 2)$, such as 1.7. Divide synthetically with a calculator:

$$
\begin{array}{r|rrrrr}
1.7 & 1 & 0 & -10 & 0 & 21 \\
 & & 1.7 & 2.89 & -12.087 & -20.5479 \\
\hline
 & 1 & 1.7 & -7.11 & -12.087 & 0.4521 \\
\end{array}
$$

Since $f(1.5) = 3.5625$, $f(1.7) = 0.4521$, and $f(2) = -3$, we know that $f(d) = 0$ is between $f(1.7)$ and $f(2)$, and the zero is in the interval $(1.7, 2)$. Arbitrarily pick a one-decimal number near the midpoint of this interval such as 1.8 and divide synthetically:

$$
\begin{array}{r|rrrrr}
1.8 & 1 & 0 & -10 & 0 & 21 \\
 & & 1.8 & 3.24 & -12.168 & -21.9 \\
\hline
 & 1 & 1.8 & -6.76 & -12.168 & -0.9 \\
\end{array}
$$

Since $f(1.7) = 2.8695$, $f(1.8) = -0.9$, and $f(2) = -3$, we conclude that $f(d) = 0$ is between $f(1.7)$ and $f(1.8)$, and a zero is in the interval $(1.7, 1.8)$. To determine whether the zero is closer to 1.7 or 1.8, divide synthetically by the midpoint 1.75:

$$
\begin{array}{r|rrrrr}
1.75 & 1 & 0 & -10 & 0 & 21 \\
 & & 1.75 & 3.0625 & -12.14 & -21.246 \\
\hline
 & 1 & 1.75 & -6.9375 & -12.14 & -0.246 \\
\end{array}
$$

Since $f(1.7) = 2.8695$, $f(1.75) = -0.246$, and $f(1.8) = -0.9$, we conclude that $f(d) = 0$ is between $f(1.7)$ and $f(1.75)$. Therefore the zero is closer to 1.7 than to 1.8. Thus the zero in the interval $[1, 2]$ is 1.7 rounded off to the nearest tenth.

Therefore f has a zero of approximately 1.7 in the interval $[1, 2]$, and by symmetry with respect to the y-axis, f also has a zero of approximately -1.7 in the interval $[-2, -1]$. ■

Exercises 3.3

Use synthetic division to find (a) the quotient and (b) the remainder of $f(x) \div g(x)$ for each of the following.

1. $f(x) = x^3 - 3x^2 + x - 5$
$g(x) = x - 5$

2. $f(x) = x^3 - 3x^2 + x - 5$
$g(x) = x - 1$

3. $f(x) = x^3 + 5x^2 - x + 3$
$g(x) = x + 1$

4. $f(x) = x^3 + 5x^2 - x + 3$
$g(x) = x + 3$

5. $f(x) = x^3 + 3x^2 - x - 3$
$g(x) = x + 3$

6. $f(x) = x^3 + 3x^2 - x - 3$
$g(x) = x - 1$

7. $f(x) = x^3 + x^2 - 2x - 2$
$g(x) = x - \sqrt{2}$

8. $f(x) = x^3 + x^2 - 3x - 3$
$g(x) = x - \sqrt{3}$

9. $f(x) = x^3 + x^2 + x + 1$
$g(x) = x - i$

10. $f(x) = x^3 + x^2 + x + 1$
$g(x) = x + i$

11. $f(x) = 2x^4 - 3x^3 - x^2 + 2x - 1$
$g(x) = x - 1$

12. $f(x) = 2x^4 - 3x^3 - x^2 + 2x - 1$
$g(x) = x - 2$

13. $f(x) = 2x^4 - 3x^3 - x^2 + 2x - 1$
$g(x) = x + 2$

14. $f(x) = 2x^4 - 3x^3 - x^2 + 2x - 1$
$g(x) = x + 1$

15. $f(x) = 2x^4 - 10x^2 - 9$
$g(x) = x - 1$

16. $f(x) = 2x^4 - 10x^2 - 9$
$g(x) = x + 1$

Use the remainder theorem and synthetic division to find $f(d)$ for each of the following.

17. For $f(x) = 4x^4 + 5x^3 - x^2 - 5x + 2$, find $f\left(\frac{1}{2}\right)$.

18. For $f(x) = 10x^4 - 7x^3 - 14x^2 - 2x + 1$, find $f\left(\frac{1}{5}\right)$.

19. For $f(x) = 2x^3 + 5x - 4$, find $f\left(\frac{3}{4}\right)$.

20. For $f(x) = x^3 + 2x - 3$, find $f\left(\frac{1}{4}\right)$.

21. For $f(x) = 2x^3 + x^2 + 2x - 3$, find $f(i)$.

22. For $f(x) = 2x^3 - x^2 + 2x + 1$, find $f(i)$.

23. For $f(x) = x^4 - 2x^3 + 2x^2 + 3$, find $f(1 + i)$.

24. For $f(x) = x^4 - 2x^3 + 2x^2 + 3$, find $f(1 - i)$.

Greathouse Construction plans to manufacture prefabricated log cabin homes. They can produce a maximum of 365 kits a year. The profit function is $P(x) = x^3 - 410x^2 + 44{,}000x - 400{,}000$, and the cost function is $C(x) = -290x^2 + 106{,}000x + 400{,}000$, where x is the number of kits produced.

25. Find the profit $P(x)$, the cost $C(x)$, and the revenue $R(x)$ when $x = 35$ and $x = 110$ house kits.

26. Find the profit $P(x)$, the cost $C(x)$, and the revenue $R(x)$ when $x = 95$ and $x = 170$ house kits.

Bowen's boats produces a small recreational sailboat. They can produce up to 150 boats a month with a cost function of $C(x) = 400 - 0.001x^2$ and a profit function of $P(x) = 0.1x^3 - 21x^2 + 1100x$, where x is the number of boats produced.

27. Find the profit $P(x)$, the cost $C(x)$, and the revenue $R(x)$ when $x = 25$ and $x = 150$ boats.

28. Find the profit $P(x)$, the cost $C(x)$, and the revenue $R(x)$ when $x = 50$ and $x = 90$ boats.

Use synthetic division and the intermediate value theorem for polynomials to show that f has one zero in the indicated interval $[a, b]$.

29. $f(x) = 2x^4 + x^3 - 25x^2 - 12x + 12$ where $[a, b] = [3, 4]$

30. $f(x) = 2x^4 + x^3 - 43x^2 - 21x + 21$ where $[a, b] = [4, 5]$

31. $f(x) = 2x^4 + x^3 - 43x^2 - 21x + 21$ where $[a, b] = [-5, -4]$

32. $f(x) = 2x^4 + x^3 - 25x^2 - 12x + 12$ where $[a, b] = [-4, -3]$

33. $f(x) = x^4 + 3x^3 - 3x^2 - 15x - 10$ where $[a, b] = [2, 3]$

34. $f(x) = 2x^4 + 10x^3 + 11x^2 - 5x - 6$ where $[a, b] = [0, 1]$

35. Find the zero for Exercise 29 in $[3, 4]$ to one decimal place.

36. Find the zero for Exercise 30 in $[4, 5]$ to one decimal place.

37. Find the quotient

$$\frac{8x^3 - x + 5}{2x - 1}$$

using synthetic division. (Multiply the numerator and denominator by the reciprocal of 2, since synthetic division requires the coefficient of x in the divisor to be 1.)

38. Find the quotient

$$\frac{8x^3 - x + 5}{2x + 1}$$

using synthetic division. (Multiply the numerator and denominator by the reciprocal of 2, since synthetic division requires the coefficient of x in the divisor to be 1.)

39. Use synthetic division to evaluate

$$\frac{2ax^4 + 2bx^3 - ax^2 - bx + a}{ax + b}$$

(Factor the denominator before synthetic division.)

40. Use synthetic division to evaluate

$$\frac{3ax^4 + 3bx^3 - 5ax^2 - 5bx - a}{ax + b}$$

Given that the functions in Exercises 41–48 are polynomials, find the polynomial of smallest degree that could be represented by the graph.

41.

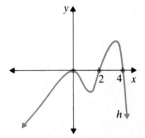

42.

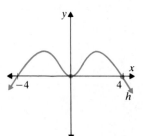

43.

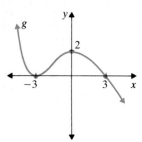

44.

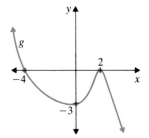

45.

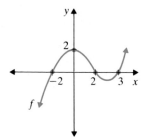

46.

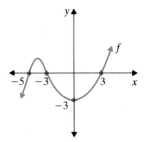

47.

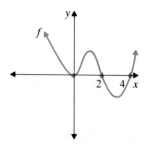

48.

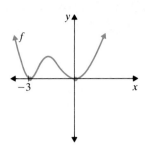

49. Show that $f(x) = ax^5 + bx^4 + cx^3 + dx^2 + ex + f$ can be written as $f(x) = (\{[(ax + b)x + c]x + d\}x + e)x + f$ by repeatedly factoring x. This second form of $f(x)$ is called **nested form** and is useful in evaluating a polynomial for a specific value of x with a calculator.

50. Calculating $f(x)$ from Exercise 49 in nested form is similar to calculating $f(x)$ with synthetic division. Show that $f(w)$ in nested form is the same as the remainder when synthetically dividing $f(x)$ by $x - w$.

3.4   Finding Polynomial Zeros

In this section and Section 3.5 we will develop techniques to systematically locate the rational zeros of a polynomial function. In addition to the new material presented here, we will use the factor theorem and synthetic division from previous sections to find rational zeros.

The Number of Zeros

We begin by finding the number of zeros for a given polynomial using the following theorems.

Theorem

The Fundamental Theorem of Algebra. If a polynomial $f(x)$ with real number (or complex number) coefficients has a degree greater than zero, then $f(x)$ has at least one complex zero.

This theorem assures us that any polynomial of degree greater than zero has at least one complex zero. We note that this theorem includes polynomials with real number coefficients or complex number coefficients. The proof of the fundamental theorem of algebra is beyond the scope of this course.

The following corollary is an immediate consequence of the fundamental theorem of algebra.

Corollary

Every polynomial $f(x)$ with degree greater than zero has at least one factor of the form $x - d$ where d is a complex number.

PROOF: From the fundamental theorem of algebra we know that $f(x)$ has at least one zero; call it d. If d is a zero, then $f(d) = 0$, and by the factor theorem, $x - d$ is a factor of $f(x)$. $\square$

In the next theorem we will consider the total number of zeros in a polynomial.

Theorem

The Number of Polynomial Zeros. If $f(x)$ is a polynomial of degree n, $n \in N$, then $f(x)$ has n zeros $d_1, d_2, d_3, \ldots, d_n$, where each d_i is a complex number. Since $d_1, d_2, d_3, \ldots, d_n$ are zeros for $f(x)$, $f(x)$ can be factored as

$$f(x) = a(x - d_1)(x - d_2)(x - d_3) \cdots (x - d_n)$$

where a is the coefficient of x^n.

PROOF: If $f(x)$ is of degree n, $n \in N$, and $n > 0$, then by the fundamental theorem of algebra we know that $f(x)$ has at least one zero d_1. We use the factor theorem to write $f(x) = (x - d_1)q_1(x)$. If the degree of $q_1(x) > 0$, then we use the fundamental theorem of algebra to conclude that $q_1(x)$ has at least one zero d_2. When d_2 is a zero for $q_1(x)$, then we use the factor theorem to write $q_1(x) = (x - d_2)q_2(x)$. By substitution we have

$$f(x) = (x - d_1)(x - d_2)q_2(x)$$

We continue this process n times until the degree of $q_i(x)$ is 1 and a is the leading coefficient of $q_i(x)$, also the leading coefficient of $f(x)$. Therefore we can write $f(x)$ in factored form as

$$f(x) = a(x - d_1)(x - d_2)(x - d_3) \cdots (x - d_n) \qquad \square$$

Using this theorem, we can find the number of zeros for any polynomial quickly because the theorem states that the number of zeros is the same as the degree of the polynomial.

It is possible that a number is a zero more than once. For example, the polynomial

$$f(x) = x^4 - 4x^3 + 16x - 16$$
$$= (x + 2)(x - 2)^3$$

has zeros of -2 and 2 with 2 as a zero three times. Since 2 is a zero of $f(x)$ three times, we call 2 **a zero of multiplicity three** for the polynomial $f(x)$. Thus the multiplicity of a zero d is the number of times $x - d$ is a factor.

EXAMPLE 1

Factor the polynomial $f(x) = x^4 + 2x^3 - 4x^2 - 2x + 3$ if 1 is a zero of multiplicity two for $f(x)$.

Solution   Since $f(x)$ has a degree of four, we know that $f(x)$ has four zeros. Knowing that 1 is a zero, we use synthetic division to divide $f(x)$ by $x - 1$:

$$
\begin{array}{r|rrrrr}
1 & 1 & 2 & -4 & -2 & 3 \\
  &   & 1 & 3 & -1 & -3 \\
\hline
  & 1 & 3 & -1 & -3 & \underline{}\,0 \\
\end{array}
$$

We call the last line of synthetic division, without the remainder, the coefficients of **the residue polynomial**. In this case the residue polynomial is $x^3 + 3x^2 - x - 3$. Thus $f(x) = (x - 1)(x^3 + 3x^2 - x - 3)$. Since 1 is a zero of multiplicity two, we divide the residue polynomial $x^3 + 3x^2 - x - 3$ synthetically by 1. (Do not go back to the original polynomial once a zero has been found. Use the residue polynomial.)

$$
\begin{array}{r|rrrr}
1 & 1 & 3 & -1 & -3 \\
  &   & 1 & 4 & 3 \\
\hline
  & 1 & 4 & 3 & \big\vert\ 0
\end{array}
$$

The residue polynomial from this division is $x^2 + 4x + 3$, providing

$$f(x) = (x - 1)(x - 1)(x^2 + 4x + 3)$$

Factoring $x^2 + 4x + 3$ into $(x + 1)(x + 3)$, we have

$$
\begin{aligned}
f(x) &= (x - 1)(x - 1)(x + 1)(x + 3) \\
     &= (x - 1)^2(x + 1)(x + 3)
\end{aligned}
$$

Thus $f(x) = x^4 + 2x^3 - 4x^2 - 2x + 3$ has the four zeros. They are 1, 1, −1, and −3. Thus $f(x)$ has zeros of 1 with multiplicity two, −1, and 3. ∎

EXAMPLE 2   Write the equation of a fourth-degree polynomial with a zero of 2 with multiplicity three and a zero of −2 with multiplicity two if $f(1) = 18$.

Solution   By the previous theorem and our discussion of multiple zeros, we can write $f(x)$ as

$$
\begin{aligned}
f(x) &= a(x - 2)^3[x - (-2)]^2 \\
     &= a(x - 2)^3(x + 2)^2
\end{aligned}
$$

Since $f(1) = 18$, we have

$$
\begin{aligned}
f(1) = a(1 - 2)^3(1 + 2)^2 &= 18 \\
a(-1)^3(3)^2 &= 18 \\
-9a &= 18 \\
a &= -2
\end{aligned}
$$

Thus

$$
\begin{aligned}
f(x) &= -2(x - 2)^3(x + 2)^2 \\
     &= -2[(x - 2)(x + 2)]^2(x - 2) \\
     &= -2(x^2 - 4)^2(x - 2) \\
     &= -2(x^4 - 8x^2 + 16)(x - 2) \\
     &= -2(x^5 - 2x^4 - 8x^3 + 16x^2 + 16x - 32) \\
     &= -2x^5 + 4x^4 + 16x^3 - 32x^2 - 32x + 64
\end{aligned}
$$
∎

Positive and Negative Zeros

Knowing that the degree of a polynomial reveals the exact number of zeros for that polynomial, we will find the number of possible positive zeros, negative zeros, and nonreal complex zeros of a polynomial with the following rule.

DECARTES' RULE OF SIGNS	For the polynomial $f(x)$ in standard form with real number coefficients the number of possible positive real zeros, negative real zeros, and nonreal complex zeros can be calculated as follows: (a) Count the sign changes in the coefficients of $f(x)$, moving from left to right. The number of *positive real zeros* is equal to the number of *sign changes in $f(x)$* or is less than this number by an even integer. (b) Count the sign changes in the coefficients of $f(-x)$, moving from left to right. The number of *negative real zeros* is equal to the number of *sign changes in $f(-x)$* or is less than this number by an even integer.

We count the sign changes in $f(x)$ or $f(-x)$ for Descartes' rule of signs by reading the polynomial in one direction, from left to right. Each time

(a) a positive coefficient is followed by a negative coefficient or
(b) a negative coefficient is followed by a positive coefficient,

we count one change in sign. We will not prove Descartes' rule of signs.

Guidelines for Using Descartes' Rule of Signs

STEP 1: First factor all common factors of x from $f(x)$, if there are any, to identify all zeros that are 0. When x is a factor of multiplicity m, then $f(x) = x^m g(x)$.

STEP 2: Next we apply Descartes' rule of signs to the polynomial $g(x)$ to find the number of possible positive real zeros.

STEP 3: Now we apply Descartes' rule of signs to the polynomial $g(-x)$ to find the number of possible negative real zeros.

STEP 4: Finally, we subtract the sum of the different combinations of possible positive and negative real zeros from the total number of zeros to find the number of possible nonreal complex zeros (of the form $a + bi$, where $b \neq 0$).

This process is illustrated in the next example.

EXAMPLE 3   Find the possible number of positive, negative, and nonreal complex zeros for the polynomial $f(x) = x^6 - 2x^5 - 7x^4 + 8x^3 + 12x^2$,

Solution   We will use the above steps to apply Descartes' rule of signs.

STEP 1: First we factor the common factor x^2 from $f(x)$. Thus

$$f(x) = x^2(x^4 - 2x^3 - 7x^2 + 8x + 120) \qquad \text{and}$$

$$f(x) = x^2 h(x) \qquad \text{where } h(x) = x^4 - 2x^3 - 7x^2 + 8x + 12$$

STEP 2: Note that the sign for the coefficients of $h(x)$ are

We count two sign changes for the coefficient of $h(x)$. Thus by part (a) of Descartes' rule of signs we know that $h(x)$ has two positive real zeros or no (0) positive real zeros. (The number of zeros is two or a number less than two by an even integer, $2 - 2 = 0$.)

STEP 3: To find the number of possible negative zeros, we use $h(-x)$. Then

$$h(-x) = (-x)^4 - 2(-x)^3 - 7(-x)^2 + 8(-x) + 12$$
$$= x^4 + 2x^3 - 7x^2 - 8x + 12$$

The signs for the coefficients of $h(-x)$ are

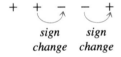

Since there are two sign changes for the coefficients of $h(-x)$, we know that $h(x)$ has two or no negative zeros. The number of zeros can be described as follows:

possible positive real zeros	2 or 0
possible negative real zeros	2 or 0

STEP 4: Since $h(x) = x^4 - 2x^3 - 7x^2 + 8x + 12$ has degree four, we know that it has four zeros. We show all possible combinations of the zeros for $h(x)$ with the following chart:

possible positive real zeros	2	2	0	0
possible negative real zeros	2	0	2	0
possible nonreal complex zeros	0	2	2	4
total zeros	4	4	4	4

Each column in this chart represents one way in which the zeros can occur for $h(x)$. Therefore the polynomial $f(x)$ has 0 as a zero of multiplicity two as well as the same possible positive and negative zeros as $h(x)$ displayed in the chart above. ■

EXAMPLE 4   Find the possible number of positive, negative, and nonreal complex zeros for the polynomial $g(x) = x^4 + 7x^3 + 17x^2 + 17x + 6$.

Solution   We will use the four steps for applying Descartes' rule of signs.

STEP 1: There is no common factor of x for each term of $g(x)$, and thus 0 is not a zero of $g(x)$.

STEP 2: We note that the signs for $g(x)$ are all positive. Thus there are no sign changes for $g(x)$, so there can be no positive real zeros for $g(x)$.

STEP 3: The signs for $g(-x)$ given by

$$g(x) = (-x)^4 + 7(-x)^3 + 17(-x)^2 + 17(-x) + 6$$
$$= x^4 - 7x^3 + 17x^2 - 17x + 6$$

are $+ - + - +$, so $g(-x)$ has four sign changes. Thus we know that $g(x)$ has four, two, or no negative zeros (four zeros or a number of zeros less than four by an even integer).

STEP 4: Since $g(x) = x^4 + 7x^3 + 17x^2 + 17x + 6$ has a degree of four, we know that it has four zeros. A chart of the possible zeros for $g(x)$ would be

possible positive real zeros	0	0	0
possible negative real zeros	4	2	0
possible nonreal complex zeros	0	2	4
total zeros	4	4	4

■

Upper and Lower Bounds

The search for polynomial zeros can be narrowed further by finding an interval $[a, b]$ that contains *all real zeros* of a polynomial. The number $b \geq 0$ of this interval is called **an upper bound**, and the number $a \leq 0$ in this interval is called **a lower bound** for the real number zeros of a polynomial. The next theorem provides us with a method of finding these upper and lower bounds for a polynomial.

BOUNDS FOR POLYNOMIAL ZEROS	For a polynomial $f(x)$ with real coefficients and a positive leading coefficient, use synthetic division to divide $f(x)$ by $x - d$.

(a) If $d > 0$ and all numbers in the quotient line (third line) of synthetic division are positive or zero, then d is an upper bound for the real zeros of $f(x)$.

(b) If $d < 0$ and the numbers in the quotient line (third line) of synthetic division alternate sign, $+ \ - \ + \ - \ + \ - \cdots$ (where zero can be either positive or negative), then d is a lower bound for the real zeros of $f(x)$.

We will not prove the above statement concerning the bounds of polynomials. From the work developed in this section it is now possible to determine

(a) the number of zeros for each polynomial,
(b) the different possible combinations of positive and negative real zeros for each polynomial, and
(c) an upper and lower bound for a polynomial's real zeros.

We will now use these techniques to find all integer zeros of a polynomial.

EXAMPLE 5   Find all integer zeros for the polynomial $h(x) = x^4 - 2x^3 - 7x^2 + 8x + 12$, and write the polynomial in factored form.

Solution   As we saw in Example 3, the fourth-degree polynomial $h(x) = x^4 - 2x^3 - 7x^2 + 8x + 12$ has four zeros with two or no positive and two or no negative real zeros. Rather than randomly picking possible integer zeros, we will first look for an integer that is an upper bound of $h(x)$. We try the integer 4, since it might be large enough to be an upper bound:

$$
\begin{array}{r|rrrrr}
4 & 1 & -2 & -7 & 8 & 12 \\
  &   & 4 & 8 & 4 & 48 \\
\hline
  & 1 & 2 & 1 & 12 & \boxed{60} \\
\end{array}
$$

Thus 4 is an upper bound, since all the numbers in the quotient are positive. (Note also that 4 is not a zero, since $h(4) = 60$.) Next, try a second positive integer less than 4. It is possible that 3 could be a smaller upper bound or a zero:

$$
\begin{array}{r|rrrrr}
3 & 1 & -2 & -7 & 8 & 12 \\
  &   & 3 & 3 & -12 & -12 \\
\hline
  & 1 & 1 & -4 & -4 & \boxed{0} \\
\end{array}
$$

Thus 3 is a zero of $h(x)$ yet not an upper bound for real zeros. At this point we use the residue polynomial $x^3 + x^2 - 4x - 4$ in future divisions. Next check to see whether 3 is a multiple zero:

$$
\begin{array}{r|rrrr}
3 & 1 & 1 & -4 & -4 \\
  &   & 3 & 12 & 24 \\
\hline
  & 1 & 4 & 8 & \boxed{20}
\end{array}
$$

We see that 3 is not a multiple zero but is an upper bound of the remaining zeros. The remaining positive integer zero could be either 1 or 2. We arbitrarily try 2:

$$
\begin{array}{r|rrrr}
2 & 1 & 1 & -4 & -4 \\
  &   & 2 & 6 & 4 \\
\hline
  & 1 & 3 & 2 & \boxed{0}
\end{array}
$$

Thus 2 is a zero of $h(x)$. We can now factor the residue polynomial $x^2 + 3x + 2$ into $(x + 2)(x + 1)$. Then $h(x)$ has zeros of -2, -1, 2, and 3, and in factored form, $h(x) = (x + 2)(x + 1)(x - 2)(x - 3)$.   ∎

EXAMPLE 6   Find all integer zeros for the polynomial $g(x) = x^4 + 7x^3 + 17x^2 + 17x + 6$, and write the polynomial in factored form.

Solution   As we saw in Example 4, the fourth-degree polynomial $g(x) = x^4 + 7x^3 + 17x^2 + 17x + 6$ has four zeros, no positive real zeros, and four, two, or no negative zeros. We begin by searching for a negative integer that is a lower bound. We try -7 to see whether it will make the coefficients of the quotient alternate signs:

$$
\begin{array}{r|rrrrr}
-7 & 1 & 7 & 17 & 17 & 6 \\
   &   & -7 & 0 & -119 & 714 \\
\hline
   & 1 & 0 & 17 & -102 & \boxed{720}
\end{array}
$$

Thus $g(-7) = 720$, and -7 is a lower bound, since the 0 in the third row can be thought of as -0. (We also note that -7 is the greatest integer that can be a lower bound, since any integer greater than -7, such as -6, would make the second number in the third line of synthetic division positive.) We now try another negative integer value. We arbitrarily pick -4:

$$
\begin{array}{r|rrrrr}
-4 & 1 & 7 & 17 & 17 & 6 \\
   &   & -4 & -12 & -20 & 12 \\
\hline
   & 1 & 3 & 5 & -3 & \boxed{18}
\end{array}
$$

Thus $g(-4) = 18$, and -4 is neither a zero nor a lower bound. We arbitrarily try another negative integer such as -3:

$$
\begin{array}{r|rrrrr}
-3 & 1 & 7 & 17 & 17 & 6 \\
   &   & -3 & -12 & -15 & -6 \\
\hline
   & 1 & 4 & 5 & 2 & \boxed{0}
\end{array}
$$

Thus -3 is a zero. Next we use the residue polynomial to see whether -3 is a multiple zero:

$$
\begin{array}{r|rrrr}
-3 & 1 & 4 & 5 & 2 \\
   &   & -3 & -3 & -6 \\
\hline
   & 1 & 1 & 2 & \boxed{-4}
\end{array}
$$

Thus -3 is not a multiple zero. We now try another negative integer such as -2:

$$
\begin{array}{r|rrrr}
-2 & 1 & 4 & 5 & 2 \\
   &   & -2 & -4 & -2 \\
\hline
   & 1 & 2 & 1 & \boxed{0}
\end{array}
$$

Thus -2 is a zero. Now we can write the residue polynomial $x^2 + 2x + 1$ in factored form as $(x + 1)(x + 1)$. Hence $g(x)$ has zeros of -1 of multiplicity 2, -2, and -3. We can write $g(x)$ in factored form as

$$g(x) = (x + 1)(x + 1)(x + 2)(x + 3)$$

or

$$g(x) = (x + 1)^2(x + 2)(x + 3) \qquad\blacksquare$$

The polynomials in Examples 5 and 6 have only integer zeros, which is not always the case with polynomial zeros. However, the methods used to find integer zeros are very similar to the methods we will use in the next section to locate rational zeros. In the next section, upper and lower bounds will become even more important in finding polynomial zeros which are rational numbers.

EXAMPLE 7   A 10×20 inch sheet of metal is used to form a candy box with an open top by cutting squares with sides of x out of each corner (Fig. 3.12). Find the size of the corner cut x such that the volume of the box is not less than 144 cubic inches.

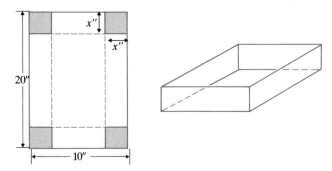

Figure 3.12

Solution   The volume

$$V = x(10 - 2x)(20 - 2x)$$
$$= 4x^3 - 60x^2 + 200x$$

When $V \geq 144$, we have

$$4x^3 - 60x^2 + 200x \geq 144$$
$$4x^3 - 60x^2 + 200x - 144 \geq 0$$
$$4(x^3 - 15x^2 + 50x - 36) \geq 0$$
$$x^3 - 15x^2 + 50x - 36 \geq 0$$

To find the solution to this inequality, we now search for the zeros of $f(x) = x^3 - 15x^2 + 50x - 36$. We narrow our search by noting that the possible values for the cut x must be less than half the shortest side, $0 < 2x < 10$, or $0 < x < 5$. We try $x = 3$, and

$$
\begin{array}{r|rrr r}
3 & 1 & -15 & 50 & -36 \\
  &   & 3 & -36 & 42 \\
\hline
  & 1 & -12 & 14 & \boxed{6} \\
\end{array}
$$

Thus $f(3) = 6$. We also know that $f(0) = -36$, and by the intermediate value theorem for polynomials there must be a zero between 0 and 3. We try $x = 1$, and

$$
\begin{array}{r|rrr r}
1 & 1 & -15 & 50 & -36 \\
  &   & 1 & -14 & 36 \\
\hline
  & 1 & -14 & 36 & \boxed{0} \\
\end{array}
$$

Thus 1 is a zero. We can now find the zeros of the residue polynomial, $x^2 - 14x + 36$, by solving $x^2 - 14x + 36 = 0$ with the quadratic formula:

$$x = \frac{-(-14) \pm \sqrt{(-14)^2 - 4(1)(36)}}{2 \cdot 1}$$

$$= \frac{14 \pm \sqrt{52}}{2} = 7 \pm \sqrt{13}$$

$$= \begin{cases} 7 + \sqrt{13} \approx 10.6 \\ 7 - \sqrt{13} \approx 3.4 \end{cases}$$

We use the sign chart to evaluate the inequality

$$f(x) = x^3 - 15x^2 + 50x - 36 \geq 0$$

which can be approximately written as

$$f(x) = (x - 1)(x - 10.6)(x - 3.4) \geq 0$$

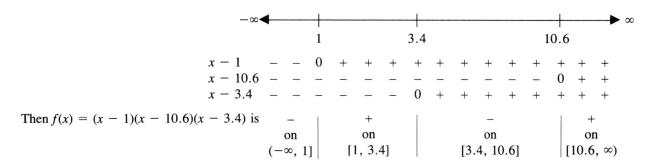

Then $f(x) = (x - 1)(x - 10.6)(x - 3.4)$ is

	$-$	$+$	$-$	$+$
	on	on	on	on
	$(-\infty, 1]$	$[1, 3.4]$	$[3.4, 10.6]$	$[10.6, \infty)$

Thus for our candy box to have a volume of at least 144 cubic inches with $0 < x < 5$, the corner cut x must be between 1 and 3.4 inches in length. ■

Exercises 3.4

State the number of zeros for each polynomial. Find the smallest interval $[a, b]$ where a is a negative integer lower bound and b is a positive integer upper bound for the zeros of each polynomial. (It is possible that these integer bounds could also be zeros.)

1. $f(x) = x^3 + 7x^2 + 7x - 15$

2. $f(x) = x^3 - 4x^2 - 4x + 16$

3. $g(x) = x^3 - x^2 - 8x + 12$

4. $g(x) = x^3 + 2x^2 - 9x - 18$

5. $h(x) = x^3 - x^2 - 16x + 16$

6. $h(x) = x^3 - 7x^2 + 15x - 9$

7. $f(x) = x^3 - 2x^2 - 7x - 4$

8. $f(x) = x^3 - 12x + 16$

9. $g(x) = x^4 - x^3 - 5x^2 + 3x + 6$

10. $g(x) = x^4 - x^3 - 8x^2 + 2x + 12$

11. $h(x) = x^4 - x^3 + 2x^2 - 4x - 8$

12. $h(x) = x^4 - 2x^3 - 7x^2 - 2x - 8$

Find the possible combinations of positive and negative real zeros. Then use the information in Exercises 1–8 to find the integers that are zeros for the following polynomials.

13. $f(x) = x^3 + 7x^2 + 7x - 15$

14. $f(x) = x^3 - 4x^2 - 4x + 16$

15. $g(x) = x^3 - x^2 - 8x + 12$

16. $g(x) = x^3 + 2x^2 - 9x - 18$

17. $h(x) = x^3 - x^2 - 16x + 16$

18. $h(x) = x^3 - 7x^2 + 15x - 9$

19. $f(x) = x^3 - 2x^2 - 7x - 4$

20. $f(x) = x^3 - 12x + 16$

Find all zeros for the following polynomials using the information from Exercises 9–12.

21. $g(x) = x^4 - x^3 - 5x^2 + 3x + 6$

22. $g(x) = x^4 - x^3 - 8x^2 + 2x + 12$

23. $h(x) = x^4 - x^3 + 2x^2 - 4x - 8$

24. $h(x) = x^4 - 2x^3 - 7x^2 - 2x - 8$

Use the information in Exercises 13–24 and the sign graph to sketch graphs of the following polynomials.

25. $f(x) = x^3 + 7x^2 + 7x - 15$

26. $f(x) = x^3 - 4x^2 - 4x + 16$

27. $g(x) = x^3 - x^2 - 8x + 12$

28. $g(x) = x^3 + 2x^2 - 9x - 18$

29. $h(x) = x^3 - x^2 - 16x + 16$

30. $h(x) = x^3 - 7x^2 + 15x - 9$

31. $f(x) = x^3 - 2x^2 - 7x - 4$

32. $f(x) = x^3 - 12x + 16$

33. $g(x) = x^4 - x^3 - 5x^2 + 3x + 6$

34. $g(x) = x^4 - x^3 - 8x^2 + 2x + 12$

35. $h(x) = x^4 - x^3 + 2x^2 - 4x - 8$

36. $h(x) = x^4 - 2x^3 - 7x^2 - 2x - 8$

37. Find a fourth-degree polynomial that has one zero of -1 with multiplicity three and another zero of 3.

38. Find a fourth-degree polynomial that has one zero of 2 with multiplicity two and another zero of -1 with multiplicity two.

39. Find a fourth-degree polynomial that has one zero of 3 with multiplicity two and another zero of -2 with multiplicity two.

40. Find a fourth-degree polynomial with one zero of -3 with multiplicity three and another zero of -1.

Solve the following inequalities.

41. $x^3 - 2x^2 \geq 24x + 12$ **42.** $x^3 - 3x^2 > 9x + 5$

43. $x^3 + 4x^2 \leq 13x - 2$ **44.** $x^3 + x^2 \leq 21x - 4$

45. A 10×20 inch sheet of metal is used to form a candy box with an open top by cutting squares with sides of x out of each corner as shown in Fig. 3.13. Find the size of the corner cut such that the volume of the box is not less than 96 cubic inches.

46. A 12×20 inch sheet of metal is used to form a candy box with an open top by cutting squares with sides of x out of each corner (Fig. 3.13). Find the size of the corner cut such that the volume of the box is not less than 252 cubic inches.

47. A funnel is constructed with a top radius 5 cm larger than its height (Fig. 3.14). Find the possible values for the height of the funnel when the volume is less than 64π cm^3. (Use the volume of the cone as an approximation of the volume of the funnel.)

48. A funnel is constructed with a top radius 5 cm less than its height (Fig. 3.14). Find the possible values for the height of the funnel when the volume is less than $\left(\frac{4}{3}\right)\pi$ cm^3. (Use the volume of the cone as an approximation of the volume of the funnel.)

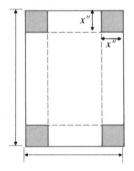
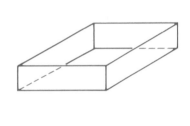

Figure 3.13

Figure 3.14

3.5   Finding Rational Zeros for Polynomials

In Section 3.4 we found the domain intervals that contained all real zeros of a polynomial. There are many numbers in these intervals that are possible zeros for a given polynomial. To make our search for the zeros of a polynomial more efficient, we will use the following theorem. This new theorem is called the rational zero theorem and will provide us with a list of all possible rational zeros for a polynomial with integer coefficients.

Theorem | *The Rational Zero Theorem.* Let the polynomial

$$f(x) = a_n x^n + a_{n-1} x^{n-1} + a_{n-2} x^{n-2} + \cdots + a_2 x^2 + a_1 x + a_0$$

be a polynomial with integer coefficients. Let $\dfrac{p}{q}$ be a rational number, where

$q > 0$ and there is no common natural number factor for p and q. If $\frac{p}{q}$ is a zero of $f(x)$, then p is an integer factor of a_0 and q is a positive integer factor of a_n.

We require q to be positive in this theorem as a matter of convenience. With q positive we will eliminate duplicate values when both p and q are negative and will keep the denominator nonzero. The proof of the rational zero theorem follows.

PROOF: If $\frac{p}{q}$ is a zero for

$$f(x) = a_n x^n + a_{n-1} x^{n-1} + \cdots + a_2 x^2 + a_1 x + a_0$$

then by substituting we have

$$f\left(\frac{p}{q}\right) = a_n\left(\frac{p}{q}\right)^n + a_{n-1}\left(\frac{p}{q}\right)^{n-1} + \cdots + a_z\left(\frac{p}{q}\right)^2 + a_1\left(\frac{p}{q}\right) + a_0 = 0$$

Next, we multiply both sides by q^n to obtain

$$a_n p^n + a_{n-1} p^{n-1} q + \cdots + a_2 p^2 q^{n-2} + a_1 p q^{n-1} + a_0 q^n = 0$$

Then we subtract $a_0 q^n$ from both sides and factor the left side:

$$a_n p^n + a_{n-1} p^{n-1} q + \cdots + a_2 p^2 q^{n-2} + a_1 p q^{n-1} = -a_0 q^n$$
$$p(a_n p^{n-1} + a_{n-1} p^{n-2} q + \cdots + a_2 p q^{n-2} + a_1 q^{n-1}) = -a_0 q^n$$

Notice that p is a factor of the left side of this equation and thus must also be a factor of the right side, $-a_0 q^n$. But since p and q have no common natural number factors, then p must be a factor of a_0. Similarly, we take the equation above,

$$a_n p^n + a_{n-1} p^{n-1} q + \cdots + a_2 p^2 q^{n-2} + a_1 p q^{n-1} + a_0 q^n = 0$$

and subtract $a_n p^n$ from both sides:

$$a_{n-1} p^{n-1} q + a_{n-2} p^{n-2} q^2 + \cdots + a_1 p q^{n-1} + a_0 q^n = -a_n p^n$$

Next, we factor the left side to produce

$$q(a_{n-1} p^{n-1} + a_{n-2} p^{n-2} q + \cdots + a_1 p q^{n-2} + a_0 q^{n-1}) = -a_n p^n$$

Since q is a factor of the left side, it must also be a factor of $-a_n p^n$. Since q and p have no common natural number factors, q must be a factor of a_n. **Therefore if the rational number $\frac{p}{q}$ in lowest terms is a zero for $f(x)$, then p is an integer factor of a_0 (the constant term) and q is a positive integer factor of a_n (the leading coefficient).** ☐

The following examples show how the rational zero theorem, along with material from previous sections, can be used to find the rational zeros of a polynomial.

EXAMPLE 1   Find all possible rational zeros for $f(x) = 4x^3 + 4x^2 - 29x + 21$.

Solution   To find all possible rational zeros $\dfrac{p}{q}$, we first find the set of all values for p. Since the possible values of p are the integer factors of $a_0 = 21$, $p \in \{\pm 1, \pm 3, \pm 7, \pm 21\}$. The set of possible values for q are the positive integer factors of $a_n = 4$, so $q \in \{1, 2, 4\}$. All the rational numbers $\dfrac{p}{q}$ that are possible zeros for $f(x)$ are given by

$$\frac{p}{q} \in \left\{ \pm 1, \pm 3, \pm 7, \pm 21, \pm \frac{1}{2}, \pm \frac{3}{2}, \pm \frac{7}{2}, \pm \frac{21}{2}, \pm \frac{1}{4}, \pm \frac{3}{4}, \pm \frac{7}{4}, \pm \frac{21}{4} \right\}$$

When we rearrange the elements of this set in order, we have

$$\frac{p}{q} \in \left\{ \pm \frac{1}{4}, \pm \frac{1}{2}, \pm \frac{3}{4}, \pm 1, \pm \frac{3}{2}, \pm \frac{7}{4}, \pm 3, \pm \frac{7}{2}, \pm \frac{21}{4}, \pm 7, \pm \frac{21}{2}, \pm 21 \right\}$$ ∎

EXAMPLE 2   Find all possible rational zeros for $f(x) = 4x^3 - 7x - 3$. Use the techniques of the previous section (the number of zeros, Descartes' rule of signs, upper and lower bounds) to find the zeros of f. Then factor f completely.

Solution   Since $f(x) = 4x^3 - 7x - 3$ is a third-degree polynomial function, we know from Section 3.4 that f has three zeros. In addition, since $f(x) = 4x^3 - 7x - 3$ has one sign change, we know that f has exactly one positive real zero. Likewise, we know that f has two or no negative real zeros because $f(-x) = -4x^3 + 7x - 3$ has two sign changes. Next, we find the set of all possible rational zeros $\dfrac{p}{q}$. If p is an integer factor of $a_0 = -3$, then $p \in \{\pm 1, \pm 3\}$. Since q is a positive integer factor of $a_n = -4$, then $q \in \{1, 2, 4\}$. Thus all possible $\dfrac{p}{q}$ are accounted for with

$$\frac{p}{q} \in \left\{ \pm 1, \pm 3, \pm \frac{1}{2}, \pm \frac{3}{2}, \pm \frac{1}{4}, \pm \frac{3}{4} \right\}$$

We arrange these set elements in order:

$$\frac{p}{q} \in \left\{ \pm \frac{1}{4}, \pm \frac{1}{2}, \pm \frac{3}{4}, \pm 1, \pm \frac{3}{2}, \pm 3 \right\}$$

We can now use synthetic division to search for the rational zeros of $f(x)$, trying only elements $\dfrac{p}{q}$ from this set above. We first look for a zero that may be an upper bound. Try 3 to see whether it is an upper bound or a zero:

$$
\begin{array}{r|rrrr}
3 & 4 & 0 & -7 & -3 \\
  &   & 12 & 36 & 87 \\
\hline
  & 4 & 12 & 29 & \boxed{84} \\
\end{array}
$$

Thus $f(3) = 84$ and 3 is an upper bound because 3 is positive and all the signs of the coefficients of the residue polynomial and the remainder are positive. Since 3 is an upper bound, we must now try a smaller value. We try 1, since it is a value for $\frac{p}{q}$, a possible zero and easy to divide:

$$\begin{array}{r} 1\,\underline{\big|}\ 4 \quad 0 \ -7 \ -3 \\ 4 \quad 4 \ -3 \\ \hline 4 \quad 4 \ -3\,\underline{\big|\,-6} \end{array}$$

Thus $f(1) = -6$, and 1 is not an upper bound. Since $f(3) = 84$ is positive and $f(1) = -6$ is negative, by the mean value theorem of polynomials there must be a zero between 1 and 3. The *only possible rational zero* in this interval is $\frac{3}{2}$. For $\frac{3}{2}$,

$$\begin{array}{r} \tfrac{3}{2}\,\underline{\big|}\ 4 \quad 0 \ -7 \ -3 \\ 6 \quad 9 \quad 3 \\ \hline 4 \quad 6 \quad 2\ \underline{\big|\,0} \end{array}$$

Thus $\frac{3}{2}$ is a zero. We can now factor the residue polynomial,

$$4x^2 + 6x + 2 = 2(2x^2 + 3x + 1)$$
$$= 2(2x + 1)(x + 1)$$

Therefore $f(x)$ has zeros -1, $-\frac{1}{2}$, and $\frac{3}{2}$. The factored polynomial is

$$f(x) = 2\left(x - \frac{3}{2}\right)(2x + 1)(x + 1)$$
$$= (2x - 3)(2x + 1)(x + 1)$$

■

EXAMPLE 3   Find all the zeros, if possible, for

$$g(x) = 2x^5 - 9x^4 + 14x^3 - 13x^2 + 12x - 4$$

Solution   Since $g(x) = 2x^5 - 9x^4 + 14x^3 - 13x^2 + 12x - 4$ is of degree 5, g has five zeros. There are five sign changes for $g(x)$, so g has five, three, or one positive real zero. The signs of $g(-x)$ are all negative, so g has no negative real zeros. Since $p \in \{\pm 1,\ \pm 2,\ \pm 4\}$ and $q \in \{1, 2\}$, all possible rational zeros, $\frac{p}{q}$, are given by

$$\frac{p}{q} \in \left\{\pm 1,\ \pm 2,\ \pm 4,\ \pm\frac{1}{2}\right\} = \left\{\pm\frac{1}{2},\ \pm 1,\ \pm 2,\ \pm 4\right\}$$

Because we have already eliminated all negative numbers as possible zeros, we can condense this set further to

$$\frac{p}{q} \in \left\{\frac{1}{2},\ 1,\ 2,\ 4\right\}$$

We will first try 4, the largest possible *rational* zero, to see whether it is a zero and/or a rational upper bound. (There could be an irrational number that is an upper bound.)

$$
\begin{array}{r|rrrrr}
4 & 2 & -9 & 14 & -13 & 12 & -4 \\
  &   & 8 & -4 & 40 & 108 & 480 \\
\hline
  & 2 & -1 & 10 & 27 & 120 & \boxed{476}
\end{array}
$$

Thus 4 is not a zero and not an upper bound for the real zeros by this test. Next, try the next smaller $\dfrac{p}{q}$ value such as 2:

$$
\begin{array}{r|rrrrrr}
2 & 2 & -9 & 14 & -13 & 12 & -4 \\
  &   & 4 & -10 & 8 & -10 & 4 \\
\hline
  & 2 & -5 & 4 & -5 & 2 & \boxed{0}
\end{array}
$$

Thus 2 is a zero. Check to see whether 2 is a zero of multiplicity two:

$$
\begin{array}{r|rrrrr}
2 & 2 & -5 & 4 & -5 & 2 \\
  &   & 4 & -2 & 4 & -2 \\
\hline
  & 2 & -1 & 2 & -1 & \boxed{0}
\end{array}
$$

Hence 2 is a zero twice. Check to see whether 2 is a zero of multiplicity three:

$$
\begin{array}{r|rrrr}
2 & 2 & -1 & 2 & -1 \\
  &   & 4 & 6 & 16 \\
\hline
  & 2 & 3 & 8 & \boxed{15}
\end{array}
$$

Thus 2 is a zero of multiplicity two. Next, consider one of the two remaining positive values of $\dfrac{p}{q}$, $\dfrac{1}{2}$ and 1, with the above residue polynomial $2x^3 - 1x^2 + 2x - 1$. (This is the residue polynomial resulting from the last zero we identified.)

$$
\begin{array}{r|rrrr}
1 & 2 & -1 & 2 & -1 \\
  &   & 2 & 2 & 4 \\
\hline
  & 2 & 1 & 4 & \boxed{3}
\end{array}
\qquad
\begin{array}{r|rrrr}
\tfrac{1}{2} & 2 & -1 & 2 & -1 \\
  &   & 1 & 0 & 1 \\
\hline
  & 2 & 0 & 2 & \boxed{0}
\end{array}
$$

Thus the rational zeros of g are 2 with multiplicity two and $\frac{1}{2}$. The remaining zeros are zeros of the residue polynomial $2x^2 + 2$:

$$
\begin{aligned}
2x^2 + 2 &= 0 \\
2x^2 &= -2 \\
x^2 &= -1 \\
x &= \pm i
\end{aligned}
$$

The five zeros of g are 2 with multiplicity two, $\frac{1}{2}$, i, and $-i$.   ■

Note in Example 3 that both i and its conjugate $-i$ are zeros for this polynomial. The next theorem will generalize this observation.

Theorem

Complex Zero Pairs. If a complex number $a + bi$, $b \neq 0$, is a zero of a polynomial, $f(x)$, with real coefficients, then the conjugate complex number $a - bi$ is also a zero of $f(x)$.

Thus complex zeros occur in conjugate pairs when the polynomial has real coefficients.

PROOF: Let $f(x)$ be the polynomial with $a + bi$ as a zero. Then $f(a + bi) = 0$ and $[x - (a + bi)]$ is a factor of $f(x)$. To prove that $f(a - bi) = 0$, we will show that $[x - (a - bi)]$ is a factor of $f(x)$. To do this, we will check to see whether

$$D(x) = [x - (a + bi)][x - (a - bi)]$$
$$= [(x - a) - bi][(x - a) + bi]$$
$$= x^2 - 2ax + (a^2 + b^2)$$

is a factor of $f(x)$. Notice that $D(x)$ has $a + bi$ and its conjugate $a - bi$ as zeros. We now divide $f(x)$ by $D(x)$ using the division algorithm. Then $f(x) = D(x)Q(x) + (Rx + S)$, where $Q(x)$ is the quotient and $Rx + S$ is the remainder. (The remainder $Rx + S$ is either the zero polynomial or a polynomial of degree 0 or 1, since its degree must be less than that of the second-degree polynomial $D(x)$.) Next we remember that $a + bi$ is a zero of both f and D, so $f(a + bi) = 0$ and $D(a + bi) = 0$. Then

$$f(a + bi) = D(a + bi)Q(a + bi) + [R \cdot (a + bi) + S]$$
$$0 = 0 \cdot Q(a + bi) + [R \cdot (a + bi) + S]$$
$$0 + 0i = (Ra + S) + Rbi$$

For two equal complex numbers the real parts are equal, $0 = (Ra + S)$, and the coefficients of i are equal, $0 = Rb$. With $Rb = 0$ and $b \neq 0$, we can conclude that $R = 0$. With $Ra + S = 0$ and $R = 0$ we can conclude that $S = 0$. Thus the remainder $Rx + S = 0$, and $D(x)$ is a factor of $f(x)$. Therefore with

$$D(x) = [x - (a + bi)][x - (a + bi)]$$

as a factor of $f(x)$, we have proven that $a + bi$ and $a - bi$ are both zeros of $f(x)$.

□

A similar relationship exists between the irrational zeros as indicated in the next theorem.

Theorem

Irrational Zero Pairs. The expression $a + \sqrt{b}$, where a and b are rational but $\sqrt{b}$ is irrational, is called a **quadratic surd**. If a quadratic surd $a + \sqrt{b}$ is a zero of a polynomial, $f(x)$, with rational coefficients, then the conjugate surd $a - \sqrt{b}$ is also a zero of $f(x)$.

Thus irrational zeros of a polynomial function occur in conjugate pairs when the polynomial has rational coefficients. The proof of this theorem is similar to the proof above and is a problem in the exercise set that follows.

EXAMPLE 4   Find a fourth-degree polynomial $g(x)$ with rational coefficients and zeros of $1 - i$ and $1 + \sqrt{2}$.

Solution   If the polynomial with rational coefficients has zeros of $1 - i$ and $1 + \sqrt{2}$, then by the previous two theorems their conjugates $1 + i$ and $1 - \sqrt{2}$ are also zeros. Thus

$$
\begin{aligned}
g(x) &= [x - (1 - i)][x - (1 + i)][x - (1 + \sqrt{2})][x - (1 - \sqrt{2})] \\
&= (x - 1 + i)(x - 1 - i)(x - 1 - \sqrt{2})(x - 1 + \sqrt{2}) \\
&= [(x - 1) + i][(x - 1) - i][(x - 1) - \sqrt{2}][(x - 1) + \sqrt{2}] \\
&= [(x - 1)^2 - i^2][(x - 1)^2 - (\sqrt{2})^2] \\
&= (x^2 - 2x + 1 + 1)(x^2 - 2x + 1 - 2) \\
&= (x^2 - 2x + 2)(x^2 - 2x - 1) \\
&= x^4 - 4x^3 + 5x^2 - 2x - 2
\end{aligned}
$$

or in general for any rational number a.

$$
g(x) = a(x^4 - 4x^3 + 5x^2 - 2x - 2)
$$

Exercises 3.5

For each of the following polynomials, find the set of all possible rational zeros $\dfrac{p}{q}$. Do not solve further for any zeros.

1. $f(x) = x^4 + 7x^2 + 6$
2. $f(x) = x^4 - 6x^3 - 10$
3. $g(x) = x^4 + 6x^3 - 15$
4. $g(x) = x^4 + 6x^2 + 21$
5. $h(x) = 2x^4 + 3x^3 - 10$
6. $h(x) = 2x^4 - 3x^2 + 6$
7. $f(x) = 3x^4 + 6x^3 - 21$
8. $f(x) = 3x^4 + 16x^2 - 4$
9. $g(x) = 4x^5 - 7x^3 + 12$
10. $g(x) = 6x^5 - 2x^3 + 15$
11. $h(x) = 6x^4 - 5x^3 - 18$
12. $h(x) = 4x^4 - 5x^3 - 18$

Write the polynomials of degree four with rational coefficients and the following zeros. Write the answer with integer coefficients.

13. $5, 1, -1, 2$
14. $2, -2, 3, 4$
15. $1, \frac{1}{2}, -3, -\frac{1}{2}$
16. $1, -\frac{1}{3}, 2, \frac{1}{3}$
17. -1 with multiplicity three and 1
18. -2 with multiplicity three and 2
19. $-1, 2, 1 - i$
20. $1, -2, 2 - i$
21. $1, -2, 2 - \sqrt{3}$
22. $-1, 2, 3 - \sqrt{2}$
23. $1 - 2i, 1 - \sqrt{2}$
24. $2 - 3i, 1 - \sqrt{3}$

For each of the following polynomials, find the set of all possible rational zeros. Then use the techniques of Descartes' rule of signs, synthetic division, and locating upper and lower bounds to find all zeros.

25. $f(x) = x^3 - 4x^2 - 4x + 16$
26. $f(x) = x^3 + 7x^2 + 7x - 15$
27. $g(x) = 4x^3 + 8x^2 - x - 2$
28. $g(x) = 9x^3 - 18x^2 - x + 2$
29. $h(x) = 4x^3 - 7x - 3$
30. $h(x) = 4x^3 - 4x^2 - 11x + 6$
31. $f(x) = 12x^4 + 20x^3 - 11x^2 - 5x + 2$
32. $f(x) = 18x^4 + 27x^3 - 20x^2 - 3x + 2$
33. $g(x) = 6x^4 - 17x^3 - 7x^2 + 13x - 3$
34. $g(x) = 12x^4 + 20x^3 - 11x^2 - 5x + 2$
35. $h(x) = x^4 - x^3 - 8x^2 + 2x + 12$

36. $h(x) = x^4 - x^3 - 5x^2 + 3x + 6$

37. $f(x) = x^4 - 2x^3 - 7x^2 - 2x - 8$

38. $f(x) = x^4 - x^3 + 2x^2 - 4x - 8$

Verify that the following polynomials have no rational zeros.

39. $f(x) = x^4 - x^3 + x^2 + 1$ **40.** $f(x) = x^4 + x^3 + x^2 + 1$

41. $g(x) = x^4 - x^3 - x^2 + 2$ **42.** $g(x) = x^4 + x^3 + x^2 + 2$

43. $h(x) = 2x^4 - x^3 - x^2 - 1$ **44.** $h(x) = 3x^4 - x^3 + x^2 + 1$

45. Show that i is a double root of $f(x) = 2x^5 - x^4 + 4x^3 - 2x^2 + 2x - 1$. Also factor $f(x)$.

46. Show that $2 + i$ and $2 - i$ are double roots of $f(x) = x^4 - 8x^3 + 26x^2 - 40x + 25$. Also factor $f(x)$.

47. For rational numbers a and b where $\sqrt{b}$ is irrational, prove that if the quadratic surd $a + \sqrt{b}$ is a zero of $f(x)$, then the conjugate surd $a - \sqrt{b}$ is also a zero of $f(x)$. Assume that $f(x)$ has rational coefficients.

3.6 Graphing Rational Functions

In Section 2.5 we worked briefly with the graphs of rational functions when we sketched the graph of a function's reciprocal. We will now sketch the graphs of more complicated rational functions. The following examples illustrate and explain the behavior of these graphs.

EXAMPLE 1   Graph $f(x) = \dfrac{x^2 - 1}{x - 1}$.

Solution   At first, $f(x)$ appears to be a rational function with the degree of the numerator greater than the degree of the denominator. By reducing the common factors we see that

$$f(x) = \frac{x^2 - 1}{x - 1} = \frac{(x - 1)(x + 1)}{(x - 1)} = x + 1 \qquad \text{for } x \neq 1$$

We notice that $x \neq 1$, since $f(1)$ is undefined ($x = 1$ makes the denominator of $f(x)$ zero.) Thus $f(x) = x + 1$ with the exception that $x \neq 1$. This leaves a hole in the graph of $f(x)$ at $x = 1$ as seen in Fig. 3.15. ■

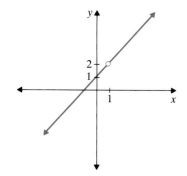

Figure 3.15

We see in Example 1 how a factor in the denominator can create a hole in the graph of a function. We note that the graph of $f(x) = \dfrac{x^2 - 1}{x - 1}$, (Fig. 3.15) does not have 1 as a domain element. As the value of x gets closer and closer to 1, the value of $f(x)$ gets closer and closer to 2.

We can also describe the behavior of f around the hole at $x = 1$ using the limit concept. Recalling the limit notation discussed in Section 2.5, we can say that

$$\lim_{x \to 1^-} f(x) = 2 \qquad \text{and} \qquad \lim_{x \to 1^+} f(x) = 2$$

(Remember that $x \to 1^-$ means that x approaches 1 from the left and $x \to 1^+$ means that x approaches 1 from the right.) We find the values of these limits by evaluating the reduced form of the rational function at $x = 1$ and/or observing the graph.

Definition | *Holes of Rational Function.* If a factor is common to the numerator and the denominator and that factor is not a factor in the denominator of the rational function in reduced form, then the graph will contain a hole. The hole will appear for those domain values of x that make the reduced factor zero.

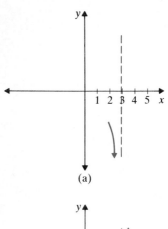

(a)

(b)

Figure 3.16

We will not always be able to reduce factors in the denominator with factors in the numerator. An example of this is the function $f(x) = \dfrac{1}{x - 3}$. Obviously, $x \neq 3$, since $x = 3$ causes the denominator of $f(x)$ to equal zero. We will now look at the graph of this function for x-values close to $x = 3$. The following values are found by using a calculator.

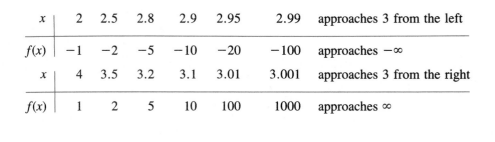

x	2	2.5	2.8	2.9	2.95	2.99	approaches 3 from the left
$f(x)$	-1	-2	-5	-10	-20	-100	approaches $-\infty$
x	4	3.5	3.2	3.1	3.01	3.001	approaches 3 from the right
$f(x)$	1	2	5	10	100	1000	approaches ∞

From the charts above of $f(x) = \dfrac{1}{x - 3}$ we can use limit notation to describe this behavior as follows:

$$\lim_{x \to 3^-} f(x) = -\infty \qquad \text{and} \qquad \lim_{x \to 3^+} f(x) = \infty$$

We graph these results in Fig. 3.16. In these graphs the dashed line at $x = 3$ is called a **vertical asymptote**. The graph of the function gets very close to a vertical asymptote but never intersects, meets, or crosses the vertical asymptote. In fact, when the vertical asymptote is $x = 3$, the number 3 is not an element of the domain of f. On each side of the asymptote the function is either increasing or decreasing, approaching ∞ or $-\infty$, as x approaches a vertical asymptote.

Definition | *Vertical Asymptotes of Rational Functions.* If g and h are polynomial functions, then the rational function

$$f(x) = \frac{g(x)}{h(x)}$$

will have a **vertical asymptote** at $x = a$ whenever $h(a) = 0$ and $g(a) \neq 0$. (As x approaches a from one side, the function f is increasing or decreasing and $\lim_{x \to a^-} f(x) = \pm\infty$ and $\lim_{x \to a^+} f(x) = \pm\infty$.)

To be able to sketch the graph of $f(x) = \dfrac{1}{x-3}$, we must also understand the behavior of the value $f(x)$ as x approaches $\pm\infty$. We find these values for $f(x)$ with a calculator.

x	10	100	1000	10,000	approaches ∞
$f(x)$	0.1428	0.0103	0.001003	0.00010003	approaches 0 as a positive number
x	-10	-100	-1000	$-10,000$	approaches $-\infty$
$f(x)$	-0.0769	-0.0097	-0.000997	-0.00009997	approaches 0 as a negative number

Thus as x approaches ∞, f is a decreasing function, and $f(x)$ approaches 0 as a positive number (Fig. 3.17a). Also as x approaches $-\infty$, $f(x)$ approaches 0 as a negative number (Fig. 3.17b). When a graph behaves this way around the line $y = 0$, the line $y = 0$ is called a **horizontal asymptote**. From the charts we see that as the value of $|x|$ gets larger and larger, the value of $f(x)$ gets closer and closer to 0. In limit notation we can describe this behavior as

$$\lim_{x\to\infty} f(x) = 0 \qquad \text{and} \qquad \lim_{x\to-\infty} f(x) = 0$$

When we combine this information about the vertical asymptote, the line $x = 3$, and the horizontal asymptote, the line $y = 0$, along with the fact that $f(0) = \dfrac{1}{0-3} = -\dfrac{1}{3}$, we can sketch the graph of $f(x) = \dfrac{1}{x-3}$ as shown in Fig. 3.17(c).

Although we could sketch the graph of the function $f(x) = \dfrac{1}{x-3}$ as the reciprocal of $y = x - 3$, this new method of analyzing the graph of a rational function will be very helpful with more complicated functions.

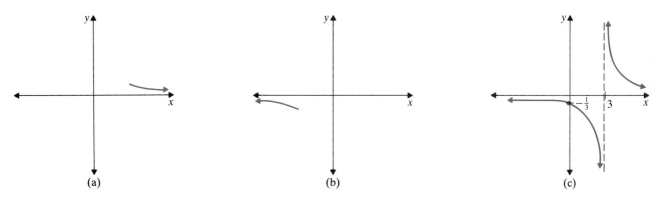

(a)   (b)   (c)

Figure 3.17

Definition

Horizontal Asymptotes of a Rational Function. For a constant real number K and the rational function $f(x) = \dfrac{g(x)}{h(x)}$, the **horizontal asymptote** of f is a line $y = K$ when the value of $f(x)$ approaches K as the value of x approaches $\pm\infty$ (or $\lim\limits_{x \to \pm\infty} f(x) = K$). It is possible for the function f to intersect the horizontal asymptote, since the asymptote best approximates the function as x approaches $\pm\infty$.

Another condition for a horizontal asymptote occurs when the degree of the numerator equals the degree of the denominator. When this happens, the horizontal asymptote is the ratio of the leading coefficients from the numerator and denominator. An example of this is the function

$$g(x) = \frac{2x^2 + x - 1}{x^2 + 1}$$

After division,

$$g(x) = 2 + \frac{x - 3}{x^2 + 1}$$

Now consider the remainder fraction. As x approaches $\pm\infty$, the value of the denominator $x^2 + 1$ is much larger than the value of the numerator $x - 3$. Thus as x approaches $\pm\infty$, the remainder fraction approaches 0. In limit notation we have

$$\lim_{x \to \infty} \frac{x - 3}{x^2 + 1} = 0 \quad \text{and} \quad \lim_{x \to -\infty} \frac{x - 3}{x^2 + 1} = 0$$

Thus as x approaches $\pm\infty$, $g(x)$ approaches $2 + 0 = 2$, and the horizontal asymptote for g is $y = 2$. This is also the ratio of the leading coefficients of the numerator and denominator, $\dfrac{2x^2}{x^2} = 2$.

The two functions above, f and g, are examples of the two types of horizontal asymptotes described in Table 3.2.

There are several advantages in writing the rational expression

$$g(x) = \frac{ax^n + \cdots}{bx^n + \cdots} \quad \text{as} \quad g(x) = \frac{a}{b} + \frac{R}{bx^n + \cdots}$$

First, the horizontal asymptote is clearly visible as $y = \dfrac{a}{b}$, since

$$\lim_{x \to \pm\infty} \frac{R}{bx^n + \cdots} = 0$$

TABLE 3.2 Horizontal Asymptotes

Conditions	Example	Horizontal Asymptote
The degree of the numerator is less than the degree of the denominator.	$f(x) = \dfrac{ax^n + \cdots}{bx^m + \cdots}$ for $n < m$	$y = 0$
The degree of the denominator equals the degree of the numerator.	$g(x) = \dfrac{ax^n + \cdots}{bx^n + \cdots}$ or by dividing (quotient form) $g(x) = \dfrac{a}{b} + \dfrac{R}{bx^n + \cdots}$	$y = \dfrac{a}{b}$

Also, when $\dfrac{R}{bx^n + \cdots} > 0$, the graph of g is above the asymptote $y = \dfrac{a}{b}$, and

when $\dfrac{R}{bx^n + \cdots} < 0$, the graph of g is below the asymptote $y = \dfrac{a}{b}$. When

$R = 0$, the graph intersects the asymptote $y = \dfrac{a}{b}$. Thus the quotient form

$$g(x) = \frac{a}{b} + \frac{R}{bx^n + \cdots}$$

can be most helpful in constructing the graph of g. We will now use these results in the examples that follow.

EXAMPLE 2   The altitude of a satellite during launch can be described by the function

$$h(t) = \frac{210.03t + 0.03}{t + 1}$$

where $h(t)$ is the altitude in miles above sea level and t is the time in seconds after launch.
(a) Find the altitude of the satellite at $t = 0$, $t = 60$, and $t = 100$ seconds.
(b) What is the highest possible orbit based on this equation?
(c) Sketch the graph of this equation for $t \geq 0$.

Solution

(a) To find the altitude at $t = 0$, 60, and 100 seconds, we substitute these values into the equation

$$h(t) = \frac{210.03t + 0.03}{t + 1}$$

For $t = 0$,

$$h(0) = \frac{(210.03)(0) + 0.03}{0 + 1} = \frac{0 + 0.03}{1} = 0.03 \text{ miles} = 158.4 \text{ feet}$$

For $t = 60$,

$$h(60) = \frac{(210.03)(60) + 0.03}{60 + 1} = \frac{12601.8 + 0.03}{61} = 206.6 \text{ miles}$$

For $t = 100$,

$$h(100) = \frac{(210.03)(100) + 0.03}{100 + 1} = \frac{21003 + 0.03}{101} = 207.95 \text{ miles}$$

(b) We divide $210t + 0.03$ by $x + 1$ and write $h(t)$ as

$$h(t) = 210.03 - \frac{210}{t + 1}$$

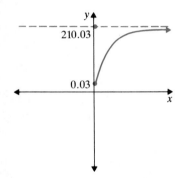

Figure 3.18

Since the remainder fraction, $-\dfrac{210}{(t + 1)}$, is negative and $\lim\limits_{t \to \infty} -\dfrac{210}{(t - 1)} = 0$, the highest possible orbit is approximated by the horizontal asymptote, the line $y = 210.03$ miles. Using the leading coefficients, we can also find the horizontal asymptote to be $y = \dfrac{210.03}{1} = 210.03$ miles. Thus the highest altitude of the satellite is less than 210.03 miles.

(c) Using this information, we graph $h(t)$ in Fig. 3.18. ■

EXAMPLE 3   Sketch the graph of

$$g(x) = \frac{2x^2 + x - 1}{x^2 + 1} = 2 + \frac{x - 3}{x^2 + 1}$$

Solution   The form

$$g(x) = \frac{2x^2 + x - 1}{x^2 + 1}$$

will help us to identify the vertical asymptotes. The vertical asymptotes occur when the denominator, $x^2 + 1$, equals zero. Since there are no real number solutions to $x^2 + 1 = 0$, there are no vertical asymptotes for $g(x)$. The quotient form

$$g(x) = 2 + \frac{x - 3}{x^2 + 1}$$

helps us to find the horizontal asymptote $y = 2$. The graph is above the asymptote

$$y = 2 \text{ for } \frac{x - 3}{x^2 + 1} > 0,$$

which is $x - 3 > 0$ (since $x^2 + 1 > 0$),

or $x > 3$, the interval $(3, \infty)$.

The graph is below the asymptote when $\frac{x - 3}{x^2 + 1} < 0,$

which is $x - 3 < 0$ (since $x^2 + 1 > 0$)

or $x < 3$, the interval $(-\infty, 3)$.

The graph intersects the asymptote when $\frac{x - 3}{x^2 + 1} = 0$, which is $x - 3 = 0$ or $x = 3$. We summarize this information in Table 3.3, which will help in graphing.

TABLE 3.3

Domain Interval	$(-\infty, 3)$	$x = 3$	$(3, \infty)$
The remainder fraction $\frac{x - 3}{x^2 + 1}$ is and $g(x)$	$-$ is below the asymptote $y = 2$	0 intersects the asymptote $y = 2$	$+$ is above the asymptote $y = 2$

The y-intercept of g is $g(0) = -1$, and the x-intercepts or zero of $g(x)$ are all x-values when

$$g(x) = \frac{2x^2 + x - 1}{x^2 + 1} = 0$$

$$2x^2 + x - 1 = 0$$

$$(2x - 1)(x + 1) = 0$$

$$x = \frac{1}{2} \quad \text{or} \quad x = -1$$

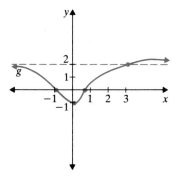

Figure 3.19

Thus the y-intercept is -1, and the x-intercepts are $\frac{1}{2}$ and -1. Using all of this information, we can sketch the graph shown in Fig. 3.19.   ∎

The steps listed below can be used as guidelines for graphing rational functions.

Guidelines for Sketching the Graph
of a Rational Function

STEP 1: Find and graph the vertical asymptotes and holes, if there are any.

STEP 2: Write the rational function in the quotient form to identify and graph the horizontal asymptote (the remainder fraction will be used in Step 4).

STEP 3: Find all zeros (the rational expression equals zero when the numerator equals zero and the denominator does not equal zero) and the y-intercept b, from the point $(0, b)$.

STEP 4: Find the domain interval in which the graph of the function is above, below, and equal to the horizontal asymptote by identifying the intervals where the remainder fraction of Step 2 is positive, negative, and zero.

STEP 5: Find intervals in which the rational function is positive and negative (sign chart).

STEP 6: Sketch in the graph to conform with all of the above information.

Using these steps, we will graph the rational function in Example 4.

EXAMPLE 4   Sketch the graph of

$$h(x) = \frac{x^2 - 1}{x^2 - 4}$$

Solution   For

$$h(x) = \frac{x^2 - 1}{x^2 - 4} = \frac{(x - 1)(x + 1)}{(x - 2)(x + 2)}$$

complete the six steps for graphing a rational function.

STEP 1: The vertical asymptotes are $x = \pm 2$, since each makes the denominator zero and makes the numerator different from zero.

STEP 2: The horizontal asymptote is $y = 1$, since upon division

$$h(x) = 1 + \frac{3}{x^2 - 4}$$

STEP 3: Finding zeros, we have $h(x) = 0$ when $x^2 - 1 = 0$ or $x = \pm 1$. Therefore $(1, 0) \in h$ and $(-1, 0) \in h$. Also, $h(0) = \frac{(0 - 1)}{(0 - 4)} = \frac{1}{4}$, so $\left(0, \frac{1}{4}\right) \in h$.

STEP 4: To find when the range values are above or below the horizontal asymptote, construct the sign chart for the remainder

$$\frac{3}{x^2 - 4} = \frac{3}{(x - 2)(x + 2)}$$

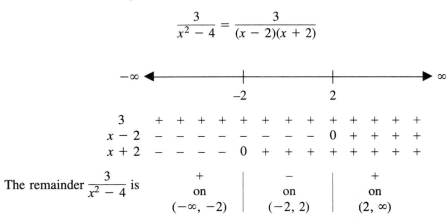

The remainder $\frac{3}{x^2 - 4}$ is

As you can see, the remainder $\frac{3}{x^2 - 4}$ is positive on the interval $(-\infty, -2) \cup (2, \infty)$. Therefore the graph of h is above the asymptote $y = 1$ in that interval. Likewise, the remainder $\frac{3}{x^2 - 4}$ is negative on the interval $(-2, 2)$, and the graph of h is below the asymptote $y = 1$ on the interval $(-2, 2)$. Since $\frac{3}{x^2 - 4} \neq 0$, there is no point of intersection between h and the horizontal asymptote.

STEP 5: To find when the range values of h are positive and negative, we construct the sign chart for

$$h(x) = \frac{(x - 1)(x + 1)}{(x - 2)(x + 2)}$$

$x - 1$

$x + 1$

$x - 2$

$x + 2$

$h(x) = (x - 1)(x + 1)(x - 2)(x + 2)$ is

STEP 6: The information in Steps 4 and 5 is combined and summarized in Table 3.4, which will help in sketching the graph.

TABLE 3.4

Domain Interval	$(-\infty, -2)$	$(-2, -1)$	$(-1, 1)$	$(1, 2)$	$(2, \infty)$
The graph is	+ above the x-axis	− below the x-axis	+ above the x-axis	− below the x-axis	+ above the x-axis
The remainder fraction $\dfrac{3}{x^2 - 4}$ is	+	−	−	−	+
The graph of h is	above the asymptote	below the asymptote	below the asymptote	below the asymptote	above the asymptote

Figure 3.20 gives us the general shape of the graph of $h(x)$. The sketch should keep the function positive and negative in the right intervals, above and below the asymptote in the right intervals, and approaching the asymptotes in a correct manner. ∎

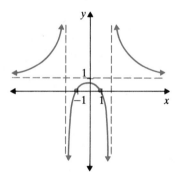

Figure 3.20

Exercises 3.6

Find the equations of all asymptotes from the graphs of each of the following rational functions.

1.

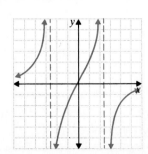

2.

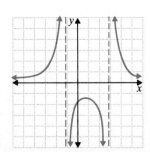

3.

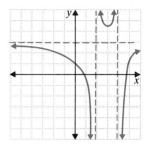

4.

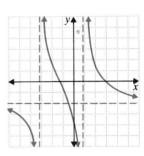

For each of the following rational functions, (a) identify the domain values of holes, the equations of vertical asymptotes, and (b) sketch the graph of each function.

5. $f(x) = \dfrac{x^2 - 9}{x + 3}$

6. $f(x) = \dfrac{x^2 - 4}{x - 2}$

7. $g(x) \dfrac{x^3 - 1}{x - 1}$

8. $g(x) = \dfrac{x^3 + 1}{x + 1}$

9. $h(x) = \dfrac{x}{x^2 - 4}$

10. $h(x) = \dfrac{x}{x^2 - 9}$

11. For $f(x)$ in Exercise 5, find (a) $\lim\limits_{x \to -3^-} f(x)$ and (b) $\lim\limits_{x \to -3^+} f(x)$.

12. For $f(x)$ in Exercise 6, find (a) $\lim\limits_{x \to 2^-} f(x)$ and (b) $\lim\limits_{x \to 2^+} f(x)$.

13. For $g(x)$ in Exercise 7, find (a) $\lim\limits_{x \to 1^-} g(x)$ and (b) $\lim\limits_{x \to 1^+} g(x)$.

14. For $g(x)$ in Exercise 8, find (a) $\lim\limits_{x \to -1^-} g(x)$ and (b) $\lim\limits_{x \to -1^+} g(x)$.

15. For $h(x)$ in Exercise 9, find (a) $\lim\limits_{x \to 2^-} h(x)$ and (b) $\lim\limits_{x \to 2^+} h(x)$.

16. For $h(x)$ in Exercise 10, find (a) $\lim\limits_{x \to 3^-} h(x)$ and (b) $\lim\limits_{x \to 3^+} h(x)$.

For each of the following rational functions, (a) find the vertical asymptotes, (b) find the horizontal asymptotes, and (c) use the other steps listed in this section to sketch the functions graph.

17. $h(x) = \dfrac{2(x^2 - 1)}{x + 1}$

18. $h(x) = \dfrac{-2(x^2 - 4)}{x - 2}$

19. $f(x) = \dfrac{x + 1}{x^2 - 9}$

20. $f(x) = \dfrac{x - 1}{x^2 - 16}$

21. $g(x) = \dfrac{x^2 + 2x + 1}{x^2 + x - 6}$

22. $g(x) = \dfrac{x^2 - 4x + 4}{x^2 - 2x - 3}$

23. $h(x) = \dfrac{x + 2}{x^2 + 2}$

24. $h(x) = \dfrac{x - 2}{x^2 + 3}$

25. The equation $H_T = \dfrac{H_S + H_M}{A}$ is used in solar design to calculate total heat gain H_T per square foot of floor area, where H_S is the heat gained through unshaded windows; H_M is the heat gained through thermal storage walls, roof ponds, and walls adjacent to an attached greenhouse; and A is the total floor area of the structure in square feet. Sketch the graph of H_T as a function of A when H_S and H_M are constants.

Martinez Management Consultants operates with a revenue equation of $R(x) = \dfrac{(2000x + 350)}{(x - 0.1)}$ and a cost equation of $C(x) = \dfrac{(1500x^2 + 299x + 16)}{(x + 0.1)^2}$ for x days of consultation.

26. Sketch the graph of $C(x)$.

27. Sketch the graph for $R(x)$.

28. Find the profit function $P(x)$ and sketch its graph.

3.7   Graphing Other Rational Functions

In this section we will use the ideas developed in the last section to graph more complicated rational functions. Specifically, we will graph rational functions with other nonvertical asymptotes. In these rational functions the degree of the numerator will be greater than the degree of the denominator. We will use the six steps from Section 3.6, which are listed below as guidelines. In Steps 2 and 4 we change from horizontal to nonvertical asymptotes. We will define nonvertical asymptotes later in this section. For now we will use horizontal asymptotes as the nonvertical asymptotes in Steps 2 and 4.

Guidelines for Sketching the Graph of a Rational Function

STEP 1: Find and graph the vertical asymptotes and holes, if there are any.

STEP 2: Write the rational function in the quotient form to identify and graph the nonvertical asymptote (the remainder fraction will be used in Step 4).

STEP 3: Find all zeros (the rational expression equals zero when the numerator equals zero and the denominator does not equal zero) and the y-intercept b, from the point $(0, b)$.

STEP 4: Find the domain interval in which the graph of the function is above, below, and equal to the nonvertical asymptote by identifying the interval where the remainder fraction of Step 2 is positive, negative, and zero.

STEP 5: Find the intervals in which the rational function is positive and negative (sign chart).

STEP 6: Sketch in the graph to conform with all of the above information.

We will now use these steps to graph the following rational function.

EXAMPLE 1   Sketch the graph of

$$g(x) = \frac{2x^2 - 18}{x^2 - 5x + 6}$$

Solution First factor completely and simplify $g(x)$:

$$g(x) = \frac{2(x - 3)(x + 3)}{(x - 3)(x - 2)} = \frac{2x + 6}{x - 2}$$

STEP 1: The vertical asymptote is $x = 2$ with a hole in the graph at $x = 3$, since $x - 3$ is a common factor in the numerator and the denominator.

STEP 2: The horizontal asymptote is $y = 2$, the ratio of the leading coefficient, since the degrees are equal and also since the quotient form is

$$g(x) = 2 + \frac{10}{x - 2}$$

STEP 3: Finding zeros, we have $g(x) = 0$ when $2x + 6 = 0$ or $x = -3$, the zero of g so that $(-3, 0) \in g$. Also, $g(0) = \frac{0 + 6}{0 - 2} = -3$ is the y-intercept, and $(0, -3) \in g$.

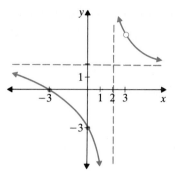

Figure 3.21

STEP 4: The remainder fraction $\dfrac{10}{x-2}$ is positive for $x - 2 > 0$ or $x > 2$. Thus the graph of g is above the asymptote $y = 2$ for x in the interval $(2, \infty)$. The remainder fraction $\dfrac{10}{x-2}$ is negative for $x - 2 < 0$ or $x < 2$. Thus the graph of g is below the asymptote $y = 2$ for x in the interval $(-\infty, 2)$. Since $\dfrac{10}{x-2} \neq 0$, then the graph of g does not intersect the asymptote $y = 2$.

STEP 5: The sign chart for the reduced form of $g(x)$ is

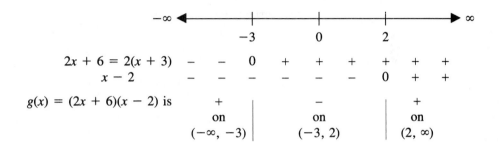

STEP 6: The information in Steps 4 and 5 is summarized in Table 3.5, which will be helpful in graphing Fig. 3.21.

TABLE 3.5

Domain Interval	$(-\infty, -3)$	$(-3, 0)$	$(0, 2)$	$(2, \infty)$
$g(x)$ is and the graph of g is	+ above the x-axis	− below the x-axis	− below the x-axis	+ above the x-axis
The remainder fraction $\dfrac{10}{x-2}$ is	−	−	−	+
The graph of g is	below the asymptote $y = 2$	below the asymptote $y = 2$	below the asymptote $y = 2$	above the asymptote $y = 2$ ∎

Other Nonvertical Asymptotes

We will now consider rational functions with nonvertical asymptotes that are not horizontal. We will first define a nonvertical asymptote.

Definition

Nonvertical Asymptotes of a Rational Function. For a function $k(x)$ and the rational function $f(x) = \dfrac{g(x)}{h(x)}$ the **nonvertical asymptote** of f is a curve $y = k(x)$ when the value of $f(x)$ approaches $k(x)$ as the value of x approaches $\pm\infty$ (or $\lim\limits_{x \to \pm\infty} f(x) = k(x)$. It is possible for the function f to intersect the nonvertical asymptote, since the asymptote best approximates the function as x approaches $\pm\infty$.

Example 2 illustrates the graph of a rational function with a nonvertical asymptote.

EXAMPLE 2   Sketch the graph of

$$f(x) = \frac{x^2 - 3}{x + 2}$$

Solution   We begin by factoring $f(x)$ completely:

$$f(x) = \frac{x^2 - 3}{x + 2} = \frac{(x - \sqrt{3})(x + \sqrt{3})}{x + 2}$$

STEP 1: We have a vertical asymptote at $x = -2$. (It is not a hole, since $x + 2$ is not a factor of the numerator.)

STEP 2: To find the nonvertical asymptote, we divide $x^2 - 3$ by $x + 2$, using synthetic division. We then have

$$\begin{array}{r} -2\,\big|\ 1 \quad 0 \ -3 \\ \underline{\quad -2 \ \ 4\quad} \\ 1 \ -2 \ \ \big|\underline{1} \end{array} \qquad \text{which is} \qquad f(x) = x - 2 + \frac{1}{x + 2}$$

As x approaches $\pm\infty$, the remainder fraction, $\dfrac{1}{x + 2}$, approaches 0. Since the remainder fraction is approaching 0, the function itself is approaching the line $y = x - 2$. This line is therefore the nonvertical asymptote. That is, as x approaches $\pm\infty$, the value of $f(x)$ will get closer and closer to the asymptote $y = x - 2$.

STEP 3: Since $f(x) = 0$ when $x^2 - 3 = 0$, the x-intercepts are $x = \pm\sqrt{3}$. We also have $f(0) = -\frac{3}{2}$, which indicates that the y-intercept is $-\frac{3}{2}$.

STEP 4: The sign of the remainder fraction $\dfrac{1}{x + 2}$ indicates when the graph of $f(x)$ is above or below the asymptote $y = x - 2$. (When the remainder fraction is zero, the graph of f will intersect the asymptote.) The form

$$f(x) = x - 2 + \frac{1}{x + 2}$$

shows that the function consists of two parts: the asymptote, $y = x - 2$,

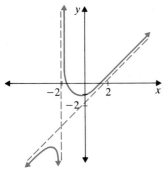

Figure 3.22

and the remainder fraction, $\dfrac{1}{x + 2}$. With this in mind you can think of this function as

$$f(x) = \text{(the nonvertical asymptote)} + \text{(remainder fraction)}$$

For $x > -2$ the remainder fraction is positive, so the graph of f lies above the line $y = x - 2$. For $x < -2$ the remainder fraction is negative, so the graph of f lies below the line $y = x - 2$. Since $\dfrac{1}{x + 2} \neq 0$, the graph of f does not intersect the asymptote $y = x - 2$.

STEP 5: The sign chart for $f(x)$ is

The information from Steps 4 and 5 is combined in Table 3.6.

TABLE 3.6

Domain Intervals	$(-\infty, -2)$	$(-2, -\sqrt{3})$	$(-\sqrt{3}, \sqrt{3})$	$(\sqrt{3}, \infty)$
$f(x)$ is	$-$	$+$	$-$	$+$
The graph of f is	below the x-axis	above the x-axis	below the x-axis	above the x-axis
The remainder fraction $\dfrac{1}{x+2}$ is	$-$	$+$	$+$	$+$
The graph of f is	below the asymptote $y = x - 2$	above the asymptote $y = x - 2$	above the asymptote $y = x - 2$	above the asymptote $y = x - 2$

STEP 6: Using the information from Steps 1–5, we sketch the graph of Fig. 3.22. ∎

As seen in Step 2 of Example 2, the graph of a rational function will have a nonvertical asymptote when the degree of the numerator is greater than or equal to the degree of the denominator. This nonvertical asymptote may be a line or a higher-degree polynomial curve. When the nonvertical asymptote is a line with nonzero slope, the asymptote is called a **slant or oblique asymptote**.

We find nonvertical asymptotes by dividing the denominator of our rational function into the numerator and writing the function as the quotient plus the remainder fraction:

$$f(x) = \text{(the nonvertical asymptote)} + \text{(remainder fraction)}$$

The quotient of that division process is the asymptote. The intervals in which the remainder fraction is positive and negative indicate where the graph of the function is above and below the asymptote. When the remainder is zero, the graph will intersect the asymptote, as with nonvertical asymptotes.

EXAMPLE 3   Sketch the graph of

$$g(x) = \frac{x^3 - x^2 - 8x + 8}{2x^2 + 2x - 12}$$

Solution   We begin by factoring $g(x)$ completely:

$$g(x) = \frac{x^3 - x^2 - 8x + 8}{2x^2 + 2x - 12} = \frac{(x - 1)(x + 2\sqrt{2})(x - 2\sqrt{2})}{2(x - 2)(x + 3)}$$

Here are the six steps used to sketch the graph of $g(x)$.

STEP 1: Since there are no factors common to the numerator and denominator of $g(x)$, the vertical asymptotes are $x = 2$ and $x = -3$.

STEP 2: To find the nonvertical asymptote, we divide:

$$
\require{enclose}
\begin{array}{r}
\frac{1}{2}x - 1 \\
2x^2 + 2x - 12 \enclose{longdiv}{x^3 - x^2 - 8x + 8} \\
\underline{x^3 + x^2 - 6x } \\
-2x^2 - 2x + 8 \\
\underline{-2x^2 - 2x + 12} \\
-4
\end{array}
$$

Thus

$$g(x) = \frac{x^3 - x^2 - 8x + 8}{2x^2 + 2x - 12}$$

$$= \frac{1}{2}x - 1 + \frac{-4}{2(x - 2)(x + 3)}$$

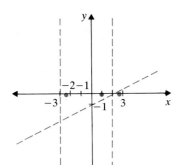

Figure 3.23(a)

The nonvertical asymptote is $y = \frac{1}{2}x - 1$, and the remainder fraction is

$$\frac{-4}{2(x-2)(x+3)}$$

STEP 3: The x-intercept occurs when $g(x) = 0$. Since $g(x) = 0$ when its numerator is zero, the x-intercepts are $x = \pm 2\sqrt{2}$ and $x = 1$. Since $g(0) = \dfrac{8}{(-12)} = -\frac{2}{3}$, the y-intercept is $-\frac{2}{3}$.

STEP 4: The remainder fraction cannot equal zero, so the graph of f does not intersect the asymptote $y = \frac{1}{2}x - 1$. The sign chart for the remainder fraction is

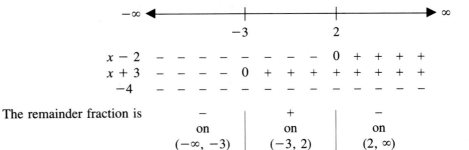

These results indicate that the graph of g is below the asymptote $y = \frac{1}{2}x - 1$ when $x < 3$ and when $x > 2$. Similarly, the graph of g lies above the asymptote when $-3 < x < 2$.

STEP 5: The sign chart for $g(x)$ is

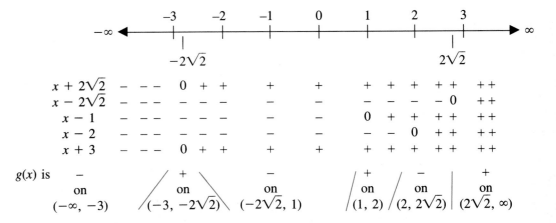

The results from Steps 4 and 5 are combined in Table 3.7.

TABLE 3.7

Domain Intervals	$(-\infty, -3)$	$(-3, -2\sqrt{2})$	$(-2\sqrt{2}, 1)$	$(1, 2)$	$(2, 2\sqrt{2})$	$(2\sqrt{2}, \infty)$
$g(x)$ is	$-$	$+$	$-$	$+$	$-$	$+$
The graph of g is	below the x-axis	above the x-axis	below the x-axis	above the x-axis	below the x-axis	above the x-axis
The remainder fraction $\dfrac{-4}{2(x-2)(x+3)}$ is	$-$	$+$	$+$	$+$	$-$	$-$
The graph of g is	below the asymptote	above the asymptote	above the asymptote	above the asymptote	below the asymptote	above the asymptote

STEP 6: We sketch the graph of g using Steps 1–5 (see Fig. 3.23).

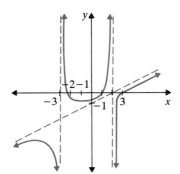

Figure 3.23(b)

There are rational functions $f(x) = \dfrac{g(x)}{h(x)}$ in which the degree of g is greater than the degree of h by more than 1. These rational functions have interesting nonvertical asymptotes as in the next example.

EXAMPLE 4   Sketch the graph of

$$h(x) = \frac{x^4 + 1}{10x^2}$$

Solution We will use the six steps for graphing rational functions.
STEP 1: Since the denominator is $10x^2$, the vertical asymptote is $x = 0$.

STEP 2: Writing $h(x)$ in quotient form we have

$$h(x) = \frac{x^4 + 1}{10x^2} = \frac{1}{10}x^2 + \frac{1}{10x^2}$$

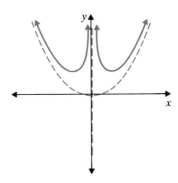

Figure 3.24

Thus the nonvertical asymptote is $y = \frac{1}{10}x^2$.

STEP 3: There are no zeros for $h(x)$, since $x^4 + 1$ has no real zeros.

STEP 4: The remainder $\frac{1}{10}x^2$ is always positive. Thus the value of $h(x)$ is always above the nonvertical asymptote $y = \frac{1}{10}x^2$.

STEP 5: Since $x^4 + 1 > 0$ and $10x^2 > 0$ for $x \neq 0$, we can conclude that $h(x) > 0$ with a vertical asymptote at $x = 0$.

STEP 6: Using the above information, we sketch the graph of $h(x)$ as shown in Fig. 3.24. ■

Exercises 3.7

Find the equations of all asymptotes for the rational functions in the following graphs.

1.

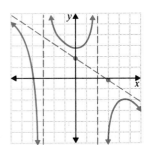

2.

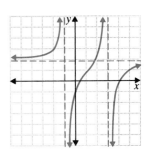

3.

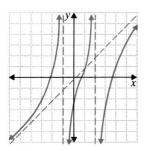

4.

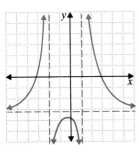

Identify all asymptotes and then sketch the graphs of each rational function.

5. $g(x) = \dfrac{3x^2}{(x^2 - 1)(x - 3)}$

6. $g(x) = \dfrac{2x^2}{(x + 1)(x^2 - 9)}$

7. $f(x) = \dfrac{2x^2 - 4x + 2}{x^2 - 9}$

8. $f(x) = \dfrac{2x^2 + 4x + 2}{x^2 - 4}$

9. $h(x) = \dfrac{(x - 1)(x + 1)}{(x^2 - 4)(x^2 - 1)}$

10. $h(x) = \dfrac{(x - 2)(x + 2)}{(x^2 - 4)(x - 3)}$

11. $g(x) = \dfrac{2x^2 + 2x - 12}{x^2 - 1}$

12. $g(x) = \dfrac{2x^2 - 2}{x^2 - x - 6}$

13. $f(x) = \dfrac{x^2}{x - 1}$

14. $f(x) = \dfrac{x^2}{x - 2}$

15. $g(x) = \dfrac{x^2 - 2x + 1}{x + 2}$

16. $g(x) = \dfrac{x^2 + 4x + 4}{x - 1}$

17. $f(x) = \dfrac{x^3 - x^2 - 2x + 2}{3(x + 1)(x - 2)}$

18. $f(x) = \dfrac{x^3 - 2x^2 - 5x + 10}{2(x + 2)(x - 3)}$

19. $g(x) = \dfrac{(x^2 - 1)(x - 2)}{x^2 + 2x}$

20. $g(x) = \dfrac{(x^2 - 4)(x + 1)}{x^2 - x}$

21. $f(x) = \dfrac{x^4 + 1}{10x^2}$

22. $f(x) = \dfrac{x^3 + 3}{10x}$

23. $g(x) = \dfrac{x^5 - 3}{20x^2}$

24. $g(x) = \dfrac{x^6 - 1}{20x^3}$

25. $k(x) = \dfrac{x^4}{3x^2 - 9}$

26. $k(x) = \dfrac{x^4 - 3x^2}{x^2 - 4}$

Ohm's Law for energy cells in series is $I = \dfrac{nE}{R_0 + nb}$, where I is the current, n is the number of cells each having an electromotive force E and an internal force of b, and R_0 is the external resistance (Fig. 3.25).

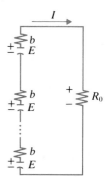

Figure 3.25

27. Sketch the graph of the current I as a function of the external resistance R_0 when n, b, and E are constants and $I > 0$. Describe the value of the current as the value of the external resistance increases.

28. Sketch the graph of the current I as a function of the number of cells n when b, E, and R_0 are constants and $I > 0$. Describe the value of the current as the value of the energy cells increases.

29. Dr. Stork's Sports Cards manufactures baseball and football trading cards. The cards can be produced according to the supply function

$$p_s(x) = \frac{5x}{x + 1}$$

and sold according to the demand function

$$p_d(x) = \frac{2x + 36}{x + 3}$$

where p_s and p_d are the price per card in cents and x is the number of cards in units of 100,000. On the basis of these equations,
a) find the largest price anticipated by the supply function for large production numbers,
b) find the smallest price anticipated by the demand function for a large availability of cards, and
c) sketch the graphs of the supply and demand function. (These graphs are called the supply and demand curves.)

30. The rational function $f(x) = \dfrac{x^3 + 1}{x - 1}$ has an asymptote of $x = 1$. What is the other asymptote for $f(x)$?

31. The rational function $g(x) = \dfrac{x^3 + 1}{x + 2}$ has an asymptote of $x = -2$. What is the other asymptote for $g(x)$?

32. Find all three asymptotes for $h(x) = \dfrac{x^5 - 4x^3 + 5}{x^2 - 4}$.

33. Find all three asymptotes for $k(x) = \dfrac{x^5 - x^3 + 2}{x^2 - 1}$.

CHAPTER SUMMARY AND REVIEW

KEY TERMS

Polynomial function

Zero of a function

The division algorithm

The remainder theorem

The factor theorem

Synthetic division

Intermediate value theorem

The fundamental theorem of algebra

The number of polynomial zeros

Descartes' rule of signs

Upper and lower bounds for polynomial zeros

The rational zero theorem

Complex zero pairs

Irrational zero pairs

Holes of rational function

Vertical asymptotes of rational functions

Horizontal asymptote of rational functions

Nonvertical asymptotes of rational functions

CHAPTER EXERCISES

Sketch the graphs of the following polynomial functions.

1. $f(x) = x^3 - 16x$

2. $g(x) = x^3 - 4x^2 - 3x + 12$

3. $f(x) = (x + 1)^2(x - 2)^2$ **4.** $g(x) = x^3 - 1$

5. $g(x) = x^2 + 3x - 4$ **6.** $g(x) = x^2 - 9$

7. $h(x) = (x^2 - 5x + 6)(x - 2)$

8. $h(x) = (x^2 - 5x - 6)(x - 3)$

9. $f(x) = (x^2 + x - 2)(x^2 - 4)$

10. $f(x) = (x^2 - 5x + 6)(x^2 - x - 6)$

11. Find a third-degree polynomial with zeros of 1, 2, and -3.

12. Write the quotient and remainder of $x^4 - x^3 + x + 5$ and $x^2 - 3$ using both forms of the division algorithm.

Use synthetic division to find the quotient of $f(x)$ and $g(x)$.

13. $f(x) = x^3 - 3x^2 + x - 5$
    $g(x) = x - 2$

14. $f(x) = x^3 + x^2 - 3x - 3$
    $g(x) = x + 3$

Use synthetic division to find the value of $f(d)$.

15. $f(x) = x^3 + 5x^2 - x + 3, d = -3$

16. $f(x) = x^3 + x^2 + x + 1, d = 1$

Find integer upper and lower bounds for the zeros of the following polynomials.

17. $f(x) = x^3 - 2x^2 + 3x - 5$

18. $f(x) = 2x^3 - x^2 + x - 5$

19. Find all possible rational zeros for $f(x) = 5x^3 - 16x^2 + 2x - 10$.

20. Find all zeros for $f(x) = x^4 - 4x^3 - 8x^2 + 12x + 15$.

Find the asymptotes for the following graphs of rational functions.

21.

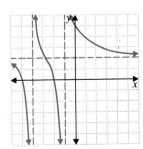

22.

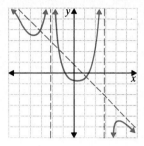

23.

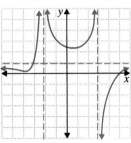

24.

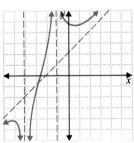

Find all asymptotes for each of the following rational functions.

25. $f(x) = \dfrac{x^2 + 5x + 6}{(x^2 - 1)(x - 3)}$ **26.** $f(x) = \dfrac{x^2 - 7x + 12}{(x^2 - 4)(x + 1)}$

27. $h(x) = \dfrac{x^2 + 3x + 1}{x - 2}$ **28.** $h(x) = \dfrac{x^2 - 2x - 3}{x - 1}$

29. $h(x) = \dfrac{x^3 + x^2 + 1}{x - 1}$ **30.** $h(x) = \dfrac{x^3 - x^2 + 1}{x + 1}$

31. $f(x) = \dfrac{2}{x^2 + 2}$ **32.** $f(x) = \dfrac{3}{x^2 + 3}$

Graph the following rational functions.

33. $f(x) = \dfrac{x - 1}{x^2 - 4}$ **34.** $g(x) = \dfrac{2x^2}{(x^2 - 1)(x - 3)}$

35. $f(x) = \dfrac{x^2 - 6x + 9}{x - 2}$ **36.** $g(x) = \dfrac{x^5 + x^3 - 2}{x^3}$

37. Write a fifth-degree polynomial with rational coefficients and zeros of -3, $-3i$, and $\sqrt{3}$.

38. Bowen's Boats produces a small recreational sailboat. They can produce up to 150 boats a month with a cost function of $C(x) = 400 - 0.001x^2$ and a revenue function of $R(X) = 0.1x^3 - 21.001x^2 + 1100x + 400$. Find the cost, revenue, and profit when 62 boats are produced and when 132 boats are produced.

Kennedy's Computer Products can produce up to 4000 computer disks a month. Under current conditions the monthly profit function is $P(x) = -0.01x^2 + 40x - 100$, and the monthly cost function is $C(x) = 0.1x + 100$, where x is the number of disks produced.

39. Find the profit $P(x)$, the cost $C(x)$, and the revenue $R(x)$ when $x = 100$ disks and when $x = 3500$ disks.

40. Find the profit $P(x)$, the cost $C(x)$, and the revenue $R(x)$ when $x = 2500$ disks and when $x = 4000$ disks.

41. When two condensers with capacity of C_1 and C_2 are used in series (Fig. 3.26), the electrical capacity C of the combination is found by

$$\frac{1}{C} = \frac{1}{C_1} + \frac{1}{C_2}$$

If C_1 is a constant value, sketch the graph of C as a function of C_2 for $C_2 > 0$.

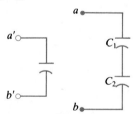

Figure 3.26

42. A 12×24 inch sheet of metal is used to form a candy box with an open top by cutting squares with sides of x out of each corner (Fig. 3.27). Find the size of the corner cut such that the volume of the box is not less than 324 in.^{3}

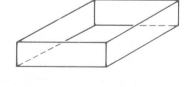

Figure 3.27

4

Exponential and Logarithmic Functions

METHADONE, AN ALTERNATIVE TO DRUG ADDICTION

Whether the addiction is to drugs, alcohol, or tobacco, the problem is self-destructive behavior. One treatment that has succeeded in restoring heroin addicts to health is maintenance on methadone. To better understand this treatment, we will consider the concentration of morphine and methadone in the bloodstream as a function of time.

The concentration of narcotics in the bloodstream and its physiological effects has some dramatic differences for nonaddicts and addicts. A graph of the concentration of morphine in the bloodstream as a function of time after an intravenous injection shows that the nonaddict benefits longer from the narcotic effects without withdrawal symptoms. The nonaddict receives these narcotic effects for up to 2 hours, while the addict receives these effects for at most 20 to 30 minutes. The addict receives narcotic effects only when the bloodstream concentration is above 1 microgram per milliliter (μg/ml), while the nonaddict continues to receive narcotic effects for a concentration above 0.05 μg/ml. The addict finds it harder and

harder to benefit from the narcotic effects. It is also interesting to notice that the concentration at which the nonaddict loses all narcotic effects is the same concentration at which the addict begins experiencing withdrawal symptoms, 0.05 μg/ml.

The oral ingestion of methadone provides narcotic effects for the nonaddict, while the addict is suspended somewhere between narcotic effects and withdrawal symptoms. The concentration of methadone does not drop off with time as quickly as that of a narcotic such as morphine. This factor is especially important to the recovering addict. Studies by Mary Jeanne Kreek and Vincent P. Dole showed that the correct daily oral dose of methadone maintains the former addict between narcotic effects and withdrawal symptoms. On this daily dosage the patient is unaffected by the sedative effects and analgesic powers of methadone and is able to function with normal vigilance and coordination. Another important result of methadone treatment is that it causes heroin injections to be ineffective, not providing

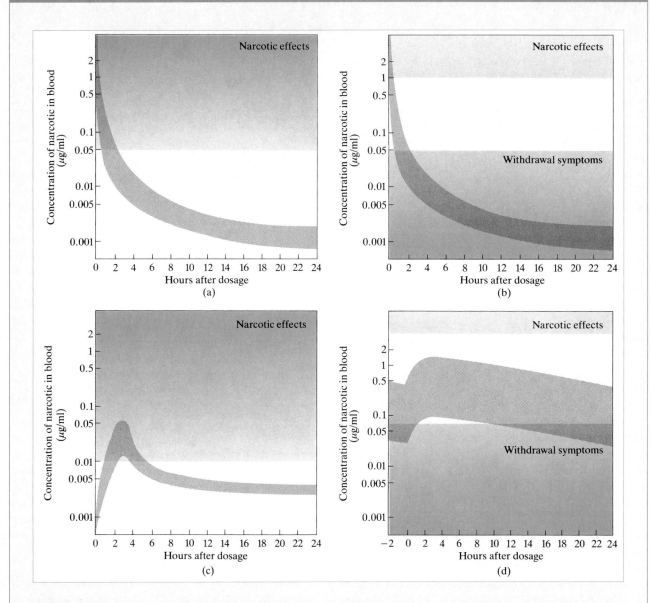

Concentration of morphine in the blood and its functional effects following an intravenous injection of 10 milligrams are shown for a nonaddict (a) and an addict (b). The concentration is approximately the same in each case, but the functional effects are quite different because of the addict's tolerance and physical dependence. The period between the disappearance of the narcotic effect and the onset of withdrawal symptoms is short for the addict. The body rapidly produces morphine from heroin after an injection. Effect and concentration of methadone are shown for naive subjects (c) and methadone-maintenance patients (d) after an oral dose. In the naive subject taking a 10-milligram dose the gradual absorption of the drug from the intestinal tract and the buffering effect provided by absorption in the liver before the methadone enters the bloodstream prevent the concentration of the drug in the blood from rising rapidly. This smooth course of action makes methadone useful for relieving pain in medical and surgical patients. In maintenance patients stabilized on an adequate daily dose (usually 50 to 100 milligrams) the concentration of methadone in the blood is kept at all times above the threshold for withdrawal symptoms. (In a few cases a higher dose may be required if the elimination of methadone is unusually fast.) The peak concentration in a stabilized patient remains well below the threshold for narcotic effects. A maintenance patient who is stabilized on a correct daily dose of methadone is functionally normal, protected from narcotic effects by his pharmacological tolerance of the drug and from withdrawal symptoms by the constant presence of methadone in the bloodstream.

any narcotic effects. Thus the methadone treatment blocks the narcotic effects of additional narcotics, and cravings for heroin can disappear.

The curves in these diagrams are graphs of a new function that we will study in this chapter. This new function is called the exponential function.

SOURCE: Vincent P. Dole, "Addictive Behavior," *Scientific American*, December 1980, p. 142, 144. Reprinted by permission.

The graphs in the example on methadone treatment are examples of the exponential function, which we will study in this chapter. We will also study another new function related to the exponential function and called the logarithmic function. We will investigate several other applications of both the exponential and logarithmic functions in this chapter.

Not many years ago, before the pocket calculator, exponential and logarithmic functions were useful in completing the basic arithmetic calculations of multiplication, division, roots, and powers such as y^x. Now we can easily make all of these calculations with a pocket calculator. In addition to this historical use, exponential functions, logarithmic functions, and their properties have always been used in calculus. In this chapter we will define and graph the exponential and logarithmic functions. We will also develop their related properties, which are used in calculations and in simplifying certain exponential and logarithmic expressions.

4.1 Exponential Functions

In this section we will define the exponential function, sketch its graph, and solve applications involving exponential functions. We start with the definition of the exponential function.

Definition | For $a > 0$, $a \neq 1$, and any real number x, the **exponential function** f with base a is defined by

$$f(x) = a^x$$

In Chapter 1 we worked with the properties of exponents for integer and rational number exponents. It seems like a big step to now allow an exponent to be any real number. We know that $f(x) = a^x$ is a function, since for every value of x, $f(x) = a^x$ is a unique number. To better understand $f(x) = a^x$, where x could be any real number (rational or irrational), consider the graph of $f(x) = 2^x$ in the next example.

EXAMPLE 1   Sketch the graph of $f(x) = 2^x$ where **(a)** x is an integer and **(b)** x is a rational number.

Solution   We will make a chart of representative ordered pairs for each part of this problem and then graph the results.

(a) For $f(x) = 2^x$ for any integer x we have

x	$f(x) = 2^x$
-3	$f(-3) = 2^{-3} = \left(\dfrac{1}{2}\right)^3 = \dfrac{1}{8}$
-2	$f(-2) = 2^{-2} = \left(\dfrac{1}{2}\right)^2 = \dfrac{1}{4}$
-1	$f(-1) = 2^{-1} = \left(\dfrac{1}{2}\right)^1 = \dfrac{1}{2}$
1	$f(1) = 2^1 = 2$
2	$f(2) = 2^2 = 4$
3	$f(3) = 2^3 = 8$

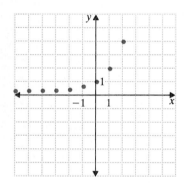

Figure 4.1

The graph is shown in Fig. 4.1.

(b) For $f(x) = 2^x$ where x is a rational number, realizing that the chart represents only a small sample of the infinite number of possible rational exponents, we have

x	$f(x) = 2^x$
$\dfrac{1}{4}$	$f\left(\dfrac{1}{4}\right) = 2^{1/4} = \sqrt[4]{2} \approx 1.1892 \ldots$
$\dfrac{1}{3}$	$f\left(\dfrac{1}{3}\right) = 2^{1/3} = \sqrt[3]{2} \approx 1.2599 \ldots$
$\dfrac{1}{2}$	$f\left(\dfrac{1}{2}\right) = 2^{1/2} = \sqrt{2} \approx 1.4142 \ldots$
$\dfrac{2}{3}$	$f\left(\dfrac{2}{3}\right) = 2^{2/3} = \sqrt[3]{2^2} \approx 1.5874 \ldots$
$\dfrac{3}{4}$	$f\left(\dfrac{3}{4}\right) = 2^{3/4} = \sqrt[4]{2^3} \approx 1.6817 \ldots$

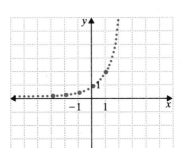

We now have a small number of the ordered pairs for $f(x) = 2^x$ with rational exponents in the interval $[0, 1]$. Notice that the values calculated indicate that $f(x) = 2^x$ is an increasing function on this interval. It is also true that f is an increasing function in all other domain intervals. We can now sketch the graph seen in Fig. 4.2. The graph in Fig. 4.2 has a domain that includes all rational numbers, yet the domain is still missing an infinite number of irrational numbers. ◾

Figure 4.2

Along with Example 1 we will consider $f(x) = 2^x$ when x is an irrational number. If $x = \pi$, then π is between two rational numbers, $3.141 < \pi < 3.142$. Thus it is consistent with the above example, since $f(x) = 2^x$ is increasing, to expect the range value for $f(\pi)$ to be between the function range values of these two rational numbers, $f(3.141) < f(\pi) < f(3.142)$. Since the values for $f(3.141)$ and $f(3.142)$ are both on the graph in Fig. 4.2, the value for $f(\pi)$ should be a blank space between the points $(3.141, f(3.141))$ and $(3.142, f(3.142))$. The values in the following chart are found with the aid of a calculator (using the y^x key) and confirm the above conclusions.

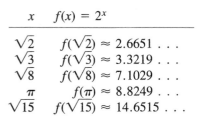

x	$f(x) = 2^x$
$\sqrt{2}$	$f(\sqrt{2}) \approx 2.6651 \ldots$
$\sqrt{3}$	$f(\sqrt{3}) \approx 3.3219 \ldots$
$\sqrt{8}$	$f(\sqrt{8}) \approx 7.1029 \ldots$
π	$f(\pi) \approx 8.8249 \ldots$
$\sqrt{15}$	$f(\sqrt{15}) \approx 14.6515 \ldots$

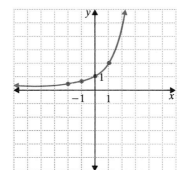

We observe that $\sqrt{2} < \sqrt{3} < \sqrt{8} < \pi < \sqrt{15}$, and likewise it follows that $f(\sqrt{2}) < f(\sqrt{3}) < f(\sqrt{8}) < f(\pi) < f(\sqrt{15})$.

The graph in Fig. 4.3 demonstrates the characteristics of the exponential function. The exponential function $f(x) = 2^x$ is increasing and positive for all x ($2^x > 0$), has a horizontal asymptote to the left (the line $y = 0$), and has a y-intercept of 1 ($f(0) = 2^0 = 1$). Adding the irrational numbers to the domain of $f(x) = a^x$ allows Fig. 4.2 to be changed from a dotted curve to the solid curve of Fig. 4.3.

Figure 4.3

EXAMPLE 2

Graph the exponential function $f(x) = \left(\frac{1}{3}\right)^x$ for any real number x.

Solution From the characteristics described above, the graph of $f(x) = \left(\frac{1}{3}\right)^x$ can be sketched by connecting selected ordered pairs of this function. The domain

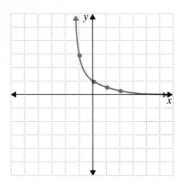

Figure 4.4

values can be integers, rational numbers, or irrational numbers. We will use integer domain values, since they are easier to calculate.

x	$f(x) = \left(\dfrac{1}{3}\right)^x = 3^{-x}$
-2	$f(-2) = \left(\dfrac{1}{3}\right)^{-2} = 3^2 = 9$
-1	$f(-1) = \left(\dfrac{1}{3}\right)^{-1} = 3$
0	$f(0) = \left(\dfrac{1}{3}\right)^0 = 1$
1	$f(1) = \dfrac{1}{3}$
2	$f(2) = \left(\dfrac{1}{3}\right)^2 = \dfrac{1}{9}$

The graph is shown in Fig. 4.4. ∎

Note that the exponential function $f(x) = \left(\frac{1}{3}\right)^x$ is decreasing and positive for all x and has a horizontal asymptote to the right (the line $y = 0$), and $f(0) = 1$. In general, we have the following graphing techniques.

The graph of the rational function $f(x) = a^x$, for $a \neq 0$ and $a \neq 1$,

(a) is positive, $f(x) = a^x > 0$,
(b) has a one-directional horizontal asymptote, the line $y = 0$,
(c) has a y-intercept of 1, since $f(0) = a^0 = 1$, and
(d) is either increasing for $1 < a$ or decreasing for $0 < a < 1$.

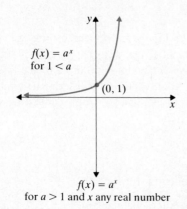

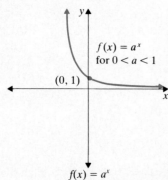

The letter e is a famous and most useful base for an exponential function. The exponential base e was named by the renowned mathematician Leonhard Euler (1707–1783). The value of the number e is the result of a calculus problem. In calculus it can be shown that e is the value of $\left[1 + \left(\dfrac{1}{x}\right)\right]^x$ as x approaches $+\infty$. Thus $y = e$ is a one-directional horizontal asymptote for $\left[1 + \left(\dfrac{1}{x}\right)\right]^x$ with a value of

$$e = 2.71828182845904523536 \ldots$$

Although the first nine digits in this decimal representation could indicate a rational number, we can see from the rest of this decimal representation that there is no obvious repeating pattern in the first 20 digits. The conclusion that the decimal representation does not repeat is correct, and e is an irrational number.

The exponential function is very helpful in calculating growth. We will consider a scientifically planted forest that starts with a volume of b_0 board feet and maintains an annual percent increase. We will let r be the decimal or fractional equivalent for the annual percent increase. Thus 21% would be written as $r = 0.21$ or $r = \frac{21}{100}$. We will now find a formula for $b(n)$, the board feet of lumber after n years of growth.

At the end of any given year the volume of lumber is the sum of the volume of lumber at the beginning of the year and r times that volume.

$$\text{After one year:} \quad b(1) = b_0 + rb_0$$
$$= b_0(1 + r)$$
$$\text{After two years:} \quad b(2) = [b_0(1 + r)] + r[b_0(1 + r)]$$
$$= [b_0(1 + r)](1 + r)$$
$$= b_0(1 + r)^2$$
$$\text{After three years:} \quad b(3) = [b_0(1 + r)^2] + r[b_0(1 + r)^2]$$
$$= [b_0(1 + r)^2] + r[b_0(1 + r)^2]$$
$$= b_0(1 + r)^3$$

In general, after n years we have

$$b(n) = b_0(1 + r)^n$$

Therefore the volume of lumber present after n years is $b(n) = b_0(1 + r)^n$, where r is the decimal or fraction equivalent of the annual percent increase in volume and b_0 is the volume of lumber at the starting time ($n = 0$).

EXAMPLE 3   If a tract of land contained 30,000 cubic feet of lumber in 1988 and is expected to show a 21.5% annual increase in volume, how much lumber will be available in 1993?

Solution   For $b(n) = b_0(1 + r)^n$, in this problem $b_0 = 30{,}000$, $n = 1993 - 1988 = 5$, and for a 21.5% annual growth rate $r = 0.215$. Thus

$$b(5) = 30{,}000(1.215)^5 \approx 79{,}433.307 \ldots$$

Therefore in 1993 there should be approximately 79,433 cubic feet of lumber. ■

The growth of trees and their production of lumber is similar to the interest formula for compounded interest. Let A equal the amount of money in an account when A_0 is invested at a rate r (r is the decimal or fraction equivalent for the percent of interest) and k the number of times the money is compounded in 1 year.

After one interest period: $A = A_0 + A_0 \dfrac{r}{k}$

$$= A_0\left(1 + \frac{r}{k}\right)$$

After two interest periods: $A = A_0\left(1 + \dfrac{r}{k}\right) + A_0\left(1 + \dfrac{r}{k}\right)\dfrac{r}{k}$

$$= A_0\left(1 + \frac{r}{k}\right)\left(1 + \frac{r}{k}\right)$$

$$= A_0\left(1 + \frac{r}{k}\right)^2$$

After three interest periods: $A = A_0\left(1 + \dfrac{r}{k}\right)^2 + A_0\left(1 + \dfrac{r}{k}\right)^2\left(\dfrac{r}{k}\right)$

$$= A_0\left(1 + \frac{r}{k}\right)^2\left(1 + \frac{r}{k}\right)$$

$$= A_0\left(1 + \frac{r}{k}\right)^3$$

Following this pattern after n interest periods, we have

$$A = A_0\left(1 + \frac{r}{k}\right)^n$$

If k is the number of times interest is compounded in one year and n is the total number of times interest is compounded in t years, then $n = tk$.

Therefore

$$A(t) = A_0\left(1 + \frac{r}{k}\right)^{kt}$$

is the formula for compounded interest when A_0 is the original amount invested, r is the annual rate of interest, k is the number of times interest is compounded a year, and t is the time in years for which the interest is earned.

EXAMPLE 4   Find the amount of money available when $10,000 is invested for 5 years and the interest of 8% is compounded **(a)** quarterly and **(b)** daily.

Solution   Use $A(t) = A_0\left(1 + \dfrac{r}{k}\right)^{kt}$ with $A_0 = 10,000$, $t = 5$, and $r = 0.08$.

(a) When $k = 4$,

$$A(5) = 10,000\left(1 + \frac{0.08}{4}\right)^{4 \cdot 5}$$

$$= 10,000(1.02)^{20}$$

$$= \$14,859.47$$

(b) For a business year of $k = 360$ days,

$$A(5) = 10,000\left(1 + \frac{0.08}{360}\right)^{360 \cdot 5}$$

$$= 10,000(1.000222)^{1800}$$

$$= \$14,917.58$$

We can change the formula $A(t) = A_0\left(1 + \dfrac{r}{k}\right)^{kt}$ to another form by substituting with $x = \dfrac{k}{r}$. Then $A(t) = A_0\left(1 + \dfrac{1}{x}\right)^{xrt}$. As we increase the number of times the interest is compounded a year (n approaches ∞), the value of $\left[1 + \left(\dfrac{1}{x}\right)\right]^x$ will approach e as mentioned earlier. This is called **continuous interest** and provides the upper bound for the principal and earned interest on A_0 at a rate r for t years:

$$A_c(t) = A_0 e^{rt}$$

EXAMPLE 5   What amount is a limit for the amount of money that could be earned, including the principal, if $10,000 were invested at 8% for 5 years (continuous interest)?

Solution   For continuous interest, $A_c(t) = A_0 e^{rt}$. Then

$$A_c(5) = 10,000 e^{(0.08)5}$$

$$= 10,000 e^{0.4}$$

$$= \$14,918.24$$

It is interesting to note that this is only $0.66 more than would be earned with daily interest over a 5-year period as calculated in Example 4.

EXAMPLE 6

An electrical circuit consists of a series of a battery of V volts, a resistor of R ohms, a capacitor of C farads, and a switch S (Fig. 4.5). The capacitor is initially uncharged, and the switch is closed at $t = 0$. The current $i(t)$ in the circuit and the voltage $v(t)$ across the capacitor are given as a function of time by

$$i(t) = \left(\frac{V}{R}\right)e^{-t/RC} \qquad \text{and} \qquad v(t) = V(1 - e^{-t/RC})$$

where $i(t)$ is measured in amperes and $v(t)$ is measured in volts. Sketch a graph of $i(t)$ and $v(t)$ for $t \geq 0$ when $V = 100$, $R = 10$, and $C = \frac{1}{10}$.

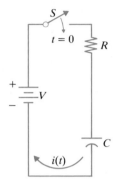

Figure 4.5

Solution   For $V = 100$, $R = 10$, and $C = \frac{1}{10}$ from the formulas above, we have $i(t) = 10e^{-t}$ and $v(t) = 100(1 - e^{-t})$.

(a) We will use the graphs of e^t and e^{-t} to graph $i(t) = 10e^{-t}$. The graph of e^t contains $(0, 1)$ and is an increasing function (the solid curve in Fig. 4.6(a)). We can then graph e^{-t} as the reciprocal of e^t (the dashed curve in Fig. 4.6(a)). Thus e^{-t} contains $(0, 1)$ and is decreasing, and $0 < e^{-t} \leq 1$. Now we can graph $i(t) = 10e^{-t}$ as the multiple of e^{-t} as seen in Fig. 4.6(b).

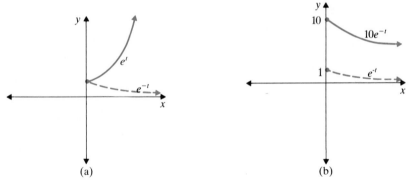

Figure 4.6

(b) To graph $y = 100(1 - e^{-t})$, we start with the graph of e^{-t} and graph its negative $-e^{-t}$ (Fig. 4.7a). Then we graph $1 - e^{-t}$ as the vertical translation of $-e^{-t}$ (Fig. 4.7b). Then we graph $v(t) = 100(1 - e^{-t})$ as the multiple of $1 - e^{-t}$.

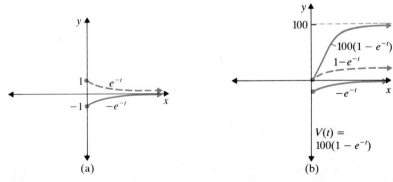

Figure 4.7

Exercises 4.1

Sketch the graph of each of the following functions.

1. $f(x) = 3^x$

2. $f(x) = 4^x$

3. $g(x) = -3^x$

4. $g(x) = -2^x$

5. $h(x) = 3^{-x}$

6. $h(x) = 4^{-x}$

7. $f(x) = \left(\frac{2}{3}\right)^x$

8. $f(x) = \left(\frac{3}{4}\right)^x$

9. $g(x) = e^x$

10. $g(x) = e^{-x}$

11. $h(x) = 1 + e^x$

12. $h(x) = 1 + e^{-x}$

13. $f(x) = 3 + e^{-x}$

14. $f(x) = 2 + e^x$

15. $g(x) = e^{x+1}$

16. $g(x) = e^{x-1}$

17. $f(x) = 2^{2-x}$

18. $f(x) = 2^{2+x}$

19. $g(x) = 2 + 3^x$

20. $g(x) = 3 + 2^x$

21. $h(x) = 3 - 2^{-x}$

22. $h(x) = 2 + 3^{-x}$

23. If a forest had 15,000 cubic feet of lumber in 1988 and its volume increased by 18% every year, how many cubic feet of lumber should be available in 1995?

24. If a forest had 8000 cubic feet of lumber in 1988 and its volume increased by 27.5% every year, how many cubic feet of lumber should be available in 1997?

25. If $4000 were invested at an annual rate of 8%, how much money would be in the account at the end of 9 years if the interest was compounded
   (a) quarterly?   **(b)** daily?   **(c)** continuously?

26. If $7000 were invested at an annual rate of 9%, how much money would be in the account at the end of 4 years if the interest was compounded
   (a) quarterly?   **(b)** daily?   **(c)** continuously?

27. The number of bacteria in a culture doubles every day. Find a function that describes the number of bacteria present as a function of time in days if there are 2000 bacteria present at time zero.

28. The number of bacteria in a culture triples every two days. Find a function that describes the number of bacteria present as a function of time in days if there are 500 bacteria present at time zero.

29. A student has a credit card bill of $5000. Each month the credit card company adds 1.4% of the unpaid balance to his bill. The student must make a minimum payment of 5% of the unpaid balance. If the student makes only the minimum payment each month and makes no new charges, find a function $A(n)$ that describes the balance owed as a function of time in months. Use your calculator to find $A(12)$, $A(24)$, and $A(36)$.

30. A student has a credit card bill of $2000. Each month the credit card company adds 1.8% of the unpaid balance to her bill. The student must make a minimum payment of 5% of the unpaid balance. If the student makes only the minimum payment each month and makes no new charges, find a function $A(n)$ that describes the balance owed as a function of time in months. Use your calculator to find $A(12)$, $A(24)$, and $A(36)$.

31. A population of 20 foxes on an uninhabited island increases by 25% every year for the first 20 years. After 20 years the fox population is too large, destroys important elements of the food chain, and decreases by 30% every year for the next 30 years. Find a piecewise-defined function that describes the size of the population as a function of time in years for this 50-year period. What are the largest and smallest fox population, and when do they occur?

32. A contagious disease is identified in a country when it is contracted by N_0 people. Unchecked, the number of cases increases by 60% each year. After 7 years a medical cure is discovered, reducing the number of cases by 30% per year. Find a piecewise-defined function that describes the number of cases as a function of time. What is the largest number of cases under these conditions, and when does this number occur?

Find **(a)** $(f \circ g)(x)$ and **(b)** $(g \circ f)(x)$ for each of the following pairs of functions.

33. $f(x) = e^x$
     $g(x) = x^2 - 2x + 3$

34. $f(x) = e^x$
     $g(x) = x^2 - 1$

35. $f(x) = |3x|$
     $g(x) = e^x$

36. $f(x) = \sqrt{x}$
     $g(x) = e^{2x}$

37. $f(x) = x(x - 3)^2$
     $g(x) = e^x$

38. $f(x) = (x - 1)^2$
     $g(x) = e^x$

Find functions f and g for the following composite functions.

39. $(f \circ g)(x) = 5e^{2x} - e^x + 1$ **40.** $(f \circ g)(x) = e^{2x} - 7e^x + 1$

41. $(f \circ g)(x) = e^{-x} + e^x$ **42.** $(f \circ g)(x) = e^x - e^{-x}$

43. $(g \circ f)(x) = (e^{2x} - 1)^2$ **44.** $(g \circ f)(x) = e^{3x-5}$

Sketch the graph for each of the following functions.

45. $g(x) = 2^{|x|}$ **46.** $g(x) = 3^{|x|}$

47. $h(x) = 3^{|2x|}$ **48.** $h(x) = 2^{|3x|}$

49. $f(x) = \dfrac{2}{1 - e^x}$ **50.** $f(x) = \dfrac{3}{e^x - 1}$

An electrical circuit consists of a series of a battery of V volts, a resistor of R ohms, a capacitor of C farads, and a switch S (Fig. 4.8). The capacitor is initially uncharged, and the switch is closed at $t = 0$. The current $i(t)$ in the circuit and the voltage $v(t)$ across the capacitor are given as a function of time by

$$i(t) = \left(\frac{V}{R}\right)e^{-t/RC} \qquad \text{and} \qquad v(t) = V(1 - e^{-t/RC})$$

where $i(t)$ is measured in amperes and $v(t)$ is measured in volts.

51. Sketch a graph of $i(t)$ and $v(t)$ for $t > 0$ when $V = 100$, $R = 10$, and $C = \frac{1}{20}$ (Fig. 4.8).

52. Sketch a graph of $i(t)$ and $v(t)$ for $t > 0$ when $V = 100$, $R = 10$, and $C = \frac{1}{30}$ (Fig. 4.8).

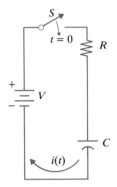

Figure 4.8

53. If $f(x) = a^x$ with $a > 0$ is an increasing function, prove that $a > 1$. (*Hint:* Compare $f(r)$ and $f(r + 1)$.)

54. If $f(x) = a^x$ with $a > 0$ is a decreasing function, prove that $a < 1$. (*Hint:* Compare $f(r)$ and $f(r + 1)$.)

55. The exponential function $f(x) = a^x$ is not defined for $a \leq 0$. What problems are created when $a \leq 0$ in calculating $f(x) = a^x$?

4.2   The Logarithmic Function

In Section 4.1 we defined and sketched the graph of the exponential function $f(x) = a^x$. The graph of $f(x) = a^x$ with $a > 0$ and $a \neq 1$ indicates that the exponential function is a one-to-one function mapping the real numbers onto the positive real numbers. As we saw in Chapter 2, any one-to-one onto function has an inverse function. The exponential function $y = a^x$ has an inverse function of $x = a^y$. Since it is difficult, if not impossible, to solve for y in the equation $x = a^y$, we will define the inverse of the exponential function as follows.

Definition | For $a > 0$, $a \neq 1$, and $x > 0$ the **logarithmic function** of x with base a is

$$y = \log_a x \qquad \text{and is equivalent to} \qquad a^y = x$$

It will be very important for you to be able to change the equation $y = \log_a x$ into $a^y = x$ as well as change the equation $a^y = x$ into $y = \log_a x$. To do this, we notice that a is the base for both the exponential and the logarithmic function.

This change between exponential and logarithmic form is demonstrated by following the arrows in this diagram:

$$\log_a x = y$$
$$a^y = x$$

We will refer to this process as the definition of the logarithmic function. Being able to use the definition of the logarithmic function is most important, as we will see in the examples in this section.

EXAMPLE 1   Write each in logarithmic form, using the definition.
(a) $3^4 = 81$ (b) $5^{-x} = 3$

Solution
(a) By the definition, $3^4 = 81$ is $\log_3 81 = 4$.
(b) By the definition, $5^{-x} = 3$ is $\log_5 3 = -x$.

EXAMPLE 2   Write each in exponential form, using the definition.
(a) $\log_3 9 = 2$,   (b) $\log_a x = r$.

Solution
(a) By the definition, $\log_3 9 = 2$ is $3^2 = 9$.
(b) By the definition, $\log_a x = r$ is $a^r = x$.

We can use logarithmic functions and the corresponding calculator keys to solve different types of functions, as we will see in the examples that follow.

EXAMPLE 3   Chemists use a pH number to determine the acidity or basicity of a substance. The pH number is determined by

$$pH = -\log_{10} [H^+]$$

where $[H^+]$ is the concentration of hydrogen ions measured in moles per liter. A pH between 1 and 7 indicates an acid, and a pH between 7 and 17 indicates a base. The closer the pH is to 7, the closer the substance is to being neutral, not acidic or basic.
(a) Find the pH if the $[H^+]$ of a 0.1% normal ammonia solution is 7.9432×10^{-12} mole per liter.
(b) Find $[H^+]$ if the pH of rhubarb is about 3.15.

Solution   We will use the equation $pH = -\log_{10} [H^+]$, the logarithmic definition, and a calculator to solve these problems.
(a) For a $[H^+]$ of 7.9432×10^{-12} mole per liter we have

$$pH = -\log_{10} [H^+]$$

$$pH = -\log_{10} (7.9432 \times 10^{-12})$$

$$pH \approx 11.1 \quad \textit{(Use the log calculator key)}$$

(This pH number was calculated with the $\log_{10}$ or log calculator key.)

(b) For rhubarb with a pH $= 3.15$ we have

$$\text{pH} = -\log_{10} [\text{H}^+]$$
$$3.15 = -\log_{10} [\text{H}^+]$$
$$-3.15 = \log_{10} [\text{H}^+] \qquad \textit{(Definition of a logarithm)}$$
$$[\text{H}^+] = 10^{-3.15} \qquad \textit{(Use the } 10^x \textit{ or the inv log calculator key)}$$
$$[\text{H}^+] \approx 7.079 \times 10^{-4} \text{ mole per liter} \qquad ■$$

We can also use the definition of the logarithmic function to solve equations containing logarithmic functions, as we will see in the following examples.

EXAMPLE 4   Solve the equation $\log_5 (4x + 1) = 2$.

Solutions By the logarithmic definition, $\log_5 (4x + 1) = 2$ is

$$5^2 = 4x + 1$$
$$25 = 4x + 1$$
$$24 = 4x$$
$$6 = x \qquad \textit{the solution}$$

As stated in the definition, the domain of the logarithmic function must be positive.

In this example the domain $4x + 1$ must be positive. Thus $4x + 1 > 0$ or $x > \frac{1}{4}$. Since the solution is $x = 6 > \frac{1}{4}$, the domain will be positive, and the solution is $x = 6$.   ■

Always check your answer to verify that it makes the domain positive for the logarithmic function in the original problem. The domain of the logarithmic function must be positive.

EXAMPLE 5   Solve the logarithmic equation $\log_2 (x + 5) = 3$.

Solution   By the definition of the logarithmic function,

$$\log_2 (x + 5) = 3$$

is

$$2^3 = (x + 5)$$

Thus

$$8 = x + 5 \qquad \text{and} \qquad x = 3$$

Notice that the domain $x + 5$ is positive when $x + 5 > 0$ or $x > -5$. Since $x = 3$ is greater than -5, the domain is positive, and the solution is $x = 3$.   ■

As we mentioned earlier, the exponential and logarithmic functions are inverse functions. Starting with the logarithmic function $y = \log_a x$, which is equivalent to $x = a^y$, and interchanging the roles of x and y, we have $y = a^x$, the exponential function. Since $f(x) = a^x$ and $f^{-1}(x) = \log_a x$ are inverse functions, we know that $f(f^{-1}(x)) = x$ and $f^{-1}(f(x)) = x$. To demonstrate this result of function composition in exponential and logarithmic form when $f(x) = a^x$ and $f^{-1}(x) = \log_a x$, we have

$$f(f^{-1}(x)) = a^{\log_a x} = x$$

and

$$f^{-1}(f(x)) = \log_a a^x = x$$

As well as being the result of the inverse function relation, these properties can also be verified by the definition of the logarithmic function,

$$a^{\log_a x} = x \qquad \text{and} \qquad \log_a a^x = x$$
$$\log_a x = \log_a x \qquad\qquad\qquad a^x = a^x$$

Therefore $f(f^{-1}(x)) = a^{\log_a x} = x$ and $f^{-1}(f(x)) = \log_a a^x = x$ are both valid results.

We can identify another property if we let $x = 0$ in $\log_a a^x = x$. Then we have

$$\log_a a^0 = 0$$

or

$$\log_a 1 = 0$$

Thus the logarithm of 1 to any base equals zero, and $\log_a 1 = 0$. We can also verify this result using the definition, since $\log_a 1 = 0$ is $a^0 = 1$.

We now have three properties that are results of the inverse function relationship between the exponential and logarithmic functions.

Logarithmic Function Properties

For $a > 0$, $a \neq 1$, and $x > 0$,

(a) $a^{\log_a x} = x$,
(b) $\log_a a^x = x$, and
(c) $\log_a 1 = 0$.

We will now use these logarithmic properties in some examples.

EXAMPLE 6

Simplify each of the following:
(a) $3^{\log_3 5}$ **(b)** $\log_7 5^0$ **(c)** $\log_c c^4$

Solution   Use the logarithmic function properties.
(a) By property (a), $3^{\log_3 5} = 5$, since 3^x and $\log_3 x$ are inverse functions in composition.
(b) By property (c), $\log_7 5^0 = \log_7 1 = 0$.
(c) By property (b), $\log_c c^4 = 4$, since $\log_c x$ and x^4 are inverse functions in composition. ◼

EXAMPLE 7

Solve these equations:
(a) $\log_a a^{2x-1} = x + 1$ **(b)** $a^{\log_a 3x-1} = x + 5$

Solution   We apply the properties of logarithmic functions.
(a) By property (b),

$$\log_a a^{2x-1} = x + 1$$

is

$$2x - 1 = x + 1 \quad \text{\textit{(Since} } \log_a a^{2x-1} = 2x - 1\text{\textit{)}}$$

and

$$x = 2$$

(b) By property (a),

$$a^{\log_a 3x-1} = x + 5$$

is

$$3x - 1 = x + 5 \quad \text{\textit{(Since} } a^{\log_a 3x-1} = 3x - 1\text{\textit{)}}$$

and

$$2x = 6$$

with

$$x = 3$$

Note that the domain $3x - 1$ is positive when $x = 3$. ◼

Exercises 4.2

Write the following problems in logarithmic form.

1. $3^5 = 243$ **2.** $2^7 = 128$

3. $10^{-2} = 0.01$ **4.** $10^{-3} = 0.001$

5. $5^x = 3$ **6.** $3^y = 5$

Write the following problems in exponential form.

7. $\log_2 32 = 5$ **8.** $\log_3 9 = 2$

9. $\log_5 x = 3$ **10.** $\log_4 x = 3$

11. $\log_3 8 = x$ **12.** $\log_3 27 = x$

Chemists use a pH number to determine the acidity or basicity of a substance. The pH number is determined by

$$pH = -\log_{10} [H^+]$$

where $[H^+]$ is the concentration of hydrogen ions measured in moles per liter. A pH between 1 and 7 indicates an acid, and a pH between 7 and 17 indicates a base. The closer the pH is to 7, the closer the substance is to being neutral, not acidic or basic.

13. Find the pH if the $[H^+]$ of a 0.1% normal boric acid solution is 6.3095×10^{-6} moles per liter.

14. Find the pH if the $[H^+]$ of a 0.01% normal hydrochloric acid solution is 7.9432×10^{-2} moles per liter.

15. Find the pH if the $[H^+]$ of cow's milk is between 5.0118×10^{-7} and 2.5118×10^{-7} moles per liter.

16. Find the pH if the $[H^+]$ of cheese is between 1.5848×10^{-5} and 1.661×10^{-3} moles per liter.

17. Find $[H^+]$ if the pH of cherries is about 3.8.

18. Find $[H^+]$ if the pH of corn is about 6.25.

19. Find $[H^+]$ if the pH of wine is between 2.8 and 3.8.

20. Find $[H^+]$ if the pH of soft drinks is between 6.5 and 4.0.

Simplify each of the following expressions.

21. $2^{\log_2 6}$ **22.** $5^{\log_5 4}$

23. $\log_3 3^5$ **24.** $\log_3 1$

25. $\log_5 1$ **26.** $\log_4 4^7$

27. $\log_x x^{y-1}$ **28.** $\log_y y^{1-x}$

29. $x^{\log_x 7}$ **30.** $y^{\log_y 4}$

Solve each of the following equations.

31. $\log_2 x = 4$ **32.** $\log_3 x = 2$

33. $\log_5 x = 3$ **34.** $\log_4 x = 3$

35. $\log_2 (x - 2) = 3$ **36.** $\log_3 (x + 7) = 2$

37. $\log_3 (x^2 - 1) = 2$ **38.** $\log_4 (x^2 + 3) = 3$

39. $\log_2 8 = x$ **40.** $\log_4 16 = x$

41. $\log_6 36 = x$ **42.** $\log_2 32 = x$

43. $\log_3 27 = x - 2$ **44.** $\log_2 16 = x + 5$

45. $\log_{10} 10^{x^2-1} = 2x - 2$ **46.** $\log_5 5^{x^2} = 4x - 4$

47. $5^{\log_5 3x-7} = x^2 - 4$

48. $10^{\log_{10} 2x+1} = x^2 - 3x - 5$

49. $\log_x (7x - 10) = 2$ **50.** $\log_x (7x - 12) = 2$

51. $\log_x (10 - 3x) = 2$ **52.** $\log_x (5x - 4) = 2$

4.3   Graphing Logarithmic Functions

In Section 4.2 we defined the logarithmic function and found some of its properties. Now we will graph the logarithmic function.

EXAMPLE 1

Graph $y = \log_a x$, using its inverse function.

Solution   The inverse of the logarithmic function $y = \log_a x$, or $a^y = x$, is the exponential function $y = a^x$. In Section 2.7 we found that a function and its inverse are symmetric with each other with respect to the line $y = x$. Using the graph of $y = a^x$ from Section 4.1 and symmetry with respect to the line $y = x$, we graph the logarithmic function as shown in Fig. 4.9. ∎

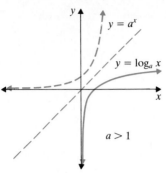

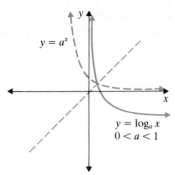

Figure 4.9   The exponential and logarithmic functions for $a > 1$ The exponential and logarithmic functions for $0 < a < 1$

The graph of the logarithmic function has certain properties, which we will now identify.

GRAPHING
LOGARITHMIC
FUNCTIONS

The logarithmic function $f(x) = \log_a x$, with $a > 0$ and $a \neq 1$,

(a) has a domain containing the positive real numbers ($x > 0$),
(b) has a vertical asymptote, the line $x = 0$,
(c) has an x-intercept of 1, $f(1) = \log_a 1 = 0$, and
(d) is an increasing function for $a > 0$ and a decreasing function for $0 < a < 1$.

In preparation for graphing we will consider the domains of the following logarithmic functions.

EXAMPLE 2   What is the domain of $y = \log_{10} (x - 2)$?

Solution   The domain of the logarithmic function must be positive. Thus for $y = \log_{10} (x - 2)$ we have

$$x - 2 > 0 \qquad \text{and}$$
$$x > 2$$

Therefore the domain of $y = \log_{10} (x - 2)$ is $\{x : x > 2\}$.   ■

EXAMPLE 3   What is the domain of $y = \log_2 x^2$?

Solution   For the domain of $y = \log_2 x^2$ to be positive we have

$$x^2 > 0$$

Therefore the domain of $y = \log_2 x^2$ is $\{x : x \text{ is a real number and } x \neq 0\}$.   ■

EXAMPLE 4   What is the domain of $y = \log_3 |x - 3|$?

Solution   The domain is positive when $|x - 3| > 0$. Thus x can be any real number except 3. Therefore the domain is $\{x : x \text{ is a real number and } x \neq 3\}$. ∎

Keeping in mind the properties of logarithmic functions, we will now sketch the graphs of logarithmic functions. We will also use the graphing techniques from Chapters 2 and 3 whenever possible to sketch these graphs.

EXAMPLE 5   Sketch the graph of $y = \log_{10}(x - 2)$.

Solution   We first notice that the graph of $y = \log_{10}(x - 2)$ is a translation of the graph $y = \log_{10} x$. Thus for $x - 2 = x - h$ we have $h = 2$. We then sketch the basic shape of $y = \log_{10} x$ with a dashed curve and translate it $h = 2$ units to the right to obtain the graph $y = \log_{10}(x - 2)$, the solid curve in Fig. 4.10. Notice that the domain is $\{x : x > 2\}$ as in Example 2. ∎

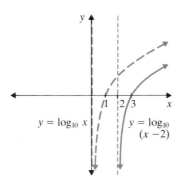

$y = \log_{10} x$ $y = \log_{10}$ $(x - 2)$

Figure 4.10

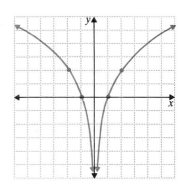

Figure 4.11

EXAMPLE 6   Sketch the graph of $f(x) = \log_2 x^2$.

Solution   To graph $f(x) = \log_2 x^2$, we notice that the domain includes all $x^2 > 0$. Thus the domain is all real numbers x with $x \neq 0$. This function also has symmetry with respect to the y-axis, since $\log_2 (-x)^2 = \log x^2$. The function has a vertical asymptote, the line $x = 0$. The points $(1, 0)$ and $(-1, 0)$ are on the graph of f, since $\log_2 (1)^2 = 0$ and $\log_2 (-1)^2 = 0$. The logarithmic base of 2 is greater than 1, and thus f is an increasing function for $x > 0$. Having sketched the graph of $f(x) = \log_2 x^2$ for $x > 0$, we sketch the graph for $x < 0$ using symmetry with respect to the y-axis (Fig. 4.11). Some other points on this graph are $(2, 2)$ and $(-2, 2)$, since from the definition of the logarithmic function, $\log_2 (\pm 2)^2 = 2$. ∎

EXAMPLE 7   Sketch the graph of $g(x) = \log_3 |x - 3|$.

Solution   As in Example 4, the graph of $g(x) = \log_3 |x - 3|$ has a domain of $\{x : x \text{ is a real number and } x \neq 3\}$. Since the values of $|x - 3|$ have symmetry about the line $x = 3$, the function $g(x) = \log_3 |x - 3|$ will also have symmetry about the line $x = 3$. We see this symmetry when $x = 2$ and $x = 4$. Both domain values have a range value of 0, since

$$\log_3 |2 - 3| = \log_3 |4 - 3| = \log_3 1 = 0$$

Thus $(2, 0)$ and $(4, 0)$ are points on the graph. Notice that for $x > 3$ we have $g(x) = \log_3 (x - 3)$. This is the graph of $y = \log_3 x$ shifted horizontally three units to the right. Using symmetry with respect to the line $x = 3$, we graph $g(x) = \log_3 |x - 3|$ for $x < 3$ (Fig. 4.12). By the definition of the logarithmic function, other points on the graph are $(0, 1)$ and $(6, 1)$, since

$$\log_3 |0 - 3| = 1 \text{ and } \log_3 |6 - 3| = 1 \qquad \blacksquare$$

Using the properties of the graphs of the exponential and logarithmic functions we can write the equations for these functions as in the following examples.

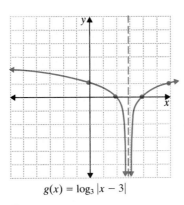

$g(x) = \log_3 |x - 3|$

Figure 4.12

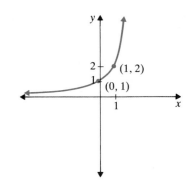

Figure 4.13

EXAMPLE 8

Identify the base and find the equation for the exponential function $y = a^x$ in Fig. 4.13.

Solution This is the graph of an exponential function of the form $y = a^x$, which contains the points $(0, 1)$ and $(1, 2)$. We will use both points in the equation $y = a^x$, since both points are on the graph. Thus for the point $(0, 1)$ we have $a^0 = 1$, a true statement. This result verifies that $(0, 1)$ is on the graph of $y = a^x$. Now substitute the point $(1, 2)$, and we have $x = 1$, $y = 2$, and $y = a^x$ resulting in

$$a^1 = 2 \quad \text{or} \quad a = 2$$

Thus the base is $a = 2$, and the equation is $y = 2^x$. $\blacksquare$

EXAMPLE 9

Identify the base and find the equation for the logarithmic function of the form $y = \log_a x$ in Fig. 4.14.

Solution From the graph we see that the graph contains the points $(1, 0)$ and $(8, 3)$. To find the base, we substitute each point into the equation $y = \log_a x$.

Using $(1, 0)$, we have $\log_a 1 = 0$ or $a^0 = 1$. This true statement verifies that $(1, 0)$ is on the graph of $y = \log_a x$. Using $(8, 3)$, we have

$$\log_a 8 = 3$$

$$a^3 = 8 \quad \text{or} \quad a = 2$$

Thus the base is $a = 2$, and the function is $y = \log_2 x$.   ■

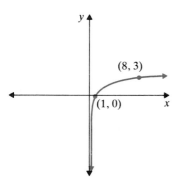

Figure 4.14   **Figure 4.15**

EXAMPLE 10   Use the results of Example 7 to find the equation of the logarithmic function in Fig. 4.15.

Solution   In Fig. 4.15 we have the graph of the logarithmic function from Example 9 (Fig. 4.14), but it is shifted eight units to the left. Thus we have a horizontal translation of $h = -8$ and $y = \log_2 (x + 8)$.   ■

Exercises 4.3

Find the domain of the following functions.

1. $f(x) = \log_2 x$

2. $f(x) = \log_3 x$

3. $g(x) = \log_3 (x + 2)$

4. $g(x) = \log_2 (x + 3)$

5. $h(x) = \log_2 (x - 3)$

6. $h(x) = \log_3 (x - 2)$

7. $f(x) = \log_3 |x|$

8. $f(x) = \log_5 |x|$

9. $h(x) = \log_3 |x + 2|$

10. $h(x) = \log_2 |x - 2|$

11. $f(x) = |\log_2 (x - 3)|$

12. $f(x) = |\log_3 (x + 2)|$

Sketch the graph for each of the following logarithmic functions.

13. $f(x) = \log_2 x$

14. $f(x) = \log_3 x$

15. $g(x) = \log_3 (x + 2)$

16. $g(x) = \log_2 (x + 3)$

17. $h(x) = \log_2 (x - 3)$

18. $h(x) = \log_3 (x - 2)$

19. $f(x) = -3 + \log_2 x$

20. $f(x) = -2 + \log_3 x$

21. $g(x) = 2 + \log_3 (x + 2)$

22. $g(x) = 3 + \log_2 (x - 3)$

23. $h(x) = 2 - \log_3 x$

24. $h(x) = 3 - \log_2 x$

25. $f(x) = |\log_3 x|$

26. $f(x) = |\log_2 x|$

27. $f(x) = \log_3 |x|$

28. $f(x) = \log_2 |x|$

29. $g(x) = \log_2 \left(\dfrac{1}{x}\right)$

30. $g(x) = \log_3 \left(\dfrac{1}{x}\right)$

31. $h(x) = \log_3 |x + 2|$

32. $h(x) = \log_2 |x - 2|$

33. $f(x) = |\log_2 (x - 3)|$

34. $f(x) = |\log_3 (x + 2)|$

Find the inverse function for each of the following functions.

35. $f(x) = \log_{10} x^3$ for $x > 0$

36. $f(x) = \log_e x^3$ for $x > 0$

37. $g(x) = \log_e (x + 1)$ for $x > -1$

38. $g(x) = \log_{10}(x - 1)$ for $x > 1$

39. $h(x) = \log_e x^{1/3}$ for $x > 0$

40. $h(x) = \log_{10} x^{1/5}$ for $x > 0$

Find the equation of the exponential functions in the following graphs.

41.

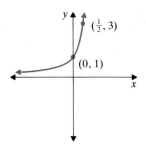

42.

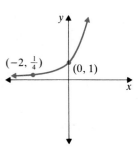

43.

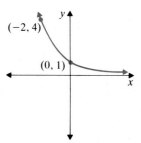

44.

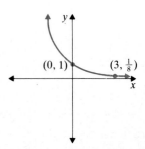

Find the equations for the logarithmic functions in the following graphs.

45.

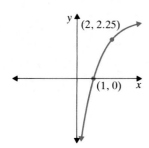

46.

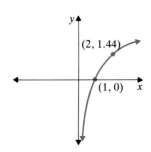

Use the results from Exercises 45 and 46 to find the equations of the following graphs.

47.

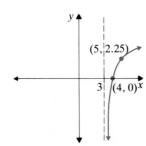

48.

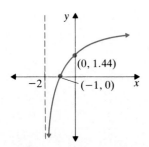

4.4   Properties of Logarithmic Functions

The definition and properties of logarithmic functions are most helpful tools in manipulating functions, especially in calculus. The following definition and properties were developed in the two preceding sections.

Definition

For $a > 0$, $a \neq 1$, and $x > 0$ the **logarithmic function** of x with base a is

$$\log_a x = y \qquad \text{and is equivalent} \qquad \text{to } a^y = x$$

Logarithmic Properties

For $a > 0$, $a \neq 1$, and $x > 0$,

   (a) $a^{\log_a x} = x$,

   (b) $\log_a a^x = x$, and

   (c) $\log_a 1 = 0$.

We will now add three other logarithmic properties that help in manipulating logarithmic functions.

Logarithmic Properties

For positive numbers a, n_1, and n_2 with $a \neq 1$,

   (d) $\log_a (n_1 \cdot n_2) = \log_a n_1 + \log_a n_2$,

   (e) $\log_a \left(\dfrac{n_1}{n_2} \right) = \log_a n_1 - \log_b n_2$, and

   (f) $\log_a (n_1)^r = r \log_a n_1$.

PROOF: To prove properties (d), (e), and (f), let

$$A = \log_a n_1 \qquad \text{and} \qquad B = \log_a n_2$$

Thus by the logarithmic definition,

$$a^A = n_1 \qquad \text{and} \qquad a^B = n_2 \qquad\qquad \square$$

PROOF OF PROPERTY (d): Using the multiplication property of equality with $a^A = n_1$ and $a^B = n_2$, we have

$$a^A \cdot a^B = n_1 \cdot n_2$$

$$a^{A+B} = n_1 \cdot n_2$$

$$\log_a (n_1 \cdot n_2) = A + B \qquad \text{(Definition of the logarithmic function)}$$

and

$$\log_a (n_1 \cdot n_2) = \log_a n_1 + \log_a n_2 \quad \text{(Substitution)}$$

Therefore the logarithm of a product equals the sum of the logarithms of the factors.   □

PROOF OF PROPERTY (e): For $a^A = n_1$ and $a^B = n_2$ and the multiplication (or division) property of equality,

$$\frac{a^A}{a^B} = \frac{n_1}{n_2}$$

$$a^{A-B} = \frac{n_1}{n_2}$$

$$\log_a \left(\frac{n_1}{n_2}\right) = A - B \qquad \text{(Definition of the logarithmic function)}$$

and

$$\log_a \left(\frac{n_1}{n_2}\right) = \log_a n_1 - \log_a n_2 \quad \text{(Substitution)}$$

Therefore the logarithm of a quotient equals the difference of the logarithm of the numerator and the logarithm of the denominator.   □

PROOF OF PROPERTY (f): Raise both sides of $a^A = n_1$ to the r power. Then

$$(a^A)^r = n_1{}^r$$

$$a^{rA} = n_1{}^r$$

$$\log_a n_1{}^r = rA \qquad \text{(Definition of the logarithmic function)}$$

$$\log_a n_1{}^r = r \log_a n_1 \quad \text{(Substitution)}$$

Therefore the logarithm of an exponential (any base) equals the product of the power and the logarithm of the exponential base.   □

We will use these properties to expand and condense logarithmic notation in the following examples.

EXAMPLE 1   Write these expressions as the sum, difference, and multiples of logarithms:

(a) $\log_7 \dfrac{37(15)^3}{25}$ and   (b) $\log_a \dfrac{\sqrt{xy}}{b^2}$

Solution   We will use the properties of logarithms to simplify these expressions.

(a) $\log_7 \dfrac{37(15)^3}{25} = \log_7 37(15)^3 - \log_7 25$ *(By property (e))*

$$= \log_7 37 + \log_7 15^3 - \log_7 5^2 \quad \text{\textit{(By property (d))}}$$

$$= \log_7 37 + 3 \log_7 15 - 2 \log_7 5 \quad \text{\textit{(By property (f))}}$$

(b) $\log_a \dfrac{\sqrt{xy}}{b^2} = \log_a \sqrt{xy} - \log_a b^2$ *(By property (e))*

$$= \log_a (xy)^{1/2} - \log_a b^2 \quad \text{\textit{(By the properties of exponents)}}$$

$$= \frac{1}{2} \log_a xy - 2 \log_a b \quad \text{\textit{(By property (f))}}$$

$$= \frac{1}{2}(\log_a x + \log_a y) - 2 \log_a b \quad \text{\textit{(By property (d))}}$$

$$= \frac{1}{2} \log_a x + \frac{1}{2} \log_a y - 2 \log_a b \quad \text{\textit{(By the distributive property)}}$$

This could also have been accomplished as follows:

$$\log_a \frac{\sqrt{xy}}{b^2} = \log_a [(\sqrt{xy})(b^{-2})]$$

$$= \log_a [(xy)^{1/2} b^{-2}]$$

$$= \log_a (xy)^{1/2} + \log_a b^{-2}$$

$$= \frac{1}{2} \log_a (xy) + (-2) \log_a b$$

$$= \frac{1}{2}(\log_a x + \log_a y) - 2 \log_a b$$

$$= \frac{1}{2} \log_a x + \frac{1}{2} \log_a y - 2 \log_a b$$

EXAMPLE 2   Condense these expressions into the logarithm of a single expression:

(a) $\dfrac{1}{2} \log_5 x - 2 \log_5 y$,   (b) $5 \log_a x - 2 \log_a 7 - \dfrac{1}{2} \log_a y$.

Solution

(a) $\dfrac{1}{2} \log_5 x - 2 \log_5 y = \log_5 x^{1/2} - \log_5 y^2$ *(By property (f))*

$$= \log_5 \left(\frac{x^{1/2}}{y^2}\right) \quad \text{\textit{(By property (e))}}$$

This could also have been accomplished as follows:

$$\frac{1}{2} \log_5 x - 2 \log_5 y = \log_5 x^{1/2} + \log_5 y^{-2} \quad \textit{(By property (f))}$$

$$= \log_5 (x^{1/2}y^{-2}) \quad \textit{(By property (d))}$$

$$= \log_5 \left(\frac{x^{1/2}}{y^2}\right)$$

(b) $5 \log_a x - 2 \log_a 7 - \dfrac{1}{2} \log_a y$

$$= 5 \log_a x - \left(2 \log_a 7 + \frac{1}{2} \log_a y\right) \quad \textit{(By the distributive property)}$$

$$= \log_a x^5 - (\log_a 7^2 + \log_a y^{1/2}) \quad \textit{(By property (f))}$$

$$= \log_a x^5 - \log_a (49y^{1/2}) \quad \textit{(By property (d))}$$

$$= \log_a \left(\frac{x^5}{49y^{1/2}}\right) \quad \textit{(By property (e))}$$

This could also have been accomplished as follows:

$$5 \log_a x - 2 \log_a 7 - \frac{1}{2} \log_a y$$

$$= \log_a x^5 + \log_a 7^{-2} + \log_a y^{-1/2} \quad \textit{(By property (f))}$$

$$= \log_a x^5(7^{-2})y^{-1/2} \quad \textit{(By property (d))}$$

$$= \log_a \frac{x^5}{7^2y^{1/2}} = \log_a \frac{x^5}{49y^{1/2}}$$

■

We will also use these logarithmic properties to solve equations.

EXAMPLE 3   Solve the equation $\log_5 (x + 2) - \log_5 (x - 1) = \log_5 2$.

Solution   By using logarithmic property (e) the equation

$$\log_5 (x + 2) - \log_5 (x - 1) = \log_5 2$$

becomes

$$\log_5 \frac{x + 2}{x - 1} = \log_5 2$$

Since the logarithms are equal, the bases are equal, and the logarithmic function is one-to-one, we conclude that the domains must also be equal. Thus

$$\frac{x + 2}{x - 1} = 2$$
$$x + 2 = 2(x - 1)$$
$$x + 2 = 2x - 2$$

and

$$4 = x \quad \textit{the solution}$$

We must check this solution $x = 4$ to verify that the domain of the original functions, $x + 2$ and $x - 1$, are both positive. For $x = 4$,

$$x + 2 = 6 > 0 \quad \text{and} \quad x - 1 = 3 > 0$$

Thus $x = 4$ is the solution. ■

Remember that the domain of the logarithmic function must be positive.

EXAMPLE 4   Solve the equation $\log_a (x - 2) - \log_a 4 = -\log_a (x + 1)$.

Solution   By using properties (e) and (f),

$$\log_a (x - 2) - \log_a 4 = -\log_a (x + 1)$$

becomes

$$\log_a \frac{(x - 2)}{4} = \log (x + 1)^{-1}$$

Then with equal logarithms and equal bases we conclude that the domains are also equal. Thus

$$\frac{x - 2}{4} = (x + 1)^{-1}$$
$$\frac{x - 2}{4} = \frac{1}{x + 1}$$
$$(x - 2)(x + 1) = 4$$
$$x^2 - x - 2 = 4$$
$$x^2 - x - 6 = 0$$
$$(x - 3)(x + 2) = 0$$

Thus the solution could be $x \in \{3, -2\}$. But we must check both solutions to see whether the domains of the logarithmic functions are positive. The solution $x = -2$ is eliminated, since it produces the logarithm of a negative number, $\log (-2 - 2) = \log (-4)$ and $\log (-2 + 1) = \log (-1)$. The domain of the logarithm must be positive; thus $x \neq -2$. Therefore the only solution is $x = 3$. ■

The variable can also be the base of the logarithmic function. Remember that the base is always a positive number and cannot equal 1.

EXAMPLE 5

Solve $\log_x (x + 2) = 2$.

Solution By definition,

$$\log_x (x + 2) = 2$$

is

$$x^2 = x + 2$$

Then

$$x^2 - x - 2 = 0$$

$$(x - 2)(x + 1) = 0$$

The possible solutions are 2 and -1. But $x \neq -1$, since x is the base of the logarithm and by definition all bases must be positive. Therefore $x = 2$ is the only solution. ■

In working with logarithmic functions there are two values for the logarithmic base that are used the most. These two bases are 10 and the irrational number e.

Definitions

The logarithmic function with base 10 is called the **common logarithm** and written as

$$\log_{10} x = \log x$$

When no base number is written for a logarithm, it is considered to be base 10. The logarithmic function with base e is called the **natural logarithm** and written as

$$\log_e x = \ln x$$

All of the previous properties for logarithms are also valid for common and natural logarithms.

Logarithmic Properties

Natural Logarithms	Common Logarithms
(a) $e^{\ln x} = x$	(a) $10^{\log x} = x$
(b) $\ln e^x = x$	(b) $\log 10^x = x$
(c) $\ln 1 = 0$	(c) $\log 1 = 0$

(d) $\ln (a \cdot b) = \ln a + \ln b$

(e) $\ln \left(\dfrac{a}{b}\right) = \ln a - \ln b$

(f) $\ln a^r = r \ln a$

(d) $\log (a \cdot b) = \log a + \log b$

(e) $\log \left(\dfrac{a}{b}\right) = \log a - \log b$

(f) $\log a^r = r \log a$

It is easy for us to evaluate common logarithms and natural logarithms with a calculator or logarithmic tables. It would be nice to be able to change other logarithmic bases into base 10 or base e. The theorem that follows provides a way to evaluate a logarithm of any base.

Theorem

The Change of Base Theorem. For positive numbers a, b, and n, with $a \neq 1$ and $b \neq 1$,

$$\log_b n = \frac{\log_a n}{\log_a b}$$

PROOF: Let $A = \log_b n$. Then, by the logarithmic definition, $b^A = n$. Since the logarithmic function is one-to-one, we take the logarithm base a of both sides of $b^A = n$:

$$\log_a b^A = \log_a n$$

$$A \log_a b = \log_a n \quad \textit{(By property (f))}$$

$$(\log_b n)(\log_a b) = \log_a n \quad \textit{(By substitution since } A = \log_b n\text{)}$$

Then, by the multiplication (or division) property of equality,

$$\log_b n = \frac{\log_a n}{\log_a b}$$

which is the desired result. $\square$

EXAMPLE 6

Find the value of these logarithms, using a calculator with both base e and base 10: **(a)** $\log_7 15$ and   **(b)** $\log_{21} 7$.

Solution   Using the change of base theorem, we have

(a) $\log_7 15 = \dfrac{\ln 15}{\ln 7} \approx \dfrac{2.7080 \ldots}{1.9459 \ldots} \approx 1.39166 \ldots$

or $\log_7 15 = \dfrac{\log 15}{\log 7} \approx \dfrac{1.17609 \ldots}{0.84509 \ldots} \approx 1.39166 \ldots$

(b) $\log_{21} 7 = \dfrac{\ln 7}{\ln 21} \approx \dfrac{1.9459 \ldots}{3.0445 \ldots} \approx 0.63915 \ldots$

or $\log_{21} 7 = \dfrac{\log 7}{\log 21} \approx \dfrac{0.8451 \ldots}{1.3222 \ldots} \approx 0.63915 \ldots$ ∎

Exercises 4.4

Use the properties of logarithms to write the following as the sum, difference, and multiples of logarithms.

1. $\log_5 \dfrac{x^2 y^3}{3}$ 　　　　　**2.** $\log_4 \dfrac{6x^2}{y^3}$

3. $\log_4 \dfrac{4^2 x^3}{\sqrt{y}}$ 　　　　**4.** $\log_5 \dfrac{5^2 \sqrt{x}}{y^3}$

5. $\log \dfrac{\sqrt{xy}}{14}$ 　　　　　**6.** $\log \dfrac{\sqrt[3]{xy}}{13}$

7. $\ln x^5 y^3 w^{-5}$ 　　　　　**8.** $\ln x^{-2} y^{1/3} w^{-3}$

9. $\ln e^x \sqrt{x}$ 　　　　　　**10.** $\log 1000 \sqrt[3]{y}$

11. $\log \dfrac{(x+2)^2}{x(x-1)}$ 　　　**12.** $\log \dfrac{\sqrt{x+1}}{(x+2)^2}$

13. $\ln \dfrac{x(x+2)}{\sqrt{(x-1)}}$ 　　　**14.** $\ln \dfrac{(x-1)^3}{x^2(x+2)}$

Condense the following expressions into the logarithm of one expression.

15. $\frac{1}{2} \log_5 y - 2 \log_5 x$ 　　　**16.** $\frac{3}{4} \log_2 x - \frac{1}{2} \log_2 5$

17. $\frac{1}{2} \log x - \log 5 + 2 \log y$ 　**18.** $5 \log x - \frac{1}{3} \log 2 + \frac{1}{2} \log y$

19. $\log x - \frac{1}{2} \log y - 3 \log w$ 　**20.** $\log y - 5 \log x - \frac{1}{3} \log w$

21. $\frac{3}{4} \ln x - \frac{1}{2} \ln y + \frac{1}{2} \ln 5$ 　**22.** $\frac{1}{4} \ln x - \frac{3}{4} \ln y + \frac{1}{4} \ln 3$

Simplify each of the following using the logarithmic properties.

23. $\log_2 \sqrt{8}$ 　　　　　　**24.** $\log_3 \sqrt{27}$

25. $\log \frac{1}{1000}$ 　　　　　**26.** $\log_2 \frac{1}{32}$

27. $\log_7 \sqrt[4]{7}$ 　　　　　**28.** $\log_5 \sqrt[3]{5}$

29. $\log_7 49$ 　　　　　　　**30.** $\log_5 125$

Use the logarithmic definition and properties to solve the following equations.

31. $\log (x + 99) - \log x = 2$

32. $\log_3 (x + 16) - \log_3 x = 2$

33. $\log (x + 1) - \log (x + 3) = -\log 3$

34. $\log (x + 3) - \log (x + 2) = -\log 2$

35. $\log (x - 1) - \log (x + 2) = \log 3$

36. $\log (x + 3) - \log (x + 1) = \log 2$

37. $\log (x - 1) = \log (5x + 3) - \log (x + 2)$

38. $\log (x - 3) = \log 4 + \log (x - 3) - \log (x + 2)$

39. $\log (x + 2) = 2 \log x - \log (x + 5)$

40. $\log (x - 3) = 2 \log x - \log (x + 4)$

The Richter scale is used to measure the magnitude of earthquakes. A seismograph plots the intensity of earthquake shocks and tremors. Half the height of a shock wave on the seismograph is the amplitude of that shock wave. The Richter scale measure for an earthquake is

$$R = \log \frac{A}{a}$$

where A is the amplitude of the largest shock wave in an earthquake and a is the amplitude of a normal minimal background shock wave, both measured in microns.

41. For the Richter scale measure above, find A as a function of a and R.

42. When $a = 1$ micron and $R = 8.3$, as in the 1906 San Francisco earthquake, what is the magnitude of A on the seismograph?

43. When $a = 1$ micron and $R = 6.3$, as in the 1971 Los Angeles earthquake, what is the magnitude of A on the seismograph?

44. When $a = 1$ micron and $R = 8.6$, as in the 1964 Alaska earthquake, what is the magnitude of A on the seismograph?

45. Find the ratio of A_1 for the 1906 San Francisco earthquake ($R = 8.3$) to A_2 for the 1971 Los Angeles earthquake ($R = 6.3$).

46. Find the ratio of A_1 for the 1906 Peru earthquake ($R = 8.9$) and A_2 for the 1906 San Francisco earthquake ($R = 8.3$).

The 1800 "prime number conjecture" of Adrien-Marie Legendre and Carl Friedrich Gauss states that the number of prime numbers less than a number x is approximately equal to $\dfrac{x}{\log x}$ for $x > 1$.

47. Using the "prime number conjecture" of Gauss and Legendre, approximate the number of prime numbers less than 10,000.

48. Using the "prime number conjecture" of Gauss and Legendre, approximate the number of prime numbers less than one million.

49. For integer values that are multiples of 10, sketch the graph for the approximation of the number of primes, $\dfrac{x}{\log x}$.

Change the scale on both axes from 1, 2, 3, 4, . . . to 10, 20, 30, 40 . . . before sketching this graph.

50. Using the "prime number conjecture" of Gauss and Legendre, approximate the number of prime numbers less than one billion.

4.5 Solving Equations Using Logarithmic Functions

In Section 4.4 we solved equations containing logarithmic functions using the logarithmic definition and properties. In this section we will solve more of these equations as well as other types of exponential and logarithmic equations. A necessary property that we used in the last section allowed us to take the logarithm of both sides of an equation.

Equality Property of Logarithmic Functions

For $a > 0$ and $a \neq 1$, $n_1 > 0$ and $n_2 > 0$,

$$n_1 = n_2 \quad \text{if and only if} \quad \log_a n_1 = \log_a n_2$$

There are two parts to this proof. We must prove that

(a) if $n_1 = n_2 > 0$, $a > 0$ and $a \neq 1$, then $\log_a n_1 = \log_a n_2$ and
(b) if $\log_a n_1 = \log_a n_2$, $a > 0$, and $a \neq 1$, then $n_1 = n_2 > 0$.

PROOF:
(a) The logarithmic function $f(x) = \log_a x$ is a function for $x > 0$. Thus $f(n_1) = f(n_1)$ uniquely. By substitution of $n_1 = n_2 > 0$ we have $f(n_1) = f(n_2)$ or $\log_a n_1 = \log_a n_2$.
(b) The logarithmic function $f(x) = \log_a x$ is one-to-one. Since $f(x)$ is one-to-one and $f(n_1) = f(n_2)$, the domain value must be equal, $n_1 = n_2$. □

We have proven both parts of this property. With this new property we can solve exponential equations as well as logarithmic equations. The examples that follow demonstrate the use of this property.

EXAMPLE 1   Solve $3^x = 2$.

Solution   Starting with $3^x = 2$, we use the equality property of logarithmic functions to obtain

$$\log 3^x = \log 2$$
$$x \log 3 = \log 2 \quad \textit{(Logarithmic property (f))}$$
$$x = \frac{\log 2}{\log 3} \approx \frac{0.301 \ldots}{0.477 \ldots} \approx 0.6309 \ldots$$

EXAMPLE 2   Solve $3^{x-1} = 2^{2x+3}$.

Solution   We first use the equality property of logarithmic functions:

$$3^{x-1} = 2^{2x+3}$$
$$\log 3^{x-1} = \log 2^{2x+3}$$
$$(x-1) \log 3 = (2x+3) \log 2$$
$$x \log 3 - \log 3 = 2x \log 2 + 3 \log 2$$
$$x \log 3 - 2x \log 2 = \log 3 + 3 \log 2$$
$$x(\log 3 - 2 \log 2) = \log 3 + 3 \log 2$$
$$x(\log 3 - \log 2^2) = \log 3 + \log 2^3$$
$$x\left(\log \frac{3}{4}\right) = \log 3(8)$$
$$x = \frac{\log (24)}{\log \left(\dfrac{3}{4}\right)} \approx \frac{1.3802 \ldots}{-0.1249 \ldots} \approx -11.047104 \ldots$$

Examples 1 and 2 were both solved with common logarithms but could also have been solved by using the natural logarithms. Values for the common logarithms above were computed with a calculator.

EXAMPLE 3   Solve $3^x - 5(3^{-x}) = 4$.

Solution   This problem will be easier to solve if the exponents are positive. To remove the negative exponent, we multiply both sides of this equation by 3^x:

$$3^x[3^x - 5(3^{-x})] = 4(3^x)$$
$$(3^x)^2 - 5(3^{-x+x}) = 4(3^x)$$
$$(3^x)^2 - 5(1) = 4(3^x)$$
$$(3^x)^2 - 4(3^x) - 5 = 0$$

This quadratic equation can be factored or solved with the quadratic formula. In this case,

$$(3^x - 5)(3^x + 1) = 0$$

and

$$3^x = 5 \quad \text{or} \quad 3^x = -1$$

We know that exponential functions are always positive, so $3^x > 0$. Thus $3^x = 5$ is the only solution. Solving $3^x = 5$ for x, we have

$$\log 3^x = \log 5$$
$$x \log 3 = \log 5$$

and

$$x = \frac{\log 5}{\log 3} \approx 1.4649 \ldots$$

■

EXAMPLE 4   Solve $y = 10^x - 10^{-x}$ for x as a function of y.

Solution   To remove the difficulty of dealing with the negative exponent, we multiply both sides of the equation by 10^x:

$$y = 10^x - 10^{-x}$$
$$10^x y = 10^x(10^x - 10^{-x})$$
$$y10^x = 10^{2x} - 1$$
$$0 = 10^{2x} - y10^x - 1$$
$$0 = (10^x)^2 - y(10^x) - 1$$

Then we solve this quadratic equation with the quadratic formula:

$$10^x = \frac{-(-y) \pm \sqrt{(-y)^2 - 4(-1)}}{2}$$
$$10^x = \frac{y \pm \sqrt{y^2 + 4}}{2}$$

Now we take the logarithm of both sides:

$$\log 10^x = \log \frac{y \pm \sqrt{y^2 + 4}}{2}$$

and

$$x = \log \frac{y \pm \sqrt{y^2 + 4}}{2}$$

Since the domain of the logarithm must be positive, then the negative square root must be eliminated from this solution, and

$$x = \log \frac{y + \sqrt{y^2 + 4}}{2}$$

∎

EXAMPLE 5   The amount of lumber a forest produces, $b(n)$, is given by $b(n) = b_0(1 + r)^n$, where n is the number of years of growth, r is the decimal or fraction equivalent of the percentage of new lumber produced each year, and b_0 is the original volume.

(a) How many years will it take a forest of 15,000 cubic feet of lumber with an annual growth rate of 17% to produce 15,000,000 cubic feet of lumber?

(b) How many years will it take any forest growing at a rate of 17% to have a volume 15 times its original volume?

Solution

(a) For $b(n) = b_0(1 + r)^n$ with $b(n) = 15{,}000{,}000$, $b_0 = 15{,}000$, and $r = 0.17$, then

$$15{,}000{,}000 = 15{,}000(1 + 0.17)^n$$

We will simplify and use the equality property of logarithms, base 10:

$$\frac{15{,}000{,}000}{15{,}000} = (1 + 0.17)^n$$

$$1000 = (1 + 0.17)^n$$

$$10^3 = (1.17)^n$$

$$\log 10^3 = \log (1.17)^n$$

$$3 = n \log (1.17)$$

$$n = \frac{3}{\log (1.17)} \approx 43.997 \ldots$$

Therefore the answer is approximately 44 years.

(b) For $b(n) = b_0(1 + r)^n$ with $b(n) = 15b_0$, and $r = 0.17$ we have $15b_0 = b_0(1 + 0.17)^n$. Next we simplify and use the equality property of logarithms:

$$\frac{15b_0}{b_0} = (1.17)^n$$

$$15 = (1.17)^n$$

$$\ln 15 = \ln (1.17)^n$$

$$\ln 15 = n \ln (1.17)$$

$$n = \frac{\ln 15}{\ln (1.17)} \approx 17.2483 \ldots$$

Therefore with an annual growth rate of 17% a forest will be 15 times larger in about 17.25 years.

∎

EXAMPLE 6   Newton's cooling law states that the rate of change in temperature of an object is proportional to the difference between its temperature and the temperature of its environment. This relationship can be written as

$$\frac{T(t) - T_E}{T_0 - T_E} = e^{-kt}$$

where $T(t)$ is the temperature as a function of time t, T_E is the temperature of the environment, T_0 is the original temperature of the object at $t = 0$, and k is a constant resulting from the cooling properties of the object. A tub of ice and water is used to cool soft drinks at a picnic. The ice and water in the tub maintain a temperature of 32°F. The soft drinks have a temperature of 70°F when they are placed in the tub, and after 10 minutes they have a temperature of 60°F. Use Newton's cooling law to solve the following:

(a) Find the temperature of the soft drinks as a function of time.

(b) How long will it take the soft drinks to cool from 70°F to 35°F?

Solution   From the given information we have $T_E = 32°F$ and $T_0 = 70°F$.

(a) To find $T(t)$, we substitute these values into Newton's cooling law:

$$\frac{T(t) - 32}{70 - 32} = e^{-kt}$$

$$T(t) - 32 = (70 - 32)e^{-kt}$$

$$T(t) - 32 = 38e^{-kt}$$

$$T(t) = 32 + 38e^{-kt}$$

Since the soft drinks are at a temperature of 60°F after 10 minutes, we can find k when $t = 10$:

$$60 = 32 + 38e^{-10k}$$

$$28 = 38e^{-10k}$$

$$\frac{28}{38} = e^{-10k}$$

$$\frac{38}{28} = e^{10k}$$

$$\ln\left(\frac{19}{14}\right) = 10k$$

$$k = \left(\frac{1}{10}\right)\ln\left(\frac{19}{14}\right) \approx 0.0305 \ldots$$

and

$$T(t) = 32 + 38e^{-0.0305t}$$

(b) For the soft drinks to reach a temperature of 35°F, we must solve for t when $T(t) = 35$. Thus

$$35 = 32 + 38e^{-(0.0305)t}$$

$$3 = 38e^{-(0.0305)t}$$

$$\frac{3}{38} = e^{-(0.0305)t}$$

$$\frac{38}{3} = e^{(0.0305)t}$$

$$\ln\left(\frac{38}{3}\right) = 0.0305t$$

$$t = \left(\frac{1}{0.0305}\right)\ln\left(\frac{38}{3}\right) \approx 83.245 \text{ minutes}$$

Thus it would take almost 84 minutes to cool the soft drinks to 35°F.   ■

Exercises 4.5

Solve the following equations, using the logarithmic function and its properties.

1. $2^{x-3} = 3$

2. $3^{2x-1} = 2$

3. $4^{2x-6} = 5$

4. $5^{x-5} = 4$

5. $2^{x-3} = 3^x$

6. $3^{x-2} = 2^x$

7. $7^x = 3^{2x-5}$

8. $5^x = 2^{2x-3}$

9. $4^x - 3(4^{-x}) = -2$

10. $2^x - 6(2^{-x}) = 5$

11. $e^x - 5e^{-x} = 4$

12. $e^x - 3e^{-x} = 2$

13. $e^{2x} - 7e^{-2x} = 6$

14. $e^{2x} - 6e^{-2x} = 1$

15. $\log x^3 = (\log x)^2$

16. $\log \sqrt{x} = (\log x)^2$

17. $x^{\sqrt[3]{\log x}} = 10^{16}$

18. $x^{\sqrt{\log x}} = 10^{27}$

Solve the following equations for x as a function of y.

19. $y = 10^x + 10^{-x}$

20. $y = 10^{-x} - 10^x$

21. $y = \dfrac{e^x - e^{-x}}{2}$

22. $y = \dfrac{e^x + e^{-x}}{2}$

Find $(f \circ g)(x)$ for the following pairs of functions.

23. $f(x) = \ln x$
  $g(x) = \sqrt{x}$

24. $f(x) = \log x$
  $g(x) = \sqrt[3]{x}$

25. $f(x) = \ln x^2$
  $g(x) = e^{x-3}$

26. $f(x) = \log x^3$
  $g(x) = 10^{5x-2}$

27. $f(x) = \log x$
  $g(x) = 10x^{3x}$

28. $f(x) = \ln x$
  $g(x) = ex^{5x}$

29. $f(x) = \log x$
  $g(x) = \sqrt[4]{10x^3}$

30. $f(x) = \ln x$
  $g(x) = \sqrt[3]{ex^2}$

The amount of lumber a forest produces, $b(n)$, is given by $b(n) = b_0(1 + r)^n$, where n is the number of years of growth, r is the decimal or fraction equivalent of the percentage of new lumber produced each year, and b_0 is the original volume.

31. A forest has a 21.7% annual growth rate (increase in volume of lumber). How many years will it take this forest to become 8 times larger than it is now?

32. A forest has a 13.5% annual growth rate (increase in volume of lumber). How many years will it take this forest to become 8 times larger than it is now?

33. A tract of land is scientifically planted so that its volume will increase by 35% each year. Since its volume is very small at planting, how long will it take the volume of wood in this forest to increase to 1000 times its original volume?

34. A tract of land is scientifically planted so that its volume will increase by 49% each year. Since its volume is very small at planting, how long will it take the volume of wood in this forest to increase to 10,000 times its original volume?

Radioactive decay is described by $y = y_0 e^{-kt}$, where y is the mass present after time t, y_0 is the amount at $t = 0$, and k is a positive constant.

35. Find t, the half-life for a substance as a function of k. (*Hint:* $y = \left(\frac{1}{2}\right)y_0$.)

36. Find k, the constant in the radioactive decay formula, for sodium-24, which has a half-life of 15 hours.

37. Find k, the constant in the radioactive decay formula, for thorium-230, which has a half-life of about 80,000 years.

38. Find k, the constant in the radioactive decay formula, for iron-60, which has a half-life of about 343,000 years.

39. If the original amount of thorium-230 was 155 units and there are now 135 units, how much time has passed? (See Exercise 37.)

40. If the original amount of sodium-24 was 155 units and there are now 25 units, how much time has passed? (See Exercise 36.)

An annuity consists of a given number of equal payments made at fixed periods of time with interest being compounded at the end of each payment period. The future value F of an annuity is

$$F = R\left[\frac{(1 + i)^n - 1}{i}\right]$$

where n is the number of payments of R dollars each and i is the interest rate for each payment period of time (i = annual interest ÷ number of payment periods per year).

41. Find the future value of an annuity consisting of 36 monthly payments of $100 each with an interest of 9%.

42. Find the future value of an annuity consisting of 24 monthly payments of $125 at an annual interest rate of 10.5%.

43. Find the number of months needed to accumulate $20,000 when $100 is deposited each month at an annual rate of 9%. (Remember that your answer is in months.)

44. Find the number of months needed to accumulate $30,000 when $125 is deposited each month at an annual rate of 8%.

45. How many months are needed for the future value F to be equal to 20,000 times the monthly investment when the annual interest is 8.5%?

46. How many months are needed for the future value F to be equal to 1,000,000 times the monthly investment when the annual interest is 9%?

Newton's cooling law states that the rate of change in temperature of an object is proportional to the difference between its temperature and the temperature of its environment. This relationship can be written as

$$\frac{T(t) - T_E}{T_0 - T_E} = e^{-kt}$$

where $T(t)$ is the temperature as a function of time t, T_E is the temperature of the environment, T_0 is the original temperature of the object at $t = 0$, and k is a constant resulting from the cooling properties of the object. Use Newton's cooling law to solve the following.

47. Tap water at 60°F is placed in an ice tray and then inserted into a 20°F freezer. After 10 minutes the temperature of the water is 45°F.
   (a) Find the temperature of the water as a function of time.
   (b) What is the temperature of the water after 15 minutes?
   (c) How long does it take the water to turn into ice?

48. A 3-inch-thick stream of molten lava is released from a volcano at a temperature of approximately 2200°F. It will cool to about 1200°F after 24 hours when the average external temperature is 60°F.
   (a) Find the temperature of the lava as a function of time.
   (b) What is the temperature of the lava after 5 days?
   (c) How long will it take the temperature of the lava to reach 70°F?

49. Bread is baked at 350°F and cools to 72°F in 50 minutes when room temperature is 70°F.
   (a) Find the temperature of the bread as a function of time.
   (b) What is the temperature of the bread after 35 minutes?
   (c) How long does it take the bread to cool from 350°F to 200°F?

CHAPTER SUMMARY AND REVIEW

KEY TERMS

Exponential functions

Logarithmic functions

Properties of logarithmic functions

Common logarithms

Natural logarithms

The change of base theorem

Equality property of logarithmic functions

CHAPTER EXERCISES

Sketch the graph of each of the following.

1. $f(x) = 4^x$

2. $g(x) = 2^{x+2}$

3. $h(x) = 3^{-x} - 2$

Write each of the following in logarithmic form.

4. $2^5 = 32$

5. $4^x = 3$

Write each of the following in exponential form.

6. $\log_3 27 = 3$

7. $\log_x 7 = 3$

Use the inverse function relationship between the exponential and logarithmic functions to evaluate the following.

8. $x^{\log_x 7}$

9. $\log_y y^{x^2-1}$

Sketch the graph of each of the following.

10. $f(x) = \log_4 x$

11. $g(x) = \log_3 |x - 3|$

12. $h(x) = \dfrac{1}{\log_4 x}$

Use the logarithmic properties to write each of the following as the sum, difference, and multiples of logarithms.

13. $\log \dfrac{x\sqrt{y}}{w^2}$

14. $\ln \dfrac{xy}{\sqrt[3]{x^2 w}}$

15. $\ln \sqrt{\dfrac{(x + 1)}{(x - 3)}}$

16. $\ln \sqrt[3]{\dfrac{x(x + 1)}{(x + 2)}}$

Use the logarithmic properties to write each of the following as the logarithm of one expression.

17. $\frac{1}{2} \ln x - 3 \ln y + 5 \ln w$

18. $\frac{3}{4} \log y - \frac{1}{4} \log x - \frac{1}{2} \log w$

19. $3 \log (x + 1) - \frac{1}{2} \log (x - 1)$

20. $\frac{1}{2} \log x - 3 \log (x + 2)$

21. $\frac{1}{4}[2 \ln x - 3 \ln (x - 1)]$

22. $\frac{1}{5}[3 \ln (x + 1) - 2 \ln x]$

Use the definition to solve each equation.

23. $\log_3 x = 4$

24. $\log_5 25 = x - 1$

Solve the following equations.

25. $3^{4x+2} = 2$

26. $\log_x (5x - 6) = 2$

27. $\log_3 (10x - 3) = 3$

28. $\log_4 (5x - 4) = 2$

29. $\log (10x) = 3$

30. $\log (2x + 50) = 2$

31. $\log (x + 1) = \log (5x - 27) - \log (x - 6)$

32. $\ln x = \frac{1}{2} \ln (2x + 3)$

33. Decibels are measured as $D = 10 \log (I/I_0)$ where I is the intensity of sound and I_0 is the intensity of sound at the threshold of hearing. Find I as a multiple of I_0 when $D = 82$.

34. Chemists use a pH number to determine the acidity or basicity of a substance. The pH number is determined by

$$\text{pH} = -\log [\text{H}^+]$$

where $[\text{H}^+]$ is the concentration of hydrogen ions measured in moles per liter. A pH between 1 and 7 indicates an acid, and a pH between 7 and 17 indicates a base. The closer the pH is to 7, the closer the substance is to being neutral, not acidic or basic. Find $[\text{H}^+]$ if the pH of drinking water is between 6.5 and 8.0.

35. A shock absorber can be made up of a spring and a dash pot as seen in Fig. 4.16. A force of F newtons is applied to the bumper and is resisted by a spring with compliance K meters per newton and a dash pot with a viscous friction B newtons per meter per second. The force F produces a bumper velocity $v(t)$ meters per second and a bumper displacement $x(t)$. The velocity and displacement are

$$v(t) = \left(\frac{F}{B}\right)e^{-(t/KB)} \qquad \text{and} \qquad x(t) = KF(1 - e^{-t/KB})$$

Sketch the graphs of the velocity and displacement functions.

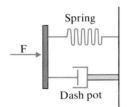

Figure 4.16

36. Newton's cooling law states that the rate of change in temperature of an object is proportional to the difference between its temperature and the temperature of its environment. This relationship can be written as

$$\frac{T(t) - T_E}{T_0 - T_E} = e^{-kt}$$

where $T(t)$ is the temperature as a function of time t, T_E is the temperature of the environment, T_0 is the original temperature of the object at $t = 0$, and k is a constant resulting from the cooling properties of the object. A cup of coffee is made with boiling (212°F) water and 10 minutes later has cooled to 102°F when room temperature is 68°F.
  (a) Find the temperature of the coffee as a function of time.
  (b) What is the temperature of the coffee 15 minutes after it is made?
  (c) How long does it take the coffee to reach 75°F?

The sales for a new line of videotape players is expected to follow the following relationship:

$$S(x) = \frac{10,000}{1 + 100e^{-x}}$$

where x is time in years.

37. Using the videotape player sales function $S(x)$ above, how many years will it take to sell 5000 units?

38. Using the videotape sales function $S(x)$, how many years will it take to sell 7000 units?

39. What is the maximum number of sales that can be expected from the videotape sales function? (*Hint*: Let x approach ∞.)

40. Using the videotape sales function $S(x)$, how many years will it take to sell 9950 units?

If the value of an auto depreciates by $\frac{1}{3}$ each year, then it would have a current value of

$$V(t) = V_N\left(1 - \frac{1}{3}\right)^{t/12}$$

where V_N is the original value of the new auto and t is the time in months.

41. If an auto sells for $8700 new and follows the above depreciation, what is its value after $2\frac{1}{2}$ years?

42. Using the auto depreciation function, find the value of an auto that originally sold for $11,500 after 4 years.

43. How long does it take for an auto to equal one third of its original value?

44. Using the auto depreciation function, how long does it take for an auto to equal half of its original price?

45. What was the net cost of an auto that was purchased when it was 2 years old and sold 5 years later? Use the depreciation function above and assume that the value of the auto new was $12,000.

46. What was the net cost of an auto that was purchased when it was 2 years old and sold 5 years later? Use the depreciation function above and assume that the value of the auto new was $8000.

Chemists use a pH number to determine the acidity or basicity of a substance. The pH number is determined by

$$\text{pH} = -\log\,[\text{H}^+]$$

where $[\text{H}^+]$ is the concentration of hydrogen ions measured in moles per liter. A pH between 1 and 7 indicates an acid, and a pH between 7 and 17 indicates a base. The closer the pH is to 7, the closer the substance is to being neutral, not acidic or basic.

47. Find $[\text{H}^+]$ if the pH of fresh white eggs is about 7.8.

48. Find $[\text{H}^+]$ if the pH of pumpkin is between 4.8 and 5.2.

49. Find the pH if the $[\text{H}^+]$ of a 0.01 normal sulfuric acid solution is 7.9432×10^{-6} moles per liter.

50. Find the pH if the $[H^+]$ of a 0.1 normal oxalic acid solution is 2.5118×10^{-2} moles per liter.

51. What are the $[H^+]$ for the extremes measured by pH values of 1 and 7?

52. Find $[H^+]$ as a function of the pH.

An electrical circuit consists of a series of a battery of V volts, a resistor of R ohms, a capacitor of C farads, and a switch S (Fig. 4.17). The capacitor is initially uncharged, and the switch is closed at $t = 0$. The current $i(t)$ in the circuit and the voltage $v(t)$ across the capacitor are given as a function of time by

$$i(t) = \left(\frac{V}{R}\right)e^{-t/RC} \qquad \text{and} \qquad v(t) = V(1 - e^{-t/RC})$$

where $i(t)$ is measured in amperes and $v(t)$ is measured in volts.

53. Find the time when the voltage across the capacitor is 25% of the voltage produced by the battery when $R = 10$ ohms and $C = \frac{1}{10}$ farad.

54. Find the time when the voltage across the capacitor is 55% of the voltage produced by the battery when $R = 10$ ohms and $C = \frac{1}{10}$ farad.

55. Find the time when the current in the circuit is 75% of the current at $t = 0$ when $R = 10$ ohms and $C = \frac{1}{10}$ farad.

56. Find the time when the current in the circuit is 45% of the current at $t = 0$ when $R = 10$ ohms and $C = \frac{1}{10}$ farad.

57. What is the relationship between the times and percents in Exercises 53 and 55?

58. What is the relationship between the times and percents in Exercises 54 and 56?

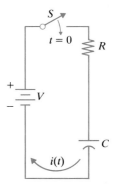

Figure 4.17

5

Basic Trigonometry

THE CARBON DIOXIDE CRISIS

Carbon dioxide, an apparently innoxious gas, may be quickly becoming a major threat to the present world order. Human activities such as the burning of fossil fuels, the razing of the world's forests, the expansion of agricultural and grazing lands, and industrial development are increasing the carbon dioxide content of the earth's atmosphere.

Since 1850, the amount of carbon dioxide in the earth's atmosphere has increased from 290 parts per million or less to a little more than 330 parts per million. About one quarter of that increase has come within the last 10 years. The graph on the following page showing the amount of measurable carbon dioxide in the atmosphere reflects the seasonal changes in the concentration of atmospheric carbon dioxide, considered the "pulse" of photosynthesis. If the current rate of increase continues, the amount of carbon dioxide in the atmosphere could double by 2020.

Carbon dioxide, just a trace gas in the atmosphere at a concentration of about 0.03% by volume, plays a critical role in controlling the earth's climate. It absorbs radiant energy at infrared wavelengths. By holding heat in this way, carbon dioxide has the potential of substantially altering climate. This phenomenon is commonly known as "the Greenhouse Effect."

Should we accelerate the development of nuclear power plants rather than those based on coal? Should we preserve forest areas instead of providing new lands for agriculture? Widespread destruction of the rain forests today could have profound long-term ecological and political effects for our world. How do we provide the increasing amounts of food that we are sure to need? The potential hazards associated with a steady increase in the carbon dioxide content of the atmosphere will have serious impacts on both national and international policies as we look at the prospect of global climatic change.

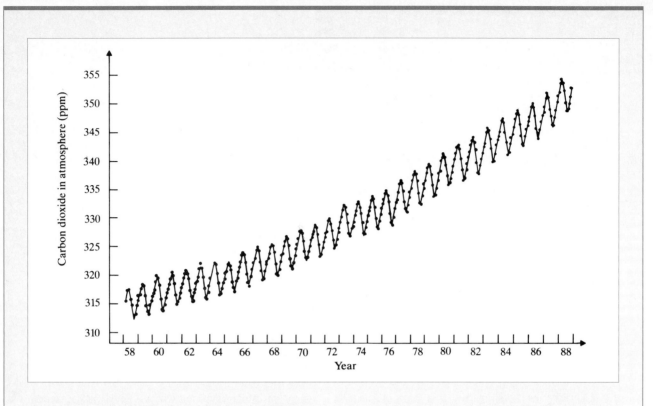

Average atmospheric carbon dioxide content has been measured since 1958 at the Mauna Loa Observatory on the island of Hawaii by Charles D. Keeling of the Scripps Institution of Oceanography. The dots indicate the monthly average concentration of carbon dioxide. Photosynthesis causes seasonal oscillations by removing carbon dioxide during the growing season in the northern hemisphere and releasing it during the fall and winter months.

SOURCE: C. D. Keeling, R. B. Bacastow, A.F. Carter, S. C. Piper, and T. P. Whorf, "A Three Dimensional Model of Atmospheric CO_2 Transport Based on Observed Winds: Observational Data and Preliminary Analysis," *Aspects of Climate Variability in the Pacific and the Western Americas*, Geophysical Monograph American Geophysical Union, vol. 55, p. 179, 1989. Reprinted by permission.

Basic trigonometry deals with the relationships between the acute angles of a right triangle and the ratio of its sides. Trigonometry has ancient origins. The Babylonian cuneiform tablet Plimpton 322 (dated around 1900–1600 B.C.), contains a trigonometric table. The ancient Egyptians empirically used trigonometry in construction and surveying, and the ancient Hindus of India developed trigonometry in their study of astronomy.

In this chapter we will define the trigonometric relations, sketch their graphs, and use them to solve related applications. The graph of the amount of carbon dioxide in the atmosphere described in the discussion of the carbon dioxide crisis is similar to some of the graphs we will sketch in this chapter.

5.1   Degrees and Radians

We first review the basic geometry of angles. Each angle has a vertex, an initial side, and a terminal side (Fig. 5.1). The angle in Fig. 5.1 can be identified as angle *ABC* or angle *CBA*. Although it is not essential, we traditionally use the Greek letters α (alpha), β (beta), γ (gamma), θ (theta), and ϕ (phi) to represent angles. The angle in Fig. 5.1 really represents two angles, angles α and β, as seen in Fig. 5.2. When two angles have the same initial side and the same terminal side, as in Fig. 5.2, the angles are called **coterminal angles**. Thus angle α and angle β in Fig. 5.2 are coterminal angles. Coterminal angles always have the same initial side and the same terminal side.

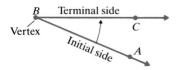

Figure 5.1

Figure 5.2

Measuring Angles

In trigonometry we measure angles with two different units of measurement, degrees and radians. The following statement defines both degree and radian measurement.

Definition

Degree measure: A central angle of a circle (an angle with its vertex at the center of the circle) has a measure of one degree, 1°, when the intercepted arc is $\frac{1}{360}$ of the circumference of the circle (angle α in (a) below).
Radian Measure: A central angle of a circle has a measure of 1 radian when the intercepted arc has a measure equal to the length of the circle's radius (angle β in (b) below).

(a)   (b)

From this definition we will calculate the radian measure of a circle. Since 1 radian occurs when the intercepted arc has a length r, the radius, we need only find the number of radii in the circumference of the circle. Thus the circle's circumference divided by the radius provides the number of radians in a circle. Therefore we have $\dfrac{2\pi r}{r} = 2\pi$ radians in a circle. Since a circle is $360°$ and also 2π radians, we conclude that

$$360° = 2\pi \text{ radians}$$
$$\mathbf{180° = \pi \text{ radians}}$$

Memorize this relationship. It makes it possible to change degrees into radians and radians into degrees. It is also helpful to remember that 1 can be written as

$$1 = \frac{180°}{1 \text{ radian}} \quad \text{and} \quad 1 = \frac{1 \text{ radian}}{180°}$$

when changing units. Changing from degrees to radians and from radians to degrees is demonstrated in the following two examples.

EXAMPLE 1   Change the following angles to radian measure:
(a) $\alpha = 90°$,   (b) $\beta = 60°$, and   (c) $\gamma = 75°$.

Solution   Remembering that $\dfrac{\pi \text{ radians}}{180°} = 1$, we have

(a) $\alpha = 90° = 90°(1) = 90°\left(\dfrac{\pi \text{ radians}}{180°}\right) = \dfrac{\pi}{2}$ radians

(b) $\beta = 60° = 60°(1) = 60°\left(\dfrac{\pi \text{ radians}}{180°}\right) = \dfrac{\pi}{3}$ radians

(c) $\gamma = 75° = 75°\left(\dfrac{\pi \text{ radians}}{180°}\right) = \dfrac{5\pi}{12}$ radians   ∎

We will usually not write $\dfrac{\pi}{2}$ radians; instead we will write $\dfrac{\pi}{2}$, without any reference to the measurement units of radians. Thus **when degrees are not indicated, you must assume that the units are radians**. The following example demonstrates the process of changing radians to degrees.

EXAMPLE 2   Change the following angles to degree measure.
(a) $\alpha = \dfrac{\pi}{6}$,   (b) $\beta = \dfrac{\pi}{4}$,   and   (c) $\theta = 1.75$

Solution   Remembering that $180° = \pi$ radians, in parts (a) and (b) we have

(a) $\alpha = \dfrac{\pi}{6} = \dfrac{\pi \text{ radians}}{6} = \dfrac{180°}{6} = 30°$

(b) $\beta = \dfrac{\pi}{4} = \dfrac{\pi \text{ radians}}{4} = \dfrac{180°}{4} = 45°$

(c) Since $\dfrac{180°}{(\pi \text{ radians})} = 1$, we have

$$\theta = 1.75 = (1.75 \text{ radians}) \cdot 1 = 1.75 \text{ radians} \left(\frac{180°}{\pi \text{ radians}} \right) \approx 100.2676°$$

The symbol $\approx$ indicates "approximately equal to" for rounded-off results.   ■

The following conclusions from Examples 1 and 2 are most important and should be memorized:

$$30° = \frac{\pi}{6}, \quad 45° = \frac{\pi}{4}, \quad 60° = \frac{\pi}{3}, \quad 90° = \frac{\pi}{2}, \quad 180° = \pi, \quad \text{and} \quad 360° = 2\pi$$

When an angle has a measure that contains a fractional part of a degree, we can write the measure in degrees with decimals or in degrees with minutes and seconds. The relation between degrees, minutes, and seconds is

$$1° = 60 \text{ minutes} = 60'$$
$$1 \text{ minute} = 60 \text{ seconds} = 60''$$

Using these different forms of degree measurement, we can write

$$35\frac{9}{25}° = 35.36° \qquad \text{with decimals}$$
$$= 35°21'36'' \qquad \text{with degrees, minutes, and seconds}$$

Measuring angles in degrees with decimals is done frequently today, now that calculators are in the common use. Angles are still measured with degrees, minutes, and seconds in areas such as navigation and astronomy, but with the development of new instruments containing digital readouts, more and more areas are converting to degrees with decimals.

Even with this change to degrees with decimals it is helpful to know how to change between degrees with decimals and degrees, minutes, and seconds.

EXAMPLE 3   Change $125.341°$ to degrees, minutes, and seconds.

Solution   Remembering that $1° = 60'$, we change units as follows:

$$125.341° = 125° + (0.341°)\left(\frac{60'}{1°}\right)$$
$$= 125° + 20.46'$$
$$= 125°20.46' \qquad \left(\textit{Substitute } 1 = \frac{60''}{1'} \right)$$
$$= 125°20' + (0.46')\frac{60''}{1'}$$
$$= 125°20' + 27.6''$$
$$= 125°20'27.6''$$
$$\approx 125°20'28''$$

■

EXAMPLE 4   Change $38°15'39''$ into decimal form.

Solution   Remembering that $1' = 60''$, we change units as follows:

$$38°15'39'' = 38°15' + (39'')\left(\frac{1'}{60''}\right)$$

$$= 38°15' + 0.65'$$

$$= 38°15.65' \left(Substitute\ 1 = \frac{1°}{60'}\right)$$

$$= 38° + (15.65')\left(\frac{1°}{60'}\right)$$

$$\approx 38° + 0.26083333° = 38.26083333°$$

$$\approx 38.26°$$

Arc Length, Linear and Angular Speed

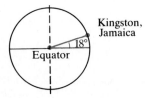

Central angle θ with
radian measure of $\theta = \frac{s}{r}$

Figure 5.3

We can also use radian measure to measure the arc of a circle. From geometry we know that the ratio of an arc length s to the circumference equals the ratio of the central angle θ in radians to the measure of the circle in radians. Thus

$$\frac{s}{2\pi r} = \frac{\theta}{2\pi} \quad (Multiply\ both\ sides\ by\ 2\pi r)$$

$$2\pi r\frac{s}{2\pi r} = 2\pi r\frac{\theta}{2\pi}$$

$$s = r\theta \quad (With\ \theta\ in\ radians)$$

Thus the arc length s is the radius times the measure of the angle in radians (Fig. 5.3). We can use this description of arc length to complete the following problem.

EXAMPLE 5   How many miles north of the equator is Kingston, Jamaica, if its latitude is 18°N and the radius of the earth is 3960 miles?

Solution   The latitude of 18°N is the central angle in Fig. 5.4, and the distance above the equator is the arc length s. We can use this information with $s = r\theta$ to find s, but we must also change 18° into radians. Doing this, we solve for s:

$$s = r\theta$$

$$s = 3960 \text{ miles } (18°)\left(\frac{\pi}{180°}\right)$$

$$s \approx 1244.07069 \text{ miles}$$

Figure 5.4   Thus Kingston, Jamaica, is about 1244 miles north of the equator.

The equation $s = r\theta$ also provides us with the relationship between the linear speed v and the angular speed ω (omega) of a rolling object. The linear speed v is $v = \dfrac{s}{t}$, the distance the object travels s for some time t. The object's angular speed ω is $\omega = \dfrac{\theta}{t}$, the amount of rotation θ in radians for some time t. The linear speed, $v = \dfrac{s}{t}$, is the "straight line" speed of a wheel, while the angular speed, $\omega = \dfrac{\theta}{t}$, is the "turning" or "spinning" speed of the same wheel.

To find the relationship between v and ω, we start with

$$s = r\theta \qquad \textit{(Divide both sides by time t)}$$

$$\frac{s}{t} = \frac{r\theta}{t}$$

$$\frac{s}{t} = r\left(\frac{\theta}{t}\right) \qquad \text{or}$$

$$v = r\omega \qquad \textit{(With } \omega \textit{ in radians per time)}$$

Thus the linear speed is the product of the radius and the angular velocity. This new relationship, $v = r\omega$, can help us to solve some interesting problems, as in the next example.

EXAMPLE 6

The rear wheel of a bicycle has a diameter of 27 inches with a fixed chain gear that is 4 inches in diameter (Fig. 5.5). (The fixed chain gear and the rear wheel always rotate together at the same rate.) Find the speed of the bicycle chain when the bicycle is traveling at a speed of 30 miles per hour.

Solution   The linear speed of the chain, v_c, is the same as the linear speed of the fixed gear. To find v_c, we must first find the angular speed ω, which is common to both the gear and the wheel. We will use the linear speed of the wheel $v_w = 30$ mi/h and $v = r\omega$. (Remember to change these units to miles and hours.)

$$v_w = r\omega$$

$$30\frac{\text{miles}}{\text{hour}} = \frac{27}{2}\text{ inches}\left(\frac{1\text{ foot}}{12\text{ inches}}\right)\left(\frac{1\text{ mile}}{5280\text{ feet}}\right)\omega$$

Solve for ω:

$$\omega = 30\left(\frac{2}{27}\right)\left(\frac{12}{1}\right)\left(\frac{5280}{1}\right)\frac{\text{radians}}{\text{hour}}$$

$$\omega \approx 140{,}800 \text{ radians per hour}$$

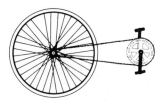

Figure 5.5

Using ω in radians per hour and $v = r\omega$, we have

$$v_c = r\omega$$

$$v_c = 2 \text{ inches} \left(\frac{1 \text{ foot}}{12 \text{ inches}}\right)\left(\frac{1 \text{ mile}}{5280 \text{ feet}}\right)\omega$$

$$v_c = 2 \text{ inches} \left(\frac{1 \text{ foot}}{12 \text{ inches}}\right)\left(\frac{1 \text{ mile}}{5280 \text{ feet}}\right)140,800 \frac{\text{radians}}{\text{hour}}$$

$$v_c \approx 4.44 \text{ miles per hour.}$$

Thus for this bicycle to have a speed of 30 miles per hour the chain speed is only about 4.44 miles per hour. ◾

Exercises 5.1

Change the following angle measurements to radians. (Round off to the nearest thousandth.)

1. 20°

2. 50°

3. 75°

4. 16°

5. 37.65°

6. 102.21°

7. 102°35′

8. 25°45′

Change the following angle measurements from radians to degrees with decimals and also to degrees, minutes, and seconds. (Round off to the nearest second.)

9. $\dfrac{5\pi}{12}$

10. $\dfrac{\pi}{7}$

11. $\dfrac{3\pi}{8}$

12. $\dfrac{11\pi}{12}$

13. 1.34

14. 2.2

15. 5.71

16. 6.54

17. The latitude of Diamond Head near Honolulu, Hawaii, is approximately 21°15′ (Fig. 5.6). How far north of the equator is Diamond Head? (Use 3960 miles for the radius of the earth.)

18. The latitude of Boston, Massachusetts, is approximately 42°24′. How far north of the equator is Boston?

19. Mt. McKinley, Alaska, at a latitude of 63° is almost directly north of Mouna Loa, Hawaii, at a latitude of 19.5°. What is the distance between the two mountains on the surface of the earth?

20. Salt Lake City, Utah, at a latitude of 40.75° is almost directly north of Phoenix, Arizona, at a latitude of 33.5°. What is the distance between the two cities on the surface of the earth?

21. The drive gear of a motor has a diameter of 1 centimeter and an angular speed of 2000 revolutions per minute (Fig. 5.7). This drive gear drives a second gear with a 4-centimeter diameter. What is the angular speed of the second gear?

22. The drive gear of a motor has a diameter of 2 centimeters and an angular speed of 2000 revolutions per minute (Fig. 5.7). This drive gear drives a second gear with a 5-centimeter diameter. What is the angular speed of the second gear?

Figure 5.6

Figure 5.7

23. A conveyor belt is to advance at 2 miles per hour (Fig. 5.8). What is the radius needed for the drive drum if the motor drives the drum at 2000 revolutions per hour?

Figure 5.8

24. A conveyor belt is to advance at 5 miles per hour. What is the radius needed for the drive drum if the motor drives the drum at 3000 revolutions per hour?

25. A satellite can send an effective signal back to earth over an angle of $\alpha = 12°$. What is the minimum height h of this satellite such that all of its signal reaches the maximum area on earth? The radius of the earth is about 3960 miles. Find the arc length s covered by the satellite signal (Fig. 5.9).

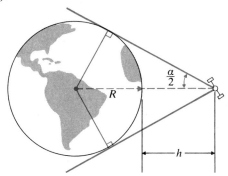

Figure 5.9

26. A satellite can send an effective signal back to earth over an angle of $\alpha = 17°$. What is the minimum height h of this satellite such that all of its signal reaches the maximum area on earth? The radius of the earth is about 3960 miles. Find the arc length s covered by the satellite signal.

27. The distance from the surface of the earth to the center of the moon is about 240,000 miles. The diameter of the moon is about 2160 miles. Find the angle subtended by the moon for an observer on earth in degrees and radians. (Use a calculator.)

28. The distance from the surface of the earth to the center of the moon is about 240,000 miles, and the diameter of the earth is about 7920 miles. Find the angle subtended by the earth for an observer on the moon in degrees and radians. (Use a calculator.)

29. The bicycle in Fig. 5.10 has a pedal gear with a 6-inch diameter, and each pedal is 6 inches from the center of the pedal gear. Use the information from Example 6 to find the angular speed of the pedal gear and the linear speed of the pedals.

30. The bicycle in Fig. 5.10 has a pedal gear with a 5-inch diameter, and each pedal is 7 inches from the center of the pedal gear. Use the information from Example 6 to find the angular speed of the pedal gear and the linear speed of the pedals.

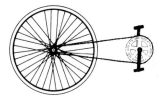

Figure 5.10

31. A child's bicycle has a 20-inch diameter rear wheel, a 2-inch diameter rear wheel fixed gear, a 6-inch diameter pedal gear, and a 5-inch pedal radius. Find the linear speed of the pedals when the bicycle's linear speed is 10 miles per hour.

32. A child's bicycle has a 24-inch diameter rear wheel, a 2-inch diameter rear wheel gear, a 6-inch diameter pedal gear, and a 4-inch pedal radius. Find the linear speed of the bicycle when the pedals' linear speed is 10 miles per hour.

33. Find the length of a belt around two pulleys when $L = 10$ inches, $R = 8$ inches, and $r = 2$ inches with $\alpha = 72.54°$ as shown in Fig. 5.11.

34. Find the length of a belt around two pulleys when $L = 3$ inches, $R = 6$ inches, and $r = 2$ inches with $\alpha = 68.68°$ as shown in Fig. 5.11.

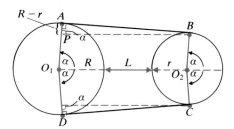

Figure 5.11

5.2 Trigonometric Ratios

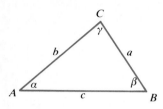

Figure 5.12

In this section we will measure angles in degrees and radians to define some special ratios called trigonometric ratios. To do this, we first review some basic geometry of triangles.

Every triangle has three sides and three angles as shown in Fig. 5.12. Notice that the lowercase letters a, b, and c represent the sides of the triangle, and the Greek letters α (alpha), β (beta), and γ (gamma) represent the angles of the triangle. Notice that in the triangle ABC, the angle with vertex A (angle CAB) is α and a is the side opposite the angle α. This relationship is true for all vertices. Side b is opposite vertex B and angle β. Likewise, side c is opposite vertex C and angle γ. Although it is not essential, it is traditional to use the Greek letters α, β, γ, as well as θ (theta) and ϕ (phi) to represent angles. Remember that the sum of the interior angles of a triangle is $180°$.

There are six trigonometric ratios or relations. They are the **sine, cosine, tangent, cosecant, secant,** and **cotangent**. The domains for these relations are angles measured in radians or degrees. When the domain is the measure of the angle β, then these trigonometric relations are written as $\sin \beta$, $\cos \beta$, $\tan \beta$, $\csc \beta$, $\sec \beta$, and $\cot \beta$. Initially, these relations will be defined as the ratios of the sides of a right triangle (Fig. 5.13). By using the right triangle, the sine, cosine, and tangent of α can be defined as

$$\sin \alpha = \frac{\text{opposite side}}{\text{hypotenuse}} = \frac{a}{c}$$

$$\cos \alpha = \frac{\text{adjacent side}}{\text{hypotenuse}} = \frac{b}{c}$$

$$\tan \alpha = \frac{\text{opposite side}}{\text{adjacent side}} = \frac{a}{b}$$

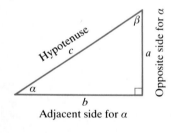

Figure 5.13

Since the opposite side and the adjacent side are relative to the angle, from Fig. 5.13 we also have

$$\sin \beta = \frac{\text{opposite side}}{\text{hypotenuse}} = \frac{b}{c} \qquad \cos \beta = \frac{\text{adjacent side}}{\text{hypotenuse}} = \frac{a}{c}$$

and

$$\tan \beta = \frac{\text{opposite side}}{\text{adjacent side}} = \frac{b}{a}$$

With the sine and cosine defined in terms of the sides of a right triangle we will define the other trigonometric relations in terms of the sine and cosine as seen here.

Definition

By using the right triangle the **right triangle trigonometric relations** are defined as

$$\sin \theta = \frac{\text{opposite side}}{\text{hypotenuse}} = \frac{b}{c} \qquad \csc \theta = \frac{1}{\sin \theta}$$

$$\cos \theta = \frac{\text{adjacent side}}{\text{hypotenuse}} = \frac{a}{c} \qquad \sec \theta = \frac{1}{\cos \theta}$$

$$\tan \theta = \frac{b}{a} = \frac{\sin \theta}{\cos \theta} \qquad \cot \theta = \frac{1}{\tan \theta} = \frac{\cos \theta}{\sin \theta}$$

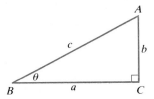

Defining the trigonometric relations with sine and cosine, as in the definition above, is most helpful. Notice that $\csc \theta$, $\sec \theta$, and $\cot \theta$ are the reciprocals of $\sin \theta$, $\cos \theta$, and $\tan \theta$, respectively. At first glance you might mistake the $\sin^{-1}$ key on the calculator for the reciprocal of the sine, but this is not the case. The reciprocal key on the calculator is the $\frac{1}{x}$ key. It is important for you to remember that

$$\sin^{-1} x \neq \frac{1}{\sin x}$$

but

$$\frac{1}{\sin x} = \csc x$$

The $\sin x$ and $\sin^{-1} x$ are inverse relations just as we saw earlier that $f(x)$ and $f^{-1}(x)$ were inverse functions. We will work with the inverse relations, $\sin^{-1} x$, $\cos^{-1} x$, and $\tan^{-1} x$, later in more detail.

We will now find the exact trigonometric values of some special angles in the following examples.

EXAMPLE 1

Find the values for the six trigonometric ratios of α and β when

$$\textbf{(a)} \ \alpha = \frac{\pi}{3} \qquad \text{and} \qquad \textbf{(b)} \ \beta = \frac{\pi}{6}.$$

Solution   Remembering that $\pi = 180°$, we know that

$$\alpha = \frac{\pi}{3} = 60° \qquad \text{and} \qquad \beta = \frac{\pi}{6} = 30°$$

We will construct a 30°–60° right triangle as shown in Fig. 5.14. First draw an equilateral triangle with all sides of length $2t$. Then sketch the altitude to the base, producing two right triangles that divide the base into two equal segments of length t. By using the Pythagorean Theorem the length of the altitude can be calculated to be $t\sqrt{3}$. Using the lengths in the 30°–60° right triangle and the definitions of the trigonometric ratios, we have the following:

(a) For $\alpha = \dfrac{\pi}{3} = 60°$,

$$\sin\frac{\pi}{3} = \frac{t\sqrt{3}}{2t} = \frac{\sqrt{3}}{2} \qquad \csc\frac{\pi}{3} = \frac{2}{\sqrt{3}} = \frac{2\sqrt{3}}{3}$$

$$\cos\frac{\pi}{3} = \frac{t}{2t} = \frac{1}{2} \qquad \sec\frac{\pi}{3} = \frac{2}{1} = 2$$

$$\tan\frac{\pi}{3} = \frac{\dfrac{\sqrt{3}}{2}}{\dfrac{1}{2}} = \sqrt{3} \qquad \cot\frac{\pi}{3} = \frac{1}{\sqrt{3}} = \frac{\sqrt{3}}{3}$$

(b) For $\beta = \dfrac{\pi}{6} = 30°$,

$$\sin\frac{\pi}{6} = \frac{t}{2t} = \frac{1}{2} \qquad \csc\frac{\pi}{6} = \frac{2}{1} = 2$$

$$\cos\frac{\pi}{6} = \frac{t\sqrt{3}}{2t} = \frac{\sqrt{3}}{2} \qquad \sec\frac{\pi}{6} = \frac{2}{\sqrt{3}} = \frac{2\sqrt{3}}{3}$$

$$\tan\frac{\pi}{6} = \frac{t}{t\sqrt{3}} = \frac{\sqrt{3}}{3} \qquad \cot\frac{\pi}{6} = \frac{\sqrt{3}}{1} = \sqrt{3}$$

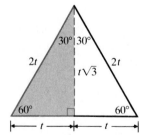

Figure 5.14

EXAMPLE 2   Find the values for the six trigonometric ratios of $\theta = \dfrac{\pi}{4}$.

Solution   For $\theta = \dfrac{\pi}{4} = 45°$ we construct a 45°–45° right triangle. To do this, draw a square with sides of length t. The diagonal divides the square into two 45°–45° right triangles (Fig. 5.15). By using the Pythagorean Theorem the diagonal has

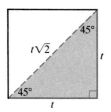

Figure 5.15

a length of $t\sqrt{2}$. With this information and the definitions of trigonometric ratios we have

$$\sin\frac{\pi}{4} = \frac{t}{t\sqrt{2}} = \frac{1}{\sqrt{2}} = \frac{\sqrt{2}}{2} \qquad \csc\frac{\pi}{4} = \sqrt{2}$$

$$\cos\frac{\pi}{4} = \frac{t}{t\sqrt{2}} = \frac{1}{\sqrt{2}} = \frac{\sqrt{2}}{2} \qquad \sec\frac{\pi}{4} = \sqrt{2}$$

$$\tan\frac{\pi}{4} = \frac{\frac{\sqrt{2}}{2}}{\frac{\sqrt{2}}{2}} = 1 \qquad\qquad \cot\frac{\pi}{4} = 1$$

Although we will not prove these results until the next section, we will now add the following trigonometric values to the results above:

$$\sin 0° = 0, \qquad \cos 0° = 1, \quad \sin 90° = 1, \qquad \cos 90° = 0$$

We condense the results in the following chart:

$\theta =$	0	$\frac{\pi}{6}$	$\frac{\pi}{4}$	$\frac{\pi}{3}$	$\frac{\pi}{2}$
$\sin\theta$	0	$\frac{1}{2}$	$\frac{\sqrt{2}}{2}$	$\frac{\sqrt{3}}{2}$	1
$\cos\theta$	1	$\frac{\sqrt{3}}{2}$	$\frac{\sqrt{2}}{2}$	$\frac{1}{2}$	0

The values in this chart are exact values and should be used rather than the approximate values given by a calculator when you use these special angles. **You will need to use *exact values* whenever possible.** Since these values are so important, it is necessary to memorize this chart. The following chart should make it easier to memorize these values.

$\theta =$	0	$\frac{\pi}{6} = 30°$	$\frac{\pi}{4} = 45°$	$\frac{\pi}{3} = 60°$	$\frac{\pi}{2} = 90°$
$\sin\theta$	$\frac{\sqrt{0}}{2} = 0$	$\frac{\sqrt{1}}{2} = \frac{1}{2}$	$\frac{\sqrt{2}}{2}$	$\frac{\sqrt{3}}{2}$	$\frac{\sqrt{4}}{2} = 1$
$\cos\theta$	$\frac{\sqrt{4}}{2} = 1$	$\frac{\sqrt{3}}{2}$	$\frac{\sqrt{2}}{2}$	$\frac{\sqrt{1}}{2} = \frac{1}{2}$	$\frac{\sqrt{0}}{2} = 0$

The patterns to note in this chart are that all denominators are 2 and that all numerators are the square root of 0, 1, 2, 3, 4 or 4, 3, 2, 1, 0 sequentially. We will now use the patterns in this chart to find the exact value, without a calculator, for the trigonometric relations of these special angles.

We can solve many practical problems by using right triangles and the trigonometric relations. When working with triangles, we use two terms to describe angles we measure, the **angle of elevation** and the **angle of depression**. The angle of elevation is the angle we measure when we look up. The angle of depression is the angle we measure when we look down. Both are shown in Fig. 5.16. Notice that the angle of elevation and the angle of depression are both measured with respect to the horizontal. Thus the angle of elevation equals the angle of depression. (If two parallel lines are cut by a transversal, the alternate interior angles are equal.)

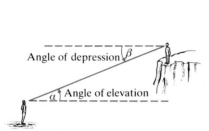

Figure 5.16

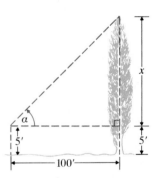

Figure 5.17

EXAMPLE 3

A field biologist measures the height of a tree by measuring the angle of elevation, angle α, from a point 100 feet from the base of the tree as shown in Fig. 5.17. Use a calculator to find the following:

(a) One year the angle α is 40°. How tall is the tree?

(b) Find the height of the tree in the next year when α is 47°.

(c) What is the percent of increase in the height of the tree during that year?

Solution   We first examine Fig. 5.17 and consider what trigonometric relation applies to the given data.

(a) When $\alpha = 40°$, we have

$$\tan 40° = \frac{x}{100}$$

$$x = 100 \tan 40°$$

$$x \approx 83.90996 \text{ feet}$$

Thus when $\alpha = 40°$, the height of the tree is approximately equal to

$$83.9 + 5 = 88.9 \text{ feet}$$

(b) When $\alpha = 47°$, we have

$$\tan 47° = \frac{x}{100}$$

so that

$$x = 100 \tan 47°$$
$$x \approx 107.236871 \text{ feet}$$

Thus when $\alpha = 47°$, the height of the tree is approximately equal to

$$107.2 + 5 = 112.2 \text{ feet}$$

(c) The percent of height increase is

$$\frac{\text{change in height}}{\text{height at beginning}} = \frac{112.23687 - 88.90996}{88.90996}$$
$$\approx 0.262365543$$
$$\approx 26.2\% \text{ increase in height} \qquad \blacksquare$$

Exercises 5.2

Find the exact values for each of the following.

1. $\sin 60°$ **2.** $\cos 30°$ **3.** $\sin 90°$ **4.** $\cos 0°$

5. $\cos 60°$ **6.** $\sin 30°$ **7.** $\cos 45°$ **8.** $\sin 45°$

9. $\cos 90°$ **10.** $\sin 0°$ **11.** $\tan 0°$ **12.** $\tan 90°$

13. $\sec 60°$ **14.** $\cos 30°$ **15.** $\tan 30°$ **16.** $\tan 60°$

17. $\sin \dfrac{\pi}{6}$ **18.** $\cos \dfrac{\pi}{2}$ **19.** $\tan \dfrac{\pi}{4}$ **20.** $\cot \dfrac{\pi}{4}$

21. $\sec \dfrac{\pi}{6}$ **22.** $\csc \dfrac{\pi}{3}$ **23.** $\csc \dfrac{\pi}{4}$ **24.** $\sec \dfrac{\pi}{3}$

25. $\cot \dfrac{\pi}{2}$ **26.** $\tan \dfrac{\pi}{3}$ **27.** $\tan \dfrac{\pi}{2}$ **28.** $\cot \dfrac{\pi}{6}$

29. $\sec \dfrac{\pi}{3}$ **30.** $\sec \dfrac{\pi}{4}$ **31.** $\csc \dfrac{\pi}{6}$ **32.** $\csc \dfrac{\pi}{2}$

Use right triangle ABC (Fig. 5.18) to solve the following triangles with **exact** values. (Do not use a calculator.)

33. For $\alpha = \dfrac{\pi}{4}$ and $a = 15$, find b and c.

34. For $\alpha = \dfrac{\pi}{3}$ and $a = 15$, find b and c.

35. For $\alpha = \dfrac{\pi}{6}$ and $a = 15$, find b and c.

36. For $\alpha = \dfrac{\pi}{3}$ and $b = 25$, find a and c.

37. For $\alpha = \dfrac{\pi}{6}$ and $c = 25$, find a and b.

38. For $\alpha = \dfrac{\pi}{4}$ and $c = 25$, find a and b.

39. For $\alpha = \dfrac{\pi}{3}$ and $c = 25$, find a and b.

40. For $\beta = \dfrac{\pi}{6}$ and $b = 18$, find a and c.

41. For $\beta = \dfrac{\pi}{6}$ and $a = 18$, find b and c.

42. For $\beta = \dfrac{\pi}{3}$ and $a = 18$, find b and c.

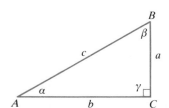

Figure 5.18

43. A field biologist measures the height of a tree by measuring the angle of elevation, angle α, from a point 100 feet from the base of the tree as shown in Fig. 5.19. Use a calculator to find the following:
  (a) Find the height if the angle α is 37°.
  (b) Find the height of the tree one year later when the angle α is 43°.
  (c) Find the percent of increase in the height of the tree during the year. (Use a calculator.)

44. A field biologist measures the height of a tree by measuring the angle of elevation, angle α, from a point 100 feet from the base of the tree as shown in Fig. 5.19. Use a calculator to find the following:
  (a) Find the height if the angle α is 44°.
  (b) Find the height of the tree one year later when the angle α is 52°.
  (c) Find the percent of increase in the height of the tree during the year.

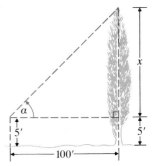

Figure 5.19

45. Two guy wires that are used to support an antenna are anchored 80 feet from the base of the antenna and make angles of 55° and 75°, respectively, with the ground as shown in Fig. 5.20. Find the length of the antenna between

the guy wires (distance CD). Also find the amount of wire needed for three such sets of guy wires. (Use a calculator and disregard any slack in the wires.)

46. Two guy wires that are used to support an antenna are anchored 50 meters from the base of the antenna and make angles of 57° and 73°, respectively, with the ground as shown in Fig. 5.20. Find the length of the antenna between the guy wires (distance CD). Also find the amount of wire needed for three such sets of guy wires. (Use a calculator and disregard any slack in the wires.)

47. To measure the distance across a canyon, a Boy Scout standing at point A sights a tree directly across the canyon at point T. He then walks 100 feet to point P in a direction perpendicular to the line AT as shown in Fig. 5.21. Using his compass, he calculates the angle APT to be 82°.
  (a) Find the distance across the canyon from point A to point T.
  (b) Find the percent of error if the angle measurement could be off by as much as $2\frac{1}{2}°$ in either direction. (Use a calculator.)

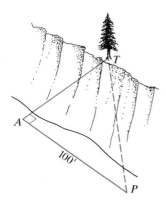

Figure 5.21

48. To measure the distance across a river, a Girl Scout standing at point A sights a tree directly across the river at point T. She then walks 100 feet to point P in a direction perpendicular to the line AT as shown in Fig. 5.21. Using her compass, she calculates the angle APT to be 38°.
  (a) Find the distance across the river from point A to point T.
  (b) Find the percent of error if the angle measurement could be off by as much as $2\frac{1}{2}°$ in either direction. (Use a calculator.)

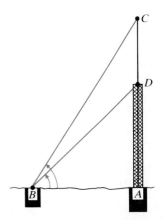

Figure 5.20

49. Show that sin x is a function for $0° \leq x \leq 90°$. To do this, use the right triangle in Fig. 5.22 to verify that the value of sin x is the same for any triangle and that for each value of x there is only one value for sin x.

50. Show that cos x is a function for $0° \leq x \leq 90°$. To do this, use the right triangle in Fig. 5.22 to verify that the value

of cos x is the same for any triangle and that for each value of x there is only one value for cos x.

51. Show that tan x is a function for $0° \leq x \leq 90°$. To do this, use the right triangle in Fig 5.23 to verify that the value of tan x is the same for any triangle and that for each value of x there is only one value for tan x.

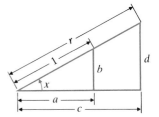

Figure 5.22

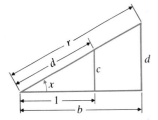

Figure 5.23

5.3   Circle Definitions of the Trigonometric Functions

In Section 5.2 we defined the trigonometric relations as the ratios of the sides of right triangles. That approach is a good starting point, but there are limitations. To apply these trigonometric relations to all angles, especially angles θ where $\theta > 90°$ as well as $\theta < 0$, we will use another approach.

An angle on the Cartesian coordinate system is in **standard position** when the vertex is at the origin and the initial side is the positive x-axis. An angle in standard position that opens counterclockwise has a positive measure, while an angle in standard position that opens clockwise has a negative measure, as shown in Fig. 5.24. We now define the sine and cosine of angles in standard position and the circle $x^2 + y^2 = r^2$.

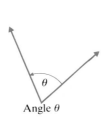

Angle θ

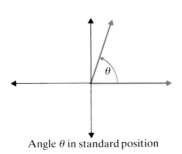

Angle θ in standard position

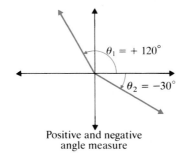

Positive and negative angle measure

Figure 5.24

Definition

Trigonometric Relations from Circles. For any angle θ in standard position that intersects the circle $x^2 + y^2 = r^2$ with radius r at the point (x, y), the trigonometric ratios are

$$\sin \theta = \frac{y}{r} \quad \text{and} \quad \cos \theta = \frac{x}{r}$$

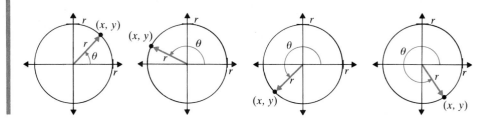

Notice that a central angle θ in quadrant *I* and the point (x, y) on the terminal side of θ provide a right triangle, where x is the adjacent side and y is the opposite side for θ. This shows how the definitions in Section 5.2 are the same as these definitions:

$$\sin \theta = \frac{\text{opposite side}}{\text{hypotenuse}} = \frac{y}{r} \quad \text{and} \quad \cos \theta = \frac{\text{adjacent side}}{\text{hypotenuse}} = \frac{x}{r}$$

The advantage of this new definition of the sine and cosine is that it is now possible for the domain to be any real number. Thus we can find $\sin \theta$ and $\cos \theta$ for any real number θ in degrees or radians. Also for any value of θ there is a unique value $\frac{y}{r}$ for $\sin \theta$ and a unique value $\frac{x}{r}$ for $\cos \theta$. **Therefore the sine and cosine relations are functions**. Since the sine and cosine are functions, we can show that the cosecant, secant, tangent, and cotangent, are also functions with a few restrictions on the domains.

This new definition for the sine and cosine functions does not change the previously stated trigonometric relationships. These relationships are called **identities**. Identities are statements that are always true for all domain values of θ where the denominator is not zero.

BASIC TRIGONOMETRIC IDENTITIES

$$\csc \theta = \frac{1}{\sin \theta} \quad \text{for } \sin \theta \neq 0 \qquad \sec \theta = \frac{1}{\cos \theta} \quad \text{for } \cos \theta \neq 0$$

$$\tan \theta = \frac{\sin \theta}{\cos \theta} \quad \text{for } \cos \theta \neq 0 \qquad \cot \theta = \frac{\cos \theta}{\sin \theta} \quad \text{for } \sin \theta \neq 0$$

These four basic trigonometric identities will help us to prove other trigonometric identities. In the following example we will use the trigonometric definition, the basic trigonometric identities, algebra, and geometry to verify another trigonometric identity.

EXAMPLE 1   Prove that $\sin^2 \theta + \cos^2 \theta = 1$.

Solution By definition, $\sin \theta = \dfrac{y}{r}$ and $\cos \theta = \dfrac{x}{r}$; then $\sin^2 \theta = \left(\dfrac{y}{r}\right)^2$ and $\cos^2 \theta = \left(\dfrac{x}{2}\right)^2$. Thus

$$\sin^2 \theta + \cos^2 \theta = \frac{y^2}{r^2} + \frac{x^2}{r^2}$$

$$= \frac{y^2 + x^2}{r^2} \qquad \textit{(Since (x, y) is a point on the circle}$$
$$\textit{with radius r, we have } x^2 + y^2 = r^2)$$

$$= \frac{r^2}{r^2}$$

$$= 1$$

Therefore we have $\sin^2 \theta + \cos^2 \theta = \dfrac{r^2}{r^2} = 1$. ∎

The identity in Example 1 and two other identities, which will be proven in the exercises at the end of this section, can now be added to the list of basic identities.

OTHER BASIC TRIGONOMETRIC IDENTITIES	$\sin^2 \theta + \cos^2 \theta = 1$ $\tan^2 \theta + 1 \quad\;\; = \sec^2 \theta$ $1 + \cot^2 \theta = \csc^2 \theta$

The trigonometric functions can also be used with function composition, as the following examples illustrate.

EXAMPLE 2   Find $(f \circ g)(x)$ and $(g \circ f)(x)$ for each pair of functions:
   (a) $f(x) = \cos x$ (b) $f(x) = \sin x$
       $g(x) = x^2$ $g(x) = x^3 - 1$

Solution   Using function composition from Section 2.6, we have
(a) For $f(x) = \cos x$ and $g(x) = x^2$,

$$(f \circ g)(x) = f(g(x)) \qquad \text{and} \qquad (g \circ f)(x) = g(f(x))$$
$$= f(x^2) \qquad\qquad\qquad\qquad = g(\cos x)$$
$$= \cos (x^2) \qquad\qquad\qquad\qquad = (\cos x)^2$$
$$= \cos^2 x$$

(b) For $f(x) = \sin x$ and $g(x) = x^3 - 1$,

$$(f \circ g)(x) = f(g(x)) \qquad \text{and} \qquad (g \circ f)(x) = g(f(x))$$
$$= f(x^3 - 1) \qquad\qquad\qquad\qquad = g(\sin x)$$
$$= \sin (x^3 - 1) \qquad\qquad\qquad\qquad = (\sin x)^3 - 1$$
$$= (\sin^3 x) - 1 \qquad \text{or}$$
$$= -1 + \sin^3 x \qquad \blacksquare$$

In many problems we need to know the exact values of certain trigonometric functions rather than an approximate answer from a calculator. The following examples show how to use this new definition to find these values.

EXAMPLE 3   Find the exact values for the sine and cosine of the following angles:

(a) π **(b)** $\dfrac{3\pi}{2}$ **(c)** 2π **(d)** $\dfrac{\pi}{2}$ **(e)** 0

Solution
(a) Remember that $\pi = 180°$ and draw a circle with this angle in standard position (Fig. 5.25a). Thus the point on the circle is $(x, y) = (-r, 0)$, and the radius is r. Then

$$\sin \pi = \frac{y}{r} = \frac{0}{r} = 0 \qquad \text{and} \qquad \cos \pi = \frac{x}{r} = \frac{-r}{r} = -1$$

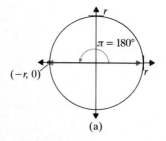

(a)

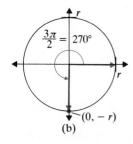

(b)

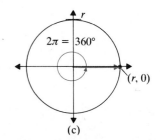

(c)

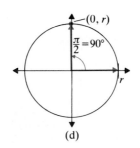

(d)

Figure 5.25

(b) First we have $\dfrac{3\pi}{2} = 270°$. Then from Fig. 5.25(b) the point $(x, y) = (0, -r)$, and the radius is r. Thus

$$\sin \frac{3\pi}{2} = \frac{y}{r} = \frac{-r}{r} = -1 \qquad \text{and} \qquad \cos \frac{3\pi}{2} = \frac{x}{r} = \frac{0}{r} = 0$$

(c) First $2\pi = 360°$. From Fig. 5.25(c) the point $(x, y) = (r, 0)$, and the radius is r. Thus

$$\sin 2\pi = \frac{y}{r} = \frac{0}{r} = 0 \qquad \text{and} \qquad \cos 2\pi = \frac{x}{r} = \frac{r}{r} = 1$$

(d) First $\dfrac{\pi}{2} = 90°$. From Fig. 5.25(d) the point $(x, y) = (0, r)$, and the radius is r. Thus

$$\sin \frac{\pi}{2} = \frac{y}{r} = \frac{r}{r} = 1 \qquad \text{and} \qquad \cos \frac{\pi}{2} = \frac{x}{r} = \frac{0}{r} = 0$$

(e) The terminal side for $\theta = 2\pi = 360°$ is coterminal with $\theta = 0°$. Thus

$$\sin 0 = \sin 360° = 0 \qquad \text{and} \qquad \cos 0° = \cos 360° = 1 \qquad \blacksquare$$

The answers from parts (d) and (e) of Example 3 prove the values that we used in the chart in Section 5.2 for sin 0°, cos 0°, sin 90°, and cos 90°.

EXAMPLE 4

Find the exact values for the six trigonometric functions of the angle α in standard position when the terminal side of α passes through the point $(3, -4)$.

Solution   When $(3, -4)$ is on the terminal side of the angle, the radius r is $r = \sqrt{(3)^2 + (-4)^2} = \sqrt{25} = 5$ (Fig. 5.26). Thus

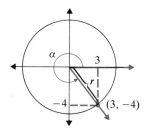

$$\sin \alpha = \frac{y}{r} = \frac{-4}{5} = -\frac{4}{5} \qquad\qquad \csc \alpha = -\frac{5}{4}$$

$$\cos \alpha = \frac{x}{r} = \frac{3}{5} \qquad\qquad \sec \alpha = \frac{5}{3}$$

$$\tan \alpha = \frac{\sin \alpha}{\cos \alpha} = \frac{-\dfrac{4}{5}}{\dfrac{3}{5}} = -\frac{4}{3} \qquad\qquad \cot \alpha = -\frac{3}{4}$$

Figure 5.26

$\blacksquare$

The next example is an interesting application of trigonometry.

EXAMPLE 5

A backpacker observes that the lightning from a thundershower emerges from the clouds with an angle of elevation of $\theta = 15°$ from the horizontal. There is a 15-second interval between seeing lightning and hearing its thunder. Find the height h of the storm and the horizontal distance x from the storm (Fig. 5.27).

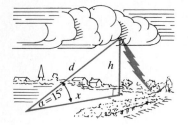

Figure 5.27

Solution The speed of sound is 333 meters per second. We let t be the time in seconds between seeing the lightning and hearing the thunder. Then the distance d to the storm is $d = 333t$. When $t = 15$ seconds,

$$d = 333(15) = 4995 \text{ meters}$$

$$= 4995 \text{ meters}\left(\frac{1 \text{ mile}}{1609 \text{ meters}}\right)$$

$$= \frac{4995}{1609} \text{ miles} \approx 3.1 \text{ miles}$$

And since $\sin \theta = \dfrac{h}{d}$ and $\cos \theta = \dfrac{x}{d}$, we have

$$h = d \sin \theta \qquad\qquad \text{and} \qquad x = d \cos \theta$$
$$h = 4995 \sin 15° \qquad\qquad\qquad\quad x = 4995 \cos 15°$$
$$h \approx 1292.8 \text{ meters} \qquad\qquad\qquad x \approx 4824.8 \text{ meters}$$

or

$$h \approx 0.8 \text{ miles} \qquad\quad \text{and} \qquad x \approx 2.998 \approx 3 \text{ miles}$$

Thus the lightning is about 3 miles away horizontally and 0.8 mile above the backpacker's present altitude. ■

We see in Example 5 that the time, $t = 15$ seconds, is equivalent to the distance, $d = 3$ miles, in $d = rt$. Thus every 5 seconds of time between seeing lightning and hearing the thunder is equal to approximately 1 mile. (*Note*: The distance formula that we used, $d = 333t$, accounts for only the speed of sound, 333 meters per second. It seems that we should also account for the speed of light, 3×10^8 meters per second. Doing this would change our equation to $d = 333.00037t$ meters per second. Since this introduces an error less than 0.00011%, far beyond the accuracy of the measurements, it is most convenient and correct for us to use only the speed of sound, 333 meters per second.)

The next example demonstrates how to find the exact value for the trigonometric functions of some obtuse angles.

EXAMPLE 6   Find the exact values for the six trigonometric functions of each angle:

$$\textbf{(a)} \;\; \alpha = \frac{3\pi}{4} \qquad \text{and} \qquad \textbf{(b)} \;\; \beta = \frac{5\pi}{3}$$

Solution

(a) Sketch the diagram where $\alpha = \dfrac{3\pi}{4}$ in standard position and solve the right triangle formed by the angle's terminal side and the x-axis (Fig. 5.28). Then

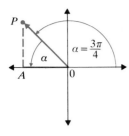

Figure 5.28

$\alpha' = \pi - \dfrac{3\pi}{4} = \dfrac{\pi}{4} = 45°$. Finding a point on the terminal side of the right triangle OAP, we arbitrarily let $OA = 1$. Then we solve for OP and AP, using α':

$$\cos \frac{\pi}{4} = \frac{OA}{OP} \qquad \text{and} \qquad \sin \frac{\pi}{4} = \frac{AP}{OP}$$

$$\frac{\sqrt{2}}{2} = \frac{1}{OP} \qquad\qquad\qquad \frac{\sqrt{2}}{2} = \frac{AP}{\sqrt{2}}$$

$$OP = \frac{2}{\sqrt{2}} = \sqrt{2} \qquad\qquad \frac{\sqrt{2}\sqrt{2}}{2} = AP$$

$$1 = AP$$

Thus with $OP = \sqrt{2}$ and $AP = 1$ we have the point $P = (-1, 1)$ on the terminal side of $\alpha = \dfrac{3\pi}{4}$ with a radius of $\sqrt{2}$. Therefore

$$\sin \frac{3\pi}{4} = \frac{y}{r} = \frac{1}{\sqrt{2}} = \frac{\sqrt{2}}{2} \qquad \csc \frac{3\pi}{4} = \sqrt{2}$$

$$\cos \frac{3\pi}{4} = \frac{x}{r} = \frac{-1}{\sqrt{2}} = -\frac{\sqrt{2}}{2} \qquad \sec \frac{3\pi}{4} = -\sqrt{2}$$

$$\tan \frac{3\pi}{4} = \frac{\dfrac{1}{\sqrt{2}}}{\dfrac{-1}{\sqrt{2}}} = -1 \qquad \cot \frac{3\pi}{4} = -1$$

(b) Sketch the diagram where $\beta = \dfrac{5\pi}{3}$ in standard position and solve the right triangle formed by the angle's terminal side and the x-axis (Fig. 5.29). Then $\beta' = 2\pi - \dfrac{5\pi}{3} = \dfrac{\pi}{3} = 60°$. We arbitrarily let $OA = 1$ and solve for OP and AP:

$$\cos \frac{\pi}{3} = \frac{OA}{OP} \qquad \text{and} \qquad \sin \frac{\pi}{3} = \frac{AP}{OP}$$

$$\frac{1}{2} = \frac{1}{OP} \qquad\qquad\qquad \frac{\sqrt{3}}{2} = \frac{AP}{2}$$

$$2 = OP \qquad\qquad\qquad \sqrt{3} = AP$$

With $OP = 2$ and $AP = \sqrt{3}$ we have the point $P = (1, -\sqrt{3})$ on the terminal side of $\beta = \dfrac{5\pi}{3}$ with a radius of 2. Therefore

$$\sin \frac{5\pi}{3} = \frac{y}{4} = \frac{-\sqrt{3}}{2} \qquad\qquad \csc \frac{5\pi}{3} = -\frac{2}{\sqrt{3}} = -\frac{2\sqrt{3}}{3}$$

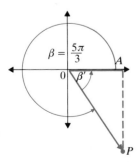

Figure 5.29

$$\cos\frac{5\pi}{3} = \frac{x}{r} = \frac{1}{2} \qquad\qquad \sec\frac{5\pi}{3} = \frac{2}{1} = 2$$

$$\tan\frac{5\pi}{3} = \frac{\dfrac{-\sqrt{3}}{2}}{\dfrac{1}{2}} = -\sqrt{3} \qquad \cot\frac{5\pi}{3} = -\frac{1}{\sqrt{3}} = -\frac{\sqrt{3}}{3}$$

■

As we see in Example 6, the acute angle between the terminal side and the x-axis is very helpful in finding the values of the trigonometric function for angles greater than $\frac{\pi}{2} = 90°$. We will now formalize these observations in a definition.

Definition

Angle θ' is called the **reference angle** for angle θ, when θ is in standard position and θ' is the acute angle between the terminal side of θ and the x-axis.

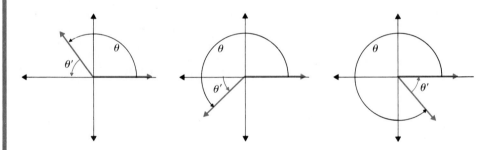

In quadrant I, $\theta' = \theta$.
In quadrant II, $\theta' = 180° - \theta$ or $\theta' = \pi - \theta$.
In quadrant III, $\theta' = \theta - 180°$ or $\theta' = \theta - \pi$.
In quadrant IV, $\theta' = 360° - \theta$ or $\theta' = 2\pi - \theta$.

From Example 6 we notice that

$$|\sin \beta| = \left|\sin\frac{5\pi}{3}\right| = \left|-\frac{\sqrt{3}}{2}\right| = \frac{\sqrt{3}}{2}$$

$$\sin \beta' = \sin\frac{\pi}{3} = \frac{\sqrt{3}}{2}$$

and for β in quadrant IV,

$$\sin \beta = -\sin \beta' = -\frac{\sqrt{3}}{2}$$

Thus the value of sin β can be found by using the product of the sin β' and $+1$ or -1, depending on the quadrant of β and the specific trigonometric function. In this case, β is in quadrant IV, where the sine is negative. Therefore

$$\sin \beta = \sin \frac{5\pi}{3} = (-1) \sin \beta' = -\frac{\sqrt{3}}{2}$$

In general, this observation can be summarized as follows:

TRIGONOMETRIC FUNCTION VALUES FOR NONACUTE ANGLES

For angle θ, find its reference θ'; then

$$\sin \theta = (\pm 1) \sin \theta' \qquad \csc \theta = (\pm 1) \csc \theta'$$
$$\cos \theta = (\pm 1) \cos \theta' \qquad \sec \theta = (\pm 1) \sec \theta'$$
$$\tan \theta = (\pm 1) \tan \theta' \qquad \cot \theta = (\pm 1) \cot \theta'$$

The factors $+1$ and -1 are based on the function and the quadrant as described in the following chart:

For θ in Quadrant	Use $(+1)$ for	Use (-1) for
I	all functions	no functions
II	$\sin \theta$, $\csc \theta$	$\cos \theta$, $\sec \theta$, $\tan \theta$, $\cot \theta$
III	$\tan \theta$, $\cot \theta$	$\sin \theta$, $\csc \theta$, $\cos \theta$, $\sec \theta$
IV	$\cos \theta$, $\sec \theta$	$\sin \theta$, $\csc \theta$, $\tan \theta$, $\cot \theta$

There are several ways to determine the factor of $+1$ or -1 when using a reference angle. Another common and effective method is to use the positive or negative values of x and y from the point (x, y) on the terminal side of angle θ in standard position. By using x, y, and the trigonometric definitions it is easy to determine the factor of $+1$ or -1 for a trigonometric function when using the reference angle.

This factor of $+1$ or -1 can also be easily remembered by Fig. 5.30 and the following mnemonic:

All   (all functions positive)
Scholars   (the sine and its reciprocal are positive)
Take   (tangent and its reciprocal are positive)
Calculus   (the cosine and its reciprocal are positive)

We will now use the concept of reference angles in an example.

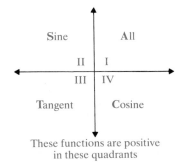

Sine   All

        II | I
        III | IV

Tangent   Cosine

These functions are positive
in these quadrants

Figure 5.30

EXAMPLE 7   Find the exact values for the six trigonometric functions of $\alpha = \dfrac{7\pi}{6}$.

Solution   For $\alpha = \dfrac{7\pi}{6}$ in quadrant IV (Fig. 5.31) we have

$$\alpha' = \frac{7\pi}{6} - \pi = \frac{7\pi}{6} - \frac{6\pi}{6} = \frac{\pi}{6}$$

Since α is in quadrant III, then only tangent and cotangent are positive. Therefore

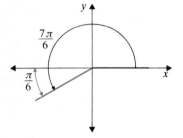

$$\sin \frac{7\pi}{6} = -\sin \frac{\pi}{6} = -\frac{1}{2} \qquad\qquad \csc \frac{7\pi}{6} = -2$$

$$\cos \frac{7\pi}{6} = -\cos \frac{\pi}{6} = -\frac{\sqrt{3}}{2} \qquad\qquad \sec \frac{7\pi}{6} = -\frac{2}{\sqrt{3}} = -\frac{2\sqrt{3}}{3}$$

$$\tan \frac{7\pi}{6} = \tan \frac{\pi}{6} = \frac{\dfrac{1}{2}}{\dfrac{\sqrt{3}}{2}} = \frac{1}{\sqrt{3}} = \frac{\sqrt{3}}{3} \qquad \cot \frac{7\pi}{6} = \sqrt{3}$$

Figure 5.31

EXAMPLE 8   Find the exact values for the six trigonometric functions of the angle β in standard position with its terminal in quadrant III perpendicular to the line $y = -\frac{3}{5}x - 3$.

Solution   Since the terminal side of β is perpendicular to the line $y = -\frac{3}{5}x - 3$, the slope of the terminal side is the negative reciprocal of the slope of the line. The slope of the line $y = -\frac{3}{5}x - 3$ is $-\frac{3}{5}$, and thus the slope of the terminal side of β is $m = \left(-\frac{5}{3}\right) = \frac{5}{3}$. As we see in Fig. 5.32, the point $(-3, -5)$ is on the terminal side of angle β, since the slope between $(0, 0)$ and $(-3, -5)$ is $\frac{5}{3}$. The radius from $(0, 0)$ to $(-3, -5)$ is $r = \sqrt{(-3)^2 + (-5)^2} = \sqrt{24} = 2\sqrt{6}$. Thus

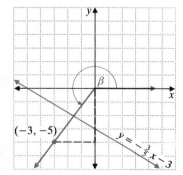

$$\sin \beta = \frac{y}{r} = \frac{-5}{2\sqrt{6}} = -\frac{5\sqrt{6}}{12} \qquad \csc \beta = -\frac{2\sqrt{6}}{5}$$

$$\cos \beta = \frac{x}{r} = \frac{-3}{2\sqrt{6}} = -\frac{3\sqrt{6}}{12} \qquad \sec \beta = -\frac{2\sqrt{6}}{3}$$

$$\tan \beta = \frac{\sin \beta}{\cos \beta} = \frac{\dfrac{-5}{2\sqrt{6}}}{\dfrac{-3}{2\sqrt{6}}} = \frac{5}{3} \qquad \cot \beta = \frac{3}{5}$$

Figure 5.32

Exercises 5.3

Find the exact value of the six trigonometric functions for each of the following angles in standard position with the given point on the terminal side.

1. $(-3, 4)$ **2.** $(-4, -3)$

3. $(4, -3)$ **4.** $(-6, 8)$

5. $(-8, -6)$ **6.** $(8, -6)$

7. $(5, -12)$ **8.** $(-12, -5)$

9. $(-5, -2)$ **10.** $(2, -5)$

11. $(-\sqrt{3}, 1)$ **12.** $(-1, \sqrt{3})$

13. A vertical pine tree on a hill has an 83-foot shadow on the downhill side of the tree (Fig. 5.33). The hill has a 10° angle of elevation, and the sun's angle of elevation is 68°. How tall is the tree?

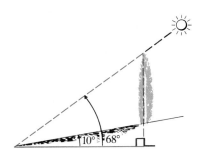

Figure 5.33

14. Find the height of the tree in Exercise 13 if the 83-foot shadow is on the uphill side of the tree as shown in Fig. 5.34.

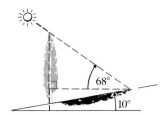

Figure 5.34

15. A 2100 pound crate is on a ramp with a 20° angle of elevation. How much of its weight, s, is being supported by the ramp and how much weight, r, is directed down the ramp as shown in Fig. 5.35?

16. A 1300 pound crate is on a ramp with a 10° angle of elevation. How much of its weight, s, is being supported by the ramp(s) and how much weight, r, is directed down the ramp as shown in Fig. 5.35?

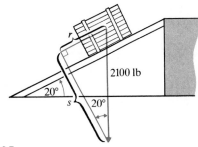

Figure 5.35

Find the exact values of $\cos \theta$, $\csc \theta$, and $\cot \theta$, using the correct reference angles (θ'), when θ is

17. 135°   **18.** 210°

19. 120°   **20.** 300°

21. $\dfrac{7\pi}{4}$ **22.** $\dfrac{5\pi}{4}$

23. $\dfrac{11\pi}{6}$ **24.** $\dfrac{5\pi}{6}$

25. $\dfrac{4\pi}{3}$ **26.** $\dfrac{7\pi}{3}$

Find the exact values of $\sin \theta$, $\sec \theta$, and $\tan \theta$, using the correct reference angles (θ'), when θ is

27. 135°   **28.** 345°

29. 570°   **30.** 495°

31. $\dfrac{17\pi}{6}$ **32.** $\dfrac{8\pi}{3}$

33. $\dfrac{9\pi}{4}$ **34.** $\dfrac{13\pi}{6}$

35. $\dfrac{11\pi}{3}$ **36.** $\dfrac{19\pi}{6}$

37. A hiker observes the time between seeing lightning and hearing thunder to be 13 seconds. The angle of elevation α of the initial point of the lightning is 10° (Fig. 5.36). Find the distance to the lightning d, the altitude h, and the horizontal distance x from the lightning.

38. A hiker observes the time between seeing lightning and hearing thunder to be 18 seconds. The angle of elevation α of the initial point of the lightning is 20° (Fig. 5.36). Find the distance to the lightning d, the altitude h, and the horizontal distance x from the lightning.

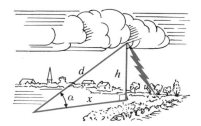

Figure 5.36

Find (a) $(f \circ g)(x)$ and (b) $(g \circ f)(x)$ for the following pairs of functions.

39. $f(x) = \sin x$
$g(x) = 2x$

40. $f(x) = \cos x$
$g(x) = 2x + 1$

41. $f(x) = \cos x$
$g(x) = x^2 + 1$

42. $f(x) = \sin x$
$g(x) = x^3$

43. $f(x) = \tan (x^2)$
$g(x) = \sqrt{x}$

44. $f(x) = \sqrt[3]{\sin x}$
$g(x) = x^3$

45. $f(x) = \sqrt{x}$
$g(x) = \cos^2 x$

46. $f(x) = \sqrt{x}$
$g(x) = \sin^2 x$

47. A unit circle (a circle of radius 1) is constructed as shown in Fig. 5.37 with a line through the point $A = (1, 0)$ perpendicular to the radius of the circle at point A. The terminal side of angle θ intersects the circle at $P = (x, y)$ and intersects the perpendicular line at $B = (1, z)$. Prove that $\sin \theta = y$, $\cos \theta = x$, $\tan \theta = z$, and $\sec \theta = w$.

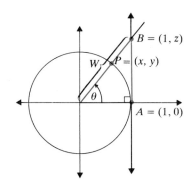

Figure 5.37

48. A unit circle is constructed as shown in Fig. 5.38 with a line through the point $C = (1, 0)$ perpendicular to the radius at point C. The terminal side of θ intersects the circle at $P = (x, y)$ and intersects the perpendicular line at $D = (z, 1)$. Prove that $\cot \theta = z$ and $\csc \theta = w$.

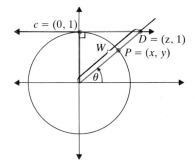

Figure 5.38

49. Use the identity $\sin^2 \alpha + \cos^2 \alpha = 1$ to prove that $\tan^2 \alpha + 1 = \sec^2 \alpha$.

50. Prove that $1 + \cot^2 \alpha = \csc^2 \alpha$ by using the identity $\sin^2 \alpha + \cos^2 \alpha = 1$.

51. Find sin α, sec α, and tan α when α is in standard position with its terminal side in quadrant III parallel to $y = 3x - 5$.

52. Find sin α, sec α, and tan α when α is in standard position with its terminal side in quadrant II parallel to $y = -3x + 5$.

53. Find cos β, csc β, and cot β when β is in standard position with its terminal side in quadrant II perpendicular to $y = 2x + 5$.

54. Find cos β, csc β, and cot β when β is in standard position with its terminal side in quadrant III perpendicular to $y = -2x + 1$.

5.4   Basic Graphs and Inverse Functions

We first defined the trigonometric functions using a right triangle. Then to enlarge the domains to all real numbers, we defined the trigonometric functions using a circle with any radius r. Now we will let $r = 1$ and define the trigonometric functions using a unit circle (a circle with a radius of 1 and center at the origin). The unit circle will make it easier to find the values of the sin θ and cos θ and also make it easier to graph the sine and cosine functions.

Definition

Unit Circle Trigonometric Functions. For any point $P = (x, y)$ on the unit circle with center at the origin,

$$\sin \theta = \frac{y}{1} = y \quad \text{and} \quad \cos \theta = \frac{x}{1} = x$$

The other four trigonometric functions, for nonzero denominators, are

$$\tan \theta = \frac{\sin \theta}{\cos \theta} \qquad \csc \theta = \frac{1}{\sin \theta}$$

$$\cot \theta = \frac{1}{\tan \theta} = \frac{\cos \theta}{\sin \theta} \qquad \sec \theta = \frac{1}{\cos \theta}$$

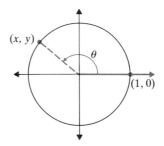

This definition will now help us to develop some identities as well as sketch the graph of the sine and cosine functions.

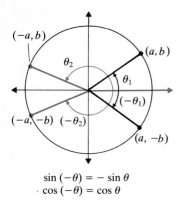

$$\sin(-\theta) = -\sin\theta$$
$$\cos(-\theta) = \cos\theta$$

Figure 5.39

Trigonometric Function of Negative Angles

We will use the definition above to derive some identities for negative angles. First, we will consider the two angles θ and $-\theta$ when the terminal side of θ is in quadrant I on the unit circle. Using Fig. 5.39, we can compile the following observations:

Quadrant	Angle	Point on Terminal Side of the Angle with $a > 0$ and $b > 0$	Value of Sine and Cosine Functions	
I	θ_1	(a, b)	$\sin\theta = b$	$\cos\theta = a$
IV	$-\theta_1$	$(a, -b)$	$\sin(-\theta) = -b$	$\cos(-\theta) = a$

For θ_1 in quadrant I and the value of the trigonometric functions above, we can conclude that

$$\sin(-\theta_1) = -\sin\theta_1 \qquad \text{and} \qquad \cos(-\theta_1) = \cos\theta_1$$

We will now complete the same process for θ_2 in quadrant II.

Using Fig. 5.39, we can compile the following observations:

Quadrant	Angle	Point on Terminal Side of the Angle with $a > 0$ and $b > 0$	Value of Sine and Cosine Functions	
II	θ_2	$(-a, b)$	$\sin\theta = b$	$\cos\theta = -a$
III	$-\theta_2$	$(-a, -b)$	$\sin(-\theta) = -b$	$\cos(-\theta) = -a$

For θ_2 in quadrant II and the value of the trigonometric functions above, we can conclude that

$$\sin(-\theta_2) = -\sin\theta_2 \qquad \text{and} \qquad \cos(-\theta_2) = \cos\theta_2$$

Notice that we have already taken care of quadrants III and IV, since

$$-(-\theta_1) = \theta_1 \qquad \text{and} \qquad -(-\theta_2) = \theta_2$$

Thus we now have two more basic identities.

NEGATIVE ANGLE
IDENTITIES

For any θ, $\theta \in R$,

$$\sin(-\theta) = -\sin\theta \qquad \text{and} \qquad \cos(-\theta) = \cos\theta$$

These new identities may remind you of the definition of odd and even functions. We will consider this idea later in the exercise set. We now add these identities to our list of basic identities.

Graphing sin θ and cos θ

The domain of each trigonometric function is a real number with units in degrees or radians, and the range is a real number. Thus we can graph the trigonometric functions on a Cartesian coordinate system with the domain, the measure of angle θ, on the horizontal axis and the real numbers of the range on the vertical axis. The trigonometric function $\sin \theta = y$ would consist of all ordered pairs (θ, y), and the trigonometric function $\cos \theta = x$ would consist of all ordered pairs (θ, x). We will now graph $\sin \theta = y$ using the unit circle in Fig. 5.40 and θ values with reference angles of $\theta' \in \left\{ 0, \dfrac{\pi}{6}, \dfrac{\pi}{4}, \dfrac{\pi}{3}, \text{ and } \dfrac{\pi}{2} \right\}$. Connecting these points, we have the graph of $\sin \theta = y$.

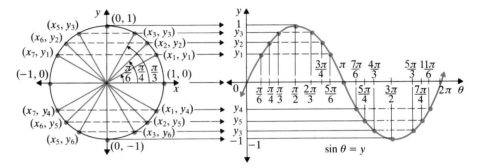

Figure 5.40

Similarly, we will graph $\cos \theta = x$ using the unit circle in Fig. 5.41 and θ values with the reference angles $\theta' \in \left\{ 0, \dfrac{\pi}{6}, \dfrac{\pi}{4}, \dfrac{\pi}{3}, \dfrac{\pi}{2} \right\}$. Connecting these points provides the graph of $\cos \theta = x$.

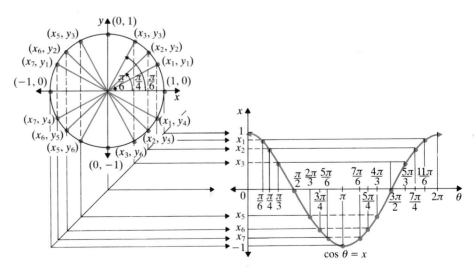

Figure 5.41

Periodic Functions

When we used the unit circle to graph the sine and cosine functions, we restricted θ to $0 \le \theta \le 2\pi$, one revolution around the circle (where 0 and 2π duplicate the same point on the unit circle). Although each additional revolution around the unit circle produces no new values for the $\sin \theta$ or $\cos \theta$, it extends the domain θ to all positive real numbers. With the identities for negative angles the domain θ can also be extended to all negative real numbers. Thus the following graphs provide a more complete picture of $y = \sin \theta$ and $y = \cos \theta$ (Figs. 5.42 and 5.43).

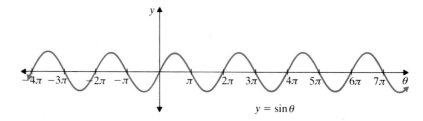

$y = \sin \theta$

Figure 5.42

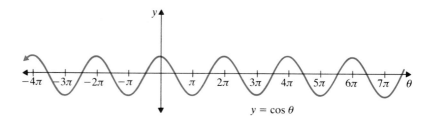

$y = \cos \theta$

Figure 5.43

It is interesting to note the repetitive nature of these graphs. The graph of the function for the domain interval of $[0, 2\pi)$ is repeated over and over every 2π interval of the domain on both graphs. This type of function is called a periodic function. We define a periodic function as follows.

Definition

The function $f(x)$ is a **periodic function** if

$$f(x + k) = f(x)$$

for every value x in the domain. The smallest positive number k that satisfies this equation is the **period** of f.

As indicated by their graphs both the sine and cosine functions have a period of 2π.

EXAMPLE 1   For $-\pi \le \theta \le 3\pi$, sketch the graph of (a) $y = \frac{1}{2} \cos \theta$ and (b) $y = 3 \sin \theta$.

Solution

(a) We first sketch the dashed graph of $y = \cos \theta$. Then we multiply each range value by $\frac{1}{2}$ to obtain the solid graph of $y = \frac{1}{2} \cos \theta$ (Fig. 5.44a).

(b) We first sketch the dashed graph of $y = \sin \theta$. Then we multiply each range value by 3 to obtain the solid graph of $y = 3 \sin \theta$ (Fig. 5.44b).

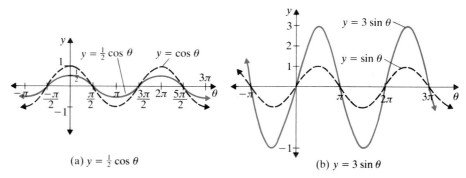

(a) $y = \frac{1}{2} \cos \theta$ (b) $y = 3 \sin \theta$

Figure 5.44

Graphing the Other Trigonometric Functions

To graph $\tan \theta$, we remember that $\tan \theta$ equals $\dfrac{\sin \theta}{\cos \theta}$ and has a vertical asymptote whenever $\cos \theta = 0$. Thus the tangent has vertical asymptotes at $\theta \in \left\{ \pm \dfrac{\pi}{2}, \right.$ $\pm \dfrac{3\pi}{2}, \pm \dfrac{5\pi}{2}, \pm \dfrac{7\pi}{2}, \ldots \left. \right\}$. It can be shown by similar triangles in Fig. 5.45 that $\tan \theta = \dfrac{y}{x} = z$ for $-\dfrac{\pi}{2} < \theta < \dfrac{\pi}{2}$. Thus as θ goes from 0 toward $\dfrac{\pi}{2}$, z is positive and $\tan \theta = z$ increases from 0 toward $+\infty$. And as θ goes from 0 toward $-\dfrac{\pi}{2}$, z is negative and $\tan \theta = z$ decreases from 0 toward $-\infty$.

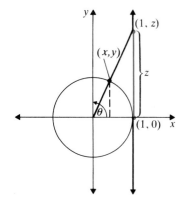

Figure 5.45

We also note that $\tan \dfrac{\pi}{4} = \dfrac{\sin \dfrac{\pi}{4}}{\cos \dfrac{\pi}{4}} = 1$ and $\tan \left(-\dfrac{\pi}{4} \right) = \dfrac{\sin \left(-\dfrac{\pi}{4} \right)}{\cos \left(-\dfrac{\pi}{4} \right)} = -1$.

Using this information, we graph one part of $\tan \theta = z$ for $-\dfrac{\pi}{2} < \theta < \dfrac{\pi}{2}$ (Fig. 5.46).

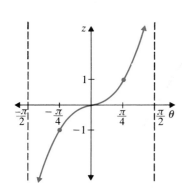

Figure 5.46

To graph the rest of the graph of tan θ, we notice the following:

(a) With $\cos\left(\dfrac{\pi}{2} + k\pi\right) = 0$ for any integer k, tan θ has a vertical asymptote

for $\theta = \dfrac{\pi}{2} + k\pi$, as noted above.

(b) With $\sin k\pi = 0$ for any integer k, $\tan k\pi = 0$.

(c) With $\sin\left(\dfrac{\pi}{4} + k\pi\right) = \cos\left(\dfrac{\pi}{4} + k\pi\right)$ for any integer k,

$\tan\left(\dfrac{\pi}{4} + k\pi\right) = 1$.

(d) With $\sin\left(-\dfrac{\pi}{4} + k\pi\right) = -\cos\left(-\dfrac{\pi}{4} + k\pi\right)$ for an integer k,

$\tan\left(-\dfrac{\pi}{4} + k\pi\right) = -1$.

We can now use this information to fill in the graph of tan θ in Fig. 5.47. You can use a calculator to confirm more points on this graph if desired. As seen in this graph and the steps above,

the period for $y = \tan \theta$ is π

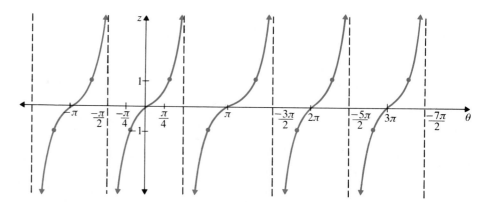

Figure 5.47

We can now graph the cosecant, secant, and cotangent functions as the reciprocals of the sine, cosine, and tangent functions, using the techniques of Section 2.5. The graphs of $y = \csc x$, $y = \sec x$, and $y = \cot x$ are presented in Fig. 5.48.

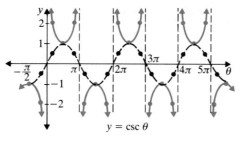

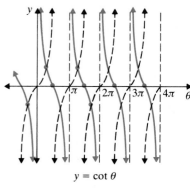

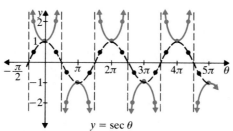

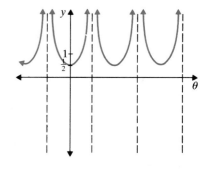

Figure 5.48

EXAMPLE 2

Sketch the graph of $y = \left| \frac{1}{2} \sec \theta \right|$.

Solution   To sketch the graph of $y = \left| \frac{1}{2} \sec \theta \right|$, we will first sketch the graph of $y = \frac{1}{2} \sec \theta$, with each range value being half the range value of $y = \sec \theta$ (Fig. 5.49). Then we take the absolute value of each range value to finish the graph of $y = \left| \frac{1}{2} \sec \theta \right|$ (Fig. 5.50).

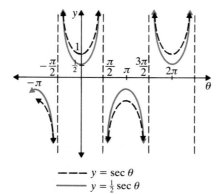

$---$ $y = \sec \theta$
$\underline{\quad\quad}$ $y = \frac{1}{2} \sec \theta$

Figure 5.49

Figure 5.50

Inverse Trigonometric Functions

As we noted before, the $\sin^{-1}$ key on a calculator represents the inverse function of the sine and not its reciprocal. As we discussed in Section 2.7, the function and its inverse function f^{-1} are defined as

$$f(f^{-1}(x)) = x = f^{-1}(f(x))$$

Thus

$$\sin(\sin^{-1} x) = x = \sin^{-1}(\sin x)$$

$$\cos(\cos^{-1} x) = x = \cos^{-1}(\cos x)$$

$$\tan(\tan^{-1} x) = x = \tan^{-1}(\tan x)$$

We will work with inverse functions in more detail in Section 6.6. At this point we will study inverse functions so that we can use a calculator to solve certain trigonometric equations. It is important to recognize that with $\sin \theta = y$ we have

$$\sin^{-1} y = \sin^{-1}(\sin \theta) \quad \textit{(Substituting } y = \sin \theta\textit{)}$$

$$\sin^{-1} y = \theta \quad\qquad \textit{(Definition of inverse functions)}$$

We can do this with the other trigonometric functions with the following results:

$$\sin^{-1} x = \theta, \quad \cos^{-1} x = \theta, \quad \text{and} \quad \tan^{-1} x = \theta$$

It is important to remember that **the value of an inverse trigonometric function is an angle**.

As we mentioned in Section 2.7, only one-to-one functions have inverse functions. This is a problem, since none of the trigonometric functions are one-to-one. To have an inverse trigonometric function, we must restrict each inverse function to a one-to-one branch or section of the trigonometric function. Thus we have $-\dfrac{\pi}{2} \le \sin^{-1} y \le \dfrac{\pi}{2}$, since the sine is one-to-one on the interval $\left[-\dfrac{\pi}{2}, \dfrac{\pi}{2}\right]$ (Fig. 5.51). Similarly, we have $0 \le \cos^{-1} y \le \pi$, since the cosine is one-to-one on the interval $[0, \pi]$ and $-\dfrac{\pi}{2} < \tan^{-1} y < \dfrac{\pi}{2}$, since the tangent is one-to-one on the interval $\left(-\dfrac{\pi}{2}, \dfrac{\pi}{2}\right)$ (Fig. 5.51).

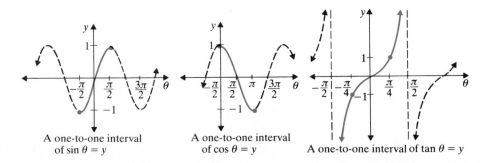

Figure 5.51

A one-to-one interval of $\sin \theta = y$

A one-to-one interval of $\cos \theta = y$

A one-to-one interval of $\tan \theta = y$

INVERSE TRIGONOMETRIC FUNCTIONS	For real numbers y and θ,

$$\sin^{-1} y = \theta \qquad \text{if and only if} \qquad y = \sin \theta \quad \text{for } -\frac{\pi}{2} \le \theta \le \frac{\pi}{2}$$

$$\cos^{-1} y = \theta \qquad \text{if and only if} \qquad y = \cos \theta \quad \text{for } 0 \le \theta \le \pi$$

and

$$\tan^{-1} y = \theta \qquad \text{if and only if} \qquad y = \tan \theta \quad \text{for } -\frac{\pi}{2} < \theta < \frac{\pi}{2}$$

These inverse functions now make it possible to find the angles in triangles when sides are given, as we will see in the following examples.

EXAMPLE 3   A camera is to take a family picture from a distance of 10 feet. Standing next to each other, the family takes up a distance of 8 feet. What is the angle α that the lens of the camera must be able to photograph (Fig. 5.52)?

Solution   First we draw in the altitude to point C and then

$$\tan \frac{\alpha}{2} = \frac{4}{10}$$

$$\frac{\alpha}{2} = \tan^{-1} \frac{4}{10} = \tan^{-1} 0.4$$

Thus

$$\frac{\alpha}{2} \approx 21.80°$$

and

$$\alpha \approx 43.6°$$ ∎

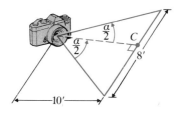

Figure 5.52

We must be careful when evaluating inverse trigonometric functions. Since both

$$\sin 43.6° = 0.689619543 \qquad \text{and} \qquad \sin 136.4° = 0.689619543$$

the solution to $\sin \alpha = 0.689619543$ could be either $\alpha = 43.6°$ or $\alpha = 136.4°$ when α is in $[0, 360°)$. In general, when α is any angle, the solution to $\sin \alpha = 0.689619543$ is $\alpha = (43.6 + 360k)°$ and $\alpha = (136.4 + 360k)°$ for any integer k. On the other hand, the solution to $\alpha = \sin^{-1} (0.689619543)$ is only one value, $\alpha = 43.6°$, since $-90° \le \sin^{-1} \theta \le 90°$. Thus $\sin \alpha = y$ has an infinite number of y values, while $\sin^{-1} y = \alpha$ has only one value for α because of the restrictions on the inverse trigonometric functions. It is convenient that these restrictions are already built into calculators.

Care must be taken in evaluating inverse trigonometric functions. The following examples demonstrate solving problems with these inverse functions.

EXAMPLE 4   Find the exact value for each of the following:

(a) $\sin^{-1}\left(-\dfrac{\sqrt{2}}{2}\right)$,   (b) $\cos^{-1}\dfrac{1}{2}$, and   (c) $\tan^{-1}(-1)$.

Solution

(a) If the value of $\sin^{-1}\left(-\dfrac{\sqrt{2}}{2}\right)$ is an angle, then

$$\sin^{-1}\left(-\frac{\sqrt{2}}{2}\right) = \beta$$

Thus $\sin\beta = -\dfrac{\sqrt{2}}{2}$. The reference angle is $\beta' = \dfrac{\pi}{4}$, and $\sin\beta$ is negative in quadrants III and IV. Therefore with

$$-\frac{\pi}{2} \leq \sin^{-1} x \leq \frac{\pi}{2}$$

we have $\beta = -\dfrac{\pi}{4}$.

(b) If the value of $\cos^{-1}\dfrac{1}{2}$ is an angle, then

$$\cos^{-1}\frac{1}{2} = \theta$$

and

$$\cos\theta = \frac{1}{2}$$

The reference angle is $\theta' = \dfrac{\pi}{3}$, and $\cos\theta$ is positive in quadrants I and IV. Therefore with

$$0 \leq \cos^{-1} x \leq \pi$$

we have $\theta = \dfrac{\pi}{3}$.

(c) If the value of $\tan^{-1}(-1)$ is an angle, then

$$\tan^{-1}(-1) = \phi$$

and

$$\tan \phi = -1$$

The reference angle is $\phi' = \dfrac{\pi}{4}$, and $\tan \phi$ is negative in quadrants II and IV. Therefore with

$$-\frac{\pi}{2} \le \tan^{-1} x \le \frac{\pi}{2}$$

we have $\phi = -\dfrac{\pi}{4}$.   ■

EXAMPLE 5   Find the value of each of the following angles using a calculator:
(a) θ when $\sin \theta = \frac{4}{5}$ and $\sec \theta < 0$,
(b) ϕ when $\cos \phi = -\frac{4}{5}$ and $\tan \phi > 0$.

Solution

(a) If $\sin \theta = \dfrac{4}{5}$, then $\sin \theta > 0$, and θ is in quadrant I or II. We also have $\sec \theta = \dfrac{1}{\cos \theta} < 0$, so θ is in quadrant II or III. Both conditions are true when θ is in quadrant II. Using a calculator, we get

$$\sin^{-1} \frac{4}{5} = \theta' \approx 53.13°$$

And since θ is in quadrant II, we have

$$\theta \approx 180 - \theta' \approx 126.87°$$

(b) If $\cos \phi = -\dfrac{4}{5}$, then $\cos \phi < 0$, and ϕ is in quadrant II or III. We also have $\tan \phi > 0$, so ϕ is in quadrant I or III. Both conditions are met when ϕ is in quadrant III. Using a calculator, we get

$$\cos^{-1}\left(-\frac{4}{5}\right) \approx 143.13°$$

and

$$\phi' \approx 180° - 143.13° = 36.87°$$

Since ϕ must be in quadrant III, we have

$$\phi \approx 180° + 36.87° = 216.87° \quad\quad ■$$

Exercises 5.4

Find the angle that is the solution to each of the following. (Use a calculator.)

1. $\sin^{-1} 0.3 = \alpha$

2. $\cos^{-1} 0.7 = \beta$

3. $\cos^{-1} (-0.2) = \theta$

4. $\sin^{-1} (-0.4) = \phi$

5. $\tan^{-1} (1.5) = \alpha$

6. $\tan^{-1} (0.3) = \beta$

7. $\csc^{-1} 5 = \theta$

8. $\sec^{-1} 7 = \phi$

Find the exact value of each of the following angles. (Do not use a calculator.)

9. $\sin^{-1} \left(-\dfrac{1}{2}\right) = \alpha$

10. $\cos^{-1} \left(-\dfrac{1}{2}\right) = \beta$

11. $\cos^{-1} \left(-\dfrac{\sqrt{3}}{2}\right) = \theta$

12. $\sin^{-1} \left(-\dfrac{\sqrt{3}}{2}\right) = \phi$

13. $\tan^{-1} (\sqrt{3}) = \alpha$

14. $\tan^{-1} \left(\dfrac{\sqrt{3}}{3}\right) = \beta$

15. $\sec^{-1} (2) = \theta$

16. $\csc^{-1} (2) = \phi$

Identify each graph as one of the following: (a) $y = \sin \theta$, (b) $y = \cos \theta$, (c) $y = \tan \theta$, (d) $y = \csc \theta$, (e) $y = \sec \theta$, and (f) $y = \cot \theta$.

17.

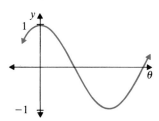

18.

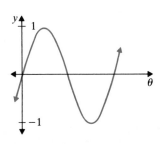

19.

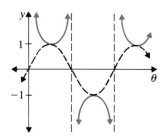

20.

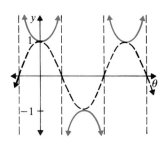

Use the graphing techniques of this section and the techniques in Section 2.3 and 2.5 for graphing multiple or reciprocal functions to sketch the graphs of the following trigonometric functions. Sketch each graph on the interval of $[-\pi, 3\pi]$.

21. $y = \sin \theta$

22. $y = \cos \theta$

23. $y = \sec \theta$

24. $y = \csc \theta$

25. $y = 3 \cos \theta$

26. $y = 2 \sin \theta$

27. $y = \frac{1}{2} \csc \theta$

28. $y = 2 \sec \theta$

29. $y = 3 \sec \theta$

30. $y = \frac{1}{3} \csc \theta$

31. $y = \tan \theta$

32. $y = \cot \theta$

33. $y = |2 \cos \theta|$

34. $y = |3 \sin \theta|$

Find the six trigonometric functions of the angles described in the following problems. (Use exact values.)

35. θ when $\sin \theta = \dfrac{4}{5}$ and $\cos \theta < 0$

36. α when $\cos \alpha = -\dfrac{4}{5}$ and $\sin \alpha < 0$

37. β when $\sin \beta = -\dfrac{\sqrt{3}}{2}$ and $\tan \beta > 0$

38. ϕ when $\cos \phi = \dfrac{1}{2}$ and $\cot \phi < 0$

39. α when $\tan \alpha = -1$ and $\sec \alpha > 0$

40. β when $\tan \beta = \sqrt{3}$ and $\csc \beta > 0$

41. Verify that $\cot (-\theta) = -\cot \theta$.

42. Verify that $\sec (-\theta) = \sec \theta$.

43. Verify that $\csc (-\theta) = -\csc \theta$.

44. Verify that $\tan (-\theta) = -\tan \theta$.

45. A photographer wants to photograph a 5280-foot bridge from a point 100 feet from one end of the bridge perpendicular to the bridge (Fig. 5.53). What angle α must the lens be able to photograph?

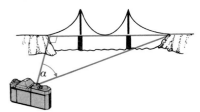

Figure 5.53

46. A portrait of a family is to be taken from a distance of 8 feet. If the width to be photographed is 12 feet, what angle α must the camera lens be able to record (Fig. 5.54)?

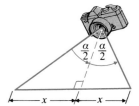

Figure 5.54

47. A new mountain expressway is to be built. One route will increase in altitude 2 miles with each 5 miles of road (Fig. 5.55).
  (a) What is the angle of elevation of the roadway?
  (b) How much more roadway must be added to change this angle by 2°?

Figure 5.55

48. Repeat Exercise 47 when the altitude must increase 1.25 miles for each 5 miles of roadway.

49. A six-foot-tall man has an 8-foot shadow. What is the angle of elevation of the sun?

50. A six-foot-tall man has a 15-foot shadow. What is the angle of elevation of the sun?

51. A communications satellite's orbit is 26,000 miles above the earth, which has a radius of about 4000 miles (Fig. 5.56).
  (a) What is the smallest angle α that the satellite would need to send a message to cover the largest portion of the earth?
  (b) What percent of the earth's circumference can one satellite cover?
  (c) How many of these satellites would it take to cover the whole circumference of the earth? (Solve this in two dimensions.)

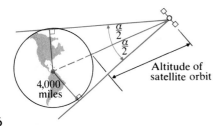

Figure 5.56

52. Repeat Exercise 51 for a disabled satellite that gained an altitude of only 16,000 miles above the earth.

53. At what angle should an aircraft climb to clear a mountain pass 5 miles away that is 8000 feet higher than its present altitude (Fig. 5.57)?

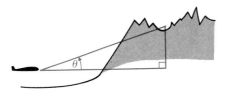

Figure 5.57

54. At what angle should an aircraft climb to clear a mountain pass 13 miles away that is 8000 feet higher than its present altitude?

55. Is $y = \sin \alpha$ an even function, an odd function, or neither? Prove your result.

56. Is $y = \cos \alpha$ an even function, an odd function, or neither? Prove your result.

5.5   Graphing Trigonometric Functions

In Section 5.4 we graphed the sine, cosine, tangent, cosecant, secant, and cotangent functions. You should be able to recognize and identify these basic graphs. In this section we will develop techniques for graphing more general trigonometric functions such as

$$y = k + a \sin b(x - h)$$
$$y = k + a \cos b(x - h)$$
$$y = k + a \tan b(x - h)$$

and their reciprocal functions. We will now see how the values of a, b, h, and k change the shape of the basic graphs developed in Section 5.4.

Change in Amplitude

In nature and technology the mathematical description of oscillations is based on the sine and cosine functions. The greatest displacement of $y = a \sin \theta$ and $y = a \cos \theta$ from the line $y = 0$ is called the amplitude and has a value equal to $|a|$. In these oscillations the period is often called the wavelength λ. For the functions $y = \sin \theta$ and $y = \cos \theta$ the amplitude is $|a| = |1| = 1$, and the wavelength is $\lambda = 2\pi$. **The amplitude of $y = a \sin x$ and $y = a \cos x$ is $|a|$.**

The amplitude provides the upper and lower boundaries for the range values of these graphs. Notice the geometric significance of the amplitudes in the graphs of $y = \sin \theta$ with amplitude of 1 and $y = 3 \cos \theta$ with amplitude $|3| = 3$ in Figs. 5.58(a) and 5.58(b).

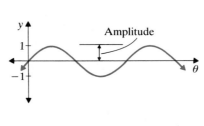   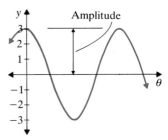

(a) $y = \sin \theta$ with amplitude of 1   (b) $y = 3\cos \theta$ with amplitude of 3

Figure 5.58

Change in Period

As we saw in Section 5.4, $\sin \theta$ and $\cos \theta$ both have a period, or wavelength λ, of 2π. What is the period for the functions $y = \sin 2\theta$ and $y = \cos \frac{1}{3}\theta$? When considering the period of $y = \sin 2\theta$, we notice the values of 2θ for each domain value θ in the following chart.

θ	0	$\dfrac{\pi}{6}$	$\dfrac{\pi}{4}$	$\dfrac{\pi}{3}$	π
2θ	0	$\dfrac{\pi}{3}$	$\dfrac{\pi}{2}$	$\dfrac{2\pi}{3}$	2π

For θ in $[0, \pi)$ we see that 2θ is in $[0, 2\pi)$, producing one period of $y = \sin 2\theta$. Thus $y = \sin 2\theta$ has a period of π, $\frac{1}{2}$ of its normal period of 2π; the new period is $\left(\frac{1}{2}\right)2\pi = \pi$. We now consider the period of $y = \cos \frac{1}{3}\theta$. For each value of θ we observe the values of $\frac{1}{3}\theta$ in the following chart.

θ	0	π	2π	3π	4π	5π	6π
$\dfrac{1}{3}\theta$	0	$\dfrac{\pi}{3}$	$\dfrac{2\pi}{3}$	π	$\dfrac{4\pi}{3}$	$\dfrac{5\pi}{3}$	2π

For θ in $[0, 6\pi)$ we see that $\frac{1}{3}\theta$ is in $[0, 2\pi)$, producing one period in $y = \cos \frac{1}{3}\theta$, three times its normal period; the new period is $3(2\pi) = 6\pi$.

From these two examples we see that

$$y = \sin 2\theta \text{ has a period of } \frac{2\pi}{2} = \pi \text{ and}$$

$$y = \cos \frac{1}{3}\theta \text{ has a period of } \frac{2\pi}{\frac{1}{3}} = 6\pi.$$

Thus the coefficient of θ changes the period. In general, we conclude that

$$\text{the new period is } \frac{\text{normal period}}{\text{the coefficient of } \theta}$$

$$y = \sin b\theta \quad \text{has a new period} = \frac{2\pi}{|b|}$$

$$y = \cos b\theta \quad \text{has a new period} = \frac{2\pi}{|b|}$$

and

$$y = \tan b\theta \quad \text{has a new period} = \frac{\pi}{|b|}$$

We will now use the amplitude and new period to sketch the graph of the trigonometric function.

EXAMPLE 1   Sketch the graph of $y = 2 \sin \frac{1}{3}\theta$ for at least one period to the right of the origin.

Solution   For $y = 2 \sin \frac{1}{3}\theta$, the amplitude is $|2| = 2$, and the new period is

$$\frac{2\pi}{1/3} = 6\pi.$$

First use the amplitude to sketch in dashed horizontal lines for the upper and lower boundaries of $y = \pm 2$. Next mark off the new period from 0 to 6π with quarter-period divisions at $\frac{3\pi}{2}$, 3π, and $\frac{9\pi}{2}$. Finally, sketch in this basic sine curve meeting these conditions (Fig. 5.59).

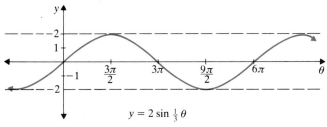

$$y = 2 \sin \tfrac{1}{3}\theta$$

Figure 5.59

Horizontal Shift

In Chapter 2 (Section 2.5) we worked with the horizontal shift or translation of functions in general. Now we will consider the translation of trigonometric functions. The horizontal shift of trigonometric functions is identical to the translation of axes for functions in Section 2.5. Using the methods of Section 2.5, we find that the graph of

$$y = \sin\left[\theta + \left(\frac{\pi}{3}\right)\right]$$

or

$$y = \sin\left[\theta - \left(-\frac{\pi}{3}\right)\right]$$

is the graph of $y = \sin \theta$ shifted $-\frac{\pi}{3}$ units $\left(\frac{\pi}{3}\right.$ units to the left, as in Fig. 5.60(a)$\left.\right)$. Notice that we rewrite the equation of $y = \sin\left[\theta + \left(\frac{\pi}{3}\right)\right]$ in the form

$$y = a \sin b(x - h)$$

where h determines the horizontal shift. In the case of $y = \sin\left[\theta - \left(-\dfrac{\pi}{3}\right)\right]$, $h = -\dfrac{\pi}{3}$, which shifts the graph of $y = \sin\theta$ a distance of $\dfrac{\pi}{3}$ to the left.

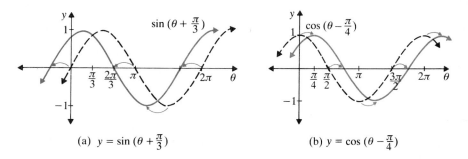

(a) $y = \sin\left(\theta + \frac{\pi}{3}\right)$

(b) $y = \cos\left(\theta - \frac{\pi}{4}\right)$

Figure 5.60

Similarly, $y = \cos\left(\theta - \dfrac{\pi}{4}\right)$ is the graph of $y = \cos\theta$ shifted $\dfrac{\pi}{4}$ units $\left(\dfrac{\pi}{4}\text{ units to the right, as in Fig. 5.60(b)}\right)$. In this case, $h = \dfrac{\pi}{4}$, which shifts the graph of $y = \cos\theta$ a distance of $\dfrac{\pi}{4}$ to the right.

In general, the graph of

$$y = \sin(\theta - h) \qquad \text{and} \qquad y = \cos(\theta - h)$$

are the graphs of $\qquad y = \sin\theta \qquad$ and $\qquad y = \cos\theta$ shifted h units to the right when $h > 0$ and h units to the left when $h < 0$. Remember that we must have the domain in the form $(\theta - h)$ to obtain the correct direction of the shift. (Do not use the subtraction sign with $-h$ when evaluating the direction of the horizontal shift.)

The graphing process is more complex when there is a change in period. For $y = \sin(b\theta - S)$ and $y = \cos(b\theta - S)$ we first factor b, the coefficient of θ, from the parentheses. Then

$$y = \sin(b\theta + k) = \sin b\left(\theta - \frac{S}{b}\right) = \sin b(\theta - h)$$

and

$$y = \cos(b\theta + k) = \cos b\left(\theta - \frac{S}{b}\right) = \cos b(\theta - h)$$

are both shifted $\dfrac{S}{b} = h$ units (to the right when $\dfrac{S}{b} = h > 0$ and to the left when $\dfrac{S}{b} = h < 0$), with a new period of $\dfrac{2\pi}{|b|}$.

> ### *Guidelines for Graphing Trigonometric Functions*
>
> Follow these steps to sketch the graph of the trigonometric functions
>
> $$y = a\ \sin b(x - h), \quad y = a\ \cos b(x - h), \quad \text{or} \quad y = a\ \tan b(x - h)$$
>
> STEP 1: For $y = a \sin \theta$ and $y = a \cos \theta$, use the amplitude $|a|$ to sketch in upper and lower boundaries of $y = \pm|a|$. For $y = a \tan \theta$, use $|a|$ to find the points $\left(\pm\dfrac{\pi}{4}, y\right)$, where $y = a \tan\left(\pm\dfrac{\pi}{4}\right) = \pm|a|$.
>
> STEP 2: Find the new period, $\dfrac{\text{normal period}}{|b|}$, and use it to mark off each quarter-period of the function. Sketch in the unshifted trigonometric function with a dashed curve.
>
> STEP 3: With the form $(\theta - h)$ the shift number is h. Shift the graph h units to the left (for $h < 0$) or h units to the right (for $h > 0$).

We will now use these steps to sketch the graphs in the following examples.

EXAMPLE 2   Sketch the graph of $y = 3 \cos\left(\theta - \dfrac{\pi}{4}\right)$

Solution   For $y = 3 \cos\left(\theta - \dfrac{\pi}{4}\right) = 3 \cos 1\left(\theta - \dfrac{\pi}{4}\right)$,

$$\text{the amplitude is } |3| = 3,$$

$$\text{the new period is } \frac{2\pi}{1} = 2\pi,$$

$$\text{and the shift is } h = \frac{\pi}{4}$$

We use this information to sketch the graph.

STEP 1: Use the amplitude of 3 to sketch dashed horizontal lines for the upper and lower boundaries of $y = \pm 3$ (Fig. 5.61a).

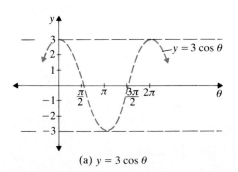

(a) $y = 3 \cos \theta$

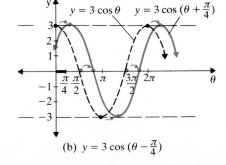

(b) $y = 3 \cos\left(\theta - \dfrac{\pi}{4}\right)$

Figure 5.61

STEP 2: Next mark off the period from 0 to 2π with quarter-period divisions at $\dfrac{\pi}{2}$, π, and $\dfrac{3\pi}{2}$. Sketch this basic curve of $y = 3 \cos \theta$ with a dashed curve (Fig. 5.61a).

STEP 3: Then the shift is $h = \dfrac{\pi}{4}$. We shift each point on the dashed curve $\dfrac{\pi}{4}$ units to the right (Fig. 5.61b). ■

EXAMPLE 3

Sketch the graph of $y = \dfrac{1}{3} \sin\left(2\theta + \dfrac{\pi}{2}\right)$.

Solution   To have this function in a more recognizable form, we rewrite it as

$$y = \frac{1}{3} \sin\left(2\theta + \frac{\pi}{2}\right)$$

$$y = \frac{1}{3} \sin 2\left(\theta + \frac{\pi}{4}\right)$$

$$y = \frac{1}{3} \sin 2\left[\theta - \left(-\frac{\pi}{4}\right)\right]$$

This simulates the form $y = a \sin b(\theta - h)$, so we can identify

the amplitude as $\left|\dfrac{1}{3}\right| = \dfrac{1}{3}$,

the new period as $\dfrac{2\pi}{2} = \pi$,

and the shift as $h = -\dfrac{\pi}{4}$.

STEP 1: First sketch dashed horizontal lines for the upper and lower bounds of $y = \pm\dfrac{1}{3}$.

STEP 2: Next mark off the period from 0 to π with quarter-period divisions at $\dfrac{\pi}{4}$, $\dfrac{\pi}{2}$, and $\dfrac{3\pi}{4}$.

STEP 3: The shift is $-\dfrac{\pi}{4}$, so shift each point $\dfrac{\pi}{4}$ units to the left. The result is the solid graph in Fig. 5.62.

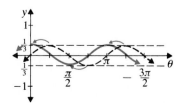

Figure 5.62 ■

EXAMPLE 4

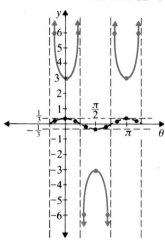

Sketch the graph of $y = 3 \csc \left(2\theta + \dfrac{\pi}{2} \right)$.

Solution   Since $y = 3 \csc \left(2\theta + \dfrac{\pi}{2} \right) = 3\left[1/\sin \left(2\theta + \dfrac{\pi}{2} \right) \right]$, we will sketch

the graph as a multiple of $\csc \left(2\theta + \dfrac{\pi}{2} \right)$. First sketch $y = \sin \left(2\theta + \dfrac{\pi}{2} \right)$ as in

Example 3. Next graph the reciprocals of range values to sketch the dashed graph

of $\csc \left(2\theta + \dfrac{\pi}{2} \right)$. Last we multiply each range value of $y = \csc \left(2\theta + \dfrac{\pi}{2} \right)$ by 3

to obtain the solid graph of $y = 3 \csc \left(2\theta + \dfrac{\pi}{2} \right)$ in Fig. 5.63.

Figure 5.63

EXAMPLE 5

Sketch the graph of $y = \dfrac{3}{2} \tan \dfrac{1}{3}\theta$.

Solution   For $y = \dfrac{3}{2} \tan \dfrac{1}{3}\theta = \dfrac{3}{2} \tan \dfrac{1}{3} (\theta + 0)$, the amplitude $a = \dfrac{3}{2}$, the new

period is $\dfrac{\pi}{1/3} = 3\pi$, and the shift is 0.

STEP 1: Use $a = \dfrac{3}{2}$ to find the values of $y = \dfrac{3}{2} \tan \left(\pm \dfrac{\pi}{4} \right)$. In this case,

$y = \dfrac{3}{2} \tan \left(\pm \dfrac{\pi}{4} \right) = \dfrac{3}{2}(\pm 1) = \pm \dfrac{3}{2}$. Two values for the angle θ with the

range values of $\pm \dfrac{3}{2}$ occur on the unshifted graph $y = \left(\dfrac{3}{2} \right) \tan \left(\dfrac{1}{3} \right)\theta$

when

$$\pm \dfrac{\pi}{4} = \dfrac{1}{3}\theta$$

and

$$\theta = \pm \dfrac{3\pi}{4}$$

Thus two points on the graph are $\left(-\dfrac{3\pi}{4}, -\dfrac{3}{2} \right)$ and $\left(\dfrac{3\pi}{4}, \dfrac{3}{2} \right)$.

STEP 2: Mark off one period of 3π centered around the origin from $-\dfrac{3\pi}{2}$ to $\dfrac{3\pi}{2}$.

Then mark quarter-period divisions at $-\dfrac{3\pi}{4}$, 0, and $\dfrac{3\pi}{4}$ (Fig. 5.64) and

vertical asymptotes at $\theta = \pm \dfrac{3\pi}{2}$. Then sketch the graph of $y = \dfrac{3}{2} \tan \dfrac{1}{3}\theta$.

STEP 3: There is no shift.

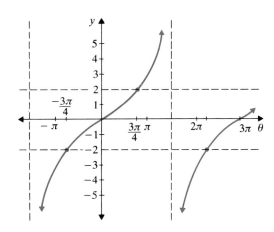

Figure 5.64

EXAMPLE 6   Sketch the graph of $y = \sin^6 \theta$.

Solution   For $y = \sin^6 \theta$ we use the amplitude of $y = \sin \theta$, which is 1; the new period, which is $\frac{2\pi}{1} = 2\pi$; and the shift, which is 0. We will use what we know about the graph of $y = \sin \theta$ and the techniques for graphing the power of a function from Section 2.5.

STEP 1: The maximum boundary is $y = 1$, and the minimum is $y = 0$, since the exponent is an even integer ($\sin^6 \theta \geq 0$ for all θ).

STEP 2: Mark off the period from 0 to 2π with quarter-periods at $\frac{\pi}{2}$, π, and $\frac{3\pi}{2}$.

Sketch in $y = \sin \theta$ with a dashed curve (Fig. 5.65).

STEP 3: There is no shift.

Now sketch in $y = \sin^6 \theta$ with a solid curve remembering that

$$-1 \leq \sin \theta \leq 1$$
$$0 \leq \sin^6 \theta \leq 1$$

and

$$\sin^6 \theta \leq |\sin \theta|$$

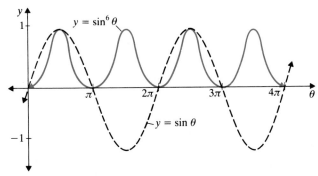

Figure 5.65

Vertical Shift

As we discovered in Chapter 2 (Section 2.5), we can have both a horizontal shift and a vertical shift when translating functions. For the function

$$y = k + a \sin b(x - h)$$

and

$$y = k + a \cos b(x - h)$$

h accounts for the horizontal shift, and k accounts for the vertical shift. As we saw earlier, the horizontal shift of the graph, h, is to the right when $h > 0$ and to the left when $h < 0$. Likewise, the vertical shift of the graph, k, is up when $k > 0$ and down when $k < 0$.

The following example shows how the vertical shift k is used in graphing.

EXAMPLE 7   Sketch the graph of $y = 1 + \cos 2\theta$.

Solution   Notice that this is the graph of $y = \cos 2\theta$ with a vertical shift of $k = 1$. Thus we first sketch the graph of
$y_1 = \cos 2\theta$ which has an
amplitude of 1,

$$\text{a new period of } \frac{2\pi}{2} = \pi,$$

and no horizontal shift.
Sketch $y_1 = \cos 2\theta$ with a dashed curve (Fig. 5.66). Then since $k = 1$, move each point on the dashed curve up one unit to complete the graph of $y = 1 + \cos 2\theta$.

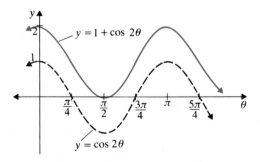

Figure 5.66

Exercises 5.5

Graph each of the following functions on the domain of $[-\pi, 3\pi]$.

1. $y = 3 \cos \alpha$

2. $y = 3 \sin \alpha$

3. $y = \sin 3\theta$

4. $y = \cos 3\theta$

5. $y = \frac{1}{2} \csc \beta$

6. $y = \frac{1}{2} \sec \beta$

7. $y = 2 \tan \alpha$

8. $y = \frac{1}{2} \cot \alpha$

9. $y = 3 \sin 2\beta$

10. $y = 3 \cos 2\beta$

11. $y = 3 \sec 2\alpha$

12. $y = \csc 2\alpha$

13. $y = \tan 2\alpha$

14. $y = \tan \frac{1}{2}\alpha$

15. $y = |\cos \frac{1}{2}\beta|$

16. $y = |\sin \frac{1}{2}\beta|$

17. $y = 2 + \cos x$

18. $y = 2 + \sin x$

19. $y = 2 - \sin x$

20. $y = 2 - \cos x$

21. $y = 1 + \sin 2x$

22. $y = 1 - \cos 2x$

Find the amplitude, period, and equation for each of the following graphs.

23.

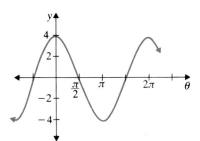

24.

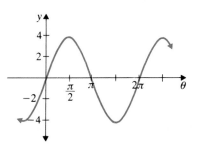

25.

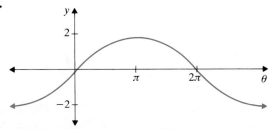

26.

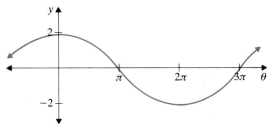

27.

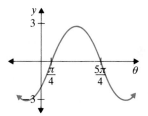

28.

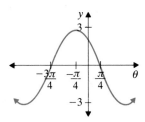

Sketch the graphs of the following functions.

29. $y = 3 \sin \left(\alpha + \dfrac{\pi}{2} \right)$

30. $y = 3 \cos \left(\alpha - \dfrac{\pi}{2} \right)$

31. $y = 2 \cos \left(\beta - \dfrac{\pi}{4} \right)$

32. $y = 2 \sin \left(\beta + \dfrac{\pi}{4} \right)$

33. $y = \sec \left(\alpha - \dfrac{\pi}{2} \right)$

34. $y = \csc \left(\alpha + \dfrac{\pi}{2} \right)$

35. $y = \tan \left(\beta - \dfrac{\pi}{3} \right)$

36. $y = \tan \left(\beta + \dfrac{\pi}{4} \right)$

37. $y = \dfrac{1}{2} \sin \left(2\alpha - \dfrac{\pi}{2} \right)$

38. $y = \dfrac{1}{2} \cos \left(2\alpha + \dfrac{\pi}{2} \right)$

39. $y = 2 \cos (3\beta + 2\pi)$

40. $y = 2 \sin \left(3\beta - \dfrac{\pi}{2} \right)$

41. $y = \dfrac{3}{4} \sin \left(\dfrac{1}{2}\alpha + \dfrac{\pi}{8} \right)$

42. $y = \dfrac{2}{3} \cos \left(\dfrac{1}{3}\alpha - \dfrac{\pi}{6} \right)$

43. $y = 2 \csc \left(2\beta - \dfrac{\pi}{2} \right)$

44. $y = 3 \sec \left(2\beta + \dfrac{\pi}{2} \right)$

45. A ferris wheel has a radius of R feet and has an angular speed of ω radians per minute (Fig. 5.67). When the ferris wheel fills up, you are halfway to the top. Place the center of the coordinate axis system at the center of the circular ferris wheel and complete the following questions.

   (a) As the ferris wheel rotates counterclockwise, find θ as a function of time t and ω.

   (b) After the ferris wheel is full, it rotates with a constant angular speed of ω. Find your height above and below the center of the ferris wheel as a function of the constant ω and the variable t for time. Graph this function.

   (c) Redo part (b) if your ride starts at the top of the ferris wheel.

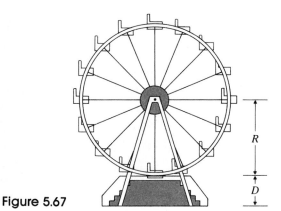

Figure 5.67

46. Repeat part (b) of Exercise 45 when your ride starts (a) at the bottom and (b) half way down the ferris wheel.

47. Sketch the graph of $y = \cos x$ and $y = \sin\left(x + \dfrac{\pi}{2}\right)$ on the same coordinate system.

48. Sketch the graph of $y = -\sin x$ and $y = \cos\left(x + \dfrac{\pi}{2}\right)$ on the same coordinate system.

49. The almost circular orbit of the space shuttle is a sine curve on the Mercator projection as shown in Fig. 5.68. The most northerly point in this orbit is a latitude of approximately 32°N, achieved 8 minutes away from Cape Canaveral, and its most southerly point in the orbit is a latitude approximately 32°S. The latitude of Cape Canaveral is 28.5°N. Find the latitude of the space shuttle as a function of time if the shuttle takes about 90 minutes to circle the earth.

   (a) Place the origin of the axes on the equator directly south of Cape Canaveral.

   (b) Place the origin of the coordinate axes at Cape Canaveral.

50. For the space shuttle in Exercise 49, find the number of miles the shuttle is above or below the plane of the equator as a function of time. The average altitude of the shuttle is 220 miles, and the radius of the earth is 3960 miles. (Use the results of Exercise 49.)

Sketch the graphs of the following functions.

51. $y = 5 \cos^3 \frac{1}{2}\alpha$ **52.** $y = 5 \sin^3 \frac{1}{2}\alpha$

53. $y = 2 \sin^4 2\beta$ **54.** $y = 3 \cos^4 3\alpha$

55. $y = 3 \cos^{10} 2\alpha$ **56.** $y = 2 \sin^{10} 2\alpha$

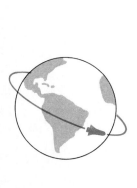

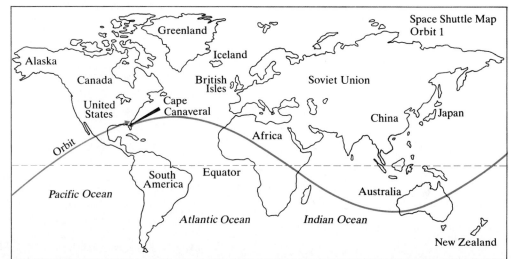

Figure 5.68

5.6   Graphing Trigonometric Sums and Products

In Section 5.5 we sketched the graphs of the basic trigonometric functions. It is interesting to note that any periodic functions can be written as sums and products of the sine and cosine functions. This observation was made by the French physicist Jean Baptiste Joseph de Fourier (1768–1830). These sums of sine and cosine functions are called Fourier series and are an important tool in calculus and physics. In this section we will sketch the graphs of some basic sums and products of trigonometric functions.

Graphing Trigonometric Sums

To sketch the graph of trigonometric sums, we will graph each part of the sum and then add the range values of these parts. We can do this by adding range values of several representative domain values. It is especially easy to add these range values when one part of the sum has a range value of zero. This method is demonstrated in the following examples.

EXAMPLE 1   Sketch the graph for $y = x + \cos \frac{3}{2}\left(x + \frac{\pi}{3}\right)$. Is this a periodic function?

Solution   We rewrite the equation $y = x + \cos \frac{3}{2}\left(x + \frac{\pi}{3}\right)$ as

$$y = x + \cos \frac{3}{2}\left[x - \left(-\frac{\pi}{3}\right)\right]$$

We now rewrite this function as the sum of two functions y_1 and y_2:

$$y = y_1 + y_2$$

where

$$y_1 = x$$

and

$$y_2 = \cos \frac{3}{2}\left[x + \left(\frac{\pi}{3}\right)\right]$$

We first graph the line $y_1 = x$, which contains the points (0, 0) and (1, 1). Since $\frac{\pi}{3}$ is approximately equal to 1 radian, the point $\left(\frac{\pi}{3}, 1\right)$ is very close to the point (1, 1) on the line $y = x$ (Fig. 5.69a). Next we graph $y_2 = \cos \frac{3}{2}\left[x - \left(-\frac{\pi}{3}\right)\right]$, with a dashed line, using its period of $\frac{2\pi}{3/2} = \frac{4\pi}{3}$ and horizontal shift of $-\frac{\pi}{3}$

$\left(\dfrac{\pi}{3} \text{ units to the left}\right)$ as shown in Fig. 5.69(a). Then the representative values of y_2 can be added to y_1 to construct the graph of

$$y = y_1 + y_2 = x + \cos \frac{3}{2}\left(x - \frac{\pi}{3}\right)$$

When y_2 is zero, the graph of y is the point on the line $y_1 = x$. Also note that the range values of y_2, the vertical line segments, were physically added to or subtracted from the range values of y_1, the line $y = x$ as shown in Fig. 5.69(b). When $y_2 > 0$, the line segments are placed on top of y_1. When $y_2 < 0$, the vertical line segments are hung under y_1.

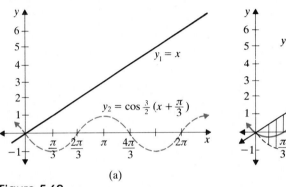

(a)   (b)

Figure 5.69

Although there is a pattern to this graph, we notice that

$$y = x + \cos \frac{3}{2}\left(x + \frac{\pi}{3}\right)$$

is *not* a periodic function, since $f(x + k) \neq f(x)$ for any $k \in R$.   ■

The graph in Example 1 is similar to the graph of seasonal changes in carbon dioxide at the beginning of the chapter. Using this pattern, we could now write an equation to model the amount of carbon dioxide in the atmosphere. We will do this later in the exercise set.

EXAMPLE 2   Sketch the graph of $y = 6 \sin \theta + 2 \sin 2\theta$ for at least one period.

Solution   To graph $y = 6 \sin \theta + 2 \sin 2\theta$, we write y as

$$y = y_1 + y_2$$

where

$$y_1 = 6 \sin \theta \quad \text{and} \quad y_2 = 2 \sin 2\theta$$

First sketch the graphs of y_1 and y_2 on the same Cartesian coordinate system. The graph of $y_1 = 6 \sin \theta$ has an amplitude of 6 with a period of 2π. Similarly, $y_2 = 2 \sin 2\theta$ has an amplitude of 2 with a period of $\dfrac{2\pi}{2} = \pi$. Since $y_1 = 6 \sin \theta$ is the more dominant of the two curves (its shape is much "larger" with amplitude larger than the amplitude of $y_2 = 2 \sin 2\theta$), take the value of y_2, a vertical segment, for every θ in the domain, and add or subtract this value to the curve $y_1 = 6 \sin \theta$. Note that whenever $y_2 = 2 \sin 2\theta = 0$, the value of y is the value of y_1. Also if $y_2 = 2 \sin 2\theta > 0$, the line segments are placed on top of $y_1 = 6 \sin \theta$. Similarly, if $y_2 = 2 \sin 2\theta < 0$, the line segments are hung under $y_1 = 6 \sin \theta$. In this way we can graphically represent the addition of y_1 to y_2 as seen in Fig. 5.70.

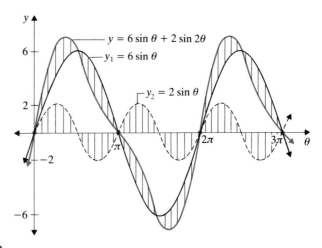

Figure 5.70

It is interesting to notice that adding y_1 to y_2 is not difficult when $y_1 = 0$ or $y_2 = 0$. We also note from Example 2 that when the period of y_1 is different from the period of y_2, the period of the sum is the least common multiple (LCM) of the two periods measured in degrees.

EXAMPLE 3   Sketch the graph of $y = 2 \sin \theta + \cos 3\theta$ for at least one period.

Solution   To graph $y = 2 \sin \theta + \cos 3\theta$, we let

$$y = y_1 + y_2$$

where

$$y_1 = 2 \sin \theta \quad \text{and} \quad y_2 = \cos 3\theta$$

Next we sketch the graph of $y_1 = 2 \sin \theta$ (with amplitude 2 and period 2π) and $y_2 = \cos 3\theta$ $\left(\text{with amplitude 1 and period } \dfrac{2\pi}{3}\right)$ on the same coordinate system. Then we add representative values of y_2 to y_1 to produce the graph of

$$y = y_1 + y_2 = 2 \sin \theta + \cos 3\theta \text{ (Fig. 5.71)}$$

Since the period of $y_1 = 2 \sin \theta$ is 2π and the period of $y_2 = \cos 3\theta$ is $\dfrac{2\pi}{3}$, the period for $y = y_1 + y_2$ is the LCM of $2\pi = 360°$ and $\dfrac{2\pi}{3} = 120°$. This LCM is $2\pi = 360°$, the period for the sum $y = 2 \sin \theta + \cos 3\theta$.

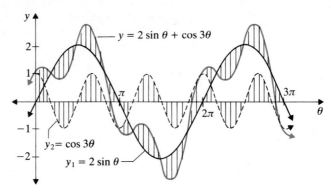

Figure 5.71

Graphing Products

To graph functions that are the products of the sine or cosine functions, we utilize the fact that the range values of $\sin x$ and $\cos x$ are bounded by ± 1. Thus we have

$$-1 \leq \sin x \leq 1 \qquad \text{and} \qquad -1 \leq \cos x \leq 1$$

Multiplying both sides by any function $f(x)$, we have

$$-|f(x)| \leq f(x) \sin x \leq |f(x)| \qquad \text{and} \qquad -|f(x)| \leq f(x) \cos x \leq |f(x)|$$

For any function $f(x)$ the multiple functions

$$y = f(x) \sin x \qquad \text{or} \qquad y = f(x) \cos x$$

are bounded by $\pm f(x)$, since $-|f(x)| \leq y \leq |f(x)|$. If $y = f(x) \sin x$ or $y = f(x) \cos x$, then y is equal to either $\pm f(x)$ or a fractional value of $\pm f(x)$. Sketching the graphs of these products can be completed with the following steps.

Guidelines for Graphing the Products
$y = f(x) \sin x$ or $y = f(x) \cos x$

STEP 1: Graph $y = \pm f(x)$, since this provides the upper and lower bounds of the range values.

STEP 2: Graph all points $(x, 0)$ where one factor of this product equals zero.

STEP 3: Graph all points $(x, f(x))$ where $\sin x = 1$ or $\cos x = 1$ and all points $(x, -f(x))$ where $\sin x = -1$ or $\cos x = -1$.

STEP 4: Plot some other approximate points on the curve as fractional values of $f(x)$ and sketch the graph.

We will now use these techniques to graph the following functions.

EXAMPLE 4   Sketch the graph of $y = (|x| + 2) \cos 3x$.

Solution   For $y = (|x| + 2) \cos 3x$ we let $f(x) = |x| + 2$.

STEP 1: We first graph the boundaries of $f(x)$ and $-f(x)$ along with $\cos 3x$ on the same coordinate system (Fig. 5.72a).

STEP 2: On this graph we mark the point $(x, 0)$ where $\cos 3x = 0$.

STEP 3: We mark the points $(x, \pm f(x))$ where $\cos 3x = \pm 1$.

STEP 4: Then we approximate points on the curve $y = f(x) \cos 3x$ by finding the product of the fraction and $f(x)$, as in Fig. 5.72(b). Thus the graph oscillates between $f(x)$ and $-f(x)$.

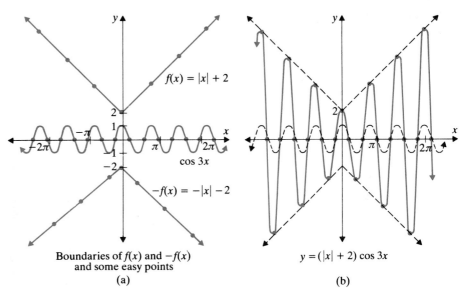

Boundaries of $f(x)$ and $-f(x)$
and some easy points
(a)

$y = (|x| + 2) \cos 3x$
(b)

Figure 5.72

Exercises 5.6

Sketch the graphs of the following sums and products.

1. $y = x + \cos 3x$
2. $y = x + \cos 3x$
3. $y = \frac{1}{3}x + \sin 2x$
4. $y = \frac{1}{3}x + \sin 2x$
5. $y = \frac{1}{2}x - \cos 3x$
6. $y = \frac{1}{2}x - \sin 3x$
7. $y = \cos 2\theta + 3 \sin \theta$
8. $y = \sin 2\theta + 3 \cos \theta$
9. $y = 3 \cos \theta - \sin \theta$
10. $y = 3 \sin \theta - \cos \theta$
11. $y = 4 \sin \theta - \cos 2\theta$
12. $y = 4 \cos \theta - \sin 2\theta$
13. $y = 2 \sin \theta + \cos \theta$
14. $y = 2 \sin \theta - \cos \theta$
15. $y = |x| \sin x$
16. $y = |x| \cos x$
17. $y = 2^x \cos x$
18. $y = 2^x \sin x$

19. $y = \frac{1}{3}x \sin x$

20. $y = \frac{1}{3}x \cos x$

21. $y = 3^{-x} \sin x$

22. $y = 3^{-x} \cos x$

23. $y = |x \cos x|$

24. $y = |x \sin x|$

25. $y = 2^{x/4} \sin x$

26. $y = 2^{x/4} \cos x$

27. $y = 2^{x/4} + \sin x$

28. $y = 2^{x/4} + \cos x$

29. $y = \sin |x|$

30. $y = \cos |x|$

31. $y = |\sin x|$

32. $y = |\cos x|$

33. $y = 3 \sin \frac{1}{2}\theta \cos \theta$

34. $y = 3 \cos \frac{1}{2}\theta \sin \theta$

35. $y = 4 \cos \theta \cos 2\theta$

36. $y = 4 \sin \theta \sin 2\theta$

37. A ferris wheel has a radius of R feet, is D feet off the ground, and has an angular speed of ω radians per minute (Fig. 5.73). When the ferris wheel fills up, you are halfway to the top. Place the center of the coordinate axis system on ground level directly below the center of the circular ferris wheel and complete the following questions:

(a) As the ferris wheel rotates, find θ as a function of time t and ω.

(b) After the ferris wheel is full, it rotates counterclockwise with an angular speed of ω. Find your height above the ground as a function of the constants ω, R and D and the variable t for time. Graph this function.

(c) Redo part (b) if your ride starts at the top of the ferris wheel.

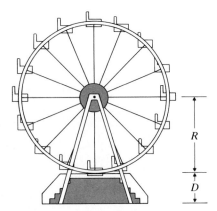

Figure 5.73

38. Complete part (b) of Exercise 37 when your ride starts (i) at the bottom and (ii) halfway down the ferris wheel.

39. Rectangle $ABCD$ with a diagonal of 1 is inscribed in rectangle $EFGH$ as shown in Fig. 5.74. Prove that **(a)** $AE = \cos \alpha \cos \beta$, **(b)** $EB = \sin \alpha \cos \beta$, **(c)** $CF = \sin \alpha \sin \beta$, and **(d)** $BF = \cos \alpha \sin \beta$.

40. If $\alpha + \beta = \dfrac{\pi}{2}$ in rectangle $ABCD$ and rectangle $EFGH$ as shown in Fig. 5.74, prove that (a) $EB = \sin^2 \alpha$ and (b) $BF = \cos^2 \alpha$.

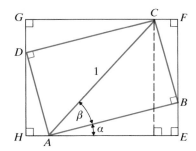

Figure 5.74

41. The graph in the introduction to Chapter 5 provides a model for the amount of carbon dioxide in the atmosphere (Fig. 5.75). Using the function $g(x)$, find the function $f(x)$ that is a rough approximation of this curve. This equation will provide the amount of carbon dioxide in the atmosphere as a function of time. (Let the domain x be the number of days after the beginning of 1958. For more detail use the graph on p. 284.)

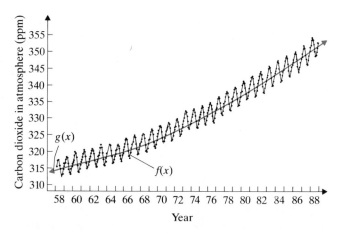

Figure 5.75

5.7   The Laws of Sines and Cosines

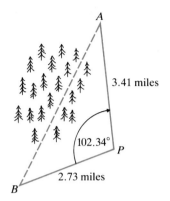

Figure 5.76

In all our previous work we solved only right triangles or triangles that could be divided into right triangles. Sometimes a problem does not contain a right triangle.

For example, an electrical line is to be installed in a straight line through a forest between points A and B. A surveyor at point P measures the distance from P to A to be 3.41 miles, the distance from P to B to be 2.73 miles, and the angle APB to be 102.34° (Fig. 5.76). What is the distance from A to B that the electrical wire must span?

As Fig. 5.76 shows, this is not a right triangle and cannot even be divided into right triangles that can be solved with the given information. We will solve this problem later in this section.

We will now develop two methods for solving triangles which are not right triangles. The first method for solving these triangles is called the **law of sines**.

THE LAW OF SINES

The ratio of each side of a triangle to the sine of the opposite angle is the same constant. For triangle ABC,

$$\frac{a}{\sin \alpha} = \frac{b}{\sin \beta} = \frac{c}{\sin \gamma}$$

or

$$\frac{\sin \alpha}{a} = \frac{\sin \beta}{b} = \frac{\sin \gamma}{c}$$

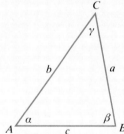

PROOF: We will use either triangle ABC with α as an acute angle (Fig. 5.77a) or triangle ABC with α as an obtuse angle (Fig. 5.77b) to prove the law of sines. These two triangles cover all possibilities.

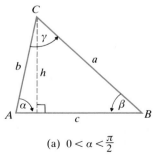

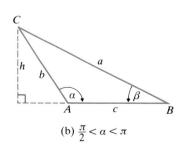

Figure 5.77

(a) $0 < \alpha < \frac{\pi}{2}$ (b) $\frac{\pi}{2} < \alpha < \pi$

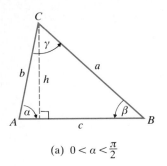

(a) $0 < \alpha < \frac{\pi}{2}$

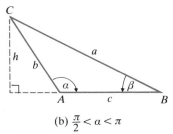

(b) $\frac{\pi}{2} < \alpha < \pi$

Figure 5.77

We let h be the altitude from vertex C as indicated in these triangles. Using the right triangles with altitude h to evaluate $\sin \alpha$ and $\sin \beta$, we have

$$\sin \alpha = \frac{h}{b} \quad \text{or} \quad h = b \sin \alpha$$

and

$$\sin \beta = \frac{h}{a} \quad \text{or} \quad h = a \sin \beta$$

Since h is the same altitude in both equations above, we substitute, and

$$b \sin \alpha = a \sin \beta \quad \textit{(Divide both sides by the product } \sin \alpha \sin \beta\textit{)}$$

$$\frac{b}{\sin \beta} = \frac{a}{\sin \alpha}$$

Divide both sides of $b \sin \alpha = a \sin \beta$ by ab, and we also have

$$\frac{\sin \alpha}{a} = \frac{\sin \beta}{b}$$

Using another altitude (from vertex A or B), we can also show that

$$\frac{b}{\sin \beta} = \frac{c}{\sin \gamma} \quad \text{and} \quad \frac{\sin \beta}{b} = \frac{\sin \gamma}{c}$$

Therefore we can conclude that

$$\frac{a}{\sin \alpha} = \frac{b}{\sin \beta} = \frac{c}{\sin \gamma} \quad \text{and} \quad \frac{\sin \alpha}{a} = \frac{\sin \beta}{b} = \frac{\sin \gamma}{c}$$

This proves the law of sines.   □

The law of sines involves two sides and the angles opposite these sides. When one of these four parts is missing, we use the law of sines to calculate the missing part. Notice that one form of the law of sines has the sides in the numerators, while the other has the angles in the numerator. It is slightly easier to use the form with the unknown in the numerator. Also remember that the sum of the interior angles of a triangle is $180°$ ($\alpha + \beta + \gamma = 180°$).

EXAMPLE 1

Solve for all sides and angles of triangle ABC when $\alpha = 33°$, $\beta = 77°$, and $a = 7.2$.

Solution   Sketch a diagram and then use the law of sines to find the other parts of triangle ABC. Since

$$\alpha + \beta + \gamma = 180°$$
$$33° + 77° + \gamma = 180°$$

we have $\gamma = 70°$. Then by the law of sines,

$$\frac{b}{\sin 77°} = \frac{7.2}{\sin 33°} \quad \text{and} \quad b = \frac{7.2 \sin 77°}{\sin 33°} \approx 12.88$$

and

$$\frac{c}{\sin 70°} = \frac{7.2}{\sin 33°} \quad \text{and} \quad c = \frac{7.2 \sin 70°}{\sin 33°} \approx 12.42$$

Therefore $\gamma = 70°$, and the sides are $b \approx 12.9$ and $c \approx 12.4$.   ■

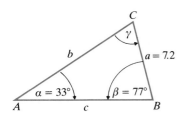

Figure 5.78

Given the right information, the law of sines can be an effective method for solving triangles. You must be aware of a problem that can occur when using the law of sines. For $\sin \theta \geq 0$, θ could be in quadrant I or II. Thus when working with $\sin \alpha = 0.629320391$, we must check the diagram to decide whether $\alpha = 39°$ or $\alpha = 180° - 39° = 141°$. **When using the law of sines, always check the answer with the diagram to determine whether the angle is acute or obtuse.**

EXAMPLE 2

Find α when $a = 120.2$, $b = 68.9$, $c = 66.6$, and $\beta = 28°$.

Solution   Sketch a triangle with the given data and solve with the law of sines.

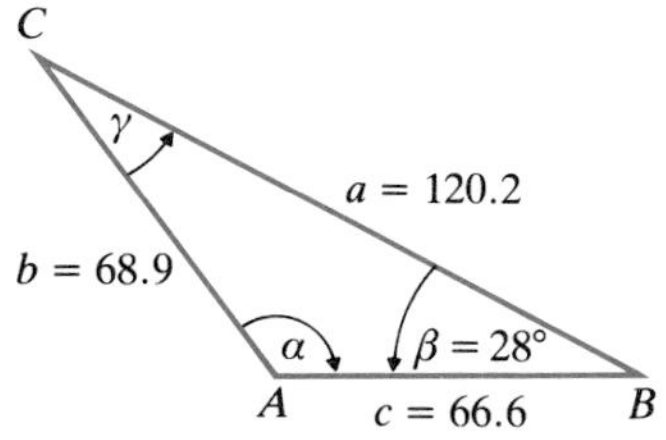

$$\frac{\sin \alpha}{120.2} = \frac{\sin 28°}{68.9} \quad \textit{(Solve with a calculator)}$$

$$\sin \alpha = \frac{120.2 \sin 28°}{68.9} \approx 0.819020055$$

and $\sin^{-1}(0.819020055) \approx 54.98681756$. From the diagram we see that angle α is obtuse. Therefore $\alpha' \approx 54.98681756$ and $\alpha = \alpha'$ or $\alpha = 180° - \alpha' \approx 125.0131825°$. As indicated in the diagram $\alpha > 54°$, and thus $\alpha \approx 125°$.   ■

Figure 5.79

As we saw in Example 2, the sketch of the triangle is most important in knowing whether the angle is acute or obtuse when using the law of sines.

Another problem can occur with the law of sines. When we have two sides and the nonincluded angle, we could have no triangle, one triangle, or even two triangles that fit the given data, as shown below.

POSSIBLE TRIANGLES
WHEN GIVEN TWO
SIDES AND THE
NONINCLUDED
ANGLE

If we are given sides a and b and angle α of triangle ABC, we have

(a) No triangle when $a < b$ and $\sin \beta > 1$:

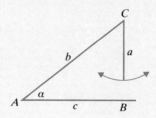

(b) One right triangle when $a < b$ and $\sin \beta = 1$:

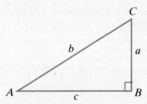

(c) One triangle when $a \geq b$ and $0 < \sin \beta < 1$:

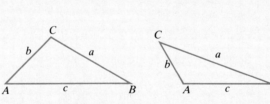

(d) Two triangles when $a < b$ and $0 < \sin \beta < 1$:

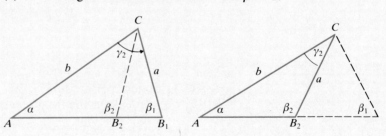

The four possibilities are easily computed with the given conditions and the law of sines. It is easy to overlook the possibility of two triangles when working with the law of sines. **When given two sides and the nonincluded angle, always consider the possibility of two triangles.**

EXAMPLE 3   Solve for side c when $\alpha = 25°$, $a = 25$, and $b = 37$.

Solution   We first sketch the triangle and notice that there is a possibility of two triangles, since $a < b$ (Fig. 5.80). Using the law of sines, we have

$$\frac{\sin \beta_1}{37} = \frac{\sin 25°}{25}$$

and

$$\sin \beta_1 = \frac{37 \sin 25°}{25} \approx 0.625475027$$

which yields $\beta_1 \approx 38.7170617°$ and $\beta_2 = 180 - \beta_1 \approx 141.2829383°$. We now use α, β_1, and β_2 to find γ_1 and γ_2. Remembering that the sum of the interior angles of a triangle is $180°$, we have

$$\gamma_1 = 180 - 25° - \beta_1 \approx 116.28293°$$
$$\gamma_2 = 180 - 25° - \beta_2 \approx 13.717061°$$

We use the law of sines again for

$$\frac{c_1}{\sin \gamma_1} = \frac{25}{\sin 25°} \text{ with } c_1 = \frac{25 \sin \gamma_1}{\sin 25°} \approx 53.039496$$

and

$$\frac{c_2}{\sin \gamma_2} = \frac{25}{\sin 25°} \text{ with } c_2 = \frac{25 \sin \gamma_2}{\sin 25°} \approx 14.027282$$

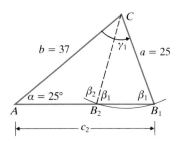

Figure 5.80

Thus the possible appropriate lengths of c are 14 and 53.   ■

The law of sines is easy to remember and calculate even though there are two problems that may arise with its use. A second method for solving triangles that are not right triangles is called the **law of cosines**.

THE LAW OF COSINES

In a triangle the square of one side is equal to the sum of the squares of the other two sides minus twice the product of these two sides and the cosine of the angle between them. For triangle ABC the law of cosines provides

$$a^2 = b^2 + c^2 - 2bc \cos \alpha$$
$$b^2 = a^2 + c^2 - 2ac \cos \beta$$
$$c^2 = a^2 + b^2 - 2ab \cos \gamma$$

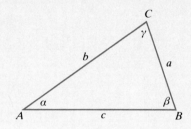

PROOF: To prove the law of cosines, solve for the distance a in either triangle in Fig. 5.81.

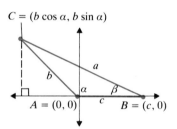 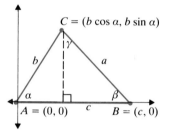

Figure 5.81

First we note that for the point $C = (x, y)$,

$$\sin \alpha = \frac{y}{b} \quad \text{and} \quad \cos \alpha = \frac{x}{b}$$

Therefore $y = b \sin \alpha$, $x = b \cos \alpha$, and point $C = (b \cos \alpha, b \sin \alpha)$ for either triangle. Now use the distance formula between $B = (c, 0)$ and $C = (b \cos \alpha, b \sin \alpha)$ to find the distance a. Then we have

$$a = \sqrt{(b \cos \alpha - c)^2 + (b \sin \alpha - 0)^2}$$
$$a^2 = (b \cos \alpha - c)^2 + (b \sin \alpha - 0)^2 \quad \textit{(Square both sides)}$$
$$a^2 = b^2 \cos^2 \alpha - 2bc \cos \alpha + c^2 + b^2 \sin^2 \alpha$$
$$a^2 = b^2(\cos^2 \alpha + \sin^2 \alpha) + c^2 - 2bc \cos \alpha$$
$$a^2 = b^2 + c^2 - 2bc \cos \alpha$$

Using the same procedure with β and γ, we can also derive the other forms of the law of cosines.   □

The law of cosines is most useful in solving triangles when three sides are known or when two sides and the included angle are known. The law of cosines has no complications or second triangles as does the law of sines. Thus acute and obtuse angles are immediately identified, and there is no possibility of two triangles.

EXAMPLE 4   Find the angles α, β, and γ of triangle ABC when $a = 8.5$, $b = 16.2$, and $c = 9.4$.

Solution   Sketch a triangle using the data in this problem (Fig. 5.82). By the law of cosines we have

$$a^2 = b^2 + c^2 - 2bc \cos \alpha$$
$$(8.5)^2 = (16.2)^2 + (9.4)^2 - 2(16.2)(9.4) \cos \alpha$$
$$\cos \alpha = \frac{(16.2)^2 + (9.4)^2 - (8.5)^2}{2(16.2)(9.4)} \approx 0.9145981$$

and

$$\alpha \approx 23.851282° \approx 23.9°$$

Likewise,

$$(16.2)^2 = (9.4)^2 + (8.5)^2 - 2(9.4)(8.5) \cos \beta$$
$$\cos \beta = \frac{(9.4)^2 + (8.5)^2 - (16.2)^2}{2(9.4)(8.5)} \approx -0.63723404$$

and

$$\beta \approx 129.58587° \approx 129.6°$$

Then $\gamma = 180 - \alpha - \beta \approx 26.562848° \approx 26.6°$. Therefore $\alpha \approx 23.9°$, $\beta \approx 129.6°$, and $\gamma \approx 26.6°$. ∎

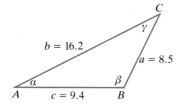

Figure 5.82

EXAMPLE 5   An electrical line is to be installed in a straight line through a forest between points A and B. A surveyor at point P measures the distance from P to A to be 3.41 miles, the distance from P to B to be 2.73 miles, and the angle APB to be 102.34°. What is the distance from A to B that the electrical wire must span?

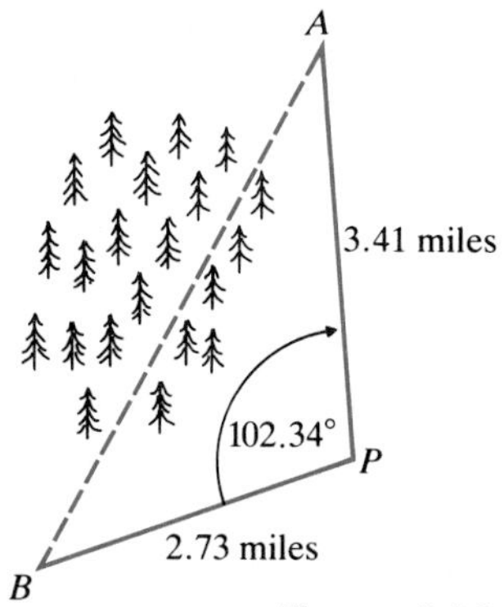

Solution   We first sketch a diagram for this data (Fig. 5.83). Next we use the law of cosines to solve for the distance AB:

$$(AB)^2 = (2.73)^2 + (3.41)^2 - 2(2.73)(3.41) \cos 102.34°$$
$$(AB)^2 \approx 23.060026$$
$$AB = 4.8020856$$

Thus the distance AB, the distance the wire must span, is a little more that 4.8 miles. ∎

Figure 5.83

EXAMPLE 6   What error is introduced if the surveyed angle is in error by $\pm 1°$?

Solution   To find the percent error, we calculate AB as in Example 5 with maximum and minimum error of the angle, using $101.34°$ and $103.34°$. Then

$$AB^2 = (2.73)^2 + (3.41)^2 - 2(2.73)(3.41)\cos 101.34°$$
$$AB^2 \approx 22.741988 \quad \text{or} \quad AB \approx 4.768856$$

and

$$AB^2 = (2.73)^2 + (3.41)^2 - 2(2.73)(3.41)\cos 103.34°$$
$$AB^2 \approx 23.376852 \quad \text{or} \quad AB \approx 4.8349614$$

Thus we find that the distance for AB could be miscalculated by $+0.0332296$ miles or -0.0328758 miles, an error of less than $4/100$ of a mile, or about $0.0332296(5280) \approx 175.5$ feet.   ■

Exercises 5.7

Solve for all parts of triangle ABC using the given parts, the law of sines, and the law of cosines. Where there is a possibility of two triangles, find all parts of both. Round off the final answers.

1. $\alpha = 41.5°$, $\beta = 62.3°$, and $b = 181.1$
2. $\alpha = 60.5°$, $\gamma = 52.3°$, and $c = 7.4$
3. $\alpha = 55.2°$, $\beta = 15.7°$, and $c = 17.6$
4. $\alpha = 115.2°$, $\beta = 16.7°$, and $c = 28.7$
5. $\alpha = 57°$, $b = 22.3$, and $c = 16.5$
6. $\beta = 121.2°$, $a = 63.2$, and $c = 42.5$
7. $a = 5.2$, $b = 7.5$, and $c = 10.3$
8. $a = 17.1$, $b = 12.7$, and $c = 10.5$
9. $a = 5.7$, $b = 4.2$, and $\beta = 37.1°$
10. $a = 5.3$, $b = 7.7$, and $\alpha = 40.75°$
11. $a = 7.2$, $b = 3.1$, and $c = 13.7$
12. $a = 17.3$, $b = 5.6$, and $c = 33.7$
13. $b = 105.8$, $c = 37.7$, and $\gamma = 10.3°$
14. $a = 171.8$, $b = 155.2$, and $\alpha \doteq 55.5°$
15. $a = 12.1$, $b = 2.7$, and $\gamma = 125°$
16. $a = 10.2$, $b = 7.8$, and $\gamma = 109°$
17. $a = 13.7$, $b = 12.4$, and $\beta = 40.1°$
18. $a = 9.4$, $c = 7.8$, and $\alpha = 42°$

19. $b = 14.3$, $c = 20.1$, and $\alpha = 17°$
20. $b = 13.7$, $c = 41.8$, and $\alpha = 107°$

21. A water pipeline is to be constructed on a straight-line path from point A through a forest to point B. A surveyor at point P measures the distance from point P to A to be 4.1 miles, the distance from P to B to be 3.3 miles, and the angle APB to be $\alpha = 126.5°$ (Fig. 5.84). What is the distance from A to B? What is the possible error (in feet) in the distance AB if the error in measuring the angle is $\pm 1°$?

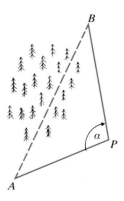

Figure 5.84

22. A water pipeline is to be constructed on a straight-line path from point A through a forest to point B. A surveyor at point P measures the distance from point P to A to be 2.7 miles, the distance from P to B to be 5.4 miles, and the angle APB to be $\alpha = 132.7°$ (Fig. 5.84). What is the distance from A to B? What is the possible error (in feet) in the distance AB if the error in measuring the angle is $\pm 1°$?

23. A tunnel is to be constructed on a straight line from point A through a mountain ridge to point B. A surveyor is at point P, where both A and B are visible, and measures the distance AP to be 2.3 miles, the distance BP to be 2.7 miles, and the angle APB to be $21.5°$. How long is the tunnel from point A to B? What is the possible error (in feet) in the length of the tunnel if the error in the measurement of the distances AP and BP is ± 0.05 mile?

24. A tunnel is to be constructed on a straight line from point A through a mountain ridge to point B. A surveyor is at point P, where both A and B are visible, and measures the distance AP to be 2.8 miles, the distance BP to be 3.4 miles, and the angle APB to be $16.75°$. How long is the tunnel from point A to B? What is the possible error (in feet) in the length of the tunnel if the error in the measurement of the distances AP and BP is ± 0.05 mile?

25. Two observers stand at points A and B, 1000 feet apart on a very long road that makes a $10°$ angle with the horizontal (Fig. 5.85). Both observers spot an aircraft at the same moment and measure the angle between the line of sight and the horizontal. From point A the angle is $37.4°$, and from point B the angle is $65.8°$. If the aircraft is between points A and B, how far is the aircraft from B and how high is it above the road?

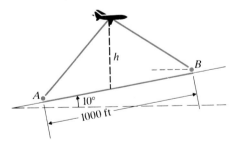

Figure 5.85

26. Two observers stand at points A and B, 1000 feet apart on a very long road that makes a $10°$ angle with the horizontal (Fig. 5.86). Both observers spot an aircraft at the same

moment and measure the angle between the line of sight and the horizontal. From point A the angle is $37.4°$, and from point B the angle is $65.8°$. If the aircraft is uphill from points A and B, how far is the aircraft from B and how high is it above the road?

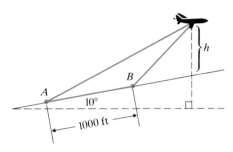

Figure 5.86

27. Prove that triangle ABC has an area of $A = \frac{1}{2}ac \sin \beta$. Find the area of a triangular piece of property with sides of 250 feet, 355 feet, and 425 feet using this formula.

28. Prove that triangle ABC has an area of $A = \frac{1}{2}bc \sin \alpha$. Find the area of a triangular piece of property with sides of 553 feet, 474 feet, and 735 feet using this formula.

29. Find the angles of the triangular piece of property in Exercise 27 with sides of 250 feet, 325 feet, and 425 feet.

30. Find the angles of the triangle in Exercise 28 with sides of 553 feet, 474 feet, and 735 feet.

31. All construction on the triangular piece of property in Exercises 27 and 29 is subject to a 10-foot setback line on all boundaries (Fig. 5.87). What is the area of the part of this property that is available for construction?

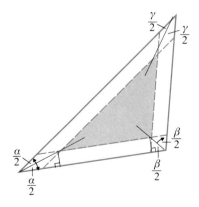

Figure 5.87

32. All construction on the triangular piece of property in Exercises 28 and 30 is subject to a 10-foot setback line on all boundaries (Fig. 5.87). What is the area of the part of this property that is available for construction?

33. A telephone pole is on a 12° incline and is tilted 9° from vertical (away from the sun), and the sun is on the uphill side of the telephone pole (Fig. 5.88). If the shadow of the telephone pole is 68.7 feet and the angle of elevation of the sun is 37.5°, how long is the telephone pole?

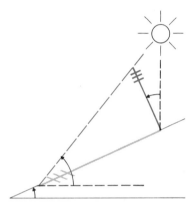

Figure 5.88

34. A telephone pole is on a 12° incline and is tilted 9° from vertical (away from the sun), and the sun is on the downhill side of the telephone pole, casting the shadow uphill. If the shadow of the telephone pole is 68.7 feet and the angle of elevation of the sun is 37.5°, how long is the telephone pole?

35. Starting at point A, an airplane travels north for 3 hours and 42 minutes at an average of 153 miles per hour and then changes its course to N78.2°E (78.2° from north toward the east) with a new average speed of 113 miles per hour for 2 hours and 18 minutes (Fig. 5.89). After these 6 hours of flying, what are the plane's distance and direction from point A?

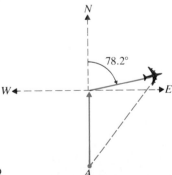

Figure 5.89

36. Starting at point A, an airplane travels south for 2 hours and 18 minutes at an average of 178 miles per hour and then changes its course to S62.7°W (62.7° from south toward the west) with a new average speed of 135 miles per hour for 2 hours and 42 minutes (Fig. 5.90). After these 5 hours of flying, what are the plane's distance and direction from point A?

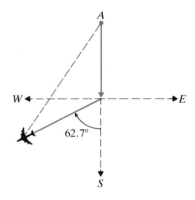

37. A lighthouse is at the top of a 15° incline. From two points A and B, 100 feet apart on the incline, the angle to the top of the lighthouse with respect to the horizontal is 32.7° at A and 56.7° at B (Fig. 5.91). Find the height of the lighthouse.

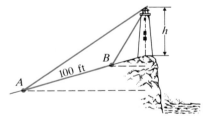

Figure 5.91

38. A lighthouse is at the top of a 12° incline. From two points A and B, 100 feet apart on the incline, the angle to the top of the lighthouse with respect to the horizontal is 25.5° at A and 62.7° at B (Fig. 5.91). Find the height of the lighthouse.

39. Rangers at two ranger stations, 16 miles apart at points A and B, spot a forest fire at point F. The angle FAB from station A is 42.7°, and the angle FBA from station B is 39.8°. How far is the fire from each ranger station?

40. Rangers at two ranger stations, 8 miles apart at points A and B, spot a forest fire at point F. The angle FAB from station A is 25.7°, and the angle FBA from station B is 82.4°. How far is the fire from each ranger station?

41. A radar-equipped aircraft records a submarine 15 miles away and an oil tanker 20 miles away. The angle between the lines of sight to the submarine and the oil tanker from the aircraft is 27.4°. What is the distance between the submarine and the oil tanker?

CHAPTER SUMMARY AND REVIEW

KEY TERMS

Degrees and radians

Arc length, $s = r\theta$ with θ in radians

Linear speed and angular speed

The trigonometric relations

Basic trigonometric identities

Trigonometric functions of 0°, 30°, 45°, 60°, and 90°

Trigonometric functions using reference angles

Negative angle identities

Inverse trigonometric functions

Graphing trigonometric functions

Horizontal shift and vertical shift

The law of sines

The law of cosines

CHAPTER EXERCISES

Change each of the following to radians.

1. 75°

2. 130°

3. 235°

4. 300°

Change each of the following to degrees.

5. $\dfrac{3\pi}{4}$

6. $\dfrac{5\pi}{6}$

7. $\dfrac{4\pi}{3}$

8. $\dfrac{7\pi}{4}$

Find the exact values for the following.

9. $\sin \dfrac{\pi}{4}$

10. $\cos \dfrac{\pi}{6}$

11. $\tan \dfrac{\pi}{3}$

12. $\sec 45°$

13. $\csc 60°$

14. $\cot 45°$

Use the given information to solve for the indicated parts of right triangle ABC (Fig. 5.92).

15. For $\alpha = 30°$ and $a = 5$, find b and c.

16. For $\alpha = 45°$ and $c = 15$, find a and b.

17. For $\alpha = 45°$ and $c = 10$, find a and b.

18. For $\alpha = 60°$ and $a = 5$, find b and c.

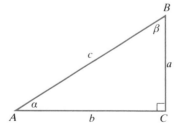

Figure 5.92

19. A vertical television antenna is mounted at the top of a 20-foot garage. From point A, 40 feet from another point directly under the antenna, the angle of elevation to the top of the antenna is 40.5° (Fig 5.93). What is the length of the antenna?

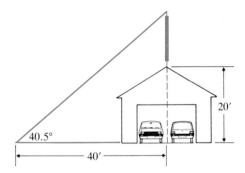

Figure 5.93

Find the exact values of the six trigonometric functions for the angle θ in standard position with the given point on the terminal side.

20. $(-3, 4)$ **21.** $(-3, -4)$

22. $(3, -4)$

Find the six trigonometric functions of each angle.

23. 135°   **24.** 300°

25. 240°   **26.** $\dfrac{5\pi}{3}$

27. $\dfrac{7\pi}{6}$ **28.** $\dfrac{5\pi}{4}$

Find $(f \circ g)(x)$ and $(g \circ f)(x)$.

29. $f(x) = 1 + \sin x$ **30.** $f(x) = \cos (x^2)$
$\quad\ g(x) = \sqrt{x}$ $\qquad\ g(x) = x - 1$

Find the exact value of each of the following inverse functions.

31. $\sin^{-1}\left(\dfrac{\sqrt{2}}{2}\right)$ **32.** $\cos^{-1}\left(-\dfrac{1}{2}\right)$

33. $\tan^{-1}(-1)$

Find the value of each inverse function using a calculator.

34. $\cos^{-1}(-0.41)$ **35.** $\sin^{-1}(0.78)$

36. $\tan^{-1}(-2)$

Find the six trigonometric functions of the indicated angle. (Use exact values.)

37. α when $\sin \alpha = -\frac{4}{5}$ and $\sec \alpha < 0$

38. α when $\sin \alpha = \frac{3}{4}$ and $\cot \alpha < 0$

39. β when $\tan \beta = -\frac{3}{4}$ and $\cos \beta < 0$

40. β when $\tan \beta = \frac{4}{5}$ and $\sin \beta < 0$

41. What is the angle of elevation of the sun when a 5'10" man has a 15'4" shadow?

42. To photograph the ground floor of a 100-foot-long building, a photographer can be only 35 feet away from the middle of the building (Fig. 5.94). Find the angle α the camera must be able to photograph.

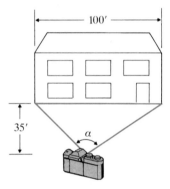

Figure 5.94

43. To photograph a 5280-foot bridge, the best viewpoint is 300 feet from one end of the bridge perpendicular to the bridge (Fig. 5.95). What angle α must the lens be able to photograph?

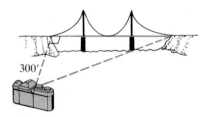

Figure 5.95

44. Ranger station A is 6.3 miles due west of ranger station B. Rangers at both stations spot a forest fire. The direction of the fire from A is N18°E, and the direction of the fire from B is N56°W. How far is the fire from each ranger station?

Graph the following functions.

45. $y = \cos 4x$ **46.** $y = \sin 4x$

47. $y = \sin 3x$ **48.** $y = \cos 3x$

49. $y = 3 \sin \left(2x - \dfrac{\pi}{3}\right)$ **50.** $y = \dfrac{1}{2} \cos \left(\dfrac{1}{3}x - \dfrac{\pi}{6}\right)$

51. $y = \sin^2 2x$ **52.** $y = \cos^3 2x$

53. $y = 1 + 2 \sin x$ **54.** $y = 2 + \cos x$

55. $y = 6 \cos 2\theta - \sin \dfrac{1}{2}\theta$ **56.** $y = \dfrac{1}{3}x + \cos 2x$

57. $y = \left|\dfrac{x}{2}\right| + \cos 2x$ **58.** $y = 3 \sin \theta \cos 2\theta$

59. $y = 6 \sin \theta + \cos 4\theta$ **60.** $y = 6 \cos \theta + \sin 4\theta$

61. $y = 2 \sin \theta + \cos 2\theta$ **62.** $y = 2 \cos 2\theta + \sin \theta$

63. $y = 2 \cos \theta - \sin \theta$ **64.** $y = 2 \sin \theta - \cos 2\theta$

65. A triangle has sides of 15, 21, and 29 meters. Find one angle and the area of this triangle.

66. Seattle, Washington, at a latitude of 47.5° is almost directly north of San Francisco, California, at a latitude of 37.5°. What is the distance between the two cities?

67. Urbana, Illinois, at a latitude of 40.2° is almost directly north of Mobile, Alabama, at a latitude of 30.6°. What is the distance between the two cities?

68. As shown in Fig. 5.96, triangle ABC is constructed with angle $A = \theta$ and AB_1 perpendicular to BB_1 with hypotenuse $\cos \theta$. Similarly, triangle AB_1B_2 is constructed with angle $A = \theta$ and AB_2 perpendicular to B_1B_2 with hypotenuse x_1. All other triangles are constructed following the same pattern. Find the values of x_1, x_2, x_3, and x_4. What is the value of x_n?

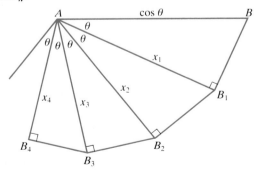

Figure 5.96

69. As shown in Fig. 5.97, triangle ABB_1 is constructed with angle $B = \theta$ and BB_1 is perpendicular to AB_1. Similarly, triangle AB_1B_2 is constructed with angle $AB_1B_2 = \theta$ and B_1B_2 is perpendicular to AB_2. The other triangles are constructed following the same pattern. Find the values of x_1, x_2, x_3, and x_4. What is the value of x_n?

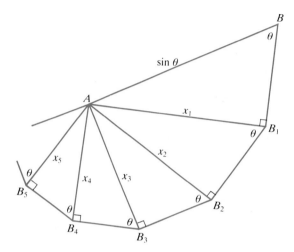

Figure 5.97

70. A stationary buoy moves up and down with the ocean waves. The ocean waves produce a motion equivalent to a sine curve with a period of 30 feet and an amplitude of 6 feet as shown in Fig. 5.98. Find the height of the buoy as a function of time. The speed v of the wave is $v = 1.424\sqrt{\lambda}$, where λ is the period of the wave in feet.

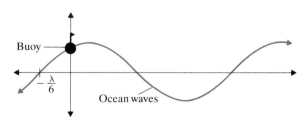

Figure 5.98

6

Analytic Trigonometry

IRIDESCENCE: LIGHT WAVES OF COLOR

It is from the beautiful Greek messenger goddess Iris, personifying the rainbow in all her glorious colors, that we get the word *iridescence*. This brilliant display of colors can be breathtaking whether seen in a butterfly's wings, peacock plumes, an abalone shell, opals, or even soap bubbles. Wherever the shimmering optical effect is seen, the process is the same: It is the interference, or interaction, of light waves as they reflect or scatter off two closely spaced surfaces. The surface may consist of a transparent film that reflects just one color, such as the emerald of a beetle's wing, or the iridescence may be caused by surfaces of tiny, regularly spaced bumps or ridges known as diffraction gratings.

Slight variations in the spacing of the surfaces may cause scattered or reflected waves of a certain color (certain wavelength) to interfere with and cancel out another color, or rays of colors may reinforce each other, making that color particularly intense. This cancellation or reinforcement of colors also depends on the angle from which the iridescence is viewed; as

an observer shifts positions, the colors seem to vanish and reappear.

In diagram (a), we see what happens when light shines on a thin patch of oil. Part of the beam reflects off the top of the oil, while part penetrates the oil and is then reflected off the bottom. When the two rays emerge and combine on the way to the observer's eye, their peaks and valleys align and reinforce each other to produce a powerful wave of light and an intense color. The alignment occurs when the difference in the distance traveled by the two rays has the proper relationship to the spacing of their peaks (the wavelength).

In diagram (b), the cancellation of color takes place as the light is reflected off a thinner patch of oil. The reflected waves are not synchronized, and the peaks and valleys line up against each other. There is no reflected color to the observer's eye.

Diagrams (c) and (d) show how iridescence is caused by the scattering of light off a surface of many closely spaced ridges (diffraction grating). Depending on the angle of the light beam,

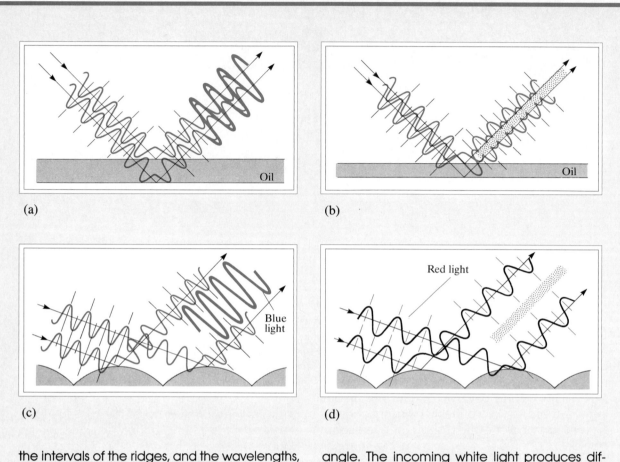

(a)

(b)

(c)

Blue light

(d)

Red light

the intervals of the ridges, and the wavelengths, the scattered rays reinforce or cancel each other. The blue light waves reinforce each other, while at the same angle the red light waves cancel each other because of their longer wavelength. At another angle the red waves would reinforce each other. This behavior accounts for the rainbow pattern that appears on an LP record or CD held in light at the proper angle. The incoming white light produces different colors at different angles.

Thus viewing the color of light as a sine wave provides a good model for understanding the phenomenon of iridescence.

SOURCE: Bruce Scheckter, "Bubbles That Bend the Mind," *Science 84*, March 1984; "The Light Fantastic," *Science 84*, May 1984, © 1984 by the AAAS. Original illustrations by Rick Farrell. Reprinted by permission.

Trigonometric functions are very useful in solving problems with triangles and periodic motion, as we saw in Chapter 5. Building on the trigonometric foundations of Chapter 5, we will now simplify trigonometric expessions, solve trigonometric equations, and prove new identities. Your ability to apply these skills successfully will be essential to your success in calculus.

6.1   Establishing Identities

We established several important identities in Chapter 5, which are listed here.

BASIC
TRIGONOMETRIC
IDENTITIES

$$\tan \theta = \frac{\sin \theta}{\cos \theta} \qquad \csc \theta = \frac{1}{\sin \theta}$$

$$\cot \theta = \frac{1}{\tan \theta} = \frac{\cos \theta}{\sin \theta} \qquad \sec \theta = \frac{1}{\cos \theta}$$

$$\sin^2 \theta + \cos^2 \theta = 1 \qquad \tan^2 \theta + 1 = \sec^2 \theta \qquad 1 + \cot^2 \theta = \csc^2 \theta$$

$$\sin (-\theta) = -\sin \theta \qquad \cos (-\theta) = \cos \theta \qquad \tan (-\theta) = -\tan \theta$$

Each of these equations is called an identity because each is always true and valid for any domain values of θ where the denominator does not equal zero. To prove these identities, we start with one side and show that it is equivalent to the other side in the following proof of $\tan (-\theta) = -\tan \theta$.

EXAMPLE 1

Prove that $\tan (-\theta) = -\tan \theta$.

Solution   We start with the left side and simplify it, using previous definitions and previously proven identities. The reason for each step is noted.

$$\tan (-\theta) = -\tan \theta$$

$$= \frac{\sin (-\theta)}{\cos (-\theta)} \qquad \left(\textit{Definition of } \tan \alpha = \frac{\sin \alpha}{\cos \alpha} \right)$$

$$= \frac{-\sin \theta}{\cos \theta} \qquad \begin{array}{l} \textit{(Identity for } \sin (-\theta)) \\ \textit{(Identity for } \cos (-\theta)) \end{array}$$

$$= -\tan \theta \qquad \textit{(Definition of } \tan \alpha)$$

We have shown that both sides are equivalent and thus have proven that

$$\tan (-\theta) = -\tan \theta.$$

In this section we will establish or prove several other identities. To prove an identity, the expression on one side of the equation must be shown to be equivalent to the expression on the other side of the equation. The following guidelines will help to provide strategies for proving new identities.

Guidelines for Proving Identities

(a) Each side of the equation is manipulated independently of the other side of the equation.

(b) Substitute with the basic identities to simplify expressions.

(c) Simplify expressions algebraically on one side by factoring, adding fractions, and other algebraic techniques.

(d) Simplify the more complicated side of the equation so that it becomes equivalent to the other side of the equation.

(e) When simplifying one side of an equation, continue to look at the other side to remember the desired form of the result.

(f) To see possible strategies for the proof, analyze the identity before beginning.

Almost all identities can be proved with the guidelines above. You will become more effective in using these guidelines as you understand the example proofs in this section and as you practice working out proofs yourself. The following examples show how you can use these guidelines.

EXAMPLE 2   Prove the identity $\sec^2 \alpha + \csc^2 \alpha = \sec^2 \alpha \csc^2 \alpha$.

Solution   To prove $\sec^2 \alpha + \csc^2 \alpha = \sec^2 \alpha \csc^2 \alpha$, we notice that the left side is a sum, while the right side is a product. It is often easier to change sums into products by addition (guideline (c) above):

$$\sec^2 \alpha + \csc^2 \alpha = \sec^2 \alpha \csc^2 \alpha$$

$$= \frac{1}{\cos^2 \alpha} + \frac{1}{\sin^2 \alpha} \qquad \textit{(Basic identities of sec } \theta = 1/\cos \theta \textit{ and csc } \theta = 1/\sin \theta\textit{)}$$

$$= \frac{1 \cdot \sin^2 \alpha}{\cos^2 \alpha \cdot \sin^2 \alpha} + \frac{1 \cdot \cos^2 \alpha}{\sin^2 \alpha \cdot \cos^2 \alpha} \qquad \textit{(Find the LCD and add the fractions)}$$

$$= \frac{\sin^2 \alpha + \cos^2 \alpha}{\cos^2 \alpha \sin^2 \alpha}$$

$$= \frac{1}{\cos^2 \alpha \sin^2 \alpha} \qquad \textit{(Substitute the basic identity } sin^2 \alpha + cos^2 \alpha = 1\textit{)}$$

$$= \frac{1}{\cos^2 \alpha} \frac{1}{\sin^2 \alpha} \qquad \textit{(Using the basic identities for sec } \theta \textit{ and csc } \theta\textit{)}$$

$$= \sec^2 \alpha \csc^2 \alpha \qquad \blacksquare$$

Notice how sums can be simplified into products in Example 2. Also notice that in this proof, all work is completed on only one side of the equation (guideline (a)). This is normal, although in some harder proofs, identities will be established by simplifying both sides *independently* to the same form.

EXAMPLE 3   Prove that

$$\frac{\cos \beta}{1 + \sin \beta} = \sec \beta - \tan \beta$$

Solution   Although the left side seems more complicated, we will first simplify the right side by changing both terms to $\sin \beta$ and $\cos \beta$ using the basic identities:

$$\frac{\cos \beta}{1 + \sin \beta} = \sec \beta - \tan \beta$$

$$= \frac{1}{\cos \beta} - \frac{\sin \beta}{\cos \beta}$$

$$= \frac{1 - \sin \beta}{\cos \beta}$$

Now we will work with the left side to complete this proof. We can get the numerator on the left to look more like the numerator on the right by multiplying by $1 - \sin \beta$:

$$\frac{\cos \beta}{1 + \sin \beta} = \sec \beta - \tan \beta$$

$$= \frac{1}{\cos \beta} - \frac{\sin \beta}{\cos \beta}$$

$$= \frac{1 - \sin \beta}{\cos \beta}$$

$$= \frac{\cos \beta}{1 + \sin \beta} \cdot \frac{1 - \sin \beta}{1 - \sin \beta} \qquad \textit{(Multiply numerator and denominator by } 1 - \sin \beta\textit{)}$$

$$= \frac{\cos \beta (1 - \sin \beta)}{1 - \sin^2 \beta} \qquad \textit{(Substitute } \cos^2 \beta \textit{ for } 1 - \sin^2 \beta\textit{, since } \sin^2 \beta + \cos^2 \beta = 1\textit{)}$$

$$= \frac{\cos \beta (1 - \sin \beta)}{\cos^2 \beta} \qquad \textit{(Now reduce the common factor of } \cos \beta \textit{ in the numerator and denominator)}$$

$$= \frac{1 - \sin \beta}{\cos \beta}$$

Now that the result on the left side is the same as the result on the right side, we have completed this proof.   ■

Notice that the product of $1 + \sin \beta$ and its conjugate $1 - \sin \beta$ produces a single function $\cos^2 \beta$:

$$(1 + \sin \beta)(1 - \sin \beta) = 1 - \sin^2 \beta = \cos^2 \beta$$

Multiplying by the conjugate is helpful in solving some identities, as we saw in the previous example.

The basic identities are always helpful, as we will see in the examples that follow.

EXAMPLE 4   Prove that

$$\frac{\sec \beta - \cos \beta}{\sec \beta + \cos \beta} = \frac{\sin^2 \beta}{1 + \cos^2 \beta}$$

Solution   We will simplify the more complicated left side by substituting the basic identities:

$$\frac{\sec \beta - \cos \beta}{\sec \beta + \cos \beta} = \frac{\sin^2 \beta}{1 + \cos^2 \beta}$$

$$= \frac{\dfrac{1}{\cos \beta} - \cos \beta}{\dfrac{1}{\cos \beta} + \cos \beta}$$

$$= \frac{\left(\dfrac{1}{\cos \beta} - \cos \beta\right) \cos \beta}{\left(\dfrac{1}{\cos \beta} + \cos \beta\right) \cos \beta} \qquad \textit{(Multiply the numerator and denominator by the LCD, cos β)}$$

$$= \frac{1 - \cos^2 \beta}{1 + \cos^2 \beta} \qquad \textit{(Substitute sin}^2 \textit{β for 1 − cos}^2 \textit{β since} \\ \sin^2 \textit{β + cos}^2 \textit{β = 1)}$$

$$= \frac{\sin^2 \beta}{1 + \cos^2 \beta}$$

■

EXAMPLE 5   Prove that

$$\frac{2 \sin^2 \alpha - 1}{(\sin \alpha + \cos \alpha)^2} = \frac{\sin \alpha - \cos \alpha}{\sin \alpha + \cos \alpha}$$

Solution   Although both sides are possible starting points, it might be easier to start with the right side. Notice that if we multiply the denominator on the right hand side by $\sin \alpha + \cos \alpha$, it will become the denominator on the left. (The factor $\sin \alpha + \cos \alpha$ is also the conjugate of the numerator on the right side.) Thus we will multiply the numerator and denominator on the right by $\sin \alpha + \cos \alpha$. At least the denominator on the right will be correct, and we will need to simplify only the numerator.

$$\frac{2 \sin^2 \alpha - 1}{(\sin \alpha + \cos \alpha)^2} = \frac{\sin \alpha - \cos \alpha}{\sin \alpha + \cos \alpha}$$

$$= \frac{(\sin \alpha - \cos \alpha)(\sin \alpha + \cos \alpha)}{(\sin \alpha + \cos \alpha)(\sin \alpha + \cos \alpha)}$$

$$= \frac{\sin^2 \alpha - \cos^2 \alpha}{(\sin \alpha + \cos \alpha)^2}$$

(Substitute $1 - \sin^2 \alpha$ for $\cos^2 \alpha$ since $\sin^2 \alpha + \cos^2 \alpha = 1$)
$$= \frac{\sin^2 \alpha - (1 - \sin^2 \alpha)}{(\sin \alpha + \cos \alpha)^2}$$

$$= \frac{2 \sin^2 \alpha - 1}{(\sin \alpha + \cos \alpha)^2} \qquad ■$$

EXAMPLE 6   Prove that $\sin^4 \beta - \cos^4 \beta = 1 - 2 \cos^2 \beta$.

Solution   Since the fourth power is more complicated than the second power, we will factor the left side of the equation and simplify:

$$\sin^4 \beta - \cos^4 \beta = 1 - 2 \cos^2 \beta$$

$= (\sin^2 \beta - \cos^2 \beta)(\sin^2 \beta + \cos^2 \beta)$ | *(Factoring)*

$= (\sin^2 \beta - \cos^2 \beta) \cdot 1$ | *(Substitution of $\sin^2 \beta + \cos^2 \beta = 1$)*

$= (1 - \cos^2 \beta) - \cos^2 \beta$ | *(Substitution of $\sin^2 \beta = 1 - \cos^2 \beta$)*

$= 1 - 2 \cos^2 \beta$ | ■

In Example 5 we forced part of an expression into the desired form. We will use this approach again to prove the following identity.

EXAMPLE 7   Prove that

$$\frac{\cos \alpha \cos \beta - \sin \alpha \sin \beta}{\sin \alpha \cos \beta + \cos \alpha \sin \beta} = \frac{1 - \tan \alpha \tan \beta}{\tan \alpha + \tan \beta}$$

Solution   We will first consider two possible methods of proof.

(a) One option is to change the right side of the equation into sines and cosines and simplify. (We will do this in Example 8.)

(b) We also notice that both numerators are differences and both denominators are sums. In addition, when we multiply the left numerator by $\dfrac{1}{\cos \alpha \cos \beta}$, it becomes $1 - \tan \alpha \tan \beta$. Using this last observation, we multiply the numerator and denominator of the left side by $\dfrac{1}{\cos \alpha \cos \beta}$. This will force both the numerator and denominator into the correct expressions.

$$\frac{\cos \alpha \cos \beta - \sin \alpha \sin \beta}{\sin \alpha \cos \beta + \cos \alpha \sin \beta} = \frac{1 - \tan \alpha \tan \beta}{\tan \alpha + \tan \beta}$$

$$= \frac{\dfrac{1}{\cos \alpha \cos \beta}(\cos \alpha \cos \beta - \sin \alpha \sin \beta)}{\dfrac{1}{\cos \alpha \cos \beta}(\sin \alpha \cos \beta + \cos \alpha \sin \beta)}$$

$$= \frac{\dfrac{\cos \alpha \cos \beta}{\cos \alpha \cos \beta} - \dfrac{\sin \alpha \sin \beta}{\cos \alpha \cos \beta}}{\dfrac{\sin \alpha \cos \beta}{\cos \alpha \cos \beta} + \dfrac{\cos \alpha \sin \beta}{\cos \alpha \cos \beta}}$$

$$= \frac{1 - \tan \alpha \tan \beta}{\tan \alpha + \tan \beta}$$

Sometimes more than one approach or strategy is available in proving an identity. The strategy in Example 7 forces part of the expression into the right form and the rest of the simplification follows.

We can also prove the identity in Example 7 by using sines and cosines, as shown in Example 8.

EXAMPLE 8   Prove that

$$\frac{\cos \alpha \cos \beta - \sin \alpha \sin \beta}{\sin \alpha \cos \beta + \cos \alpha \sin \beta} = \frac{1 - \tan \alpha \tan \beta}{\tan \alpha + \tan \beta}$$

by simplifying the left side with sines and cosines.

Solution   Work on the right side of the equation with the basic identities:

$$\frac{\cos \alpha \cos \beta - \sin \alpha \sin \beta}{\sin \alpha \cos \beta + \cos \alpha \sin \beta} = \frac{1 - \tan \alpha \tan \beta}{\tan \alpha + \tan \beta}$$

$$= \frac{1 - \dfrac{\sin \alpha}{\cos \alpha}\dfrac{\sin \beta}{\cos \beta}}{\dfrac{\sin \alpha}{\cos \alpha} - \dfrac{\sin \beta}{\cos \beta}}$$

(Multiply the numerator and denominator by $\cos \alpha \cos \beta$)

$$= \frac{\cos \alpha \cos \beta \left(1 - \dfrac{\sin \alpha}{\cos \alpha}\dfrac{\sin \beta}{\cos \beta}\right)}{\cos \alpha \cos \beta \left(\dfrac{\sin \alpha}{\cos \alpha} - \dfrac{\sin \beta}{\cos \beta}\right)}$$

$$= \frac{\cos \alpha \cos \beta - \sin \alpha \sin \beta}{\sin \alpha \cos \beta + \cos \alpha \sin \beta}$$

As shown in Examples 7 and 8, the same identity can be proved in at least two different ways. The following example shows how we can factor to prove an identity.

EXAMPLE 9   Prove that

$$\frac{\sin^2 \theta}{1 + 5 \cos \theta + 4 \cos^2 \theta} = \frac{1 - \cos \theta}{1 + 4 \cos \theta}$$

Solution   Begin on the left side, since it appears to be a little more complicated than the right side. Note that on the left side we can factor the denominator and change the numerator from sine to cosine:

$$\frac{\sin^2 \theta}{1 + 5 \cos \theta + 4 \cos^2 \theta} = \frac{1 - \cos \theta}{1 + 4 \cos \theta}$$

$$= \frac{1 - \cos^2 \theta}{1 + 5 \cos \theta + 4 \cos^2 \theta} \qquad \textit{(Substitute } 1 - \cos^2 \theta \textit{ for } \sin^2 \theta\textit{)}$$

$$= \frac{(1 - \cos \theta)(1 + \cos \theta)}{(1 + 4 \cos \theta)(1 + \cos \theta)} \qquad \begin{array}{l}\textit{(Factor both numerator and denominator)} \\ \textit{(Reduce the common factor)}\end{array}$$

$$= \frac{1 - \cos \theta}{1 + 4 \cos \theta}$$

Not all equations are identities. An equation that is not an identity is not true for all possible domain values; one or more domain values make the equation false. Any value that makes the equation false is called a *counterexample*.

EXAMPLE 10   Is the equation

$$\sin^2 \alpha + \cos^2 \alpha = (\sin \alpha + \cos \alpha)^2$$

true or false for all α? If it is true, prove it as an identity. If it is false, find a counterexample.

Solution   This equation appears to be false, since the binomial squared on the right should have a middle term that is not on the left side. To find a counterexample, we substitute a value for α. For example let $\alpha = \frac{\pi}{2}$. Then

$$\sin^2 \frac{\pi}{2} + \cos^2 \frac{\pi}{2} = \left(\sin \frac{\pi}{2} + \cos \frac{\pi}{2} \right)^2$$

is

$$1^2 + 0^2 = (1 + 0)^2$$

or

$$1 = 1, \qquad \text{a true statement}$$

This true result means that $\alpha = \dfrac{\pi}{2}$ is not a counterexample. On the other hand, this true result does not prove this statement true.

We now try another value for α. We let $\alpha = \dfrac{\pi}{4}$ and

$$\sin^2 \frac{\pi}{4} + \cos^2 \frac{\pi}{4} = \left(\sin \frac{\pi}{4} + \cos \frac{\pi}{4} \right)^2$$

$$\left(\frac{\sqrt{2}}{2} \right)^2 + \left(\frac{\sqrt{2}}{2} \right)^2 = \left(\frac{\sqrt{2}}{2} + \frac{\sqrt{2}}{2} \right)^2$$

$$\frac{2}{4} + \frac{2}{4} = \left(\frac{2\sqrt{2}}{2} \right)^2$$

$$1 = 2, \qquad \text{a false statement}$$

Therefore $\alpha = \dfrac{\pi}{4}$ is a counterexample, and $\sin^2 \alpha + \cos^2 \alpha = (\sin \alpha + \cos \alpha)^2$ is not an identity. ■

As we saw in Example 10, you cannot prove an identity true with one domain value that yields a true statement. But you can prove that a statement is not an identity by finding one counterexample.

Exercises 6.1

Prove the following identities.

1. $\dfrac{\cot \theta}{\csc \theta} = \cos \theta$

2. $\dfrac{\tan \theta}{\sec \theta} = \sin \theta$

3. $\dfrac{\cos \theta - \tan \theta}{\sin \theta} = \cot \theta - \sec \theta$

4. $\dfrac{\sin \theta - \cot \theta}{\cos \theta} = \tan \theta - \csc \theta$

5. $\dfrac{\sin \theta}{\csc \theta} + \dfrac{\cos \theta}{\sec \theta} = 1$

6. $\dfrac{\tan^2 \theta}{\sec^2 \theta} + \dfrac{\cot^2 \theta}{\csc^2 \theta} = 1$

7. $\tan (-\theta) \cos \theta = -\sin \theta$

8. $\cot (-\theta) \sin \theta = -\cos \theta$

9. $\cos^2 \theta + \cos^2 \theta \sin^2 \theta + \sin^4 \theta = 1$

10. $\sin^2 \theta + \sin^2 \theta \cos^2 \theta + \cos^4 \theta = 1$

11. $\sec^2 \theta - \csc^2 \theta = \tan^2 \theta - \cot^2 \theta$

12. $\sec^2 \theta + \csc^2 \theta = 2 + \tan^2 \theta + \cot^2 \theta$

13. $\dfrac{\sin^2 \theta}{1 + 2 \cos \theta + \cos^2 \theta} = \dfrac{1 - \cos \theta}{1 + \cos \theta}$

14. $\dfrac{\sin^2 \theta}{1 - 2 \cos \theta + \cos^2 \theta} = \dfrac{1 + \cos \theta}{1 - \cos \theta}$

15. $\dfrac{\cos^2 \theta}{(1 - \sin \theta)^2} = \dfrac{1 + \sin \theta}{1 - \sin \theta}$

16. $\dfrac{\cos^2 \theta}{1 + \sin \theta - 2 \sin^2 \theta} = \dfrac{1 + \sin \theta}{1 + 2 \sin \theta}$

17. $\dfrac{\cot x}{\csc x + 1} = \dfrac{\csc x - 1}{\cot x}$

18. $\dfrac{1 + \sec x}{\sin x + \tan x} = \csc x$

19. $(\sin \alpha \cos \beta - \cos \alpha \sin \beta)^2$
$\qquad\qquad + (\cos \alpha \cos \beta + \sin \alpha \sin \beta)^2 = 1$

20. $(\sin \alpha \cos \beta + \cos \alpha \sin \beta)^2$
$\qquad\qquad + (\cos \alpha \cos \beta - \sin \alpha \sin \beta)^2 = 1$

21. $\dfrac{\sin^3 x + \cos^3 x}{\sin x + \cos x} = 1 - \sin x \cos x$

22. $\dfrac{\sin^3 x - \cos^3 x}{\sin x - \cos x} = 1 + \sin x \cos x$

23. $\sin^4 \theta - \cos^4 \theta = 1 - 2 \cos^2 \theta$

24. $\sin^4 \theta - \cos^4 \theta = 2 \sin^2 \theta - 1$

25. $\sin^6 \theta + \cos^6 \theta = 1 - 3 \sin^2 \theta \cos^2 \theta$

26. $\sin^6 \theta - \cos^6 \theta = (1 - 2 \cos^2 \theta)(\sin^2 \theta + \cos^4 \theta)$

27. $\dfrac{\tan x - \tan y}{1 + \tan x \tan y} = \dfrac{\cot y - \cot x}{1 + \cot x \cot y}$

28. $\dfrac{\csc \theta}{1 + \csc \theta} - \dfrac{\csc \theta}{1 - \csc \theta} = 2 \sec^2 \theta$

29. $\dfrac{\sin \theta}{1 + \cos \theta} + \dfrac{1 + \cos \theta}{\sin \theta} = 2 \csc \theta$

30. $\dfrac{\cos \theta}{1 - \tan \theta} + \dfrac{\sin \theta}{1 - \cot \theta} = \cos \theta + \sin \theta$

31. $\ln |\csc \alpha| = -\ln |\sin \alpha|$

32. $\log |\tan \alpha| = -\log |\cot \alpha|$

33. $\ln |\csc \alpha - \cot \alpha| = -\ln |\csc \alpha + \cot \alpha|$

34. $\ln |\sec \beta - \tan \beta| = -\ln |\sec \beta + \tan \beta|$

35. $\cos^4 \alpha + 1 - \sin^4 \alpha = 2 \cos^2 \alpha$

36. $(1 - \tan^2 \beta)^2 = \sec^4 \beta - 4 \tan^2 \beta$

Substitute the given value for u in each of the following expressions and simplify the resulting value of y for $0 \le \theta \le \dfrac{\pi}{2}$.

37. $y = \dfrac{u}{\sqrt{1 - u^2}}$

where $u = \sin \theta$

38. $y = \dfrac{u}{\sqrt{1 - u^2}}$

where $u = \cos \theta$

39. $y = \dfrac{1}{\sqrt{1 + u^2}}$

where $u = \tan \theta$

40. $y = \dfrac{1 + u^2}{\sqrt{4 + u^2}}$

where $u = 2 \tan \theta$

41. $y = \dfrac{u}{\sqrt{9 + u^2}}$

where $u = 3 \tan \theta$

42. $y = \dfrac{2u}{\sqrt{(4/9) - u^2}}$

where $u = \frac{2}{3} \sin \theta$

Mark each of the following as true or false. Explain your answer. If true, prove the identity. If false, find a counterexample.

43. $(\sin \alpha + \cos \alpha)^2 = 1$

44. $\sin^2 \theta + \cos^2 \theta = (\sin \theta + \cos \theta)^2$

45. $\cos 2\theta = 2 \cos \theta$ **46.** $\sin 2\theta = 2 \sin \theta$

47. $\sin (\alpha + \beta) = \sin \alpha + \sin \beta$

48. $\cos (\alpha + \beta) = \cos \alpha + \cos \beta$

49. $\cos (\alpha - \beta) = \cos \alpha - \cos \beta$

50. $\sin (\alpha - \beta) = \sin \alpha - \sin \beta$

51. $\cos \theta = \sqrt{1 - \sin^2 \theta}$ **52.** $\sin \theta = \sqrt{1 - \cos^2 \theta}$

6.2   Trigonometric Equations

An equation that is not an identity may be true for some domain values. These equations are said to be conditionally true. Much of the work in algebra is centered around solving these conditional equations. We will now use algebraic techniques and inverse trigonometric functions to solve trigonometric equations.

EXAMPLE 1   Solve $\sin \alpha \cos \alpha - 3 \sin \alpha = 0$ for α in $[0, 2\pi)$.

Solution   We will factor the common factor $\sin \alpha$ and solve:

$$\sin \alpha \cos \alpha - 3 \sin \alpha = 0$$

$$\sin \alpha (\cos \alpha - 3) = 0$$

Then

$$\sin \alpha = 0 \qquad \text{or} \qquad \cos \alpha - 3 = 0$$

$$\alpha = 0, \pi \qquad\qquad\qquad \cos \alpha = 3$$

$$\text{No solution, since } -1 \le \cos \alpha \le 1$$

Therefore the solution to this equation is $\alpha \in \{0, \pi\}$.   ■

EXAMPLE 2

Solve $4 \cos \beta \sin \beta - 2 \cos \beta - 2 \sin \beta + 1 = 0$ for β in $[0, 2\pi)$.

Solution For these four terms we will factor by grouping. (We will group two terms at a time and factor the greatest common factor from each pair.)

$$4 \cos \beta \sin \beta - 2 \cos \beta - 2 \sin \beta + 1 = 0$$
$$(4 \cos \beta \sin \beta - 2 \cos \beta) - (2 \sin \beta - 1) = 0$$
$$2 \cos \beta \, (\mathbf{2 \sin \beta - 1}) - 1 \, (\mathbf{2 \sin \beta - 1}) = 0$$
$$(2 \sin \beta - 1)(2 \cos \beta - 1) = 0$$

Thus

$$2 \sin \beta - 1 = 0 \qquad \text{or} \qquad 2 \cos \beta - 1 = 0$$
$$\sin \beta = \frac{1}{2} \qquad\qquad\qquad \cos \beta = \frac{1}{2}$$
$$\beta = \frac{\pi}{6}, \frac{5\pi}{6} \qquad\qquad\qquad \beta = \frac{\pi}{3}, \frac{5\pi}{3}$$

Therefore $\beta \in \left\{ \dfrac{\pi}{6}, \dfrac{\pi}{3}, \dfrac{5\pi}{6}, \dfrac{5\pi}{3} \right\}$.   ■

EXAMPLE 3

Solve $2 \sin^2 2\theta - 1 = 0$ for θ in $[0, 360°)$.

Solution For $2 \sin^2 2\theta - 1 = 0$,

$$\sin^2 2\theta = \frac{1}{2} \quad \textit{(Now use the square root method)}$$
$$\sin 2\theta = \pm\frac{1}{\sqrt{2}} = \pm\frac{\sqrt{2}}{2}$$

Since the angle θ is $0 \le \theta < 360°$, we must have the angle 2θ such that $0 \le 2\theta \le 720°$. In solving for a multiple of an angle the length of the period must be multiplied by the same factor. Thus for

$$\sin 2\theta = \pm\frac{\sqrt{2}}{2}$$

2θ is in $[0, 720°)$ for θ to be in $[0, 180°)$. The reference angle for 2θ is $45°$, and 2θ is in every quadrant, since the $\sin 2\theta$ is both positive and negative. We also note that with a reference angle of $45°$ these 2θ solutions are $90°$ apart. Thus we can start with $45°$ and add $90°$ to get the next angle up to $720°$. Doing this, we have

$$2\theta \in \{45°, 135°, 225°, 315°, 405°, 495°, 585°, 675°\}$$

and dividing by 2, we have

$$\theta \in \{22.5°, 67.5°, 112.5°, 157.5°, 202.5°, 247.5°, 292.5°, 337.5°\} \quad ■$$

Notice in Examples 1 and 2 that we solved for β in radians, since the interval was $[0, 2\pi)$. On the other hand, in Example 3 we used degrees, since θ was in the interval $[0, 360°)$.

Sometimes we will not restrict solutions to the interval $[0, 2\pi)$ or $[0, 360°)$. In these cases the solution set must include all solutions in the interval $(-\infty, \infty)$. These solutions would include all solutions in the interval $[0, 2\pi)$ or $[0, 360°)$ as well as all integer multiples of 2π or $360°$ added to the solutions in the interval $[0, 2\pi)$ or $[0, 360°)$. Thus when the solution α is not restricted to $[0, 2\pi)$, $\alpha = \dfrac{\pi}{4}$ in $[0, 2\pi)$ becomes

$$\alpha = \frac{\pi}{4} + 2k\pi \qquad \text{where } k \text{ is an integer}$$

Notice how this is demonstrated in the next example.

EXAMPLE 4   Find all solutions to $2 \sin^2 \alpha - 1 = 0$, where α is in radians.

Solution   We notice that we are using radians and α is not restricted to $[0, 2\pi)$. Thus

$$2 \sin^2 \alpha - 1 = 0$$

$$\sin \alpha = \frac{1}{2} \quad \textit{(Now use the square root method)}$$

$$\sin \alpha = \pm\frac{1}{\sqrt{2}} = \pm\frac{\sqrt{2}}{2}$$

Thus for $\alpha' = \dfrac{\pi}{4}$ we have

$$\alpha \in \left\{\frac{\pi}{4}, \frac{3\pi}{4}, \frac{5\pi}{4}, \frac{7\pi}{4}\right\} \qquad \text{for } \alpha \text{ in } [0, 2\pi)$$

But this problem asks for *all* values of α, which include the above answers and $2k\pi$ added to each answer where k is an integer. Thus

$$\alpha \in \left\{\frac{\pi}{4} + 2k\pi, \frac{3\pi}{4} + 2k\pi, \frac{5\pi}{4} + 2k\pi, \frac{7\pi}{4} + 2k\pi\right\} \qquad \text{for any integer } k$$

Fortunately, these answers are only $\dfrac{\pi}{2}$ apart, and the above answer can be simplified and written as

$$\alpha = \frac{\pi}{4} + \frac{k\pi}{2} \qquad \text{for any integer } k \qquad\blacksquare$$

As well as using algebraic techniques, we can also use identities in solving equations. Identities can change the functions in equations, so they are compatible for factoring as we will see in the following examples.

EXAMPLE 5   Solve $2 \cos^2 \alpha + 3 \sin \alpha = 3$ for α in $[0, 2\pi)$.

Solution   Because this equation involves sines and cosines, we will substitute $1 - \sin^2 \alpha$ for $\cos^2 \alpha$, since $\sin^2 \alpha + \cos^2 \alpha = 1$. Then we will be working only with the sine function.

$$2 \cos^2 \alpha + 3 \sin \alpha = 3$$
$$2(1 - \sin^2 \alpha) + 3 \sin \alpha = 3$$
$$2 - 2 \sin^2 \alpha + 3 \sin \alpha = 3 \qquad \textit{(Change to standard form and simplify)}$$
$$0 = 2 \sin^2 \alpha - 3 \sin \alpha + 3 - 2$$
$$0 = 2 \sin^2 \alpha - 3 \sin \alpha + 1 \quad \textit{(Now factor and solve)}$$
$$0 = (2 \sin \alpha - 1)(\sin \alpha - 1)$$

Thus

$$2 \sin \alpha - 1 = 0 \qquad \text{or} \qquad \sin \alpha - 1 = 0$$
$$\sin \alpha = \frac{1}{2} \qquad \text{or} \qquad \sin \alpha = 1$$

and

$$\alpha = \frac{\pi}{6}, \frac{5\pi}{6} \qquad \text{or} \qquad \alpha = \frac{\pi}{2}$$

$$\text{Therefore } \alpha \in \left\{ \frac{\pi}{6}, \frac{\pi}{2}, \frac{5\pi}{6} \right\}.$$

When you cannot factor a quadratic equation, you can always use the quadratic formula and a calculator as in the following example.

EXAMPLE 6   Solve $\cos^2 \beta + \sin \beta = 0$ for β in $[0, 360°)$.

Solution   First substitute $1 - \sin^2 \beta$ for $\cos^2 \beta$, since $\sin^2 \alpha + \cos^2 \alpha = 1$, and the equation will contain only sines.

$$\cos^2 \beta + \sin \beta = 0$$
$$(1 - \sin^2 \beta) + \sin \beta = 0 \qquad \textit{(Change to standard form)}$$
$$0 = \sin^2 \beta - \sin \beta - 1 \quad \begin{array}{l}\textit{(Since this does not factor,}\\ \textit{solve with the quadratic formula)}\end{array}$$
$$\sin \beta = \frac{-(-1) \pm \sqrt{(-1)^2 - 4(1)(-1)}}{2(1)}$$
$$= \frac{1 \pm \sqrt{1 + 4}}{2} = \frac{1 \pm \sqrt{5}}{2}$$

Thus we have

$$\sin \beta = \frac{1 + \sqrt{5}}{2} \qquad \text{or} \qquad \sin \beta = \frac{1 - \sqrt{5}}{2}$$

Since $\sin \beta = \dfrac{1 + \sqrt{5}}{2} \approx 1.6180339 > 1$ and no angle exists for $\sin \beta = \dfrac{1 + \sqrt{5}}{2}$, the remaining solution is

$$\sin \beta = \frac{1 - \sqrt{5}}{2} \approx -0.61803398$$

with a reference angle of $\beta' \approx 38.1727°$. For this β' we have $\beta \approx 180° + 38.2°$ or $\beta \approx 360° - 38.2°$. Therefore the solution to the equation is $\beta \in \{218.2°, 321.8°\}$.

■

As we saw in these examples, we must use algebraic techniques, basic identities, and inverse trigonometric functions to solve trigonometric equations.

Exercises 6.2

Find the solutions to the following equations in the interval $[0, 360°)$.

1. $(\sin \alpha - 1)(2 \cos \alpha + 1) = 0$
2. $(\cos \alpha + 1)(2 \sin \alpha - 1) = 0$
3. $\sin \beta \, (2 \cos \beta - \sqrt{3}) = 0$
4. $\cos \beta \, (2 \sin \beta + \sqrt{3}) = 0$
5. $\sin^2 \alpha - 2 \sin \alpha + 1 = 0$
6. $\cos^2 \alpha + 2 \cos \alpha + 1 = 0$
7. $\cos^2 \beta - 1 = 0$
8. $\sin^2 \beta - 1 = 0$

Find the solutions to the following equations in the interval $[0, 2\pi)$.

9. $\cos^2 \alpha - 4 \cos \alpha + 3 = 0$
10. $\sin^2 \alpha - \sin \alpha - 2 = 0$
11. $4 \sin^2 \beta - 2 = 0$
12. $4 \cos^2 \beta - 2 = 0$
13. $\tan \alpha \, (2 \cos \alpha - 1) - (2 \cos \alpha - 1) = 0$
14. $2 \sin \alpha \, (\tan \alpha + 1) - (\tan \alpha + 1) = 0$
15. $\sin^2 2\beta - 1 = 0$
16. $\cos^2 2\beta - 1 = 0$
17. $\cos^2 2\alpha - 1 = 0$
18. $\sin^2 2\alpha - 1 = 0$

Find *all* solutions, in radians, to the following equations.

19. $\sin \alpha \tan \alpha = 0$
20. $\cot \alpha \cos \alpha = 0$
21. $(2 \sin \beta - \sqrt{3})(2\cos \beta + \sqrt{3}) = 0$
22. $(2 \sin \beta + \sqrt{2})(2 \cos \beta - \sqrt{2}) = 0$
23. $\cot^2 \alpha \cos \alpha - \cos \alpha = 0$

24. $\tan^2 \alpha \sin \alpha - \sin \alpha = 0$
25. $\sin^2 \beta - 1 = 0$
26. $\cos^2 \beta - 1 = 0$
27. $4 \cos \alpha \sin \alpha - 2 \cos \alpha + 2 \sin \alpha = 1$
28. $\sin^2 2\beta + 2 \sin 2\beta = 3$
29. $\cos^2 2\beta - 3 \cos 2\beta = -2$
30. $4 \sin \alpha \cos \alpha - 2 \sin \alpha + 2 \cos \alpha = 1$

Find the zeros (*x*-intercepts) for each of the following in the interval $[0, 2\pi)$.

31. $y = \sin^2 x + \cos x + 1$
32. $y = \cos^2 x - \sin x + 1$
33. $y = 3 \tan^2 x - \sec^2 x - 1$
34. $y = 3 \sec^2 x - \tan^2 x - 1$
35. $y = \cos^2 x + 2 \sin x - 2$
36. $y = 2 \cos^2 x - \sin x - 1$
37. $y = 2 \sin^2 x - \cos x - 1$
38. $y = \sin^2 x + 2 \cos x - 2$

Find the zeros (*x*-intercepts) for each of the following using a calculator in the interval $[0, 360°)$.

39. $y = \sin^2 x + 5 \sin x - 3$
40. $y = 2 \cos^2 x + 4 \cos x + 1$
41. $y = 2 \cos^2 x - 10 \cos x + 9$
42. $y = 2 \sin^2 x - 10 \sin x + 5$
43. $y = 3 \sin^2 x - 5 \sin x + 1$
44. $y = 3 \cos^2 x - 6 \cos x + 1$

6.3   Trigonometric Functions of Sums and Differences

We will now establish some identities with more complicated domains, such as sin $(\alpha + \beta)$, cos $(\alpha + \beta)$, and tan $(\alpha + \beta)$. We will first find an identity for cos $(\alpha - \beta)$, using some geometry and the distance formula. Remember that for any point generated by angle θ on the unit circle we have sin $\theta = y$ and cos $\theta = x$. Therefore this point on the unit circle generated by the angle θ has coordinates of $(x, y) = (\cos \theta, \sin \theta)$ as shown in Fig. 6.1.

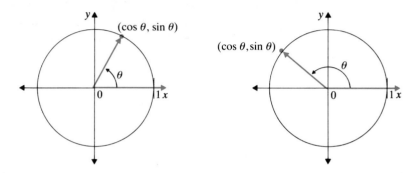

The angle θ intersects the unit circle at the point $(\cos \theta, \sin \theta)$

Figure 6.1

We now consider angles α and β in standard position as in Fig. 6.2(a). Then we notice that the angle $\alpha - \beta$ is the angle P_1OP_2. We also place the same angle $\alpha - \beta$ in standard position as in Fig. 6.2(b) and rename it angle P_4OP_3. Now we use the coordinates of the ordered P_1 and P_2 in Fig. 6.2(a) to find the distance d_1 from P_1 to P_2:

$$d_1 = \sqrt{(\cos \alpha - \cos \beta)^2 + (\sin \alpha - \sin \beta)^2}$$
$$
\begin{aligned}
d_1{}^2 &= (\cos \alpha - \cos \beta)^2 + (\sin \alpha - \sin \beta)^2 \\
&= \cos^2 \alpha - 2 \cos \alpha \cos \beta + \cos^2 \beta + \sin^2 \alpha - 2 \sin \alpha \sin \beta + \sin^2 \beta \\
&= (\cos^2 \alpha + \sin^2 \alpha) + (\cos^2 \beta + \sin^2 \beta) - 2 \cos \alpha \cos \beta - 2 \sin \alpha \sin \beta \\
&= 1 + 1 - 2 \cos \alpha \cos \beta - 2 \sin \alpha \sin \beta
\end{aligned}
$$

Thus $d_1{}^2 = 2 - 2 \cos \alpha \cos \beta - 2 \sin \alpha \sin \beta$.

Similarly, we use the coordinates of the points P_3 and P_4 to find the distance d_2 from P_3 to P_4:

$$
\begin{aligned}
d_2 &= \sqrt{[\cos (\alpha - \beta) - 1]^2 + [\sin (\alpha - \beta) - 0]^2} \\
d_2{}^2 &= [\cos (\alpha - \beta) - 1]^2 + [\sin (\alpha - \beta)]^2 \\
&= \cos^2 (\alpha - \beta) - 2 \cos (\alpha - \beta) + 1 + \sin^2 (\alpha - \beta) \\
&= [\cos^2 (\alpha - \beta) + \sin^2 (\alpha - \beta)] - 2 \cos (\alpha - \beta) + 1 \\
&= 1 - 2 \cos (\alpha - \beta) + 1
\end{aligned}
$$

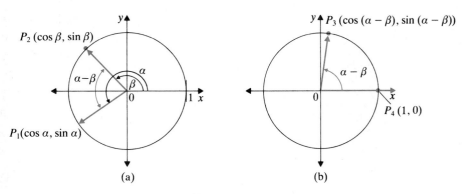

Figure 6.2

Thus $d_2{}^2 = 2 - 2 \cos (\alpha - \beta)$. Since d_1 and d_2 are both chords for the same angle, $\alpha - \beta$, on the unit circle, we have $d_1 = d_2$ and $d_1{}^2 = d_2{}^2$. Substituting these results and simplifying, we have

$$d_1{}^2 = d_2{}^2$$
$$2 - 2 \cos \alpha \cos \beta - 2 \sin \alpha \sin \beta = 2 - 2 \cos (\alpha - \beta)$$
$$-2 \cos \alpha \cos \beta - 2 \sin \alpha \sin \beta = -2 \cos (\alpha - \beta)$$
$$2 \cos (\alpha - \beta) = 2 \cos \alpha \cos \beta + 2 \sin \alpha \sin \beta$$
$$\cos (\alpha - \beta) = \cos \alpha \cos \beta + \sin \alpha \sin \beta$$

This result, $\cos (\alpha - \beta) = \cos \alpha \cos \beta + \sin \alpha \sin \beta$, is an important identity. We can now utilize this identity whenever it might be helpful, as in the following example.

EXAMPLE 1   Prove that $\cos (\alpha + \beta) = \cos \alpha \cos \beta - \sin \alpha \sin \beta$.

Solution   Remembering that

$$\sin (-\theta) = -\sin \theta \quad \text{and} \quad \cos (-\theta) = \cos \theta$$

we can write

$$\cos (\alpha + \beta) = \cos [\alpha - (-\beta)] \qquad \textit{(Use the identity for}$$
$$= \cos [\alpha - (-\beta)] \qquad \qquad \textit{cos } (\alpha + \beta))$$
$$= \cos \alpha \cos (-\beta) + \sin \alpha \sin (-\beta) \quad \textit{(Use the identities for}$$
$$\qquad \qquad \qquad \qquad \qquad \qquad \qquad \textit{cos } (-\beta) \textit{ and sin } (-\beta))$$
$$= \cos \alpha \cos \beta + \sin \alpha (-\sin \beta)$$
$$= \cos \alpha \cos \beta - \sin \alpha \sin \beta$$

Therefore $\cos (\alpha + \beta) = \cos \alpha \cos \beta - \sin \alpha \sin \beta$, another important identity.

■

To find identities for $\sin(\alpha + \beta)$ and $\sin(\alpha - \beta)$, it is necessary to first prove that

(a) $\cos\left(\dfrac{\pi}{2} - \theta\right) = \sin\theta$ and

(b) $\sin\left(\dfrac{\pi}{2} - \theta\right) = \cos\theta.$

PROOF:

(a) Using the identity $\cos(\alpha - \beta) = \cos\alpha\cos\beta + \sin\alpha\sin\beta$, we have

$$\cos\left(\frac{\pi}{2} - \theta\right) = \cos\frac{\pi}{2}\cos\theta + \sin\frac{\pi}{2}\sin\theta$$

$$= 0 \cdot \cos\theta + 1 \cdot \sin\theta$$

$$= \sin\theta$$

(b) To prove

$$\sin\left(\frac{\pi}{2} - \theta\right) = \cos\theta$$

we let $\alpha = \dfrac{\pi}{2} - \theta$ and thus $\theta = \dfrac{\pi}{2} - \alpha.$ Then we have

$$\cos\theta = \cos\left(\frac{\pi}{2} - \alpha\right)$$

$$= \cos\frac{\pi}{2}\cos\alpha + \sin\frac{\pi}{2}\sin\alpha$$

$$= 0 \cdot \cos\alpha + 1 \cdot \sin\alpha$$

$$= \sin\alpha$$

$$= \sin\left(\frac{\pi}{2} - \theta\right)$$

Therefore $\cos\left(\dfrac{\pi}{2} - \theta\right) = \sin\theta$ and $\sin\left(\dfrac{\pi}{2} - \theta\right) = \cos\theta.$ □

These identities are called cofunction identities and are listed below. The $\tan\left(\dfrac{\pi}{2} - \theta\right)$ identity will be proved in the exercise set of this section.

COFUNCTION IDENTITIES		
$\cos\left(\dfrac{\pi}{2} - \theta\right) = \sin\theta$	or	$\cos(90° - \theta) = \sin\theta$
$\sin\left(\dfrac{\pi}{2} - \theta\right) = \cos\theta$	or	$\sin(90° - \theta) = \cos\theta$
$\tan\left(\dfrac{\pi}{2} - \theta\right) = \cot\theta$	or	$\tan(90° - \theta) = \cot\theta$

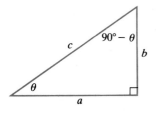

Figure 6.3

Although we proved these identities for any θ, we can also verify these identities when θ and $90° - \theta$ are acute angles in a right triangle. As we can see in Fig. 6.3, we have the following cofunction identities:

$$\sin \theta = \frac{b}{c} = \cos (90° - \theta)$$

$$\cos \theta = \frac{a}{c} = \sin (90° - \theta)$$

$$\cot \theta = \frac{a}{b} = \tan (90° - \theta)$$

With these cofunction identities and the identities for $\cos (\alpha - \beta)$ and $\cos (\alpha + \beta)$ we can now establish identities for $\sin (\alpha + \beta)$ and $\sin (\alpha - \beta)$. Thus

$$\sin (\alpha + \beta) = \cos \left[\frac{\pi}{2} - (\alpha + \beta) \right] \qquad \textit{(Cofunction identity)}$$

$$= \cos \left(\frac{\pi}{2} - \alpha - \beta \right)$$

$$= \cos \left[\left(\frac{\pi}{2} - \alpha \right) - \beta \right] \qquad \textit{(Now use the cos $(\alpha - \beta)$ identity)}$$

$$= \cos \left(\frac{\pi}{2} - \alpha \right) \cos \beta + \sin \left(\frac{\pi}{2} - \alpha \right) \sin \beta$$

$$= \sin \alpha \cos \beta + \cos \alpha \sin \beta \qquad \textit{(Cofunction identities)}$$

Therefore $\sin (\alpha + \beta) = \sin \alpha \cos \beta + \cos \alpha \sin \beta$. Similarly,

$$\sin (\alpha - \beta) = \sin [\alpha + (-\beta)] \qquad \textit{(Use the sin $(\alpha + \beta)$ identity)}$$

$$= \sin \alpha \cos (-\beta) + \cos \alpha \sin (-\beta) \qquad \textit{Use the cos $(-\theta)$ and}$$

$$= \sin \alpha \cos \beta - \cos \alpha \sin \beta \qquad \textit{sin $(-\theta)$ identities)}$$

Therefore $\sin (\alpha - \beta) = \sin \alpha \cos \beta - \cos \alpha \sin \beta$.

The conclusions for the trigonometric functions of sums and differences are summarized below. The proof of the tangent function identities is left for the exercises at the end of this section.

FUNCTIONS OF SUMS AND DIFFERENCES

$$\sin (\alpha + \beta) = \sin \alpha \cos \beta + \cos \alpha \sin \beta$$

$$\sin (\alpha - \beta) = \sin \alpha \cos \beta - \cos \alpha \sin \beta$$

$$\cos (\alpha + \beta) = \cos \alpha \cos \beta - \sin \alpha \sin \beta$$

$$\cos (\alpha - \beta) = \cos \alpha \cos \beta + \sin \alpha \sin \beta$$

$$\tan (\alpha + \beta) = \frac{\tan \alpha + \tan \beta}{1 - \tan \alpha \tan \beta} \qquad \tan (\alpha - \beta) = \frac{\tan \alpha - \tan \beta}{1 + \tan \alpha \tan \beta}$$

These identities along with the cofunction identities can be used to prove other identities and solve related equations. The following examples show several ways in which we can use these new identities.

EXAMPLE 2   Find the exact value of **(a)** sin 225° and **(b)** tan 15°.

Solution   To find exact values for these functions, we need to find ways to write the domain values with angles of 35°, 45°, 60°, 90°, and 180°.

(a)
$$\sin 225° = \sin(180° + 45°)$$
$$= \sin 180° \cos 45° + \cos 180° \sin 45°$$
$$= (0)\frac{\sqrt{2}}{2} + (-1)\frac{\sqrt{2}}{2}$$

Therefore

$$\sin 225° = -\frac{\sqrt{2}}{2}$$

(b)
$$\tan 15° = \tan(45° - 30°)$$
$$= \frac{\tan 45° - \tan 30°}{1 + \tan 45° \tan 30°}$$
$$= \frac{1 - \dfrac{1}{\sqrt{3}}}{1 + 1 \cdot \dfrac{1}{\sqrt{3}}} \cdot \frac{\sqrt{3}}{\sqrt{3}}$$
$$= \frac{\sqrt{3} - 1}{\sqrt{3} + 1} = \frac{(\sqrt{3} - 1)(\sqrt{3} - 1)}{(\sqrt{3} + 1)(\sqrt{3} - 1)}$$
$$= \frac{(\sqrt{3})^2 - 2\sqrt{3} + 1}{(\sqrt{3})^2 - 1^2} = \frac{3 - 2\sqrt{3} + 1}{3 - 1}$$
$$= \frac{4 - 2\sqrt{3}}{2} = \frac{2(2 - \sqrt{3})}{2} = 2 - \sqrt{3}$$

Therefore the exact value is tan 15° = $2 - \sqrt{3}$.   ■

EXAMPLE 3   Write cos 18° sin 42° + sin 18° cos 42° as a single trigonometric function and evaluate the result. Find the exact value.

Solution   Using the appropriate identity for the functions of a sum or difference, we have

$$\cos 18° \sin 42° + \sin 18° \cos 42° = \sin(18° + 42°)$$
$$= \sin 60°$$
$$= \frac{\sqrt{3}}{2}$$ ■

EXAMPLE 4   If α and β are *acute* angles with $\sin \alpha = \frac{2}{3}$ and $\tan \beta = \frac{3}{4}$, find the exact values for $\sin(\alpha + \beta)$, $\cos(\alpha + \beta)$, and the quadrant of $(\alpha + \beta)$ in standard position.

Solution   First we use the information of $\sin \alpha = \frac{2}{3}$ (Fig. 6.4a) and $\tan \beta = \frac{3}{4}$ (Fig. 6.4b), to draw triangles for *acute* angles α and β. Next we use the Pythagorean Theorem to find the third side of each triangle. From these diagrams we have

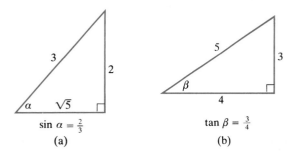

$$\sin \alpha = \frac{2}{3}$$
(a)

$$\tan \beta = \frac{3}{4}$$
(b)

Figure 6.4

$$\sin \alpha = \frac{2}{3} \qquad \sin \beta = \frac{3}{5}$$

$$\cos \alpha = \frac{\sqrt{5}}{3} \qquad \cos \beta = \frac{4}{5}$$

To find the quadrant of $\alpha + \beta$, we evaluate $\sin(\alpha + \beta)$ and $\cos(\alpha + \beta)$:

$$\sin(\alpha + \beta) = \sin \alpha \cos \beta + \cos \alpha \sin \beta$$

$$= \frac{2}{3} \cdot \frac{4}{5} + \frac{\sqrt{5}}{3} \cdot \frac{3}{5}$$

$$= \frac{8}{15} + \frac{\sqrt{5}}{5}$$

$$\approx 0.98054692$$

and

$$\cos(\alpha + \beta) = \cos \alpha \cos \beta - \sin \alpha \sin \beta$$

$$= \frac{\sqrt{5}}{3} \cdot \frac{4}{5} - \frac{2}{3} \cdot \frac{3}{5}$$

$$= \frac{4\sqrt{5}}{15} - \frac{2}{5}$$

$$\approx 0.19628479$$

Since $\sin(\alpha + \beta) > 0$ and $\cos(\alpha + \beta) > 0$, then $\alpha + \beta$ is in quadrant I. Quadrant I is the only quadrant where $\sin(\alpha + \beta)$ and $\cos(\alpha + \beta)$ are both positive.   ■

EXAMPLE 5   Prove that $\sin 4\theta = 4 \sin \theta \cos \theta - 8 \sin^3 \theta \cos \theta$.

Solution   The left side is a little more complicated, since its domain is a multiple angle. Simplify $\sin 4\theta$ by using the identity for $\sin (\alpha + \beta)$.

$$\sin 4\theta = 4 \sin \theta \cos \theta - 8 \sin^3 \theta \cos \theta$$

$= \sin (2\theta + 2\theta)$	*(Now use the sin $(\alpha + \beta)$ identity)*
$= \sin 2\theta \cos 2\theta + \cos 2\theta \sin 2\theta$	*(Add like terms)*
$= 2 \sin 2\theta \cos 2\theta$	
$= 2 \sin (\theta + \theta) \cos (\theta + \theta)$	*(Now use the sin $(\alpha + \beta)$ and cos $(\alpha + \beta)$ identities)*
$= 2(\sin \theta \cos \theta + \cos \theta \sin \theta)$ $\cdot (\cos \theta \cos \theta - \sin \theta \sin \theta)$	*(Add like terms and multiply terms)*
$= 2(2 \sin \theta \cos \theta)(\cos^2 \theta - \sin^2 \theta)$	
$= (4 \sin \theta \cos \theta)(1 - \sin^2 \theta - \sin^2 \theta)$	*(Substitute $1 - \sin^2 \theta$ for $\cos^2 \theta$ and add like terms)*
$= (4 \sin \theta \cos \theta)(1 - 2 \sin^2 \theta)$	*(Multiply)*
$= 4 \sin \theta \cos \theta - 8 \sin^3 \theta \cos \theta$	∎

EXAMPLE 6   Solve $\cos 2x \cos 3x + \sin 2x \sin 3x = 0$ for x in $[0, 2\pi)$.

Solution   We will use the identity for $\cos (\alpha - \beta)$ to simplify and then solve the resulting equation:

$$\cos 2x \cos 3x + \sin 2x \sin 3x = 0 \quad \textit{(Use the cos } (\alpha - \beta) \textit{ identity)}$$
$$\cos (2x - 3x) = 0$$
$$\cos (-x) = 0 \quad \textit{(Use the cos } (-x) \textit{ identity)}$$
$$\cos x = 0$$
$$x = \frac{\pi}{2} \quad \text{and} \quad \frac{3\pi}{2} \qquad ∎$$

As we have seen in these examples, the trigonometric functions of the sums and differences of angles are very helpful in solving many different types of problems.

Exercises 6.3

Find the exact value of each of the following, using the trigonometric functions of sums and differences.

1. $\sin 75°$
2. $\cos 75°$
3. $\cos 105°$
4. $\sin 105°$
5. $\sin 15°$
6. $\cos 15°$
7. $\tan 75°$
8. $\tan 105°$
9. $\cos 165°$
10. $\sin 165°$
11. $\sin 150°$
12. $\cos 150°$

Write each of the following in terms of a single trigonometric function.

13. $\cos 17° \cos 52° - \sin 17° \sin 52°$

14. $\cos 17° \sin 52° - \sin 17° \cos 52°$

15. $\cos 81° \cos 71° + \sin 81° \sin 71°$

16. $\cos 81° \sin 71° + \sin 81° \cos 71°$

17. $\sin \dfrac{\pi}{8} \cos \dfrac{\pi}{8} + \cos \dfrac{\pi}{8} \sin \dfrac{\pi}{8}$

18. $\cos \dfrac{\pi}{4} \cos \dfrac{\pi}{4} - \sin \dfrac{\pi}{4} \sin \dfrac{\pi}{4}$

19. $\sin 3 \cos 2 + \cos 3 \sin 2$ **20.** $\cos 3 \cos 2 - \sin 3 \sin 2$

21. If α and β are acute angles with $\cos \alpha = \frac{3}{5}$ and $\sin \beta = \frac{1}{3}$, find the exact values of $\cos(\alpha + \beta)$ and $\sin(\alpha + \beta)$ and the quadrant of $(\alpha + \beta)$.

22. If α and β are acute angles with $\sin \alpha = \frac{3}{5}$ and $\tan \beta = \frac{3}{4}$, find the exact values of $\cos(\alpha + \beta)$ and $\sin(\alpha + \beta)$ and the quadrant of $(\alpha + \beta)$.

23. If α and β are acute angles with $\tan \alpha = \frac{1}{2}$ and $\cos \beta = \frac{3}{5}$, find the exact values of $\cos(\alpha + \beta)$ and $\sin(\alpha + \beta)$ and the quadrant of $(\alpha + \beta)$.

24. If α and β are acute angles with $\sin \alpha = \frac{1}{2}$ and $\cos \beta = \frac{4}{5}$, find the exact values of $\cos(\alpha + \beta)$ and $\sin(\alpha + \beta)$ and the quadrant of $(\alpha + \beta)$.

Prove the following identities.

25. $\tan \left(\dfrac{\pi}{2} - \theta \right) = \cot \theta$ **26.** $\cot \left(\dfrac{\pi}{2} - \theta \right) = \tan \theta$

27. $\tan(\alpha + \beta) = \dfrac{\tan \alpha + \tan \beta}{1 - \tan \alpha \tan \beta}$

28. $\tan(\alpha - \beta) = \dfrac{\tan \alpha - \tan \beta}{1 + \tan \alpha \tan \beta}$

29. $\cos \left(\theta + \dfrac{\pi}{2} \right) = -\sin \theta$ **30.** $\sin \left(\theta + \dfrac{\pi}{2} \right) = \cos \theta$

31. $\sin 3\theta = 3 \sin \theta - 4 \sin^3 \theta$

32. $\cos 3\theta = 4 \cos^3 \theta - 3 \cos \theta$

33. $\sin(\alpha + \beta) \sin(\alpha - \beta) = \cos^2 \beta - \cos^2 \alpha$

34. $\sin(\alpha + \beta) \sin(\alpha - \beta) = \sin^2 \alpha - \sin^2 \beta$

35. $\cos(\alpha + \beta) \cos(\alpha - \beta) = \cos^2 \beta - \sin^2 \alpha$

36. $\cos(\alpha + \beta) \cos(\alpha - \beta) = \cos^2 \alpha - \sin^2 \beta$

37. $\cos \alpha + \sin \alpha = \sqrt{2} \cos \left(\dfrac{\pi}{4} - \alpha \right) = \sqrt{2} \sin \left(\dfrac{\pi}{4} + \alpha \right)$

38. $\cos \alpha - \sin \alpha = \sqrt{2} \cos \left(\dfrac{\pi}{4} + \alpha \right) = \sqrt{2} \sin \left(\dfrac{\pi}{4} - \alpha \right)$

39. $\tan \alpha + \tan \beta = \dfrac{\sin(\alpha + \beta)}{\cos \alpha \cos \beta}$

40. $\tan \alpha - \tan \beta = \dfrac{\sin(\alpha - \beta)}{\cos \alpha \cos \beta}$

Solve the following equations for α in $[0, 2\pi)$.

41. $2 \sin(5\alpha) \cos(3\alpha) - 2 \cos(5\alpha) \sin(3\alpha) = 1$

42. $\sin(5\alpha) \cos(3\alpha) + \cos(5\alpha) \sin(3\alpha) = 1$

43. $\cos(2\alpha) \cos(3\alpha) - \sin(2\alpha) \sin(3\alpha) + 1 = 0$

44. $2 \cos(2\alpha) \cos(3\alpha) + 2 \sin(2\alpha) \cos(3\alpha) + 1 = 0$

45. Use the law of sines and triangle ABC, as shown in Fig. 6.5 to prove that $\sin(\alpha + \beta) = \sin \alpha \cos \beta + \cos \alpha \sin \beta$. (*Hint*: Find AD and BD in terms of α and β. Start with $\sin(\alpha + \beta) = \sin \gamma$.) What are the restrictions on α and β in this proof?

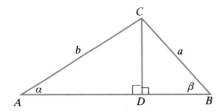

Figure 6.5

A streetlight is suspended by two cables of unequal length with angles α and β as shown in Fig. 6.6. The hanging force of the streetlight caused by gravity is F. If we ignore any sag in the cables, the tension in the cables, S_1 and S_2, can be calculated as $S_1 = \dfrac{F \cos \beta}{\sin(\alpha + \beta)}$ and $S_2 = \dfrac{F \cos \alpha}{\sin(\alpha + \beta)}$.

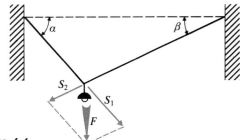

Figure 6.6

46. Solve $S_1 = \dfrac{F \cos \beta}{\sin(\alpha + \beta)}$ for β.

47. Solve $S_2 = \dfrac{F \cos \alpha}{\sin(\alpha + \beta)}$ for α.

To find the height of the cliff FC shown in Fig. 6.7, a surveyor of height $h = AD$ looks down at the reflection E of the top of the cliff in a puddle and carefully measures the angles α and β.

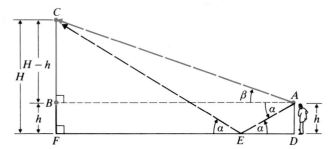

Figure 6.7

48. Find H as a function of h, α, and β. Specifically, prove that

$$H = h\left(\frac{1 + \tan \beta \cot \alpha}{1 - \tan \beta \cot \alpha}\right) = h\left(\frac{\tan \alpha + \tan \beta}{\tan \alpha - \tan \beta}\right)$$

(*Hints*: Let $CB = H - h$. Find FE and ED. Substitute into $AB = FE + ED$ and solve for H.)

49. Using the results of Exercise 48, prove that

$$H = h\left[\frac{\sin(\alpha + \beta)}{\sin(\alpha - \beta)}\right]$$

6.4 Trigonometric Functions of Double Angles and Half Angles

The identities of Section 6.3 are the basic foundation for the identities in this section. These new identities will have domains of double angles and half angles. Double-angle and half-angle identities are very important in the manipulation of trigonometric functions in calculus. The double-angle identities are listed here.

DOUBLE-ANGLE IDENTITIES	
$\sin 2\theta = 2 \sin \theta \cos \theta$	$\cos 2\theta = \cos^2 \theta - \sin^2 \theta \qquad \tan 2\theta = \dfrac{2 \tan \theta}{1 - \tan^2 \theta}$
	$\cos 2\theta = 2 \cos^2 \theta - 1$
	$\cos 2\theta = 1 - 2 \sin^2 \theta$

We will now prove these identities using the identities of sums.

PROOF: Using previous identities, we have

(a) $\sin 2\theta = \sin(\theta + \theta)$ (*Use the* $\sin(\alpha + \beta)$ *identity*)

$= \sin \theta \cos \theta + \sin \theta \cos \theta$ (*Add like terms*)

$= \mathbf{2 \sin \theta \cos \theta}$

(b) $\cos 2\theta = \cos (\theta + \theta)$ *(Use the cos $(\alpha + \beta)$ identity)*

$\qquad\quad\ = \cos \theta \cos \theta - \sin \theta \sin \theta$ *(Multiply)*

$\qquad\quad\ = \mathbf{\cos^2 \theta - \sin^2 \theta}$

or

$\qquad\quad\ = (1 - \sin^2 \theta) - \sin^2 \theta$ *(Substitute $1 - \sin^2 \theta$ for $\cos^2 \theta$)*

$\qquad\quad\ = \mathbf{1 - 2 \sin^2 \theta}$ *(Subtract like terms)*

or

$\qquad\quad\ = 1 - 2(1 - \cos^2 \theta)$ *(Substitute $1 - \cos^2 \theta$ for $\sin^2 \theta$)*

$\qquad\quad\ = 1 - 2 + 2 \cos^2 \theta$

$\qquad\quad\ = -1 + 2 \cos^2 \theta$

$\qquad\quad\ = \mathbf{2 \cos^2 \theta - 1}$

(c) $\tan 2\theta = \tan (\theta + \theta)$

$\qquad\quad\ = \dfrac{\tan \theta + \tan \theta}{1 - \tan \theta \tan \theta}$ *(Use the tan $(\alpha + \beta)$ identity and simplify)*

$\qquad\quad\ = \mathbf{\dfrac{2 \tan \theta}{1 - \tan^2 \theta}}$ □

It is important to notice that there are three equivalent forms for cos 2θ. Each form of cos 2θ is important. Sometimes one form will be more helpful than the other forms, depending on the situation.

We can now use these double-angle identities to solve equations or verify other identities.

EXAMPLE 1   Solve $2 \sin 2\theta - \sin \theta = 0$ for θ in $[0, 2\pi)$.

Solution   We use the double-angle identity for the sine and then try to factor the result:

$$2 \sin 2\theta - \sin \theta = 0 \quad \text{\textit{(Use the sin 2θ identity)}}$$

$$2 \sin \theta \cos \theta - \sin \theta = 0 \quad \text{\textit{(Factor)}}$$

$$\sin \theta (2 \cos \theta - 1) = 0$$

then

$$\sin \theta = 0 \qquad \text{or} \qquad 2 \cos \theta - 1 = 0$$

$$\theta = 0,\ \pi \qquad\qquad\qquad \cos \theta = \frac{1}{2}$$

$$\theta = \frac{\pi}{3},\ \frac{5\pi}{3}$$

Therefore $\theta \in \left\{0,\ \dfrac{\pi}{3},\ \pi,\ \dfrac{5\pi}{3}\right\}$.   ■

EXAMPLE 2   Prove that $\cos 4\theta = 8 \cos^4 \theta - 8 \cos^2 \theta + 1$.

Solution   We use the double-angle identities to simplify $\cos 4\theta$, since it has a more complicated domain:

$$\cos 4\theta = 8 \cos^4 \theta - 8 \cos^2 \theta + 1$$

$= \cos 2(2\theta)$ *(Use the cos 2α identity)*

$= \cos^2 2\theta - \sin^2 2\theta$

$= (\cos 2\theta)^2 - (\sin 2\theta)^2$ *(Use the cos 2α and sin 2α identities)*

$= (2 \cos^2 \theta - 1)^2 - (2 \sin \theta \cos \theta)^2$ *(Multiply)*

$= 4 \cos^4 \theta - 4 \cos^2 \theta + 1$
$\qquad - 4 \sin^2 \theta \cos^2 \theta$ *(Substitute $1 - \cos^2 \theta$ for $\sin^2 \theta$)*

$= 4 \cos^4 \theta - 4 \cos^2 \theta + 1$
$\qquad - 4(1 - \cos^2 \theta) \cos^2 \theta$ *(Simplify)*

$= 4 \cos^4 \theta - 4 \cos^2 2\theta + 1$
$\qquad - 4(\cos^2 \theta - \cos^4 \theta)$

$= 4 \cos^4 \theta - 4 \cos^2 \theta + 1$
$\qquad - 4 \cos^2 \theta + 4 \cos^4 \theta$

$= 8 \cos^4 \theta - 8 \cos^2 \theta + 1$ ∎

EXAMPLE 3   For the acute angle α with $\sin \alpha = \dfrac{5}{8}$, find $\sin 2\alpha$, $\cos 2\alpha$, and the quadrant of 2α.

Solution   First we make a triangle for the acute angle α (Fig. 6.8) and solve for the third side, using the Pythagorean Theorem. Then

$$\sin \alpha = \frac{5}{8} \qquad \text{and} \qquad \cos \alpha = \frac{\sqrt{39}}{8}$$

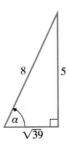

Figure 6.8

We can now evaluate $\sin 2\alpha$ and $\cos 2\alpha$. Thus

$$\sin 2\alpha = 2 \sin \alpha \cos \alpha$$

$$= 2\left(\frac{5}{8}\right)\left(\frac{\sqrt{39}}{8}\right) = \frac{5\sqrt{39}}{32}$$

and

$$\cos 2\alpha = \cos^2 \alpha - \sin^2 \alpha$$

$$= \left(\frac{\sqrt{39}}{8}\right)^2 - \left(\frac{5}{8}\right)^2 = \frac{39}{64} - \frac{25}{64}$$

$$= \frac{39 - 25}{64} = \frac{14}{64} = \frac{7}{32}$$

Since α is an acute angle, $\sin 2\alpha$ is positive, and $\cos 2\alpha$ is positive, we conclude that 2α is in quadrant I. ■

The next two identities are a result of the double-angle identities and are an important bridge to the half-angle identities. We will now prove that

$$\sin^2 \theta = \frac{1 - \cos 2\theta}{2} \qquad \text{and} \qquad \cos^2 \theta = \frac{1 + \cos 2\theta}{2}$$

PROOF: We will start with the double-angle identities:

(a)
$$\cos 2\theta = 1 - 2 \sin^2 \theta \quad \text{(Solve for } \sin^2 \theta)$$
$$\cos 2\theta - 1 = -2 \sin^2 \theta$$
$$\frac{\cos 2\theta - 1}{-2} = \sin^2 \theta$$

and thus

$$\boldsymbol{\sin^2 \theta = \frac{1 - \cos 2\theta}{2}}$$

(b) Similarly,
$$\cos 2\theta = 2 \cos^2 \theta - 1 \quad \text{(Solve for } \cos^2 \theta)$$
$$1 + \cos 2\theta = 2 \cos^2 \theta$$
$$\frac{1 + \cos 2\theta}{2} = \cos^2 \theta$$

and thus

$$\boldsymbol{\cos^2 \theta = \frac{1 + \cos 2\theta}{2}}$$

□

Having proved these preliminary identities, we need only substitute $\frac{\alpha}{2}$ for θ to establish the half-angle identities. When

$$\frac{\alpha}{2} = \theta \quad \text{we have} \quad \alpha = 2\theta$$

Making these substitutions, we have the following half-angle identities.

HALF-ANGLE
IDENTITIES

For an angle α,

$$\sin \frac{\alpha}{2} = \pm \sqrt{\frac{1 - \cos \alpha}{2}} \qquad \cos \frac{\alpha}{2} = \pm \sqrt{\frac{1 + \cos \alpha}{2}}$$

$$\tan \frac{\alpha}{2} = \pm \sqrt{\frac{1 - \cos \alpha}{1 + \cos \alpha}} = \frac{1 - \cos \alpha}{\sin \alpha} = \frac{\sin \alpha}{1 + \cos \alpha}$$

where the plus or minus sign is determined by the quadrant of $\frac{\alpha}{2}$ and the specific trigonometric function.

We can now use these half-angle identities to solve the following examples.

EXAMPLE 4   Find the exact value of (a) $\sin \frac{\pi}{12}$ and (b) $\cos 67\frac{1}{2}°$.

Solution   We will use the half-angle identities to find these values.

(a)
$$\sin \frac{\pi}{12} = \sin \frac{1}{2}\left(\frac{\pi}{6}\right)$$

$$= +\sqrt{\frac{1 - \cos \frac{\pi}{6}}{2}} = \sqrt{\frac{1 - \frac{\sqrt{3}}{2}}{2}}$$

$$= \sqrt{\frac{\left(1 - \frac{\sqrt{3}}{2}\right) \cdot 2}{2 \cdot 2}} = \frac{1}{2}\sqrt{2 - \sqrt{3}}$$

(b)
$$\cos 67\frac{1}{2}° = \cos \frac{135°}{2}$$

$$= +\sqrt{\frac{1 + \cos 135°}{2}} = \sqrt{\frac{1 - \frac{\sqrt{2}}{2}}{2}}$$

$$= \sqrt{\frac{\left(1 - \frac{\sqrt{2}}{2}\right)2}{2 \cdot 2}} = \frac{1}{2}\sqrt{2 - \sqrt{2}}$$

EXAMPLE 5   Solve $\sin^2\left(\dfrac{\theta}{2}\right) - \cos^2\theta = 0$ for all values of θ in radians.

Solution   Using a half-angle identity, we have

$$\sin^2\left(\frac{\theta}{2}\right) - \cos^2\theta = 0 \qquad \text{(Substitute } 1 - \cos^2\theta \text{ for } \sin^2\theta)$$

$$\left(\pm\sqrt{\frac{1 - \cos\theta}{2}}\right)^2 - \cos^2\theta = 0 \qquad \text{(Square and multiply both sides by 2)}$$

$$1 - \cos\theta - 2\cos^2\theta = 0$$

$$0 = 2\cos^2\theta + \cos\theta - 1 \qquad \text{(Factor)}$$

$$0 = (2\cos\theta - 1)(\cos\theta + 1)$$

$$\cos\theta = \frac{1}{2} \qquad \text{or} \qquad \cos\theta = -1$$

$$\theta = \frac{\pi}{3}, \frac{5\pi}{3} \qquad \text{or} \qquad \theta = \pi$$

For θ in $[0, 2\pi)$ we have $\theta = \dfrac{\pi}{3}, \dfrac{5\pi}{3}$, or $\theta = \pi$. Thus for any real number θ,

$$\theta \in \left\{x : x = \frac{\pi}{3} + 2k\pi, x = \frac{5\pi}{3} + 2k\pi, \text{ or } x = k\pi \text{ for any integer } k\right\}. \qquad \blacksquare$$

The identities in this section are used extensively in calculus. If you memorize these identities, you will find calculus easier.

Exercises 6.4

Use the half-angle formulas to find the exact values of each of the following.

1. $\sin 22\frac{1}{2}^\circ$

2. $\cos 15^\circ$

3. $\cos\dfrac{\pi}{12}$

4. $\sin\dfrac{\pi}{8}$

5. $\tan\dfrac{\pi}{8}$

6. $\tan\dfrac{\pi}{12}$

7. $\sin\dfrac{11\pi}{12}$

8. $\cos\dfrac{11\pi}{12}$

9. $\cos\dfrac{7\pi}{8}$

10. $\sin\dfrac{7\pi}{8}$

11. For the acute angle α with $\cos\alpha = \frac{5}{8}$, find $\sin 2\alpha$, $\cos 2\alpha$, and the quadrant of 2α.

12. For the acute angle β with $\sin\beta = \frac{3}{5}$, find $\sin 2\beta$, $\cos 2\beta$, and the quadrant of 2β.

13. For the acute angle α with $\tan\alpha = \frac{4}{3}$, find $\sin 2\alpha$, $\cos 2\alpha$, and the quadrant of 2α.

14. For the acute angle β with $\tan\beta = \frac{12}{5}$, find $\sin 2\beta$, $\cos 2\beta$, and the quadrant of 2β.

15. For the acute angle α with $\sin\alpha = \dfrac{\sqrt{35}}{18}$, find $\sin 2\alpha$, $\cos 2\alpha$, and the quadrant of 2α.

16. For the acute angle β with $\cos\beta = \frac{12}{13}$, find $\sin 2\beta$, $\cos 2\beta$, and the quadrant of 2β.

Solve the following equations for the unknown angle in $[0, 2\pi)$.

17. $\sin 2\theta - \cos \theta = 0$

18. $2 \sin 2\theta - \cos \theta = 0$

19. $2 \sin 2\theta - \sin \theta = 0$

20. $2 \cos 2\theta \sin 2\theta = 1$

21. $4 \sin 2\theta \cos 2\theta = 1$

22. $8 \cos 2\theta \sin 2\theta = 1$

23. $\cos 2\theta + \sin^2 \theta = 1$

24. $\cos 2\theta - \cos^2 \theta = 1$

25. $\cos 2\theta + \cos \theta - 1 = 0$

26. $\cos 2\theta - \cos \theta - 1 = 0$

27. $\cos 2\theta + \cos \theta = 14$

28. $\cos 2\theta + 2 \cos \theta = 11$

29. $\cos 2\theta - \sin \theta = 0$

30. $\cos 2\theta - 3 \sin \theta = -1$

Prove each of the following identities.

31. $\sin \alpha = 2 \sin \dfrac{\alpha}{2} \cos \dfrac{\alpha}{2}$

32. $\cos \alpha = \cos^2 \dfrac{\alpha}{2} + \sin^2 \dfrac{\alpha}{2}$

33. $\tan \dfrac{\beta}{2} = \pm\sqrt{\dfrac{1 - \cos \beta}{1 + \cos \beta}} = \dfrac{1 - \cos \beta}{\sin \beta}$

34. $\sin \beta = \dfrac{2 \tan \dfrac{\beta}{2}}{1 + \tan^2 \dfrac{\beta}{2}}$

35. $\cos 3\beta = 4 \cos^3 \beta - 3 \cos \beta$

36. $\sin 3\beta = 3 \sin \beta - 4 \sin^3 \beta$

37. $\sin^4 \alpha = \frac{1}{8}(\cos 4\alpha - 4 \cos 2\alpha + 3)$

38. $\cos^4 \alpha = \frac{1}{8}(\cos 4\alpha + 4 \cos 2\alpha + 3)$

Use an identity to write each of the following as a single trigonometric function and sketch the graph of each.

39. $y = 2 \sin x \cos x$

40. $y = \sin^2 x - \cos^2 x$

41. $y = \cos^2 x - \sin^2 x$

42. $y = 3 \sin x \cos x$

43. $y = \dfrac{1 - \cos x}{2}$

44. $y = \dfrac{1 + \cos x}{2}$

45. Find an identity containing irrational expressions with the following steps:
   (a) Find the cos 15° using the half-angle identity.
   (b) Find the cos 15° using the identity for cos $(\alpha - \beta)$.
   (c) Does the result of part (a) equal the results of part (b) above? Why?

46. Find an identity containing irrational expressions with the following steps:
   (a) Find the sin 15° using the half-angle identity.
   (b) Find the sin 15° using the identity for sin $(\alpha - \beta)$.
   (c) Does the result of part (a) equal the results of part (b) above? Why?

6.5   Product and Factor Identities

We can easily prove the following product identities using the identities for the trigonometric functions of sums and differences. We call these new identities the **product identities**, since one side of each is a product of two trigonometric functions.

PRODUCT IDENTITIES	
	$2 \sin \alpha \cos \beta = \sin (\alpha + \beta) + \sin (\alpha - \beta)$
	$2 \cos \alpha \sin \beta = \sin (\alpha + \beta) - \sin (\alpha - \beta)$
	$2 \cos \alpha \cos \beta = \cos (\alpha + \beta) + \cos (\alpha - \beta)$
	$2 \sin \alpha \sin \beta = \cos (\alpha - \beta) - \cos (\alpha + \beta)$

PROOF: We will use the identities for trigonometric functions of sums and differences to prove these product identities.

(a) $2 \sin \alpha \cos \beta = \sin (\alpha + \beta) + \sin (\alpha - \beta)$

$\qquad = (\sin \alpha \cos \beta + \cos \alpha \sin \beta)$

$\qquad\qquad\qquad\qquad + (\sin \alpha \cos \beta - \cos \alpha \sin \beta)$

$\qquad = \sin \alpha \cos \beta + \sin \alpha \cos \beta + \cos \alpha \sin \beta - \cos \alpha \sin \beta$

$\qquad = 2 \sin \alpha \cos \beta$

Therefore $2 \sin \alpha \cos \beta = \sin (\alpha + \beta) + \sin (\alpha - \beta)$.

(b) $2 \cos \alpha \sin \beta = \sin (\alpha + \beta) - \sin (\alpha - \beta)$

$\qquad = (\sin \alpha \cos \beta + \cos \alpha \sin \beta)$

$\qquad\qquad\qquad\qquad - (\sin \alpha \cos \beta - \cos \alpha \sin \beta)$

$\qquad = \sin \alpha \cos \beta + \cos \alpha \sin \beta - \sin \alpha \cos \beta + \cos \alpha \sin \beta$

$\qquad = 2 \cos \alpha \sin \beta$

Therefore $2 \cos \alpha \sin \beta = \sin (\alpha + \beta) - \sin (\alpha - \beta)$.

(c) The proofs of the last two parts will be in the exercises at the end of this section.   □

We will use these product identities to change a product into a sum or difference as in the following example.

EXAMPLE 1   Write each of the following as a sum or difference:
(a) $2 \cos 5\theta \sin 3\theta$ and **(b)** $\sin \theta \sin 3\theta$.

Solution   We will use the product identities.
(a) Thus the identity

$$2 \cos \alpha \sin \beta = \sin (\alpha + \beta) - \sin (\alpha - \beta)$$

becomes

$$2 \cos 5\theta \sin 3\theta = \sin (5\theta + 3\theta) - \sin (5\theta - 3\theta)$$
$$= \sin 8\theta - \sin 2\theta$$

(b) The identity

$$2 \sin \alpha \sin \beta = \cos (\alpha - \beta) - \cos (\alpha + \beta)$$

becomes

$$\sin \theta \sin 3\theta = \frac{1}{2}[\cos (\theta - 3\theta) - \cos (\theta + 3\theta)]$$

$$= \frac{1}{2}[\cos (-2\theta) - \cos 4\theta]$$

$$= \frac{1}{2}(\cos 2\theta - \cos 4\theta)$$

$$= \frac{1}{2}\cos 2\theta - \frac{1}{2}\cos 4\theta$$

Sometimes we also need to change a sum or difference into a product. To do this, we use the following factoring identities.

FACTORING
IDENTITIES

$$\sin a + \sin b = 2 \sin \frac{a+b}{2} \cos \frac{a-b}{2}$$

$$\sin a - \sin b = 2 \cos \frac{a+b}{2} \sin \frac{a-b}{2}$$

$$\cos a + \cos b = 2 \cos \frac{a+b}{2} \cos \frac{a-b}{2}$$

$$\cos b - \cos a = 2 \sin \frac{a+b}{2} \sin \frac{a-b}{2}$$

PROOF: To prove these identities, we start with the product identities and let

$$\alpha = \frac{a+b}{2} \quad \text{and} \quad \beta = \frac{a-b}{2}$$

With these values, $\alpha + \beta = a$ and $\alpha - \beta = b$. Substituting these values into the product formula, we will prove the first two factoring identities.

(a) We substitute the values for α and β above into the identity

$$2 \sin \alpha \cos \beta = \sin (\alpha + \beta) + \sin (\alpha - \beta)$$

and

$$2 \sin \frac{a+b}{2} \cos \frac{a-b}{2} = \sin a + \sin b$$

Therefore

$$\mathbf{\sin a + \sin b = 2 \sin \frac{a+b}{2} \cos \frac{a-b}{2}}$$

(b) We substitute the values for α and β above into the identity

$$2 \cos \alpha \sin \beta = \sin (\alpha + \beta) - \sin (\alpha - \beta)$$

and

$$2 \cos \frac{a+b}{2} \sin \frac{a-b}{2} = \sin a - \sin b$$

Therefore

$$\mathbf{\sin a - \sin b = 2 \cos \frac{a+b}{2} \sin \frac{a-b}{2}}$$

The proofs for the last two factoring identities are part of the exercises at the end of this section.   □

When using the factoring identities, notice that the left side of the last identity has cos b first, followed by cos a, as the difference cos b − cos a:

$$\cos b - \cos a = 2 \sin \frac{a + b}{2} \sin \frac{a - b}{2}$$

This order is distinctively different from that of the other three identities. Be careful when using this identity.

EXAMPLE 2   Write each sum or difference as a product:
(a) cos 5θ + cos 3θ and **(b)** sin 5θ − sin 3θ.

Solution   We will use the factoring identities.
(a) The identity

$$\cos a + \cos b = 2 \cos \frac{a + b}{2} \cos \frac{a - b}{2}$$

becomes

$$\cos 5\theta + \cos 3\theta = 2 \cos \frac{5\theta + 3\theta}{2} \cos \frac{5\theta - 3\theta}{2}$$
$$= 2 \cos \frac{8\theta}{2} \cos \frac{2\theta}{2}$$
$$= 2 \cos 4\theta \cos \theta$$

(b) The identity

$$\sin a - \sin b = 2 \cos \frac{a + b}{2} \sin \frac{a - b}{2}$$

becomes

$$\sin 5\theta - \sin 3\theta = 2 \cos \frac{5\theta + 3\theta}{2} \sin \frac{5\theta - 3\theta}{2}$$
$$= 2 \cos \frac{8\theta}{2} \sin \frac{2\theta}{2}$$
$$= 2 \cos 4\theta \sin \theta$$

EXAMPLE 3   Prove that

$$\frac{\sin 7\theta + \sin 5\theta}{\cos 7\theta + \cos 5\theta} = \tan 6\theta$$

Solution We will use the factoring identities to rewrite the left side of the equation and simplify:

$$\frac{\sin 7\theta + \sin 5\theta}{\cos 7\theta + \cos 5\theta} = \tan 6\theta$$

$$= \frac{2 \sin \dfrac{12\theta}{2} \cos \dfrac{2\theta}{2}}{2 \cos \dfrac{12\theta}{2} \cos \dfrac{2\theta}{2}} \quad \textit{(Use the factoring theorem)}$$

$$\textit{(Simplify)}$$

$$= \frac{2 \sin 6\theta \cos \theta}{2 \cos 6\theta \cos \theta} \quad \textit{(Reduce and use the definition of tan 6}\theta\textit{)}$$

$$= \tan 6\theta \qquad \blacksquare$$

There is also an identity for reducing the sum of the multiples of sine and cosine functions with the same angle to a single trigonometric function. We can use this identity to simplify some trigonometric sums. This will be helpful in both graphing and solving equations.

REDUCTION IDENTITY

For real numbers a and b, angles θ and ϕ, and $A = \sqrt{a^2 + b^2}$,

$$a \sin \theta + b \cos \theta = A \sin (\theta + \phi)$$

where $\sin \phi = \dfrac{b}{A}$, $\cos \phi = \dfrac{a}{A}$, and $\tan \phi = \dfrac{b}{a}$.

PROOF: First sketch angle ϕ with the point (a, b) on the terminal side of angle ϕ (Fig. 6.9). Then the radius from the origin to point (a, b) is $A = \sqrt{a^2 + b^2}$, and we have $\sin \phi = \dfrac{b}{A}$ and $\cos \phi = \dfrac{a}{A}$. With this information we can prove the required identity

$$a \sin \theta + b \cos \theta = A \sin (\theta + \phi)$$

$$\textit{(Identity for sin } (\alpha + \beta)) \qquad = A(\sin \theta \cos \phi + \cos \theta \sin \phi)$$

$$\textit{(Substitution of given above)} \qquad = A\left[(\sin \theta)\left(\frac{a}{A}\right) + (\cos \theta)\left(\frac{b}{A}\right) \right]$$

$$\textit{(Simplify)} \qquad = a \sin \theta + b \cos \theta$$

Therefore $a \sin \theta + b \cos \theta = A \sin (\theta + \phi)$, where $A = \sqrt{a^2 + b^2}$, $\sin \phi = \dfrac{b}{A}$, $\cos \phi = \dfrac{a}{A}$, and $\tan \phi = \dfrac{b}{a}$. $\qquad \square$

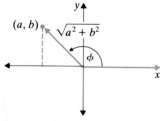

Figure 6.9

The following examples show how we can use the reduction identity to solve some equations and sketch the graph of this type of sum.

EXAMPLE 4   Solve the equation $2 \sin \alpha + \sqrt{2} \cos \alpha = 2$ using the reduction identity for α in $[0, 360°)$.

Solution   For $2 \sin \alpha + \sqrt{2} \cos \alpha$ we have $A = \sqrt{4 + 2} = \sqrt{6}$, $\sin \phi = \dfrac{\sqrt{2}}{A} > 0$, and $\cos \phi = \dfrac{2}{A} > 0$. Thus ϕ is in quadrant I, $\tan \phi = \dfrac{b}{a} = \dfrac{\sqrt{2}}{2}$, and

$$\phi = \tan^{-1} \frac{\sqrt{2}}{2} \approx 35.3°$$

Using the reduction identity, we have

$$2 \sin \alpha + \sqrt{2} \cos \alpha = 2$$
$$\sqrt{4 + 2} \sin (\alpha + 35.3°) = 2$$
$$\sin (\alpha + 35.3°) = \frac{2}{\sqrt{6}}$$

and

$$(\alpha + 35.3°) \in \{54.7°, 125.3°\}$$

Therefore $\alpha \in \{54.7° - 35.3°, 125.3° - 35.3°\}$, which simplifies to $\alpha \in \{19.4°, 90°\}$.   ■

EXAMPLE 5   Use the reduction identity to simplify and graph $y = \frac{3}{2}\sin x + 2 \cos x$.

Solution   First compute ϕ using $A = \sqrt{\left(\dfrac{3}{2}\right)^2 + 2^2} > 0$, $\sin \phi = \dfrac{2}{A} > 0$, $\cos \phi = \dfrac{3}{A} > 0$, and

$$\tan \phi = \frac{b}{a} = \frac{2}{\frac{3}{2}} = \frac{4}{3}$$

With $\sin \phi$ and $\cos \phi$ both positive we have ϕ in quadrant I. Thus $\phi = \tan^{-1} \frac{4}{3} \approx 53.1°$. Now we simplify A and

$$A = \sqrt{a^2 + b^2} = \sqrt{\left(\frac{3}{2}\right)^2 + 2^2} = \sqrt{\frac{9}{4} + 4} = \sqrt{\frac{25}{4}} = \frac{5}{2}$$

We combine this information, and

$$y = \frac{3}{2} \sin x + 2 \cos x$$

becomes

$$y = \frac{5}{2} \sin (x - 53.1°)$$

To sketch the graph, the amplitude is $a = \frac{5}{2}$, the period is 2π, and the shift is $+53.1°$ (53.1° to the right) as shown in Fig. 6.10.

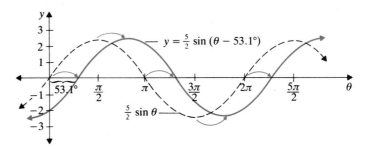

Figure 6.10

The factoring identities discussed earlier in this section can also be used to simplify the graphing of a trigonometric sum, as in the next example.

EXAMPLE 6   Use the factoring identity to simplify and graph $y = \sin 3x - \sin x$.

Solution   When we use the factoring identity, we have

$$y = \sin 3x - \sin x$$
$$= 2 \cos \frac{3x + x}{2} \sin \frac{3x - x}{2}$$
$$= 2 \cos 2x \sin x$$

and

$$y = 2 \sin x \cos 2x$$

We will sketch the graph of this equation as a product of two functions using the techniques we developed in Section 5.6. In using these techniques we let $f(x) = 2 \sin x$ and then

$$-1 \le \cos 2x \le 1$$
$$f(x) \cdot (-1) \le f(x) \cos 2x \le f(x) \cdot 1$$
$$-f(x) \le y \le f(x)$$
$$-2 \sin x \le y \le 2 \sin x$$

We graph $f(x) = 2 \sin x$ and $\cos 2x$ on the same coordinate system. Then we sketch the graph of y as a positive or negative fraction of $f(x)$ based on the value of $\cos 2x$. It may be helpful to review these graphing techniques at the end of Section 5.6. The graph of $y = 2 \sin x \cos 2x$ has been sketched in Fig. 6.11.

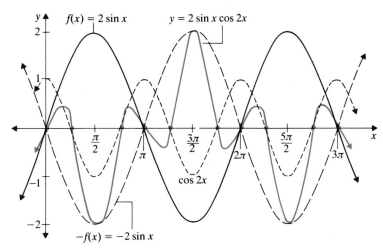

Figure 6.11

Exercises 6.5

Write each of the following products as sums or differences.

1. $2 \cos \alpha \sin 5\alpha$

2. $2 \cos 3\alpha \cos \alpha$

3. $\sin 4\beta \cos 2\beta$

4. $\sin 4\beta \sin 2\beta$

5. $\sin 5\alpha \sin (-3\alpha)$

6. $\cos 3\alpha \sin (-2\alpha)$

7. $\cos 7\beta \cos (-3\beta)$

8. $\sin 4\beta \cos (-2\beta)$

Write each of the following sums and differences as products.

9. $\sin 2\alpha + \sin 5\alpha$

10. $\cos 2\alpha + \sin 5\alpha$

11. $\cos 5\beta - \cos 2\beta$

12. $\sin 5\beta - \sin 2\beta$

13. $\cos 3\alpha + \cos \alpha$

14. $\cos 2\alpha - \cos 5\alpha$

15. $\sin 7\beta - \sin 5\beta$

16. $\sin 3\alpha + \sin 2\alpha$

Establish each of the following identities.

17. $2 \cos \alpha \cos \beta = \cos (\alpha + \beta) + \cos (\alpha - \beta)$

18. $2 \sin \alpha \sin \beta = \cos (\alpha - \beta) - \cos (\alpha + \beta)$

19. $\cos b - \cos a = 2 \sin \dfrac{a + b}{2} \sin \dfrac{a - b}{2}$

20. $\cos a + \cos b = 2 \cos \dfrac{a + b}{2} \cos \dfrac{a - b}{2}$

21. $\sin (\alpha + \beta) \sin (\alpha - \beta) = \sin^2 \alpha - \sin^2 \beta$

22. $\sin (\alpha + \beta) \sin (\alpha - \beta) = \cos^2 \beta - \cos^2 \alpha$

23. $\cos (\alpha + \beta) \cos (\alpha - \beta) = \cos^2 \alpha - \sin^2 \beta$

24. $\cos (\alpha + \beta) \cos (\alpha - \beta) = \cos^2 \beta - \sin^2 \alpha$

25. $\dfrac{\sin 5x + \sin 3x}{\cos 5x + \cos 3x} = \tan 4x$

26. $\dfrac{\cos 3x - \cos x}{\sin 3x - \sin x} = \tan 2x$

27. $\dfrac{\cos 3\beta - \cos \beta}{\cos 3\beta + \cos \beta} = -\tan 2\beta \tan \beta$

28. $\dfrac{\sin 5\beta + \sin 3\beta}{\cos 3\beta - \cos 5\beta} = -\cot \beta$

29. $\sin 6\alpha + \sin 4\alpha + \sin 2\alpha = 4 \cos \alpha \cos 2\alpha \sin 3\alpha$

30. $2 \cos 3\beta \sin 5\beta + \sin 4\beta + \sin 6\beta = 4 \cos \beta \cos 2\beta \sin 5\beta$

31. $\sin 2\beta + \sin 4\beta + 2 \cos \beta \sin 3\beta = 4 \cos \beta \sin 3\beta$

32. $\sin 2\alpha + \sin 4\alpha + 2 \cos \alpha \sin 3\alpha = 2 \sin 4\alpha + 2 \sin 2\alpha$

Use the reduction identity to solve the following equations for the angle in the interval $[0, 360°)$. (Use a calculator.)

33. $\sin \beta + \cos \beta = 1$

34. $\sin \beta + \cos \beta = \sqrt{3}$

35. $2 \sin \alpha - 3 \cos \alpha = 2$

36. $3 \sin \alpha - 2 \cos \alpha = 2$

37. $3 \sin \beta + 4 \cos \beta = 3$

38. $4 \sin \beta + 3 \cos \beta = 2$

Use the reduction identity or the factoring identities to simplify and then sketch the graph of each of the following.

39. $y = \sin x - \sqrt{3} \cos x$

40. $y = \sqrt{3} \sin x + \cos x$

41. $y = 3 \sin x + 4 \cos x$

42. $y = -4 \sin x + 3 \cos x$

43. $y = \sqrt{5} \sin x + 2 \cos x$

44. $y = 2 \sin x + \sqrt{5} \cos x$

45. $y = \sin 3x + \sin x$

46. $y = \cos 3x + \cos x$

6.6   Inverse Trigonometric Functions

We first worked with the definitions of the inverse trigonometric functions in Section 5.4. Remembering that only one-to-one functions have inverse functions, we limit the domains of the trigonometric functions and work with a one-to-one interval of each. The solid section in each graph in Fig. 6.12 is the interval we use for sine, cosine, and tangent to keep the functions one-to-one and thus make it possible to have a one-to-one function. The following definition also delineates these one-to-one intervals.

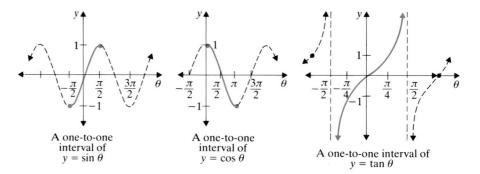

Figure 6.12

A one-to-one interval of $y = \sin \theta$

A one-to-one interval of $y = \cos \theta$

A one-to-one interval of $y = \tan \theta$

INVERSE TRIGONOMETRIC FUNCTIONS

The inverse of the sine function is $\theta = \sin^{-1} x$, where

$$-1 \leq x \leq 1 \qquad \text{and} \qquad -\frac{\pi}{2} \leq \theta \leq \frac{\pi}{2}$$

The inverse of the cosine function is $\theta = \cos^{-1} x$, where

$$-1 \leq x \leq 1 \qquad \text{and} \qquad 0 \leq \theta \leq \pi$$

The inverse of the tangent function is $\theta = \tan^{-1} x$, where

$$x \text{ is any real number} \qquad \text{and} \qquad -\frac{\pi}{2} < \theta < \frac{\pi}{2}$$

Although the inverse trigonometric function could be defined on any one-to-one interval of a trigonometric function, we choose these intervals because they are close to the origin. We will also use the one-to-one intervals of the other trigonometric functions based on the intervals of their reciprocals. Thus we have

$$\csc^{-1} x = \theta \quad \text{is} \quad \csc \theta = x \quad \text{when} \quad -\frac{\pi}{2} < \theta < \frac{\pi}{2}$$

$$\sec^{-1} x = \theta \quad \text{is} \quad \sec \theta = x \quad \text{when} \quad 0 < \theta < \pi$$

$$\tan^{-1} x = \theta \quad \text{is} \quad \tan \theta = x \quad \text{when} \quad -\frac{\pi}{2} < \theta < \frac{\pi}{2}$$

We will now consider the graphs of these inverse trigonometric functions. As we discovered in Section 2.7, a function and its inverse functions have symmetry with respect to the line $y = x$. Thus we can sketch the graphs of $\sin^{-1} x = \theta$, $\cos^{-1} x = \theta$, and $\tan^{-1} x = \theta$ using the graphs of $\sin \theta = x$, $\cos \theta = x$, and $\tan \theta = x$ and symmetry with respect to the line $y = x$ (or in this case, $\theta = x$).

Thus the inverse trigonometric functions $\sin^{-1} x = \theta$, $\cos^{-1} x = \theta$, and $\tan^{-1} x = \theta$ are symmetrical to the trigonometric functions $x = \sin \theta$, $x = \cos \theta$, and $x = \tan \theta$ (on the correct domain intervals) with respect to the line $x = \theta$. Using this technique, we sketch the graphs of the inverse trigonometric functions in Fig. 6.13.

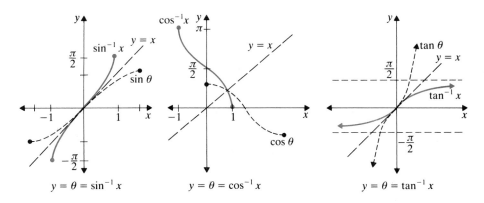

Figure 6.13

$y = \theta = \sin^{-1} x$ \qquad $y = \theta = \cos^{-1} x$ \qquad $y = \theta = \tan^{-1} x$

It is important to remember that the value of an inverse trigonometric function is an angle. We can consider these angles a part of a right triangle or as an angle in standard form. Representing the value of an inverse trigonometric function as an angle will be most useful as demonstrated in the following examples.

EXAMPLE 1   Find $\sin \theta$, $\cos \theta$, and $\tan \theta$ when $\sin^{-1} \dfrac{\sqrt{2}}{2} = \theta$.

Solution   When $\sin^{-1} \dfrac{\sqrt{2}}{2} = \theta$, we also have

$$\sin \theta = \frac{\sqrt{2}}{2} \qquad \text{with} \qquad -\frac{\pi}{2} < \theta < \frac{\pi}{2}$$

Thus we have $\theta = \dfrac{\pi}{4}$. Now we can calculate the following:

$$\sin \theta = \sin \frac{\pi}{4} = \frac{\sqrt{2}}{2}$$

$$\cos \theta = \cos \frac{\pi}{4} = \frac{\sqrt{2}}{2}$$

$$\tan \theta = \tan \frac{\pi}{4} = 1$$

EXAMPLE 2　Sketch a diagram of the angle $\alpha = \sin^{-1}(1 - a)$.

Solution　We remember that $\alpha = \sin^{-1}(1 - a)$

is also $\sin \alpha = (1 - a)$ when $-\dfrac{\pi}{2} \le \alpha \le \dfrac{\pi}{2}$.

Then we construct the diagrams in Fig. 6.14.

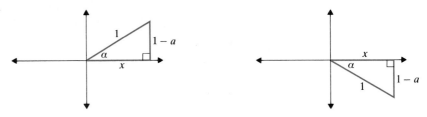

Figure 6.14

EXAMPLE 3　Find $\sin \alpha$, $\cos \alpha$, and $\tan \alpha$ when $\alpha = \sin^{-1}(1 - a)$.

Solution　Since $\alpha = \sin^{-1}(1 - a)$ is the inverse trigonometric function in Example 2, we will use the diagrams in Fig. 6.14. First calculate x, using the Pythagorean Theorem:

$$x^2 + (1 - a)^2 = 1^2$$
$$x^2 = 1 - (1 - a)^2$$
$$x^2 = 1 - (1 - 2a + a^2)$$
$$x^2 = 1 - 1 + 2a - a^2$$
$$x = \sqrt{2a - a^2}$$

Since α is in quadrant I or IV, x must be the positive square root. Therefore,

$$\sin \alpha = 1 - a$$
$$\cos \alpha = \frac{\sqrt{2a - a^2}}{1} = \sqrt{2a - a^2}$$
$$\tan \alpha = \frac{1 - a}{\sqrt{2a - a^2}} = \frac{(1 - a)\sqrt{2a - a^2}}{2a(1 - a)}$$
$$= \frac{\sqrt{2a - a^2}}{2a}$$

Note that $2a - a^2 \ge 0$, since we have $x = \sqrt{2a - a^2}$. This restricts the value of a. To find a, we solve $2a - a^2 \ge 0$ by factoring the left side and using a sign chart (Section 1.5) as shown here.

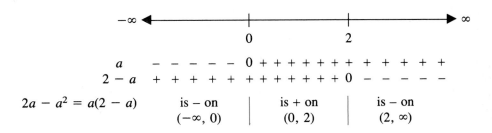

$2a - a^2 = a(2 - a)$ is − on   is + on   is − on
                        $(-\infty, 0)$ $(0, 2)$ $(2, \infty)$

Therefore in Examples 2 and 3, a is restricted such that $0 \le a \le 2$.   ■

Remembering that the value of an inverse trigonometric function is an angle, we can use the identities of this chapter when working with the composition of trigonometric functions and their inverses. The following examples demonstrate this.

EXAMPLE 4   Evaluate $\sin\left[\tan^{-1} 1 - \cos^{-1}\left(-\dfrac{\sqrt{3}}{2}\right)\right]$.

Solution   Since the value of inverse trigonometric functions are angles, we let

$\alpha = \tan^{-1} 1$ and $\beta = \cos^{-1}\left(-\dfrac{\sqrt{3}}{2}\right)$.

Then we have $\tan \alpha = 1$ with $-\dfrac{\pi}{2} \le \alpha \le \dfrac{\pi}{2}$ and $\alpha = \dfrac{\pi}{4}$.

Similarly, with $\cos \beta = -\dfrac{\sqrt{3}}{2}$ for $0 \le \beta \le \pi$ we have $\beta' = \dfrac{\pi}{6}$ and $\beta = \dfrac{5\pi}{6}$

With this information and the identity for $\sin(\alpha - \beta)$ we have

$$\sin\left[\tan^{-1} 1 - \cos^{-1}\left(-\dfrac{\sqrt{3}}{2}\right)\right] = \sin(\alpha - \beta)$$

$$= \sin \alpha \cos \beta - \cos \alpha \sin \beta$$

$$= \sin\left(\dfrac{\pi}{4}\right)\cos\left(\dfrac{5\pi}{6}\right) - \cos\left(\dfrac{\pi}{4}\right)\sin\left(\dfrac{5\pi}{6}\right)$$

$$= \dfrac{\sqrt{2}}{2}\left(-\dfrac{\sqrt{3}}{2}\right) - \dfrac{\sqrt{2}}{2}\left(\dfrac{1}{2}\right)$$

$$= \dfrac{-\sqrt{6} - \sqrt{2}}{4}$$

■

EXAMPLE 5   Evaluate $\sin(2 \sin^{-1}\sqrt{1 - x^2})$.

Solution   Since the value of an inverse trigonometric function is an angle, we let β be defined as $\beta = \sin^{-1}\sqrt{1 - x^2}$. Thus we have

$$\sin \beta = \sqrt{1 - x^2} \quad \text{and} \quad 0 \le \beta \le \frac{\pi}{2}$$

Sketch the triangles for β and solve for the unknown side (Fig. 6.15). From the diagram and the third side of the triangles we have

$$\sin \beta = \sqrt{1 - x^2}$$
$$\cos \beta = \pm|x|$$
$$\tan \beta = \pm\frac{\sqrt{1 - x^2}}{|x|}$$

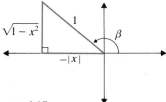
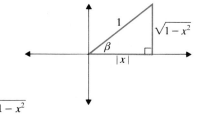

Figure 6.15 $\beta = \sin^{-1}\sqrt{1 - x^2}$

This information along with the double-angle identity yields

$$\begin{aligned} \sin(2 \sin^{-1}\sqrt{1 - x^2}) &= \sin 2\beta \\ &= 2 \sin \beta \cos \beta \\ &= 2(\sqrt{1 - x^2})(\pm x) \\ &= \pm 2x\sqrt{1 - x^2} \end{aligned}$$ ■

Although the skills in this section are not difficult, they require a good understanding of trigonometry. The following exercises will help you to develop your proficiency in using these skills.

Exercises 6.6

Find the exact values of each of the following functions.

1. $\sin^{-1}\dfrac{\sqrt{3}}{2}$

2. $\sin^{-1}\dfrac{1}{2}$

3. $\cos^{-1}\dfrac{\sqrt{3}}{2}$

4. $\cos^{-1}\dfrac{1}{2}$

5. $\tan^{-1}(-1)$

6. $\tan^{-1} 1$

7. $\sin^{-1}\left(-\dfrac{\sqrt{2}}{2}\right)$

8. $\cos^{-1}\left(-\dfrac{1}{2}\right)$

9. $\cos^{-1}\left(-\dfrac{\sqrt{2}}{2}\right)$

10. $\sin^{-1}\left(-\dfrac{1}{2}\right)$

11. $\cos^{-1} 0$

12. $\tan^{-1} 0$

13. $\sin\left(2 \cos^{-1} \dfrac{\sqrt{3}}{2}\right)$

14. $\cos\left(2 \sin^{-1} \dfrac{\sqrt{3}}{2}\right)$

15. $\tan\left[\sin^{-1}\left(-\dfrac{\sqrt{2}}{2}\right)\right]$

16. $\tan\left[\cos^{-1}\left(-\dfrac{\sqrt{3}}{2}\right)\right]$

17. $\sin(-\tan^{-1} 1)$

18. $\cos[\tan^{-1}(-1)]$

19. $\cos\left[\sin^{-1}\left(-\dfrac{\sqrt{2}}{2}\right) + \cos^{-1}\left(\dfrac{\sqrt{3}}{2}\right)\right]$

20. $\sin\left[\sin^{-1}\left(\dfrac{\sqrt{2}}{2}\right) + \cos^{-1}\left(-\dfrac{\sqrt{3}}{2}\right)\right]$

21. $\sin\left[\sin^{-1}\left(-\dfrac{1}{2}\right) - \cos^{-1}\left(\dfrac{\sqrt{2}}{2}\right)\right]$

22. $\cos\left[\sin\left(-\dfrac{\sqrt{2}}{2}\right) - \cos\left(\dfrac{1}{2}\right)\right]$

Find the other trigonometric functions (sine, cosine, tangent, secant, cosecant, and cotangent) of each of the following angles.

23. $\alpha = \cos^{-1} x$

24. $\alpha = \sin^{-1} x$

25. $\beta = \cos^{-1} \dfrac{1}{x}$

26. $\beta = \sin^{-1} \dfrac{1}{x}$

27. $\alpha = \sin^{-1} \sqrt{x^2 - 1}$

28. $\alpha = \cos\sqrt{x^2 - 1}$

29. $\beta = \tan^{-1} \dfrac{1}{x}$

30. $\beta = \cot^{-1} \dfrac{1}{x}$

31. $\alpha = \sec^{-1} \sqrt{1 - x^2}$

32. $\alpha = \csc^{-1} \sqrt{1 - x^2}$

33. $\beta = \cos^{-1} 3u$

34. $\beta = \sin^{-1} \dfrac{1}{3u}$

35. $\alpha = \tan^{-1} \dfrac{\sqrt{x^2 - 1}}{x}$

36. $\alpha = \tan^{-1} \sqrt{x^2 - 1}$

Use the trigonometric identities in evaluating each of the following expressions.

37. $\sin(2 \cos^{-1} x)$

38. $\sin(2 \sin^{-1} x)$

39. $\cos(2 \cos^{-1} x)$

40. $\cos(2 \sin^{-1} x)$

41. $\sin(\sin^{-1} x + \cos^{-1} x)$

42. $\sin(\cos^{-1} x - \sin^{-1} x)$

43. $\cos(\sin^{-1} x + \cos^{-1} \sqrt{1 - x^2})$

44. $\cos(\cos^{-1} x - \sin^{-1} \sqrt{1 - x^2})$

45. $\sin\left(\dfrac{1}{2} \sin^{-1} x\right)$

46. $\cos\left(\dfrac{1}{2} \cos^{-1} x\right)$

47. $\tan(2 \cos^{-1} x)$

48. $\tan(2 \sin^{-1} x)$

49. $\sin(2 \tan^{-1} x)$

50. $\cos(2 \tan^{-1} x)$

Prove the following identities.

51. $\cos(\sin^{-1} x + \cos^{-1} x) = 0$

52. $\cos(\sin^{-1} x - \cos^{-1} x) = 2x\sqrt{1 - x^2}$

53. $2 \cos^2\left(\dfrac{1}{2}\cos^{-1} x\right) = 1 + x$

54. $2 \sin^2\left(\dfrac{1}{2}\cos^{-1} x\right) = 1 - x$

55. $\cos(2 \sin^{-1} \sqrt{x^2 - 1}) = 3 - 2x^2$

56. $\sin(2 \cos^{-1} \sqrt{x^2 - 1}) = 2x\sqrt{x^2 - 1}$

57. Find the domain and range of (**a**) $y = \cos^{-1} 2x$ and (**b**) $y = 2 \cos^{-1} x$.

58. Find the domain and range of (**a**) $y = \sin^{-1} 2x$ and (**b**) $y = 2 \sin^{-1} x$.

59. Simplify and sketch the graphs of (**a**) $y = \sin(\cos^{-1} x)$ and (**b**) $y = \sin(\sin^{-1} x)$.

60. Simplify and sketch the graphs of (**a**) $y = \cos(\sin^{-1} x)$ and (**b**) $y = \cos(\cos^{-1} x)$.

CHAPTER SUMMARY AND REVIEW

KEY TERMS

Proving trigonometric identities

The basic trigonometric identities

Solving trigonometric equations

Trigonometric cofunction identities

Trigonometric functions of sums and differences

Double-angle identities

Half-angle identities

Product identities

Factoring identities

Reduction identity

Inverse trigonometric functions

Chapter 6

CHAPTER EXERCISES

Find the exact values of

1. $\sin 15°$

2. $\cos 15°$

3. $\cos 22\frac{1}{2}°$

4. $\sin 22\frac{1}{2}°$

For the acute angles α and β, where $\tan \alpha = \frac{3}{4}$ and $\sin \beta = \frac{5}{13}$, find exact values for each of the following.

5. $\sin (\alpha + \beta)$

6. $\sin (\alpha - \beta)$

7. $\cos (\alpha - \beta)$

8. $\cos (\alpha + \beta)$

9. $\sin 2\alpha$

10. $\cos 2\alpha$

11. $\tan 2\alpha$

12. $\sin 2\beta$

13. $\cos 2\beta$

14. $\tan 2\beta$

15. $\sin \dfrac{\alpha}{2}$

16. $\cos \dfrac{\alpha}{2}$

17. $\cos \dfrac{\beta}{2}$

18. $\sin \dfrac{\beta}{2}$

19. $\cos \left(\dfrac{\pi}{2} - \alpha\right)$

20. $\cos \left(\dfrac{\pi}{2} - \beta\right)$

21. $\sin \left(\dfrac{\pi}{2} - \beta\right)$

22. $\sin \left(\dfrac{\pi}{2} - \alpha\right)$

23. $\tan \left(\dfrac{\pi}{2} - \alpha\right)$

24. $\tan \left(\dfrac{\pi}{2} - \beta\right)$

25. $\tan (\alpha + \beta)$

26. $\tan (\alpha - \beta)$

27. $\cot 2\beta$

28. $\cot 2\alpha$

Prove each of the following identities.

29. $\tan^2 \theta = -1 + \dfrac{\sec \theta}{\cos \theta}$

30. $\sec \theta - 1 = \dfrac{\tan^2 \theta}{1 + \sec \theta}$

31. $\dfrac{\cos^4 \theta - \sin^4 \theta}{\cos 2\theta} = 1$

32. $\sin \left(\dfrac{\pi}{6} + \theta\right) - \cos \left(\dfrac{\pi}{3} + \theta\right) = \sqrt{3} \sin \theta$

33. $\sin 4\alpha = 4 \sin \alpha \cos \alpha - 8 \sin^3 \alpha \cos \alpha$

34. $\tan 4x = \dfrac{4 \tan x - 4 \tan^3 x}{1 - 6 \tan^2 x + \tan^4 x}$

35. $\tan \left(\dfrac{\pi}{4} - \theta\right) = \dfrac{1 - \tan x}{1 + \tan x}$

36. $\cos^2 \dfrac{\theta}{2} = \dfrac{1}{2}(1 + \cos \theta)$

37. $\sec^2 \dfrac{\theta}{2} = \dfrac{2}{1 + \cos \theta}$

38. $\left(\cos \dfrac{\alpha}{2} - \sin \dfrac{\alpha}{2}\right)^2 = 1 - \sin \alpha$

39. $\dfrac{\cos 3\alpha + \cos 5\alpha}{\cos 3\alpha - \cos 5\alpha} = \cot \alpha \cot 4\alpha$

40. $\dfrac{\sin 2\alpha + \sin 6\alpha}{\sin 2\alpha - \sin 6\alpha} = -\tan 4\alpha \cot 2\alpha$

Solve each of the following equations for angles in $[0, 2\pi)$.

41. $4 \sin^2 \alpha = 1$

42. $4 \sin \alpha \cos \alpha - 2 \sin \alpha + 2 \cos \alpha = 2$

43. $\cos^2 \alpha - 2 \cos \alpha = 3$

44. $2 \cos (5\alpha) \cos (3\alpha) - 2 \sin (5\alpha) \sin (3\alpha) = -1$

45. $\cos 2\theta + 2 \cos \theta = 0$

46. $6 \sin \theta \cos 3\theta + 6 \cos \theta \sin 3\theta = 6$

47. $\sin 2\theta + \cos \theta = 0$

48. $\sin^2 \theta - \cos^2 \theta + 3 \sin \theta = -3$

Write each as a single trigonometric function.

49. $y = 6 \sin 2\theta \cos 2\theta$

50. $y = 2 \cos^2 \alpha - 2 \sin^2 \alpha$

51. $y = \dfrac{1 + \cos \beta}{4}$

52. $y = 3 \sin \alpha - 4 \cos \alpha$

Find the six trigonometric functions for each of the following angles.

53. $\alpha = \cos^{-1} a$

54. $\alpha = \sin^{-1} \dfrac{1}{a}$

55. $\beta = \csc^{-1} 2a$

56. $\beta = \sec^{-1} \dfrac{1}{a}$

Evaluate each of the following.

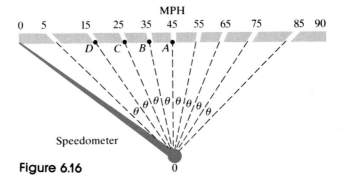

Figure 6.16

57. $\cos (2 \cos^{-1} a)$ **58.** $\sin \left(2 \sin^{-1} \dfrac{1}{a}\right)$

59. $\sin (2 \tan^{-1} a)$ **60.** $\cos (\sin^{-1} a)$

61. An automobile speedometer is designed so that the speed indicator bar rotates through the same angle θ for each 10-mile-per-hour change of speed as shown in Fig. 6.16. If $OA = x$, $AC = x$, and $AB = 1$, find the numerical value of BC and OA.

Other Applications
of Trigonometry

THE WAGGLE DANCE OF DIRECTION AND DISTANCE

The celebrated waggle dance of the honeybee is one of the most sophisticated of all animal communication systems. It was first decoded by Karl van Frisch in 1945. When a worker bee returns to the nest with the news of a new food source or a new nest site, she communicates the location by performing this special dance. The pattern that she dances is a figure 8 repeated many times in the midst of her sister worker bees. The most informative part of the dance is the straight run (the middle of the figure 8), which represents a miniature version of the flight from the hive to the target. It points directly at the target if the bee is on the horizontal outside the hive (the sun provides the orientation); if she is inside the darkened hive on the vertical surface, the straight run points at an appropriate angle using gravity as the temporary orientation (see diagram on next page).

The straight run also provides information on the distance of the target from the hive. The farther away the goal lies, the longer the straight run lasts; a straight run lasting a second indicates a target about 500 meters away, while a run

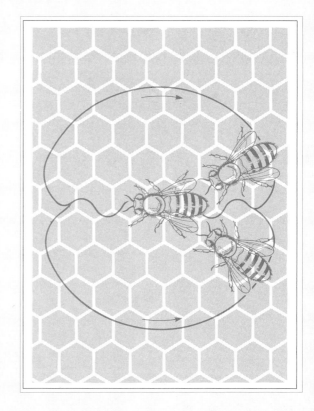

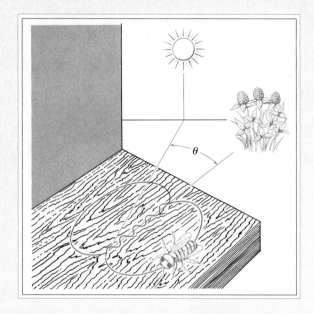

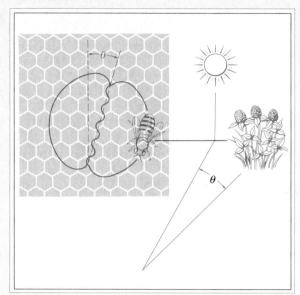

lasting 2 seconds indicates a target 2 kilometers away.

The other workers touch the waggle dancer with their antennae many times during her dance. Within minutes, some of the workers begin to leave the nest and search for the tar-

get. Amazingly, the great majority land very near the intended object.

In Chapters 5 and 6 we defined the basic trigonometric functions, developed many trigonometric identities, solved trigonometric equations, and found trigonometric solutions to applied problems. The introduction to this chapter demonstrates some of the material we will cover in this chapter. The trigonometry in this chapter will add important skills to your mathematical repertoire that you will use in calculus.

7.1   Polar Coordinates

In all previous work in this book we used the Cartesian (or rectangular) coordinate system to graph relations and functions. Because of this major reliance on the Cartesian coordinate system, we can now graph difficult equations with variables x and y, such as $x^2 + y^2 = 1$, or variables y and θ, such as $y = \sin \theta$. However, not all equations can be graphed easily in the Cartesian coordinate system. We will now consider another important coordinate system called **polar coordinates**.

We will define polar coordinates and then use this coordinate system with analytic geometry. In the Cartesian coordinate system, a point P is denoted by the ordered pair (x, y) in which x is the coordinate on the x-axis and y is the coordinate on the y-axis associated with the point (Fig. 7.1). In the polar coordinate system the x- and y-axes are replaced by a ray that starts at the origin, $(0, 0)$, and extends to the right on the positive x-axis. This ray is called the **polar axis**. The starting point of this ray, the point $(0, 0)$, is called the **pole**. In polar coordinates a point P is denoted by the ordered pair (r, θ), in which r is the distance from the pole to P and θ is the angle from the polar axis associated with P (Fig. 7.2). This new polar coordinate system is very similar to the method used by the honeybee in gathering honey.

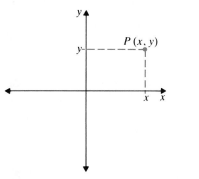

Figure 7.1

Figure 7.2

The angle θ, with vertex at the pole, has the polar axis as its initial side and has a positive value when measured counterclockwise and a negative value when measured clockwise. To be able to change from Cartesian coordinates to polar coordinates and back again, it is helpful to visualize both representations together and observe the following conclusions.

POLAR
COORDINATES

The point (x, y) in rectangular coordinates can be written as (r, θ) in polar coordinates, where

(a) $\sin \theta = \dfrac{y}{r}$ so that $y = r \sin \theta$

(b) $\cos \theta = \dfrac{x}{r}$ so that $x = r \cos \theta$

(c) $x^2 + y^2 = r^2$

(d) $\tan \theta = \dfrac{y}{x}$.

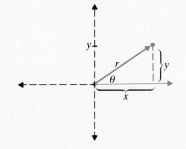

We will now use these conclusions to change from Cartesian to polar coordinates and vice versa.

EXAMPLE 1

For the point $(r, \theta) = (2, 120°)$,
(a) graph this point in polar coordinates, and
(b) find this point in rectangular coordinates.

Solution

(a) To graph the point $(2, 120°)$ in polar coordinates, we start with a polar coordinate system and sketch a $120°$ angle in standard position (vertex at the pole, initial side on the polar axis, and terminal side opening in a counterclockwise direction). Then measure 2 units from the vertex on the terminal side (Fig. 7.3).

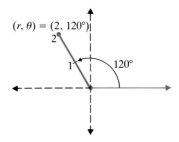

Figure 7.3

(b) Using the polar coordinate equations and $(r, \theta) = (2, 120°)$, we have

$$x = r \cos \theta \qquad \text{and} \qquad y = r \sin \theta$$
$$x = 2 \cos 120° \qquad\qquad y = 2 \sin 120°$$
$$x = 2\left(-\frac{1}{2}\right) \qquad\qquad y = 2\left(\frac{\sqrt{3}}{2}\right)$$
$$x = -1 \qquad\qquad y = \sqrt{3}$$

Thus the point $(r, \theta) = (2, 120°)$ has Cartesian coordinates of $(x, y) = (-1, \sqrt{3})$. ■

We can also change equations in Cartesian coordinates to equivalent equations in polar coordinates.

EXAMPLE 2

Change each of the following equations from Cartesian coordinates to polar coordinates:
(a) $x^2 + y^2 = 4$, **(b)** $x = 4$, and **(c)** $y = x^2 + y^2$.

Solution We will use the conversion equations for polar coordinates listed above.
(a) Since $x^2 + y^2 = r^2$, the circle $x^2 + y^2 = 4$
$$\text{becomes } r^2 = 4.$$

Both equations, $x^2 + y^2 = 4$ and $r^2 = 4$, are equations for the circle with center at $(0, 0)$ and radius of $r = 2$. (Both equations can also be written as $x^2 + y^2 = 2^2$ and $r^2 = 2^2$.)

(b) Since $x = r \cos \theta$, the equation of the line $x = 4$
$$\text{becomes } r \cos \theta = 4.$$

In polar coordinates, equations are usually written with r isolated on one side of the equation. In this case,

$$r \cos \theta = 4$$

is

$$r = \frac{4}{\cos \theta} \quad \text{or} \quad r = 4 \sec \theta$$

Therefore $r = 4 \sec \theta$ is the equation of the line $x = 4$.

(c) We start with $y = x^2 + y^2$ and substitute with $r^2 = x^2 + y^2$ and $y = r \sin \theta$. Then $y = x^2 + y^2$ becomes

$$r \sin \theta = r^2 \qquad \textit{(Write in standard form)}$$
$$0 = r^2 - r \sin \theta \quad \textit{(Solve for r)}$$
$$0 = r(r - \sin \theta)$$
$$r = 0 \qquad \text{and} \qquad r = \sin \theta$$

The solution $r = 0$ is the pole that is also included in $r = \sin \theta$. Thus the equation of the circle $x^2 + y^2 = y$ in Cartesian coordinates is $r = \sin \theta$ in polar coordinates. ■

Equations in polar form are written with r isolated on one side. The equations from the example above, $r^2 = 4$, $r = 4 \sec \theta$, and $r = \sin \theta$, demonstrate this form.

One problem with polar coordinates is that there is always more than one way to name a point with polar coordinates. The same point can be named in several ways, using coterminal angles. For example,

$$P = (r, \theta) = (r, \theta + 2\pi) = (r, \theta - 2\pi)$$

and in general, $P = (r, \theta) = (r, \theta + 2k\pi)$, where k is an integer. Thus we can rewrite the coordinates of a given point (r, θ) by adding integer multiples of 2π to the angle θ.

We can also use a negative as well as a positive radius when describing a point in polar coordinates. A positive radius starts at the pole and extends out the same direction as the terminal side of the given angle. A negative radius starts at the pole and uses the same angle as the terminal side of the given angle but extends in the direction opposite to the terminal side. Notice how the points $(2, 120°)$ and $(-2, 120°)$ in Fig. 7.4 demonstrate this. Thus $(r, \theta) = (-r, \theta + \pi)$ as seen in Fig. 7.5.

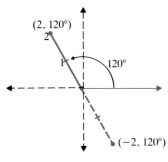

Figure 7.4

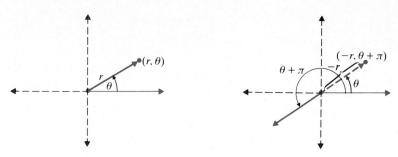

Two order pairs for the same point
$(r, \theta) = (-r, \theta + \pi)$

Figure 7.5

In working with polar coordinates it is important to remember that

(a) it is possible to work with either a positive or a negative radius and

(b) each point may be named with more than one ordered pair.

The following example demonstrates these points.

EXAMPLE 3

Graph each of the following points in polar coordinates and then find another ordered pair for each point:

(a) $(3, 135°)$, (b) $(-2, 30°)$, and (c) $\left(-3, -\dfrac{\pi}{3}\right)$.

Solution

(a) To graph $(3, 135°)$, first locate the angle $\theta = 135°$ (with a $45°$ reference angle with the former x-axis). Then on that angle, mark the point 3 units from the pole (Fig. 7.6). Two other ordered pairs for this point are $(3, 135° + 360°) = (3, 495°)$ and $(-3, 135° + 180°) = (-3, 315°)$.

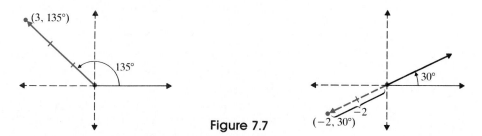

Figure 7.6   **Figure 7.7**

(b) To graph $(-2, 30°)$, first locate the angle $\theta = 30°$. Then from the pole, mark off two units in a negative direction as seen in Fig. 7.7. Two other ordered pairs for this point are $(-2, 30° + 360°) = (-2, 390°)$ and $(2, 30° + 180°) = (2, 210°)$.

(c) To graph $\left(-3, -\dfrac{\pi}{3}\right)$, first locate the angle $\theta = -\dfrac{\pi}{3}$ and then mark the radius 3 units in the negative direction (Fig. 7.8). Two other ordered pairs for this point are $\left(-3, 2\pi - \dfrac{\pi}{3}\right) = \left(-3, \dfrac{5\pi}{3}\right)$ and $\left(3, \pi - \dfrac{\pi}{3}\right) = \left(3, \dfrac{2\pi}{3}\right)$.

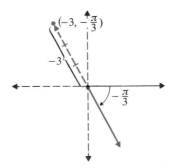

Figure 7.8

As well as graphing points, we can also graph equations. Sometimes it will be easier to graph the equation in polar form; other times it will be easier to graph the equivalent equation in Cartesian coordinate form. The following example demonstrates this.

EXAMPLE 4   Graph each of the following:

(a) $r = 5$, (b) $\theta = \dfrac{\pi}{6}$, and (c) $r = 2 \csc \theta$.

Solution

(a) Solutions to the equation $r = 5$ are all ordered pairs of the form $(5, \theta)$, where θ can be any angle. Therefore $r = 5$ is the circle with center at the pole and a radius of 5 (Fig. 7.9).

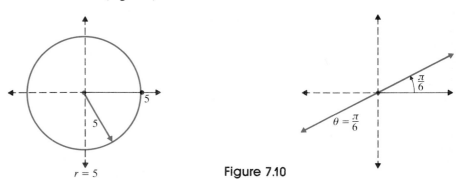

Figure 7.9 $r = 5$ **Figure 7.10**

(b) Solutions to the equation $\theta = \dfrac{\pi}{6}$ are all ordered pairs of the form $\left(r, \dfrac{\pi}{6}\right)$, where r can be any and all real numbers. Therefore $\theta = \dfrac{\pi}{6}$ is a straight line through the origin, taking into account both positive and negative values for r (Fig. 7.10).

(c) The equation $r = 2 \csc \theta$ is easier to graph in Cartesian coordinates. Changing to Cartesian coordinates, we have

$$r = 2 \csc \theta \qquad \left(\textit{Substitute } csc \ \theta = \frac{1}{\sin \theta}\right)$$

$$r = \frac{2}{\sin \theta} \qquad \textit{(Multiply both sides by sin } \theta)$$

$$\sin \theta \ r = \frac{2}{\sin \theta} \sin \theta \qquad \textit{(Simplify)}$$

$$r \sin \theta = 2 \qquad \textit{(Substitute } y = r \sin \theta)$$

$$y = 2$$

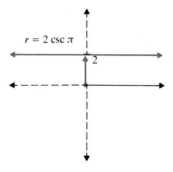

Figure 7.11

Therefore $y = 2$ is equivalent to $r = 2 \csc \theta$ and is much easier to graph (Fig. 7.11). ∎

Although we will do much more with polar coordinates in the next section, we will now start working with these basic foundations of polar coordinates.

Exercises 7.1

Graph each of the following points on a polar coordinate system and then find two additional ordered pairs for each point.

1. $(2, 60°)$

2. $(3, 45°)$

3. $(3, 150°)$

4. $(2, 120°)$

5. $(-2, 45°)$

6. $(-3, 60°)$

7. $\left(3, -\frac{5\pi}{6}\right)$

8. $\left(2, -\frac{3\pi}{4}\right)$

9. $\left(4, -\frac{\pi}{4}\right)$

10. $\left(5, -\frac{\pi}{6}\right)$

11. $\left(-5, -\dfrac{\pi}{6}\right)$ **12.** $\left(-4, -\dfrac{\pi}{4}\right)$

Change the following points from polar coordinates to Cartesian coordinates.

13. $(2, 60°)$ **14.** $(3, 45°)$

15. $(3, 150°)$ **16.** $(2, 120°)$

17. $(-2, 45°)$ **18.** $(-3, 60°)$

19. $\left(3, -\dfrac{5\pi}{6}\right)$ **20.** $\left(2, -\dfrac{3\pi}{4}\right)$

Graph each of the following pairs of polar equations and find the points of intersection in polar form.

21. $r = 3$ and $\theta = 45°$ **22.** $r = 2$ and $\theta = 135°$

23. $\theta = \dfrac{\pi}{6}$ and $\theta = \dfrac{\pi}{3}$ **24.** $\theta = \dfrac{\pi}{2}$ and $\theta = \dfrac{\pi}{4}$

25. $r = 5$ and $\theta = \dfrac{3\pi}{2}$ **26.** $r = 1$ and $\theta = \dfrac{5\pi}{4}$

27. $r = 2\frac{1}{2}$ and $r = 4$ **28.** $r = 3$ and $r = 5$

Change each of the following equations into equivalent equations in polar coordinates. (Do not graph.)

29. $x^2 + y^2 = 7$ **30.** $x^2 + y^2 = 16$

31. $x^2 - y^2 = 4$ **32.** $x^2 - y^2 = 1$

33. $x = 3y$ **34.** $5x = y$

35. $xy = 3$ **36.** $xy = -4$

37. $x = 5$ **38.** $y = 3$

39. $(x - 1)^2 + y^2 = 1$ **40.** $x^2 + (y - 1)^2 = 1$

41. $x^2 + (y + 1)^2 = 1$ **42.** $(x + 1)^2 + y^2 = 1$

Change each of the following equations into their Cartesian coordinate equivalents. (Do not graph.)

43. $r = 3 \sec \theta$ **44.** $r = 5 \csc \theta$

45. $r^2 = 3r \cos \theta$ **46.** $r^2 = 2r \sin \theta$

47. $r^2 = 1 - r \sin \theta$ **48.** $r^2 = 3 - r \cos \theta$

49. $r = \dfrac{3}{1 - \cos \theta}$ **50.** $r = \dfrac{2}{1 = \sin \theta}$

51. $r^2 = 2\theta$ **52.** $r^2 = 3\theta$

53. $r = \dfrac{2}{1 + \sin \theta}$ **54.** $r = \dfrac{3}{1 + \cos \theta}$

55. $2r = r \sin \theta + 3$ **56.** $r = r \cos \theta + 2$

57. $r = \sin 2\theta$ **58.** $r = \tan \theta$

59. $r = \cos \theta$ **60.** $r = \sin \theta$

61. $r^2 \sin^2 \theta = 4$ **62.** $r^2 \cos^2 \theta = 1$

7.2   Polar Graphing

In Section 7.1 we worked with some polar equations that were easy to graph. For example, we can easily graph $\theta = \dfrac{3\pi}{4}$ as a line through the pole and $r = 3$ as a circle with center at the pole and a radius of 3.

We will now sketch the graphs of other polar equations by analyzing the trigonometric function in the equation. To do this, we will note how the radius r changes for each quarter period of the trigonometric function in the equation. This method relies on the fact that in each quarter period

(a) the sine and cosine functions are only increasing or decreasing, and

(b) the sine and cosine functions are positive or negative.

The following examples demonstrate this technique.

EXAMPLE 1   Sketch the graph of $r = 1 + \cos \theta$.

Solution   We will first sketch the graph of $r = 1 + \cos \theta$ on the Cartesian coordinates as in Section 5.5 (Fig. 7.12). To do this, we shift all the points on the graph $y = \cos \theta$ vertically, up one unit. Remember that in this graph, θ is the

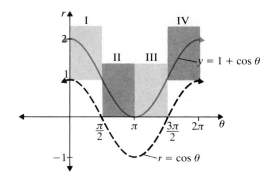

Figure 7.12

horizontal axis and r is the vertical axis. Now we will make a chart that shows the changes in θ and r for each of the four quarter periods denoted by I, II, III, and IV.

Quarter Period	Changes in θ	Changes in r
I	$0 \longrightarrow \dfrac{\pi}{2}$	$2 \longrightarrow 1$
II	$\dfrac{\pi}{2} \longrightarrow \pi$	$1 \longrightarrow 0$
III	$\pi \longrightarrow \dfrac{3\pi}{2}$	$0 \longrightarrow 1$
IV	$\dfrac{3\pi}{2} \longrightarrow 2\pi$	$1 \longrightarrow 2$

Looking at the first quarter period (I), we see that

as θ increases from 0 to $\dfrac{\pi}{2}$

the radius r decreases from 2 to 1.

We also see that in the second quarter period (II),

as θ increases from $\dfrac{\pi}{2}$ to π,

the radius r decreases from 1 to 0.

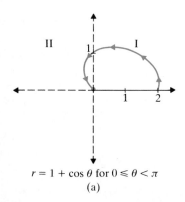

$r = 1 + \cos \theta$ for $0 \le \theta < \pi$
(a)

Figure 7.13a

We will now use these changes to sketch the graph in polar coordinates for θ from 0 to $\dfrac{\pi}{2}$ and from $\dfrac{\pi}{2}$ to π as shown in Fig. 7.13(a).

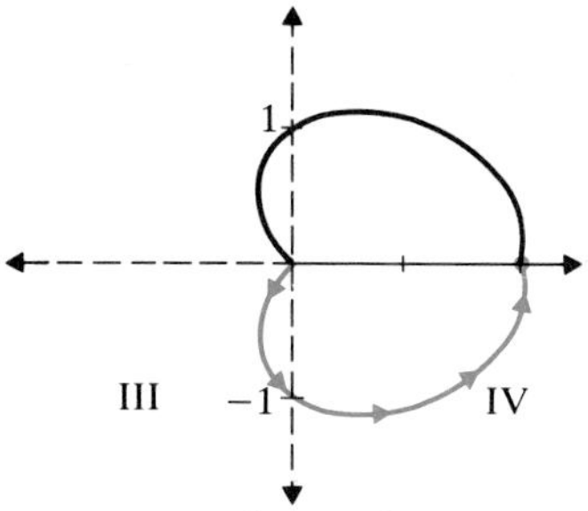

$r = 1 + \cos \theta$
(b)

Figure 7.13b

We follow the same procedure for the next two quarter periods. In these quarter periods we see that

as θ increases from π to $\frac{3\pi}{2}$,

the radius r increases from 0 to 1,

and

as θ increases from $\frac{3\pi}{2}$ to 2π,

the radius r increases from 1 to 2.

We now use this information to finish the sketch of the graph in Fig. 7.13(b). This curve is called a **cardioid** because of its heartlike shape.   ■

Although you can obtain more accuracy in these graphs by plotting more points (ordered pairs each containing a radius and an angle) in each interval, the graph in Fig. 7.13 provides the basic shape of the curve. You will find that it is more helpful to work with quarter periods than plotting many points. Using the quarter-period method provides an overall picture of the graph fairly quickly with the confidence that you did not miss a part of the graph. Note that I, II, III, and IV are used here to denote quarter periods, not quadrants.

As we found symmetry helpful when graphing in Cartesian coordinates, we will also find symmetry helpful when graphing in polar coordinates. The following symmetry tests will help us to know when we have symmetry in polar coordinates.

SYMMETRY TESTS

When each of the following substitutions is made and the resulting equation is equivalent to the original, then the graph of the equation has the indicated symmetry.

Substitution	Resulting Symmetry
$(r, -\theta)$ for (r, θ)	Symmetry about the polar axis (formerly x-axis symmetry)
$(-r, \theta)$ for (r, θ)	Symmetry with respect to the pole (origin symmetry)
$(r, \pi + \theta)$ for (r, θ)	Symmetry with respect to the pole (origin symmetry)
$(r, \pi - \theta)$ for (r, θ)	Symmetry about the line $\theta = \frac{\pi}{2}$ (formerly y-axis symmetry)

It is interesting to note that the failure of these substitutions to produce an equivalent equation does not necessarily rule out the indicated symmetry; it just does not confirm that symmetry.

In Example 1, $r = 1 + \cos\theta$ has symmetry with respect to the polar axis (formerly the x-axis), since

$$r = 1 + \cos(-\theta) = 1 + \cos\theta \quad \text{is} \quad r = 1 + \cos\theta$$

which is equivalent to the original equation. This symmetry is also observed in Fig. 7.13(b). Failure of the symmetry tests does not rule out symmetry, but a valid test does guarantee symmetry.

We will now use symmetry and the quarter-period method to sketch the graphs of equations in polar coordinates.

EXAMPLE 2   Sketch the graph of $r = \sin 2\theta$

Solution

(a) *Symmetry*: Notice that this curve has symmetry with respect to the line $\theta = \frac{\pi}{2}$ (formerly the y-axis), since substituting $(r, \pi - \theta)$ for (r, θ) provides an equivalent equation. To confirm this, we substitute $(r, \pi - \theta)$ for (r, θ):

$$r = \sin 2\theta \qquad \text{(Substitute } \pi - \theta \text{ for } \theta)$$
$$r = \sin 2(\pi - \theta) \qquad \text{(Use the double-angle identity)}$$
$$r = 2\sin(\pi - \theta)\cos(\pi - \theta)$$
$$r = 2\cos\theta\sin\theta \qquad \text{(Use the double-angle identity)}$$
$$r = \sin 2\theta$$

Thus the result is equivalent to the original equation, and this graph has symmetry with respect to the line $\theta = \frac{\pi}{2}$.

(b) *Quarter-period method*: To graph $r = \sin 2\theta$, we first sketch the graph of $y = \sin 2\theta$ in Cartesian coordinates and then analyze the behavior of the radius in each quarter period. The period of this function is $\frac{2\pi}{2} = \pi$ (Fig. 7.14). Now

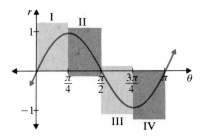

Figure 7.14

we will consider the changes in θ and r for each quarter period of $r = \sin 2\theta$ in the chart below.

Quarter Period	Changes in θ	Changes in r
I	$0 \longrightarrow \dfrac{\pi}{4}$	$0 \longrightarrow 1$
II	$\dfrac{\pi}{4} \longrightarrow \dfrac{\pi}{2}$	$1 \longrightarrow 0$
III	$\dfrac{\pi}{2} \longrightarrow \dfrac{3\pi}{4}$	$0 \longrightarrow -1$
IV	$\dfrac{3\pi}{4} \longrightarrow \pi$	$-1 \longrightarrow 0$

We notice that

as θ increases from 0 to $\dfrac{\pi}{4}$,
the value of r increases from 0 to 1.

As we see in Fig. 7.14, the change in r is larger (faster) when θ is close to 0, and smaller (slower) for θ close to $\dfrac{\pi}{4}$. Similarly,

as θ increases from $\dfrac{\pi}{4}$ to $\dfrac{\pi}{2}$,
the value of r decreases from 1 to 0.

These observations are shown in Fig. 7.15(a). Now we will consider the next two quarter periods. Notice that

as θ increases from $\dfrac{\pi}{2}$ to $\dfrac{3\pi}{4}$,
r is negative and decreasing from 0 to -1,

and

as θ increases from $\dfrac{3\pi}{4}$ to π,
r is negative and increases from -1 to 0.

We add this to the parts previously graphed with the results shown in Fig. 7.15(b).

After the fourth quarter period the sketch of the graph repeats the pattern of the first four quarter periods (I, II, III, and IV) as θ increases from π to 2π, since $\sin 2\theta$ has a period of π. The portion of the graph from π to 2π can also be sketched by using symmetry, established earlier, to the line $\theta = \pi$ (formerly the y-axis). This particular equation $r = \sin 2\theta$ is a **four-leaf rose**, and the name is reinforced by the shape of its graph (Fig. 7.15c). ∎

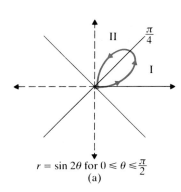

$r = \sin 2\theta$ for $0 \le \theta \le \dfrac{\pi}{2}$
(a)

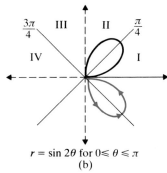

$r = \sin 2\theta$ for $0 \le \theta \le \pi$
(b)

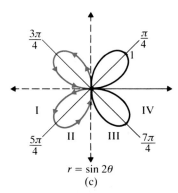

$r = \sin 2\theta$
(c)

Figure 7.15

The graphs of $r = a\ \sin\ n\theta$ or $r = a\ \cos\ n\theta$ for any integer $n > 1$ will produce a multileaf rose with n leaves when n is odd and $2n$ leaves when n is even. For $n = 1$, both equations are circles.

This quarter-period method of graphing provides a technique for sketching the graph of a relation in polar form based on the unique behavior of trigonometric functions during each quarter period. This method relies on the fact that in each quarter period

(a) the sine and cosine functions are only increasing or decreasing, and
(b) the sine and cosine functions are usually positive or negative.

To gain some confidence in the more exact shape of the curve, you might want to find more points, using a calculator. But in general this is not necessary. The quarter-period method is a conceptual method of graphing that is far superior. When graphing with this conceptual method, always graph all angles θ in the interval from 0 to 2π.

The more you sketch graphs using this method, the easier it will be for you to graph in polar coordinates. The following examples demonstrate graphing more complicated curves in polar coordinates.

EXAMPLE 3   Sketch the graph of $r^2 = 4\ \sin\ 2\theta$.

Solution

(a) *Symmetry*: Because of the r^2, the substitution of $(-r, \theta)$ for (r, θ) produces an equivalent equation. Thus $r^2 = 4\ \sin\ 2\theta$ has symmetry with respect to the pole (origin symmetry).
(b) *Quarter-period method*: Next, consider the graph of $y = r^2 = 4\ \sin\ 2\theta$ in Cartesian coordinates (Fig. 7.16). Then find the range of r^2 and r for each

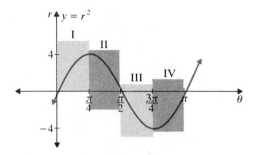

$y = r^2 = \sin\ 2\theta$ in cartesian co-ordinates

Figure 7.16

quarter period. We will now make a chart of θ, r^2, and r for each quarter period. (Remember that when you take the square root, you will have both the positive and the negative square roots.)

Quarter Period	Changes in θ	Changes in r^2	Changes in r
I	$0 \longrightarrow \dfrac{\pi}{4}$	$0 \longrightarrow 4$	$0 \longrightarrow 2$ $0 \longrightarrow -2$
II	$\dfrac{\pi}{4} \longrightarrow \dfrac{\pi}{2}$	$4 \longrightarrow 0$	$2 \longrightarrow 0$ $-2 \longrightarrow 0$
III	$\dfrac{\pi}{2} \longrightarrow \dfrac{3\pi}{4}$	$0 \longrightarrow -4$	not a real number
IV	$\dfrac{3\pi}{4} \longrightarrow \pi$	$-4 \longrightarrow 0$	not a real number

Thus

as θ increases from 0 to $\dfrac{\pi}{4}$,

the value of r increases from 0 to 2
and decreases from 0 to -2.

Likewise, as

θ increases from $\dfrac{\pi}{4}$ to $\dfrac{\pi}{2}$,

the value of r decreases from   2 to 0
and increases from -2 to 0

as sketched in Fig. 7.17(a).

Now we consider the third and fourth quarter periods. Since r must be a real number and $r^2 < 0$, there are no values for r that can be graphed when $\dfrac{\pi}{2} < \theta < \pi$ (Fig. 7.17a). Since $r^2 = 4 \sin 2\theta$ has a period of π, the patterns for the intervals I, II, III, and IV are repeated as θ increases from π to 2π. This produces the graph in Fig. 7.17(b). Notice how this graph contains the origin symmetry mentioned earlier. This graph is called a **lemniscate**.

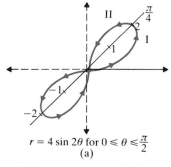

$r = 4 \sin 2\theta$ for $0 \le \theta \le \dfrac{\pi}{2}$
(a)

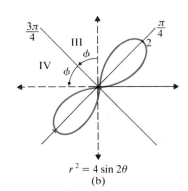

$r^2 = 4 \sin 2\theta$
(b)

Figure 7.17

The graphs of $r^2 = a^2 \cos 2\theta$ and $r^2 = a^2 \sin 2\theta$ are also **lemniscates** in different positions or orientations with their centers at the pole.

EXAMPLE 4 Sketch the graph of $r = 1 - 2 \cos \theta$.

Solution Notice that this equation has polar axis symmetry, since $\cos (-\theta) = \cos \theta$. To graph $r = 1 - 2 \cos \theta$ in polar coordinates, we first sketch $y = 1 - 2 \cos \theta$ in Cartesian coordinates (Fig. 7.18). Now we examine the changes of θ and r in each quarter period.

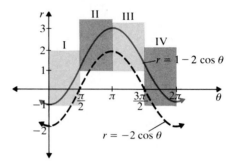

Figure 7.18

Quarter Period	Changes in θ	Changes in r
I	$0 \longrightarrow \dfrac{\pi}{2}$	$-1 \longrightarrow 1$
II	$\dfrac{\pi}{2} \longrightarrow \pi$	$1 \longrightarrow 3$
III	$\pi \longrightarrow \dfrac{3\pi}{2}$	$3 \longrightarrow 1$
IV	$\dfrac{3\pi}{2} \longrightarrow 2\pi$	$1 \longrightarrow -1$

We note that in quarter periods I and IV the radius is both positive and negative. It will help us to sketch the graph if we find the values of θ where $r = 0$. To find these points, solve the equation $r = 1 - 2 \cos \theta = 0$. Then

$$1 - 2 \cos \theta = 0$$
$$2 \cos \theta = 1$$
$$\cos \theta = \frac{1}{2}$$

and

$$\theta = \frac{\pi}{3} \quad \text{and} \quad \frac{5\pi}{3}$$

Combining this with the information above, we have

Quarter Period	Changes in θ	Changes in r
Ia	$0 \longrightarrow \dfrac{\pi}{3}$	$-1 \longrightarrow 0$
Ib	$\dfrac{\pi}{3} \longrightarrow \dfrac{\pi}{2}$	$0 \longrightarrow 1$
II	$\dfrac{\pi}{2} \longrightarrow \pi$	$1 \longrightarrow 3$
III	$\pi \longrightarrow \dfrac{3\pi}{2}$	$3 \longrightarrow 1$
IVa	$\dfrac{3\pi}{2} \longrightarrow \dfrac{5\pi}{3}$	$1 \longrightarrow 0$
IVb	$\dfrac{5\pi}{3} \longrightarrow 2\pi$	$0 \longrightarrow -1$

We now use this information to sketch the graph of $r = 1 - 2 \cos \theta$ step by step in Fig. 7.19. This curve is called a **limaçon**.

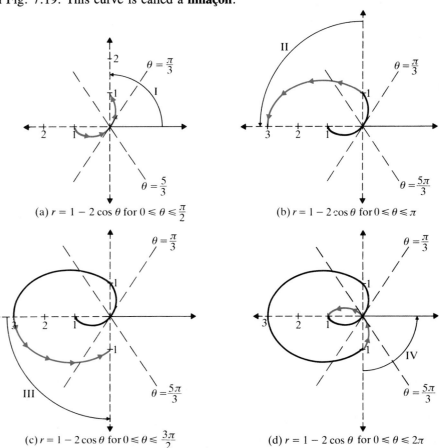

(a) $r = 1 - 2 \cos \theta$ for $0 \leqslant \theta \leqslant \dfrac{\pi}{2}$

(b) $r = 1 - 2 \cos \theta$ for $0 \leqslant \theta \leqslant \pi$

(c) $r = 1 - 2 \cos \theta$ for $0 \leqslant \theta \leqslant \dfrac{3\pi}{2}$

(d) $r = 1 - 2 \cos \theta$ for $0 \leqslant \theta \leqslant 2\pi$

Figure 7.19

Equations of the form $r = a \pm b \cos \theta$ and $r = a \pm b \sin \theta$ will be a **limaçon** when $a \neq b$ and a cardioid when $a = b$ (Example 1). The limaçon with $a < b$ goes through the pole and contains an internal loop. On the other hand, the limaçon with $a > b$ does not go through the pole and does not have an internal loop but does contain a dimple. We will sketch the graph of some of these in the exercises at the end of this section.

Some polar equations such as $r = 2 \sec \theta$ are still easier to graph when changed into Cartesian coordinates.

EXAMPLE 5   Sketch the graph of $r = 2 \sec \theta$.

Solution   Changing to the Cartesian coordinate equivalent, we have

$$r = 2 \sec \theta$$

$$r = \frac{2}{\cos \theta}$$

$$r \cos \theta = 2$$

$$x = 2$$

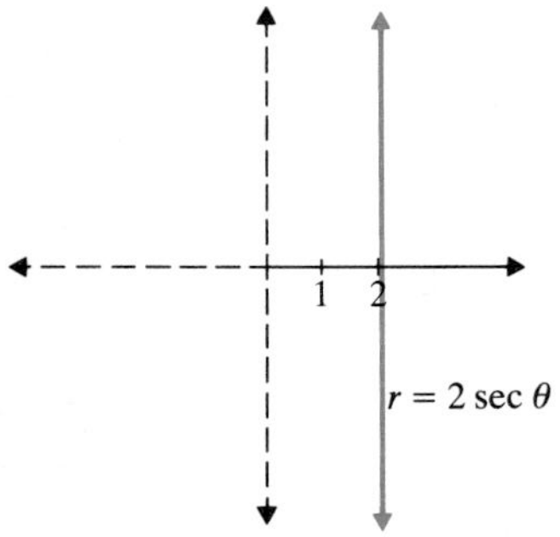

We can easily sketch the graph once $r = 2 \sec \theta$ is seen to be equivalent to $x = 2$ (Fig. 7.20). ∎

$r = 2 \sec \theta$

Figure 7.20

As demonstrated in this section, we can sketch the graph of a polar equation by

(a) using the appropriate symmetry, and
(b) graphing the radius in each quarter period of the basic function from 0 to 2π, or
(c) using the equivalent equation in Cartesian coordinates.

Exercises 7.2

For each of the following polar equations, **(a)** determine the symmetry (if any), **(b)** sketch the graph, and **(c)** name each curve (using the new names in this section when they apply).

1. $r = 2 - 2 \sin \theta$

2. $r = 2 - 2 \cos \theta$

3. $r = 3 + 3 \sin \theta$

4. $r = 3 + 3 \cos \theta$

5. $r = 2 \cos \theta$

6. $r = 3 \sin \theta$

7. $r = 2 \cos 2\theta$

8. $r = 2 \sin 2\theta$

9. $r = \sin 3\theta$

10. $r = \cos 3\theta$

11. $r = \sin 4\theta$

12. $r = \cos 4\theta$

13. $r^2 = \cos 2\theta$

14. $r^2 = \sin 2\theta$

15. $r^2 = \sin \theta$

16. $r^2 = \cos \theta$

17. $r = 3 + 2 \sin \theta$

18. $r = 2 + \sin \theta$

19. $r = 2 + \cos \theta$

20. $r = 3 + 2 \cos \theta$

Sketch the graphs of the following in polar coordinates. For spirals, make a chart of ordered pairs (r, θ) with radian measure for θ.

21. $r = \theta$ for $\theta \geq 0$
    (the spiral of Archimedes)

22. $r = e^\theta$ for $\theta \geq 0$
    (logarithmic spiral)

23. $r = 5 \csc \theta$

24. $r = -3 \sec \theta$

25. $r = 2 + 3 \sin \theta$

26. $r = 2 + 3 \cos \theta$

27. $r = 2 - 3 \cos \theta$

28. $r = 2 - 3 \sin \theta$

29. $r = \dfrac{2}{\theta}$ for $\theta > 0$
    (the reciprocal spiral)

30. $r = \cos\left(\theta - \dfrac{\pi}{4}\right)$

31. $r = \sin\left(\theta - \dfrac{\pi}{4}\right)$

32. $r = \sin\left(\theta + \dfrac{\pi}{3}\right)$

33. $r = \cos\left(\theta + \dfrac{\pi}{3}\right)$

34. $r = \cos\left(\theta - \dfrac{\pi}{4}\right)$

35. $r = \cos 2\left(\theta - \dfrac{\pi}{4}\right)$

36. $r = \sin 2\left(\theta - \dfrac{\pi}{4}\right)$

37. Prove that $r = a \sin \theta$ and $r = a \cos \theta$ are the equations of circles.

7.3 Vectors

The force of the wind on the sail of a yacht, the thrust of an airplane engine, and the weight of a lake against the wall of a spillway are all examples of a force that has a specific strength and direction. We will use vectors to represent these directed forces. Thus a vector will be used to represent the *magnitude* and *direction* of a force.

Definition | A **vector** is a directed line segment (a segment starting at an initial point and directed to a terminal point). Each vector has **(a)** a magnitude (length) and **(b)** a direction (angle).

In Fig. 7.21, v is a vector with an initial point of (x_1, y_1) and a terminal point of (x_2, y_2). By the Pythagorean Theorem we see that the **magnitude** or length of vector v is

$$|v| = \sqrt{(x_2 - x_1)^2 + (y_2 - y_1)^2}$$

The angle θ is the angle in standard position associated with the direction from (x_1, y_1) to (x_2, y_2) as seen in Fig. 7.21.

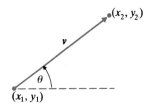

Figure 7.21

We can use the graphical representation of vectors to add and subtract vectors as seen in Fig. 7.22(a). The parallelograms in Fig. 7.22(a) show the ways in which vectors *v* and *w* are added and subtracted graphically. We can use right triangles, the law of sines, and the law of cosines to solve many practical vector problems. In this section we will use the horizontal and vertical components of a vector to solve vector problems. This new method, working with the horizontal and vertical components, will be most helpful but requires that we define two new vectors as follows.

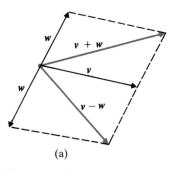

(a)

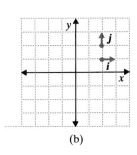

(b)

Figure 7.23

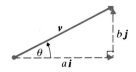

Figure 7.23

Definition

i is the unit vector parallel to and in the same direction as the positive *x*-axis ($|i| = 1$) (Fig. 7.22(a)).
j is the unit vector parallel to and in the same direction as the positive *y*-axis ($|j| = 1$) (Fig. 7.22(b)).

It is important to note that the vector *i* is not the same as the imaginary number

i. The vector *i* has a length of 1, an angle of $\theta = 0°$, and $i \neq \sqrt{-1}$. The vector *j* has a length of 1 and an angle of 90°.

For vector *v* in Fig. 7.23 a right triangle with sides *a* and *b* can be constructed where *a* is horizontal and *b* is vertical. Then the representation

$$v = ai + bj$$

uniquely represents the vector *v* where *ai* and *bj* are the horizontal and vertical components of vector *v*, respectively. Using θ and some basic trigonometry, we have

$$a = |v| \cos \theta$$
$$b = |v| \sin \theta$$

and thus

$$v = ai + bj = |v| \cos \theta i + |v| \sin \theta j$$

Notice that in the form $v = (|v| \cos \theta)i + (|v| \sin \theta)j$ the angle θ is in standard position.

We can easily add vectors, subtract vectors, and multiply a vector by a constant once we have written the vectors in this $ai + bj$ form.

Vector Properties

For real numbers a, b, c, and d,

Addition: $(ai + bj) + (ci + dj) = (a + c)i + (b + d)j$

Subtraction: $(ai + bj) - (ci + dj) = (a - c)i + (b - d)j$

Real number multiple of a vector: $c(ai + bj) = aci + bcj$

Equal vectors: $ai + bj = ci + dj$ if and only if $a = c$ and $b = d$

Vector magnitude: $|ai + bj| = \sqrt{a^2 + b^2}$

We proved the equation for vector magnitude earlier, using the Pythagorean Theorem. The proofs of the other properties are easily completed by using the associative, commutative, and distributive properties of real numbers, but we will not complete them here.

Sometimes an ordered pair notation is used to denote a vector. In this notation the horizontal component of the vector is written first and the vertical component second, and the parentheses are replaced by special brackets as seen here:

$$v = ai + bj = \langle a, b \rangle$$

We now have several ways to write a vector.

VECTOR NOTATION

A vector v can be represented as

$$v = ai + bj$$
$$v = \langle a, b \rangle$$

and

$$v = |v| \cos \theta\, i + |v| \sin \theta\, j$$

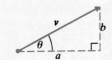

where $|v| = \sqrt{a^2 + b^2}$ and θ is in standard position.

Also note that the multiple of a vector can be written by using any of the vector forms. For a real number c and $v = ai + bj$ we have

$$cv = c(ai + bj) = cai + cbj$$
$$cv = c\langle a, b \rangle = \langle ca, cb \rangle$$

and

$$cv = (c|v| \cos \theta)i + (c|v| \sin \theta)j$$

These representations of a vector will be useful in solving vector problems.

EXAMPLE 1

Write the vector v in the $ai + bj$ form and the $\langle a, b \rangle$ form when v is the vector from (3, 2) to (6, 3).

Solution   We first sketch the vector v and find its horizontal and vertical components (Fig. 7.24). Thus

$$a = 6 - 3 = 3$$
$$b = 3 - 2 = 1$$

and

$$v = 3i + j = \langle 3, 1 \rangle$$

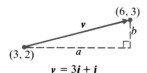

$$v = 3i + j$$

Figure 7.24

EXAMPLE 2

Write the vector w in the ai and bj form and the $\langle a, b \rangle$ form when w has a magnitude of 6 ($|w| = 6$) and an angle of $\theta = 60°$ with the horizontal.

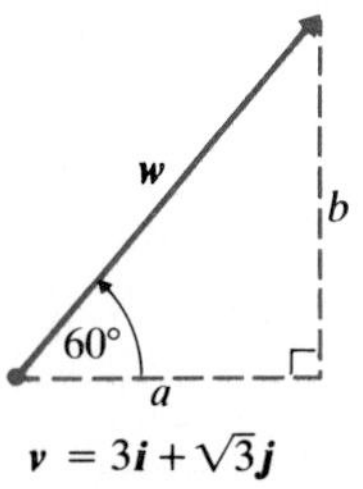

Solution   We first sketch the vector w and find its horizontal and vertical components (Fig. 7.25). Thus

$$|w| = 6$$
$$a = |w| \cos 60° = 6\left(\frac{1}{2}\right) = 3$$
$$b = |w| \sin 60° = 6\left(\frac{\sqrt{3}}{2}\right) = 3\sqrt{3}$$

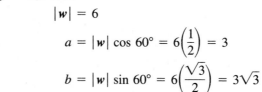

$$v = 3i + \sqrt{3}j$$

Figure 7.25   and thus

$$w = 3i + 3\sqrt{3}j = \langle 3, 3\sqrt{3} \rangle.$$

As we saw in Example 2, we can have a vector without knowing the coordinates of its endpoints. This is possible because vectors are not "tied down" to points but are what we call "free-floating" forces. Vectors have magnitude and a direction but are free to be moved to any convenient position. This freedom to move a vector around is helpful in solving applied problems.

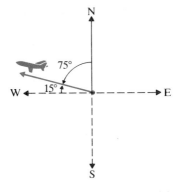

Figure 7.26

There are different ways to describe the direction of a force. In the past, we used the four points of a compass (N for north, E for east, S for south, and W for west). Thus the direction of an airplane could be described as N75°W as in Fig. 7.26. By this method this direction could also be named as S105°W or W15°N. This method can be confusing and misread; it is being replaced by writing the direction as a compass angle from north between 0° and 360° in a clockwise direction (Fig. 7.27). A direction written in this way with a compass is called a **heading** or an **azimuth**.

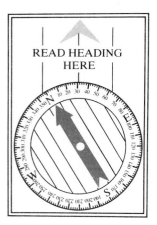

Figure 7.27

It will be helpful to use the form $v = |v| \cos \theta i + |v| \sin \theta j$ in solving word problems. But in working with the form $v = (|v| \cos \theta)i + (|v| \sin \theta)j$ the angle θ must be in standard position. Be careful because standard position and a heading are very different. Remember that **a heading or azimuth is always measured from north in a clockwise direction.**

EXAMPLE 3

An airplane is flying at 180 miles per hour with a heading of 15°. During the flight the wind is 60 miles per hour from the southwest with a heading of 47°. What are the azimuth and ground speed of this airplane?

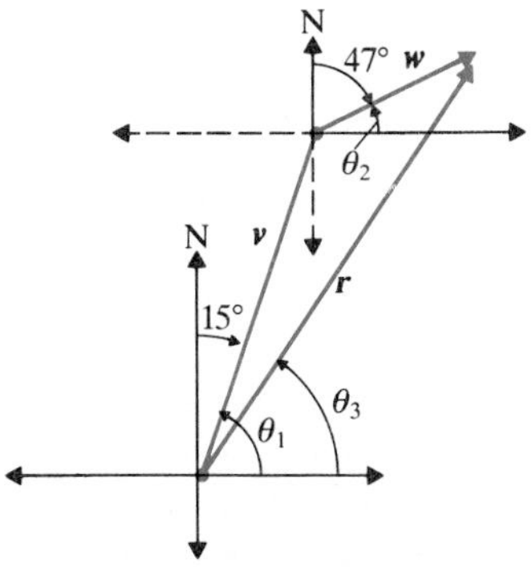

Solution   We let vectors v and w represent the airplane and the wind, respectively. We will find the resulting vector r by adding v and w. Thus $r = v + w$. Figure 7.28 is the sketch associated with this problem. To make this sketch, we start at the origin and sketch v. Then at the terminal point of v we sketch w. The resultant vector $r = v + w$ is the vector from the initial point of v to the terminal point of w. We make this sketch using vector addition and the fact that vectors are free floating.

To calculate r, we change the vector v of the airplane and the vector w of the wind into $ai + bj$ form. We first calculate θ_1 and θ_2:

$$\theta_1 = 90° - 15° = 75°$$
$$\theta_2 = 90° - 47° = 43°$$

Figure 7.28

Then the airplane vector is

$$v = 180 \cos 75°i + 180 \sin 75°j$$
$$\approx 46.59i + 173.87j$$

and the wind vector is

$$w = 60 \cos 43°i + 60 \sin 43°j$$
$$\approx 43.88i + 40.92j$$

Thus r, the resultant vector, is calculated by

$$r = v + w$$
$$\approx (46.59i + 173.87j) + (43.88i + 40.92j)$$
$$\approx 90.47i + 214.79j$$

The ground speed of the airplane is the magnitude of the resultant vector r:

$$|r| = \sqrt{(233.07)^2 + (214.79)^2}$$
$$\approx 233.07 \text{ miles per hour}$$

To find θ_3, we use the vertical and horizontal components of r:

$$\tan \theta_3 = \frac{\text{vertical change}}{\text{horizontal change}} \approx \frac{214.79}{90.47}$$
$$\theta_3 \approx 67.16°$$

and the azimuth of the r is $90° - 67.16° \approx 22.84°$ ■

EXAMPLE 4   An ocean tanker's destination is 20 nautical miles away with a heading of 15°. To get to this destination, the tanker must navigate around the end of the island. To do this, it will travel for 30 nautical miles at a heading of 340° and then turn and head directly for the destination point. What are the distance and heading of the second leg of this trip?

Solution   We first sketch the diagram for this problem (Fig. 7.29a). We let r be the vector from the first position to the destination and v_1 and v_2 be the first and second parts of the trip. Using angles in standard position for r, v_1, and v_2 as in Fig. 7.29(b), we calculate r, v_1, and v_2 in $ai + bj$ form:

$$r = 20 \cos 75°i + 20 \sin 75°j$$
$$v_1 = 30 \cos 110°i + 30 \sin 110°j$$

For

$$r = v_1 + v_2$$

we have

$$v_2 = r - v_1$$

Thus

$$v_2 = (20 \cos 75°\mathbf{i} + 20 \sin 75°\mathbf{j}) - (30 \cos 110°\mathbf{i} + 30 \sin 110°\mathbf{j})$$
$$= (20 \cos 75° - 30 \cos 110°)\mathbf{i} + (20 \sin 75° - 30 \sin 110°)\mathbf{j}$$
$$\approx 15.44\mathbf{i} - 8.87\mathbf{j}$$

The distance of v_2 is

$$|v_2| \approx \sqrt{(15.44)^2 + (8.87)^2} \approx 17.81$$

and θ_3 is

$$\tan^{-1}\left(\frac{-8.87}{15.44}\right) \approx -29.88°$$

in standard position. Therefore the second leg of this trip is 17.8 miles with a heading of $90° + 29.88° = 119.88°$. ∎

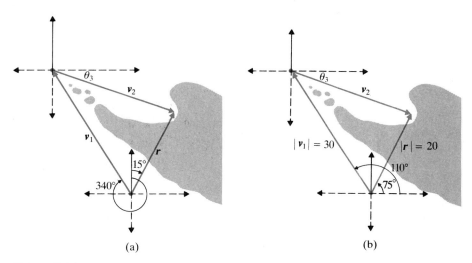

(a)   (b)

Figure 7.29

Vectors have other operations besides addition, subtraction, and multiplication by a real number. One of these operations is called the **dot product** (*scalar product* or *inner product*). We will now define the dot product.

Definition

The **dot product** of vectors

$$v = a\mathbf{i} + b\mathbf{j} \text{ and } w = c\mathbf{i} + d\mathbf{j}$$

is written as $v \cdot w$ and defined as

$$v \cdot w = ac + bd$$

It is important to notice that the dot product of two vectors $v \cdot w$ is not a vector but a real number (or scalar). With the dot product we can find the angle between two vectors.

ANGLE BETWEEN
VECTORS

For the angle θ between the vectors v and w,

$$v \cdot w = |v|\,|w|\cos\theta \qquad \text{or} \qquad \cos\theta = \frac{v \cdot w}{|v|\,|w|} \qquad \text{for } 0 \le \theta \le 180°$$

PROOF: To begin, we let

$$v = ai + bj$$
$$w = ci + dj$$

as in Fig. 7.30. For these values, $v \cdot w = ac + bd$ and

$$r = v - w = (a - c)i + (b - d)j$$

Next we solve for $\cos\theta$, using the law of cosines:

$$|r|^2 = |v|^2 + |w|^2 - 2|v|\,|w|\cos\theta$$

Now we substitute the magnitude of v and simplify:

$$(\sqrt{(a - c)^2 + (b - d)^2})^2 = |\sqrt{a^2 + b^2}|^2 + |\sqrt{c^2 + d^2}|^2 - 2|v|\,|w|\cos\theta$$

$$(a - c)^2 + (b - d)^2 = a^2 + b^2 + c^2 + d^2 - 2|v|\,|w|\cos\theta$$

$$a^2 - 2ac + c^2 + b^2 - 2bd + d^2 = a^2 + b^2 + c^2 + d^2 - 2|v|\,|w|\cos\theta$$

$$-2ac - 2bd = -2|v|\,|w|\cos\theta \qquad \textit{(Subtract } a^2, b^2, \textit{ and } c^2$$
$$\textit{from both sides and simplify)}$$

$$ac + bd = |v|\,|w|\cos\theta \qquad \textit{(Substitute } v \cdot w \textit{ for } ac + bd\textit{)}$$

$$v \cdot w = |v|\,|w|\cos\theta$$

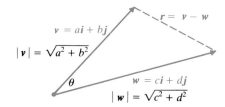

Figure 7.30

or

$$\cos\theta = \frac{v \cdot w}{|v|\,|w|}$$

☐

Now we can use the dot product to find the angle between two vectors.

EXAMPLE 5 Find the angle θ between the following pairs of vectors:

(a) $v = 3i - 5j$ (b) $r = 3i - 2j$
     $w = 4i + 2j$ $s = 2i + 3j$

Solution We will use the information concerning the dot product from both the definition and the previous theorem to find the angle between these vectors.

(a) To find θ, first evaluate $|v|$, $|w|$, and $v \cdot w$:

$$|v| = \sqrt{3^2 + (-5)^2} = \sqrt{9 + 25} = \sqrt{34}$$
$$|w| = \sqrt{4^2 + 2^2} = \sqrt{16 + 4} = \sqrt{20}$$
$$v \cdot w = 3(4) + (-5)2 = 12 - 10 = 2$$

Then

$$\cos \theta = \frac{v \cdot w}{|v|\,|w|} = \frac{2}{\sqrt{20}\sqrt{34}} = \frac{2}{2\sqrt{5}\sqrt{34}}$$

$$= \frac{1}{\sqrt{5 \cdot 34}} \approx 0.076696$$

and

$$\theta \approx 85.6°$$

(b) To find θ, first evaluate $|r|$, $|s|$, and $r \cdot s$:

$$|r| = \sqrt{3^2 + (-2)^2} = \sqrt{13}$$
$$|s| = \sqrt{2^2 + 3^2} = \sqrt{13}$$
$$r \cdot s = 3(2) + (-2)3 = 0$$

Then

$$\cos \theta = \frac{r \cdot s}{|r|\,|s|} = \frac{0}{\sqrt{13}\sqrt{13}} = 0$$

and

$$\theta = 90°.$$

Thus the vectors r and s are perpendicular. ∎

EXAMPLE 6 Find $v \cdot w$ when $|v| = 6$, $|w| = 3$, and θ, the angle between v and w, is $72°$.

Solution Since

$$\cos \theta = \frac{v \cdot w}{|v|\,|w|}$$

we have

$$v \cdot w = |v|\,|w| \cos \theta$$

and by substituting,

$$v \cdot w = 6 \cdot 3 \cdot \cos 72°$$
$$\approx 5.56$$

The dot product obeys many of the patterns and properties of real numbers. We will prove some of these properties in the exercises that follow this section. The work in this section is a good introduction for future work with the dot product and vectors that you will encounter in calculus.

Exercises 7.3

Write each of the following vectors v in $ai + bi$ form and as $\langle a, b \rangle$ when v is the vector from point P to point Q.

1. $P = (2, 5)$, $Q = (1, -3)$

2. $P = (3, 7)$, $Q = (-3, 1)$

3. $P = (-2, -5)$, $Q = (3, 16)$

4. $P = (2, 7)$, $Q = (-5, -3)$

5. $P = \left(\frac{1}{2}, \frac{3}{4}\right)$, $Q = \left(-\frac{1}{2}, -7\right)$

6. $P = \left(-\frac{1}{3}, 6\right)$, $Q = \left(\frac{2}{3}, \frac{1}{2}\right)$

Write each of the following vectors w in $ai + bj$ form and as $\langle a, b \rangle$ when the magnitude and angle are given for w:

7. $|w| = 3$, $\theta = 30°$

8. $|w| = 7$, $\theta = 45°$

9. $|w| = \frac{1}{2}$, $\theta = -\frac{\pi}{4}$

10. $|w| = \sqrt{3}$, $\theta = \frac{\pi}{3}$

11. $|w| = 3\frac{1}{2}$, $\theta = \frac{2\pi}{3}$

12. $|w| = 7$, $\theta = -\frac{3\pi}{4}$

For the following pairs of vectors, find $v + w$, $v - w$, cv, and $v \cdot w$.

13. $v = 3i - 2j$
$w = -2i + 7j$
$c = -4$

14. $v = 5i - 7j$
$w = -4i + 2j$
$c = 2$

15. $v = \frac{1}{2}i + \frac{1}{3}j$
$w = -\frac{1}{2}i + \frac{2}{3}j$
$c = 6$

16. $v = \frac{1}{4}i + 3j$
$w = -4i - \frac{2}{3}j$
$c = 6$

17. $v = \langle -3, 9 \rangle$
$w = \langle -\frac{1}{3}, -\frac{2}{3} \rangle$
$c = \frac{1}{9}$

18. $v = \langle -4, 2 \rangle$
$w = \langle \frac{1}{4}, -\frac{3}{4} \rangle$
$c = \frac{1}{2}$

19. A man and his son try to push a crate diagonally across the garage floor (Fig. 7.31). The father pushes on the west side of the crate with a force of 40 pounds, and the son pushes on the south side with a force of 15 pounds. What are the azimuth for the movement of the box and the force in that direction?

Figure 7.31

20. Repeat Exercise 19 when the father and son push with forces of 75 pounds and 35 pounds, respectively.

21. An airplane flies at 305 miles per hour with a heading of 312°. During the flight the wind is 72 miles per hour from the southwest with a heading of 27°. What are the azimuth and ground speed of the flight of this airplane?

22. An airplane flies at 387 miles per hour with a heading of 17°. During the flight the wind is 52 miles per hour from the northeast with a heading of 321°. What are the azimuth and ground speed of the flight of this airplane?

23. An oil tanker wants to reach a port 750 nautical miles away at an azimuth of 48°. To avoid the end of an island, the tanker travels at an azimuth of 35° for 715 nautical miles and then turns and heads to port. What are the azimuth and distance of the last part of this voyage?

24. An oil tanker wants to reach a port 150 nautical miles away at an azimuth of 52°. To avoid the end of an island, the tanker travels at an azimuth of 41° for 205 nautical miles and then turns and heads to port. What are the azimuth and distance of the last part of this voyage?

25. A father and son enter an orienteering competition. The instructions direct the following moves:
(a) heading 42.5°, distance 350 feet
(b) heading 245°, distance 425 feet
(c) heading 125°, distance 175 feet
After these three instructions, what should be the heading and distance to the starting point?

26. A father and daughter enter an orienteering competition. The instructions direct the following moves:
(a) heading 165°, distance 175 feet
(b) heading 49°, distance 350 feet
(c) heading 259°, distance 300 feet
After these three instructions, what should be the heading and distance to the starting point?

27. An airplane flies for 3.75 hours with a heading of 123° and then changes course and flies for 1.62 hours with a heading of 214°. The airplane's average air speed is 315 miles per hour, and the wind's average speed is 32 miles per hour from the southwest with a heading of 21°. What are the heading and distance from the end of the second part of this trip back to the starting point?

28. An airplane flies for 1.92 hours with a heading of 16° and then changes course and flies for 4.15 hours with a heading of 285°. The airplane's average air speed is 417 miles per hour, and the wind's average speed is 43 miles per hour from the northwest with a heading of 131°. What are the heading and distance from the end of the second part of this trip back to the starting point?

29. Using the vectors for wind and boat speed of a sailing ship (Fig. 7.32) and the law of cosines, find the boat speed v and the speed made good v_{mg} when the apparent wind speed is 30 knots, the true wind speed is 20 knots, and the true wind angle is 37°.

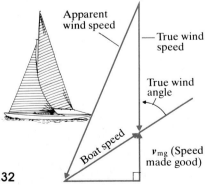

Figure 7.32

30. Using the vectors for wind and boat speed of a sailing ship (Fig. 7.32) and the law of cosines, find the boat speed v and the speed made good v_{mg} when the apparent wind speed is 40 knots, the true wind speed is 28 knots, and the true wind angle is 20°.

31. Prove that the length of the projection (or right angle shadow) of vector v onto vector w is $\dfrac{v \cdot w}{|w|}$, as shown in Fig. 7.33.

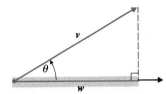

Figure 7.33

32. Prove that the length of the projection (or right angle shadow) of vector w onto vector v is $\dfrac{v \cdot w}{|v|}$, as shown in Fig. 7.34.

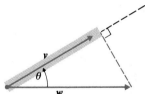

Figure 7.34

33. Find the length of the projection of $v = 5i - 2j$ onto $w = 7i - 3j$. (See Exercises 31 and 32.)

34. Find the length of the projection of $w = 7i - 3j$ onto $v = 5i - 2j$. (See Exercises 31 and 32.)

35. For any vector $v = ai + bj$, prove that $v \cdot v = |v|^2$.

36. For any vector $v = ai + bj$, prove that $0 \cdot v = 0$.

37. For vectors v and w, prove that $v \cdot w = w \cdot v$ (the commutative property of the dot product).

38. For vectors u, v, and w, prove that $u \cdot (v + w) = u \cdot v + u \cdot w$ (the distributive property for the dot product).

39. For nonzero vectors v and w, prove that $v \cdot w = 0$ if and only if v and w are perpendicular vectors (or orthogonal vectors).

40. Prove that $u = bi - aj$ and $w = -bi + aj$ are each perpendicular to $v = ai + bj$.

41. Find two vectors orthogonal to $v = 7i - 2j$. (See Exercise 40.) Use the dot product to verify your results.

42. Find two vectors orthogonal to $w = 3i + 12j$. (See Exercise 40.) Use the dot product to verify your result.

43. When $v = ai + bj$ and $w = \dfrac{v}{|v|}$, prove that $|w| = 1$.

44. Find the unit vector in the same direction as $v = 7i - 2j$. (See Exercise 43.)

45. Find the unit vector in the same direction as $w = 3i + 12j$. (See Exercise 43.)

7.4   Trigonometric Form of a Complex Number and de Moivre's Theorem

Complex numbers, which we studied earlier, have been a part of algebra since Raffaele Bombelli's 1572 book, which established a consistent theory of complex numbers. A century later the theory of complex numbers was advanced by the work of three great mathematicians, Johann Bernoulli (1667–1748), Leonhard Euler (1707–1783), and Karl Friedrich Gauss (1777–1855).

The geometric representation for complex numbers, attributed to Caspard Wessel (1745–1818) and Jean Robert Argand (1768–1822), became widely accepted as a result of its use by Karl Friedrich Gauss. This coordinate system used to graph complex numbers is called the Gaussian plane. We can represent a complex number in the Gaussian plane by slightly modifying the Cartesian coordinate system. This modification changes the vertical axis into the i-axis, where $i = \sqrt{-1}$, so that the point (a, b) represents the complex number $a + bi$ (Fig. 7.35). The complex numbers and the Gaussian plane are essential foundations for the more advanced topics of the theory of complex variables and complex analysis.

Be careful not to confuse the i in $a + bi$, where $i = \sqrt{-1}$, with the horizontal unit vector $\boldsymbol{i}$. The context in which you find i will often help to differentiate between the complex number i and the unit vector $\boldsymbol{i}$. It is interesting to note that the graphical representation of a complex number is very similar to the graphical depiction of a

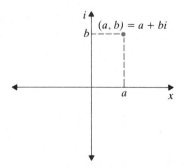

Figure 7.35

vector. Thus as we saw earlier, the addition, subtraction, and real number multiplication of complex numbers is very similar to the addition and real number multiplication of vectors, as you can see in the equations below and Figs. 7.36(a) and 7.36(b).

$$(a + bi) + (c + di) = (a + c) + (b + d)i$$
$$c(a + bi) = ac + bci$$

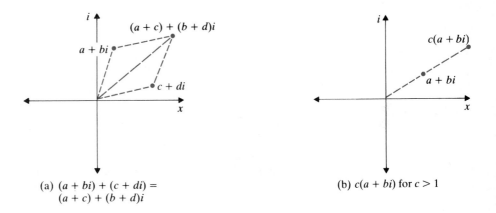

(a) $(a + bi) + (c + di) =$
$(a + c) + (b + d)i$

(b) $c(a + bi)$ for $c > 1$

Figure 7.36

Also, as we stated earlier, complex numbers are equal when the real parts are equal and the coefficients of i are equal. Symbolically stated, we have

$$a + bi = c + di \quad \text{if and only if} \quad a = c \quad \text{and} \quad b = d$$

This expression $|a + bi|$ is the **absolute value** (or *modulus*) of the complex number $a + bi$, and

$$|a + bi| = \sqrt{a^2 + b^2}$$

We note that $|a + bi|$ is the distance from the origin to the complex number in the Gaussian plane.

With this graphical representation of $a + bi$ we can represent a complex number in terms of its absolute value $|a + bi| = \sqrt{a^2 + b^2}$, the modulus of the complex number, and its angle θ, the **argument** of the complex number (Fig. 7.37). Using trigonometric functions, we can rewrite $a + bi$ in a new form as follows:

$$a = \sqrt{a^2 + b^2} \cos \theta \quad \text{and} \quad b = \sqrt{a^2 + b^2} \sin \theta$$

then, by substitution,

$$a + bi = (\sqrt{a^2 + b^2} \cos \theta) + (\sqrt{a^2 + b^2} \sin \theta)i$$
$$= \sqrt{a^2 + b^2} (\cos \theta + i \sin \theta)$$

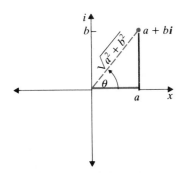

Figure 7.37

This new form is called the **trigonometric form** of a complex number.

TRIGONOMETRIC
FORM OF A
COMPLEX NUMBER

For real numbers a and b with $i = \sqrt{-1}$ the trigonometric form of the complex number $a + bi$ is

$$a + bi = r(\cos \theta + i \sin \theta)$$

where

$$r = \sqrt{a^2 + b^2}$$
$$a = r \cos \theta, \qquad b = r \sin \theta$$

and

$$\tan \theta = \frac{b}{a}$$

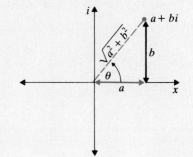

We will find this trigonometric form most helpful in evaluating large powers of complex numbers such as $(2 - 2i)^{200}$, and in evaluating the roots of complex numbers such as $(2 - 2i)^{1/4}$. Before attempting these more advanced problems it is important that we take time to change complex numbers into trigonometric form.

EXAMPLE 1

Change each of the following complex numbers to trigonometric form:
(a) $2 - 2i$, **(b)** $-\sqrt{3} + i$, and **(c)** $-5 - 2i$.

Solution

(a) We sketch a graph of the complex number $2 - 2i$ (Fig. 7.38). Then we calculate $r = |a + bi|$ and θ for $2 - 2i$. Thus

$$r = |2 - 2i| = \sqrt{2^2 + (-2)^2} = \sqrt{8} = 2\sqrt{2}$$

and $\tan \theta = -\frac{2}{2} = -1$ with

$$\theta = \tan^{-1}\left(\frac{-2}{2}\right) = \tan^{-1}(-1) = -\frac{\pi}{4}$$

so that

$$\theta = -\frac{\pi}{4} \quad \text{or} \quad \frac{7\pi}{4}$$

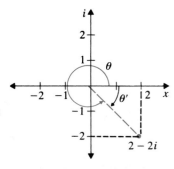

Figure 7.38

Therefore

$$2 - 2i = 2\sqrt{2}\left[\cos\left(-\frac{\pi}{4}\right) + i \sin\left(-\frac{\pi}{4}\right)\right]$$

or

$$2 - 2i = 2\sqrt{2}\left(\cos\frac{7\pi}{4} + i \sin\frac{7\pi}{4}\right)$$

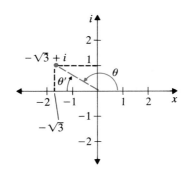

Figure 7.39

(b) We sketch a graph of the complex number $-\sqrt{3} + i$ (Fig. 7.39). Then we calculate $r = |a + bi|$ and θ for $-\sqrt{3} + i$. Thus

$$r = |-\sqrt{3} + i| = \sqrt{(-\sqrt{3})^2 + (1)^2} = \sqrt{4} = 2$$

and

$$\theta = \tan^{-1}\left(\frac{1}{-\sqrt{3}}\right)$$

so that

$$\theta' = \frac{\pi}{6} \quad \text{and} \quad \theta = \frac{5\pi}{6}$$

Therefore

$$-\sqrt{3} + i = 2\left(\cos\frac{5\pi}{6} + i\sin\frac{5\pi}{6}\right)$$

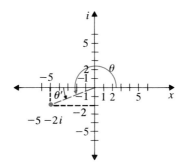

Figure 7.40

(c) We sketch a graph of the complex number $-5 - 2i$ (Fig. 7.40). Then we calculate $r = |a + bi|$ and θ for $-5 - 2i$. Thus

$$r = |-5 - 2i| = \sqrt{(-5)^2 + (-2)^2} = \sqrt{29}$$

and

$$\tan\theta = \frac{-2}{-5} = \frac{2}{5} = 0.4$$

so that

$$\theta = \tan^{-1} 0.4 \quad \text{or} \quad \theta' \approx 21.8°$$

with $\theta \approx 201.8°$. Therefore

$$-5 - 2i = \sqrt{29}\,[\cos(\tan^{-1} 0.4) + i\sin(\tan^{-1} 0.4)]$$

or

$$-5 - 2i \approx \sqrt{29}\,(\cos 201.8° + i\sin 201.8°) \qquad \blacksquare$$

We will also change a complex number in trigonometric form into the $a + bi$ form in the following example.

EXAMPLE 2   Change the following complex numbers into $a + bi$ form:

(a) $5\left(\cos\frac{\pi}{6} + i\sin\frac{\pi}{6}\right)$, **(b)** $\sqrt{3}\left[\cos\left(-\frac{\pi}{3}\right) + i\sin\left(-\frac{\pi}{3}\right)\right]$, and

(c) $18(\cos 176° - i\sin 176°)$.

Solution   To change a complex number from trigonometric form into $a + bi$ form, we evaluate the trigonometric functions and then use the distributive property. (Use exact values for trigonometric functions whenever possible.)

(a) $5\left(\cos \dfrac{\pi}{6} + i \sin \dfrac{\pi}{6}\right) = 5\left(\dfrac{\sqrt{3}}{2} + i\dfrac{1}{2}\right)$

$$= \dfrac{5\sqrt{3}}{2} + \dfrac{5}{2}i$$

(b) $\sqrt{3}\left[\cos\left(-\dfrac{\pi}{3}\right) + i \sin\left(-\dfrac{\pi}{3}\right)\right] = \sqrt{3}\left[\dfrac{1}{2} + i\left(-\dfrac{\sqrt{3}}{2}\right)\right]$

$$= \dfrac{\sqrt{3}}{2} - \dfrac{\sqrt{3}\sqrt{3}}{2}i$$

$$= \dfrac{\sqrt{3}}{2} - \dfrac{3}{2}i$$

(c) $18(\cos 176° - i \sin 176°) \approx 18(-0.99756 - 0.06976i)$

$$\approx -17.956 - 1.256i$$ ■

Now that we can write a complex number in trigonometric form, $r (\cos \theta + i \sin \theta)$, as well as the $a + bi$ form, we can state and prove the following theorem for multiplying and dividing complex numbers.

Theorem
Multiplying and
Dividing Complex
Numbers

For complex numbers

$$z_1 = r_1(\cos \theta_1 + i \sin \theta_1) \quad \text{and} \quad z_2 = r_2(\cos \theta_2 + i \sin \theta_2)$$

then

$$z_1 \cdot z_2 = r_1 r_2[\cos (\theta_1 + \theta_2) + i \sin (\theta_1 + \theta_2)]$$

and

$$\dfrac{z_1}{z_2} = \dfrac{r_1}{r_2}[\cos (\theta 1 - \theta_2) + i \sin (\theta_1 - \theta_2)] \quad \text{for} \quad z_2 \neq 0$$

Thus to multiply complex numbers in trigonometric form, we multiply the values of r and add the angles. To divide, we divide the values of r and subtract the angle in the denominator from the angle in the numerator. The following proof verifies this theorem.

PROOF:

(a) For

$$z_1 = r_1(\cos \theta_1 + i \sin \theta_1) \qquad \text{and} \qquad z_2 = r_2(\cos \theta_2 + i \sin \theta_2)$$

we have

$$\begin{aligned}
z_1 \cdot z_2 &= [r_1(\cos \theta_1 + i \sin \theta_1)][r_2(\cos \theta_2 + i \sin \theta_2)] \\
&= r_1 r_2 (\cos \theta_1 + i \sin \theta_1)(\cos \theta_2 + i \sin \theta_2) \\
&= r_1 r_2 [\cos \theta_1 \cos \theta_2 + (\cos \theta_1 \sin \theta_2 + \sin \theta_1 \cos \theta_2)i + i^2 \sin \theta_1 \sin \theta_2] \\
&= r_1 r_2 [(\cos \theta_1 \cos \theta_2 - \sin \theta_1 \sin \theta_2) + (\cos \theta_1 \sin \theta_2 + \sin \theta_1 \cos \theta_2)i] \\
&= r_1 r_2 [\cos (\theta_1 + \theta_2) + i \sin (\theta_1 + \theta_2)]. \quad \textit{(Using the cos } (\alpha + \beta) \\
& \textit{and sin } (\alpha + \beta) \textit{ identities)}
\end{aligned}$$

(b) Using the result of part (a) and the fact that $z_2 \neq 0$, we have

$$\begin{aligned}
\frac{z_1}{z_2} &= \frac{r_1(\cos \theta_1 + i \sin \theta_1)}{r_2(\cos \theta_2 + i \sin \theta_2)} \\
&= \frac{r_1(\cos \theta_1 + i \sin \theta_1)[\cos (-\theta_2) + i \sin (-\theta_2)]}{r_2(\cos \theta_2 + i \sin \theta_2)[\cos (-\theta_2) + i \sin (-\theta_2)]} \quad \textit{(Using part (a) above} \\
&\phantom{= \frac{r_1(\cos \theta_1 + i \sin \theta_1)[\cos (-\theta_2)]}{r_2}} \textit{and simplifying)} \\
&= \frac{r_1[\cos (\theta_1 - \theta_2) + i \sin (\theta_1 - \theta_2)]}{r_2[\cos (\theta_2 - \theta_2) + i \sin (\theta_2 - \theta_2)]} \\
&= \frac{r_1[\cos (\theta_1 - \theta_2) + i \sin (\theta_1 - \theta_2)]}{r_2[\cos (0) + i \sin (0)]} \\
&= \frac{r_1[\cos (\theta_1 - \theta_2) + i \sin (\theta_1 - \theta_2)]}{r_2(1 + 0i)} \\
&= \frac{r_1}{r_2}[\cos (\theta_1 - \theta_2) + i \sin (\theta_1 - \theta_2)] \qquad \qquad \square
\end{aligned}$$

We can now use this theorem to multiply and divide complex numbers in trigonometric form.

EXAMPLE 3   For $z_1 = \sqrt{3} + i$ and $z_2 = -1 - i\sqrt{3}$ (Fig. 7.41), find **(a)** $z_1 \cdot z_2$ and **(b)** $z_1 \div z_2$.

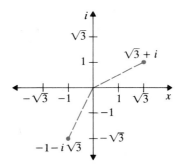

Figure 7.41

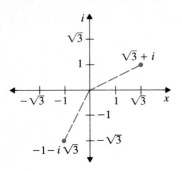

Figure 7.41

Solution We first change each complex number to trigonometric form. For $z_1 = \sqrt{3} + i$,

$$r = \sqrt{(\sqrt{3})^2 + 1^2} = \sqrt{4} = 2$$

and $\tan \theta = \dfrac{1}{\sqrt{3}}$, so $\theta = \dfrac{\pi}{6}$. Thus

$$z_1 = 2\left(\cos \frac{\pi}{6} + i \sin \frac{\pi}{6}\right)$$

Likewise, for $z_2 = -1 - i\sqrt{3}$ we have

$$r = \sqrt{(-1)^2 + (-\sqrt{3})^2} = \sqrt{4} = 2$$

and $\tan \theta = \dfrac{-\sqrt{3}}{-1} = \sqrt{3}$; so $\theta' = \dfrac{\pi}{3}$ and $\theta = \dfrac{4\pi}{3}$. Thus

$$z_2 = 2\left[\cos\left(\frac{4\pi}{3}\right) + i \sin\left(\frac{4\pi}{3}\right)\right]$$

(a) Then

$$z_1 \cdot z_2 = \left[2\left(\cos \frac{\pi}{6} + i \sin \frac{\pi}{6}\right)\right]\left[2\left(\cos \frac{4\pi}{3} + i \sin \frac{4\pi}{3}\right)\right]$$

$$= 2 \cdot 2\left[\cos\left(\frac{\pi}{6} + \frac{4\pi}{3}\right) + i \sin\left(\frac{\pi}{6} + \frac{4\pi}{3}\right)\right]$$

$$= 4\left(\cos \frac{9\pi}{6} + i \sin \frac{9\pi}{6}\right)$$

$$= 4\left(\cos \frac{3\pi}{2} + i \sin \frac{3\pi}{2}\right)$$

$$= 4(0 - i)$$

$$= -4i$$

(b)

$$\frac{z_1}{z_2} = \frac{2\left(\cos \dfrac{\pi}{6} + i \sin \dfrac{\pi}{6}\right)}{2\left(\cos \dfrac{4\pi}{3} + i \sin \dfrac{4\pi}{3}\right)}$$

$$= \frac{2}{2}\left[\cos\left(\frac{\pi}{6} - \frac{4\pi}{3}\right) + i \sin\left(\frac{\pi}{6} - \frac{4\pi}{3}\right)\right]$$

$$= 1\left[\cos\left(-\frac{7\pi}{6}\right) + i \sin\left(-\frac{7\pi}{6}\right)\right] \quad \textit{(Use coterminal angles)}$$

$$= \cos \frac{5\pi}{6} + i \sin \frac{5\pi}{6}$$

$$= -\frac{\sqrt{3}}{2} + \frac{1}{2}i = \frac{1}{2}(-\sqrt{3} + i)$$

The theorem above is important for multiplying and dividing complex numbers and will help us prove the following more important theorem.

de Moivre's Theorem

For the complex number $z = r(\cos \theta + i \sin \theta)$ and any natural number n, the nth power of z is

$$z^n = r^n (\cos n\theta + i \sin n\theta)$$

PROOF: Using the previous theorem for multiplying complex numbers and the factored form of z^n, we have

$$
\begin{aligned}
z^n &= [r(\cos \theta + i \sin \theta)]^n \\
&= r^n(\cos \theta + i \sin \theta)^n \quad \text{(Multiply these factors one at a time)} \\
&= r^n(\cos \theta + i \sin \theta)(\cos \theta + i \sin \theta)(\cos \theta + i \sin \theta)^{n-2} \\
&= r^n(\cos 2\theta + i \sin 2\theta)(\cos \theta + i \sin \theta)^{n-2} \\
&= r^n(\cos 2\theta + i \sin 2\theta)(\cos \theta + i \sin \theta)(\cos \theta + i \sin \theta)^{n-3} \\
&= r^n(\cos 3\theta + i \sin 3\theta)(\cos \theta + i \sin \theta)^{n-3} \\
&= r^n(\cos 3\theta + i \sin 3\theta)(\cos \theta + i \sin \theta)(\cos \theta + i \sin \theta)^{n-3} \\
&= r^n(\cos 4\theta + i \sin 4\theta)(\cos \theta + i \sin \theta)^{n-4} \\
&= r^n(\cos 5\theta + i \sin 5\theta)(\cos \theta + i \sin \theta)^{n-5}
\end{aligned}
$$

We continue this process until all n factors of $\cos \theta + i \sin \theta$ have been multiplied. Continuing the pattern above, we conclude that

$$z^n = r^n[\cos (n\theta) + i \sin (n\theta)]$$

We will prove de Moivre's Theorem more precisely later on, using mathematical induction. □

Thus with de Moivre's Theorem we can easily evaluate a complex number to any natural number exponent. The following example demonstrates this process.

EXAMPLE 4   Evaluate **(a)** $(2 - 2i)^{16}$, **(b)** $(-1 + i\sqrt{3})^{63}$, and **(c)** $\left(-\dfrac{\sqrt{2}}{2} - \dfrac{\sqrt{2}}{2}i\right)^{73}$.

Solution   We will first change each complex number to trigonometric form and then use de Moivre's Theorem.

(a) Changing $2 - 2i$ into trigonometric form, we have

$$r = \sqrt{2^2 + (-2)^2} = \sqrt{8} = 2\sqrt{2}$$

and $\tan \theta = \dfrac{-2}{2} = -1$, so $\theta = -\dfrac{\pi}{4}$ (Fig. 7.42). Thus

$$2 - 2i = 2\sqrt{2}\left[\cos\left(-\frac{\pi}{4}\right) + i \sin\left(-\frac{\pi}{4}\right)\right]$$

$$= 2 \cdot 2^{1/2}\left[\cos\left(-\frac{\pi}{4}\right) + i \sin\left(-\frac{\pi}{4}\right)\right]$$

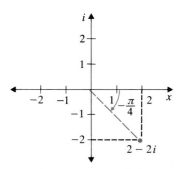

Figure 7.42

Then we use de Moivre's Theorem:

$$(2 - 2i)^{16} = \left[2 \cdot 2^{1/2}\left(\cos\left(-\frac{\pi}{4}\right) + i \sin\left(-\frac{\pi}{4}\right)\right)\right]^{16}$$

$$= 2^{16} \cdot 2^{16/2}\left[\cos\left(-\frac{16\pi}{4}\right) + i \sin\left(-\frac{16\pi}{4}\right)\right]$$

$$= 2^{16}2^{8}[\cos(-4\pi) + i \sin(-4\pi)]$$

$$= 2^{24}[\cos(-2 \cdot 2\pi) + i \sin(-2 \cdot 2\pi)] \quad \textit{(Substitute}$$
$$\textit{coterminal angles)}$$

$$= 2^{24}(\cos 0 + i \sin 0)$$

$$= 2^{24}(1 + 0i)$$

$$= 2^{24} = 16{,}777{,}216$$

(b) We first change $-1 + i\sqrt{3}$ to trigonometric form:

$$r = \sqrt{(-1)^2 + (\sqrt{3})^2} = \sqrt{4} = 2$$

and

$$\tan \theta = \frac{\sqrt{3}}{-1} = -\sqrt{3}$$

with $\theta' = \dfrac{\pi}{3}$ so that

$$\theta = \frac{3\pi}{3} - \frac{\pi}{3} = \frac{2\pi}{3}$$

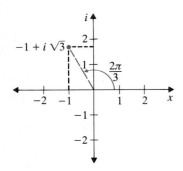

Figure 7.43

as seen in Fig. 7.43. Thus

$$-1 + i\sqrt{3} = 2\left(\cos\frac{2\pi}{3} + i \sin\frac{2\pi}{3}\right)$$

and

$$(-1 + i\sqrt{3})^{63} = \left[2\left(\cos\frac{2\pi}{3} + i \sin\frac{2\pi}{3}\right)\right]^{63}$$

$$= 2^{63}\left[\cos\left(\frac{63 \cdot 2\pi}{3}\right) + i \sin\left(\frac{63 \cdot 2\pi}{3}\right)\right]$$

$$= 2^{63}[\cos(21 \cdot 2\pi) + i \sin(21 \cdot 2\pi)]$$

$$= 2^{63}(\cos 0 + i \sin 0)$$

$$= 2^{63}(1 + 0i)$$

$$= 2^{63}$$

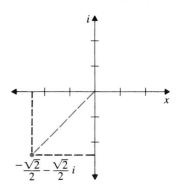

Figure 7.44

(c) We first change $-\dfrac{\sqrt{2}}{2} - \dfrac{\sqrt{2}}{2}i$ (Fig. 7.44) to trigonometric form:

$$r = \sqrt{\left(\frac{\sqrt{2}}{2}\right)^2 + \left(\frac{\sqrt{2}}{2}\right)^2} = \sqrt{\frac{4}{4}} = 1$$

and

$$\tan\theta = \frac{-\sqrt{2}}{-\sqrt{2}} = 1$$

with

$$\theta' = \frac{\pi}{4} \quad \text{and} \quad \theta = \frac{5\pi}{4}$$

Thus using de Moivre's Theorem, we have

$$\left(-\frac{\sqrt{2}}{2} - \frac{\sqrt{2}}{2}i\right)^{73} = \left[\cos\left(\frac{5\pi}{4}\right) + i\sin\left(\frac{5\pi}{4}\right)\right]^{73}$$

$$= \cos\left(\frac{73 \cdot 5\pi}{4}\right) + i\sin\left(\frac{73 \cdot 5\pi}{4}\right)$$

$$= \cos\left(\frac{365}{4}\pi\right) + i\sin\left(\frac{365}{4}\pi\right)$$

$$= \cos\left(91\frac{1}{4}\pi\right) + i\sin\left(91\frac{1}{4}\pi\right) \qquad \textit{(Since } 91\frac{1}{4}\pi = 92\pi - \frac{3}{4}\pi$$
$$\textit{and } 92\pi = 46(2\pi), \textit{ we}$$
$$\textit{know } 91\frac{1}{4}\pi \textit{ is}$$
$$\textit{coterminal with } -3\pi/4\textit{)}$$

$$= \cos\left(-\frac{3\pi}{4}\right) + i\sin\left(-\frac{3\pi}{4}\right) \qquad \textit{(Substitute coterminal}$$
$$\textit{angles again)}$$

$$= \cos\left(\frac{5\pi}{4}\right) + i\sin\left(\frac{5\pi}{4}\right)$$

$$= -\frac{\sqrt{2}}{2} - \frac{\sqrt{2}}{2}i$$

As we saw in Example 4, the trigonometric form is most helpful when evaluating a complex number with a natural number exponent.

Exercises 7.4

Change each of the following numbers to trigonometric form.

1. $-5 + 5i$

2. $3 - 3i$

3. $\sqrt{3} - i$

4. $1 - \sqrt{3}i$

5. $2 - 2\sqrt{3}i$

6. $3\sqrt{3} - 3i$

7. $1 - i$

8. $-1 + i$

9. $3 + 5i$

10. $5 + 3i$

11. $5i$

12. $-3i$

13. 7

14. 5

15. $-2i$

16. i

For the following pairs of complex numbers, find $z_1 \cdot z_2$, $z_1 \div z_2$, $z_1{}^4$, and $z_2{}^{12}$.

17. $z_1 = -5 + 5i$
$z_2 = \sqrt{3} - i$

18. $z_1 = 3 - 3i$
$z_2 = 1 - \sqrt{3}i$

19. $z_1 = 2 - 2\sqrt{3}i$
$z_2 = 1 - i$

20. $z_1 = 3\sqrt{3} - 3i$
$z_2 = -1 + i$

21. $z_1 = 3 + 5i$
$z_2 = 3 - 5i$

22. $z_1 = 5 + 3i$
$z_2 = -5 + 3i$

Evaluate each of the following.

23. $(\sqrt{2} - \sqrt{2}i)^{164}$

24. $(-\sqrt{2} + \sqrt{2}i)^{100}$

25. $(\sqrt{3} - i)^{93}$

26. $(1 - \sqrt{3}i)^{300}$

27. $(-1 - i)^{32}$

28. $(-1 - i)^{16}$

29. Prove that the reciprocal and the conjugate of $z = \cos \theta + i \sin \theta$ are the same number. Also find the reciprocal of $z = r(\cos \theta + i \sin \theta)$ in trigonometric form and $a + bi$ form.

30. Prove that $[r(\cos \theta - i \sin \theta)]^n = r^n(\cos n\theta - i \sin n\theta)$.

31. Evaluate $(\cos \theta + i \sin \theta)^2$ **(a)** as a binomial squared and **(b)** as an example of de Moivre's Theorem. Set the two results equal to establish the identities for $\cos 2\theta$ and $\sin 2\theta$.

32. Evaluate $(\cos \theta + i \sin \theta)^3$ **(a)** as a binomial cubed and **(b)** as an example of de Moivre's Theorem. Set the two results equal to find $\cos 3\theta$ as a function of $\cos \theta$ and $\sin 3\theta$ as a function of $\sin \theta$.

33. Use the method in Exercise 31 to find $\cos 4\theta$ as a function of $\cos \theta$ and $\sin 4\theta$ as a function of $\sin \theta$.

7.5 Complex Roots with Rational Exponents

In Section 7.4 we developed and used de Moivre's Theorem,

$$z^n = [r(\cos \theta + i \sin \theta)]^n = r^n(\cos n\theta + i \sin n\theta)$$

where n is a natural number. In this section we will extend de Moivre's Theorem to include rational numbers exponents for complex numbers. To do this, we first establish a few preliminary properties.

Theorem
Special Powers of
Complex Numbers

For $z = r(\cos \theta + i \sin \theta)$ and $z \neq 0$,

(a) $z^0 = 1$,

(b) $z^{-1} = r^{-1}(\cos \theta - i \sin \theta)$, and

(c) $z^{-n} = r^{-n}[\cos (n\theta) - i \sin (n\theta)]$, where n is a natural number.

PROOF:

(a) For $z = r(\cos \theta + i \sin \theta) \neq 0$ and $r \neq 0$ we have

$$\frac{z}{z} = \frac{r(\cos \theta + i \sin \theta)}{r(\cos \theta + i \sin \theta)}$$

$$z^{1-1} = \frac{r}{r}[\cos (\theta - \theta) + i \sin (\theta - \theta)]$$

$$z^0 = 1(\cos 0 + i \sin 0)$$

$$z^0 = 1(1 + i \cdot 0)$$

$$z^0 = 1 \cdot 1$$

Therefore

$$z^0 = 1$$

(b) For $z = r(\cos \theta + i \sin \theta)$ and $z \neq 0$ we have

$$z^{-1} = \frac{1}{z}$$

$$= \frac{1}{r(\cos \theta + i \sin \theta)} \qquad \textit{(Substitute cos 0 + i sin 0 for 1,}$$
$$\textit{since 1 = cos 0 + i sin 0)}$$

$$= \frac{\cos 0 + i \sin 0}{r(\cos \theta + i \sin \theta)} \qquad \textit{(Divide the complex numbers)}$$

$$= \frac{1}{r}[\cos (0 - \theta) + i \sin (0 - \theta)]$$

$$= r^{-1}[\cos(-\theta) + i \sin (-\theta)] \qquad \textit{(Use the identities for}$$
$$= r^{-1}(\cos \theta - i \sin \theta) \qquad\qquad \textit{cos } (-\theta) \textit{ and sin } (-\theta))$$

(c) For $z = r(\cos \theta + i \sin \theta)$ and $z \neq 0$ we have

$$z^{-n} = (z^n)^{-1} \qquad\qquad \textit{(Use de Moivre's Theorem)}$$

$$= \{r^n[\cos (n\theta) + i \sin (n\theta)]\}^{-1} \qquad \textit{(Use part (b) above)}$$

$$= r^{-n}[\cos (n\theta) - i \sin (n\theta)] \qquad\qquad\qquad \square$$

Therefore we can use de Moivre's Theorem from Section 7.4 and the theorem in this section to evaluate

$$z^n = [r(\cos \theta + i \sin \theta)]^n$$

when the exponent n is any integer. We will now use these properties to evaluate the complex numbers in an example.

EXAMPLE 1   For $z = 5\left(\cos \dfrac{\pi}{4} + i \sin \dfrac{\pi}{4}\right)$, find each of the following:

(a) z^{-1},   (b) z^{-4}, and   (c) z^{-17}.

Solution   Using property (c) of the theorem above and de Moivre's Theorem, we have

(a)
$$z^{-1} = \left[5\left(\cos \frac{\pi}{4} + i \sin \frac{\pi}{4}\right)\right]^{-1}$$

$$= 5^{-1}\left(\cos \frac{\pi}{4} - i \sin \frac{\pi}{4}\right)$$

$$= \frac{1}{5}\left(\frac{\sqrt{2}}{2} - i\frac{\sqrt{2}}{2}\right)$$

$$= \frac{1}{10}(\sqrt{2} - i\sqrt{2})$$

$$= \frac{\sqrt{2}}{10}(1 - i)$$

(b)
$$z^{-4} = \left[5\left(\cos \frac{\pi}{4} + i \sin \frac{\pi}{4}\right)\right]^{-4}$$

$$= 5^{-4}\left[\cos \left(\frac{4\pi}{4}\right) - i \sin \left(\frac{4\pi}{4}\right)\right]$$

$$= 5^{-4}(\cos \pi - i \sin \pi)$$

$$= \frac{1}{5^4}(-1 - 0)$$

$$= -\frac{1}{5^4} = -\frac{1}{625}$$

(c)
$$z^{-17} = \left[5\left(\cos \frac{\pi}{4} + i \sin \frac{\pi}{5}\right)\right]^{-17}$$

$$= 5^{-17}\left[\cos \left(\frac{17\pi}{4}\right) - i \sin \left(\frac{17\pi}{4}\right)\right]$$

$$= 5^{-17}\left[\cos \left(4\frac{1}{4}\right)\pi - i \sin \left(4\frac{1}{4}\right)\pi\right]$$

$$= 5^{-17}\left(\cos \frac{\pi}{4} - i \sin \frac{\pi}{4}\right)$$

$$= 5^{-17}\left(\frac{\sqrt{2}}{2} - i\frac{\sqrt{2}}{2}\right)$$

$$= \frac{\sqrt{2}}{2 \cdot 5^{17}}(1 - i)$$

■

Now we consider $z^{1/n}$, where n is a natural number. The following nth root theorem states the formula for $z^{1/n}$.

The *n*th Root Theorem

For $z = r(\cos \theta + i \sin \theta)$ with $z \neq 0$ and any natural number n, the n different nth roots of z are

$$z^{1/n} = r^{1/n}\left[\cos\left(\frac{\theta + 2k\pi}{n}\right) + i \sin\left(\frac{\theta + 2k\pi}{n}\right)\right]$$

where $k \in \{0, 1, 2, 3, \ldots, (n-1)\}$ or

$$z^{1/n} = r^n\left[\cos\left(\frac{\theta + 360°k}{n}\right) + i \sin\left(\frac{\theta + 360°k}{n}\right)\right]$$

where $k \in \{0, 1, 2, 3, \ldots, (n-1)\}$.

We will use this theorem now but save the proof until the end of this section. The results of this theorem are consistent with the pattern of de Moivre's Theorem with the exception of the angle. This specific angle is necessary to account for the periodicity of the trigonometric functions, which produce n different roots (or zeros). Earlier, we noticed that for θ such that

$$0 \leq \frac{\theta}{n} < 2\pi$$

it was necessary for

$$0 \leq \theta < 2\pi n$$

The more general solution to a trigonometric equation $\theta + 2k\pi$, where k is an integer, is also bounded by 0 and $2\pi n$. Thus we have

$$0 \leq \theta + 2k\pi < 2\pi n$$

and

$$0 \leq \frac{\theta + 2k\pi}{n} < 2\pi$$

To stay between 0 and 2π, we must limit k so that the angles are

$$\frac{\theta + 2k\pi}{n} \qquad \text{where} \qquad k \in \{0, 1, 2, 3, \ldots, (n-1)\}$$

as stated in the nth root theorem. This representation is described in more detail in the proof at the end of this section.

We will now use the nth root theorem in the following problems.

EXAMPLE 2

Find **(a)** the three values of $z^{1/3}$ when $z = 8(\cos 270° + i \sin 270°)$ and **(b)** the five values of $(-3 - 3i)^{1/5}$.

Solution   Using the nth root theorem, we have

(a) $z^{1/3} = [8(\cos 270° + i \sin 270°)]^{1/3}$

$= [2^3(\cos 270° + i \sin 270°)]^{1/3}$

$= (2^3)^{1/3}\left[\cos\left(\dfrac{270° + 360°k}{3}\right) + i \sin\left(\dfrac{270° + 360°k}{3}\right)\right]$

$= 2[\cos(90° + 120°k) + i \sin(90° + 120°k)]$ where $k \in \{0, 1, 2\}$

Now we evaluate $z^{1/3}$ for each value of k:

$z^{1/3} = \begin{cases} 2[\cos(90°) + i \sin(90°)] & \text{for}\quad k = 0 \\ 2[\cos(90° + 120°) + i \sin(90° + 120°)] & \text{for}\quad k = 1 \\ 2[\cos(90° + 240°) + i \sin(90° + 240°)] & \text{for}\quad k = 2 \end{cases}$

$= \begin{cases} 2[\cos(90°) + i \sin(90°)] \\ 2[\cos(210°) + i \sin(210°)] \\ 2[\cos(330°) + i \sin(330°)] \end{cases}$

$= \begin{cases} 2(0 + i) = 2i \\ 2\left(-\dfrac{\sqrt{3}}{2} - \dfrac{1}{2}i\right) = -\sqrt{3} - i \\ 2\left(\dfrac{\sqrt{3}}{2} - \dfrac{1}{2}i\right) = \sqrt{3} - i \end{cases}$

(b) We first change $-3 - 3i$ into trigonometric form:

$$r = \sqrt{(-3)^2 + (-3)^2} = \sqrt{18} = 18^{1/2}$$

and $\tan \theta = \dfrac{-3}{-3} = 1$ so that $\theta' = \dfrac{\pi}{4}$ and $\theta = \dfrac{5\pi}{4}$. Thus

$$-3 - 3i = 18^{1/2}\left(\cos\frac{5\pi}{4} + i \sin\frac{5\pi}{4}\right).$$

Using the nth root theorem, we have

$$(-3 - 3i)^{1/5} = \left[18^{1/2}\left(\cos\frac{5\pi}{4} + i \sin\frac{5\pi}{4}\right)\right]^{1/5}$$

$$= (18^{1/2})^{1/5}\left[\cos\frac{1}{5}\left(\frac{5\pi}{4} + 2k\pi\right) + i \sin\frac{1}{5}\left(\frac{5\pi}{4} + 2k\pi\right)\right]$$

$$= 18^{1/10}\left[\cos\left(\frac{\pi}{4} + \frac{2k\pi}{5}\right) + i \sin\left(\frac{\pi}{4} + \frac{2k\pi}{5}\right)\right]$$

$$= 18^{1/10}[\cos(45° + 72k°) + i \sin(45° + 72k°)]$$

where $k \in \{0, 1, 2, 3, 4\}$.

Now evaluating for each value of k, we have

$$(-3 - 3i)^5 = \begin{cases} 18^{1/10}(\cos 45° + i \sin 45°) & \text{for} \quad k = 0 \\ 18^{1/10}(\cos 117° + i \sin 117°) & \text{for} \quad k = 1 \\ 18^{1/10}(\cos 189° + i \sin 189°) & \text{for} \quad k = 2 \\ 18^{1/10}(\cos 261° + i \sin 261°) & \text{for} \quad k = 3 \\ 18^{1/10}(\cos 333° + i \sin 333°) & \text{for} \quad k = 4 \end{cases}$$ ■

EXAMPLE 3

Find the two square roots of $2 - 2i$.

Solution We first change $2 - 2i$ into trigonometric form and then evaluate $(2 - 2i)^{1/2}$:

$$r = \sqrt{2^2 + (-2)^2} = \sqrt{8} = 2\sqrt{2} = 2 \cdot 2^{1/2} = 2^{3/2}$$

and $\tan \theta = \dfrac{-2}{2} = -1$ so that $\theta' = \dfrac{\pi}{4}$ and $\theta = -\dfrac{\pi}{4}$. Thus

$$2 - 2i = 2^{3/2}\left[\cos\left(-\frac{\pi}{4}\right) + i \sin\left(-\frac{\pi}{4}\right)\right]$$

and

$$(2 - 2i)^{1/2} = (2^{3/2})^{1/2}\left[\cos\frac{1}{2}\left(-\frac{\pi}{4} + 2k\pi\right) + i \sin\frac{1}{2}\left(-\frac{\pi}{4} + 2k\pi\right)\right] \text{ for } k \in \{0, 1\}$$

$$= 2^{3/4}\left[\cos\left(-\frac{\pi}{8} + k\pi\right) + i \sin\left(-\frac{\pi}{8} + k\pi\right)\right]$$

$$= \begin{cases} 2^{3/4}\left[\cos\left(-\dfrac{\pi}{8}\right) + i \sin\left(-\dfrac{\pi}{8}\right)\right] \\ 2^{3/4}\left[\cos\left(\dfrac{7\pi}{8}\right) + i \sin\left(\dfrac{7\pi}{8}\right)\right] \end{cases}$$ ■

We will now consider the graphical behavior of these complex roots. In the following example we will graph the solutions of prior examples.

EXAMPLE 4

Graph **(a)** the square roots of $2 - 2i$ (Example 3), **(b)** the three solutions to $z^{1/3}$ when $z = 8(\cos 270° + i \sin 270°)$ (Example 2a), and **(c)** the five solutions to $(-3 - 3i)^{1/5}$ (Example 2b).

Solution

(a)

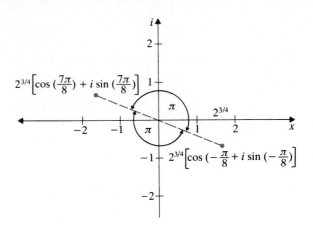

(b)

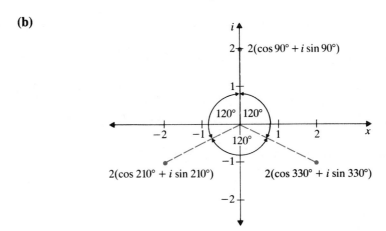

(c)

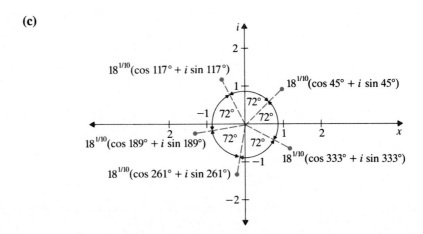

We note that in any graph in Example 4 the angle between each pair of adjacent complex nth roots (zeros) is the same and equal to $2\pi/n = 360°/n$. Thus in part (a), each root is $2\pi/2 = \pi$ or $180°$ apart; in part (b), each root is $360°/3 = 120°$ apart; and in part (c), each root is $2\pi/5 = 72°$ apart. This interesting geometric pattern is the result of $2k\pi/n$, for $k \in \{1, 2, 3, \ldots, (n - 1)\}$, being added to the angle θ in the nth root theorem. Thus the angle of each new root is $2\pi/n$ more than the previous angle.

We will now use the nth root theorem to solve some other problems.

EXAMPLE 5   Find all seven solutions for $x^7 - 1 = 0$.

Solution   We first solve for x:

$$x^7 - 1 = 0$$

$$x^7 = -1$$

$$x = (-1)^{1/7} \qquad \textit{(Substitute trigonometric form of } -1 = 1(\cos \pi + i \sin \pi))$$

$$x = [1(\cos \pi + i \sin \pi)]^{1/7} \quad \textit{(Use the nth root theorem)}$$

$$x = 1^{1/7}\left[\cos \frac{1}{7}(\pi + 2k\pi) + i \sin \frac{1}{7}(\pi + 2k\pi)\right]$$

$$x = \cos \frac{(1 + 2k)\pi}{7} + i \sin \frac{(1 + 2k)\pi}{7} \qquad \text{for} \quad k \in \{0, 1, 2, 3, 4, 5, 6\}$$

∎

We can now use the nth root theorem along with de Moivre's Theorem to simplify a complex number with a rational exponent. For $z^{p/q}$ we can calculate $(z^p)^{1/q}$ with $q > 0$ by first using de Moivre's Theorem and then the nth root theorem. When $p/q < 0$, the property for z^{-n} is also needed to complete the computation.

The next example shows how to simplify complex numbers with rational exponents.

EXAMPLE 6   Evaluate $z^{2/3}$ when

$$z = 8\left(\cos \frac{3\pi}{4} + i \sin \frac{3\pi}{4}\right)$$

Solution

$$z^{2/3} = (z^2)^{1/3}$$

$$= \left[8^2 \left(\cos \frac{2 \cdot 3\pi}{4} + i \sin \frac{2 \cdot 3\pi}{4} \right) \right]^{1/3}$$

$$= \left[64 \left(\cos \frac{3\pi}{2} + i \sin \frac{3\pi}{2} \right) \right]^{1/3}$$

$$= 64^{1/3} \left[\cos \frac{1}{3} \left(\frac{3\pi}{2} + 2k\pi \right) + i \sin \frac{1}{3} \left(\frac{3\pi}{2} + 2k\pi \right) \right]$$

$$= 4 \left[\cos \left(\frac{\pi}{2} + \frac{2k\pi}{3} \right) + i \sin \left(\frac{\pi}{2} + \frac{2k\pi}{3} \right) \right] \quad \text{for} \quad k \in \{0, 1, 2\}$$

or

$$= 4[\cos (90 + 120k)° + i \sin (90 + 120k)°] \quad \text{for} \quad k \in \{0, 1, 2\}$$

■

We will now restate and prove the nth root theorem.

The nth Root Theorem

For $z = r(\cos \theta + i \sin \theta)$ with $z \neq 0$ and any natural number n, the n different nth roots of z are

$$z^{1/n} = r^{1/n} \left[\cos \left(\frac{\theta + 2k\pi}{n} \right) + i \sin \left(\frac{\theta + 2k\pi}{n} \right) \right]$$

where $k \in \{0, 1, 2, 3, \ldots, (n - 1)\}$ or

$$z^{1/n} = r^n \left[\cos \left(\frac{\theta + 360°k}{n} \right) + i \sin \left(\frac{\theta + 360°k}{n} \right) \right]$$

where $k \in \{0, 1, 2, 3, \ldots, (n - 1)\}$.

PROOF: To find $z^{1/n}$, we let $z^{1/n} = w$,

$$z = r(\cos \theta + i \sin \theta), \quad \text{and} \quad w = p(\cos \alpha + i \sin \alpha)$$

In this proof we will show that $w = z^{1/n}$ when

(a) $p = r^{1/n}$ and

(b) $\alpha = \dfrac{\theta + 2k\pi}{n}$ where $n \in \{0, 1, 2, \ldots, n - 2, n - 1\}$.

To start, we can write $z^{1/n} = w$ as

$$z = w^n$$ *(Substitute and use de Moivre's Theorem)*

$$r(\cos\theta + i\sin\theta) = [p(\cos\alpha + i\sin\alpha)]^n$$

$$r(\cos\theta + i\sin\theta) = p^n[\cos(n\alpha) + i\sin(n\alpha)] \tag{1}$$

$$\frac{r(\cos\theta + i\sin\theta)}{p^n[\cos(n\alpha) + i\sin(n\alpha)]} = \frac{p^n[\cos(n\alpha) + i\sin(n\alpha)]}{p^n[\cos(n\alpha) + i\sin(n\alpha)]}$$ *(Divide both sides by $p^n[\cos(n\alpha) + i\sin(n\alpha)]$)*

$$\left(\frac{r}{p^n}\right)[\cos(\theta - n\alpha) + i\sin(\theta - n\alpha)] = 1$$

$$\left(\frac{r}{p^n}\right)[\cos(\theta - n\alpha) + i\sin(\theta - n\alpha)] = 1 + 0i \tag{2}$$

(a) We will now establish that $r^{1/n} = p$. Since these are two equal complex numbers, the i coefficients in equation (2) are equal. Thus

$$\left(\frac{r}{p^n}\right)\sin(\theta - n\alpha) = 0$$

and with $r \neq 0$ we have $\theta - n\alpha = k\pi$, where k is an integer. Using this angle value for the equal real parts of equation (2), we have

$$\left(\frac{r}{p^n}\right)\cos(\theta - n\alpha) = 1$$

$$\left(\frac{r}{p^n}\right)\cos(k\pi) = 1$$

$$\left(\frac{r}{p^n}\right)(\pm 1) = 1$$

$$r = \pm p^n$$

Since $r > 0$, we have $r = p^n$ and $r^{1/n} = p$.

(b) Now we will establish that $\alpha = \dfrac{\theta + 2k\pi}{n}$, where $n \in \{0, 1, 2, \ldots, n-2, n-1\}$. To do this, we use equation (1) above and set the real parts equal and also set the coefficients of i equal. We have

$$r\cos\theta = p^n\cos n\alpha \quad\quad \text{and} \quad\quad r\sin\theta = p^n\sin\alpha$$

Since $r = p^n \neq 0$, from part (a) we have

$$\cos\theta = \cos n\alpha \quad\quad \text{and} \quad\quad \sin\theta = \sin n\alpha$$

both of which have a solution of

$$\theta + 2k\pi = n\alpha \quad\quad \text{where} \quad k \text{ is an integer}$$

Dividing by n, we have

$$\frac{\theta + 2k\pi}{n} = \alpha \quad\quad \text{where} \quad k \text{ is an integer}$$

Since we are solving for α in the interval $[0, 2\pi)$, we will need to restrict the values of k. When $|k| = n$, the value of α is

$$\alpha = \frac{\theta + 2n\pi}{n} = \frac{\theta}{n} + 2\pi$$

Thus when $k = n$, the value of α is greater than 2π. Since any positive integer k less than n would be in the desired interval, we know that k is an integer such that $0 \leq k \leq n - 1$. We are also finding the n zeros or roots, so we know there must be n of zeros and we have $k \in \{0, 1, 2, \ldots, n - 2, n - 1\}$. Therefore $\alpha = \frac{\theta + 2k\pi}{n}$, where $k \in \{0, 1, 2, \ldots, n - 2, n - 1\}$.

We have now finished both parts of the proof, and

$$z^{1/n} = r^{1/n}\left[\cos\left(\frac{\theta + 2k\pi}{n}\right) + i \sin\left(\frac{\theta + 2k\pi}{n}\right)\right]$$

where $k \in \{0, 1, 2, 3, \ldots, (n - 1)\}$.   □

As we have seen in this section, the nth root theorem provides both algebraic and geometric patterns for the roots of complex numbers.

Exercises 7.5

Evaluate z^0, z^{-1}, z^{-2}, and the two values of $z^{1/2}$ for each of the following.

1. $z = \sqrt{3} + i$
2. $z = 1 - i$
3. $z = 4 - 4i$
4. $z = 1 - \sqrt{3}i$
5. $z = 2 - 2\sqrt{3}i$
6. $z = -\sqrt{3} + i$
7. $z = 3 + 4i$
8. $z = 12 - 5i$
9. $z = -5 + 12i$
10. $z = -5 - 12i$
11. $z = 3 - i$
12. $z = -5 + 2i$

Find all of the indicated roots.

13. the cube roots of 8
14. the cube roots of 1
15. the cube roots of i
16. the cube roots of $27i$
17. the square roots of $4i$
18. the square roots of i
19. the square roots of $-i$
20. the square roots of $-16i$
21. the fifth roots of i
22. the fifth roots of -32
23. the fourth roots of -64
24. the fourth roots of $-i$
25. the sixth roots of $-i$
26. the sixth roots of -1

27. the square roots of $\sqrt{2} - \sqrt{2}i$
28. the square roots of $\sqrt{3} - i$
29. the cube roots of $5 + 12i$
30. the cube roots of $3 + 4i$

Find and graph all solutions for each of the following equations.

31. $x^3 - i = 0$
32. $x^4 - i = 0$
33. $x^4 - 1 = 0$
34. $x^3 + 1 = 0$
35. $x^2 - i = 0$
36. $x^2 + 4i = 0$
37. $x^5 - 1 = 0$
38. $x^5 - i = 0$

Use de Moivre's Theorem and the nth root theorem to evaluate each of the following complex numbers with rational exponents.

39. $(2 + 2i)^{2/3}$
40. $(\sqrt{3} + i)^{2/3}$
41. $(\sqrt{3} - i)^{5/3}$
42. $(2 + 2i)^{7/4}$
43. $(3 - 5i)^{2/3}$
44. $(5 - 3i)^{3/4}$
45. $(5 - 12i)^{7/2}$
46. $(-12 + 5i)^{5/3}$

Find the other roots of z when z_1 is one of the indicated roots for each of the following.

47. $z_1 = 2(\cos 15° + i \sin 15°)$ is one of the cube roots of z.

48. $z_1 = 3(\cos 20° + i \sin 20°)$ is one of the cube roots of z.

49. $z_1 = \cos \dfrac{\pi}{3} + i \sin \dfrac{\pi}{3}$ is one of the fourth roots of z.

50. $z_1 = \cos \dfrac{\pi}{6} + i \sin \dfrac{\pi}{6}$ is one of the fourth roots of z.

51. $z_1 = 3(\cos 20° + i \sin 20°)$ is one of the square roots of z.

52. $z_1 = 5(\cos 15° + i \sin 15°)$ is one of the square roots of z.

53. $z_1 = 4\left(\cos \dfrac{\pi}{4} + i \sin \dfrac{\pi}{4}\right)$ is one of the cube roots of z.

54. $z_1 = 3\left(\cos \dfrac{\pi}{2} + i \sin \dfrac{\pi}{2}\right)$ is one of the cube roots of z.

55. Find z for Exercises 47 and 49.

56. Find z for Exercises 48 and 50.

57. Find z for Exercises 51 and 53.

58. Find z for Exercises 52 and 54.

CHAPTER SUMMARY AND REVIEW

KEY TERMS

Polar coordinates

Graphing polar equations

Symmetry tests with polar coordinates

Vectors

Vector properties

Vector notation

Definition of the dot product

Trigonometric form of a complex number

Multiplying and dividing complex numbers

de Moivre's Theorem

The nth root theorem

CHAPTER EXERCISES

Change each of the following points from Cartesian coordinates to polar coordinates.

1. $(-\sqrt{2}, \sqrt{2})$ **2.** $(\sqrt{3}, -1)$

3. $(5, 4)$ **4.** $(6, -8)$

Change each of the following points from polar coordinates to Cartesian coordinates.

5. $(3, -45°)$ **6.** $(2, -30°)$

7. $(-2, 120°)$ **8.** $(-1, 225°)$

Change the following equations into equivalent equations in polar coordinates.

9. $x^2 + y^2 = 8$ **10.** $x^2 - y^2 = 9$

11. $x = 3y$ **12.** $y = 17$

Change each of the following equations to an equation in Cartesian coordinates.

13. $r = 4 \csc \theta$ **14.** $r = \frac{1}{2} \sec \theta$

15. $r^2 = 2 - r \cos \theta$ **16.** $r^2 = 2 - r \sin \theta$

Sketch the graph of each of the following polar equations.

17. $r = 1 - \cos \theta$ **18.** $r = 3 - 3 \sin \theta$

19. $r = 3 \sin 2\theta$ **20.** $r = 1 - \sin \theta$

21. $r^2 = 4 \sin 2\theta$ **22.** $r = 2 + \sin \theta$

23. $r = 3 \sec \theta$ **24.** $r^2 = 4 \cos \theta$

25. $r = 2 \cos \left(\theta - \dfrac{\pi}{4} \right)$ **26.** $r = 3 \sin \left(\theta + \dfrac{\pi}{3} \right)$

Find $v + w$, $v - w$, cv, and $v \cdot w$ for each of the following.

27. $v = 3i - 4j$
$w = 2i + 3j$
$c = -2$

28. $v = 5i + 6j$
$w = -3i + 2j$
$c = \frac{1}{3}$

29. $v = \langle -1, 7 \rangle$
$w = \langle -7, -1 \rangle$
$c = 3$

30. $v = \langle 5, -1 \rangle$
$w = \langle -5, 1 \rangle$
$c = -2$

31. Find the angle θ between v and w in **(a)** Exercise 27 and **(b)** Exercise 29.

32. Find the angle θ between v and w in **(a)** Exercise 28 and **(b)** Exercise 30.

33. An airplane flies for $\frac{3}{4}$ of an hour with a heading of 62° and then changes course and flies for 1 hour with a heading of 338°. The airplane's average speed is 275 miles per hour, and the wind averages 43 miles per hour from the west with a heading of 83°. What are the heading and distance from the end of the second part of this trip back to the starting point?

34. Two Girl Scouts participate in orienteering competition. The instructions are
(a) heading 15°, distance 110 feet,
(b) heading 125°, distance 300 feet, and
(c) heading 275°, distance 290 feet.
After these instructions are completed, what are the heading and distance back to the starting point?

Change each of the following complex numbers to trigonometric form and find $z \cdot w$, $\dfrac{z}{w}$, z^3, and $w^{1/3}$.

35. $z = 2 - 2i$
$w = \sqrt{3} - i$

36. $z = \sqrt{2} + \sqrt{2}i$
$w = -\sqrt{3} + i$

Evaluate each of the following.

37. $(-2 + 2i)^5$

38. $(-\sqrt{3} - i)^6$

39. $(\sqrt{2} + \sqrt{2}i)^{-2}$

40. $(\sqrt{3} + i)^{1/3}$

41. $(1 - i)^{1/4}$

42. $(5 - 12i)^{3/2}$

43. $(3 - 4i)^{2/3}$

44. $(1 + i)^{3/2}$

45. Find the six sixth roots of $(1 - i)$. Graph these roots.

46. Find the three cube roots of $-8i$. Graph these roots.

Solve and graph the solutions of the following equations.

47. $x^4 - i = 0$

48. $x^5 - 32 = 0$

49. $x^3 + 125 = 0$

50. $x^4 + 64i = 0$

8

Solving Systems of Equations

The Continuation Method

In this chapter we will study several classical methods for solving systems of linear and non-linear equations. When large or more difficult systems of equations are encountered, finding the solution, though not impossible, may often be very time-consuming. Even computer solutions of these more difficult systems use what are called hit-or-miss local methods that rely on an interative modification of initial estimates. Reliability is compromised when there are multiple solutions or a poor initial choice is made.

In response to these problems, Dr. Alexander Morgan, a senior research scientist in the Mathematics Department of the General Motors Research Laboratories, has developed the "continuation method." The continuation method is a global method of solution. Global methods do not require an initial estimate of the solution. The continuation method is not only global but is also exhaustive, since it generates convergence to all solutions. The proof of this convergence rests on principles from the area of mathematics called differential topology.

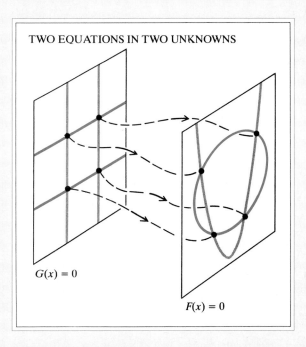

TWO EQUATIONS IN TWO UNKNOWNS

$G(x) = 0$

$F(x) = 0$

The two pairs of parallel lines of $G(x) = 0$ evolve into the parabola and ellipse of $F(x) = 0$.

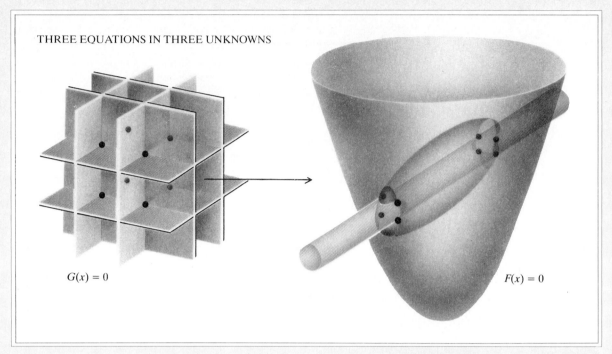

THREE EQUATIONS IN THREE UNKNOWNS

$G(x) = 0$

$F(x) = 0$

The three pairs of parallel planes of $G(x) = 0$ evolve into the paraboloid, ellipsoid and cylinder of $F(x) = 0$.

Here simply is how the continuation method works. To solve the system $F(x) = 0$, we first generate a simpler system $G(x) = 0$ that we can solve and evolve back into $F(x) = 0$. Dr. Morgan has developed a rapid and reliable method for finding the simpler $G(x) = 0$, using a theorem established by Garcia and Zangwill and some ideas from algebraic geometry (homogeneous coordinates and complex projective space). The result of Dr. Morgan's efforts is a practical numerical method based on mathematical principles with innate reliability.

The transition from a simple $G(x) = 0$ to the final $F(x) = 0$ can be shown graphically. The simplicity of $G(x) = 0$ is reflected in its linear structure (seen as lines and planes). The more complicated nonlinear $F(x) = 0$ is seen in the curvature of the final shapes in each figure.

The immediate application at General Motors is in mechanical design. The continuation method finds all eight solutions to three quadratic equations in a few tenths of a second. This is fast enough for computer-aided design on a moment-to-moment basis. (Algorithms based on this method are critical to GMSOLID, an interactive design system that models the geometric characteristics of automotive parts.)

Other uses are also foreseen. As Dr. Morgan states, "Continuation methods, although well-known to mathematicians are not widely used in science and engineering. Acoustics, kinematics, and non-linear circuit design are just a few fields that could benefit immediately. I expect to see a much greater use of this tool in the future." (See *Solving Polynomial Equations* by Alexander Morgan, © 1987 by Prentice Hall.)

SOURCE: "The Continuation Method," General Motors Research and Development advertising, Detroit, Michigan. Reprinted by permission.

Often, solutions to the practical problems of business, economics, science, and engineering are based on several considerations or criteria. These criteria can often be represented with more than one equation in two or more variables. More than one equation with more than one variable is called a **system of equations**. The solution to a system of equations is the solution common to all equations in the system.

As an example, we will consider the following business application. A toy company has just completed a test market survey that indicates the following mutually existing conditions:

(a) The revenue in dollars received by the company increases with the number of toys produced each month and can be approximated by the following equation: Revenue $= 150 + e^{x/100}$.

(b) The demand in dollars for this toy decreases with the number of toys produced each month and can be approximated by the following equation: Demand $= 5000 - e^{x/100}$.

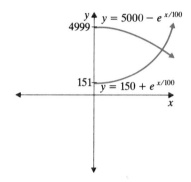

Figure 8.1

Since both revenue and demand are dollar values, we substitute y for revenue and demand in these equations. Thus these production criteria lead to the following system of equations:

$$y = 150 + e^{x/100}$$
$$y = 5000 - e^{x/100}.$$

Sketching the graph of the two equations in this system of equations (Fig. 8.1), we see that revenues increase with increased production and demand decreases with increased production. We also note that the point where the two graphs intersect is the point at which demand dollars equal revenue dollars. This point where the two curves intersect is called the *solution to the system*, since as the point of intersection, it makes both equations true.

The solution of this system of equations will provide the production level of x toys that will maximize revenue and not exceed product demand. We will solve this example later in this section.

This chapter covers some of the classical ways to find the solution to a system of equations.

8.1   Solution by Substitution and Elimination

In this section we will review some methods for solving systems of equations and expand the use of these methods to new and more challenging systems of equations.

When solving a system of equations, we can have three types of solutions:

(a) a finite number of common solutions,
(b) an infinite number of common solutions, and
(c) no common solution.

Table 8.1 delineates these solutions, with their names and characteristics after algebraic solution.

TABLE 8.1

Type of Solution	Name of This Type	Algebraic Results when Solving a System
A finite set of common solutions	A consistent system	One point or a finite number of unique points
An infinite set of common solutions	A dependent system	A true statement such as $0 = 0$
No common solutions, $\emptyset$	An inconsistent system	A false statement such as $0 = 7$ or $x = 5i$ (since all variables must be real numbers)

Solution by Substitution

To solve a system of equations by substitution, follow these guidelines.

Guidelines for Substitution

To solve a system of equations using substitution (with two equations and two unknowns),

(a) solve one equation for one variable;
(b) substitute this expression from (a) for that variable in the other equation and solve for the value of the remaining variable;
(c) substitute this value from (b) into the initial equation from (a) to find the value of the first variable;
(d) then state the solution.
(e) The solution(s) can be checked in both equations.

Using these guidelines, we will solve the revenue-demand problem from the beginning of this chapter.

EXAMPLE 1   Find the best value for monthly production x that will maximize the revenue $y = 150 + e^{x/100}$ without overproducing for the demand in dollars $y = 5000 - e^{x/100}$.

Solution   As we saw earlier in this chapter, we will need to find the solution to the system

$$y = 150 + e^{x/100}$$
$$y = 5000 - e^{x/100}$$

Since the first equation is solved for y, we substitute this expression into the second equation for y and solve:

$$150 + e^{x/100} = 5000 - e^{x/100} \qquad \textit{(Use the addition property of equality)}$$
$$2e^{x/100} = 4850 \qquad \textit{(Use the multiplication property of equality)}$$
$$e^{x/100} = 2425 \qquad \textit{(Use logarithmic form)}$$
$$\frac{x}{100} = \ln 2425 \qquad \textit{(Use the multiplication property of equality)}$$
$$x = 100 \ln 2425$$

Substituting, we have

$$y \approx 150 + e^{\ln 2425} = 2575$$

Thus monthly production should be targeted at 779 toys with a monthly income of $2575. ∎

The system solved in Example 1 is a nonlinear system, since the equations are not linear equations. Example 2 is an example of a linear system, since both equations are linear.

EXAMPLE 2   Solve the system

$$x - y = 5$$
$$3x - 2y = 20$$

Solution   For this system we solve the first equation for x:

$$x = y + 5$$

Substituting this value into the second equation, we have

$$3x - 2y = 20 \qquad \textit{(Substitute } y + 5 \textit{ for } x\textit{)}$$
$$3(y + 5) - 2y = 20 \qquad \textit{(Distributive property)}$$
$$3y + 15 - 2y = 20 \qquad \textit{(Adding like terms)}$$
$$y + 15 = 20 \qquad \textit{(Addition property of equality)}$$
$$y = 5$$

Now substituting $y = 5$ into $x = y + 5$, we have $x = 5 + 5 = 10$. Therefore the two lines in the system

$$x - y = 5$$
$$3x - 2y = 20$$

have one common point of intersection, $(x, y) = (10, 5)$. This is a consistent system. ∎

EXAMPLE 3   Solve the system

$$\log_5 x = y + 2$$
$$\log_5 x = y^2$$

Solution   Since $\log_5 x$ is in both equations, substitute $\log_5 x = y + 2$ into the second equation. Then

$$\log_5 x = y^2 \qquad \qquad \textit{(Substituting } y + 2 \text{ for } \log_5 x)$$
$$y + 2 = y^2 \qquad \qquad \textit{(Addition property of equality)}$$
$$0 = y^2 - y - 2 \qquad \textit{(Factor)}$$
$$0 = (y - 2)(y + 1) \qquad \textit{(The zero factor theorem)}$$
$$y = 2 \quad \text{or} \quad y = -1$$

Now substitute these values for y into the first equation,

$$\log_5 x = y + 2$$

$y = 2$	$y = -1$	*(Substitution)*
$\log_5 x = 2 + 2$	$\log_5 x = -1 + 2$	*(Adding and subtracting like terms)*
$\log_5 x = 4$	$\log_5 x = 1$	*(Change to exponential form)*
$5^4 = x$	$5^1 = x$	*(Simplifying)*
$625 = x$	$5 = x$	

This is a consistent system with solutions of $(625, 2)$ and $(5, -1)$. ∎

EXAMPLE 4   Solve the system

$$x^2 + y^2 = 1$$
$$4x^2 + 9y^2 = 1$$

Solution   We take the first equation and solve for x^2, then substitute this expression into the second equation.
    The first equation:

$$x^2 + y^2 = 1 \qquad \textit{(Addition property of equality)}$$
$$x^2 = 1 - y^2 \qquad \textit{(Now use the second equation)}$$

The second equation:

$$4x^2 + 9y^2 = 1 \qquad \textit{(Substitute } 1 - y^2 \textit{ for } x^2\textit{)}$$
$$4(1 - y^2) + 9y^2 = 1 \qquad \textit{(Distributive property)}$$
$$4 - 4y^2 + 9y^2 = 1 \qquad \textit{(Subtract like terms)}$$
$$4 + 5y^2 = 1 \qquad \textit{(Addition property of equality)}$$
$$5y^2 = -3 \qquad \textit{(Division property of equality)}$$
$$y^2 = -\frac{3}{5} \qquad \textit{(Square root method)}$$
$$y = \pm\sqrt{-\frac{3}{5}} = \pm i\sqrt{\frac{3}{5}}$$

Since y is not a real number, there is no solution, and this is an inconsistent system.

■

Solution by Elimination

We will now solve similar systems of equations by a method called elimination.

Guidelines for Elimination

To solve a system of equations by elimination (with two equations and two variables),

(a) add convenient multiples of the two equations to eliminate one variable from the system;
(b) solve this equation to find one variable;
(c) substitute this value into either equation to find the other variable;
(d) then state the solution.
(e) The solution(s) will check in both equations.

EXAMPLE 5   Use elimination to solve

$$x^2 + y^2 = 2$$
$$2x^2 + 3y^2 = 6$$

Solution   We multiply both sides of the first equation by -2 and add the result to the second equation. Notice how this eliminates the x^2 when the two equations are added.

$$x^2 + y^2 = 2 \xrightarrow{(-2)} -2x^2 - 2y^2 = -4 \qquad \textit{(Multiplication property of equality)}$$
$$2x^2 + 3y^2 = 6 \longrightarrow \underline{2x^2 + 3y^2 = 6} \qquad \textit{(Addition property of equality)}$$
$$y^2 = 2$$
$$y = \pm\sqrt{2} \qquad \textit{(Square root method)}$$

Then, substituting $y = \pm\sqrt{2}$ into $x^2 + y^2 = 2$, we have

$$x^2 + (\pm\sqrt{2})^2 = 2$$
$$x^2 + 2 = 2$$
$$x^2 = 0$$
$$x = 0$$

Therefore the solution of this consistent system of equations is $(0, \sqrt{2})$ and $(0, -\sqrt{2})$.   ■

EXAMPLE 6   Find the solution to

$$\log x^2 = y + 3$$
$$\log x = y - 1$$

Solution   We multiply the second equation by -1 and add equations to eliminate y:

$$\log x^2 = y + 3 \quad\longrightarrow\quad \log x^2 = y + 3$$
$$\log x = y - 1 \quad\xrightarrow{(-1)}\quad -\log x = -y + 1 \qquad \textit{(Addition property of equality)}$$
$$\overline{\log x^2 - \log x = 4} \qquad \textit{(Properties of logarithms)}$$
$$\log \frac{x^2}{x} = 4$$
$$\log x = 4 \qquad \textit{(Reducing)}$$
$$x = 10^4 \qquad \textit{(Exponential form)}$$

Then substituting into $\log x = y - 1$, we have

$$\log 10^4 = y - 1 \qquad \textit{(Properties of logarithms)}$$
$$4 \log 10 = y - 1 \qquad \textit{($log_{10}10 = 1$)}$$
$$4 = y - 1 \qquad \textit{(Addition property of equality)}$$
$$5 = y$$

Therefore the solution to this consistent system is $(10^4, 5) = (10000, 5)$.   ■

As we saw in the first example of this chapter, the solution to a system of equations in two variables is the intersection of the two curves represented by the two equations in the system. Although this intersection could be found by graphing both equations, a graphical solution is often less accurate or impractical in comparison to the precision of the solution found by substitution or elimination.

EXAMPLE 7   Find the solution to

$$x^2 - y^2 = 5$$
$$x^2 - y^2 = 4$$

Solution We multiply the second equation by -1 and add equations:

$$\begin{array}{rcl} x^2 - y^2 = 5 & \longrightarrow & x^2 - y^2 = 5 \\ x^2 - y^2 = 4 & \xrightarrow{(-1)} & -x^2 + y^2 = -4 \\ \hline & & 0 = 1 \end{array}$$

But since $0 \neq 1$, we have a false statement of $0 = 1$. Thus this is an inconsistent system with no solution.   ■

Sometimes it is easier to use elimination, and other times it is easier to use substitution. Elimination works well when the powers of each variable are the same in both equations. This makes it easier to add and subtract, since we are adding and subtracting like terms. Substitution is best when the powers of the variables in one equation are not the same as the powers of the variables in the second equation.

EXAMPLE 8   Find the solution to

$$x^2 + y^2 = 4$$
$$x - y = 2$$

Solution Since the variables in the first equation are different powers than those in the second equation, this system can be best solved by substitution. From the second equation $x - y = 2$ we have $x = y + 2$.

$$x^2 + y^2 = 4 \quad \textit{(Start with the first equation and substitute } y + 2 \textit{ for } x)$$
$$(y + 2)^2 + y^2 = 4 \quad \textit{(Simplify)}$$
$$y^2 + 4y + 4 + y^2 = 4$$
$$2y^2 + 4y + 4 = 4$$
$$2y^2 + 4y = 0 \quad \textit{(Factor)}$$
$$2y(y + 2) = 0 \quad \textit{(The zero factor theorem)}$$
$$y = 0 \quad \text{or} \quad y = -2$$

Thus for $x = y + 2$ when $y = 0$, $x = 2$ and when $y = -2$, $x = 0$. Therefore this consistent system has solutions of $(2, 0)$ and $(0, -2)$.   ■

EXAMPLE 9   Find the solution to

$$3x - y = 5$$
$$9x - 3y = 15$$

Solution Multiply both sides of the first equation by -3 and add equations:

$$\begin{array}{rcl} 3x - y = 5 & \xrightarrow{(-3)} & -9x + 3y = -15 \\ 9x - 3y = 15 & \longrightarrow & 9x - 3y = 15 \\ \hline & & 0 = 0 \end{array}$$

Since $0 = 0$ is true, this is a dependent system with an infinite number of solution points all on the same line. In fact, both of these equations are the equations for the same line and have a common solution of $\{(x, y) : y = 3x - 5\}$. ■

EXAMPLE 10

A space shuttle places a satellite into an orbit that is the same as the shuttle's orbit. After 1 minute the shuttle and satellite are 440 kilometers apart, measured along their orbital path. The satellite's rate of speed is 30 kilometers per minute slower than $\frac{1}{11}$ of the shuttle's rate (in the same direction). Find the rate of the satellite and the shuttle.

Solution   Since word problems are much easier to complete with two variables, let $x =$ the shuttle's rate of speed in kilometers per minute and $y =$ the satellite's rate of speed in kilometers per minute. Then,

	Rate ·	Time	= Distance
Shuttle	x	1 minute	d_1
Satellite	y	1 minute	d_2

or

$$x \cdot 1 = d_1$$
$$y \cdot 1 = d_2$$

The difference in the distance after 1 minute is $d_1 - d_2 = 440$ or

$$x - y = 440$$

The satellite's rate of speed is 30 kilometers per minute slower than $\frac{1}{11}$ of the shuttle's rate (in the same direction), so

$$y = \frac{1}{11}x - 30$$

Thus we will solve the system

$$x - y = 440$$
$$y = \frac{1}{11}x - 30$$

If we solve by substitution, the equation $x - y = 440$ becomes

$$x - \left(\frac{1}{11}x - 30\right) = 440 \qquad \textit{(Multiply both sides by 11)}$$

$$11x - 11\left(\frac{1}{11}x - 30\right) = (440)11 \qquad \textit{(Distribute and simplify)}$$

$$11x - x + 330 = 4840$$

$$10x = 4510 \qquad \textit{(Divide both sides by 10)}$$

$$x = 451 \qquad \text{and} \qquad y = \frac{1}{11}x - 30 = 11$$

Thus the shuttle's rate of speed is 451 kilometers per minute, and the satellite's rate of speed is 11 kilometers per minute. ∎

Exercises 8.1

Solve each of the following by substitution and name the type of system (consistent, dependent, or inconsistent).

1. $y = 2x - 5$
$x + y = 7$

2. $y = \frac{1}{2}x + 3$
$3x + 2y = 10$

3. $y = \frac{1}{2}x + 3x - 2y = -6$

4. $y = -\frac{1}{2}x + 4x + 2y = 8$

5. $\log x = 2y - 1$
$\log x = y^2$

6. $\log x = 6y - 7$
$\log x = y^2 + 2$

7. $x + y = 5$
$\quad\; y = -x - 7$

8. $4x - 2y = 11$
$\quad\; y = 2x - 3$

Solve each of the following by elimination and name the type of system (consistent, dependent, or inconsistent).

9. $x + y = 17$
$x - y = 15$

10. $x - y = 7$
$x + y = 11$

11. $x^2 + y^2 = 7$
$2x^2 + y^2 = 11$

12. $x^2 + y^2 = 7$
$2x^2 - y^2 = 5$

13. $x^2 + y^2 = 4$
$x^2 + 4y^2 = 4$

14. $x^2 + y^2 = 9$
$9x^2 + 4y^2 = 36$

15. $x^2 + y^2 = 1$
$9x^2 + 4y^2 = 36$

16. $x^2 + y^2 = 4$
$5x^2 + 6y^2 = 30$

17. $3x - 2y = 5$
$5x + 3y = 2$

18. $2x - 5y = 1$
$3x + y = 0$

19. $\frac{1}{3}x + \frac{1}{4}y = \frac{1}{3}$
$\frac{1}{2}x - \frac{1}{3}y = \frac{4}{3}$

20. $\frac{1}{2}x - \frac{1}{3}y = 1$
$\frac{1}{4}x - \frac{2}{3}y = 1$

Solve the following by either substitution or elimination.

21. $x^2 + y^2 = 2$
$x^2 - y = 0$

22. $x^2 - y^2 = 3$
$x^2 - 2y = 0$

23. $\log x = y$
$\log x^2 = y^2 - 1$

24. $\log x = 3y - 4$
$\log x^2 = y^2 + 1$

25. $x^2 + y^2 = 19$
$x^2 + 25y^2 = 25$

26. $x^2 - y^2 = -27$
$x^2 + 25y^2 = 25$

27. $x^2 + y^2 = 4$
$x^2 - y = 2$

28. $x^2 + y^2 = 4$
$x^2 + y = 2$

29. $x^2 - y^2 = 4$
$x^2 + 4y^2 = 4$

30. $x^2 - y^2 = 4$
$x^2 + y^2 = 4$

31. $x^3 + y = 0$
$x^2 - y = 0$

32. $x^3 - y = 2$
$x^2 - y = 2$

33. $y = (x - 2)^2$
$x + y = 2$

34. $y = (x - 2)^2$
$x - 4y = 2$

35. $y = 4(x - 2)^2$
$y = (x - 2)^3$

36. $y = (x - 2)^2$
$y = (x - 2)^3$

37. $2x + y + 3z = 4$
$x - y - z = 0$
$x - y - 3z = 2$

38. $x - 2y + z = 3$
$3x + y + 3z = 2$
$2x + y - 2z = 1$

39. A space shuttle places a satellite into an orbit that is the same as the shuttle's orbit. After 1 minute the shuttle and satellite are 290 miles apart. The shuttle's orbital rate of speed is 20 miles per minute less than 11 times the rate of speed of the satellite (both are in the same direction). Find the rates of speed of the shuttle and the satellite.

40. Solve Exercise 39 if the satellite and shuttle are in the same orbit, moving in opposite directions.

41. How many liters of 10% acid and 20% acid must be added together to produce a 25-liter solution of 18% acid?

42. How many liters of 25% acid and 50% acid must be added together to produce a 30-liter solution of 40% acid?

43. A factory produces telephoto and wide-angle lenses for a certain camera. The total number of lenses produced is 500 lenses per week. If orders demand two telephoto lenses for every wide-angle lens, how many of each are produced each week?

44. A factory wants to make three times as many toy blocks as toy trains. The total production of blocks and trains is 1500 units per week. How many of each are to be produced in a week?

8.2 Solution by the Matrix Method

We can solve systems of equations with only two variables straightforwardly by substitution or elimination as we saw in Section 8.1. But when a system involves three or more variables with three or more equations, substitution and elimination can be complicated in both the organization and direction of the solution.

In this section we will use a streamlined version of the elimination method called **Gauss's algorithm**. As we did in synthetic division, we will streamline our notation by not writing the variables. Essentially, this method uses the same techniques that we used previously in the elimination method.

Guidelines for Elimination by Gauss's Algorithm

In using Gauss's algorithm it is permissible to

(a) switch the order of equations,
(b) multiply (or divide) both sides of an equation by a nonzero number, and
(c) add to one equation a nonzero multiple of another equation (to eliminate variables).

Elimination by Gauss's algorithm organizes the search for the solution to a system of equations. As with the elimination method, Gauss's algorithm can be used when the terms in one equation are like terms with the corresponding terms in the other equations in the system. Most often we will use this method on linear systems. Ideally, the results of Gauss's algorithm is a system of equations with each variable remaining in only one equation.

STRATEGIES FOR
APPLYING
GAUSS'S ALGORITHM

To simplify a system of equations with Gauss's algorithm, use the guidelines above to complete the following steps:

(a) Select an equation in which the first variable will be saved. (It will be easier to work with if the coefficient of the first variable is 1.) Then switch the order of equations so that this is the first equation.
(b) Use this selected equation to eliminate the first variable from the other equations by elimination (adding or subtracting multiples of the first equation to the other equations).
(c) Leave the first equation alone and use the remaining equations to continue this process, starting with the second variable in the second equation.

These guidelines and strategies might seem abstract in theory, but the following examples will help you to understand the process better. The next example will help to clarify the techniques of elimination by Gauss's algorithm as described above. We will use some special notation to keep track of our steps. The expression E_1 will be used to indicate "equation 1," and the expression $E_1 - 3E_2$ indicates "from equation 1 subtract 3 times equation 2."

EXAMPLE 1

Use elimination and Gauss's algorithm to solve

$$x - y + 2z = 9$$
$$x + 2y - z = 6$$
$$2x + y + 2z = 18$$

Solution   We save the first variable in the first equation. Then we use multiples of the first equation to eliminate the first variable from the other equations.

$$
\begin{array}{lll}
x - y + 2z = 9 & & x - y + 2z = 9 \\
x + 2y - z = 6 & \xrightarrow{\ E_2 - E_1\ } & 0 + 3y - 3z = -3 \\
2x + y + 2z = 18 & \xrightarrow{\ E_3 - 2E_1\ } & 0 + 3y - 2z = 0
\end{array}
$$

Next we eliminate the $3y$ in E_3, using the $3y$ in E_2:

$$
\begin{array}{lll}
x - y + 2z = 9 & & x - y + 2z = 9 \\
0 + 3y - 3z = -3 & \xrightarrow{\ \frac{1}{3}E_2\ } & 0 + y - z = -1 \\
\xrightarrow{\ E_3 - E_2\ } \; 0 + 0 + z = 3 & & 0 + 0 + z = 3
\end{array}
$$

Then we multiplied both sides of E_3 by $\frac{1}{3}$ to simplify the coefficients in E_3. Although E_3 with $z = 3$ could be used with substitution to find y and then x, we can also continue Gauss's algorithm to find the solution.

$$
\begin{array}{lll}
x - y + 2z = 9 & \xrightarrow{\ E_1 - 2E_3\ } \; x - y + 0 = 3 & \xrightarrow{\ E_1 + E_2\ } \; x + 0 + 0 = 5 \\
0 + y - z = -1 & \xrightarrow{\ E_2 + E_3\ } \; 0 + y + 0 = 2 & \phantom{\xrightarrow{\ E_1 + E_2\ }} \; 0 + y + 0 = 2 \\
0 + 0 + z = 3 & \phantom{\xrightarrow{\ E_2 + E_3\ }} \; 0 + 0 + z = 3 & \phantom{\xrightarrow{\ E_1 + E_2\ }} \; 0 + 0 + z = 3
\end{array}
$$

Thus as we see in the last system, the solution is $x = 5$, $y = 2$, and $z = 3$ or $(x, y, z) = (5, 2, 3)$. ■

As in synthetic division, we can also simplify the way in which we write this algorithm. This will shorten the amount of writing. We use a precise form called the **matrix of the system**, or the **augmented matrix**. The augmented matrix keeps the coefficient and constants in the same positions without writing the variable or the equalities. Thus

$$
\begin{array}{lc}
\begin{array}{l}
x - y + 2z = 9 \\
x + 2y - z = 6 \\
2x + y + 2z = 18
\end{array}
& \text{becomes} \quad
\left[
\begin{array}{rrr|r}
1 & -1 & 2 & 9 \\
1 & 2 & -1 & 6 \\
2 & 1 & 2 & 18
\end{array}
\right]
\end{array}
$$

This matrix has both rows and columns, which are numbered as noted here:

$$
\begin{array}{cc}
 & \text{\textit{Column 2}} \quad \text{\textit{Column 3}} \\
 & \text{\textit{Column 1}} \quad\downarrow\qquad\downarrow\quad \text{\textit{Column 4}} \\
 & \qquad\qquad\downarrow\qquad\quad\qquad\downarrow \\
\begin{array}{l}
\text{\textit{Row 1}} \longrightarrow \\
\text{\textit{Row 2}} \longrightarrow \\
\text{\textit{Row 3}} \longrightarrow
\end{array}
&
\left[
\begin{array}{ccc|c}
1 & -1 & 2 & 9 \\
1 & 2 & -1 & 6 \\
2 & 1 & 2 & 18
\end{array}
\right]
\end{array}
$$

Each position in the matrix has a precise meaning. Row 1 is equivalent to equation 1, column 2 contains the coefficients for the y variables, and there is always an equality understood between column 3 and column 4. (The vertical line separates the coefficients from the constants.) The techniques for elimination are the same, and we can now restate them in terms of rows and columns.

In solving a system of equations in matrix form using Gauss's algorithm it is permissible to

(a) **switch the order of any two rows,**
(b) **multiply (or divide) a row by a nonzero number, and**
(c) **add to one row a nonzero multiple of another row.**

The strategy for using Gauss's algorithm also remains the same.

EXAMPLE 2   Using the matrix notation with Gauss's algorithm, solve

$$
\begin{aligned}
x - y + 2z &= 9 \\
x + 2y - z &= 6 \\
2x + y + 2z &= 18
\end{aligned}
$$

Solution   We change to a matrix for the system and proceed with elimination, using Gauss's algorithm:

$$
\begin{aligned}
x - y + 2z &= 9 \\
x + 2y - z &= 6 \qquad\text{becomes} \\
2x + y + 2z &= 18
\end{aligned}
\qquad
\left[
\begin{array}{ccc|c}
1 & -1 & 2 & 9 \\
1 & 2 & -1 & 6 \\
2 & 1 & 2 & 18
\end{array}
\right]
$$

Then

$$
\left[
\begin{array}{ccc|c}
1 & -1 & 2 & 9 \\
1 & 2 & -1 & 6 \\
2 & 1 & 2 & 18
\end{array}
\right]
\begin{array}{c}
\longrightarrow \\
R_2 - R_1 \\
R_3 - 2R_1
\end{array}
\left[
\begin{array}{ccc|c}
1 & -1 & 2 & 9 \\
0 & 3 & -3 & -3 \\
0 & 3 & -2 & 0
\end{array}
\right]
$$

$$
\begin{array}{c}
\longrightarrow \\
R_3 - R_2
\end{array}
\left[
\begin{array}{ccc|c}
1 & -1 & 2 & 9 \\
0 & 3 & -3 & -3 \\
0 & 0 & 1 & 3
\end{array}
\right]
\begin{array}{c}
\longrightarrow \\
\frac{1}{3}R_2
\end{array}
\left[
\begin{array}{ccc|c}
1 & -1 & 2 & 9 \\
0 & 1 & -1 & -1 \\
0 & 0 & 1 & 3
\end{array}
\right]
$$

$$
\begin{array}{c}
R_1 - 2R_3 \\
\longrightarrow \quad R_2 + R_3
\end{array}
\left[
\begin{array}{ccc|c}
1 & -1 & 0 & 3 \\
0 & 1 & 0 & 2 \\
0 & 0 & 1 & 3
\end{array}
\right]
\begin{array}{c}
R_1 + R_2 \\
\longrightarrow
\end{array}
\left[
\begin{array}{ccc|c}
1 & 0 & 0 & 5 \\
0 & 1 & 0 & 2 \\
0 & 0 & 1 & 3
\end{array}
\right]
$$

Thus the first equation yields $x = 5$, the second equation yields $y = 2$, and the third equation yields $z = 3$. Therefore the solution is $(x, y, z) = (5, 2, 3)$. ∎

This matrix method with elimination by Gauss's algorithm is an excellent way to solve any linear (all variables to the first degree) system of equations. The example above shows the solution of a consistent system. As before, an inconsistent system will have a false equation appearing in the solution process (such as $0 = 7$). We will consider dependent systems in the next section.

EXAMPLE 3   Use the matrix method to solve

$$\begin{aligned} x + y - 2z &= 7 \\ x + y + z &= 4 \\ 2x + 2y - z &= 10 \end{aligned}$$

Solution   Change to the matrix of the system and solve.

$$\left[\begin{array}{ccc|c} 1 & 1 & -2 & 7 \\ 1 & 1 & 1 & 4 \\ 2 & 2 & -1 & 10 \end{array}\right] \longrightarrow \begin{array}{c} \\ R_2 - R_1 \\ R_3 - 2R_1 \end{array} \left[\begin{array}{ccc|c} 1 & 1 & -2 & 7 \\ 0 & 0 & 3 & -3 \\ 0 & 0 & 3 & -4 \end{array}\right]$$

$$\longrightarrow \begin{array}{c} \\ \\ R_3 - R_2 \end{array} \left[\begin{array}{ccc|c} 1 & 1 & -2 & 7 \\ 0 & 0 & 3 & -3 \\ 0 & 0 & 0 & -1 \end{array}\right]$$

Row 3 indicates that $0 = -1$, but $0 \neq -1$. Therefore there is no common solution, and the system is inconsistent. ∎

In solving systems of equations with the matrix method we must be careful to complete only one operation at a time. Once a row (equation) has been changed, we cannot reintroduce its previous value by adding it to another row. Once a row has been changed, the new changed equation must be used in future manipulations. Notice how these mistakes have been avoided in these examples when two changes are made at the same time. Also, we never switch the order of columns, since this would change the order of the variables. Using the matrix method with elimination and Gauss's algorithm can be a very effective method for solving a large system of equations. It might not be shorter than elimination or substitution, but it is more organized and often easier to check. Neat work will be essential in avoiding mistakes and checking computations.

We can also solve some nonlinear systems with the matrix method when the terms in one equation are like terms with the corresponding terms in the other equations. The next example demonstrates this process.

EXAMPLE 4   Use the matrix method to solve

$$\frac{1}{x} + \frac{2}{y} + \frac{3}{z} = 17$$

$$\frac{1}{x} - \frac{2}{y} + \frac{1}{z} = 13$$

$$\frac{2}{x} + \frac{1}{y} - \frac{1}{z} = -2$$

Solution   Although this is not a linear system, we can still solve this sytem with the matrix method, since we need only the coefficients and constants for the matrix method. We must remember that the columns stand for $1/x$, $1/y$, and $1/z$, respectively. Thus we have

$$\begin{bmatrix} 1 & 2 & 3 & | & 17 \\ 1 & -2 & 1 & | & 13 \\ 2 & 1 & -1 & | & -2 \end{bmatrix} \xrightarrow[R_3 - 2R_1]{R_2 - R_1} \begin{bmatrix} 1 & 2 & 3 & | & 17 \\ 0 & -4 & -2 & | & -4 \\ 0 & -3 & -7 & | & -36 \end{bmatrix} \longrightarrow$$

$$R_3 - R_2 \begin{bmatrix} 1 & 2 & 3 & | & 17 \\ 0 & 1 & -5 & | & -32 \\ 0 & -3 & -7 & | & -36 \end{bmatrix} \xrightarrow{R_3 + 3R_2} \begin{bmatrix} 1 & 2 & 3 & | & 17 \\ 0 & 1 & -5 & | & -32 \\ 0 & 0 & -22 & | & -132 \end{bmatrix}$$

$$\xrightarrow{-\frac{1}{22}R_3} \begin{bmatrix} 1 & 2 & 3 & | & 17 \\ 0 & 1 & -5 & | & -32 \\ 0 & 0 & 1 & | & 6 \end{bmatrix} \xrightarrow{R_2 + 5R_3} \begin{bmatrix} 1 & 2 & 3 & | & 17 \\ 0 & 1 & 0 & | & -2 \\ 0 & 0 & 1 & | & 6 \end{bmatrix}$$

$$\xrightarrow{R_1 - 3R_3} \begin{bmatrix} 1 & 2 & 0 & | & -1 \\ 0 & 1 & 0 & | & -2 \\ 0 & 0 & 1 & | & 6 \end{bmatrix} \xrightarrow{R_1 - 2R_2} \begin{bmatrix} 1 & 0 & 0 & | & 3 \\ 0 & 1 & 0 & | & -2 \\ 0 & 0 & 1 & | & 6 \end{bmatrix}$$

Therefore the last system provides

$$\frac{1}{x} = 3, \qquad \frac{1}{y} = -2, \qquad \text{and} \qquad \frac{1}{z} = 6$$

and the solution is

$$(x, y, z) = \left(\frac{1}{3}, -\frac{1}{2}, \frac{1}{6}\right)$$ ∎

Systems of equations have many applications. The next example shows how to write the equation of a circle through three given points.

EXAMPLE 5   Find the equation of the circle through the points $(0, 2)$, $(1, 3)$, and $(7, -5)$.

Solution   We use the general equation of a circle

$$x^2 + y^2 + Ax + By + C = 0$$

and substitute in the three given points.

Points	$x^2 + y^2 + Ax + By + C = 0$		
$(0, 2)$	$0^2 + 2^2 + A(0) + B(2) + C = 0$	or	$2B + C = -4$
$(1, 3)$	$1^2 + 3^2 + A(1) + B(3) + C = 0$	or	$A + 3B + C = -10$
$(7, -5)$	$7^2 + (-5)^2 + A(7) + B(-5) + C = 0$	or	$7A - 5B + C = -74$

Next we solve this system of equations in variables A, B, and C:

$$\begin{bmatrix} 0 & 2 & 1 & | & -4 \\ 1 & 3 & 1 & | & -10 \\ 7 & -5 & 1 & | & -74 \end{bmatrix} \xrightarrow{\begin{subarray}{l} R_1 \\ R_2 \end{subarray}} \begin{bmatrix} 1 & 3 & 1 & | & -10 \\ 0 & 2 & 1 & | & -4 \\ 7 & -5 & 1 & | & -74 \end{bmatrix} \longrightarrow$$

$$\xrightarrow[R_3 - 7R_1]{} \begin{bmatrix} 1 & 3 & 1 & | & -10 \\ 0 & 2 & 1 & | & -4 \\ 0 & -26 & -6 & | & -4 \end{bmatrix} \xrightarrow[R_3 + 13R_2]{} \begin{bmatrix} 1 & 3 & 1 & | & -10 \\ 0 & 2 & 1 & | & -4 \\ 0 & 0 & 7 & | & -56 \end{bmatrix}$$

$$\xrightarrow[-\frac{1}{7}R_3]{} \begin{bmatrix} 1 & 3 & 1 & | & -10 \\ 0 & 2 & 1 & | & -4 \\ 0 & 0 & 1 & | & -8 \end{bmatrix} \xrightarrow[\begin{subarray}{l} R_1 - R_3 \\ R_2 - R_3 \end{subarray}]{} \begin{bmatrix} 1 & 3 & 0 & | & -2 \\ 0 & 2 & 0 & | & 4 \\ 0 & 0 & 1 & | & -8 \end{bmatrix}$$

$$\xrightarrow[\frac{1}{2}R_2]{} \begin{bmatrix} 1 & 3 & 0 & | & -2 \\ 0 & 1 & 0 & | & 2 \\ 0 & 0 & 1 & | & -8 \end{bmatrix} \xrightarrow[R_1 - 3R_2]{} \begin{bmatrix} 1 & 0 & 0 & | & -8 \\ 0 & 1 & 0 & | & 2 \\ 0 & 0 & 1 & | & -8 \end{bmatrix}$$

Thus $A = -8$, $B = 2$, and $C = -8$, and the equation of the circle through $(0, 2)$, $(1, 3)$, and $(7, -5)$ is $x^2 + y^2 - 8x + 2y - 8 = 0$.   ∎

As we complete problems with the matrix method, it is important to remember the strategy as indicated in this diagram below. (The "—" below in the matrices stands for any real number.)

$$\overset{\textit{Start}}{\begin{bmatrix} - & - & - & | & - \\ - & - & - & | & - \\ - & - & - & | & - \end{bmatrix}} \xrightarrow{\textit{Matrix Method}} \overset{\textit{Finish}}{\begin{bmatrix} 1 & 0 & 0 & | & - \\ 0 & 1 & 0 & | & - \\ 0 & 0 & 1 & | & - \end{bmatrix}}$$

Thus the strategy is to use the techniques of Gauss's algorithm to obtain 1's in the diagonal starting in the upper left-hand corner. Next obtain 0's below the 1's in the diagonal and then 0's above the 1's in the diagonal. This will leave the values of each variable in the last column on the right. Keeping this strategy in mind will help in using the matrix method.

Exercises 8.2

Solve the following systems by the matrix method, using elimination by Gauss's algorithm.

1. $x + y - 2z = -3$
$2x - 2y + z = 1$,
$x + 3y - z = 4$

2. $x + y - 2z = 11$
$2x - 2y + z = -1$
$x + 3y - z = 18$

3. $x + y - z = 8$
$5x + y + z = 6$
$2x - y - 3z = 18$

4. $x + y - z = 1$
$5x + y + z = -3$
$2x - y - 3z = -15$

5. $x + 2y - z = 7$
$x - 3y + 2z = -14$
$2x + y + z = -1$

6. $x + 2y - z = -1$
$x - 3y + 2z = 3$
$2x + y + z = 2$

7. $x + 3y - z = -7$
$x - y + 3z = 13$
$3x - 2y + z = 20$

8. $x + 3y - z = 18$
$x - y + 3z = -10$
$3x - 2y + z = -2$

9. $x - 2y + 2z = 2$
$x + y - z = 5$
$2x - y + z = 6$

10. $x - 2y + 3z = 5$
$x + y - z = 4$
$2x - y + 2z = 10$

11. $\dfrac{3}{x} + \dfrac{2}{y} - \dfrac{1}{z} = 9$
$\dfrac{2}{x} - \dfrac{1}{y} + \dfrac{1}{z} = 2$
$\dfrac{2}{x} + \dfrac{3}{y} + \dfrac{2}{z} = 5$

12. $\dfrac{3}{x} + \dfrac{2}{y} - \dfrac{1}{z} = 5$
$\dfrac{2}{x} - \dfrac{1}{y} + \dfrac{1}{z} = 9$
$\dfrac{2}{x} + \dfrac{3}{y} + \dfrac{2}{z} = 7$

13. $x + 2y - z = 5$
$2x - y + 2z = 7$
$3x + y + z = 10$

14. $x - 3y + 2z = 9$
$2x + 2y - z = 3$
$3x - y + z = 7$

15. $\dfrac{1}{x} - \dfrac{2}{y} + \dfrac{3}{z} = 3$
$\dfrac{2}{x} + \dfrac{1}{y} + \dfrac{5}{z} = 8$
$\dfrac{3}{x} - \dfrac{1}{y} - \dfrac{3}{z} = -22$

16. $\dfrac{2}{x} - \dfrac{1}{y} - \dfrac{1}{z} = 7$
$\dfrac{3}{x} + \dfrac{5}{y} + \dfrac{1}{z} = -10$
$\dfrac{4}{x} - \dfrac{3}{y} + \dfrac{2}{z} = 4$

Write the equation of the circle through the three indicated points.

17. (0, 4), (1, −1), and (5, 5).

18. (0, 5), (2, −3), and (5, 2).

19. (3, 4), (5, 0), and (6, 3).

20. (−2, 0), (0, 0), and (2, 4).

21. (2, 0), (0, 4), and (−4, 2).

22. (−4, 2), (2, 4), and (4, −2).

8.3 Solving Dependent Systems of Equations

As we saw in earlier work, some systems of equations have an infinite number of solutions and are called dependent systems. In a dependent system of two equations and two unknowns, both equations are the same curve. The following example demonstrates.

EXAMPLE 1 Solve

$$x - y = 3$$
$$2x - 2y = 6$$

Solution Solving this system with the matrix method, we have

$$\begin{bmatrix} 1 & -1 & | & 3 \\ 2 & -2 & | & 6 \end{bmatrix} \xrightarrow[R_2 - 2R_1]{} \begin{bmatrix} 1 & -1 & | & 3 \\ 0 & 0 & | & 0 \end{bmatrix}$$

Row 2 is the equation $0 = 0$, a true statement. As we remember, a true statement like this means that we have a dependent system with an infinite number of solutions. In this case the solution is the set of all points on the line $x - y = 3$.   ■

We notice that in this dependent system there is a "dependent" relationship between x and y based on the equation $x - y = 3$. We can write the solution to Example 1 in several ways. One way is to use set notation such as

$$\{(x, y) : y = x - 3\}$$

We can also write the solution using ordered pairs. If we let y equal any real number c, we have

$$y = c \quad \text{and} \quad x - c = 3$$

or

$$y = c \quad \text{and} \quad x = c + 3$$

Thus we can write the solution to Example 1 as all points $(c + 3, c)$ where c is a real number or $\{(x, y) : x = c + 3, y = c, \text{ and } c \in R\}$. This method of representing the infinite number of points on a line will be very helpful in writing the solution of dependent systems, especially when there are more than two variables.

EXAMPLE 2   Solve the following system and write the solution as a set of ordered triples if the system is dependent:

$$\begin{aligned} x + y - 2z &= 1 \\ 2x + 3y - 3z &= 9 \\ -x \quad\quad + 3z &= 6 \end{aligned}$$

Solution   Using the matrix method, we have

$$\begin{bmatrix} 1 & 1 & -2 & | & 1 \\ 2 & 3 & -3 & | & 9 \\ -1 & 0 & 3 & | & 6 \end{bmatrix} \xrightarrow[R_3 + R_1]{R_2 - 2R_1} \begin{bmatrix} 1 & 1 & -2 & | & 1 \\ 0 & 1 & 1 & | & 7 \\ 0 & 1 & 1 & | & 7 \end{bmatrix}$$

$$\xrightarrow[R_3 - R_2]{} \begin{bmatrix} 1 & 1 & -2 & | & 1 \\ 0 & 1 & 1 & | & 7 \\ 0 & 0 & 0 & | & 0 \end{bmatrix}$$

This is a dependent system, since there are only two equations and three variables. We use the second equation and arbitrarily let z equal any real number c. With this and equation 2 we have

$$z = c \quad \text{and} \quad y + z = 7$$
$$y + c = 7$$
$$y = 7 - c$$

Next we use $y = 7 - c$, $z = c$, and equation 1 to find x in terms of c. Thus

$$x + y - 2z = 1$$
$$x + (7 - c) - 2(c) = 1$$
$$x + 7 - c - 2c = 1$$
$$x + 7 - 3c = 1$$
$$x = 3c - 6$$

Therefore the solution to this system is the set of ordered triples on the line represented by $\{(3c - 6, 7 - c, c) : c \in R\}$.   ■

Although it might seem awkward to represent the infinite number of points on a line as a set of ordered pairs or ordered triples using c, it is an efficient method for representing the equation of a line especially in three-dimensional space as we will see in Chapter 9.

A **homogeneous system** of equations is a system in which all equations in standard form have a constant of 0. All homogeneous systems have the point $(0, 0, 0)$ in the solution set. The point $(0, 0, 0)$ is called the **trivial solution** for any homogeneous system. Sometimes the solution to a homogeneous system is dependent and includes the point $(0, 0, 0)$. At other times the solution is only the point $(0, 0, 0)$. A system such as

$$x + y - z = 0$$
$$x + 2y + z = 0$$
$$x - y - 2z = 0$$

is homogeneous, since each equation has a constant of 0. By substitution we can see that $(x, y, z) = (0, 0, 0)$ is part of the solution set. There may also be an infinite number of other points in the solution. The best method for determining the solution would be the matrix method.

EXAMPLE 3   Solve

$$x + y - z = 0$$
$$x + 2y + z = 0$$
$$x - y - 2z = 0$$

Solution   Using the matrix method, we have

$$\begin{bmatrix} 1 & 1 & -1 & | & 0 \\ 1 & 2 & 1 & | & 0 \\ 1 & -1 & -2 & | & 0 \end{bmatrix} \xrightarrow[R_3 - R_1]{R_2 - R_1} \begin{bmatrix} 1 & 1 & -1 & | & 0 \\ 0 & 1 & 2 & | & 0 \\ 0 & -2 & -1 & | & 0 \end{bmatrix}$$

$$\xrightarrow[R_3 + 2R_2]{} \begin{bmatrix} 1 & 1 & -1 & | & 0 \\ 0 & 1 & 2 & | & 0 \\ 0 & 0 & 3 & | & 0 \end{bmatrix} \xrightarrow[\frac{1}{3}R_3]{} \begin{bmatrix} 1 & 1 & -1 & | & 0 \\ 0 & 1 & 2 & | & 0 \\ 0 & 0 & 1 & | & 0 \end{bmatrix}$$

$$\begin{array}{cc} \begin{array}{c} R_1 + R_3 \\ \longrightarrow \quad R_2 - 2R_3 \end{array} & \left[\begin{array}{ccc|c} 1 & 1 & 0 & 0 \\ 0 & 1 & 0 & 0 \\ 0 & 0 & 1 & 0 \end{array}\right] \end{array} \quad \begin{array}{cc} \begin{array}{c} R_1 - R_2 \\ \longrightarrow \end{array} & \left[\begin{array}{ccc|c} 1 & 0 & 0 & 0 \\ 0 & 1 & 0 & 0 \\ 0 & 0 & 1 & 0 \end{array}\right] \end{array}$$

Therefore the solution is only the one point $(x, y, z) = (0, 0, 0)$. ■

EXAMPLE 4 Solve

$$\begin{aligned} x + y - z &= 0 \\ x - 3y + 2z &= 0 \\ 2x - 2y + z &= 0 \end{aligned}$$

Solution Although this homogeneous equation has at least the solution of $(0, 0, 0)$, it could also be a dependent system with an infinite number of solutions. Solving with the matrix method, we have

$$\left[\begin{array}{ccc|c} 1 & 1 & -1 & 0 \\ 1 & -3 & 2 & 0 \\ 2 & -2 & 1 & 0 \end{array}\right] \longrightarrow \begin{array}{c} R_2 - R_1 \\ R_3 - 2R_1 \end{array} \left[\begin{array}{ccc|c} 1 & 1 & -1 & 0 \\ 0 & -4 & 3 & 0 \\ 0 & -4 & 3 & 0 \end{array}\right]$$

$$\longrightarrow \begin{array}{c} \\ R_3 - R_2 \end{array} \left[\begin{array}{ccc|c} 1 & 1 & -1 & 0 \\ 0 & -4 & 3 & 0 \\ 0 & 0 & 0 & 0 \end{array}\right]$$

which is

$$\begin{aligned} x + y - z &= 0 \\ -4y + 3z &= 0 \\ 0 &= 0 \end{aligned}$$

Since this is a dependent system ($0 = 0$ is a true statement), we let $z = c$ and solve for y and x in terms of c. If $z = c$ where c is any real number, then

$$-4y + 3z = 0 \quad \text{and} \quad y = \frac{3}{4}c$$

For $x + y - z = 0$ we have $x + \frac{3}{4}c - (c) = 0$ and $x = \frac{1}{4}c$.

Therefore the solution to this system is $\left\{\left(\frac{1}{4}c, \frac{3}{4}c, c\right) : c \in R\right\}$. ■

EXAMPLE 5 Three grades of chocolate are mixed to form a 60-pound mixture worth $70. If the three grades of chocolate cost $1 per pound, $2 per pound, and $3 per pound, find the mixture that would have natural number values for each grade of chocolate.

Solution We let the amounts of each grade of chocolate be the variables x, y and z. Then

$$\begin{aligned} x + y + z &= 60 \\ x + 2y + 3z &= 70 \end{aligned}$$

This is a dependent system, since we have three variables and only two equations.

Subtracting equations, we have

$$y + 2z = 10$$

Thus we let $z = c$ where c is any real number, and solving for y in terms of c, we have $y = 10 - 2c$. For x we start with $x + y + z = 60$ and substitute for z and y:

$$x + y + z = 60$$
$$x + (10 - 2c) + c = 60$$
$$x + 10 - c = 60$$
$$x = 50 + c$$

Since each variable must be a natural number, we must have

$$z = c > 0$$
$$y = 10 - 2c > 0 \quad \text{which is} \quad 10 > 2c \quad \text{or} \quad 5 > c \quad \text{and}$$
$$z = 50 + c > 0 \quad \text{which is} \quad c > -50$$

Taking the intersection of these solution sets, we have $0 < c < 5$. Thus for $c = 1, 2, 3,$ or 4 and natural number values for x, y, and z we have the following possible solutions:

$z = c$	$y = 10 - 2c$	$x = 50 + c$
1	8	51
2	6	52
3	4	53
4	2	54

These are the different numbers of pounds of each grade of chocolate that satisfy the original conditions in the new mixture. ■

Exercises 8.3

Find all solutions to the following systems of equations.

1. $x + y - z = 3$
$2x - 3y + 2z = 5$
$3x - 2y + z = 8$

2. $2x + y - 2z = 5$
$x - 2y + z = 2$
$3x - y - z = 7$

3. $x + 2y - 3z = 2$
$x - 3y + 4z = 2$
$2x - y + z = 4$

4. $2x - y + 3z = 5$
$x - 4y + 4z = 3$
$x + 3y - z = 2$

5. $x + y - z = 3$
$x + y + z = 3$
$x + y \quad = 3$

6. $x + 2y - 3z = 5$
$x - 2y + z = 1$
$x \quad - z = 3$

7. $2x + 3y - z = 7$
$x - y + 2z = -2$
$x + 4y - 3z = 9$

8. $2x - 3y + 2z = 5$
$x + 2y + z = -1$
$x - 5y + z = 6$

9. $3x - y + z = -2$
   $2x + y + 3z = 5$
   $x - 2y - 2z = -7$

10. $x - 3y + 4z = 6$
    $2x + 5y - 3z = -1$
    $3x + 2y + z = 5$

11. $x - 3y + 5z = 0$
    $2x - 2y - z = 0$
    $x + y - 6z = 0$

12. $2x - y + z = 0$
    $5x + 2y - 2z = 0$
    $x + y - z = 0$

13. $x - 2y + 3z = 0$
    $2x - y + z = 0$
    $x + y - 2z = 0$

14. $2x - y + 3z = 0$
    $x + y + z = 0$
    $x - 2y + 2z = 0$

15. $x - 2y + 3z = 0$
    $x + y - 2z = 0$
    $2x - y + 2z = 0$

16. $x - 2y + 2z = 0$
    $x + y + z = 0$
    $2x - y + 2z = 0$

17. $x + y + z = 0$
    $x - 2y + 3z = 0$

18. $x + y + 3z = 0$
    $x - 2y + 2z = 0$

19. $x + 2y - z = 0$
    $5x - y + 2z = 0$
    $x - y + z = 0$

20. $x - 2y + z = 0$
    $2x - y + 3z = 0$
    $x - y - 2z = 0$

21. $x + 2y - z = 0$
    $5x - y + 2z = 0$
    $6x + y + z = 0$

22. $x - 2y + z = 0$
    $2x - y + 3z = 0$
    $x + y + 2z = 0$

23. $5x + 2y - 2z = 3$
    $4x - y + 3z = 5$
    $x + 3y - 5z = -2$

24. $5x - 3y + 2z = 1$
    $2x + 4y - z = 0$
    $3x - 7y + 3z = 1$

25. $5x - 2y + 3z = 0$
    $x - y - z = 0$
    $6x - 3y + 2z = 0$

26. $x + 2y - 3z = 0$
    $2x - 3y + 2z = 0$
    $3x - y - z = 0$

27. $x - y + 3z = 0$
    $y + 2z = 0$

28. $2x - y + 2z = 0$
    $3x - z = 0$

29. $x - 5y + z = -1$
    $x + 2y + z = 3$
    $2x - 3y + 2z = 2$

30. $x - 3y - 2z = -3$
    $2x - 2y - z = 5$
    $x + y + z = 8$

31. $x + 2y + 3z = 5$
    $y - 5z = 0$

32. $x + y + z = 5$
    $2x - 3y + 2z = 0$
    $x - 4y + z = -5$

33. $2x - 3y + z = 1$
    $4x - 6y + 2z = 3$
    $6x - 9y + 3z = 12$

34. $x - 2y + 3z = -1$
    $2x - 4y + 6z = 0$
    $3x - 6y + 9z = 2$

35. The sum of three numbers is 15. The first number added to twice the second less three times the third is 75. Find all triples that are solutions, and also give one specific solution.

36. The sum of three numbers is 10. The first number less the second number is then added to twice the third number with a result of 4. Find all triples that are solutions, and also give one specific solution.

37. Three grades of chocolate are mixed to make 15 pounds of a mixture costing $40. If the three grades of chocolate cost $1 per pound, $2 per pound, and $3 per pound, find the mixtures that would have natural number values for each grade of chocolate.

38. Three grades of chocolate are mixed to make 20 pounds of a mixture costing $50. If the three grades of chocolate cost $1 per pound, $2 per pound, and $3 per pound, find the mixtures that would have natural number values for each grade of chocolate.

8.4 Evaluating Determinants and Their Properties

In this section we will define and evaluate a *determinant*. We will also discuss in detail the methods for evaluating determinants. In Section 8.5 we will use determinants as a tool to solve systems of equations.

Definition | A **determinant** is a function that maps a square matrix to a real number. Thus the domain of the determinant is a square matrix, and the range of the determinant is a unique real number.

A determinant is distinctly different from a matrix in the following ways:

(a) A matrix can be a rectangular or square array, while a determinant must always have a square array as the domain and a real number as the range value.

(b) A rectangular matrix can represent an entire system of equations, while a square determinant has a real number value.

(c) A matrix array is enclosed with square brackets and a determinant array is enclosed with straight brackets.

A Matrix

$$\begin{bmatrix} a_{11} & a_{12} & a_{13} & a_{14} \\ a_{21} & a_{22} & a_{23} & a_{24} \\ a_{31} & a_{32} & a_{33} & a_{34} \end{bmatrix}$$

A Determinant

$$\begin{vmatrix} a_{11} & a_{12} & a_{13} \\ a_{21} & a_{22} & a_{23} \\ a_{31} & a_{32} & a_{33} \end{vmatrix}$$

This matrix is a 3×4 (read "3 by 4") matrix with three rows and four columns. The determinant above is a 3×3 (read "3 by 3") determinant with three rows and three columns. Each entry in the array can be identified by its row and column. The entry a_{12} is read "a sub one, two" where the 1 indicates that the entry is in the first row and the 2 indicates that the entry is in the second column. In general, a_{ij} is the entry in both row i and column j.

We will now evaluate a 2×2 and 3×3 determinant.

EVALUATING 2 X 2 AND 3 X 3 DETERMINANTS

The determinant of

$$\begin{bmatrix} a_{11} & a_{12} \\ a_{21} & a_{22} \end{bmatrix} \quad \text{is} \quad \begin{vmatrix} a_{11} & a_{12} \\ a_{21} & a_{22} \end{vmatrix} = a_{11}a_{22} - a_{21}a_{12}$$

The determinant of

$$\begin{bmatrix} a_{11} & a_{12} & a_{13} \\ a_{21} & a_{22} & a_{23} \\ a_{31} & a_{32} & a_{33} \end{bmatrix} \quad \text{is} \quad \begin{vmatrix} a_{11} & a_{12} & a_{13} \\ a_{21} & a_{22} & a_{23} \\ a_{31} & a_{32} & a_{33} \end{vmatrix} = D$$

where

$$D = a_{11}a_{22}a_{33} - a_{11}a_{32}a_{23} - a_{12}a_{21}a_{33} + a_{12}a_{31}a_{23} + a_{13}a_{21}a_{32} - a_{13}a_{31}a_{22}$$

Although the evaluation of a 2×2 determinant should be memorized, the evaluation of a 3×3 determinant need not be memorized. The more general pattern used to evaluate a 3×3 or larger determinant will be explained next. This method can always be used to evaluate any determinant.

EVALUATING 3 X 3
AND LARGER
DETERMINANTS

1. A determinant can be evaluated using any row or column. The following show how to evaluate a determinant using (a) the first-row and (b) the third column:

(a) First-row evaluation:

$$\begin{vmatrix} a_{11} & a_{12} & a_{13} \\ a_{21} & a_{22} & a_{23} \\ a_{31} & a_{32} & a_{33} \end{vmatrix} = (-1)^{1+1}a_{11}M_{11} + (-1)^{1+2}a_{12}M_{12} + (-1)^{1+3}a_{13}M_{13}$$

(b) Third-column evaluation:

$$\begin{vmatrix} a_{11} & a_{12} & a_{13} \\ a_{21} & a_{22} & a_{23} \\ a_{31} & a_{32} & a_{33} \end{vmatrix} = (-1)^{1+3}a_{13}M_{13} + (-1)^{2+3}a_{23}M_{23} + (-1)^{3+3}a_{33}M_{33}$$

(Notice how the subscripts for a_{ij} are also used with the power of -1, $(-1)^{i+j}$.)

2. The factor $(-1)^{i+j}a_{ij}$ is called the cofactor, and the factor M_{ij} is called the minor of the term (this is the cofactor and minor for the i-row and j-column). The minor M_{ij} is determined by blocking out the i-row and j-column and making a determinant out of the remaining array.

We will demonstrate with the determinant

$$\begin{vmatrix} a_{11} & a_{12} & a_{13} \\ a_{21} & a_{22} & a_{23} \\ a_{31} & a_{32} & a_{33} \end{vmatrix}$$

(a) The minor M_{12} is found by blocking out row 1 and column 2 and making a determinant from the remaining array. Thus

$$\begin{vmatrix} a_{11} & a_{12} & a_{13} \\ a_{21} & a_{22} & a_{23} \\ a_{31} & a_{32} & a_{33} \end{vmatrix} \quad \text{and} \quad M_{12} = \begin{vmatrix} a_{21} & a_{23} \\ a_{31} & a_{33} \end{vmatrix} = a_{21}a_{33} - a_{31}a_{23}$$

(b) The minor M_{23} is found by blocking out row 2 and column 3 and making a determinant from the remaining array. Thus

$$\begin{vmatrix} a_{11} & a_{12} & a_{13} \\ a_{21} & a_{22} & a_{23} \\ a_{31} & a_{32} & a_{33} \end{vmatrix} \quad \text{and} \quad M_{23} = \begin{vmatrix} a_{11} & a_{12} \\ a_{31} & a_{32} \end{vmatrix} = a_{11}a_{32} - a_{31}a_{12}$$

3. The patterns in parts 1 and 2 are used several times to evaluate a determinant.

Notice how these steps are used to evaluate the following determinants.

EXAMPLE 1   Evaluate the following determinant, using the first row:

$$\begin{vmatrix} a_{11} & a_{12} & a_{13} \\ a_{21} & a_{22} & a_{23} \\ a_{31} & a_{32} & a_{33} \end{vmatrix}$$

Solution

$$\begin{vmatrix} a_{11} & a_{12} & a_{13} \\ a_{21} & a_{22} & a_{23} \\ a_{31} & a_{32} & a_{33} \end{vmatrix} = (-1)^{1+1}a_{11}M_{11} + (-1)^{1+2}a_{12}M_{12} + (-1)^{1+3}a_{13}M_{13}$$

$$= (-1)^2 a_{11} \begin{vmatrix} a_{22} & a_{23} \\ a_{32} & a_{33} \end{vmatrix} + (-1)^3 a_{12} \begin{vmatrix} a_{21} & a_{23} \\ a_{31} & a_{33} \end{vmatrix} + (-1)^4 a_{13} \begin{vmatrix} a_{21} & a_{22} \\ a_{31} & a_{32} \end{vmatrix}$$

$$= a_{11}(a_{22}a_{33} - a_{32}a_{23}) - a_{12}(a_{21}a_{33} - a_{31}a_{23}) + a_{13}(a_{21}a_{32} - a_{31}a_{22})$$

$$= a_{11}a_{22}a_{33} - a_{11}a_{32}a_{23} - a_{12}a_{21}a_{33} + a_{12}a_{31}a_{23} + a_{13}a_{21}a_{32} - a_{13}a_{31}a_{22} \quad \blacksquare$$

Fortunately, we will find this method of evaluation much easier when working with numerical entries rather than the a_{ij} entries of the general determinant.

EXAMPLE 2   Evaluate this determinant, using **(a)** row 2 and **(b)** column 2.

$$\begin{vmatrix} 1 & 2 & 3 \\ 5 & 0 & 6 \\ 7 & 0 & 4 \end{vmatrix}$$

Solution
(a) Using row 2,

$$\begin{vmatrix} 1 & 2 & 3 \\ 5 & 0 & 6 \\ 7 & 0 & 4 \end{vmatrix} = (-1)^{2+1}(5)M_{21} + (-1)^{2+2}(0)M_{22} + (-1)^{2+3}(6)M_{23}$$

$$= (-1)^3(5) \begin{vmatrix} 2 & 3 \\ 0 & 4 \end{vmatrix} + (-1)^4(0) \begin{vmatrix} 1 & 3 \\ 7 & 4 \end{vmatrix} + (-1)^5(6) \begin{vmatrix} 1 & 2 \\ 7 & 0 \end{vmatrix}$$

$$= (-1)(5)(8 - 0) + (1)(0)(4 - 21) + (-1)(6)(0 - 14)$$

$$= -5(8) + 0 - 6(-14)$$

$$= -40 + 84 = 44$$

(b) Using column 2,

$$\begin{vmatrix} 1 & 2 & 3 \\ 5 & 0 & 6 \\ 7 & 0 & 4 \end{vmatrix} = (-1)^{1+2}(2)M_{12} + (-1)^{2+2}(0)M_{22} + (-1)^{3+2}(0)M_{32}$$

$$= (-1)(2) \begin{vmatrix} 5 & 6 \\ 7 & 4 \end{vmatrix} + 0 + 0$$

$$= (-2)(20 - 42) = (-2)(-22) = 44 \quad \blacksquare$$

Notice that the solutions of Example 2(a) and Example 2(b) both have the same value. This reinforces the earlier statement that a determinant has a unique value. No matter which row or column is used to evaluate the determinant, the value of the determinant is the same. Also notice that Example 2(b) was easier to evaluate because of the two zeros in the second column. Whenever possible, we will use the row or column that contains the most zeros.

EXAMPLE 3   Solve

$$\begin{vmatrix} x & x \\ x & 3 \end{vmatrix} = 2$$

Solution   We will evaluate the determinant and solve the resulting equation. Thus

$$\begin{vmatrix} x & x \\ x & 3 \end{vmatrix} = 2$$

is

$$3x - x^2 = 2$$
$$0 = x^2 - 3x + 2$$
$$0 = (x - 2)(x - 1)$$

Therefore $x \in \{1, 2\}$. Thus there are two determinants of this form that equal 2:

$$\begin{vmatrix} 1 & 1 \\ 1 & 3 \end{vmatrix} \quad \text{and} \quad \begin{vmatrix} 2 & 2 \\ 2 & 3 \end{vmatrix} \qquad \blacksquare$$

As we saw in Example 2, the more zeros there are in a row or column, the easier it is to evaluate a determinant. The following properties make it possible for us to reduce elements of a determinant to zeros.

DETERMINANT PROPERTIES

(a) If two rows are interchanged, the determinant changes sign (and if two columns are interchanged, the determinant changes sign). Thus when rows 1 and 2 are switched,

$$\begin{vmatrix} 1 & 2 & 3 \\ 4 & 5 & 6 \\ 7 & 8 & 9 \end{vmatrix} = - \begin{vmatrix} 4 & 5 & 6 \\ 1 & 2 & 3 \\ 7 & 8 & 9 \end{vmatrix}$$

(b) If one row (or column) is multiplied by a constant k, then the value of the determinant is also multiplied by that same constant k. Thus

$$k \begin{vmatrix} 1 & 2 & 3 \\ 4 & 5 & 6 \\ 7 & 8 & 9 \end{vmatrix} = \begin{vmatrix} 1 & 2 & 3 \\ 4k & 5k & 6k \\ 7 & 8 & 9 \end{vmatrix} \quad \text{or} \quad \begin{vmatrix} 1 & 2 & 3k \\ 4 & 5 & 6k \\ 7 & 8 & 9k \end{vmatrix}$$

This property also allows for factoring a row or a column:

$$\begin{vmatrix} 5 & 10 & 15 \\ 1 & 6 & 2 \\ 3 & 8 & 4 \end{vmatrix} = 5\begin{vmatrix} 1 & 2 & 3 \\ 1 & 6 & 2 \\ 3 & 8 & 4 \end{vmatrix} = 5 \cdot 2\begin{vmatrix} 1 & 1 & 3 \\ 1 & 3 & 2 \\ 3 & 4 & 4 \end{vmatrix} = 10\begin{vmatrix} 1 & 1 & 3 \\ 1 & 3 & 2 \\ 3 & 4 & 4 \end{vmatrix}$$

(c) The value of a determinant does not change if a nonzero multiple of one row (or column) is added to another row (or column):

$$\begin{vmatrix} 1 & 2 & 3 \\ 4 & 5 & 6 \\ 7 & 8 & 9 \end{vmatrix} = \begin{matrix} \\ R_2 - 4R_1 \\ R_3 - 7R_1 \end{matrix} \begin{vmatrix} 1 & 2 & 3 \\ 0 & -3 & -6 \\ 0 & -6 & -12 \end{vmatrix}$$

or

$$C_2 - 2C_1 \qquad C_3 - 3C_1$$

$$\begin{vmatrix} 1 & 2 & 3 \\ 4 & 5 & 6 \\ 7 & 8 & 9 \end{vmatrix} = \begin{vmatrix} 1 & 0 & 0 \\ 4 & -3 & -6 \\ 7 & -6 & -12 \end{vmatrix}$$

The preceding properties of determinants, especially the third, are similar to the properties for matrices, but are also distinctly different. A common mistake is to put $R_1 - R_2$ in for row 2 without compensating for the -1 multiple of R_2. If $R_1 - R_2$ replaces row 2, then a factor of -1 must also be placed in front of the determinant by property (b). On the other hand, $R_1 - R_2$ can be a replacement for R_1 with no other changes or new factors, using property (c). Notice how these properties are used to simplify the determinants in the following examples.

EXAMPLE 4   Evaluate determinant A, when

$$A = \begin{vmatrix} 1 & 2 & 3 & 4 \\ 2 & 0 & 1 & 2 \\ 6 & 2 & 3 & 1 \\ 1 & 5 & 0 & 1 \end{vmatrix}$$

Solution   Since this is such a large determinant (4×4), we will use the properties of determinants to simplify. The two best options are to get two more zeros in row 2 or column 3. Working for zeros in column 3, we proceed as follows:

$$A = \begin{vmatrix} 1 & 2 & 3 & 4 \\ 2 & 0 & 1 & 2 \\ 6 & 2 & 3 & 1 \\ 1 & 5 & 0 & 1 \end{vmatrix} = \begin{matrix} R_1 - 3R_2 \\ \\ R_3 - 3R_2 \\ \\ \end{matrix} \begin{vmatrix} -5 & 2 & 0 & -2 \\ 2 & 0 & 1 & 2 \\ 0 & 2 & 0 & -5 \\ 1 & 5 & 0 & 1 \end{vmatrix}$$

Expanding using column 3, we have

$$A = 0 + (-1)^{2+3}(1)\begin{vmatrix} -5 & 2 & -2 \\ 0 & 2 & -5 \\ 1 & 5 & 1 \end{vmatrix} + 0 + 0$$

$$= (-1)\begin{vmatrix} -5 & 2 & -2 \\ 0 & 2 & -5 \\ 1 & 5 & 1 \end{vmatrix} = \begin{array}{c} R_1 + 5R_3 \\ \\ (-1) \end{array}\begin{vmatrix} 0 & 27 & 3 \\ 0 & 2 & -5 \\ 1 & 5 & 1 \end{vmatrix}$$

Then expanding about column 1,

$$A = (-1)\left[0 + 0 + (-1)^{3+1}\begin{vmatrix} 27 & 3 \\ 2 & -5 \end{vmatrix}\right]$$

$$= (-1)(1)[(27)(-5) - (2)(3)] = 141$$

■

EXAMPLE 5   Show that the following equation is true:

$$\begin{vmatrix} 1 & 1 & 1 \\ x & y & z \\ x^2 & y^2 & z^2 \end{vmatrix} = (x - y)(y - z)(z - x)$$

Solution   First we simplify, using the properties of determinants:

$$\begin{array}{c} C_1 - C_2 \\ \downarrow \end{array}$$

$$\begin{vmatrix} 1 & 1 & 1 \\ x & y & z \\ x^2 & y^2 & z^2 \end{vmatrix} = \begin{vmatrix} 0 & 1 & 1 \\ (x - y) & y & z \\ (x^2 - y^2) & y^2 & z^2 \end{vmatrix}$$

$$= \begin{vmatrix} (x - y) \cdot 0 & 1 & 1 \\ (x - y) \cdot 1 & y & z \\ (x - y)(x + y) & y^2 & z^2 \end{vmatrix} = (x - y)\begin{vmatrix} 0 & 1 & 1 \\ 1 & y & z \\ (x + y) & y^2 & z^2 \end{vmatrix}$$

$$\begin{array}{c} C_2 - C_3 \\ \downarrow \end{array}$$

$$= (x - y)\begin{vmatrix} 0 & 0 & 1 \\ 1 & (y - z) & z \\ x + y & (y^2 - z^2) & z^2 \end{vmatrix}$$

$$= (x - y)(y - z)\begin{vmatrix} 0 & 0 & 1 \\ 1 & 1 & z \\ (x + y) & (y + z) & z^2 \end{vmatrix}$$

Now evaluating this determinant around the first row, we have

$$= (x - y)(y - z)\left[0 + 0 + (-1)^{1+3}(1)\begin{vmatrix} 1 & 1 \\ (x + y) & (y + z) \end{vmatrix}\right]$$

$$= (x - y)(y - z)(1)[(y + z) - (x + y)]$$

$$= (x - y)(y - z)(y + z - x - y)$$

$$= (x - y)(y - z)(z - x)$$

Exercises 8.4

Evaluate the following determinants.

1. $\begin{vmatrix} 1 & 2 \\ 3 & 4 \end{vmatrix}$

2. $\begin{vmatrix} 3 & 4 \\ 5 & 6 \end{vmatrix}$

3. $\begin{vmatrix} 1 & 5 \\ 2 & -7 \end{vmatrix}$

4. $\begin{vmatrix} -1 & 2 \\ 4 & 5 \end{vmatrix}$

5. $\begin{vmatrix} -3 & -7 \\ 2 & -1 \end{vmatrix}$

6. $\begin{vmatrix} -2 & 1 \\ 4 & 3 \end{vmatrix}$

7. $\begin{vmatrix} 1 & 2 & 3 \\ 5 & 0 & 1 \\ 2 & 4 & 6 \end{vmatrix}$

8. $\begin{vmatrix} 3 & 5 & 7 \\ 9 & 12 & 0 \\ 2 & 3 & 1 \end{vmatrix}$

9. $\begin{vmatrix} 7 & 2 & 0 \\ 3 & 6 & 1 \\ 2 & 4 & 0 \end{vmatrix}$

10. $\begin{vmatrix} 2 & -1 & 5 \\ 0 & 3 & 0 \\ 1 & -5 & 0 \end{vmatrix}$

11. $\begin{vmatrix} 3 & 6 & 1 \\ 1 & -2 & 4 \\ 0 & 0 & 5 \end{vmatrix}$

12. $\begin{vmatrix} 3 & 0 & 5 \\ 4 & 0 & 2 \\ -1 & 2 & 1 \end{vmatrix}$

Solve the following.

13. $\begin{vmatrix} x & 2 \\ 1 & x \end{vmatrix} = x$

14. $\begin{vmatrix} 3 & x \\ x & 2 \end{vmatrix} = x$

15. $\begin{vmatrix} x & 2x \\ 1 & x \end{vmatrix} = 1$

16. $\begin{vmatrix} x & x \\ 3 & x \end{vmatrix} = 4$

17. $\begin{vmatrix} x & 4 \\ 1 & x \end{vmatrix} = 0$

18. $\begin{vmatrix} x & 3 \\ 3 & x \end{vmatrix} = 0$

19. $\begin{vmatrix} x & 2x \\ 3 & x \end{vmatrix} = 0$

20. $\begin{vmatrix} 2 & x \\ x & 5x \end{vmatrix} = 0$

Evaluate each of the following determinants. Use the determinant properties to create zeros to make simplification easier.

21. $\begin{vmatrix} -1 & 3 & 4 \\ 1 & 2 & -1 \\ 3 & 6 & 5 \end{vmatrix}$

22. $\begin{vmatrix} -1 & 4 & 3 \\ 1 & 6 & 2 \\ 2 & 5 & 4 \end{vmatrix}$

23. $\begin{vmatrix} 5 & 4 & 3 \\ 1 & -1 & 1 \\ 3 & 4 & 5 \end{vmatrix}$

24. $\begin{vmatrix} 6 & 1 & 2 \\ 4 & 2 & 3 \\ 2 & -2 & 2 \end{vmatrix}$

25. $\begin{vmatrix} 1 & 7 & 3 & 5 \\ 1 & 3 & 0 & 4 \\ 2 & 4 & 1 & 2 \\ 3 & 2 & -1 & 0 \end{vmatrix}$

26. $\begin{vmatrix} 1 & 5 & 4 & 2 \\ -1 & 3 & 2 & 1 \\ 0 & -1 & 3 & 5 \\ 2 & 0 & 2 & 7 \end{vmatrix}$

27. Prove that if a determinant has one row or column of all zeros, then the determinant equals zero.

28. Prove that if two columns of a determinant are identical, then the determinant equals zero.

29. Prove that if two rows of a determinant are identical, then the determinant equals zero.

30. Show that

$$\begin{vmatrix} 1 & x & x^2 \\ 1 & y & y^2 \\ 1 & z & z^2 \end{vmatrix} = (z - x)(z - y)(y - x)$$

31. Show that

$$\begin{vmatrix} 1 & x & x^3 \\ 1 & y & y^3 \\ 1 & z & z^3 \end{vmatrix} = (z - x)(y - z)(y - x)(x + y + z)$$

32. Show that

$$\begin{vmatrix} 1 & 1 & 1 \\ x & y & z \\ yz & xz & xy \end{vmatrix} = (z - x)(z - y)(y - x)$$

33. Show that

$$\begin{vmatrix} x & y & z \\ x^2 & y^2 & z^2 \\ yz & xz & xy \end{vmatrix} = (z - x)(z - y)(y - x)(xy + xz + yz)$$

34. Show that

$$\begin{vmatrix} x & x & x & x \\ x & y & y & y \\ x & y & z & z \\ x & y & z & w \end{vmatrix} = x(y - x)(z - y)(w - z)$$

35. Show that

$$\begin{vmatrix} 1 & 1 & 1 & 1 & 1 \\ 1 & 0 & 0 & 0 & x \\ y & 1 & 0 & 0 & y \\ z & z & 1 & 0 & z \\ w & w & w & 1 & w \end{vmatrix} = (x - 1)(y - 1)(z - 1)(w - 1)$$

8.5 Cramer's Rule

In this section we will solve systems of equations using determinants. This is the last method that we will use to solve a system of equations, and it is called **Cramer's Rule**. Although it takes the same amount of time as the other methods, Cramer's Rule is very mechanical and can minimize calculation error.

We will first derive Cramer's Rule for a general system of equations in two variables using the matrix method:

$$a_{11}x + a_{12}y = k_1$$
$$a_{21}x + a_{22}y = k_2$$

$$\begin{bmatrix} a_{11} & a_{12} & | & k_1 \\ a_{21} & a_{22} & | & k_2 \end{bmatrix} \xrightarrow[-R_2 + \frac{a_{22}}{a_{12}}R_1]{} \begin{bmatrix} a_{11} & a_{12} & | & k_1 \\ -a_{21} + \frac{a_{11}a_{22}}{a_{12}} & -a_{22} + \frac{a_{12}a_{22}}{a_{12}} & | & -k_2 + \frac{k_1 a_{22}}{a_{12}} \end{bmatrix}$$

$$\xrightarrow{} \begin{bmatrix} a_{11} & a_{12} & | & k_1 \\ \frac{-a_{21}a_{12} + a_{11}a_{22}}{a_{12}} & 0 & | & \frac{-k_2 a_{12} + k_1 a_{22}}{a_{12}} \end{bmatrix}$$

From this we have

$$\frac{-a_{21}a_{12} + a_{11}a_{22}}{a_{12}}x = \frac{-k_2 a_{12} + k_1 a_{22}}{a_{12}}$$

$$(a_{11}a_{22} - a_{21}a_{12})x = k_1 a_{22} - k_2 a_{12}$$

$$x = \frac{k_1 a_{22} - k_2 a_{12}}{a_{11}a_{22} - a_{21}a_{12}}$$

$$x = \frac{\begin{vmatrix} k_1 & a_{12} \\ k_2 & a_{22} \end{vmatrix}}{\begin{vmatrix} a_{11} & a_{12} \\ a_{21} & a_{22} \end{vmatrix}}$$

As we see here, the determinants provide a much nicer way to write this result. By a similar method we can also show that

$$y = \frac{\begin{vmatrix} a_{11} & k_1 \\ a_{21} & k_2 \end{vmatrix}}{\begin{vmatrix} a_{11} & a_{12} \\ a_{21} & a_{22} \end{vmatrix}}$$

We can now combine these conclusions in writing Cramer's Rule for a system with two variables.

CRAMER'S RULE (FOR TWO EQUATIONS WITH TWO VARIABLES)

For the system

$$a_{11}x + a_{12}y = k_1$$
$$a_{21}x + a_{22}y = k_2$$

the values of the variables x and y are

$$x = \frac{D_x}{D} \quad \text{and} \quad y = \frac{D_y}{D}$$

where

$$D = \begin{vmatrix} a_{11} & a_{12} \\ a_{21} & a_{22} \end{vmatrix} \neq 0, \quad D_x = \begin{vmatrix} k_1 & a_{12} \\ k_2 & a_{22} \end{vmatrix}, \quad \text{and} \quad D_y = \begin{vmatrix} a_{11} & k_1 \\ a_{21} & k_2 \end{vmatrix}$$

It is helpful to note that the denominator D in Cramer's Rule is the determinant of the coefficients of the variables. Also notice that D_x is D with the constants k_1 and k_2 replacing the x-column coefficients a_{11} and a_{21}. Likewise, D_y is D with the constants k_1 and k_2 replacing the y-column coefficients a_{12} and a_{22}. Noticing these patterns will make it easier to set up the determinants for Cramer's Rule.

EXAMPLE 1   Using Cramer's rule, solve

$$2x - 3y = 5$$
$$3x - 7y = 2$$

Solution   For this system we first determine D, D_x, and D_y:

$$D = \begin{vmatrix} 2 & -3 \\ 3 & -7 \end{vmatrix} = -14 - (-9) = -14 + 9 = -5$$

$$D_x = \begin{vmatrix} 5 & -3 \\ 2 & -7 \end{vmatrix} = -35 - (-6) = -35 + 6 = -29$$

$$D_y = \begin{vmatrix} 2 & 5 \\ 3 & 2 \end{vmatrix} = 4 - 15 = -11$$

Therefore

$$x = \frac{D_x}{D} = \frac{-29}{-5} = \frac{29}{5} \quad \text{and} \quad y = \frac{D_y}{D} = \frac{-11}{-5} = \frac{11}{5}$$

and the point of intersection is $\left(\frac{29}{5}, \frac{11}{5}\right)$.   ■

Cramer's Rule for a system of equations with three equations and three variables follows the same basic patterns described above for a system with two variables.

CRAMER'S RULE (FOR THREE EQUATIONS WITH THREE VARIABLES)

For the system

$$a_{11}x + a_{12}y + a_{13}z = k_1$$
$$a_{21}x + a_{22}y + a_{23}z = k_2$$
$$a_{31}x + a_{32}y + a_{33}z = k_3$$

the values of the variables x, y and z are $x = \dfrac{D_x}{D}$, $y = \dfrac{D_y}{D}$, and $z = \dfrac{D_z}{D}$, where

$$D = \begin{vmatrix} a_{11} & a_{12} & a_{13} \\ a_{21} & a_{22} & a_{23} \\ a_{31} & a_{32} & a_{33} \end{vmatrix} \neq 0$$

$$D_x = \begin{vmatrix} k_1 & a_{12} & a_{13} \\ k_2 & a_{22} & a_{23} \\ k_3 & a_{32} & a_{33} \end{vmatrix}, \quad D_y = \begin{vmatrix} a_{11} & k_1 & a_{13} \\ a_{21} & k_2 & a_{23} \\ a_{21} & k_3 & a_{33} \end{vmatrix}, \quad \text{and} \quad D_z = \begin{vmatrix} a_{11} & a_{12} & k_1 \\ a_{21} & a_{22} & k_2 \\ a_{31} & a_{32} & k_3 \end{vmatrix}$$

We can now use Cramer's Rule to solve the following examples.

EXAMPLE 2   Using Cramer's Rule, solve

$$2x + 3y + z = 5$$
$$x - y + z = 6$$
$$2x - 2y + 3z = 2$$

Solution   We first evaluate D, D_x, D_y, and D_z:

$$D = \begin{vmatrix} 2 & 3 & 1 \\ 1 & -1 & 1 \\ 2 & -2 & 3 \end{vmatrix} = \begin{matrix} \\ R_2 - R_1 \\ R_3 - 3R_1 \end{matrix} \begin{vmatrix} 2 & 3 & 1 \\ -1 & -4 & 0 \\ -4 & -11 & 0 \end{vmatrix}$$

$$= (-1)^{1+3}(1) \begin{vmatrix} -1 & -4 \\ -4 & -11 \end{vmatrix} = 1(11 - 16) = -5$$

$$D_x = \begin{vmatrix} 5 & 3 & 1 \\ 6 & -1 & 1 \\ 2 & -2 & 3 \end{vmatrix} = \begin{matrix} \\ R_2 - R_1 \\ R_3 - 3R_1 \end{matrix} \begin{vmatrix} 5 & 3 & 1 \\ 1 & -4 & 0 \\ -13 & -11 & 0 \end{vmatrix}$$

$$= (-1)^{1+3}(1) \begin{vmatrix} 1 & -4 \\ -13 & -11 \end{vmatrix} = (1)(-11 - 52) = -63$$

$$D_y = \begin{vmatrix} 2 & 5 & 1 \\ 1 & 6 & 1 \\ 2 & 2 & 3 \end{vmatrix} = \begin{matrix} \\ R_2 - R_1 \\ R_3 - 3R_1 \end{matrix} \begin{vmatrix} 2 & 5 & 1 \\ -1 & 1 & 0 \\ -4 & -13 & 0 \end{vmatrix}$$

$$= (-1)^{1+3}(1) \begin{vmatrix} -1 & 1 \\ -4 & -13 \end{vmatrix} = (1)(13 + 4) = 17$$

$$D_z = \begin{vmatrix} 2 & 3 & 5 \\ 1 & -1 & 6 \\ 2 & -2 & 2 \end{vmatrix} = \begin{matrix} R_1 - 2R_2 \\ * \\ R_3 - 2R_2 \end{matrix} \begin{vmatrix} 0 & 5 & -7 \\ 1 & -1 & 6 \\ 0 & 0 & -10 \end{vmatrix}$$

$$= (-1)^{2+1}(1) \begin{vmatrix} 5 & -7 \\ 0 & -10 \end{vmatrix} = (-1)(1)(-50 + 0) = 50$$

*If $R_1 - 2R_2$ were used to replace R_2, a factor of $-\frac{1}{2}$ would have to be placed in front of the determinant's value to compensate (property (b) for determinants in Section 8.4).

Using these values for D, D_x, D_y, and D_z, we have

$$x = \frac{D_x}{D} = \frac{-63}{-5} = \frac{63}{5}$$

$$y = \frac{D_y}{D} = \frac{17}{-5} = -\frac{17}{5}$$

$$z = \frac{D_z}{D} = \frac{50}{-5} = -10$$

and the point of intersection is

$$(x, y, z) = \left(\frac{63}{5}, -\frac{17}{5}, -10 \right)$$ ■

Again, though Cramer's Rule is not any shorter than the matrix method, it is much more mechanical to complete, especially when we work with systems that have more than two variables. Cramer's Rule also isolates an error to a single part of the solution, a specific determinant, rather than influencing the validity of all parts of the solution as in the matrix method. Cramer's Rule also makes it much easier to find the solutions to systems with fractional values in the solution.

EXAMPLE 3   A canoeist paddles 2 kilometers upstream in 1 hour and then makes the return trip in 20 minutes. If the rate of paddling and the current are constant values, find each.

Solution   We let

$$p = \text{the constant rate of paddling}$$
$$c = \text{the constant rate of the current}$$

	Rate ·	Time	= Distance
Upstream	$p - c$	1	2
Downstream	$p + c$	$\dfrac{1}{3}$	2

Then we have the system

$$
\begin{array}{ll}
p - c = 2 & \\
\dfrac{1}{3}(p + c) = 2 &
\end{array}
\quad \text{or} \quad
\begin{array}{l}
p - c = 2 \\
p + c = 6
\end{array}
$$

Solving by Cramer's Rule, we have

$$D = \begin{vmatrix} 1 & -1 \\ 1 & 1 \end{vmatrix} = 1 - (-1) = 1 + 1 = 2$$

$$D_p = \begin{vmatrix} 2 & -1 \\ 6 & 1 \end{vmatrix} = 2 - (-6) = 2 + 6 = 8$$

$$D_c = \begin{vmatrix} 1 & 2 \\ 1 & 6 \end{vmatrix} = 6 - (2) = 6 - 2 = 4$$

Thus

$$p = \frac{D_p}{D} = \frac{8}{2} = 4 \quad \text{and} \quad c = \frac{D_c}{D} = \frac{4}{2} = 2$$

Therefore the rate of paddling is 4 kilometers per hour, and the rate of the current is 2 kilometers per hour.   ■

We can always use Cramer's Rule to solve a system of equations with like terms except when the denominator $D = 0$. In these cases, in which the denominator D is found to equal zero, we will use the matrix method to determine whether the system is inconsistent or dependent. If it is dependent, the solution cannot be found by Cramer's Rule but can be found by the matrix method.

Exercises 8.5

Solve the following systems of equations by Cramer's Rule. If it is not possible to use Cramer's Rule, solve with the matrix method.

1. $5x + 7y = 2$
$4x - 5y = 7$

2. $3x - 5y = 4$
$7x + 3y = 10$

3. $3x + 2y = 0$
$7x - 3y = 0$

4. $5x + 4y = 0$
$2x + 5y = 0$

5. $3x + 2y + z = 5$
$x + y - 3z = 0$
$2x - y + z = 0$

6. $2x - y + 3z = 0$
$5x - 3y + z = 2$
$x + 5y - 2z = 0$

7. $x + 5y - 2z = 0$
$3x - 2y + z = 0$
$x + 3y + 3z = 4$

8. $x - y + 5z = 0$
$2x - y + 6z = 0$
$x + 3y - z = 5$

9. $x - 3y + 4z = 6$
$2x + 5y - 3z = -1$
$3x + 2y + z = 5$

10. $3x - y + z = -2$
$2x + y + 3z = 5$
$x - 2y - 2z = -7$

11. $3x + 2y - z = 5$
$2x - y + z = 9$
$2x + 3y + 2z = 7$

12. $3x + 2y - z = 9$
$2x - y + z = 2$
$2x + 3y + 2z = 5$

13. $x - 3y + 2z = 9$
$2x + 2y - z = 3$
$3x - y + z = 7$

14. $x + 2y - z = 5$
$2x - y + 2z = 7$
$3x + y + z = 10$

15. $x + y - z = 1$
$5x + y + z = -3$
$2x - y - 3z = -15$

16. $x + y - z = 8$
$5x + y + z = 6$
$2x - y - 3z = 18$

17. $x - 2y + z = 0$
$2x - y + 3z = 0$
$x - y - 2z = 0$

18. $x + 2y - z = 0$
$5x - y + 2z = 0$
$x - y + z = 0$

19. $x + 5y - 2z = -5$
$2x + 3y - z = 4$
$x + 2y + 3z = 2$

20. $3x - 2y - z = 4$
$x - 3y + 2z = 5$
$2x + y - 3z = 9$

21. A canoeist paddled upstream 1 mile in half an hour and then paddled the return trip in 10 minutes. If the rate the canoeist paddles and the rate of the current are constant values, find both.

22. A rowboat is rowed upstream 2 miles in 45 minutes and then rowed the return trip in 15 minutes. If the rate of the rowing and the rate of the current are constant values, find each.

23. A boy's pocket contains pennies, nickels, and quarters with a total face value of $1.80. The total number of coins is 20, and there are twice as many nickels as pennies. Find the number of each type of coin.

24. A girl's purse contains nickels, dimes, and quarters with a total face value of $4.00. The number of dimes is twice the number of nickels. Find the number of each type of coin.

25. Three grades of chocolate are combined to make a 30-pound mixture worth $40. The three grades of chocolate are worth $1 per pound, $2 per pound, and $3 per pound, and there is one more pound of $2 chocolate than there is $3 chocolate. Find the amount of each grade of chocolate in this mixture.

8.6   Partial Fractions

We can also use systems of equations when working with rational expressions, as in this section.

In the past we added rational expressions using a common denominator, as seen here:

$$\frac{5}{x + 1} + \frac{2}{x - 3} + \frac{1}{x + 2}$$

$$= \frac{5(x - 3)(x + 2)}{(x + 1)(x + 2)(x - 3)} + \frac{2(x + 1)(x + 2)}{(x + 1)(x + 2)(x - 3)} + \frac{1(x + 1)(x - 3)}{(x + 1)(x + 2)(x - 3)}$$

$$= \frac{8x^2 - x - 29}{(x + 1)(x + 2)(x - 3)}$$

It is a little more difficult to take the same resulting rational expression and break it into a sum of fractions. This process is called **partial fraction decomposition** and is demonstrated here:

$$\frac{8x^2 - x - 29}{(x + 1)(x + 2)(x - 3)} = \frac{5}{x + 1} + \frac{2}{x - 3} + \frac{1}{x + 2}$$

The way to verify this result is to add these fractions with a common denominator and see whether the sum on the right side equals the rational expression on the left side of the equation. In this section we will develop a method for finding this partial fraction decomposition of a rational expression. Partial fractions have an important use in calculus. Some rational expressions cannot be integrated in calculus, except by approximation, unless we first use partial fraction decomposition.

The following principles provide the guidelines for partial fraction decomposition.

PARTIAL FRACTION DECOMPOSITION

For a rational expression $\dfrac{p(x)}{q(x)}$ where the polynomials $p(x)$ and $q(x)$ have no common factors and the degree of $p(x)$ is less than the degree of $q(x)$,

(a) if $q(x)$ has a factor of $(x - c)^n$, then the partial fraction decomposition of $\dfrac{p(x)}{q(x)}$ contains the terms

$$\frac{A_1}{(x - c)} + \frac{A_2}{(x - c)^2} + \frac{A_3}{(x - c)^3} + \frac{A_4}{(x - c)^4} + \cdots + \frac{A_n}{(x - c)^n}$$

where c and A_i are real numbers; and

(b) if $q(x)$ has a factor of $(x^2 + bx + d)^m$, where $x^2 + bx + d$ has only nonreal complex zeros, then the partial fraction decomposition of $\dfrac{p(x)}{q(x)}$ contains

$$\frac{B_1 x + C_1}{x^2 + bx + d} + \frac{B_2 x + C_2}{(x^2 + bx + d)^2} + \cdots + \frac{B_m x + C_m}{(x^2 + bx + d)^m}$$

We apply both parts of this partial fraction decomposition to every factor in the denominator $q(x)$ of the rational expression $\dfrac{p(x)}{q(x)}$. Thus the first step in partial fraction decomposition is to factor the denominator. Every denominator can be factored into factors of $(x - c)^n$, where c is a real number, and $(x^2 + bx + d)^m$, where $x^2 + bx + d$ has only nonreal complex zeros. Thus the first step is to find these factors of $q(x)$.

EXAMPLE 1   Rewrite

$$\frac{3x^2 + 8x + 6}{(x + 1)^3}$$

using partial fraction decomposition for $x \neq -1$.

Solution   The denominator is already factored, indicating that the partial fraction decomposition looks like

$$\frac{3x^2 + 8x + 6}{(x + 1)^3} = \frac{A}{(x + 1)} + \frac{B}{(x + 1)^2} + \frac{C}{(x + 1)^3}$$

where A, B, and C are real numbers. Next, we multiply both sides by the lowest common denominator, $\text{LCD} = (x + 1)^3$, and

$$(x + 1)^3 \frac{3x^2 + 8x + 6}{(x + 1)^3} = (x + 1)^3 \left[\frac{A}{(x + 1)} + \frac{B}{(x + 1)^2} + \frac{C}{(x + 1)^3} \right],$$

$$3x^2 + 8x + 6 = A(x + 1)^2 + B(x + 1) + C$$
$$= Ax^2 + 2Ax + A + Bx + B + C$$
$$= (A)x^2 + (2A + B)x + (A + B + C)$$

Thus equality occurs when the like terms on the left equal the like terms on the right. Then the coefficients of like terms are equal. Therefore

$3x^2 = Ax^2$ or $3 = A$

$8x = (2A + B)x$ or $8 = 2A + B$ *(With A = 3 we have 8 = 6 + B or 2 = B)*

$6 = A + B + C$ *(With A = 3 and B = 2 we have*
                                6 = 3 + 2 + C, 1 = C)

Therefore

$$\frac{3x^2 + 8x + 6}{(x + 1)^3} = \frac{3}{x + 1} + \frac{2}{(x + 1)^2} + \frac{1}{(x + 1)^3}$$ ■

Notice that this solution involves the system of equations

$$A \qquad\qquad = 3$$
$$2A + B \qquad = 8$$
$$A + B + C = 6$$

which could have been solved by any method in this chapter.

EXAMPLE 2   Rewrite

$$\frac{2x^2 + 5x + 5}{(x + 1)(x^2 + x + 1)}$$

using partial fraction decomposition for $x \neq -1$.

Solution   Since the denominator is factored and $x^2 + x + 1$ has only nonreal complex zeros, by the properties of partial fraction decomposition we have

$$\frac{2x^2 + 5x + 5}{(x + 1)(x^2 + x + 1)} = \frac{A}{x + 1} + \frac{Bx + C}{x^2 + x + 1}$$

We multiply both sides of the LCD $= (x + 1)(x^2 + x + 1)$ and

$$2x^2 + 5x + 5 = A(x^2 + x + 1) + (Bx + C)(x + 1)$$

Simplifying, we have

$$2x^2 + 5x + 5 = Ax^2 + Ax + A + Bx^2 + Cx + Bx + C$$
$$2x^2 + 5x + 5 = (A + B)x^2 + (A + C + B)x + (A + C)$$

For equality the like terms on both sides are equal. Thus we have

$$2x^2 = (A + B)x^2 \qquad \text{or} \qquad A + B \qquad = 2$$
$$5x = (A + C + B)x \qquad \text{or} \qquad A + B + C = 5$$
$$5 = (A + C) \qquad \text{or} \qquad A \quad + C = 5$$

Solving this system by substitution and elimination, we get $C = 3$, $A = 2$, and $B = 0$. Therefore

$$\frac{2x + 5x + 5}{(x + 1)(x^2 + x + 1)} = \frac{2}{x + 1} + \frac{3}{x^2 + x + 1}$$ ■

Very often, we can simplify this process using this shortcut: After both sides have been multiplied by the LCD, substitute for x the values that are zeros for the LCD. This shortcut is demonstrated in the following example.

EXAMPLE 3   Rewrite

$$\frac{6x^2 - x - 17}{(x + 2)(x^2 - 1)}$$

using partial fraction decomposition.

Solution   We first factor the denominator to $(x + 2)(x - 1)(x + 1)$ and set up the partial fraction decomposition.

$$\frac{6x^2 - x - 17}{(x + 2)(x^2 - 1)} = \frac{A}{x + 2} + \frac{B}{x - 1} + \frac{C}{x + 1}$$

Then multiplying both sides by the LCD $= (x + 2)(x - 1)(x + 1)$, we have

$$6x^2 - x - 17 = A(x - 1)(x + 1) + B(x + 2)(x + 1) + C(x + 2)(x - 1)$$

Now we use the shortcut and substitute for x the zeros for the LCD. When $x = -2$,

$$6(-2)^2 - (-2) - 17 = A(-2 - 1)(-2 + 1) + 0 + 0$$
$$24 + 2 - 17 = A(-3)(-1)$$
$$9 = 3A$$
$$3 = A$$

When $x = 1$,

$$6(1)^2 - (1) - 17 = 0 + B(1 + 2)(1 + 1) + 0$$
$$6 - 1 - 17 = B(3)(2)$$
$$-12 = 6B$$
$$-2 = B$$

And when $x = -1$,

$$6(-1)^2 - (-1) - 17 = 0 + 0 + C(-1 + 2)(-1 - 1)$$
$$6 + 1 - 17 = C(1)(-2)$$
$$-10 = -2C$$
$$5 = C$$

Therefore

$$\frac{6x^2 - x - 17}{(x + 2)(x^2 - 1)} = \frac{3}{x + 2} + \frac{-2}{x - 1} + \frac{5}{x + 1} \qquad \text{for} \qquad x \notin \{-2, \pm 1\} \quad \blacksquare$$

EXAMPLE 4   Rewrite

$$\frac{7x^3 + 21x^2 + 20x + 5}{(x^2 + 2x + 1)(x^2 + 3x + 3)}$$

using partial fractions.

Solution   We factor the denominator and set up the partial fraction decomposition:

$$\frac{7x^3 + 21x^2 + 20x + 5}{(x^2 + 2x + 1)(x^2 + 3x + 3)} = \frac{A}{x + 1} + \frac{B}{(x + 1)^2} + \frac{Cx + D}{x^2 + 3x + 3}$$

Multiply both sides by the LCD $= (x + 1)^2(x^2 + 3x + 1)$:

$$7x^3 + 21x^2 + 20x + 5 =$$
$$A(x + 1)(x^2 + 3x + 3) + B(x^2 + 3x + 3) + (Cx + D)(x + 1)^2$$

When $x = -1$,

$$7(-1)^3 + 21(-1)^2 + 20(-1) + 5 = 0 + B[(-1)^2 + 3(-1) + 3] + 0$$
$$-7 + 21 - 20 + 5 = B(1 - 3 + 3)$$
$$-1 = B$$

Then, since there are no other zeros for the LCD, work with

$7x^3 + 21x^2 + 20x + 5$
$= A(x + 1)(x^2 + 3x + 3) - (x^2 + 3x + 3) + (Cx + D)(x + 1)^2$
$= Ax^3 + 4Ax^2 + 6Ax + 3A - x^2 - 3x - 3 + Cx^3 + (2C + D)x^2 + (2D + C)x + D$
$= (A + C)x^3 + (4A - 1 + 2C + D)x^2 + (6A - 3 + 2D + C)x + (3A - 3 + D)$

Thus for equality the like terms on the left must equal the corresponding like terms on the right side of the equality. Thus

$$7x^3 = (A + C)x^3 \qquad \text{or} \qquad A + C = 7$$
$$21x^2 = (4A - 1 + 2C + D)x^2 \qquad \text{or} \qquad 4A + 2C + D = 22$$
$$20x = (6A - 3 + 2D + C)x \qquad \text{or} \qquad 6A + C + 2D = 23$$
$$5 = 3A - 3 + D \qquad \text{or} \qquad 3A + D = 8$$

We will solve the system of equations created by the first three equations above. The denominator for Cramer's Rule is

$$Denominator = \begin{vmatrix} 1 & 1 & 0 \\ 4 & 2 & 1 \\ 6 & 1 & 2 \end{vmatrix} = 1$$

and we have

$$A = \frac{\begin{vmatrix} 7 & 1 & 0 \\ 22 & 2 & 1 \\ 23 & 1 & 2 \end{vmatrix}}{-1} = \frac{0}{1} = 0 \qquad C = \frac{\begin{vmatrix} 1 & 7 & 0 \\ 4 & 22 & 1 \\ 6 & 23 & 2 \end{vmatrix}}{-1} = \frac{7}{1} = 7$$

$$D = \frac{\begin{vmatrix} 1 & 1 & 7 \\ 4 & 2 & 22 \\ 6 & 1 & 23 \end{vmatrix}}{-1} = \frac{8}{1} = 8$$

Therefore

$$\frac{7x^3 + 21x^2 + 20x + 5}{(x^2 + 2x + 1)(x^2 + 3x + 3)} = \frac{-1}{(x + 1)^2} + \frac{7x + 8}{x^2 + 3x + 3} \qquad \text{for } x = -1 \qquad \blacksquare$$

If a rational expression $\frac{p(x)}{q(x)}$ has the degree of $p(x)$ greater or equal to the degree of $q(x)$, we cannot immediately use partial fraction decomposition. To change this rational expression into a sum, we first divide $p(x)$ by $q(x)$ and use partial fraction decomposition on the rational expression containing the remainder. This process is demonstrated in the next example.

EXAMPLE 5   Set up the following rational expression as a sum, but do not complete the partial fraction decomposition of

$$\frac{x^4 + 5x^2 - 3}{x^2 + 5x + 6}$$

Solution   Since the degree of the numerator is greater than the degree of the denominator, we rewrite the rational expression by dividing the denominator into the numerator. Thus

$$\frac{x^4 + 5x^2 - 3}{x^2 + 5x + 6} = x^2 - 5x + 24 + \frac{-90x - 141}{x^2 + 5x + 6}$$

We factor the denominator $x^2 + 5x + 6$ to $(x + 2)(x + 3)$ and use partial fraction decomposition on the remainder fraction:

$$\frac{-90x - 141}{(x + 2)(x + 3)} = \frac{A}{(x + 2)} + \frac{B}{(x + 3)}$$

Combining this with the results of division, we then have the partial fraction decomposition of the original rational expression:

$$\frac{x^4 + 5x^2 - 3}{(x + 2)(x + 3)} = x^2 - 5x + 24 + \frac{A}{(x + 2)} + \frac{B}{(x + 3)} \quad \text{for } x \notin \{-2, -3\} \blacksquare$$

Exercises 8.6

Set the following problems up for solution with partial fractions. Do not solve them; just set them up for solution.

1. $\dfrac{x + 5}{(x + 1)(x - 3)}$

2. $\dfrac{x + 2}{(x + 1)(x + 5)}$

3. $\dfrac{x - 3}{x^2 - 5x + 6}$

4. $\dfrac{x + 1}{x^2 + 5x + 6}$

5. $\dfrac{x^2 - 2x + 1}{(x + 1)^2(x - 2)}$

6. $\dfrac{x^2 - x + 2}{(x + 3)(x - 2)^2}$

7. $\dfrac{x^2 - 5x + 7}{(x + 2)^3(x - 1)^2}$

8. $\dfrac{x^2 - 7x + 5}{(x - 1)^3(x - 2)^2}$

9. $\dfrac{x - 15}{(x^2 - 5x + 1)(x + 2)}$

10. $\dfrac{x - 17}{(x + 1)(x^2 - x + 5)}$

11. $\dfrac{x^3 + 2x^2 + x - 5}{(x^2 + x + 6)^2(x + 2)^2}$

12. $\dfrac{x^6 + 2x^3 - x + 1}{(x^2 - x + 7)^2(x + 1)^3}$

15. $\dfrac{7x + 6}{x^2 - 4}$

16. $\dfrac{4x - 6}{x^2 - 9}$

17. $\dfrac{2x^2 - x + 4}{(x - 1)^3}$

18. $\dfrac{5x^2 + 9x + 7}{(x + 1)^3}$

19. $\dfrac{3x^2 + 4x + 1}{(x^2 + x + 1)(x - 2)}$

20. $\dfrac{x^2 + 5x + 2}{(x^2 + 3x + 3)(x + 1)}$

21. $\dfrac{-x^2 + 7x - 10}{x^3 + 2x^2 - 4x - 8}$

22. $\dfrac{6x^2 - 4x - 30}{x^3 - 3x^2 - 9x + 27}$

23. $\dfrac{5x^2 - 4x + 49}{x^2 - x + 9}$

24. $\dfrac{5x^2 - 12x + 5}{x^3 - 3x^2 + x}$

25. $\dfrac{3x^2 + 11x}{x^3 - 7x - 6}$

26. $\dfrac{8x^2 + 9x - 25}{x^3 - 7x + 6}$

27. $\dfrac{x^3 + 3x^2 + 3x - 1}{x^2 - 1}$

Rewrite the following rational expressions, using partial fraction decomposition.

13. $\dfrac{4x + 3}{(x + 2)(x - 3)}$

14. $\dfrac{3x + 9}{(x + 5)(x - 1)}$

28. $\dfrac{x^4 + x^3 - 4x^2 - 5x - 6}{x^2 - 4}$

8.7   Systems of Inequalities and Linear Programming

Linear programming is a method for solving practical problems that involve finding a maximum or minimum value. To find these maximum and minimum values, we must first solve a system of linear inequalities. Using the solutions of the system of linear inequalities, we can then evaluate quantities such as profit, cost, and loss to find the maximum profit, minimum cost, or even minimum loss. The following problem provides a good example of a linear programming problem.

Midnight Auto Supply has a limited amount of space to store tires and oil drums. The total amount of both that can be stored is 600 units. Because of customer demand, two units of tires must be stored for every unit of oil, the total of both being greater than or equal to 400 units at the beginning of each month. Also, the number of units of tires must be greater than 100, and the number of units of oil must be greater than 100. Under these restrictions, find the minimum cost of stocking the warehouse when the tires cost $100 per unit and the oil costs $80 per unit. Also find the maximum profit when the profit is $80 per tire unit and $120 per oil unit.

Although this problem seems overwhelmingly complicated, we will solve it later on in this section (Example 4).

To solve this and other linear programming problems, we must be able to graph inequalities and solve systems of inequalities. The solution of a system of linear inequalities is the major part of the solution in linear programming problems.

Systems of Inequalities

We graph systems of inequalities much as we graphed regions in Section 2.4. The following steps delineate this process.

GRAPHING SYSTEMS OF INEQUALITIES	Complete the following steps to graph a system of inequalities. STEP 1: Graph each boundary as if the inequality were an equality. (Use a solid curve if the equality is included and a dashed curve if the equality is not included.) STEP 2: Shade the region on one side of each boundary where a sample point makes the inequality true. (Try representative points to verify.) STEP 3: Find the intersection of all inequality regions on the graph.

We will now use these steps to complete the following examples.

EXAMPLE 1   Find the solution to

$$x^2 + y^2 \le 9$$
$$y < -3x + 1$$

Solution   Follow the steps above.

STEP 1: Find the boundary for each inequality.

(a) $x^2 + y^2 = 9$ is a circle with center at $(0, 0)$ and a radius of 3. The circle is a solid boundary line, since the equality is included in $x^2 + y^2 \le 9$.

(b) $y = -3x + 1$ is a line with slope $m = -3$ through $(0, 1)$. The line is dashed, since the equality is not included in $y < -3x + 1$ (see Fig. 8.2(a)).

STEP 2: Test a point for each inequality not on the boundary for that inequality.

(a) For $x^2 + y^2 \le 9$, substitute $(0, 0)$. Thus

$$0^2 + 0^2 \le 9$$
$$0 \le 9$$

is true, and we shade the interior of the circle.

(b) For $y < -3x + 1$, substitute $(0, 0)$. Thus

$$0 < 0 + 1$$
$$0 < 1$$

is true, and we shade the region below the dashed line (see Fig. 8.2(b)).

STEP 3: We now indicate in one diagram the intersection of the two inequalities (see Fig. 8.2(c)).

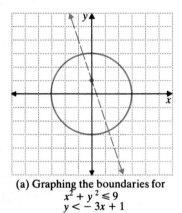

(a) Graphing the boundaries for
$x^2 + y^2 \le 9$
$y < -3x + 1$

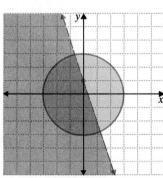

(b) Graphing the regions for
$x^2 + y^2 \le 9$
$y < -3x + 1$

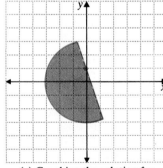

(c) Graphing the solution for
$x^2 + y^2 \le 9$
$y < -3x + 1$

Figure 8.2

EXAMPLE 2   Find the solution to the linear system of inequalities

$$2x + y \le 6$$
$$x + y \ge 2$$
$$1 \le x \le 2$$
$$y \le 3$$

Solution   We will use the steps for graphing systems of inequalities.

STEP 1: We graph the boundary line for each inequality (Fig. 8.3(a)).

STEP 2: We substitute the point (0, 0) to identify the shaded side of each inequality (Fig. 8.3(b)).

STEP 3: We find the intersection of these regions (Fig. 8.3(c)).

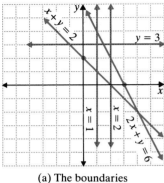

(a) The boundaries

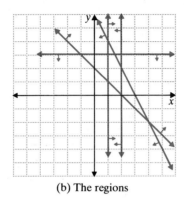

(b) The regions

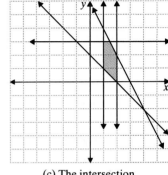

(c) The intersection

Figure 8.3

The solution to this linear system of inequalities in Example 2 is called a **convex polygon**. It is a polygon because its sides are line segments joined at their endpoints. It is convex because the segment joining any two interior points is entirely on the interior (inside) of the polygon (Fig. 8.4).

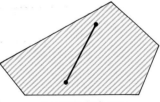
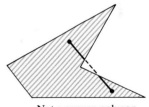

Figure 8.4   A convex polygon   Not a convex polygon

Using complex polygons, we will now solve linear programming problems.

Linear Programming

The concept of linear programming has application in many areas, especially in business problems. A linear programming problem in two variables has the following two parts:

(a) a system of linear inequalities with a convex polygon solution and

(b) a linear function in two variables that needs to be maximized or minimized within the conditions of the convex polygon.

Since the function in part (b) is a linear function, it will have a slope at any point and in any direction. Thus in solving a linear programming problem the maximums and minimums *always* occur at the vertices (corners) of the convex polygon. This is easier to see in a three-dimensional graph of a linear programming problem. The following examples demonstrate linear programming problems and their solutions.

EXAMPLE 3

For the conditions of the linear systems of inequalities

$$2x + y \le 6$$
$$x + y \ge 2$$
$$1 \le x \le 2$$
$$y \le 3$$

find the maximum for $P = 5x - 3y$.

Solution   Since this is the same system of inequalities as in Example 2 (see Fig. 8.3(c) on preceding page), we know that the systems solution is a convex polygon. Next we find the vertices for the convex polygon by substitution. The vertices of the convex polygon are $(1, 1)$, $(1, 3)$, $(2, 0)$, $(2, 2)$, and $\left(\frac{3}{2}, 3\right)$. We then evaluate $P = 5x - 3y$ at each vertex to find the maximum value of P.

Vertices	$P = 5x - 3y$
$(1, 1)$	$P = 5(1) - 3(1) = 5 - 3 = 2$
$(1, 3)$	$P = 5(1) - 3(3) = 5 - 9 = -4$
$(2, 0)$	$P = 5(2) - 3(0) = 10 - 0 = 10$
$(2, 2)$	$P = 5(2) - 3(2) = 10 - 6 = 4$
$\left(\frac{3}{2}, 3\right)$	$P = 5\left(\frac{3}{2}\right) - 3(3) = \frac{15}{2} - 9 = -\frac{3}{2}$

Therefore the maximum value for P is 10 and occurs at the point $(2, 0)$. We rely on the fact that the maximum value of P occurs at the corners of the convex polygon, since $P = 5x - 3y$ is a linear function.   ■

EXAMPLE 4

Midnight Auto Supply has a limited amount of space to store tires and oil drums. The total amount of both that can be stored is 600 units. Because of customer demand, two units of tires must be stored for every unit of oil, the total of both being greater than or equal to 400 units at the beginning of each month. Also, the number of units of tires must be greater than 100, and the number of units of oil must be greater than 100. Under these restrictions, find the minimum cost of stocking the warehouse when the tires cost $100 per unit and the oil costs $80 per unit. Also find the maximum profit when the profit is $80 per tire unit and $120 per oil unit.

Solution We let

$$x = \text{the number of tire units}$$
$$y = \text{the number of oil drum units}$$

For the sum less than 600 we have

$$x + y \leq 600$$

Because of customer demand, we have

Also, the minimum levels of each are

$$x \geq 100$$
$$y \geq 100$$

Next, draw the graph for this system of inequalities and find the vertices of the convex polynomial (Fig. 8.5). Now we calculate the cost equation $C = 100x + 80y$ and the profit equation $P = 80x + 120y$ at each vertex.

Vertices	$C = 100x + 80y$	$P = 80x + 120y$
(100, 200)	$C = 100(100) + 80(200) = 26{,}000$	$P = 80(100) + 120(200) = 32{,}000$
(100, 500)	$C = 100(100) + 80(500) = 50{,}000$	$P = 80(100) + 120(500) = 68{,}000$
(500, 100)	$C = 100(500) + 80(100) = 58{,}000$	$P = 80(500) + 120(100) = 52{,}000$
(150, 100)	$C = 100(150) + 80(100) = 23{,}000$	$P = 80(150) + 120(100) = 24{,}000$

Thus the minimum cost is $23,000 at (150, 100), 150 units of tires and 100 units of oil. Also, the maximum profit is $68,000 at (100, 500), 100 units of tires and 500 units of oil. This information would then give the company guidance in ordering inventory and running the business. ■

Exercises 8.7

Graph the solution to each of the following systems of inequalities.

1. $x^2 + y^2 < 16$
$x + y \leq 1$

2. $x^2 + y^2 < 4$
$x + y \geq 0$

3. $x^2 + y^2 \geq 9$
$y \geq x^3$

4. $x^2 + y^2 \geq 16$
$y \leq x^3$

5. $y \geq x^2$
$y \leq x^3$

6. $y \geq x^2$
$y \geq x^3$

7. $y \geq x^2$
$x - y > 3$

8. $y \geq x^2$
$2x - y > 2$

Solve the following linear programming problems. First solve the system of inequalities and find the vertices of the convex polygon. Then evaluate the maximum or minimum equation.

9. $y \geq x - 1$
$y \geq -3x + 7$
$x + y \leq 5$
Find the maximum of
$P = 30x + 50y - 10$.

10. $x + y \leq 5$
$y \geq x - 1$
$5x + y \geq 5$
Find the maximum of
$P = 15x - 5y + 3$.

11. $y \geq x - 1$
$$x + y \leq 5$$
$$y \geq 1$$
$$x \geq 1$$
Find the minimum of
$C = 30x - 5y + 10.$

12. $x + y \leq 5$
$$y \geq 2$$
$$y \leq 2x + 2$$
Find the minimum of
$C = 20x + 15y - 5.$

13. $3x + y \leq 11$
$$y \leq x + 3$$
$$x \geq 1$$
$$y \geq 2$$
Find the maximum of
$P = 30x - 15y + 25.$

14. $3x + y \leq 11$
$$y \leq x + 3$$
$$x + y \geq 5$$
Find the maximum of
$P = 25x + 10y - 10.$

15. The Pocket Rocket Toy Company makes two toys: robots and spaceships. Owing to limitations in equipment, the company cannot produce more than 400 robots per week and cannot produce more than 600 spaceships per week. The total number of toys produced each week cannot exceed 800. How many robots and spaceships should be produced to maximize profits when each robot has a profit of $3.00 and each spaceship has a profit of $2.00?

16. The Greathouse Construction Company is designing a commercial building with an exterior of mirrored windows and aluminum panels. There can be only a maximum of 300 windows and a maximum of 500 panels. The total number of windows and panels is 700. If the aluminum panels cost $300 each and the mirrored windows cost $350 each, find the number of panels and windows that minimizes the cost.

17. The VCR Electronics Company produces standard and deluxe video recorders. The company must produce at least 200 standard recorders and must also produce at least 100 deluxe recorders each month. Because of customer demand, the number of standard recorders added to twice the number of deluxe recorders cannot exceed 800 units per week. If the profit on the standard recorder is $75 and the profit on the deluxe recorder is $90, how many standard and deluxe models should be produced to maximize profits?

18. The Short Circuit Electronics Company produces a standard and a deluxe satellite-receiving antenna dish. Production cannot exceed 600 standard and 300 deluxe antennas each month. Total production of both antennas is limited to 800 units per month. If profit on the standard is $1000 and on the deluxe $2000, how many of each should be produced to maximize profits?

19. Bowen's Boat Builders makes yachts and sailboats. In production the raw materials and labor are used as indicated by the following table.

	Units of Materials	Units of Labor
Yachts	5	8
Sailboats	2	4

Owing to costs, the company has only 40 units of raw materials and 72 units of labor available each month. Each month at least two yachts and seven sailboats must be produced. If the yachts provide a profit of $50,000 and the sailboats a profit of $15,000, how many of each should be produced to maximize profits?

CHAPTER SUMMARY AND REVIEW

KEY TERMS

Types of systems of equations
    A consistent system
    A dependent system
    An inconsistent system
Solving systems of equations
    By substitution
    By elimination

    By the matrix method
    By Cramer's Rule
Homogeneous systems
Evaluating determinants
Partial fraction decomposition
Graphing systems of inequalities
Linear programming

CHAPTER EXERCISES

Solve by substitution.

1. $\log_5 y = x^2$
$\log_5 y = 2 + x$

2. $x^2 + y^2 = 3$
$y = x + 7$

Solve by elimination.

3. $5x - 3y = 2$
$x - 2y = -1$

4. $x - 3y = 3$
$\frac{1}{3}x - y = 1$

Solve by the matrix method (elimination by Gauss's algorithm).

5. $x + y - z = 8$
$2x - 4y + z = 1$
$3x - 5y + 2z = 3$

6. $x + y + z = 3$
$x - 3y + z = 1$
$x - y + z = 2$

7. $x + y + z = 0$
$5x - 3y + z = 0$
$3x - y + z = 0$

Evaluate these determinants (use the definitions as well as the properties).

8. $\begin{vmatrix} 5 & 3 \\ 4 & -1 \end{vmatrix}$

9. $\begin{vmatrix} 2 & 4 & 6 \\ 1 & 3 & 0 \\ 2 & 7 & 5 \end{vmatrix}$

10. $\begin{vmatrix} 3 & 4 & 5 \\ 1 & 2 & 3 \\ -1 & 1 & -2 \end{vmatrix}$

Solve with Cramer's Rule.

11. $x + y - z = 5$
$2x - 3y + 2z = 0$
$x - y + z = 0$

12. $x + y + z = 3$
$x - 2y - z = 2$
$2x + y + 2z = 4$

Solve.

13. $2x + 3y - 4z = 0$
$5x - 2y + z = 0$
$x + y - z = 0$

14. $x + y + z = 0$
$5x - y - z = 0$
$x - 5y + 2z = 0$

Rewrite each, using partial fraction decomposition.

15. $\dfrac{3x^2 + 8x + 4}{(x - 1)(x^2 + x + 3)}$

16. $\dfrac{8x^2 - 11x - 15}{x^3 - 2x^2 - 3x}$

17. $\dfrac{-2x^3 + 5x^2 + 9x + 6}{x^3 + 2x^2 - x - 2}$

18. Graph this system of inequalities:

$$x^2 + y^2 \geq 4$$
$$x + y \geq -2$$
$$x \leq 4$$

19. For

$$x + 2y \leq 10$$
$$x + y \geq 3$$
$$1 \leq x \leq 4$$
$$0 \leq y \leq 4$$

find the maximum and minimum for $H = 3x - 4y + 15$.

20. The Graduate Diploma Company sells Bachelor's and Master's "replacement" diplomas. Before federal prosecution the units of materials and labor followed the table below:

	Units of Materials	Units of Labor
Bachelor's	1	2
Master's	2	3

There were 5000 units of material available and 8000 units of labor available each month. Minimum production for Bachelor's diplomas was 500, and minimum production for Master's diplomas was 1000 per month. If the profit was $20 for Bachelor's diplomas and $30 for Master's diplomas, what production levels would have produced maximum profits?

Analytic Geometry

CONIC SECTION SURGERY

"Kidney Stones Miraculously Dissolved" could have been the headlines in February 1980 when Christian Chaussy of the Department of Urology at the University of Munich began dissolving kidney stones in humans without surgery. This new method, which partially displaces "therapy with the knife," is called extracorporeal shock wave lithotripsy, or ESWL.

This space-age treatment uses the classical Greek geometry of the ellipse and the modern electronics of shock waves. In ESWL, shock waves are emitted from an electrode at one focus of a semiellipsoid (one-half of a three-dimensional ellipse). The shock waves then reflect off the sides of the semiellipsoid and are focused at the ellipsoid's second focus. After about 2000 shock waves the kidney stone positioned at the ellipsoid's second focus is pulverized into particles that are small enough to pass through the urinary tract.

Schematic representation of pressure transmission of shock wave exposure. The density of the screen compares with the approximate pressure density of the transmitting pressure as it moves through the tissue.

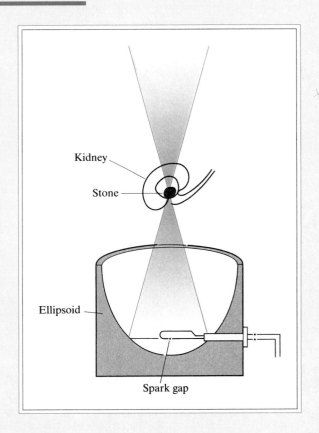

Kidney

Stone

Ellipsoid

Spark gap

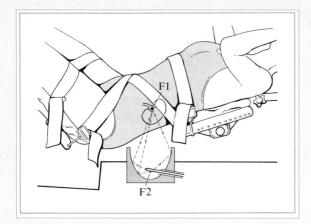

Patient on patient support with schematic drawing of the ellipsoid with spark gap (F1) and stone in focal area F2 (left).

The semiellipsoid used in extracorporeal shock wave lithotripsy (below).

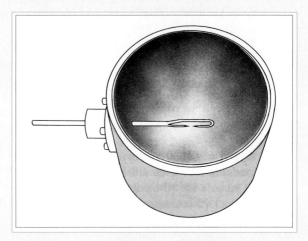

Two major advantages of ESWL are recovery time and safety. Recovery time with ESWL is limited to an average of 3.5 days in the hospital and immediate return to work, while recovery time of traditional surgery is 14 to 21 days. ESWL also avoids the 3% morbidity rate of traditional surgery. In the first six years of ESWL, more than 100,000 cases were treated internationally (Munich, Stuttgart, Sapporo, UCLA) with a success rate of 99% and higher.

ESWL is an important application of the ellipse and the ellipsoid, which we will study in this chapter. We will also study other conic sections and some of their applications.

SOURCES: Dornier Medical Systems, Inc., Marietta, Georgia; Christian Chaussy, *Extracorporeal Shock Wave Lithotripsy*, S. Karger (AG Basel 1986). Reprinted by permission.

9.1   Conic Sections: The Circle and Parabola

Analytic geometry is one spark that helped ignite the mathematical discovery of the European Renaissance (ca. 1300–1600 A.D.). During this period, European scholars studied the classical geometry of the ancient Greeks (ca. 600–400 B.C.) with the use of algebraic variables, borrowed from the Arabs and introduced into European mathematics during the Moorish occupation of Spain (ca. 711–1056 A.D.). Combining algebra with geometry produced an algebraic way of describing and manipulating geometric shapes. This mathematical method is called analytic geometry and was a necessary prerequisite for the invention and development of calculus in Europe (beginning ca. 1665 A.D.). It was René Descartes who published the first analytic geometry book in 1637. We have already done a lot of analytic geometry in this book. We will now work with the conic sections studied by both the ancient Greeks and the scholars of the European Renaissance.

Conic Sections

The surface called a **cone** is formed by two lines that intersect at a given point with θ as the angle θ between the lines. The first line is the fixed or stationary line, θ is the angle between the lines, and the second line is rotated around the fixed line, keeping the same point of intersection and angle θ with the fixed line. The set of all points which the rotating line passes through is a cone (Fig. 9.1(b)).

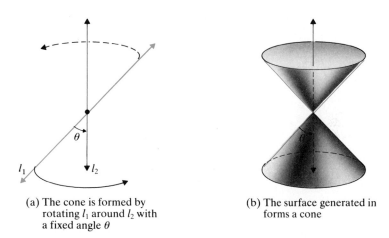

(a) The cone is formed by
rotating l_1 around l_2 with
a fixed angle θ

(b) The surface generated in
forms a cone

Figure 9.1

The intersection of a plane (a flat unending surface) and a cone create several classical geometric shapes. These shapes (the circle, parabola, ellipse, and hyperbola) are called **conic sections** (Fig. 9.2).

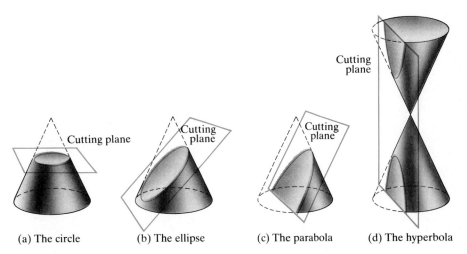

(a) The circle   (b) The ellipse   (c) The parabola   (d) The hyperbola

Figure 9.2

The Circle

Earlier, we worked with the equation of a circle. We can now describe a circle geometrically either as a cross section of a cone or in terms of distance. In fact, all conic sections can be defined by using distances and geometric cross sections of a cone.

Definition | A **circle** is the set of points in a plane a given distance r from a given point (h, k). The equation for the circle in the x-y plane with center (h, k) and radius r is $(x - h)^2 + (y - k)^2 = r^2$.

Squaring the binomials in $(x - h)^2 + (y - k)^2 = r^2$, the equation of the circle can be written as

$$x^2 + y^2 + (-2hx) + (-2ky) + (h^2 + k^2 - r^2) = 0$$

This is similar to the general form of the equation of a circle that we will use in the next example:

$$x^2 + y^2 + Ax + By + C = 0$$

We will now write the equation of a circle through three given points.

EXAMPLE 1   Find the center of the circle containing the points $(0, 5)$, $(0, -3)$, and $(5, 0)$.

Solution   We substitute these three points into the general equation of a circle $x^2 + y^2 + Ax + By + C = 0$ as follows:

$$x^2 + y^2 + Ax + By + C = 0$$

For $(0, 5)$	$0^2 + 5^2 + A(0) + B(5) + C = 0$	and $\quad 5B + C = -25$
For $(0, -3)$	$0^2 + (-3)^2 + A(0) + B(-3) + C = 0$	and $-3B + C = \quad -9$
For $(5, 0)$	$5^2 + 0^2 + A(5) + B(0) + C = 0$	and $\quad 5A + C = -25$

Solving this system of equations as we did in Chapter 8, we have $A = -2$, $B = -2$, and $C = -15$. Thus the circle has an equation of $x^2 + y^2 - 2x - 2y - 15 = 0$. To find the center of this circle, we complete the square, so we can write the equation of the circle in the same form as the definition above:

$$x^2 - 2x \quad + y^2 - 2y \quad = 15$$
$$x^2 - 2x + 1 + y^2 - 2y + 1 = 15 + 1 + 1 \qquad \textit{(Half the coefficient of the}$$
$$\textit{middle term squared is added}$$
$$\textit{to both sides)}$$
$$(x - 1)^2 \quad + (y - 1)^2 \quad = 17$$

Thus this circle has a radius of $\sqrt{17}$ and a center of $(1, 1)$.   ■

Example 1 demonstrates a technique called **curve fitting**, which means finding the equation of a specific type of curve, in this case a circle, through given points.

The Parabola

Another conic section, the parabola, can be described geometrically as a conic section (Fig. 9.2), or it can be defined in terms of distances as we see here.

Definition

The points on a **parabola** are the set of all points in a plane equidistant from a given point and a given line. The given point is called the **focus**, and the given line is called the **directrix**.

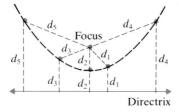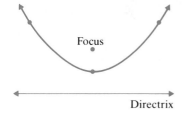

Points on the parabola are equidistant from the focus and the directrix
$(d_1 = d_1, d_2 = d_2, d_3 = d_3, d_4 = d_4, d_5 = d_5)$

The vertex is the point on the parabola halfway between the focus and directrix. The line containing the vertex and the focus is a line of symmetry for the graph of the parabola. This line of symmetry is called the **axis of symmetry** for the parabola. To find the equation of a parabola, we use the distance definition above with the diagrams in Fig. 9.3(a). To find the equation of the parabola, we square d_1 and d_2 and then simplify the equation $d_1^2 = d_2^2$:

$$d_1 = \sqrt{(x - 0)^2 + (y - p)^2} = \sqrt{x^2 + y^2 - 2py + p^2}$$
$$d_2 = \sqrt{(x - x)^2 + [y - (-p)]^2} = \sqrt{0^2 + y^2 + 2py + p^2}$$

Then

$$d_1^2 = d_2^2$$

is

$$x^2 + y^2 - 2py + p^2 = y^2 + 2py + p^2$$
$$x^2 - 2py = 2py$$
$$x^2 = 4py$$

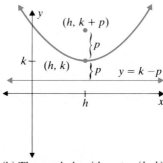

(a) The parabola with focus $(0, p)$ and directrix $y = -p$ has an equation $x^2 = 4py$

(b) The parabola with vertex (h, k), focus $(h, k + p)$, and directrix $y = k - p$ has an equation $(x - h)^2 = 4p(y - k)$

Figure 9.3

where p is the distance from the vertex to the focus and the focus is $(0, 0)$. When $p > 0$, the parabola opens up as in Fig. 9.3(a); when $p < 0$, the parabola would open down. Using the methods of translation in Chapter 2, we can translate the parabola with vertex at $(0, 0)$ so that its vertex is at (h, k), resulting in the more general equation of

$$(x - h)^2 = 4p(y - k)$$ *(See Fig. 9.3(b) on preceding page)*

The parabolas that open up or down, like those in Fig. 9.3, have equations in the form of $x^2 = 4py$. On the other hand, parabolas that open to the right or to the left have equations in the form of $y^2 = 4px$. Now we will describe the equation of a parabola in more detail.

EQUATIONS OF A PARABOLA

A parabola with vertex (h, k) and a distance p from the vertex to the focus has an equation of

(a) $(x - h)^2 = 4p(y - k)$ with a vertical axis of symmetry, $x = h$, which opens up for $p > 0$ or opens down for $p < 0$, or

(b) $(y - k)^2 = 4p(x - h)$ with a horizontal axis of symmetry, $y = k$, which opens to the right for $p > 0$ or opens to the left for $p < 0$.

We derived the equation of a parabola in part (a) with the work above. The derivation of the equation of a parabola in part (b) is similar. We will now use the equation of a parabola stated above to solve some examples.

EXAMPLE 2   Write the equation of the parabola with directrix of $x = 5$ and focus of $(1, 4)$. Sketch the graph of this parabola.

Solution   The vertex (h, k) is halfway between the focus and the directrix $x = 5$ (Fig. 9.4). As we see in the graph of this information, the vertex is $(h, k) =$

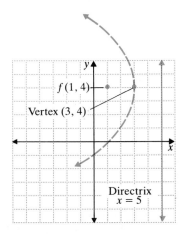

Figure 9.4

(3, 4). With the axis of symmetry horizontal the basic equation is $(y - k)^2 = 4p(x - h)$. The distance p is the directed distance from the vertex to the focus and $p = -2$. Substituting $p = -2$ and $(h, k) = (3, 4)$, we have

$$(y - 4)^2 = 4(-2)(x - 3)$$
$$(y - 4)^2 = -8(x - 3)$$ ∎

EXAMPLE 3

Find the focus and directrix of the parabola with vertex of (2, 1), with an axis of symmetry of $x = 2$, and with (5, 4) as a point on the parabola.

Solution   With a vertex of (2, 1) and a vertical axis of symmetry we know that the basic equation of this parabola is $(x - 2)^2 = 4p(y - 1)$ (Fig. 9.5). To find the value of p, we substitute a point on the parabola, (5, 4), into the basic equation:

$$(x - 2)^2 = 4p(y - 1)$$
$$(5 - 3)^2 = 4p(4 - 1)$$
$$4 = 12p$$
$$p = \frac{4}{12} = \frac{1}{3}$$

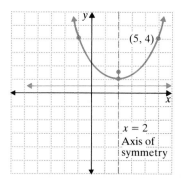

Figure 9.5

With $p = \frac{1}{3}$, we have $(x - 2)^2 = \left(\frac{4}{3}\right)(y - 1)$. Therefore the focus is $\left(2, 1 + \frac{1}{3}\right)$ $= \left(2, \frac{4}{3}\right)$, and the directrix is $y = 1 - \frac{1}{3} = \frac{2}{3}$. ∎

EXAMPLE 4

Find the vertex and focus of the parabola $y^2 - 12x - 4y = 56$.

Solution   To find the vertex and focus, we complete the square to change the equation into the basic parabolic form. Thus

$$y^2 - 12x - 4y = 56$$
$$y^2 - 4y = 12x + 56$$
$$y^2 - 4y + 4 = 12x + 56 + 4 \qquad \textit{(Half the coefficient of the middle}$$
$$\textit{term squared is added to both sides)}$$

$$\cdot \frac{1}{2} \quad SQ$$

$$(y - 2)^2 = 12x + 60$$
$$(y - 2)^2 = 12(x + 5)$$
$$(y - 2)^2 = 4(3)(x + 5)$$

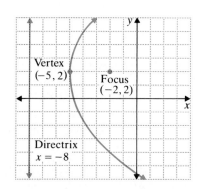

Figure 9.6

Thus $p = 3$, and the vertex is $(h, k) = (-5, 2)$. With $(y - 2)^2$ in the basic equation the axis of symmetry is horizontal, and the focus is $(-5 + 3, 2) = (-2, 2)$ (Fig. 9.6). ∎

The focus of a parabola has an important practical application. When a parabolic mirror is pointed at the sun, the parallel rays of the sun all reflect through the focus of the parabolic mirror (Fig. 9.7). When a parabolic satellite antenna dish is pointed in the direction of a satellite, the nearly parallel signals from the satellite reflect through the focus of this parabolic dish (Fig. 9.7). When the bulb of a flashlight is placed at the focus of the mirrored parabolic surface of a flashlight, the light from the bulb is reflected out from the flashlight in parallel rays (Fig. 9.8). We will consider some of these applications of the parabola in the exercise set that follows.

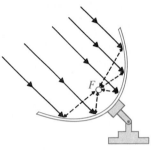

Figure 9.7

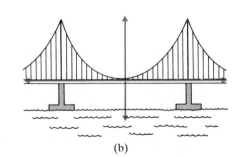

Figure 9.8

EXAMPLE 5   The cable on a suspension bridge carries the weight of the bridge and is the shape of a parabola as shown in Fig. 9.9(a). The distance between the piers is 2600 feet, the piers rise 1400 feet above the water, and the roadway is 400 feet above the water. Find the equation for the suspension cable when the origin is on the roadway at the midpoint between piers.

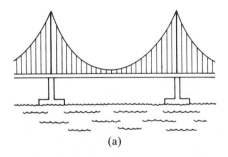

Figure 9.9

(a)   (b)

Solution   Placing the origin on the midpoint of the roadway between the piers, we have Fig. 9.9(b). With the vertex at the origin the basic equation is $x^2 = 4py$ through points $(-1300, 1200)$ and $(1300, 1200)$. Substitute one point into this equation to find p:

$$x^2 = 4py$$
$$(1300)^2 = 4p(1000)$$
$$p = \frac{(1300)^2}{4(1000)}$$
$$p = 422.5$$

Therefore the equation of the suspension cable is $x^2 = 4(422.5)y$, which is

$$x^2 = 1690y \quad \text{for} \quad -1300 \le x \le 1300$$

The limitations for x are a result of the distance between the piers. ∎

Exercises 9.1

Solve for the center and radius and graph the following circles.

1. $(x - 3)^2 + (y + 2)^2 = 4$

2. $(x + 4)^2 + (y - 2)^2 = 9$

3. $x^2 + y^2 + 2x - 2y = 7$

4. $x^2 + y^2 + 6x - 2y = 6$

5. $x^2 + y^2 - 6x - 4y = 12$

6. $x^2 + y^2 + 2x - 4y = 30$

7. $x^2 + y^2 - 4x - 2y = 20$

8. $x^2 + y^2 + 6x + 2y = 21$

9. Find the center and radius of the circle through the points $(0, 5)$, $(5, 0)$, and $(-2, 0)$.

10. Find the center and radius of the circle through the points $(0, 6)$, $(0, -2)$, and $(2, 0)$.

Find the equation of each parabola with the given information.

11. Focus at $(2, 2)$
Directrix of $x = 0$

12. Focus at $(1, 3)$
Directrix of $x = 3$

13. Focus at $(1, 3)$
Directrix of $y = 5$

14. Focus at $(2, 2)$
Directrix of $y = 0$

15. Focus at $(3, 2)$
Vertex at $(-1, 2)$

16. Focus at $(-1, 2)$
Vertex at $(-1, -1)$

17. Focus at $(-1, 2)$
Vertex at $(-1, 5)$

18. Focus at $(3, 2)$
Vertex at $(5, 2)$

19. Focus at $(2, 2)$
Directrix of $y = -4$

20. Focus at $(1, 2)$
Directrix of $x = 5$

21. Focus at $(1, 2)$
Directrix of $x = -3$

22. Focus at $(2, 2)$
Directrix of $y = -1$

23. Find the focus of the parabola with its vertex at $(2, -1)$, with a vertical axis of symmetry, and passing through the point $(5, 0)$.

24. Find the focus of the parabola with its vertex at $(2, -1)$, with a horizontal axis symmetry, and passing through the point $(0, 4)$.

Find the vertex and focus and graph each of the following parabolas.

25. $x^2 - 2x - 8y = 15$

26. $x^2 + 6x + 8y = -1$

27. $y^2 + 8x + 6y = 7$

28. $y^2 - 4y - 12x = 8$

29. $x^2 - 4x - 2y = 2$

30. $y^2 - 2x + 2y = -5$

31. $y^2 - 12x - 6y = -33$

32. $x^2 - 8x - 20y = 4$

33. A satellite antenna dish has a diameter of 6 feet and a depth of 1 foot (Fig. 9.10). How far from the vertex must the receiver be placed to be at the focus?

34. A satellite antenna dish has a diameter of 6 feet and a depth of 2 feet (Fig. 9.10). How far from the vertex must the receiver be placed to be at the focus?

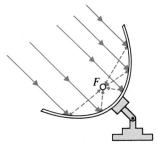

Figure 9.10

35. A flashlight has a radius of 4 inches and a depth of 3 inches (Fig. 9.11). How far from the vertex must the center of the bulb be placed to be at the focus?

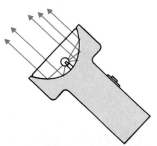

Figure 9.11

36. A flashlight has a radius of 2 inches and a depth of 1 inch (Fig. 9.11). How far from the vertex must the center of the bulb be placed to be at the focus?

37. A parabolic trough is made so that it will turn facing the sun all day (Fig. 9.12). This trough is 6 feet wide at the top and 3 feet deep. How far above the vertex of this parabola must the water pipe be placed so the center of the pipe is at the focus of the parabola?

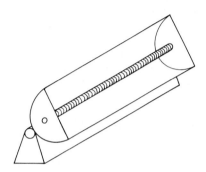

Figure 9.12

38. A parabolic trough is made so that it will turn facing the sun all day (Fig. 9.12). This trough is 6 feet wide at the top and 4 feet deep. How far above the vertex of this parabola must the water pipe be placed so the center of the pipe is at the focus of the parabola?

39. The distance a rock falls in a well is $s = \left(\frac{1}{2}\right)gt^2$, where gravity is $g = 32$ ft/sec^2. The distance the sound travels from the bottom of the well back to the top is $s = v(T - t)$, where the velocity of sound is $v = 1092.9$ ft/sec, T is the total time, and t is time for the rock to hit the bottom. If a rock is dropped into the well, and the sound of hitting the bottom returns 6 seconds later (Fig. 9.13), then the distance the rock falls $s = \left(\frac{1}{2}\right)(32)t^2$ equals the distance the sound must return $s = (1097.9)(6 - t)$. Set these equations equal and solve for t and s. Solving for s will yield the depth of the well.

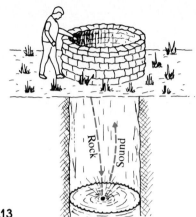

Figure 9.13

40. A rock is dropped into a well, and the sound returns 10 seconds later. How deep is the well? (See Exercise 39.)

41. *Parabola string construction:* Use a straightedge, right triangle, and length of string as shown in Fig. 9.14. The string must be the same length as the distance AB and is attached to the paper (and drawing board) with a pin at point F and to the triangle at point A. A pencil is placed on the paper at point P and held against the triangle to keep the string taut while the triangle slides freely from left to right. What distances must be equal to guarantee that the curve drawn is a parabola? Prove that these distances are equal.

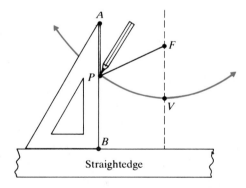

Figure 9.14 From *The VNR Concise Encyclopedia of Mathematics*, courtesy of Bibliographisches Institut Leipzig.

9.2   Conic Sections: The Ellipse

The extracorporeal shock wave lithotripter treatment is described at the beginning of Chapter 9. With this treatment, modern medicine uses the shape of the ellipse to crush kidney stones without surgery.

As we mentioned in Section 9.15 the ellipse is a conic section and can be visualized geometrically at the intersection of a plane and a cone. We will now define the ellipse, using distances.

Definition | An **ellipse** is the set of points (x, y) in a plane for which the sum of the distances from (x, y) to the two fixed points (the two foci, f_1 and f_2) is a constant.

This definition provides a nice method for drawing an ellipse. Hammer two nails into a board. Then tie one end of a string to one nail and the other end of the string to the other nail. The nails are the foci, and the string is the constant distance. Use a pencil to draw an ellipse, keeping the string taut (Fig. 9.15).

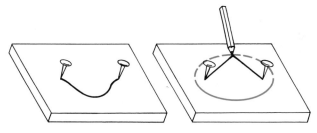

Figure 9.15 From *The VNR Concise Encyclopedia of Mathematics*, courtesy of Bibliographisches Institut Leipzig.

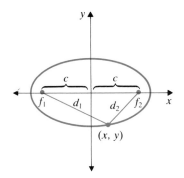

The ellipse with focii f_1 at $(-c, 0)$ and f_2 at $(c, 0)$. The point (x, y) is any point on the ellipse

Figure 9.16

The distance definition above can be used to write the equation of an ellipse. To do this, we will find the distances between (x, y) and the two foci, set the sum of these distances equal to the constant $2a$, and simplify (Fig. 9.16). Thus we have

$$d_1 = \sqrt{[x - (-c)]^2 + (y - 0)^2} = \sqrt{(x + c)^2 + y^2}$$
$$d_2 = \sqrt{(x - c)^2 + (y - 0)^2} \quad = \sqrt{(x - c)^2 + y^2}$$

and

$$d_1 + d_2 = 2a$$
$$\sqrt{(x + c)^2 + y^2} + \sqrt{(x - c)^2 + y^2} = 2a$$
$$\sqrt{(x + c)^2 + y^2} = 2a - \sqrt{(x - c)^2 + y^2}$$

Squaring both sides of this equation, we have

$$(x + c)^2 + y^2 = 4a^2 - 4a\sqrt{(x - c)^2 + y^2} + (x - c)^2 + y^2$$
$$(x + c)^2 - (x - c)^2 - 4a^2 = -4a\sqrt{(x - c)^2 + y^2}$$
$$x^2 + 2xc + c^2 - x^2 + 2xc - c^2 - 4a^2 = -4a\sqrt{(x - c)^2 + y^2}$$
$$4xc - 4a^2 = -4a\sqrt{(x - c)^2 + y^2}$$
$$xc - a^2 = -a\sqrt{(x - c)^2 + y^2}$$

Squaring both sides again, we eliminate the square root, and

$$(xc - a^2)^2 = (-a\sqrt{(x - c)^2 + y^2})^2$$

$$x^2c^2 - 2xca^2 + a^4 = a^2[(x - c)^2 + y^2]$$

$$x^2c^2 - 2a^2xc + a^4 = a^2(x^2 - 2xc + c^2 + y^2)$$

$$x^2c^2 - 2a^2xc + a^4 = a^2x^2 - 2a^2xc + a^2c^2 + a^2y^2$$

$$x^2c^2 + a^4 = a^2x^2 + a^2c^2 + a^2y^2$$

$$a^4 - a^2c^2 = a^2x^2 - c^2x^2 + a^2y^2$$

$$a^2(a^2 - c^2) = (a^2 - c^2)x^2 + a^2y^2$$

Dividing both sides by $a^2(a^2 - c^2)$, we have

$$\frac{a^2(a^2 - c^2)}{a^2(a^2 - c^2)} = \frac{(a^2 - c^2)x^2}{a^2(a^2 - c^2)} + \frac{a^2y^2}{a^2(a^2 - c^2)}$$

$$1 = \frac{x^2}{a^2} + \frac{y^2}{(a^2 - c^2)}$$

If we let $b^2 = a^2 - c^2$ and use the reflexive property, we have

$$\frac{x^2}{a^2} + \frac{y^2}{b^2} = 1$$

which is the equation of the given ellipse.

Notice that with the equation of the ellipse in this form we can easily identify the x and y intercepts of $(\pm a, 0)$ and $(0, \pm b)$. The line $y = 0$ through $(a, 0)$ and $(-a, 0)$ as well as the line $x = 0$ through $(0, b)$ and $(0, -b)$ are both axes of symmetry. Each of the line segments from $(a, 0)$ to $(-a, 0)$ and the line segment from $(0, b)$ to $(0, -b)$ is either the major or the minor axis. The longer of the two segments is the **major axis**, and the shorter of the two segments is the **minor axis** (Fig. 9.17). One-half the major axis is called the **semimajor axis**, and one-half the minor axis is called the **semiminor axis**.

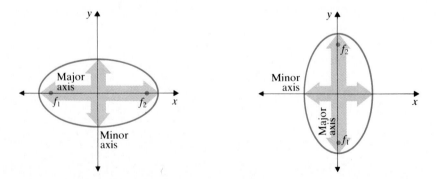

Figure 9.17

The foci are always points on the major axis, and in these nontranslated ellipses are at either $(\pm c, 0)$ or $(0, \pm c)$. We also notice (using the definition and Fig. 9.18) that the relationship between a, b, and c is the Pythagorean relationship

$b^2 + c^2 = a^2$. This relationship can also be remembered as

$$\text{(semiminor axis)}^2 + \left(\tfrac{1}{2} \text{ the distance between the foci}\right)^2 = \text{(semimajor axis)}^2$$

It is also easy to see the relationship between a, b, and c in the graph of an ellipse (Fig. 9.18). The semimajor axis will always be the hypotenuse of the Pythagorean triangle in an ellipse.

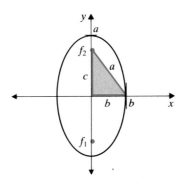

Figure 9.18

EQUATION OF AN ELLIPSE	The equation of an ellipse with the intersection of the major and minor axes at the origin $(0, 0)$ is either $$\frac{x^2}{a^2} + \frac{y^2}{b^2} = 1 \qquad \text{or} \qquad \frac{x^2}{b^2} + \frac{y^2}{a^2} = 1$$ For $a > b$, a is the length of the semimajor axis, b is the length of semiminor axis, c is one half the distance between the foci, and $$a^2 = b^2 + c^2$$ When the major and minor axes intersect at (h, k), then the translated equation of an ellipse is either $$\frac{(x - h)^2}{a^2} + \frac{(y - k)^2}{b^2} = 1 \qquad \text{or} \qquad \frac{(x - h)^2}{b^2} + \frac{(y - k)^2}{a^2} = 1$$ with the same meanings for a, b, and c as above.

We can now use these equations to sketch the graph of an ellipse as well as to find its vertices and foci. The following examples demonstrate.

EXAMPLE 1

Find the foci and graph the ellipse $16x^2 + 9y^2 + 64x + 18y - 71 = 0$.

Solution First change this equation into the standard ellipse form by completing the square:

$$16x^2 + 19y^2 + 64x + 18y + 71 = 0$$
$$16x^2 - 64x\ + 9y^2 + 18y\quad\ = 71$$
$$16(x^2 - 4x\quad) + 9(y^2 + 2y\quad\ ) = 71$$
$$16(x^2 - 4x + 4) + 9(y^2 + 2y + 1) = 71 + 16 \cdot 4 + 9 \cdot 1 \qquad \textit{(Half the coeffi-}$$

cient of the middle squared)

$$\cdot \tfrac{1}{2}\quad SQ \qquad \cdot \tfrac{1}{2}\quad SQ$$

$$16(x - 2)^2 + 9(y + 1)^2 = 144$$
$$\frac{16(x - 2)^2}{144} + \frac{9(y + 1)^2}{144} = \frac{144}{144}$$
$$\frac{(x - 2)^2}{9} + \frac{(y + 1)^2}{16} = 1$$

Thus the major and minor axes intersect at $(2, -1)$. The semimajor axis is 4, with the major axis parallel to the y-axis. The semiminor axis is 3, with the minor axis parallel to the x-axis. The major axis is on the line $x = 2$ and using the Pythagorean triangle,

$$c^2 = a^2 - b^2 = 16 - 9 = 7$$

Thus the foci are $(2, -1 + \sqrt{7})$ and $(2, -1 - \sqrt{7})$.   ■

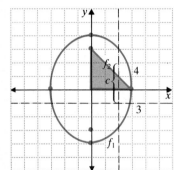

Figure 9.19

EXAMPLE 2

Find the equation of the ellipse with vertices of $(0, \pm4)$ and foci of $(\pm3, 0)$.

Solution Since the foci and vertices are symmetrical about the origin, the major and minor axes also intersect at $(0, 0)$. The major axis is on the x-axis so that $c = 3$ and $b = 4$. Thus $a^2 = 3^2 + 4^2 = 25$ and $a = \pm5$. Therefore the equation of this ellipse is

$$\frac{x^2}{5^2} + \frac{y^2}{4^2} = 1 \qquad \text{or} \qquad \frac{x^2}{25} + \frac{y^2}{16} = 1$$

    ■

It is interesting for us to note that the equation of a circle,

$$x^2 + y^2 = r^2 \qquad \text{written as} \qquad \frac{x^2}{r^2} + \frac{y^2}{r^2} = 1$$

is similar to the equation of an ellipse,

$$\frac{x^2}{a^2} + \frac{y^2}{b^2} = 1$$

Thus when $a = b$, the equation of an ellipse is the equation of a circle.

One interesting application is the elliptical orbits of satellites, planets, and comets. In these orbits the ends of the major axis are the points of the orbit that are the shortest and longest distances from a given focus. The orbit position where the distance to the focus is the shortest is called the **perigee** for an orbit about the earth and the **perihelion** for an orbit about the sun. The orbit position where the distance to the focus is the longest is called the **apogee** for an orbit around the earth and the **aphelion** for an orbit around the sun. If the orbit is around the sun (or the earth), the sun (or the earth) is a focus of the elliptical orbit. We will use this information to write the equation of an orbit in the following example.

EXAMPLE 3   Find the equation of Halley's comet, with the sun as the origin of the axis system, knowing that the perihelion is 0.5868 AU (1 AU = 93,000,000 miles) and the aphelion is 35.3002 AU.

Solution   To write the equation of this orbit, we must find a (the length of the semimajor axis), b (the length of the semiminor axis), and c (half the distance between the foci). From the diagram (Fig. 9.20) we can find a, b, and c.

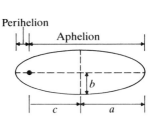

Figure 9.20

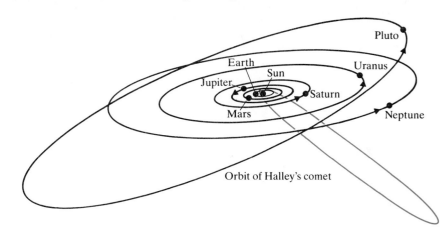

$$a = \frac{1}{2} \text{ (perihelion + aphelion)} \qquad c = a - \text{(perihelion)}$$

$$= \frac{1}{2} (0.5868 + 35.3002) \qquad\qquad = 17.9435 - 0.5868$$

$$= 17.9435 \text{ AU} \qquad\qquad\qquad = 17.3567 \text{ AU}$$

$$b = \sqrt{a^2 - c^2}$$

$$= \sqrt{(17.9435)^2 - (17.3567)^2}$$

$$\approx 4.55128 \text{ AU}$$

The equation for this elliptical orbit is

$$\frac{(x - c)^2}{a^2} + \frac{y^2}{b^2} = 1$$

Using the values that we have for a, b, and c, we have

$$\frac{(x - 17.3567)^2}{(17.9435)^2} + \frac{y^2}{(4.55128)^2} = 1$$

which is the equation of the orbit of Halley's comet around the sun with the sun at the origin. ■

Exercises 9.2

Find the equation of each ellipse with the given information.

1. The vertices are $(0, \pm 2)$ and $(\pm 3, 0)$

2. The vertices are $(0, \pm 3)$ and $(\pm 2, 0)$

3. The vertices are $(\pm 4, 0)$ and $(0, \pm 5)$

4. The vertices are $(\pm 3, 0)$ and $(0, \pm 1)$

5. The vertices are $(2, 3)$, $(10, 3)$, $(6, 1)$, and $(6, 5)$

6. The vertices are $(0, 3)$, $(4, 1)$, $(4, 5)$, and $(8, 3)$

7. The vertices are $(6, 0)$, $(6, 6)$, $(5, 3)$, and $(7, 3)$

8. The vertices are $(1, 3)$, $(7, 3)$, $(4, 2)$, and $(4, 4)$

9.

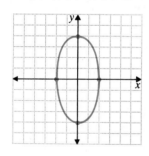

10.

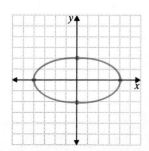

11.

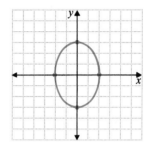

12.

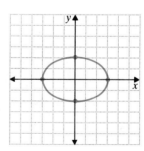

13.

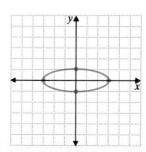

14.

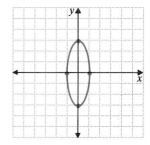

Sketch the graph and find the foci for each ellipse.

15. $\dfrac{x^2}{16} + \dfrac{y^2}{9} = 1$ **16.** $\dfrac{x^2}{9} + \dfrac{y^2}{16} = 1$

17. $\dfrac{x^2}{4} + \dfrac{y^2}{9} = 1$ **18.** $\dfrac{x^2}{16} + \dfrac{y^2}{4} = 1$

19. $\dfrac{(x-2)^2}{4} + \dfrac{(y+2)^2}{9} = 1$

20. $\dfrac{(x-3)^2}{9} + \dfrac{(y+1)^2}{4} = 1$

21. $9x^2 + y^2 - 9 = 0$

22. $x^2 + 4y^2 - 4 = 0$

23. $4x^2 + 25y^2 = 100$

24. $4x^2 + 36y^2 = 144$

25. $4x^2 + 9y^2 - 16x - 18y = 11$

26. $16x^2 + 9y^2 + 32x - 54y = 47$

27. $9x^2 + 4y^2 - 54x + 16y + 61 = 0$

28. $9x^2 + 4y^2 + 18x - 24y + 9 = 0$

Write the equation of each ellipse with the given information.

29. Foci at $(0, \pm 3)$ and a minor axis of length 8

30. Foci at $(\pm 4, 0)$ and a minor axis of length 6

31. Foci at $(\pm 3, 2)$ and a major axis of length 10

32. Foci at $(1, \pm 3)$ and a major axis of length 10

33. Foci at $(-1, 2)$ and $(5, 2)$ and a minor axis of length 6

34. Foci at $(1, 0)$ and $(1, 4)$ and a major axis of length 10

35. A steel bridge structure is half an ellipse. Find the equation for the bridge if it is 100 feet long and 20 feet above the roadway at its highest point.

36. A stone bridge is constructed with an arch that is half of an ellipse (Fig. 9.21). If this ellipse has a major axis of 50 feet and half the minor axis is 20 feet, find the equation for the arch of this stone bridge.

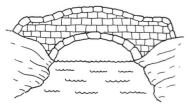

Figure 9.21

37. The cross section of the tank on a tanker truck is an ellipse (Fig. 9.22). Find the equation for this cross section if the tank is 10 feet wide and 8 feet high.

Figure 9.22

38. The cross section of the tank on a tanker truck is an ellipse (Fig. 9.22).Find the equation for this cross section if the tank is 10 feet wide and 6 feet high.

39. The Comas Sola comet has an elliptical path with the sun as one focus and a period of 8.5 years. Write the equation for the path of this comet when half the major axis is 4.18 AU and half the minor axis is 2.41 AU. Find the aphelion and perihelion of the comet in AU and miles. (1 AU is about 93,000,000 miles.)

40. Whipple's comet has an elliptical path with the sun as a focus and a period of 7.47 years. Write the equation for the path of Whipple's comet when half the major axis is 3.821 AU and half the minor axis is 3.577 AU. What are the aphelion and perihelion of the comet in AU and miles? How long does it take light from the sun to reach the comet at the aphelion and perihelion? (See Exercise 39.)

41. The Roman Colosseum was built in the shape of an ellipse (Fig. 9.23). If the length and width of this ellipse are 620

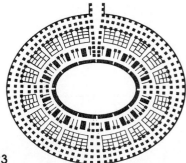

Figure 9.23

feet and 513 feet, write the equation of the ellipse. The arena floor of the Colosseum is an ellipse with a length of 287 feet and a width of 180 feet. Find the length of a center divider if it were built between the foci.

42. If the floor of the Greek Olympic Stadium is the shape of an ellipse with a length of 210 meters and a width of 32 meters, find the equation of this ellipse. Also find the length of the center divider that could be built between the foci.

43. Show that the distance from one focus to a major axis vertex and back to the other focus equals 2a for the ellipse in Fig. 9.18.

44. Find the equation of a space shuttle with a minimum altitude of 165 miles and a maximum altitude of 210 miles (let the center of the earth be the origin of the axes system). What are the apogee and perigee for this orbit? (Use 3960 miles as the radius of the earth.)

45. Find the equation of a communications satellite with a minimum altitude of 18,000 miles and a maximum altitude of 22,000 miles. (Let the center of the earth be the origin of the axes system.) What are the apogee and perigee for this orbit?

9.3   Conic Sections: The Hyperbola

As we saw in the first section of this chapter, the hyperbola is a conic section and, as such, can be visualized geometrically as an intersection of a plane and a cone. This intersection has two parts or branches as it cuts both the top and bottom parts of the cone. The hyperbola can also be described by using distances, as seen in this definition.

Definition | A **hyperbola** is the set of points (x, y) in a plane for which the differences of the distances from (x, y) to the two fixed points (the two foci, f_1 and f_2) is a constant.

We now use this definition to write the equation of a hyperbola. We will find the distances between (x, y) and the two foci, set their difference equal to the constant $2a$, and simplify (Fig. 9.24):

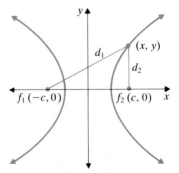

Figure 9.24

$$d_1 = \sqrt{[x - (-c)]^2 + (y - 0)^2} = \sqrt{(x + c)^2 + y^2}$$
$$d_2 = \sqrt{(x - c)^2 + (y - 0)^2} = \sqrt{(x - c)^2 + y^2}$$

Thus

$$d_1 - d_2 = 2a$$
$$\sqrt{(x + c)^2 + y^2} - \sqrt{(x - c)^2 + y^2} = 2a$$
$$\sqrt{(x + c)^2 + y^2} = 2a + \sqrt{(x - c)^2 + y^2}$$

We now square both sides of this equation:

$$(x + c)^2 + y^2 = 4a^2 + 4a\sqrt{(x - c)^2 + y^2} + (x - c)^2 + y^2$$
$$(x + c)^2 = 4a^2 + 4a\sqrt{(x - c)^2 + y^2} + (x - c)^2$$
$$x^2 + 2xc + c^2 = 4a^2 + 4a\sqrt{(x - c)^2 + y^2} + x^2 - 2xc + c^2$$
$$4xc - 4a^2 = 4a\sqrt{(x - c)^2 + y^2}$$
$$xc - a^2 = a\sqrt{(x - c)^2 + y^2}$$

We square both sides again to eliminate the square root:

$$(xc - a^2)^2 = a^2[(x - c)^2 + y^2]$$
$$x^2c^2 - 2a^2cx + a^4 = a^2x^2 - 2a^2cx + a^2c^2 + a^2y^2$$
$$x^2c^2 + a^4 = a^2x^2 + a^2c^2 + a^2y^2$$
$$a^4 - a^2c^2 = a^2x^2 - x^2c^2 + a^2y^2$$
$$a^2(a^2 - c^2) = (a^2 - c^2)x^2 + a^2y^2$$

Now we divide both sides by $a^2(a^2 - c^2)$:

$$\frac{a^2(a^2 - c^2)}{a^2(a^2 - c^2)} = \frac{(a^2 - c^2)x^2}{a^2(a^2 - c^2)} + \frac{a^2y^2}{a^2(a^2 - c^2)}$$
$$1 = \frac{x^2}{a^2} + \frac{y^2}{(a^2 - c^2)}$$

In the case of the hyperbola we have $a^2 - c^2 = -b^2$. (We will see the reason for this later in the geometric relationship between a, b, and c in the hyperbola.) Substituting $-b^2$ for $a^2 - c^2$ produces

$$\frac{x^2}{a^2} - \frac{y^2}{b^2} = 1$$

We notice that from this equation when $y = 0$ the vertices are $(\pm a, 0)$, but when $x = 0$, there are no points on the graph as $y = \pm bi$, which are not real numbers. It is interesting that for this type of hyperbola, y is not a real number for any x such that $-a < x < a$. We call the line segment from $(-a, 0)$ to $(a, 0)$ the **major axis**, and we call the line segment from $(0, -b)$ to $(0, b)$ the **conjugate axis**.

The hyperbola has two asymptotes. To identify these asymptotes, we change the equation

$$\frac{x^2}{a^2} - \frac{y^2}{b^2} = 1 \qquad \text{to} \qquad \frac{x^2}{a^2} = 1 + \frac{y^2}{b^2}$$

For some fixed values of a and b, as x and y get very large, the unit 1 will be insignificant. Thus as x and y approach $\pm\infty$, we have the following limit:

$$\lim_{x \to \pm\infty} \frac{x^2}{a^2} = \lim_{y \to \pm\infty} \left[1 + \left(\frac{y^2}{b^2}\right) \right] = \lim_{y \to \pm\infty} \left(\frac{y^2}{b^2}\right)$$

This limit shows that as x and y get larger without bound, the quantities $\frac{x^2}{a^2}$ and $\frac{y^2}{b^2}$ will get very close in value. Thus as x and y approach $\pm\infty$,

$$\frac{x^2}{a^2} \approx \frac{y^2}{b^2}$$

$$y^2 \approx \frac{b^2}{a^2} x^2$$

which is

$$y \approx \pm\frac{b}{a}x$$

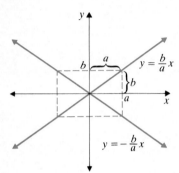

Therefore the lines $y = \pm\frac{b}{a}x$ are the asymptotes (Fig. 9.25) for the hyperbola

$$\frac{x^2}{a^2} - \frac{y^2}{b^2} = 1$$

The rectangle that crosses the x-axis at $x = \pm a$ and crosses the y-axis at $y = \pm b$ can be called an **asymptote box**. This asymptote box is helpful in graphing hyperbolas.

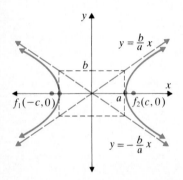

Figure 9.25

Guidelines for Graphing Hyperbolas

To sketch the graph of a hyperbola,

  (a) use the values of a and b with their respective axes to make an asymptote box, and sketch in the asymptotes through the opposite corners of the asymptote box,

  (b) find and graph the vertices by making the negative term zero [for $\dfrac{x^2}{4} - \dfrac{y^2}{2} = 1$, let $y = 0$ and the vertices are $(\pm 2, 0)$], and

  (c) sketch in the curve starting at each vertex, gently curving out toward the asymptotes.

We will now use this information to graph the hyperbolas in the following examples.

EXAMPLE 1   Graph the following hyperbolas:

(a) $\dfrac{x^2}{4} - \dfrac{y^2}{9} = 1$ and   (b) $\dfrac{y^2}{4} - \dfrac{x^2}{16} = 1$

Solution   We use the guidelines to graph each hyperbola.

(a) For $\dfrac{x^2}{4} - \dfrac{y^2}{9} = 1$ the asymptote box crosses the x-axis at $x = \pm 2$ and crosses the y-axis at $y = \pm 3$. We use these points to construct the asymptote box and sketch in the two asymptotes through opposite corners of the asymptote box. Then to find the vertices, we let $y = 0$. Thus $\dfrac{x^2}{4} + 0 = 1$, and the vertices are $(\pm 2, 0)$. We now sketch in this hyperbola (Fig. 9.26(a)).

(b) For $\dfrac{y^2}{4} - \dfrac{x^2}{16} = 1$ the asymptote box crosses the y-axis at $y = \pm 2$ and the x-axis at $x = \pm 4$. We use these values to sketch the asymptotes. To find the vertices, we let $x = 0$ and the vertices are $(0, \pm 2)$. We now sketch in this hyperbola (Fig. 9.26(b)).   ■

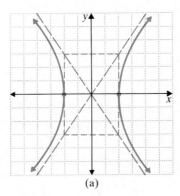
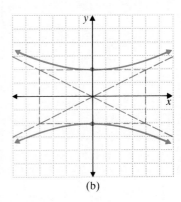

(a)   (b)

Figure 9.26

As we see in Example 1, the hyperbola

$$\frac{x^2}{a^2} - \frac{y^2}{b^2} = 1 \qquad \text{opens on the } x\text{-axis}$$

and the hyperbola

$$\frac{y^2}{b^2} - \frac{x^2}{a^2} = 1 \qquad \text{opens on the } y\text{-axis.}$$

The intersection of the two axes of symmetry for the two hyperbolas above is the origin $(0, 0)$. If this intersection of the axes of symmetry were moved to (h, k), by our earlier work with translation the equations of these hyperbolas would be

$$\frac{(x - h)^2}{a^2} - \frac{(y - k)^2}{b^2} = 1 \qquad \text{which opens on a horizontal axis,}$$

and

$$\frac{(y - k)^2}{b^2} - \frac{(x - h)^2}{a^2} = 1 \qquad \text{which opens on a vertical axis.}$$

We will now summarize the equations of a hyperbola as follows.

EQUATION OF A HYPERBOLA

The equations of a hyperbola with the intersection of the major and conjugate axes at the origin $(0, 0)$ are either

$$\frac{x^2}{a^2} - \frac{y^2}{b^2} = 1 \qquad \text{or} \qquad \frac{y^2}{b^2} - \frac{x^2}{a^2} = 1$$

The a or b with the first term (the positive term) is the length of the semimajor axis, the a or b with the second term (the negative term) is the length of the semiconjugate axis, c is one half the distance between the foci, and $c^2 = a^2 + b^2$. When the major and conjugate axes intersect at (h, k), the translated equation of an ellipse is either

$$\frac{(x - h)^2}{a^2} - \frac{(y - k)^2}{b^2} = 1 \qquad \text{or} \qquad \frac{(y - k)^2}{a^2} - \frac{(x - h)^2}{b^2} = 1$$

with the same meanings for a, b, and c above.

We can easily sketch the graph of these translated hyperbolas by using the previously mentioned guidelines around the point (h, k) rather than the origin.

The distance c from the intersection of the axes of symmetry to the focus of a hyperbola is half the length of the diagonal of the asymptote box (Fig. 9.27). We also see in this diagram that the Pythagorean relationship for the hyperbola is always $c^2 = a^2 + b^2$.

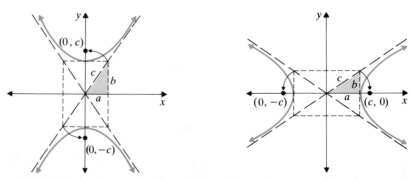

Figure 9.27

EXAMPLE 2   Sketch the graph of $4x^2 - y^2 - 16x + 2y + 11 = 0$ and find the foci.

Solution   To graph $4x^2 - y^2 - 16x + 2y + 11 = 0$, we complete the square to get the equation in the correct form:

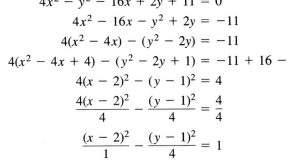

$$4x^2 - y^2 - 16x + 2y + 11 = 0$$
$$4x^2 - 16x - y^2 + 2y = -11$$
$$4(x^2 - 4x) - (y^2 - 2y) = -11$$
$$4(x^2 - 4x + 4) - (y^2 - 2y + 1) = -11 + 16 - 1$$
$$4(x - 2)^2 - (y - 1)^2 = 4$$
$$\frac{4(x - 2)^2}{4} - \frac{(y - 1)^2}{4} = \frac{4}{4}$$
$$\frac{(x - 2)^2}{1} - \frac{(y - 1)^2}{4} = 1$$

To graph this translated hyperbola, we locate the intersection of the axes of symmetry at $(2, 1)$. From this point we mark out the asymptote box for $a = \pm 1$ and $b = \pm 2$ and draw the asymptotes. Next we locate the vertices by solving for x when $y = 1$:

$$\frac{(x - 2)^2}{1} - \frac{0}{4} = 1$$
$$(x - 2)^2 = 1$$
$$\sqrt{(x - 2)^2} = \pm\sqrt{1}$$
$$x - 2 = \pm 1$$
$$x = 2 \pm 1 = 3 \quad \text{or} \quad 1$$

Thus the vertices are $(1, 1)$ and $(3, 1)$. Using the Pythagorean relationship with $a = 1$ and $b = 2$, we calculate c:

$$c^2 = a^2 + b^2 = 2^2 + 1^2 = 5$$
$$c = \pm\sqrt{5}$$

Therefore the foci are $(2 - \sqrt{5}, 1)$ and $(2 + \sqrt{5}, 1)$.   ■

Figure 9.28

Exercises 9.3

Sketch the graphs of the following hyperbolas by first graphing the asymptotes and vertices of each. Also find the foci for each hyperbola.

1. $\dfrac{x^2}{4} - \dfrac{y^2}{4} = 1$

2. $\dfrac{x^2}{9} - \dfrac{y^2}{9} = 1$

3. $\dfrac{x^2}{9} - \dfrac{y^2}{4} = 1$

4. $\dfrac{x^2}{16} - \dfrac{y^2}{9} = 1$

5. $\dfrac{y^2}{9} - \dfrac{x^2}{16} = 1$

6. $\dfrac{y^2}{9} - \dfrac{x^2}{1} = 1$

7. $\dfrac{(x - 1)^2}{4} - \dfrac{(y - 2)^2}{9} = 1$

8. $\dfrac{(x-2)^2}{9} - \dfrac{(y-1)^2}{4} = 1$

9. $\dfrac{(y-2)^2}{1} - \dfrac{(x+2)^2}{9} = 1$

10. $\dfrac{(y+1)^2}{4} - \dfrac{(x-3)^2}{3} = 1$

11. $\dfrac{x^2}{5} - \dfrac{y^2}{8} = -1$ **12.** $\dfrac{x^2}{7} - \dfrac{y^2}{10} = 1$

13. $\dfrac{(x-3)^2}{10} - \dfrac{(y+2)^2}{15} = 1$ **14.** $\dfrac{(y-2)^2}{7} - \dfrac{(x+3)^2}{9} = 1$

Find the equation of each hyperbola with the given information:

15. Vertices of $(\pm 3, 0)$ and a focus of $(4, 0)$

16. Vertices of $(0, \pm 3)$ and a focus of $(0, 2)$

17. Vertices of $(0, \pm 2)$ and a focus of $(0, 3)$

18. Vertices of $(\pm 2, 0)$ and a focus of $(4, 0)$

19. Vertices of $(\pm 2, 3)$ and a focus of $(5, 3)$

20. Vertices of $(\pm 3, 1)$ and a focus of $(5, 1)$

21. Vertices of $(2, 2)$ and $(2, 0)$ and a focus of $(2, -1)$

22. Vertices of $(-2, -1)$ and $(-2, 3)$ and a focus of $(-2, -3)$

23.

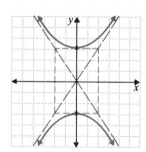

24.

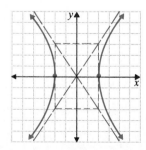

25.

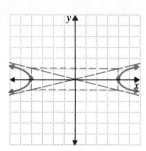

26.

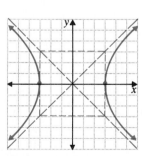

27.

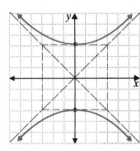

28.

Graph the following hyperbolas.

29. $x^2 - 4y^2 = 16$

30. $9y^2 - 4x^2 = 36$

31. $9x^2 - 4y^2 + 36x + 8y = 4$

32. $x^2 - 9y^2 - 2x - 36y = 44$

33. $4y^2 - 9x^2 + 36x + 16y = 56$

34. $x^2 - 4y^2 - 2x + 24y = 31$

35. $5y^2 - 3y^2 = 15$

36. $6y^2 - 5x^2 = 30$

37. A guidance system on a submarine is designed to measure the distance of the submarine from two sending sonar buoys (Fig. 9.29). To navigate through a special mine field, the guidance system will navigate a course so that the difference of the distances from the submarine to the buoys is less than or equal to half a mile. Sketch the graph of the region in which the submarine can travel safely if the buoys are at (1 mile, 0) and (−1 mile, 0). Find an inequality describing where the submarine can safely travel.

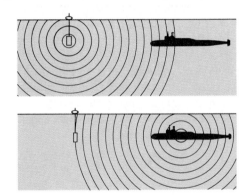

Figure 9.29

38. The guidance system in Exercise 37 can also be used to avoid a mine field. In this case the difference of the distances from the submarine to the two buoys is greater than or equal to 1 mile to avoid the mine field. Sketch the graph of the region that is the mine field if the buoys are at (0, 1 mile) and (0, −1 mile). Find an inequality describing where the submarine can safely travel.

39. Show that the distance from one focus of a hyperbola to a vertex, subtracted from the distance from that same vertex back to the other focus, has an absolute value of $2a$, for $a > 0$, using Fig. 9.25. Find an inequality describing where the submarine can safely travel.

40. Sketch a circle with center f_2 and radius $2k$ and a fixed point f_1 on the exterior of this circle. Show that the set of the centers of all circles that pass through f_1 and are tangent to the given circle (with center of f_2) is one branch of a hyperbola (Fig. 9.30).

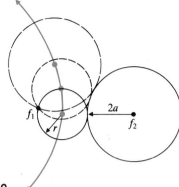

Figure 9.30

41. *Hyperbola string construction:* A straightedge is constructed so that it will rotate freely around the point f_1. The straightedge has a length of x. A string attached to the paper (and drawing board) with a pin at point f_2 and at point A on the straightedge has a length $\ell = x - 2a$. A pencil is placed on the paper so that it is held against the straightedge, keeping the string taut (Fig. 9.31). As the straightedge moves up and down, is the curve drawn a part of a hyperbola? Verify this conclusion.

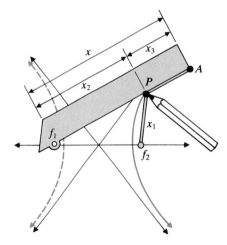

Figure 9.31 Courtesy of Bibliographisches Institut Leipzig.

42. A point, a line, and two intersecting lines can also be considered trivial conic sections. Sketch the intersection of a plane and a cone that results in (a) a point, (b) a line, and (c) two intersecting lines.

43. Graph the trivial conic section (Exercise 42) produced by the equation $x^2 - y^2 - 4x + 2y + 3 = 0$ by completing the square.

44. Graph the trivial conic section (Exercise 42) produced by the equation $x^2 - y^2 - 2x - 6y - 8 = 0$ by completing the square.

45. Graph the trivial conic section (Exercise 42) produced by the equation $x^2 - y^2 + 6x + 4y + 5 = 0$ by completing the square.

9.4 Rotation of Conic Sections

In the previous sections of this chapter we first studied conic sections with the intersection of the axes of symmetry (vertex for the parabola) at the origin $(0, 0)$. Then we studied translated equations in which the intersection of the axes of symmetry (vertex for the parabola) were at any point (h, k). In this section we will work with the equations of conic sections that are rotated around the intersection of the axes of symmetry. Figure 9.32(a) shows an ellipse that has been translated. Notice that the axes of symmetry have been translated to the point (h, k) and the axes of symmetry are parallel the axes of the untranslated ellipse. Figure 9.32(b) shows an ellipse that has been rotated. Notice that the axes of symmetry still intersect at the origin $(0, 0)$. Then in Figure 9.32(c) the ellipse has been first translated, with axes of symmetry intersecting at (h, k), and then rotated.

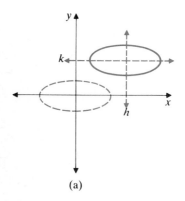

(a)

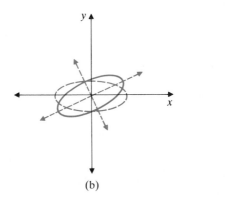

(b)

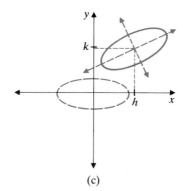

(c)

Figure 9.32

To develop the skills needed to work with rotated conic sections, it is important to see the relationship between the coordinates of a point with respect to the nonrotated x-y coordinate system and the coordinates of the same point with respect

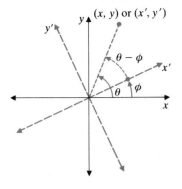

Figure 9.33

to the rotated x'-y' axis system as shown in Fig. 9.33. Notice that angle ϕ is the angle of rotation between the x-axis and the x'-axis. Angle θ is the angle between the x-axis and the line containing $(0, 0)$ and P, while the angle $\theta - \phi$ is the angle between the x'-axis and line containing $(0, 0)$ and P.

Using perpendiculars to the x- and y-axes in Fig. 9.34(a), we have $(x, y) = (r \cos \theta, r \sin \theta)$. Likewise, using perpendiculars to the x'- and y'- axes (Fig. 9.34b), we have $(x', y') = (r \cos (\theta - \phi), r \sin (\theta - \phi))$. Thus $x = r \cos \theta$, $y = r \sin \theta$,

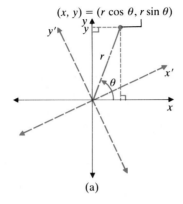

(a)

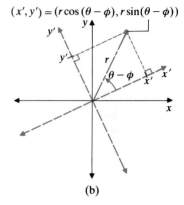

(b)

Figure 9.34

$$x' = r \cos (\theta - \phi) \qquad \text{and} \qquad y' = r \sin (\theta - \phi)$$

Then expanding x' and y', we have

$$x' = r \cos (\theta - \phi)$$
$$x' = r(\cos \theta \cos \phi + \sin \theta \sin \phi)$$
$$x' = (r \cos \theta) \cos \phi + (r \sin \theta) \sin \phi$$

Substituting $x = r \cos \theta$ and $y = r \sin \theta$, we have

$$x' = x \cos \phi + y \sin \phi$$

Similarly,

$$y' = r \sin (\theta - \phi)$$
$$y' = r(\sin \theta \cos \phi - \cos \theta \sin \phi)$$
$$y' = (r \sin \theta) \cos \phi - (r \cos \theta) \sin \phi$$

Substituting $x = r \cos \theta$ and $y = r \sin \theta$, we have

$$y' = y \cos \phi - x \sin \phi$$

Thus with ϕ as the angle of rotation for the axes, we have

$$x' = x \cos \phi + y \sin \phi$$
$$y' = y \cos \phi - x \sin \phi$$

Now this system can be solved for x and y. Since ϕ will be a specific value, we can solve the system

$$(\cos \phi)x + (\sin \phi)y = x'$$
$$(-\sin \phi)x + (\cos \phi)y = y'$$

using Cramer's Rule. Then

$$x = \frac{\begin{vmatrix} x' & \sin \phi \\ y' & \cos \phi \end{vmatrix}}{\begin{vmatrix} \cos \phi & \sin \phi \\ -\sin \phi & \cos \phi \end{vmatrix}} = \frac{x' \cos \phi - y' \sin \phi}{\cos^2 \phi - (-\sin^2 \phi)}$$

$$= \frac{x' \cos \phi - y' \sin \phi}{\cos^2 \phi + \sin^2 \phi}$$

$$= \frac{x' \cos \phi - y' \sin \phi}{1}$$

$$= x' \cos \phi - y' \sin \phi$$

and

$$y = \frac{\begin{vmatrix} \cos \phi & x' \\ -\sin \phi & y' \end{vmatrix}}{\begin{vmatrix} \cos \phi & \sin \phi \\ -\sin \phi & \cos \phi \end{vmatrix}} = \frac{y' \cos \phi - (-x' \sin \phi)}{\cos^2 \phi + \sin^2 \phi}$$

$$= \frac{y' \cos \phi + x' \sin \phi}{1}$$

$$= y' \cos \phi + x' \sin \phi$$

Therefore $x = x' \cos \phi - y' \sin \phi$ and $y = x' \sin \phi + y' \cos \phi$.

EQUIVALENCE EQUATIONS FOR ROTATIONS OF AXES

When ϕ is the angle of rotation of the axes of symmetry,

$$x = x' \cos \phi - y' \sin \phi \qquad \text{and} \qquad x' = x \cos \phi + y \sin \phi$$
$$y = x' \sin \phi + y' \cos \phi \qquad\qquad\qquad y' = -x \sin \phi + y \cos \phi$$

These equivalence formulas will be helpful when we change from the x'-y' coordinates to the x-y coordinates with a known angle of rotation ϕ. We can also use these formulas to change an equation in x-y into its x'-y' equivalent.

Now that we have established the equivalence equations, we can use them to find the value of the angle or rotation, ϕ.

The general quadratic equation is

$$Ax^2 + Bxy + Cy^2 + Dx + Ey + F = 0$$

where A, B, C, D, E, and F are real numbers. With the rotated axes, this general equation becomes

$$A'(x')^2 + B'x'y' + C'(y')^2 + D'x' + E'y' + F' = 0$$

where A', B', C', D', E', and F' are real numbers and x'-y' are the rotated axes. The Bxy and $B'x'y'$ terms are awkward, since they do not appear in any of the standard conic forms of the previous sections. Thus it will be convenient and necessary to find the angle of rotation ϕ such that $B' = 0$. To do this, we substitute the equivalence equations into

$$Ax^2 + Bxy + Cy^2 + Dx + Ey + F = 0$$

simplify, and regroup into the form

$$A'(x')^2 + B'x'y' + C'(y')^2 + D'x' + E'y' + F' = 0$$

This work is not very difficult and is not included here. The results of this process follow.

COEFFICIENTS OF THE
ROTATED EQUATION

When the conic section

$$Ax^2 + Bxy + Cy^2 + Dx + Ey + F = 0$$

is rotated with an angle of rotation of measure ϕ, the rotated equation is

$$A'(x')^2 + B'x'y' + C'(y')^2 + D'x' + E'y' + F' = 0$$

where the coefficients are

$$A' = A \cos^2 \phi + B \cos \phi \sin \phi + C \sin^2 \phi$$
$$B' = B(\cos^2 \phi - \sin^2 \phi) + 2(C - A) \sin \phi \cos \phi$$
$$C' = A \sin^2 \phi - B \sin \phi \cos \phi + C \cos^2 \phi$$
$$D' = D \cos \phi + E \sin \phi$$
$$E' = -D \sin \phi + E \cos \phi$$
$$F' = F$$

We notice that these results provide the coefficients of the rotated equation as a function of the coefficients of the nonrotated equation and the angle of rotation,

ϕ. To find the angle ϕ that eliminates the term $B'x'y'$, we solve $B' = 0$. Thus for $B' = 0$,

$$B(\cos^2 \phi - \sin^2 \phi) + 2(C - A) \sin \phi \cos \phi = 0$$

$$B(\cos^2 \phi - \sin^2 \phi) = -2(C - A) \sin \phi \cos \phi$$

$$B(\cos^2 \phi - \sin^2 \phi) = 2(A - C) \sin \phi \cos \phi$$

$$\frac{\cos^2 \phi - \sin^2 \phi}{2 \sin \phi \cos \phi} = \frac{A - C}{B}$$

$$\frac{\cos 2\phi}{\sin 2\phi} = \frac{A - C}{B}$$

$$\cot 2\phi = \frac{A - C}{B}$$

Note that when $B = 0$ the denominator is zero, and the $\cot 2\phi$ is undefined for $2\phi = k\pi$, where k is any integer. Traditionally, the angle of rotation has been defined by the cotangent function. It can just as easily be defined by

$$\tan 2\phi = \frac{B}{A - C}$$

Both equations have different domains because of the restrictions on their respective inverse functions but provide the same graph of the conic section. When

$$2\phi = \cos^{-1} \frac{A - C}{B} \qquad \text{then} \qquad 0 < 2\phi < \pi \qquad \text{or} \qquad 0 < \phi < \frac{\pi}{2}$$

and when

$$2\phi = \tan^{-1} \frac{B}{A - C} \qquad \text{then} \qquad -\frac{\pi}{2} < 2\phi < \frac{\pi}{2} \qquad \text{or} \qquad -\frac{\pi}{4} < \phi < \frac{\pi}{4}$$

Using the tangent function formula makes it possible to use a calculator to find ϕ.

It is also possible to find the exact values for the $\cos \theta$ and the $\sin \theta$, but we must use the cotangent formula for the angle of rotation with a triangle and the identities

$$\cos \theta = \sqrt{\frac{1 + \cos 2\theta}{2}} \qquad \text{and} \qquad \sin \theta = \sqrt{\frac{1 - \cos 2\theta}{2}}$$

Since $0 < \theta < \frac{\pi}{2}$ with the cotangent, then only the positive square root needs to be considered in the above formulas. We can find the values for A', B', C', D', E', and F' using either exact values with the identities above or approximate values with a calculator.

Once we find the value of θ, we substitute this value into the equations for the coefficients of the rotated equation. These coefficients provide the equation with the rotated axes. Then we can complete the square and sketch the graph of the conic section on the rotated axes. This process is demonstrated in the following examples, using a calculator in Example 1 and exact values in Example 3.

EXAMPLE 1   Sketch the graph of

$$13x^2 + 10xy + 13y^2 + 62.23x + 39.6y + 8 = 0$$

Solution   We first solve for the angle of rotation, θ:

$$\tan 2\theta = \frac{10}{13 - 13} = \frac{10}{0}$$

is undefined. Thus $2\theta = 90°$ and $\theta = 45°$. We next calculate the coefficients of rotation to be

$$A' = 13 \cos^2 45° + 10 \cos 45° \sin 45° + 13 \sin^2 45° = 18$$
$$B' = 0 \quad \text{as we planned with this choice of } \theta$$
$$C' = 13 \sin^2 45° - 10 \sin 45° \cos 45° + 13 \cos^2 45° = 8$$
$$D' = \quad 62.23 \cos 45° + 39.6 \sin 45° \approx 72$$
$$E' = -62.23 \sin 45° + 39.6 \cos 45° \approx -16$$
$$F' = F = 8$$

Thus the equation of this conic section on the axes of rotation with $\theta = 45°$ is

$$18(x')^2 + 8(y')^2 + 72x' - 16y' + 8 = 0$$

We can now complete the square to make graphing easier:

$$18(x')^2 + 72x' \quad + 8(y')^2 - 16y' \quad = -8$$
$$18[(x')^2 + 4x' \quad] + 8[(y')^2 - 2y' \quad] \quad = -8$$
$$18[(x')^2 + 4x' + 4] + 8[(y')^2 - 2y' + 1] \quad = -8 + 8 + 72$$
$$18(x' + 2)^2 + 8(y' - 1)^2 \quad = 72$$
$$\frac{18(x' + 2)^2}{72} + \frac{8(y' - 1)^2}{72} \quad = \frac{72}{72}$$
$$\frac{(x' + 2)^2}{4} + \frac{(y' - 1)^2}{9} \quad = 1$$

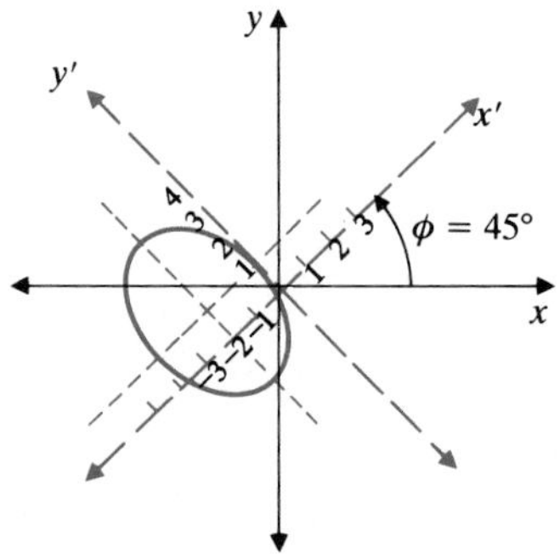

Figure 9.35

To sketch this ellipse, we sketch in the rotated x'-y' axes with dashed lines and $\theta = 45°$. The intersection of the axes of symmetry for this ellipse is at $(-2, 1)$ on the rotated x'-y' axes. Using the techniques of previous sections, we sketch the graph of this ellipse on the rotated axes (Fig. 9.35).   ■

It is also possible to find the foci of conic sections as seen in the next example.

EXAMPLE 2

Find the foci of the ellipse in Example 1.

Solution The foci are easily found with respect to the x'-y' axes. Using Fig. 9.36,

$$c^2 + 2^2 = 3^2$$
$$c^2 = 9 - 4 = 5$$
$$c = \pm\sqrt{5}$$

Figure 9.36

Thus on the x'-y' axes the foci are $(-2, 1 + \sqrt{5})$ and $(-2, 1, -\sqrt{5})$. By using the equivalence equations for rotation of axes the coordinates for the foci can be expressed in terms of the x-y axes as follows:

$$x = -2 \cos 45° - (1 - \sqrt{5}) \sin 45° \approx -0.54$$
$$y = -2 \sin 45° + (1 - \sqrt{5}) \cos 45° \approx -2.29$$

and

$$x = -2 \cos 45° - (1 + \sqrt{5}) \sin 45° \approx -3.70$$
$$y = -2 \sin 45° + (1 + \sqrt{5}) \cos 45° \approx 0.87$$

Thus the foci of the ellipse $13x^2 + 10xy + 13y^2 + 56.56x + 45.24y + 8 = 0$ are $(-0.54, -2.29)$ and $(-3.70, 0.87)$. ■

EXAMPLE 3

Sketch the graph of $x^2 + 4xy + 4y^2 - 2x + y - 9 = 0$ using exact values and the cotangent.

Solution First find the values for $\cos 2\phi$:

$$\cot 2\phi = \frac{A - C}{B} = \frac{1 - 4}{4}$$

$$\cot 2\phi = \frac{-3}{4}$$

Using the values in Fig. 9.37, $\cos 2\phi = -\frac{3}{5}$. Then

$$\cos \phi = \sqrt{\frac{1 + \cos 2\phi}{2}} \qquad \text{and} \qquad \sin \phi = \sqrt{\frac{1 - \cos 2\phi}{2}}$$

$$= \sqrt{\frac{1 - \frac{3}{5}}{2}} \qquad\qquad = \sqrt{\frac{1 + \frac{3}{5}}{2}}$$

$$= \sqrt{\frac{\left(1 - \frac{3}{5}\right)5}{(2)\,5}} \qquad\qquad = \sqrt{\frac{\left(1 + \frac{3}{5}\right)5}{(2)\,5}}$$

$$= \sqrt{\frac{5 - 3}{10}} \qquad\qquad = \sqrt{\frac{5 + 3}{10}}$$

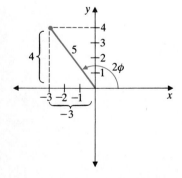

Figure 9.37

$$= \sqrt{\frac{2}{10}} \qquad\qquad = \sqrt{\frac{8}{10}}$$

$$= \sqrt{\frac{1}{5}} \qquad\qquad = \sqrt{\frac{4}{5}}$$

$$= \frac{1}{\sqrt{5}} \qquad\qquad = \frac{2}{\sqrt{5}}$$

We can now substitute these values into the formulas for the coefficients of the rotated equation. Then we have

$$A' = 1\left(\frac{1}{\sqrt{5}}\right)^2 + 4\left(\frac{1}{\sqrt{5}}\right)\left(\frac{2}{\sqrt{5}}\right) + 4\left(\frac{2}{\sqrt{5}}\right)^2$$

$$= \frac{1}{5} + \frac{8}{5} + \frac{16}{5} = \frac{25}{5} = 5$$

$$B' = 0 \qquad \text{by our choice of } \phi$$

$$C' = 1\left(\frac{2}{\sqrt{5}}\right)^2 - 4\left(\frac{2}{\sqrt{5}}\right)\left(\frac{1}{\sqrt{5}}\right) + 4\left(\frac{1}{\sqrt{5}}\right)^2$$

$$= \frac{4}{5} - \frac{8}{5} + \frac{4}{5} = 0$$

$$D' = -2\left(\frac{1}{\sqrt{5}}\right) + 1\left(\frac{2}{\sqrt{5}}\right) = 0$$

$$E' = 2\left(\frac{2}{\sqrt{5}}\right) + 1\left(\frac{1}{\sqrt{5}}\right) = \frac{5}{\sqrt{5}} = \sqrt{5}$$

$$F' = F = -9$$

Thus

$$5(x')^2 + 0x'y' + 0(y')^2 + 0(x') + \sqrt{5}y' - 9 = 0$$
$$5(x')^2 + \sqrt{5}y' + 9 = 0$$
$$5(x')^2 = -\sqrt{5}\left[y' - \left(\frac{9}{5}\right)\sqrt{5}\right]$$
$$(x')^2 = -\frac{\sqrt{5}}{5}\left[y' - \left(\frac{9}{5}\right)\sqrt{5}\right]$$

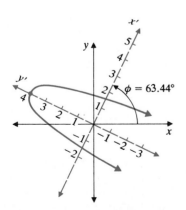

Figure 9.38

and rounding off, we have

$$(x')^2 \approx -0.447(y' - 4)$$

Thus the vertex of the parabola is at $(0, 4)$ on the $x'-y'$ axes, opening in the negative direction on the y' axis (Fig. 9.38). To sketch this graph, we must approximate the angle of rotation. Thus

$$\phi = \tfrac{1}{2}\cot^{-1}\left(-\tfrac{3}{4}\right) \text{ and } \tfrac{1}{2}\tan^{-1}\left(-\tfrac{4}{3}\right) \approx -26.56°$$

To be between $0°$ and $90°$, $\phi \approx 63.44°$. Now sketch this parabola. ∎

In Example 3 we used exact values, but we can also complete Example 3 using the tangent function and a calculator. The process is similar but not exactly the same.

EXAMPLE 4   Sketch the graph of $x^2 + 4xy + 4y^2 - 2x + y - 9 = 0$ using the tangent to find ϕ and calculator values.

Solution   We first find ϕ using the tangent:

$$\tan 2\phi = \frac{B}{A - C} = \frac{4}{1 - 4}$$

$$\tan 2\phi = -\frac{4}{3}$$

$$2\phi = \tan^{-1}\left(-\frac{4}{3}\right)$$

$$2\phi = -53.13°$$

$$\phi = -26.56°$$

Using a calculator, we evaluate the coefficients of the rotated equation:

$$A' = 1 \cos^2 \phi + 4 \cos \phi \sin \phi + 4 \sin^2 \phi \approx 0.000000039 \approx 0$$
$$B' = 0 \qquad \text{by our choice of } \phi$$
$$C' = 1 \sin^2 \phi - 4 \cos \phi \sin \phi + 4 \cos^2 \phi \approx 4.999999961 \approx 5$$
$$D' = -2 \cos \phi + \sin \phi \approx -2.236$$
$$E' = 2 \sin \phi + \cos \phi \approx 0.000197131 \approx 0$$
$$F' = F = -9$$

Therefore by using these approximate values and $\phi = -26.56°$ the rotated equation is

$$5(y')^2 - 2.236x' = 9$$
$$5(y')^2 = 2.236x' + 9$$
$$5(y')^2 = 2.236(x' + 4)$$
$$(y')^2 = 0.447(x + 4)$$

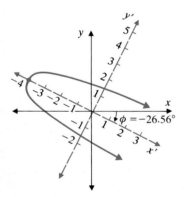

Figure 9.39

Thus this parabola has a vertex at $(-4, 0)$ and opens in the positive direction on the x-axis (Fig. 9.39). ■

It is interesting to note that in Examples 3 and 4 the angles of rotation and the rotated equations are different but the graphs are the same.

Thus it is possible to sketch the graph of a rotated conic section with either the tangent or cotangent, using calculator values or exact values as demonstrated in these examples above.

Exercises 9.4

For each of the following equations,
(a) find the angle of rotation ϕ (using cotangent or tangent),
(b) find the coefficients of the rotated equation (A', C', D', E', and F'), and
(c) then sketch the graph of each.

1. $xy = 8$ 2. $xy = 14$

3. $xy + 3\sqrt{2}x + 4\sqrt{2}y = 1$

4. $xy - 7\sqrt{2}x + 2\sqrt{2}y = 1$

5. $5x^2 + 2\sqrt{3}xy + 3y^2 = 18$

6. $9x^2 + 2\sqrt{3}xy + 7y^2 = 90$

7. $x^2 - 8xy + 7y^2 - 9\sqrt{5}x = 9$

8. $3x^2 - 4xy + 6y^2 + 9\sqrt{5}x + \frac{11}{28} = 0$

9. $2x^2 - 6xy + 10y^2 - 7\sqrt{10}x + 5\sqrt{10}y = 64$

10. $x^2 + 6xy + 9y^2 + 6\sqrt{10}x - 2\sqrt{10}y = 0$

11. $2x^2 - 4xy - y^2 + 9x - 9y = 0$

12. $x^2 - 8xy + 7y^2 - 9x - \frac{4}{9} = 0$

13. $3x^2 - 4xy + 6y^2 + 9y + \frac{18}{5} = 0$

14. $x^2 + 6xy + 9y^2 - 3x - y = 0$

15. $x^2 + 6xy + 9y^2 - 2y = 0$

16. $2x^2 - 6xy - y^2 + 3x - 2y = \frac{59}{3}$

17. Use the formulas for A', C', D', and E' to show that $A + C = A' + C'$.

18. Use the formulas for A', C', D', and E' to show that $B^2 - 4AC = (B')^2 - 4A'C'$.

19. Using the results of Ex. 18, show that the equation $Ax^2 + Bxy + Cy^2 + Dx + Ey + F = 0$ is
(a) a parabola if $B^2 - 4AC = 0$,
(b) an ellipse if $B^2 - 4AC < 0$, and
(c) a hyperbola if $B^2 - 4AC > 0$.

20. When the circle $ax^2 + ay^2 + dx + ex + f = 0$ is rotated some nonzero angle ϕ, the rotated equation is

$$A'(x')^2 + B'x'y' + C'(y')^2 + D'x' + E'y' + F' = 0$$

Show that $A' = a$, $B' = 0$, and $C' = a$.

9.5   Conic Sections in Polar Coordinates

All of the conic sections that we studied earlier in this chapter can be represented with polar coordinates. Using polar coordinates will make it much easier to rotate conic sections as we did with rectangular coordinates in Section 9.4. In this section we will develop the polar coordinate equivalent equations for conic sections. Then we will be able to easily rotate these polar equations.

In the following examples we develop the equations of conic sections when a focus is at the pole (or origin).

EXAMPLE 1   Find the equation of an ellipse in polar coordinates with one focus at the pole and the other focus on the line $\theta = 0$ (Fig. 9.40, see next page).

Solution   We remember that in an ellipse,
(a) the distance between the two foci F_1 and F_2 is $2c$,
(b) the distance from F_1 to P and back to F_2 is $2a$, and
(c) the square of the length of the semimajor axis equals the sum of the squares of the length of the semiminor axis and half the distance between the foci.

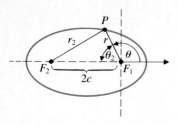

Figure 9.40

Thus $a^2 = b^2 + c^2$, and from Fig. 9.40 we see that $r + r_2 = 2a$. Using the law of cosines, and θ_2 in Fig. 9.40,

$$r_2{}^2 = r^2 + (2c)^2 - 2r(2c) \cos \theta_2$$

$$(2a - r)^2 = r^2 + 4c^2 - 4rc \cos \theta_2 \qquad \textit{(Substitute } 2a - r \textit{ for } r_2.)$$

$$4a^2 - 4ar + r^2 = r^2 + 4c^2 - 4rc \cos \theta_2 \qquad \textit{(Multiply and simplify)}.$$

$$4a^2 - 4ar = 4c^2 - 4rc \cos \theta_2 \qquad \textit{(Solve for } r).$$

$$-4ar + 4rc \cos \theta_2 = 4c^2 - 4a^2$$

$$-4r(a - c \cos \theta_2) = 4(c^2 - a^2)$$

$$-r(a - c \cos \theta_2) = c^2 - a^2$$

$$-ar\left(1 - \frac{c}{a} \cos \theta_2\right) = c^2 - a^2$$

$$r\left(1 - \frac{c}{a} \cos \theta_2\right) = \frac{c^2 - a^2}{-a}$$

$$r\left(1 - \frac{c}{a} \cos \theta_2\right) = \frac{a^2 - c^2}{a}$$

$$r\left(1 - \frac{c}{a} \cos \theta_2\right) = \frac{b^2}{a}$$

$$r = \frac{\dfrac{b^2}{a}}{1 - \dfrac{c}{a} \cos \theta_2}$$

(If $1 - \dfrac{c}{a} \cos \theta_2 = 0$, then $a^2 - c^2 = b^2 = 0$. For this ellipse, $b \neq 0$ and thus $1 - \dfrac{c}{a} \cos \theta_2 \neq 0$)

Since θ and θ_2 are supplementary angles, $\cos \theta_2 = -\cos \theta$. The eccentricity ε is $\varepsilon = \dfrac{c}{a}$. Using these substitutions, we have

$$r = \frac{\dfrac{b^2}{a}}{1 + \varepsilon \cos \theta}$$

which is an equation for an ellipse with one focus at the pole. If F_2 were to the right of F_1 as in Fig. 9.41, the derivation above would have been started with θ and resulted in

$$r = \frac{\dfrac{b^2}{a}}{1 - \varepsilon \cos \theta}$$

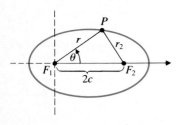

Figure 9.41

Thus we can conclude that the equation of an ellipse with one focus at the pole and the other focus on the line $\theta = 0$ is

$$r = \frac{\dfrac{b^2}{a}}{1 \pm \varepsilon \cos \theta}$$

with either plus or minus, depending on the position of the focus that is not at the pole.   ■

We note that the eccentricity of the ellipse is

$$\varepsilon = \frac{c}{a} = \frac{\dfrac{1}{2} \text{ the distance between the foci}}{\text{the length of the semimajor axis}}$$

and for the ellipse $0 < c < a$. Therefore for an ellipse we have $0 < \varepsilon < 1$. When $\varepsilon \approx 0$, the ellipse is close to a circle, since $c \approx 0$. Thus the closer ε is to 0, the more circular the ellipse will be.

EXAMPLE 2   Find the equation of a hyperbola in polar coordinates with one focus at the pole and the other focus on the line $\theta = 0$ (Fig. 9.42).

Solution   We remember that in a hyperbola,
(a) the distance between F_1 and F_2 is $2c$,
(b) the absolute value of the difference of the distance from F_2 to P and P to F_1 is $2a$, and
(c) the length of the semimajor axis squared plus half the length of the conjugate axis squared equals half the distance between the foci squared. From Fig. 9.42

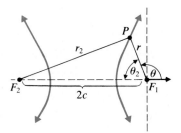

Figure 9.42

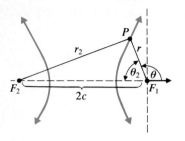

Figure 9.42

we have $r_2 - r = 2a$ and $a^2 + b^2 = c^2$. Using the law of cosines and θ_2 in Fig. 9.42, we have

$$r_2{}^2 = r^2 + (2c)^2 - 2(r)(2c) \cos \theta_2 \quad \textit{(Substitute } r + 2a \textit{ for } r_2.)$$

$$(r + 2a)^2 = r^2 + 4c^2 - 4rc \cos \theta_2 \quad \textit{(Multiply and simplify.)}$$

$$r^2 + 4ra + 4a^2 = r^2 + 4c^2 - 4rc \cos \theta_2$$

$$4ra + 4a^2 = 4c^2 - 4rc \cos \theta_2 \quad \textit{(Solve for } r.)$$

$$4ra + 4rc \cos \theta_2 = 4c^2 - 4a^2$$

$$4r(a + c \cos \theta_2) = 4(c^2 - a^2)$$

$$4ar\left(1 + \frac{c}{a} \cos \theta_2\right) = 4(c^2 - a^2)$$

$$r\left(1 + \frac{c}{a} \cos \theta_2\right) = \frac{c^2 - a^2}{a} \qquad \textit{(If } 1 + \frac{c}{a}\cos\theta_2 = 0, \textit{ then}$$

$$r\left(1 + \frac{c}{a} \cos \theta_2\right) = \frac{b^2}{a} \qquad \begin{array}{l}c^2 - a^2 = b^2 = 0.\\ \textit{For a hyperbola, } b \neq 0\end{array}$$

$$r = \frac{\dfrac{b^2}{a}}{1 + \dfrac{c}{a} \cos \theta_2} \qquad \begin{array}{l}\textit{and thus } 1 + \frac{c}{a}\cos\theta_2 \neq 0.)\end{array}$$

$$r = \frac{\dfrac{b^2}{a}}{1 + \varepsilon \cos \theta_2}$$

where $\frac{c}{a} = \varepsilon$ is the eccentricity. Since θ_2 and θ are supplementary, $\cos \theta_2 = -\cos \theta$, and

$$r = \frac{\dfrac{b^2}{a}}{1 - \varepsilon \cos \theta}$$

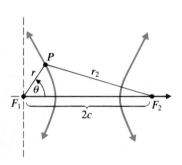

Figure 9.43

When the second focus, F_2, is placed to the right of the pole on the line $\theta = 0$ (Fig. 9.43), we complete the above calculations with θ instead of θ_2 with the following results:

$$r = \frac{\dfrac{b^2}{a}}{1 + \varepsilon \cos \theta}$$

Therefore a hyperbola with one focus at the pole and the other focus on the line $\theta = 0$ is

$$r = \frac{\dfrac{b^2}{a}}{1 \pm \varepsilon \cos \theta}$$

with either the plus or minus sign depending on the position of the focus that is not at the pole. This result for the ellipse and hyperbola appear to be identical. ∎

As we saw earlier, $c > a$ in a hyperbola, and thus $c/a = \varepsilon > 1$ for a hyperbola. Now we can combine the results of Examples 1 and 2 to conclude that the equation

$$r = \frac{\dfrac{b^2}{a}}{1 \pm \varepsilon \cos \theta}$$

is an ellipse for $0 < \varepsilon < 1$ and a hyperbola for $\varepsilon > 1$. As we will see in the next example, the conic section is a parabola when $\varepsilon = 1$.

EXAMPLE 3   Find the equation of the parabola in polar coordinates with the focus at the pole and the line $\theta = 0$ as the axis of symmetry.

Solution   This equation must be generated for the parabola opening **(a)** to the right and **(b)** to the left.

(a) For the parabola in Fig. 9.44(a) the distance for P_1 to P_2 must be the same as r. Also the distance from F to the directrix is $2p$. We also note that the distance from P_1 to P_2 is $2p + d$, where $d = r \cos \theta$. Then

$$r = 2p + d$$
$$r = 2p + r \cos \theta$$
$$r - r \cos \theta = 2p$$
$$r(1 - \cos \theta) = 2p \qquad (\textit{1 } - \cos \theta \neq 0 \textit{ since } p \neq \theta.)$$
$$r = \frac{2p}{1 - \cos \theta}$$

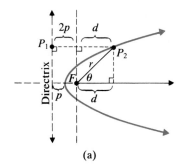

(a)

(b) Similarly, for the parabola in Fig. 9.44(b) we have

$$r = 2p + d$$
$$r = 2p + r \cos \theta_2$$
$$r - r \cos \theta_2 = 2p \qquad (\textit{1 } - \cos \theta \neq 0 \textit{ since } p \neq \theta.)$$
$$r(1 - \cos \theta_2) = 2p$$
$$r = \frac{2p}{1 - \cos \theta_2}$$

Since θ and θ_2 are supplementary angles, we have $\cos \theta_2 = -\cos \theta$, and

$$r = \frac{2p}{1 + \cos \theta}$$

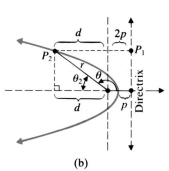

(b)

Figure 9.44

Therefore the equation of a parabola with focus at the pole and symmetry to the line $\theta = 0$ is

$$r = \frac{2p}{1 \pm \cos \theta}$$

where plus and minus signs determine the direction in which the parabola opens. ∎

We note that this form is similar to the form of the equation of the ellipse and hyperbola, as predicted, but with eccentricity $\varepsilon = 1$.

CONIC SECTIONS (WITH FOCUS AT THE POLE)

A polar equation in the form

$$r = \frac{D}{1 \pm \varepsilon \cos \theta} \qquad \text{or} \qquad r = \frac{D}{1 \pm \varepsilon \sin \theta}$$

is

(a) an ellipse with $D = \dfrac{b^2}{a}$ when $0 < \varepsilon < 1$ for $\varepsilon = \dfrac{c}{a}$,

(b) a hyperbola with $D = \dfrac{b^2}{a}$ when $\varepsilon > 1$ for $\varepsilon = \dfrac{c}{a}$, and

(c) a parabola with $D = 2p$ when $\varepsilon = 1$.

We have proved these results for

$$r = \frac{D}{1 \pm \varepsilon \cos \theta}$$

in Examples 1, 2, and 3. The proofs of

$$r = \frac{D}{1 \pm \varepsilon \sin \theta}$$

which are conic sections with foci and axes of symmetry on the line $\theta = \dfrac{\pi}{2}$, are part of this section's exercise set.

EXAMPLE 4   On February 20, 1962, John Glenn made the United States' first manned flight around the earth for three orbits on Freedom 7 with an apogee of 162 miles and a

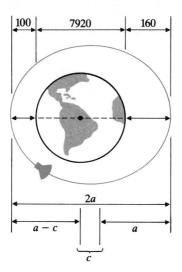

Figure 9.45

perigee of 100 miles. Find the equation of the flight in polar form if the center of the earth is one focus. Let the center of the earth be the pole of the polar coordinate system.

Solution   Using 3960 miles for the radius of the earth, we have a major axis of $a = \frac{1}{2}(162 + 3960 + 3960 + 100) = 4091$ miles, and half the distance between the focus of $c = a - 100 - 3960 = 31$ miles. Since this orbit is an ellipse, we have $a^2 = b^2 + c^2$. Thus $b = \sqrt{a^2 - c^2} \approx 4090.88$ miles. Therefore with $D = \frac{b^2}{a} = 4090.76$ and $\varepsilon = \frac{c}{a} = 0.0076$ we have

$$r = \frac{4090.6}{1 \pm 0.0076 \cos \theta} \quad \text{or} \quad r = \frac{4090.76}{1 \pm 0.0076 \sin \theta} \qquad \blacksquare$$

We will also write the equation of a conic section in polar coordinates when the axes of symmetry intersect at the pole and are symmetrical to the lines $\theta = 0$ and $\theta = \frac{\pi}{2}$ (nonrotated position).

CONIC SECTIONS
$\Big($WITH SYMMETRY
TO THE LINES $\theta = 0$
AND $\theta = \dfrac{\pi}{2}\Big)$

A polar equation of the form

(a) $r^2 = \dfrac{b^2}{1 - \varepsilon^2 \cos^2 \theta}$ is an ellipse with eccentricity $\varepsilon = \dfrac{c}{a}$, $0 < \varepsilon < 1$, and half the minor axis of length b, and

(b) $r^2 = \dfrac{b^2}{\varepsilon^2 \cos^2 \theta - 1}$ is a hyperbola with eccentricity $\varepsilon = \dfrac{c}{a}$ and half the conjugate axis of length b.

The proofs of these two polar equations are part of the exercise set at the end of this section. We will now graph conic sections using these polar equations. We first identify the conic section from the form of the polar equation and its eccentricity. Then we plot the points (r, θ) for $\theta \in \left\{0, \frac{\pi}{2}, \pi, \frac{3\pi}{2}\right\}$ that makes the equation true. The following examples demonstrate this process.

EXAMPLE 5   Sketch the graphs of each of the following equations:

(a) $r = \dfrac{5}{1 + \dfrac{5}{3} \cos \theta}$, (b) $r = \dfrac{9}{5 - 3 \cos \theta}$, and (c) $r = \dfrac{3}{3 + 3 \sin \theta}$.

Solution

(a) The equation

$$r = \frac{5}{1 + \frac{5}{3} \cos \theta}$$

is of the form

$$r = \frac{D}{1 \pm \varepsilon \cos \theta}$$

with an eccentricity of $\varepsilon = \frac{5}{3} > 1$. Thus this is a hyperbola with one focus at the pole. To find its shape, we find the values of r when $\theta \in \left\{ 0, \frac{\pi}{2}, \pi, \frac{3\pi}{2} \right\}$. We use these ordered pairs to sketch the graph, knowing that it is a hyperbola (Fig. 9.46).

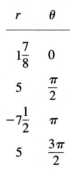

r	θ
$1\frac{7}{8}$	0
5	$\frac{\pi}{2}$
$-7\frac{1}{2}$	π
5	$\frac{3\pi}{2}$

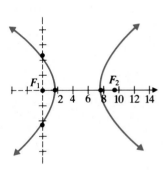

Figure 9.46

(b) To change

$$r = \frac{9}{5 - 3 \cos \theta}$$

into a form similar to those above, multiply the numerator and denominator by $\frac{1}{5}$. Thus

$$r = \frac{9 \left(\frac{1}{5} \right)}{(5 - 3 \cos \theta) \left(\frac{1}{5} \right)}$$

$$r = \frac{\frac{9}{5}}{1 - \frac{3}{5} \cos \theta}$$

This provides the proper form in the denominator of $1 - \frac{3}{5}\cos\theta$ with an eccentricity of $\varepsilon = \frac{3}{5}$. Thus

$$r = \frac{\dfrac{9}{5}}{1 - \dfrac{3}{5}\cos\theta}$$

is an ellipse with one focus at the pole. Find the points $(r,\ \theta)$ for $\theta \in \left\{0, \dfrac{\pi}{2},\ \pi,\ \dfrac{3\pi}{2}\right\}$ and sketch in the ellipse (Fig. 9.47).

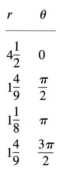

r	θ
$4\frac{1}{2}$	0
$1\frac{4}{9}$	$\dfrac{\pi}{2}$
$1\frac{1}{8}$	π
$1\frac{4}{9}$	$\dfrac{3\pi}{2}$

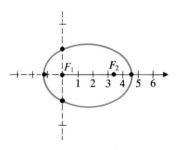

Figure 9.47

(c) The equation

$$r = \frac{3}{3 + 3\sin\theta}$$

is

$$r = \frac{3\left(\dfrac{1}{3}\right)}{(3 + 3\sin\theta)\left(\dfrac{1}{3}\right)} \quad \text{or}$$

$$r = \frac{1}{1 + 1\sin\theta}$$

Thus this is a parabola, since $\varepsilon = 1$. This form also indicates that the focus is at the pole. Now we find values of r when $\theta \in \left\{0, \dfrac{\pi}{2},\ \pi,\ \dfrac{3\pi}{2}\right\}$. We can sketch the graph of this parabola (Fig. 9.48).

r	θ
1	0
$\dfrac{1}{2}$	$\dfrac{\pi}{2}$
1	π
undefined	$\dfrac{3\pi}{2}$

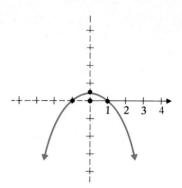

Figure 9.48

EXAMPLE 6   Sketch the graph of

$$r^2 = \frac{16}{1 - 0.7 \cos^2 \theta}$$

Solution   With r^2 and $\cos^2 \theta$ this can be identified as an ellipse or hyperbola with the lines $\theta = 0$ and $\theta = \dfrac{\pi}{2}$ as axes of symmetry. This form,

$$r^2 = \frac{16}{1 - 0.7 \cos^2 \theta}$$

with $\varepsilon^2 = 0.7$, is an ellipse. Then find r for $\theta \in \left\{0, \dfrac{\pi}{2}, \pi, \dfrac{3\pi}{2}\right\}$ and sketch the graph of this ellipse with the symmetry mentioned above (Fig. 9.49).

r	θ
± 7.3	0
± 4	$\dfrac{\pi}{2}$
± 7.3	π
± 4	$\dfrac{3\pi}{2}$

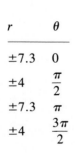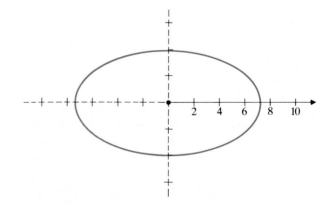

Figure 9.49

Now that we have these special forms for conic sections in polar form, it is not difficult for us to graph conic sections in polar coordinates. Polar coordinates are most convenient when we need to rotate the axes of conic sections.

ROTATION (WITH POLAR COORDINATES)	The graph of $r = f(\theta - \phi)$ is the graph of $r = f(\theta)$ rotated ϕ units (degrees or radians). The rotation is clockwise when $\phi > 0$ and counterclockwise when $\phi < 0$.

This rotation with polar coordinates is much easier than rotation in rectangular coordinates. The following example demonstrates this ease of rotation with polar coordinates.

EXAMPLE 7   Sketch the graph of

(a) $r = \dfrac{3}{3 + 3 \sin (\theta - 2.5)}$ and   (b) $r^2 = \dfrac{16}{1 - 0.7 \cos^2 (\theta + 1)}$.

Solution
(a) We notice that the graph of

$$r = \frac{3}{3 + 3 \sin (\theta - 2.5)}$$

is the graph of

$$r = \frac{3}{3 + 3 \sin \theta}$$

from Example 5(c), rotated $\phi = 2.5 = 143.2°$ clockwise (Fig. 9.50).

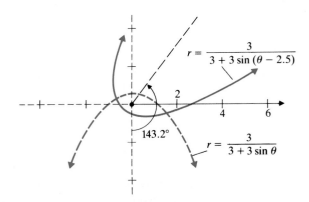

Figure 9.50

(b) Notice that the graph of

$$r^2 = \frac{16}{1 - 0.7 \cos^2 (\theta + 1)}$$

is the graph of

$$r^2 = \frac{16}{1 - 0.7 \cos \theta}$$

from Example 6, rotated $\phi = -1 = -57.2°$ ($57.2°$ counterclockwise) (Fig. 9.51).

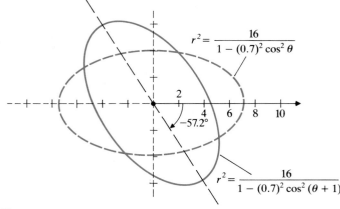

Figure 9.51

As we see in Example 7, one advantage of using polar coordinates is that it makes rotation of axes much easier.

Exercises 9.5

Sketch the graphs of the following conic sections.

1. $r = \dfrac{16}{4 - 3 \cos \theta}$

2. $r = \dfrac{8}{4 + 5 \cos \theta}$

3. $r = \dfrac{12}{4 + 5 \sin \theta}$

4. $r = \dfrac{12}{4 - 3 \sin \theta}$

5. $r = \dfrac{8}{2 + 2 \sin \theta}$

6. $r = \dfrac{9}{3 - 3 \cos \theta}$

7. $r = \dfrac{140}{5 + 13 \cos \theta}$

8. $r = \dfrac{10}{5 - 5 \sin \theta}$

9. $r = \dfrac{18}{6 - 6 \cos \theta}$

10. $r = \dfrac{26}{13 - 12 \cos \theta}$

11. $r = \dfrac{143}{13 + 5 \sin \theta}$

12. $r = \dfrac{24}{12 + 13 \sin \theta}$

13. $r^2 = \dfrac{144}{25 \cos^2 \theta - 16}$

14. $r^2 = \dfrac{225}{25 - 16 \cos^2 \theta}$

15. $r^2 = \dfrac{225}{25 - 9 \cos^2 \theta}$

16. $r^2 = \dfrac{144}{25 \cos^2 \theta - 9}$

17. $r = \dfrac{16}{4 - 3\cos\left(\theta - \dfrac{\pi}{4}\right)}$

18. $r = \dfrac{8}{4 + 5\cos\left(\theta - \dfrac{\pi}{3}\right)}$

19. $r = \dfrac{12}{4 + 5\sin\left(\theta + \dfrac{\pi}{3}\right)}$

20. $r = \dfrac{12}{4 - 3\sin\left(\theta + \dfrac{\pi}{4}\right)}$

21. $r = \dfrac{8}{2 + 2\sin(\theta + 1)}$

22. $r = \dfrac{9}{3 - 3\cos(\theta - 1)}$

23. $r = \dfrac{143}{13 + 5\sin\left(\theta - \dfrac{\pi}{6}\right)}$

24. $r = \dfrac{24}{12 + 13\sin\left(\theta + \dfrac{\pi}{6}\right)}$

25. $r^2 = \dfrac{144}{[25\cos^2(\theta - 1)] - 16}$

26. $r^2 = \dfrac{225}{25 - 16\cos^2(\theta + 1)}$

27. $r^2 = \dfrac{225}{25 - 9\cos^2\left(\theta + \dfrac{3}{4}\right)}$

28. $r^2 = \dfrac{144}{\left[25\cos^2\left(\theta - \dfrac{3}{4}\right)\right] - 9}$

Find the solutions to the following systems of equations.

29. $r = \dfrac{16}{4 - 3\cos\theta}$

$r = \dfrac{140}{5 + 13\cos\theta}$

30. $r = \dfrac{8}{4 + 5\cos\theta}$

$r = \dfrac{9}{3 - 3\cos\theta}$

31. $r = \dfrac{12}{4 + 5\sin\theta}$

$r = \dfrac{8}{2 + 2\sin\theta}$

32. $r = \dfrac{12}{4 - 3\sin\theta}$

$r = \dfrac{10}{5 - 5\sin\theta}$

33. Find the equation for the orbit of Halley's comet in polar form with the sun at the pole, knowing that the eccentricity is $\varepsilon = 0.9673$ and the semimajor axis is $a = 11.9435$ AU (Fig. 9.52).

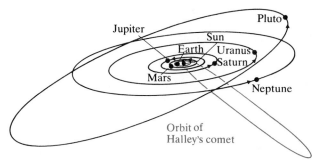

Figure 9.52

34. Find the equation for the orbit of the Schwassmann-Wachmann 1 comet in polar form with the sun at the pole, knowing that the eccentricity is $\varepsilon = 0.105$ and the semimajor axis is $a = 6.087$ AU.

35. Find the equation for the orbit of the planet Saturn in polar form with the sun at the pole, knowing that its period is 29.46 years, its eccentricity is $\varepsilon = 0.0056$, and the semimajor axis is $a = 9.55$ AU.

36. Find the equation for the orbit of the planet Mercury in polar form with the sun at the pole, knowing that its period is 0.24 years, its eccentricity is $\varepsilon = 0.2056$, and the semimajor axis is $a = 0.387$ AU.

37. Which planet, Saturn or Mercury, has a "rounder," more circular orbit? Why?

38. Which comet, Halley's comet or the Schwassmann-Wachmann 1 comet, has a "rounder," more circular orbit? Why?

39. Show that $r^2 = \dfrac{b^2}{(1 - \varepsilon^2\cos^2\theta)}$ is the polar equivalent of the equation of the ellipse $\dfrac{x^2}{a^2} + \dfrac{y^2}{b^2} = 1$, for $0 < \varepsilon < 1$ where ε is the eccentricity.

40. Show that $r^2 = \dfrac{b^2}{(\varepsilon^2\cos^2\theta - 1)}$ is the polar equivalent of the equation of the ellipse $\dfrac{x^2}{a^2} - \dfrac{y^2}{b^2} = 1$, for $\varepsilon > 1$ where ε is the eccentricity.

41. Show that the equation of the parabola with focus at the pole and vertex at $\left(p, \dfrac{\pi}{2}\right)$ or $\left(p, \dfrac{3\pi}{2}\right)$ is $r = \dfrac{D}{(1 \pm \sin\theta)}$.

42. Show that the equation of the ellipse with one focus at the pole and the other focus at $\left(2c, \frac{\pi}{2}\right)$ or $\left(2c, \frac{3\pi}{2}\right)$ is $r = \dfrac{D}{(1 \pm \varepsilon \sin \theta)}$ with $0 < \varepsilon < 1$, where ε is the eccentricity.

43. Show that the equation for the hyperbola with one focus at the pole and the other focus at $\left(2c, \frac{\pi}{2}\right)$ or $\left(2c, \frac{3\pi}{2}\right)$ is $r = \dfrac{D}{(1 \pm \varepsilon \sin \theta)}$ with the eccentricity $\varepsilon > 1$.

9.6 Parametric Equations

Earlier in this book, we wrote a function and a relation with a single equation that described the relationship between the abscissa (x-value) and the ordinate (y-value). Some examples of this are the equations $y = 2x - 3$, $x^2 + y^2 = 4$, $x^2 - y^2 = 4$, $xy = 4$, and $y = \sin^2 x$.

We will now describe the relationship between the abscissa and the ordinate with a third variable. This use of a third variable to define a curve is called the **parametric form** or **equations** of a curve.

Definition

Let f and g be functions for real number values of t. A relation consisting of ordered pairs (x, y) where

$$x = f(t) \qquad \text{and} \qquad y = g(t)$$

is in **parametric form** with a parameter t.

Parametric form has several applications. One practical application is that the parameter t could be time. Thus parametric form provides a method of describing the position (x, y) on a curve as a function of time. Another advantage occurs when curves that are not functions (such as circles, ellipses, and hyperbolas) can be defined in parametric form as functions of t.

To better understand parametric form, it will be helpful to graph the ordered pairs of a function in parametric form for several values of t.

EXAMPLE 1

Sketch the graph of the curve where $x = t + 1$ and $y = -t^2 + 2t + 2$, for $0 \le t \le 3$.

Solution One way to sketch this curve is to find the values of x and y for representative values of t in the interval $[0, 3]$. We fill in the following chart for $t \in \{0, 1, 2, 3\}$ and then sketch the graph using the ordered pairs (x, y). Since

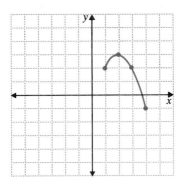

Figure 9.53

these values represent only a few of the infinite real numbers t between 0 and 3, we can sketch a smooth curve connecting these points on the parabola (Fig. 9.53).

t	$x = t + 1$	$y = -t^2 + 2t + 2$	(x, y)
0	$0 + 1 = 1$	$-(0)^2 + 2(0) + 2 = 2$	$(1, 2)$
1	$1 + 1 = 2$	$-(1)^2 + 2(1) + 2 = -1 + 4 = 3$	$(2, 3)$
2	$2 + 1 = 3$	$-(2)^2 + 2(2) + 2 = -4 + 6 = 2$	$(3, 2)$
3	$3 + 1 = 4$	$-(3)^2 + 2(3) + 2 = -9 + 8 = -1$	$(4, -1)$

It is also convenient to write vectors with parametric form where $v = f(t)i + g(t)j$. The following example illustrates this application.

EXAMPLE 2 Sketch the curve for $v = (1 + 3t)i + (-1 + 5t)j$ for $-1 \le t \le 1$ when the tail of v is fixed at the origin.

Solution Consider $v = (1 + 3t)i + (-1 + 5t)j$ to be $v = f(t)i + g(t)j$. Now make a chart for t, $f(t) = 1 + 3t = x$, $g(t) = -1 + 5t = y$, and $(f(t), g(t)) = (x, y)$. The graph is shown in Fig. 9.54.

t	$f(t) = 1 + 3t$	$g(t) = -1 + 5t$	$(f(t), g(t))$
-1	$1 + 3(-1) = -2$	$-1 + 5(-1) = -6$	$(-2, -6)$
0	$1 + 3(0) = 1$	$-1 + 5(0) = -1$	$(1, -1)$
1	$1 + 3(1) = 4$	$-1 + 5(1) = 4$	$(4, 4)$

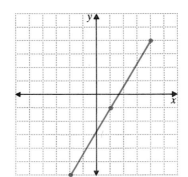

Figure 9.54

It might seem a bit presumptuous to accept the line segment that we graph in Example 2 without trying more values for t. To avoid this, we will combine the parametric forms $x = f(t)$ and $y = g(t)$ to produce one equation in x and y. Depending on the functions f and g, this combining can be accomplished through substitution, trigonometric identities, and other algebraic techniques. The following examples illustrate this process. We must be careful to find the correct intervals for x and y in this combined form.

EXAMPLE 3 Find y as a function of x when the parametric form is $x = 1 + 3t$ and $y = -1 + 5t$ for $-1 \le t \le 1$.

Solution For $x = 1 + 3t$, $y = -1 + 5t$, and $-1 \le t \le 1$ we will solve one equation for t and substitute into the other. Thus

$$x = 1 + 3t$$
$$x - 1 = 3t$$
$$\frac{x - 1}{3} = t$$

Next we substitute this value for t into $y = 1 + 5t$ and have

$$y = -1 + 5\left(\frac{x-1}{3}\right) \quad \text{(Multiply both sides by 3 and simplify)}$$
$$3y = -3 + 5(x - 1)$$
$$3y = -3 + 5x - 5$$
$$0 = 5x - 3y - 8$$

Since this is a straight line, the intervals for x and y are determined by the endpoint at $t = -1$ and $t = 1$. For $x = 1 + 3t$ and $-1 \le t \le 1$ we have $-2 \le x \le 4$. Likewise, for $y = -1 + 5t$ and $-1 \le t \le 1$ we have $-6 \le x \le 4$. Therefore the solution is $0 = 5x - 3y - 8$ for $-6 \le x \le 4$ and $-2 \le y \le 4$. Notice that this is the same line segment that we graphed earlier in Example 2.   ■

EXAMPLE 4   Find y as a function of x when the parametric form is $x = t + 1$ and $y = -t^2 + 2t + 2$ for $0 \le t \le 3$.

Solution   For $x = t + 1$, $y = -t^2 + 2t + 2$, and $0 \le t \le 3$ we substitute, as in Example 3. Then $x = t + 1$ is $x - 1 = t$; substituting, we have

$$y = -t^2 + 2t + 2$$
$$y = -(x - 1)^2 + 2(x - 1) + 2$$
$$y = -x^2 + 2x - 1 + 2x - 2 + 2$$
$$y = -x^2 + 4x - 1$$

We can then complete the square to obtain the standard form of a parabola:

$$y + 1 = -(x^2 - 4x)$$
$$y + 1 - 4 = -(x - 2)^2$$
$$y - 3 = -(x - 2)^2$$

Thus we have a parabola with vertex at $(2, 3)$ opening downward. Since this is the same curve that we sketched in Example 1, the domain and range intervals can be found easily on the graph. Thus $y - 3 = -(x - 2)^2$ for $1 \le x \le 4$ and $-1 \le y \le 3$. Notice that since this curve is a parabola, the intervals for x and y are not necessarily at the endpoints where $t = 0$ and $t = 3$.   ■

It is usually easier to sketch the graph when the parametric equations are combined into a single equation in x and y. The next example shows other ways in which we can change the parametric form into a single equation.

EXAMPLE 5   Find y as a function of x when the parametric form is $x = \sin t$, $y = \cos t$, for $0 \le t \le \frac{\pi}{4}$.

Solution   To combine $x = \sin t$ and $y = \cos t$, we remember that

$$\sin^2 t + \cos^2 t = 1$$

Thus

$$x^2 = \sin^2 t$$
$$y^2 = \cos^2 t$$

and

$$x^2 + y^2 = \sin^2 t + \cos^2 t$$

or

$$x^2 + y^2 = 1$$

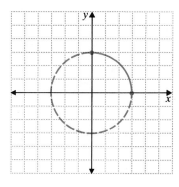

Thus the curve is the portion of the circle where $0 \le t \le \dfrac{\pi}{4}$ (Fig. 9.55). For these values of t and $x = \sin \theta$ we have $0 \le x \le 1$. Likewise, for $y = \cos \theta$ we have $0 \le y \le 1$. Thus the curve represented by $x = \sin t$ and $y = \cos t$ for $0 \le t \le \dfrac{\pi}{4}$ is the portion of the circle in Quadrant I.   ∎

Figure 9.55

EXAMPLE 6   Find y as a function of x when the parametric form is $x = t + \dfrac{1}{t}$, $y = t - \dfrac{1}{t}$, for $1 \le t$.

Solution   Combining $x = t + \dfrac{1}{t}$ and $y = t - \dfrac{1}{t}$ for $1 \le t$ is a little more complicated than in the previous example. We will first find $x - y$ and $x + y$.

For

$$x = t + \frac{1}{t}$$

$$y = t - \frac{1}{t}$$

we have

$$x + y = 2t$$

$$x - y = \frac{2}{t}$$

(Finding $x - y$ and $x + y$ is suggested by the similarity of the right-hand sides of the equations for x and y.) Now we will calculate $(x - y)(x + y)$. Thus

$$(x + y)(x - y) = 2t\left(\frac{2}{t}\right) \qquad \text{or}$$
$$x^2 - y^2 = 4$$

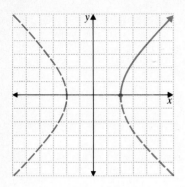

Figure 9.56

which is a hyperbola. For $1 \le t$, we have $x = t + \dfrac{1}{t} \ge 2$ and $y > 0$. Therefore the curve is the portion of this hyperbola in Quadrant I (Fig. 9.56). ■

Some curves are more easily defined by parametric equations. A good example of this is the cycloid. A **cycloid** is formed by all positions that one point on a rolling wheel passes through. Figure 9.57 illustrates a cycloid with P at the end of the radius of the circle. We will develop the parametric equations for this cycloid using geometry and trigonometry in the following example.

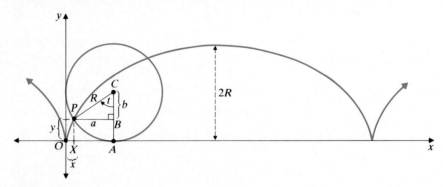

Figure 9.57

EXAMPLE 7

Find the equations of the cycloid (Fig. 9.57) in parametric form.

Solution   As circle C rolls the arc length from O to A, $x + a$ is the arc length of the circle from P to A, Rt. Thus $x + a = Rt$ and

$$x = Rt - a$$

Since $a = R \sin t$, we have

$$x = Rt - R \sin t$$

Likewise, to find y, we notice that

$$y + b = R \qquad \text{or} \qquad y = R - b$$

Since $b = R \cos t$, we have

$$y = R - R \cos t$$

Therefore the cycloid in parametric form is

$$x = R(t - \sin t)$$
$$y = R(1 - \cos t)$$

■

The graph of the cycloid (Fig. 9.57) can be sketched by using the parametric equations $x = R(t - \sin t)$ and $y = R(1 - \cos t)$ with various values of t. The domain t can be time or any real number.

As we have seen in this section, we can write the equation of a curve in parametric form as $x = f(t)$ and $y = g(t)$, where f and g are functions of some real number parameter t. Thus we can even write a relation that is not a function in terms of functions f and g with parametric form.

Sometimes the parametric form can be combined into a single equation in x and y, which may make graphing easier. There are also some parametric equations that cannot be easily combined into one equation. Parametric form is a useful tool in representing curves.

Exercises 9.6

Graph each of the following parametric equations and combine the equations to form one equation in x and y.

1. $x = t - 2$
   $y = 2 - t$
   for $2 \le t \le 6$

2. $x = 5 - t$
   $y = t - 3$
   for $5 \le t \le 10$

3. $x = (t - 2)^2$
   $y = t - 2$
   for $2 \le t \le 4$

4. $x = (t - 1)^3$
   $y = (t - 1)^2$
   for $0 \le t \le 3$

5. $x = \cos 2t$
   $y = \sin 2t$
   for $\dfrac{\pi}{4} \le t \le \dfrac{\pi}{2}$

6. $x = 2 \sin t$
   $y = 3 \cos t$
   for $0 \le t \le \pi$

7. $x = 3 \cos t$
   $y = 4 \sin t$
   for $\pi \le t \le \dfrac{3\pi}{2}$

8. $x = 5 \sin t$
   $y = 5 \cos t$
   for $\dfrac{\pi}{2} \le t \le \pi$

9. $x = t + \dfrac{9}{t}$
   $y = t - \dfrac{9}{t}$

10. $x = t - \dfrac{4}{t}$
   $y = t + \dfrac{4}{t}$

11. $x = \cos 2t$
   $y = \sin t$

12. $x = \cos 2t$
   $y = \cos t$

13. $x = t^2$
   $y = t - 2$
   for $-1 \le t \le 2$

14. $x = 3t + 2$
   $y = t^2$
   for $-2 \le t \le 1$

Sketch the curve for these vectors when the tail is fixed at the origin.

15. $v = ti + (\sqrt{1 + t^2})j$

16. $w = (\sqrt{1 - t^2})i + (t - 3)j$

17. $w = (\csc t)i + (\cot t)j$

18. $v = (\tan t)i + (\sec t)j$

19. $v = 2^t i + 2^{t+1} j$

20. $w = 3^{t+2} i + 3^{t-1} j$

21. Find and graph the parametric equations for the cycloid formed when P is at the midpoint of the radius of the wheel (Fig. 9.58).

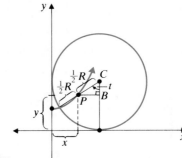

Figure 9.58

22. Find and graph the parametric equations for the cycloid when point P is a distance $\frac{3}{2}R$ from the center of the wheel (Fig. 9.59).

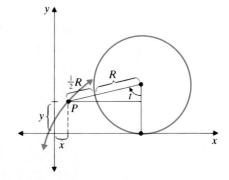

Figure 9.59

23. Verify that the orbit of the moon around the earth, given by $x = c + a \sin\left(\dfrac{27\pi}{365}\right)t$, $y = b \cos\left(\dfrac{27\pi}{365}\right)t$, and $0 \le t \le 365$, is an ellipse with the earth at one focus. For the moon in these equations the semimajor axis is $a = 2.3887 \times 10^5$ miles, the semiminor axis is $b = 2.3851 \times 10^5$ miles, one half the distance between the foci is $c = 0.1312 \times 10^5$ miles, and t is time measured in days. (The moon makes approximately 13 orbits of the earth in a year.)

24. Verify that the orbit of the earth around the sun, given by $x = C + A \sin\left(\dfrac{2\pi}{365}\right)t$, $y = B \cos\left(\dfrac{2\pi}{365}\right)t$, and $0 \le t \le 365$, is an ellipse with the earth at one focus. For the earth in these equations the semimajor axis is $A = 9.2961 \times 10^7$ miles, the semiminor axis is $B = 9.2948 \times 10^7$ miles, one half the distance between the foci is $C = 0.1552 \times 10^7$ miles, and t is time measured in days.

25. Using the equations from Exs. 23 and 24, find the equation for the cycloidlike orbit of the earth's moon about the sun with the sun at the origin (Fig. 9.60).

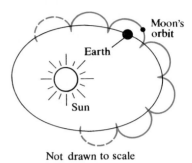

Not drawn to scale

Figure 9.60

9.7 Three-Dimensional Graphs: Planes and Lines

To this point we have worked primarily with two-dimensional geometry. We have simplified many applications with two-dimensional models. However, three-dimensional space and three-dimensional geometry are often necessary for more general solutions of these applications. The parabolic satellite receiving dish discussed in Section 9.1 can be viewed as a parabolic cross section in two dimensions, but it is more correctly seen as a parabolic solid in three dimensions. In this section we will begin working with three-dimensional analytic geometry.

In three-dimensional space, each point is named by an ordered triple of real numbers (x, y, z). This ordered triple is associated with a three-axis coordinate system. To determine this coordinate system, start with a fixed point O as the origin. Through the origin, visualize three mutually perpendicular lines. These three lines are usually labeled the x-axis, the y-axis, and the z-axis. Being mutually perpendicular, each axis is perpendicular to the other two axes. Coordinates on each axis are positive in one direction from the origin and negative in the other direction from the origin. The positive orientation usually follows the right-hand rule of thumb illustrated in Fig. 9.61.

Naming and graphing points in three dimensions is similar to naming and graphing points in two dimensions. Thus the point (a, b, c) is the point where the coordinate on the x-axis is a, the coordinate on the y-axis is b, and the coordinate

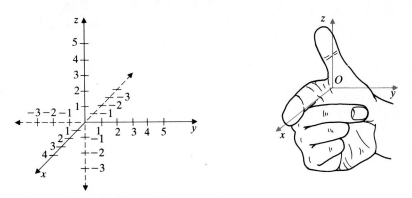

Figure 9.61

on the z-axis is c. To graph this point, start at the origin and move a units on the x-axis, then b units parallel to the y-axis, and then c units parallel to the z-axis. Thus the point (a, b, c) is at the corner of the box created by a units on the x-axis, b units on the y-axis, and c units on the z-axis as in Fig. 9.62.

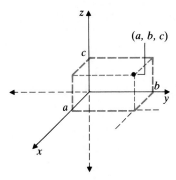

Figure 9.62

We note that the point $(x, y, 0)$ is any point in the x-y plane where $z = 0$. Likewise, the point $(x, 0, z)$ is any point in the x-z plane where $y = 0$, and $(0, y, z)$ is any point in the y-z plane where $x = 0$ (Fig. 9.63).

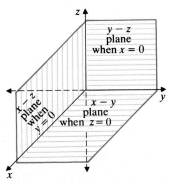

Figure 9.63

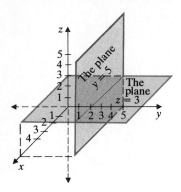

Figure 9.64

There are other planes that we can also easily identify. The equation $z = 3$ describes all points $(x, y, 3)$ in three dimensions for which x and y are real numbers. This is the x-y plane with $z = 3$ for every point (Fig. 9.64). Similarly, $y = 5$ or the points $(x, 5, z)$ represent the x-z plane with $y = 5$ (Fig. 9.64).

We will now sketch the graph of a plane that is not parallel to the x-y plane, the y-z plane, or the x-z plane by sketching the plane's given intersection with the three planes $x = 0$, $y = 0$, and $z = 0$ as in Fig. 9.63.

Guidelines for Graphing Planes

To sketch the graph of a plane:

STEP 1: Let $x = 0$ and sketch the plane's intersection with the y-z plane $x = 0$ (the y-z plane through the origin).

STEP 2: Let $y = 0$ and sketch the plane's intersection with the x-z plane $y = 0$ (the x-z plane through the origin).

STEP 3: Let $z = 0$ and sketch the plane's intersection with the x-y plane $z = 0$ (the x-y plane through the origin).

STEP 4: Shade in a portion of the plane to indicate the plane's position.

We will demonstrate this process in the following examples.

EXAMPLE 1　Sketch the graphs of the planes **(a)** $6x + 3y + 4z = 12$ and **(b)** $y + z = 3$.

Solution　We will sketch the graphs of these planes using the above guidelines.
(a) To sketch the graph of $6x + 3y + 4z = 12$:

STEP 1: Let $x = 0$ and sketch $3y + 4z = 12$ in the y-z plane through the origin (Fig. 9.65a).

STEP 2: Let $y = 0$ and sketch $6x + 4z = 12$ in the x-z plane through the origin (Fig. 9.65a).

STEP 3: Let $z = 0$ and sketch $6x + 3y = 12$ in the x-y plane through the origin (Fig. 9.65a).

STEP 4: Shade in a portion of the plane to indicate its position (Fig. 9.65b).

Figure 9.65

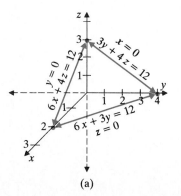

(a)

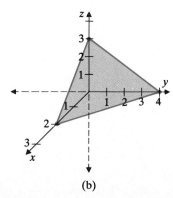

(b)

(b) To sketch the graph of $y + z = 3$:

STEP 1: Let $x = 0$ and sketch $y + z = 3$ in the y-z plane through the origin (Fig. 9.66a).

STEP 2: Let $y = 0$ and sketch $z = 3$ in the x-z plane through the origin (Fig. 9.66a).

STEP 3: Let $z = 0$ and sketch $y = 3$ in the x-z plane through the origin (Fig. 9.66a).

STEP 4: Shade in a portion of the plane to indicate its position (Fig. 9.66b).

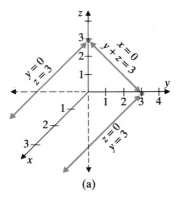
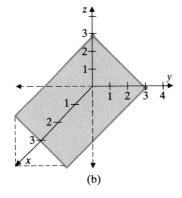

(a)   (b)

Figure 9.66   ◼

Although it is impossible to graph the whole plane, since a plane is a flat unending surface, we will shade in a portion of the plane to indicate the position of the plane.

Sometimes we will want to graph a line that is the intersection of two planes as in the next example.

EXAMPLE 2   Sketch the planes $6x + 3y + 4z = 12$ and $y + z = 3$ and the line of their intersection.

Solution   Since these are the two planes from Example 1, we will sketch them on the same coordinate system. From Fig. 9.67 it appears that the intersection of the two planes is the line L.

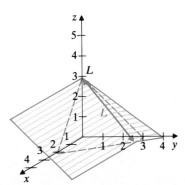

Figure 9.67   ◼

As we saw in Section 8.3, the planes in Example 2 are a dependent system of linear equations. Using the techniques of Chapter 8, we can write the equation of the solution, a line, in parametric form as follows. For

$$6x + 3y + 4z = 12$$
$$y + z = 3$$

we let $z = t$, then $y = 3 - t$. Solving for x in

$$6x + 3y + 4z = 12$$

we have

$$6x + 3(3 - t) + 4(t) = 12$$
$$6x = 3 - t$$
$$x = \frac{3 - t}{6}$$

Therefore the line L in Fig. 9.67 is represented by the parametric equations $x = \frac{3 - t}{6}$, $y = 3 - t$, and $z = t$ or, more concisely, it is represented as the set of ordered triples

$$\left(\frac{3 - t}{6}, 3 - t, t\right) \qquad \text{where } t \text{ is any real number}$$

As we see here, the parametric form is necessary and convenient when we write the equation of a three-dimensional line.

EXAMPLE 3   Sketch the graph of the line represented by $\left(1 - \frac{t}{3}, 5 - \frac{5t}{3}, t\right)$ for any real number t.

Solution   To sketch the line represented by $\left(1 - \frac{t}{3}, 5 - \frac{5t}{3}, t\right)$, we pick two values for t to find two points on this line.

t	$x = 1 - \dfrac{t}{3}$	$y = 5 - \dfrac{5t}{3}$	$z = t$	(x, y, z)
0	1	5	0	(1, 5, 0)
3	0	0	3	(0, 0, 3)

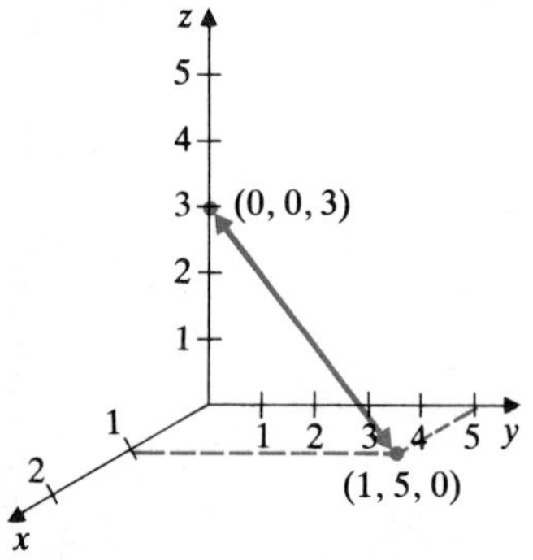

(In this case we conveniently choose the values of 0 and 3 for t, since each is evenly divisible by the 3 in the parametric equations.) We now plot and connect the points (1, 5, 0) and (0, 0, 3) (Fig. 9.68). Because of the lack of true perspective in sketching a three-dimensional graph on two-dimensional paper, it is often easier to carefully plot only two points when graphing a line.   ■

Figure 9.68

Now we can graph points, lines, and planes in three dimensions. We will conclude this section with the distance formula extended to three-dimensional space.

DISTANCE FORMULA	The distance between (x_1, y_1, z_1) and (x_2, y_2, z_2) is $$d = \sqrt{(x_2 - x_1)^2 + (y_2 - y_1)^2 + (z_2 - z_1)^2}$$

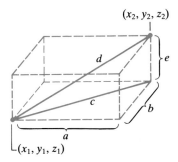

Figure 9.69

PROOF: Using Fig. 9.69, we notice that $a = |y_2 - y_1|$, $b = |x_2 - x_1|$, and $e = |z_2 - z_1|$. Using the Pythagorean Theorem, we have $c^2 = b^2 + a^2$ and $d^2 = c^2 + e^2$. By substituting,

$$d^2 = (b^2 + a^2) + e^2$$
$$d = \sqrt{b^2 + a^2 + e^2}$$

Thus

$$d = \sqrt{(|x_2 - x_1|)^2 + (|y_2 - y_1|)^2 + (|z_2 - z_1|)^2}$$

Since the square keeps the value positive, we no longer need the absolute values, and

$$d = \sqrt{(x_2 - x_1)^2 + (y_2 - y_1)^2 + (z_2 - z_1)^2} \qquad \square$$

We can now find distances in three-dimensional space as well as sketch graphs in three-dimensional space.

EXAMPLE 4

Plot the points $P_1(3, -2, 1)$ and $P_2(-1, 3, 2)$ and calculate the distance between P_1 and P_2.

Solution Using the x, y, and z coordinate system we plot the points P_1 and P_2 (Fig. 9.70). Next we calculate the distance between P_1 and P_2 with the distance formula:

$$d = \sqrt{[3 - (-1)]^2 + (-2 - 3)^2 + (1 - 2)^2}$$
$$= \sqrt{4^2 + 5^2 + (-1)^2}$$
$$= \sqrt{16 + 25 + 1}$$
$$= \sqrt{42}$$

This is the same value as if the distance were calculated as

$$d = \sqrt{(-1 - 3)^2 + [3 - (-2)]^2 + (2 - 1)^2} \qquad \blacksquare$$

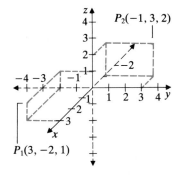

Figure 9.70

With these basics it is now possible to graph points, lines, and planes and to find distances in three-dimensional space.

Exercises 9.7

Find the distance between the following pairs of points.

1. (3, 5, 2), (5, 1, 4)

2. (3, 1, 4), (1, 4, 3)

3. (3, 0, 2), (0, 4, 1)

4. (1, 0, 5), (0, 2, 1)

5. (4, 1, 0), (1, −1, 3)

6. (2, 0, 3), (4, −2, 0)

7. (2, −1, 0), (1, −1, −2)

8. (4, 1, 0), (3, −2, −1)

Sketch the graphs of the following planes.

9. $2x + 3y + z = 6$

10. $2x + y + 3z = 6$

11. $2x + y + 2z = 8$

12. $4x + 2y + z = 8$

13. $x - y + 2z = 6$

14. $x + y - 2z = 8$

15. $2x + 3y + z = 8$

16. $x + 2y + 3z = 9$

17. $2x + 3y = 6$

18. $2y + z = 6$

19. $3y + z = 9$

20. $2x + z = 8$

21. $y = -4$

22. $z = 3$

23. $2x - y - z = 6$

24. $2x + 3y - z = -6$

Sketch the following pairs of planes and their lines of intersection.

25. $4x + 3y + 6z = 12$
    $2x + y + z = 4$

26. $x + y + z = 3$
    $7x + 5y + 3z = 15$

27. $x + 2y + z = 4$
    $3x + 4y + 3z = 12$

28. $3x + 6y + 2z = 6$
    $x + y + z = 2$

29. $x + y + z = 2$
    $y + 3z = 3$

30. $x + y + z = 3$
    $2x + z = 4$

31. $3x + 4y + 6z = 12$
    $2x + z = 4$

32. $2x + 6y + 3z = 6$
    $y + 3z = 3$

33. Find the parametric form for the line of intersection in Exercises 25, 29, and 31.

34. Find the parametric form for the line of intersection in Exercises 26, 30, and 32.

35. Midnight Auto Supply sells tires and oil with a cost of $100 per unit of tires and a cost of $80 per unit of oil. Sketch a three-dimensional graph of the cost function $C(x, y)$.

36. Midnight Auto Supply sells tires and oil with a profit of $80 per unit of tires and a profit of $120 per unit of oil. Sketch a three-dimensional graph of the profit function $P(x, y)$.

37. The Pocket Rocket Toy Company makes toy robots and spaceships. Sketch the three-dimensional graph of the profit function $P(x, y)$ if the profit for each robot is $3.00 and the profit for each spaceship is $2.00.

38. The Greathouse Construction Company designs buildings with exteriors of mirrored windows and aluminum panels. Sketch the three-dimensional graph of the cost function $C(x, y)$ if the cost of each mirrored window is $350 and the cost of each aluminum panel is $300.

39. The VCR Electronics Company produces a standard model with a profit of $75 and a deluxe model with a profit of $90. Sketch the three-dimensional graph of the profit function $P(x, y)$.

40. Bowen's Boat Builders manufacture yachts and sailboats. Sketch the three-dimensional graph of the profit when the profit is $50,000 per yacht and $15,000 per sailboat.

9.8   Three-Dimensional Graphs: Quadratic Surfaces

We will now extend the three-dimensional graphing methods of Section 9.7 to graphing three-dimensional surfaces. The surfaces that we will graph in this section will be quadratic in that their cross sections are conics. Some three-dimensional surfaces are not difficult to perceive and sketch.

*Guidelines for Sketching Cylinders Parallel
to the Coordinate Axes*

1. Sketch all three coordinate axes *very lightly*.

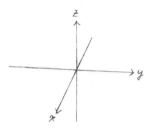

2. Sketch the trace of the cylinder in the coordinate plane of the two variables that appear in the cylinder's equation. Sketch *very lightly*.

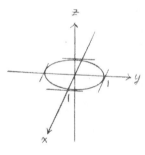

3. Sketch traces in parallel planes on either side (again, lightly).

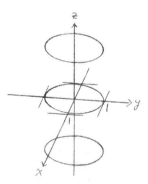

4. Add parallel outer edges to give the shape definition.

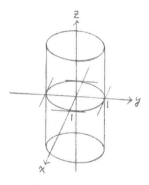

5. If more definition is required, darken the parts of the lines that are exposed to view. Leave the hidden parts light. Use line breaks when you can.

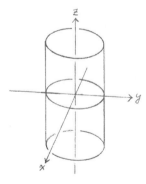

SOURCE: *Calculus*, 7th edition, by Thomas/Finney © 1988 by Addison-Wesley. Reprinted by permission.

In the following examples we will sketch several kinds of three-dimensional surfaces.

EXAMPLE 1

Sketch the graph of
(a) $x^2 + y^2 = 4$ and (b) $x^2 + y^2 = 4$ for $0 \le z \le 3$.

Solution   We notice that both of these surfaces have the same equation and shape in any x-y plane $z = c$ for any real number c.

(a) The equation of this surface, $x^2 + y^2 = 2^2$, is the equation of a circle with radius 2 and center (0, 0) for every value of z. Thus the graph is an infinitely long circular cylinder (Fig. 9.71).

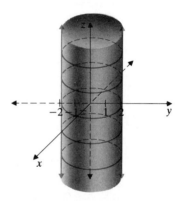

Figure 9.71

(b) The equation $x^2 + y^2 = 4$ for $0 \le z \le 3$ is the equation of the cylinder in part (a) with the restriction that $0 \le z \le 3$. Thus this graph is the circular cylinder with radius 2 and height 3 as seen in Fig. 9.72. We notice that the cylinder $x^2 + y^2 = 4$ for $0 \le z \le 3$ does not include the top and bottom but only the side surface of the cylinder.   ■

In Example 1 we sketched the graph by placing the graph of the circle, $x^2 + y^2 = 4$, in various x-y planes above and below the origin. Similarly, we will graph more complicated surfaces by allowing one variable to take on specific values and sketching the resulting two-dimensional image. The following example illustrates this process.

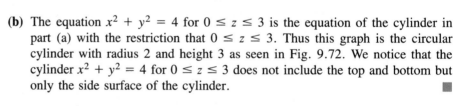

Figure 9.72

EXAMPLE 2

Sketch the graph of $x^2 + y^2 + z^2 = 4$.

Solution   We first look for conic sections where one variable is a constant, and then put together a composite picture. For $x^2 + y^2 + z^2 = 4$:

(a) We notice that $x^2 + y^2 = 4 - z^2$ is a circle for $-2 < z < 2$ and a point for $z = \pm 2$. Thus for $z = 0$ the image in the x-y plane through the origin is the circle $x^2 + y^2 = 4$. As $|z|$ gets larger, with $|z| < 2$, the equation is still a circle with the radius decreasing as z increases to ± 2 (Fig. 9.73a).

(b) Next we notice that $x^2 + z^2 = 4 - y^2$ also graphs circles, this time in the x-z plane for $-2 < y < 2$ (Fig. 9.73b).

(c) Circles also occur in the y-z plane, since $y^2 + z^2 = 4 - x^2$. Therefore the surface $x^2 + y^2 + z^2 = 4$ must be the **sphere**, as seen in Fig. 9.73(c). It is easier to visualize this surface if only one x-y cross section, one x-z cross section, and one y-z cross section are sketched in the graph. ∎

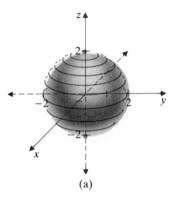

(a)

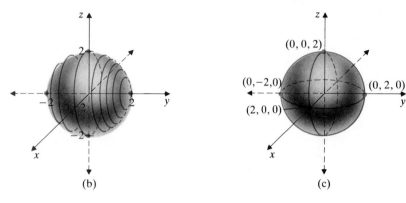

(b)   (c)

Figure 9.73

EXAMPLE 3   Sketch the graph of $36x^2 + 9y^2 + 16z^2 = 144$.

Solution   We rewrite $36x^2 + 9y^2 + 16z^2 = 144$ by dividing both sides by 144. This provides the more recognizable form

$$\frac{x^2}{4} + \frac{y^2}{16} + \frac{z^2}{9} = 1$$

(a) For different values of z,

$$\frac{x^2}{4} + \frac{y^2}{16} = 1 - \frac{z^2}{9}$$

is an ellipse when $-3 < z < 3$. This ellipse has its largest axes when $z = 0$. As $|z|$ gets larger, the axes of the ellipse get smaller. When $|z| = \pm 3$, the ellipse degenerates to a point.

(b) For different values of y,

$$\frac{x^2}{4} + \frac{z^2}{9} = 1 - \frac{y^2}{16}$$

is an ellipse when $-4 < y < 4$. As $|y|$ gets larger, the axes of the ellipse get smaller. When $y = \pm 2$, the ellipse degenerates to a point.

(c) For different values of x,

$$\frac{y^2}{16} + \frac{z^2}{9} = 1 - \frac{x^2}{4}$$

is also an ellipse when $-2 < x < 2$. As $|x|$ gets larger, the axes of the ellipse get smaller. When $x = \pm 2$, the ellipse degenerates to a point.

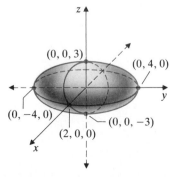

Figure 9.74

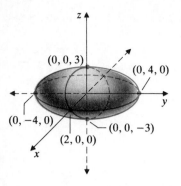

Figure 9.74

(d) These three cross sections are represented in Fig. 9.74 with one image for each of the *x-y*, *y-z*, and *x-z* planes. This surface is called an **ellipsoid**.   ◼

In general, the equation for a sphere with center at the origin as in Fig. 9.73 is $x^2 + y^2 + z^2 = r^2$ and is a direct result of the distance formula for three dimensions of Section 9.7. Also, the general equation of an ellipsoid with axes of symmetry intersecting at the origin is

$$\frac{x^2}{a^2} + \frac{y^2}{b^2} + \frac{z^2}{c^2} = 1$$

where *a*, *b*, and *c* are not all equal. Now that these graphing techniques have been introduced, we can use them to sketch the graphs of other three-dimensional surfaces.

EXAMPLE 4

Sketch the graph of $36x^2 + 16y^2 - 9z^2 = 144$.

Solution   We first divide both sides of $36x^2 + 16y^2 - 9z^2 = 144$ by 144. Then we have

$$\frac{x^2}{4} + \frac{y^2}{9} - \frac{z^2}{16} = 1$$

(a) For any value of *z*,

$$\frac{x^2}{4} + \frac{y^2}{9} = 1 + \frac{z^2}{16}$$

is an ellipse. In this case the major and minor axes of this ellipse are shortest when $z = 0$. Then as $|z|$ gets larger, the axes of the ellipse get larger. Thus any *x-y* cross section is an ellipse.

(b) For values of *y* such that $-3 < y < 3$ we see that

$$\frac{x^2}{4} - \frac{z^2}{16} = 1 - \frac{y^2}{9}$$

is a hyperbola.

(c) For values of *x* such that $-2 < x < 2$,

$$\frac{y^2}{9} - \frac{z^2}{16} = 1 - \frac{x^2}{4}$$

is also a hyperbola.

(d) These three cross sections are depicted in Fig. 9.75 to provide a graph for

$$\frac{x^2}{4} + \frac{y^2}{9} - \frac{z^2}{16} = 1$$

This surface is a **hyperboloid of one sheet**.   ◼

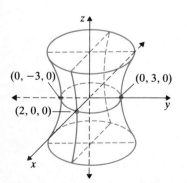

Figure 9.75

When we make a slight change in the equation of Example 4, we have another hyperboloid as in the next example.

EXAMPLE 5   Sketch the graph of $36x^2 - 16y^2 + 9z^2 = -144$.

Solution   We first divide both sides of $36x^2 - 16y^2 + 9z^2 = -144$ by -144. Then

$$-\frac{x^2}{4} + \frac{y^2}{9} - \frac{z^2}{16} = 1$$

(a) For any value of z,

$$-\frac{x^2}{4} + \frac{y^2}{9} = 1 + \frac{z^2}{16}$$

is a hyperbola.

(b) For values of y,

$$-\frac{x^2}{4} - \frac{z^2}{16} = 1 - \frac{y^2}{9}$$

is an ellipse for $|y| > 3$.

(c) For any value of x,

$$\frac{y^2}{9} - \frac{z^2}{16} = 1 + \frac{x^2}{4}$$

is a hyperbola.

(d) We sketch these cross-sectional images in Fig. 9.76 to provide the graph of $36x^2 - 16y^2 + 9z^2 = -144$.

This surface is a **hyperboloid of two sheets**.

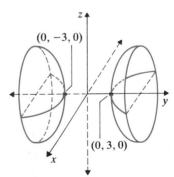

Figure 9.76

Thus there are two hyperboloid surfaces, the hyperboloid of one sheet (one surface) (Fig. 9.75), which is the design of some cooling towers for nuclear power plants, and the hyperboloid of two sheets (two surfaces), which has two parts (Fig. 9.76).

In this section, by sketching the image of a surface in the x-y plane, the y-z plane, and the x-z plane we were able to sketch a graphical representation of these three-dimensional surfaces.

Guidelines for Sketching Quadratic Surfaces

$$x^2 + \frac{y^2}{4} + z^2 = 1 \qquad\qquad z = 4 - x^2 - y^2$$

1. Lightly sketch the three coordinate axes.

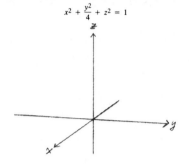

2. Decide on a scale and mark the intercepts on the axes.

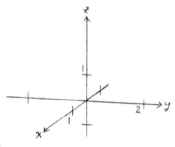

3. Sketch cross sections in the coordinate planes and in a few parallel planes, but don't clutter the picture.

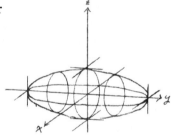

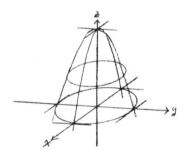

4. If more is required, darken the parts exposed to view. Leave the rest light. Use line breaks when you can.

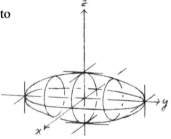

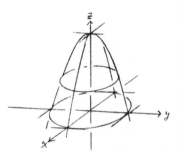

Like the parabola, the ellipse and the hyperbola have special properties when reflecting light and sound, as can be seen in Fig. 9.77. Understanding the reflecting properties of these conic sections and quadratic surfaces made possible the devel-

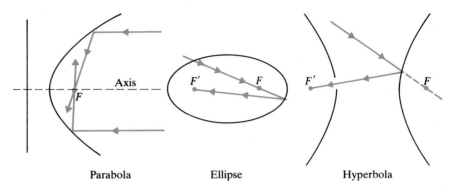

Figure 9.77

opment of the telescope. In Fig. 9.78 we can see how the reflecting properties of quadratic surfaces focus an image in the Newtonian (1668), Gregorian (1663), Cassegrain (1672), and Maksutov telescopes.

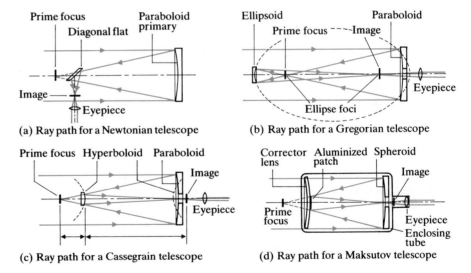

Figure 9.78

It is interesting to see in these telescope diagrams how the image converges at the focus of the conic section. Notice that in the Gregorian telescope the focus of the paraboloid is also one focus of the ellipsoid, and in the Cassegrain telescope the focus of the paraboloid is one focus of the hyperboloid. Thus the quadratic surfaces and their reflecting properties are an important part of telescope design and development.

Exercises 9.8

Sketch each of the following three-dimensional surfaces.

1. $x^2 - y^2 = 4$

2. $y = -x^2$

3. $x = -y^2$

4. $4x^2 - y^2 = 4$

5. $x^2 + y^2 - 2y = 3$

6. $x^2 + 4x + y^2 = 0$

7. $x^2 + y^2 + z^2 = 9$

8. $x^2 + y^2 + z^2 = 1$

9. $x^2 + y^2 - z^2 = 4$

10. $x^2 - y^2 + z^2 = 4$

11. $x^2 - y^2 - z^2 = 9$

12. $-x^2 + y^2 - z^2 = 9$

13. $\dfrac{x^2}{25} + \dfrac{y^2}{16} + \dfrac{z^2}{9} = 1$

14. $\dfrac{x^2}{9} + \dfrac{y^2}{4} + \dfrac{z^2}{16} = 1$

15. $\dfrac{x^2}{9} - \dfrac{y^2}{16} + \dfrac{z^2}{9} = 1$

16. $\dfrac{x^2}{9} - \dfrac{y^2}{4} - \dfrac{z^2}{16} = -1$

17. $\dfrac{x^2}{9} - \dfrac{y^2}{4} - \dfrac{z^2}{16} = 1$

18. $\dfrac{x^2}{9} - \dfrac{y^2}{16} - \dfrac{z^2}{4} = 1$

19. $x^2 + z^2 = 4y$

20. $x^2 + y^2 = 4z$

21. $4x^2 + 9y^2 = 36z$

22. $4x^2 + 9z^2 = 36y$

23. $\dfrac{x^2}{25} + \dfrac{y^2}{16} + \dfrac{z^2}{9} = 1$
for $0 \le z \le 3$

24. $\dfrac{x^2}{16} + \dfrac{y^2}{25} + \dfrac{z^2}{4} = 1$
for $-2 \le z \le 0$

25. Sketch $2y^2 + 3z^2 - x^2 = 0$. Verify that this surface is a cone by showing that the cross sections in the *x-y* plane, *y-z* plane, and *x-z* plane are the proper conic sections.

26. Find the equation of the sphere with radius 5 and center (0, 0, 0).

27. Find the equation of the sphere with radius 3 and center (2, 3, 4).

28. Find the equation of a satellite antenna dish in the shape of a paraboloid with a diameter of 6 feet and a depth of 1 foot (Fig. 9.79).

29. Find the equation of a satellite antenna dish in the shape of a paraboloid with a diameter of 6 feet and a depth of 2 feet (Fig. 9.79).

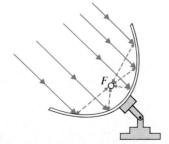

Figure 9.79

30. Find the equation of a flashlight's parabolic reflecting surface that has a radius of 4 inches and a depth of 3 inches (Fig. 9.80).

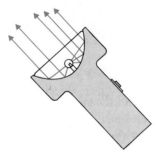

Figure 9.80

31. Find the equation of a flashlight's parabolic reflecting surface that has a radius of 2 inches and a depth of 1 inch (Fig. 9.80).

32. A parabolic trough is made so that it will turn facing the sun all day. This trough is 4 feet wide at the top, 3 feet deep, and 10 feet long (Fig. 9.81). Find the equation of this parabolic trough.

33. A parabolic trough is made so that it will turn facing the sun all day. This trough is 6 feet wide at the top, 4 feet deep, and 8 feet long (Fig. 9.81). Find the equation of this parabolic trough.

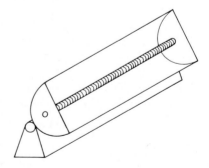

Figure 9.81

34. The lithotripter described at the beginning of this chapter is a semiellipsoid (Fig. 9.82). If the distance from the focus that produces the shock wave to the closest end of the ellipsoid is 2.14 cm and the eccentricity of the ellipse is

$\varepsilon = 0.7$, find the equation of the ellipsoid and the distance between the targeted kidney stone and the electrode emitting the shock wave.

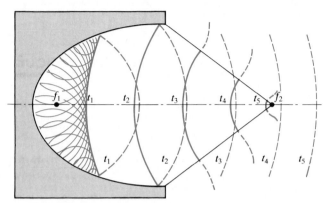

Time diagram of a shock wave expanding from the first focus f_1 of the semi ellipsoid and directed to the second focus f_2 at the kidney stone

Figure 9.82

35. The lithotripter described at the beginning of this chapter is a semiellipsoid (Fig. 9.82). If the distance from the focus that produces the shock wave to the closest end of the ellipsoid is 2.7 cm and the eccentricity of the ellipse is $\varepsilon = 0.65$, find the equation of the ellipsoid and the distance between the targeted kidney stone and the electrode emitting the shock wave.

36. Write the equation of the hyperboloid of one sheet:

$$\frac{x^2}{a^2} + \frac{y^2}{b^2} + \frac{z^2}{c^2} = 1$$

when the intersection of the axes of symmetry is translated to $(2, -1, 4)$.

37. Write the equation of the hyperboloid of two sheets:

$$\frac{x^2}{a^2} - \frac{y^2}{b^2} + \frac{z^2}{c^2} = 1$$

when the intersection of the axes of symmetry is translated to $(3, 4, -2)$.

Sketch the graphs of each of the following surfaces.

38. $x^2 + z^2 - 4x + 2z = 4y - 3$

39. $x^2 + y^2 - 6x + 4y = 4z + 5$

40. $x^2 + y^2 + z^2 - 2x + 4y - 6z + 10 = 0$

41. $x^2 + y^2 + z^2 - 4x + 6y + 4z + 13 = 0$

42. A solid has silhouettes from the front, side, and top that are a square, equilateral triangle, and a circle (Fig. 9.83). Sketch a three-dimensional diagram of this solid.

Figure 9.83

CHAPTER SUMMARY AND REVIEW

KEY TERMS

Conic sections:
  Circle
  Parabola
  Ellipse
  Hyperbola
Rotation of axes

Conic sections in polar form
Rotation with polar coordinates
Parametric form
Three-dimensional graphing

CHAPTER EXERCISES

Write the equation of the conic section in each of the following graphs.

1.

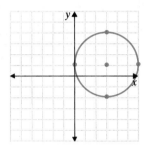

2.

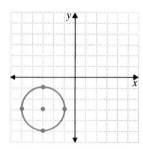

3.

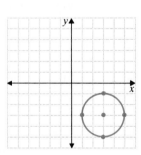

4.

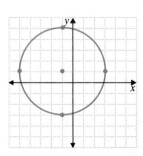

5.

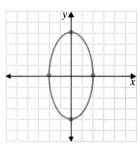

6.

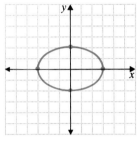

7.

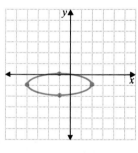

8.

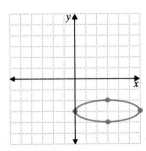

9.

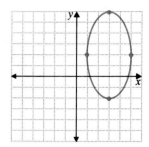

10.

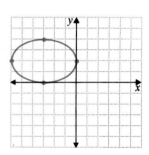

11.

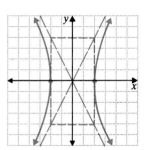

12.

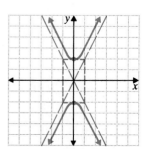

13.

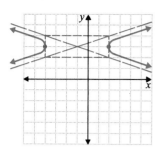

14.

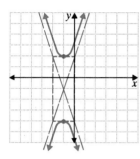

Graph each of the following. Find the ordered pairs for the center of each circle, the vertex and focus of each parabola, and the foci of each ellipse and hyperbola.

15. $(x - 3)^2 + (y + 1)^2 = 9$

16. $x^2 + y^2 - 4x + 4y + 4 = 0$

17. $x^2 - 4x - 12y + 28 = 0$

18. $4x^2 - 9y^2 = 36$

19. $9x^2 + 4y^2 = 36$

20. $4x^2 - y^2 - 8x - 4y = 4$

21. $4x^2 + y^2 - 8x + 4y = 8$

22. $xy = 4$

23. $r = \dfrac{6}{2 - \cos \theta}$

24. $r = \dfrac{6}{1 - \sin \theta}$

25. $r = \dfrac{6}{1 + 2 \sin \theta}$

26. $r = \dfrac{6}{1 - \sin (\theta - 45°)}$

27. $r^2 = \dfrac{9}{1 - 0.64 \cos^2 \theta}$

28. $r^2 = \dfrac{9}{1 - 0.64 \cos^2 (\theta + 60°)}$

29. $x = 2 \sin t$
$y = 3 \cos t$
for $\dfrac{\pi}{2} \le t \le \pi$

30. $x = t^2$
$y = t - 1$
for $-1 \le t \le 3$

31. Write the equation for the cross section of a parabolic mirror with a span of 10 feet and a depth of 2 feet. How far is the focus from the vertex?

32. The cable on a suspension bridge carries the weight of the bridge and is the shape of a parabola as shown in Fig. 9.84. The distance between the piers is 3200 feet, the piers rise 1600 feet above the water, and the roadway is 400 feet above the water. Find the equation for the suspension cable when the origin is on the roadway at the midpoint between the piers.

Figure 9.84

33. A stone arch bridge spans a 40-foot river (Fig. 9.85). Find an equation for the bottom arch of the bridge if the maximum height of the bottom arch is 12 feet above the water. How high is the arch above the water 5 feet from each end?

Figure 9.85

34. Find the equation of the orbit of Mars about the sun, with the origin of the axis system at the sun, if half the major axis is 1.5234 AU and half the minor axis is 0.1516 AU. What is the eccentricity of this orbit?

35. Find the equation of the orbit of Pluto about the sun, with the origin of the axis system at the sun, if half the major axis is 38.8 AU and half the minor axis is 37.5 AU. What is the eccentricity of this orbit?

36. A guidance system on a submarine is designed to measure the distance of the submarine from two sending sonar buoys (Fig. 9.86). To navigate through a special mine field, the guidance system will navigate a course so that the difference of the distances from the submarine to the buoys is less than or equal to 1 mile. Sketch the graph of the region in which the submarine can travel safely if the buoys are at $(0, 2 \text{ miles})$ and $(0, -2 \text{ miles})$.

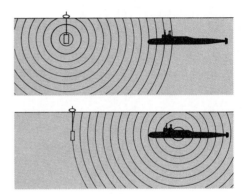

Figure 9.86

Find the angle of rotation and identify each of the following conic sections using the value of $B^2 - 4AC$.

37. $x^2 + xy + y^2 = 1$

38. $x^2 + 3xy + y^2 = 5$

39. $3x^2 + 2xy + 3y^2 = 19$

40. $2x^2 + 4xy - y^2 - 2x + 3y = 6$

Sketch the graph of each of the following in three dimensions.

41. $(25, t - 1, t)$

42. $2x + y + 3z = 6$

43. $x^2 + z^2 = 4$
for $0 \le z \le 4$

44. $\dfrac{x^2}{4} + y^2 + \dfrac{z^2}{9} = 1$

45. $4x^2 - 2y^2 + z^2 = 8$

46. Sketch the graph that is the line of intersection of the plane $2x + 3y + z = 6$ and $6x + z = 6$.

Sequences and Sums

FRACTALS

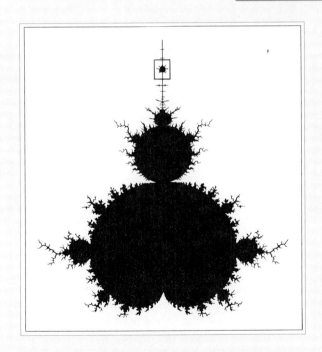

This is the Mandelbrot set. The boxed portion is enlarged in the following diagram.

Fractals or fractal geometry is one of the newest areas of mathematical development. We often work with ideal and smooth shaped curves in mathematics. Fractals, on the other hand, have been described as dealing with the competition and coexistence of chaos and order.

It was about 1964 when Benoit B. Mandelbrot began an intensive study of the ideas that make up fractals and not until 1975 that he consolidated those ideas enough to write a book on fractal geometry. As Mandelbrot wrote, "I coined the term fractal in order to be able to give a title to my first essay on this topic. But I stopped short of giving a mathematical definition, because I felt this notion—like a good wine—demanding a bit of aging before being 'bottled.'" Explaining the basic characteristics of fractals, he continued, "The shapes I was investigating, called fractals in my mind, all shared the property of being 'rough but self-similar.'" This observation describes the fairly chaotic shapes that were continually repeating the same pattern in a larger or smaller scale.

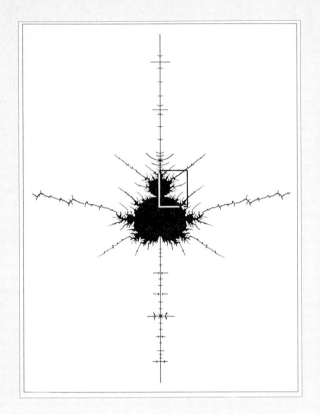

The boxed portion of this fractal is further enlarged at right.

SOURCE: H. O. Peitgen and P. H. Richter, *The Beauty of Fractals* (Heidelberg: Springer-Verlag, © 1986). Reprinted by permission.

As Mandelbrot stated, "To study it (fractals) was important because of the innumerable occurrences of self-similarity which I keep finding in nature, and it was feasible precisely because of self-similarity." Besides their artistic value, fractals are also useful in modeling nature, growth, linguistics, transmission line noise, turbulence in fluid flow, nonlinear oscillations in chemicals or electric networks, and even the translation of the normal rhythm of the heart to life-threatening fibrillation. Many other biological as well as physical phenomena appear to follow fractal patterns.

Fractals can involve functions with domains of complex numbers graphed in the Gaussian plane as well as many of the new ideas that we will work with in this chapter such as sequences, iterative (or recursive) functions, and patterns that continue on forever.

The Mandelbrot set is just one of many regular fractals. In this chapter we will work with von Koch's snowflake and Sierpinski's carpet, which are regular fractals. But there are many other types of regular and irregular fractals. Fractals provide a large field for development, future study, and application.

Historically, sequences of numbers as well as the sum of a sequence have been useful tools in finding good approximations of areas, volumes, and special numbers such as π. We have early evidence of the study of sequences and their sums from the Golden Age of Greece in the paradoxes of Zeno of Elea (ca. 490–430 B.C.) and the accurate approximations of areas, volumes, and the value of π calculated by Archimedes (ca. 287–212 B.C.). During the European Renaissance the study of sequences and their sums was continued and formalized in the development of our present-day calculus. In this chapter we will study several applications of both infinite and finite sequences and their sums.

10.1   Sequences and Summation Notation

In common conversation and usage we often use the words "sequence" and "series" interchangeably. In a mathematical context the word **sequence** denotes a set of elements arranged in a specific order, and the term **series** denotes a special type of sequence (each term of the series is a specific finite sum of a sequence). We will primarily work with sequences in this chapter.

A sequence can be written in terms of a_n where $n \in N$, the set of natural numbers, as

$$a_1, a_2, a_3, a_4, a_5, a_6, \cdots, a_n, \cdots$$

and is called an **infinite sequence**. We will now define an infinite sequence using function notation.

Definition | A function f with domain N is called an **infinite sequence**. The function value $f(n)$ is called the **general term** or ***n*th term** of the sequence.

Thus an infinite sequence could be written as

$$f(1), f(2), f(3), f(4), \cdots, f(n), \cdots$$

and we will usually refer to the sequence as

$$a_1, a_2, a_3, a_4, \cdots, a_n, \cdots$$

With the description above we will now consider some sequences.

EXAMPLE 1   Find the first five terms and the *n*th term in the sequence when $f(x) = x + \dfrac{1}{x}$ with $x \in N$.

Solution   To find the first five terms, we evaluate $f(x)$ for $x = 1$, then $x = 2$, and so forth until $x = 5$. Since

$$f(x) = x + \frac{1}{x}$$

we have

$$f(1) = 1 + \frac{1}{1} = 2$$

$$f(2) = 2 + \frac{1}{2} = 2\frac{1}{2} = \frac{5}{2}$$

$$f(3) = 3 + \frac{1}{3} = 3\frac{1}{3} = \frac{10}{3}$$

$$f(4) = 4 + \frac{1}{4} = 4\frac{1}{4} = \frac{17}{4}$$

$$f(5) = 5 + \frac{1}{5} = 5\frac{1}{5} = \frac{26}{5}$$

and in general

$$f(n) = n + \frac{1}{n} = \frac{n^2 + 1}{n}$$

Thus we can write this sequence as either

$$2, \; 2\frac{1}{2}, \; 3\frac{1}{3}, \; 4\frac{1}{4}, \; 5\frac{1}{5}, \; \cdots, \; n + \frac{1}{n}, \; \cdots$$

or

$$2, \; \frac{5}{2}, \; \frac{10}{3}, \; \frac{17}{4}, \; \frac{26}{5}, \; \cdots, \; \frac{n^2 + 1}{n}, \; \cdots$$

■

The following example is an interesting sequence with a most interesting nth term.

EXAMPLE 2   Find the fiftieth term, the two hundredth term, the five hundredth term, and the thousandth term of $f(x) = \left(1 + \frac{1}{x} \right)^x$. (Use your calculator and round off the answer to three decimal places.)

Solution   We must first evaluate $f(x) = \left(1 + \frac{1}{x} \right)^x$ for $x = 50$, $x = 200$, $x = 500$, and $x = 1000$. Then

$$f(50) = \left(1 + \frac{1}{50}\right)^{50} = (1.02)^{50} \approx 2.692$$

$$f(200) = \left(1 + \frac{1}{200}\right)^{200} = (1.005)^{200} \approx 2.712$$

$$f(500) = \left(1 + \frac{1}{500}\right)^{500} = (1.002)^{500} \approx 2.716$$

$$f(1000) = \left(1 + \frac{1}{1000}\right)^{1000} = (1.001)^{1000} \approx 2.717$$ ■

This is an interesting sequence. As the domain values get larger and larger, the sequence values get closer to the value of $e = 2.718\ldots$. As we saw in Chapter 4, we again have

$$\lim_{n \to \infty} \left(1 + \frac{1}{n}\right)^n = e$$

Another interesting sequence is the sequence $1, 1, 2, 3, 5, 8, 13, 21, 34, \cdots$ encountered by Fibonacci. Also known as Leonardo of Pisa (ca. 1180–1250), he noticed this sequence while working on a problem concerning the offspring of rabbits. In this sequence the first and second terms are 1, and each successive term is the sum of the previous two terms. Thus this **Fibonacci sequence** could be written as

$$a_1 = a_2 = 1 \quad \text{and} \quad a_n = a_{n-1} + a_{n-2} \quad \text{for } n > 2$$

since a_{n-1} and a_{n-2} are the two terms that precede a_n. We call this description of sequence a **recursive formula** for the sequence. The following example is another sequence defined by a recursive formula.

EXAMPLE 3   Find the first five terms for the sequence where $a_1 = \frac{1}{3}$ and $a_n = 3a_{n-1}$.

Solution   We notice that in this recursive formula a term is defined as 3 times the previous term, $a_n = 3a_{n-1}$, and that the first term is $a_1 = \frac{1}{3}$. The first five terms in the sequence can be developed as follows:

$$a_1 = \frac{1}{3}$$

$$a_2 = 3 \cdot a_1 = 3 \cdot \frac{1}{3} = 1$$

$$a_3 = 3 \cdot a_2 = 3 \cdot 1 = 3$$

$$a_4 = 3 \cdot a_3 = 3 \cdot 3 = 9$$

$$a_5 = 3 \cdot a_4 = 3 \cdot 9 = 27$$

Thus this is the sequence $\frac{1}{3}, 1, 3, 9, 27, \cdots$.   ■

Sometimes it is helpful to find the nth term even though the sequence is defined with a recursive formula.

EXAMPLE 4   Find the nth term of the sequence $a_1 = \frac{1}{3}$ and $a_n = 3a_{n-1}$ of Example 3.

Solution   The sequence in Example 3 is

$$a_1 = \frac{1}{3} \qquad \text{or} \qquad f(1) = \frac{1}{3}$$

$$a_2 = 1 \qquad\qquad f(2) = 1$$

$$a_3 = 3 \qquad\qquad f(3) = 3$$

$$a_4 = 9 \qquad\qquad f(4) = 9$$

$$a_5 = 27 \qquad\qquad f(5) = 27$$

Using the recursive formula, we notice that this sequence involves powers of 3 and thus

$$f(1) = \frac{1}{3} = 3^{-1}$$

$$f(2) = 3 \cdot \frac{1}{3} = 3 \cdot 3^{-1} = 3^0$$

$$f(3) = 3 \cdot 1 = 3 \cdot 3^0 = 3^1$$

$$f(4) = 3 \cdot 3 = 3^2$$

$$f(5) = 3 \cdot 3 \cdot 3 = 3^3$$

The power of 3 is always 2 less than the domain element x of $f(x)$ in this sequence. Therefore the nth term is $f(n) = 3^{n-2}$. ∎

This process of discovering the nth term can be a guessing game at times, but as in the previous example, it will always be important to involve the domain element in the resulting pattern.

It is also common to find the sum of the elements in a sequence. To shorten the writing when working with the sum of the terms in a sequence, we will use the uppercase Greek letter sigma, Σ, to indicate the addition of terms.

SUMMATION NOTATION

The symbol Σ is used to indicate the addition of the terms $a_1, a_2, a_3, \cdots, a_n$ as follows:

$$\sum_{i=1}^{n} a_i = a_1 + a_2 + a_3 + \cdots + a_n$$

The letter i is an integer and is used as an index for counting. (This is *not* the i used with complex numbers.) The equation under the sigma, $i = 1$, indicates the first value used for i in the first term of the sum. Then successive integers are substituted for i until the value of i reaches the value above (on top of) the sigma. This summation notation provides a nice compact notation for adding terms in a sequence.

EXAMPLE 5   Evaluate the following sums:

(a) $\displaystyle\sum_{i=1}^{5} 3^{i-2}$ (b) $\displaystyle\sum_{i=3}^{7} 3^{i-4}$

Solution

(a) We evaluate this sum of terms created as i takes on the integers from 1 to 5. Then

$$\sum_{i=1}^{5} 3^{i-2} = 3^{1-2} + 3^{2-2} + 3^{3-2} + 3^{4-2} + 3^{5-2}$$

$$= 3^{-1} + 3^0 + 3^1 + 3^2 + 3^3$$

$$= \frac{1}{3} + 1 + 3 + 9 + 27$$

$$= \frac{1}{3} + 40$$

$$= 40\frac{1}{3} \quad \text{or} \quad \frac{121}{3}$$

(b) We evaluate this sum of terms created as i takes on the integers from 3 to 7:

$$\sum_{i=3}^{7} 3^{i-4} = 3^{3-4} + 3^{4-4} + 3^{5-4} + 3^{6-4} + 3^{7-4}$$

$$= 3^{-1} + 3^0 + 3^1 + 3^2 + 3^3$$

$$= 40\frac{1}{3} \quad \text{or} \quad \frac{121}{3}$$

the same as in part (a) above.   ∎

We notice that in Example 5(a) and Example 5(b) the indices and general terms of the summation notation are different, yet the terms in the sum as well as the resulting sums are identical. Thus we can manipulate the summation notation and still have the same terms and the same sum.

EXAMPLE 6   Evaluate $\displaystyle\sum_{i=1}^{5} 3$.

Solution   We notice that i does not appear in this summation notation, but we must still have five elements in this sum. Therefore

$$\sum_{i=1}^{5} 3 = 3 + 3 + 3 + 3 + 3$$

$$= 3 \cdot 5$$

$$= 15$$

We can produce the same sum by keeping track of the number of terms using i as follows:

$$\sum_{i=1}^{5} 3 = \sum_{i=1}^{5} 3(1^i)$$

$$= 3(1^1) + 3(1^2) + 3(1^3) + 3(1^4) + 3(1^5)$$

$$= 3 + 3 + 3 + 3 + 3$$

$$= 15$$ ■

Powers of -1 are frequently used in summation notation when the sign of the terms changes as in the following example.

EXAMPLE 7   Evaluate $\displaystyle\sum_{i=1}^{5} 3(-1)^i$.

Solution   We will find each term and then add:

$$\sum_{i=1}^{5} 3(-1)^i = 3(-1)^1 + 3(-1)^2 + 3(-1)^3 + 3(-1)^4 + 3(-1)^5$$

$$= -3 + 3 - 3 + 3 - 3$$

$$= -3$$ ■

Exercises 10.1

Find the first five terms and the tenth term for each sequence
that has the following nth term.

1. $a_n = 5 - n$ 2. $a_n = n - 3$ 7. $a_n = 3$ 8. $a_n = (-1)^n$

3. $f(n) = 2(n - 1)$ 4. $f(n) = 3(1 - n)$

5. $a_n = (-1)^n 3^{n-1}$ 6. $a_n = (-1)^{n-1} 2^n$ 9. $f(n) = \dfrac{n^2}{2^n - 1}$ 10. $f(n) = \dfrac{n}{n^2 - 1}$

Find the first five terms of the sequences defined by the following recursive formulas.

11. $a_1 = -2, a_n = 5 - 2a_{n-1}$

12. $a_1 = 3, a_n = 2a_{n-1} - 6$

13. $a_1 = 3, a_n = \frac{1}{3}a_{n-1}$

14. $a_1 = 2, a_n = \frac{1}{4}a_{n-1}$

15. $a_1 = -1, a_n = (a_{n-1})^2$

16. $a_1 = -2, a_n = (a_{n-1})^{-1}$

17. $a_1 = 2, a_n = (a_{n-1})^{1/(n-1)}$

18. $a_1 = -3, a_n = (a_{n-1})^{n-1}$

Evaluate each of the following sums.

19. $\sum_{i=1}^{5} (2i - 5)$

20. $\sum_{i=1}^{5} (3i - 6)$

21. $\sum_{i=6}^{9} (3 - 5i)$

22. $\sum_{i=4}^{7} (1 - 3i)$

23. $\sum_{i=0}^{4} (-3i)^2$

24. $\sum_{i=0}^{3} (5 - i)^2$

25. $\sum_{i=1}^{4} 2(3i)$

26. $\sum_{i=1}^{4} 3(2i)$

27. $\sum_{i=1}^{75} (-1)^{i+1}$

28. $\sum_{i=2}^{88} (-1)^i$

29. $\sum_{i=3}^{18} 5$

30. $\sum_{i=2}^{21} 3$

31. Find the sequence $a_1, a_2, a_3, \cdots, a_n, \cdots$ for the hypotenuse of the right triangles in Fig. 10.1.

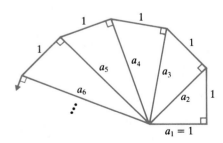

Figure 10.1

32. Find the sequence $a_1, a_2, a_3, \cdots, a_n, \cdots$ for the hypotenuse of the right triangles in Fig. 10.2.

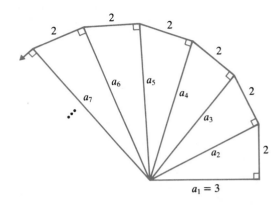

Figure 10.2

Find the nth term for each of the following sequences and write the expression for the sum of the first 50 terms in summation notation.

33. 2, 4, 6, 8, 10, $\cdots$

34. 3, 6, 9, 12, 15, $\cdots$

35. 1, 8, 27, 64, 125, $\cdots$

36. 4, 9, 16, 25, 36, $\cdots$

37. 1, $-\frac{1}{2}, \frac{1}{4}, -\frac{1}{8}, \frac{1}{16}, \cdots$

38. 1, $-\frac{1}{2}, \frac{1}{3}, -\frac{1}{4}, \frac{1}{5}, \cdots$

39. $\frac{1}{4}, \frac{2}{9}, \frac{3}{16}, \frac{4}{25}, \frac{5}{36}, \cdots$

40. $\frac{1}{3}, \frac{2}{5}, \frac{3}{7}, \frac{4}{9}, \frac{5}{11}, \cdots$

41. 1, $x, \dfrac{x^2}{2}, \dfrac{x^3}{3}, \dfrac{x^4}{4}, \cdots$

42. $\frac{1}{2}, \frac{1}{6}, \frac{1}{12}, \frac{1}{20}, \cdots$

43. Show that

$$\sum_{i=1}^{n} (a_i + b_i) = \sum_{i=1}^{n} a_i + \sum_{i=1}^{n} b_i$$

44. Show that

$$\sum_{i=1}^{n} (a_i - b_i) = \sum_{i=1}^{n} a_i - \sum_{i=1}^{n} b_i$$

45. Show that

$$\sum_{i=1}^{n} ca_i = c \sum_{i=1}^{n} a_i$$

46. Find the sequence $a_1, a_2, a_3, \cdots, a_n, \cdots$ of the long leg of the right triangle in Fig. 10.3. Also find $\lim\limits_{n \to \infty} a_n$.

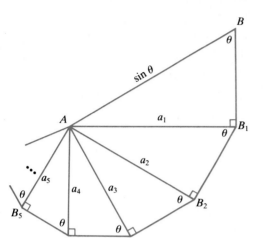

Figure 10.3

47. Find the sequence $a_1, a_2, a_3, \cdots, a_n, \cdots$ of the long leg of the right triangle in Fig. 10.4. Also find $\lim\limits_{n \to \infty} a_n$.

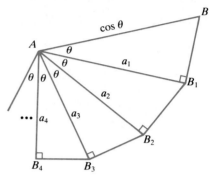

Figure 10.4

48. The Fibonacci rations are $\frac{1}{1}, \frac{1}{2}, \frac{2}{3}, \frac{3}{5}, \frac{5}{7}, \cdots, f(n) \cdots$. Find a recursive formula for this sequence. Use your calculator to verify that $\lim\limits_{n \to \infty} \dfrac{1}{f(n)} \approx \dfrac{1 + \sqrt{5}}{2}$, the **golden ratio**.

10.2 Mathematical Induction

In the last section we worked with the sums of finite sequences. The sums of some of these sequences fit predictable patterns that can be summarized by using algebraic formulas. One example is

$$\sum_{i=1}^{n} i = \frac{n}{2}(n + 1)$$

To see this result, we will first consider some examples for different integral values of n. We let (a) $n = 1$, (b) $n = 2$, (c) $n = 3$, and (d) $n = 4$.

(a) For $n = 1$ we have

$$\sum_{i=1}^{1} i = 1 \quad \text{and} \quad \frac{n}{2}(n + 1) = \frac{1}{2}(1 + 1) = 1$$

which is the desired conclusion

$$\sum_{i=1}^{n} i = \frac{n}{2}(n + 1) \quad \text{for } n = 1$$

(b) For $n = 2$ we have

$$\sum_{i=1}^{2} i = 1 + 2 = 3 \quad \text{and} \quad \frac{n}{2}(n + 1) = \frac{2}{2}(2 + 1) = 3$$

which is the desired conclusion

$$\sum_{i=1}^{n} i = \frac{n}{2}(n + 1) \quad \text{for } n = 2$$

(c) For $n = 3$ we have

$$\sum_{i=1}^{3} i = 1 + 2 + 3 = 6 \quad \text{and} \quad \frac{n}{2}(n + 1) = \frac{3}{2}(3 + 1) = 6$$

which is the desired conclusion

$$\sum_{i=1}^{n} i = \frac{n}{2}(n + 1) \quad \text{for } n = 3$$

(d) For $n = 4$ we have

$$\sum_{i=1}^{4} i = 1 + 2 + 3 + 4 = 10 \quad \text{and} \quad \frac{n}{2}(n + 1) = \frac{4}{2}(4 + 1) = 10$$

which is the desired conclusion

$$\sum_{i=1}^{n} i = \frac{n}{2}(n + 1) \quad \text{for } n = 4$$

As we can see in this process, the formula holds for $n \in \{1, 2, 3, 4\}$. It is also conceivable that this formula,

$$\sum_{i=1}^{n} i = \frac{n}{2}(n + 1)$$

is always true. Proving that this formula is true by trying every value for n is an impossible task. To prove this and other statements that are true for an infinite set of integers, we will use the following principle of mathematical induction.

MATHEMATICAL INDUCTION	Let a statement $P(n)$ be either true for every natural number n or not true for every natural number n. If (a) $P(a)$ is true for some $a \in N$, and (b) the truth of $P(k)$ implies the truth of $P(k + 1)$ for all k, $k \geq a$, then $P(n)$ is true for all natural numbers n where $n \geq a$.

The two parts of mathematical induction can be characterized as the

(a) "**firstness**"—there must be a first case $n = a$ that is true—and
(b) "**nextness**"—if $P(n)$ is valid for $n = k$, then the next case $n = k + 1$ can be shown to be true as a result of the truth of the statement when $n = k$.

This process of mathematical induction is much like the line of standing dominos in Fig. 10.5. We know that the whole line of dominos will fall if the first domino, "firstness," is knocked down and if every domino is placed close enough to the preceding domino so that any domino falling causes the next domino to fall, "nextness."

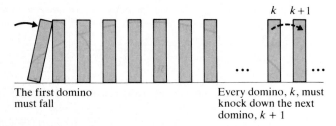

The first domino must fall

Every domino, k, must knock down the next domino, $k + 1$

Figure 10.5

With mathematical induction the first case is usually, but not necessarily, $n = a = 1$. We usually verify "nextness" by assuming that $P(k)$ is true and showing that $P(k + 1)$ is an algebraic result of $P(k)$. The following example will demonstrate the use of mathematical induction.

EXAMPLE 1   Prove that

$$\sum_{i=1}^{n} i = \frac{n}{2}(n + 1)$$

by mathematical induction.

Solution   We will prove this statement with the two steps of mathematical induction, "firstness" and "nextness."

(a) *"Firstness"*: As was shown earlier

$$\sum_{i=1}^{1} i = 1 \quad \text{and} \quad \frac{n}{2}(n + 1) = \frac{1}{2}(1 + 1) = 1$$

are equal. Thus

$$\sum_{i=1}^{n} i = \frac{n}{2}(n + 1) \quad \text{for } n = 1$$

This proves the statement true for a first case.

(b) *"Nextness"*: Assuming that this statement is true for $n = k$, we have

$$\sum_{i=1}^{k} i = \frac{k}{2}(k + 1)$$

We must then show that

$$\sum_{i=1}^{k+1} i = \frac{k + 1}{2}(k + 1 + 1) \qquad \text{or} \qquad \sum_{i=1}^{k+1} i = \frac{k + 1}{2}(k + 2)$$

Starting with $n = k$, we will add the next term, $k + 1$, to both sides so that the left side is correct. Next we will simplify the right side to see whether it yields the desired result.

Assuming that this statement is true for $n = k$, we have

$$\sum_{i=1}^{k} i = \frac{k}{2}(k + 1) \qquad \textit{(Add k + 1 to both sides)}$$

$$\sum_{i=1}^{k} i + (k + 1) = \frac{k}{2}(k + 1) + (k + 1)$$

$$\sum_{i=1}^{k+1} i = \frac{k(k + 1)}{2} + \frac{2(k + 1)}{2}$$

$$= \frac{k(k + 1) + 2(k + 1)}{2} \qquad \textit{(Factor the common factor (k + 1))}$$

$$= \frac{(k + 1)(k + 2)}{2}$$

Thus when $n = k$ and

$$\sum_{i=1}^{k} i = \frac{k}{2}(k + 1)$$

is true, we can show using only algebra that it is true for $n = k + 1$,

$$\sum_{i=1}^{k+1} i = \frac{k + 1}{2}(k + 2) \qquad\qquad \blacksquare$$

In Example 1 we proved that there was at least one case that is true and that the next case is also always true. Thus $P(n)$ is true for $n \in \{1, 2, 3, 4, \cdots\}$. The concept of "nextness" makes it possible to show that $P(n)$ is true for every $n \in N$, an infinite number of cases.

We will use the "firstness" and "nextness" of mathematical induction again in the next example.

EXAMPLE 2
Prove that

$$\sum_{i=1}^{n} i^2 = \frac{n(n + 1)(2n + 1)}{6}$$

Solution

(a) *"Firstness"*: For $n = 1$,

$$\sum_{i=1}^{1} i^2 = 1^2 = 1$$

and

$$\frac{n(n + 1)(2n + 1)}{6} = \frac{1(1 + 1)(2 \cdot 1 + 1)}{6} = \frac{1(2)(3)}{6} = 1$$

Thus

$$\sum_{i=1}^{n} i^2 = \frac{n(n + 1)(2n + 1)}{6} \qquad \text{for } n = 1$$

(b) *"Nextness"*: We assume that this statement is true when $n = k$, and add the next term, $(k + 1)^2$, to both sides:

$$\sum_{i=1}^{k} i^2 = \frac{k(k + 1)(2k + 1)}{6}$$

$$\sum_{i=1}^{k} i^2 + (k + 1)^2 = \frac{k(k + 1)(2k + 1)}{6} + (k + 1)^2$$

$$\sum_{i=1}^{k+1} i^2 = \frac{k(k + 1)(2k + 1)}{6} + \frac{6(k + 1)^2}{6}$$

$$= \frac{k(k + 1)(2k + 1) + 6(k + 1)^2}{6} \qquad \textit{(Factor the common factor } (k + 1))$$

$$= \frac{(k + 1)[k(2k + 1) + 6(k + 1)]}{6}$$

$$= \frac{(k + 1)(2k^2 + k + 6k + 6)}{6}$$

$$= \frac{(k + 1)(2k^2 + 7k + 6)}{6}$$

$$= \frac{(k + 1)(k + 2)(2k + 3)}{6}$$

This is the correct result if it is the same answer that we get when substituting $n = k + 1$ into the original problem. We check this and find that

$$\sum_{i=1}^{k+1} i^2 = \frac{(k + 1)(k + 1 + 1)[2(k + 1) + 1]}{6}$$

$$= \frac{(k + 1)(k + 2)(2k + 3)}{6}$$

The result from substitution is the same as the result produced above with $n = k$ and some algebra. **It is usually very helpful to substitute and establish the desired result as a goal before doing the algebra in the "nextness" part of an induction proof.**

■

We can also use the properties of summation notation that we proved in the last exercise set,

$$\sum_{i=1}^{n} (a_i + b_i) = \sum_{i=1}^{n} a_i + \sum_{i=1}^{n} b_i$$

$$\sum_{i=1}^{n} (a_i - b_i) = \sum_{i=1}^{n} a_i - \sum_{i=1}^{n} b_i \quad \text{and}$$

$$\sum_{i=1}^{n} ca_i = c \sum_{i=1}^{n} a_i,$$

with the results of Examples 1 and 2 to simplify more complicated summation problems, as we will see in the next example.

EXAMPLE 3   Find the value of

$$\sum_{i=1}^{n} (i + 1)^2$$

in terms of n.

Solution   To evaluate the sum, we first square $(i + 1)^2$ and use the summation notation properties:

$$\sum_{i=1}^{n} (i + 1)^2 = \sum_{i=1}^{n} (i^2 + 2i + 1)$$

$$= \sum_{i=1}^{n} i^2 + 2 \sum_{i=1}^{n} i + \sum_{i=1}^{n} 1$$

Now we substitute the results from Examples 1 and 2:

$$\sum_{i=1}^{n} (i + 1)^2 = \frac{n(n + 1)(2n + 1)}{6} + \frac{n}{2}(n + 1) + 1 \cdot n$$

$$= \frac{n(n + 1)(2n + 1)}{6} + \frac{3n(n + 1)}{6} + \frac{6n}{6}$$

$$= \frac{n(n + 1)(2n + 1) + 3n(n + 1) + 6n}{6}$$

$$= \frac{n[(n + 1)(2n + 1) + 3(n + 1) + 6]}{6}$$

$$= \frac{n(n^2 + 3n + 1 + 3n + 3 + 6)}{6}$$

$$= \frac{n(n^2 + 6n + 10)}{6}$$

Therefore

$$\sum_{i=1}^{n} (i + 1)^2 = \frac{n(n^2 + 6n + 10)}{6}$$

■

We can also use mathematical induction when proving inequalities as seen in the next example.

EXAMPLE 4   Prove that $n < 2^n$ for $n \geq 1$.

Solution   Even though this statement does not include a sum, we will prove it using mathematical induction.

(a) "*Firstness*": We let $n = 1$ and $2^n = 2^1 = 2$. Thus $1 < 2$ and $n < 2^n$ when $n = 1$.

(b) "*Nextness*": We assume that $k < 2^k$ is true and will next show that this implies that $k + 1 < 2^{k+1}$. Starting with $k < 2^k$, we multiply both sides by 2. Then we consider the results. We have

$$k < 2^k$$
$$2 \cdot k < 2 \cdot 2^k$$
$$2k < 2^{k+1}$$
$$k + k < 2^{k+1}$$

Now since $k \geq 1$, we have $k + 1 \leq k + k$. Using the transitive property with $k + 1 \leq k + k$ and $k + k < 2^{k+1}$, we can conclude that

$$k + 1 < 2^{k+1}$$ ∎

In inequalities the transitive property of inequality makes it possible to substitute a convenient smaller value for the smaller side of the inequality and/or substitute a convenient larger value for the larger side of the inequality.

Some of the theorems we used earlier in this textbook are proven by using mathematical induction and will be included in the exercises for this section. Proofs by mathematical induction will also be included in the remaining sections of this chapter. Mathematical induction is a most useful method for proving some statements that are true for an infinite number of integer cases.

Exercises 10.2

Prove the following statements.

1. $\displaystyle\sum_{i=1}^{n} 2i = n(n + 1)$

2. $\displaystyle\sum_{i=1}^{n} (2i - 1) = n^2$

7. $\displaystyle\sum_{i=1}^{n} 5^i = \frac{5}{4}(5^n - 1)$

8. $\displaystyle\sum_{i=1}^{n} 3^{-i} = \frac{1}{2}\left(1 - \frac{1}{3^n}\right)$

3. $\displaystyle\sum_{i=1}^{n} i^3 = \frac{n^2(n + 1)^2}{4}$

4. $\displaystyle\sum_{i=1}^{n} \frac{i}{2} = \frac{n(n + 1)}{4}$

9. $\displaystyle\sum_{i=1}^{n} 2^{-i} = 1 - \frac{1}{2^n}$

5. $\displaystyle\sum_{i=1}^{n} \frac{1}{i(i + 1)} = \frac{n}{n + 1}$

6. $\displaystyle\sum_{i=1}^{n} (3i - 2) = \frac{n(3n - 1)}{2}$

10. $\displaystyle\sum_{i=1}^{n} i(i + 1) = \frac{n(n + 1)(n + 2)}{3}$

Evaluate the following sums.

11. $\displaystyle\sum_{i=3}^{n} i^2$ **12.** $\displaystyle\sum_{i=3}^{n} i$

13. $\displaystyle\sum_{i=3}^{n} (i + 2)^2$ **14.** $\displaystyle\sum_{i=3}^{n} (2i + 1)^2$

15. $\displaystyle\sum_{i=3}^{n} (2i - 1)^2$ **16.** $\displaystyle\sum_{i=3}^{n} (i - 2)^2$

17. $\displaystyle\sum_{i=1}^{n} (i + 1)^3$ **18.** $\displaystyle\sum_{i=1}^{n} (i - 2)^3$

Prove the following inequalities.

19. $1 + 2n \le 3^n$ **20.** $(1 + x)^n \ge 1 + nx$

21. $\displaystyle\sum_{i=1}^{n} i < \frac{1}{8}(2n + 1)^2$ **22.** $2^n > n^2$ for $n > 4$

23. $2^n \ge 2n$ **24.** $3^n \ge 3n$

25. Use previously proven statements. Show that

$$\text{(a)} \qquad \sum_{i=1}^{n} i^3 = \left(\sum_{i=1}^{n} i\right)^2$$

and then prove that

$$\text{(b)} \qquad \sum_{i=1}^{n-1} i^3 < \frac{n^4}{4} < \sum_{i=1}^{n} i^3$$

26. Prove that $3^{2n} - 1$ is divisible by 8.

27. Prove that $n^2 + n$ is an even integer.

28. Prove that the sum of the angles in a convex polygon having $n + 2$ sides is $n(180°)$.

29. Prove de Moivre's Theorem: For $n \in N$ and $z = r(\cos \theta + i \sin \theta)$, z^n is $z^n = r^n(\cos n\theta + i \sin n\theta)$.

30. Use the fundamental theorem of algebra ("A polynomial with real number coefficients and degree greater than zero has at least one factor of the form $x - d$, where d is a complex number") to prove the following theorem: If $f(x)$ is a polynomial of degree n, $n \in N$, then $f(x)$ has n zeros, $d_1, d_2, d_3, \cdots, d_n$, where each d_i is a complex number.

31. Prove that $\sin (\theta + n\pi) = (-1)^n \sin \theta$.

32. Prove that $\cos (\theta + n\pi) = (-1)^n \cos \theta$.

33. Prove that $(a - b)$ is a factor of $a^n - b^n$ for $n \in N$. (*Hint:* $a^{k+1} - b^{k+1} = a^k(a - b) + (a^k - b^k)b$.)

34. Prove that $(a + b)$ is a factor of $a^{2n-1} + b^{2n-1}$ for $n \in N$.

In 1904 the Swedish mathematician Hele von Koch presented the following snowflake problem. We start with an equilateral triangle with sides of 1 unit (Fig. 10.6a). Each side is divided into three equal parts, and an equilateral triangle is built on the middle third of each side as seen in case 2. We continue this method of construction to create the snowflake in case 3. Using this process, we can create case 1, case 2, case 3, case 4, $\cdots$, case n, $\cdots$ (Fig. 10.6b).

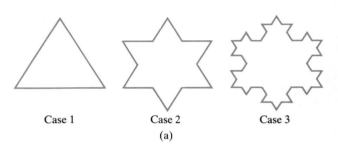

Case 1   Case 2   Case 3
(a)

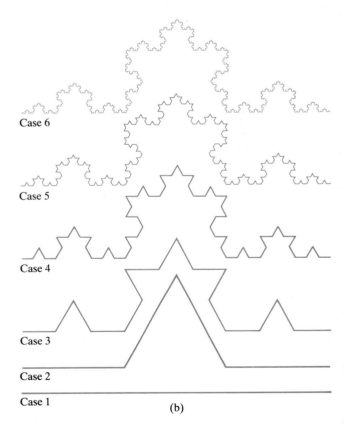

Case 6

Case 5

Case 4

Case 3

Case 2

Case 1   (b)

Figure 10.6

35. For the snowflake problem described above, find the sequence $s_1, s_2, s_3, s_4, s_5, \cdots$, where s_n is the length of the segments in the sides of snowflake in case n. Find a formula for s_n and prove it by mathematical induction. Also find $\lim\limits_{n \to \infty} s_n$.

36. For the snowflake problem described above, find the sequence $N_1, N_2, N_3, N_4, N_5, \cdots$, where N_n is the number of segments in the perimeter of the snowflake in case n.

Find a formula for N_n and prove it by mathematical induction.

37. Using the results of Problems 35 and 36, find the sequence $p_1, p_2, p_3, p_4, p_5, \cdots$, where p_n is the perimeter of the snowflake in case n. Find a formula for s_n. Is $\lim\limits_{n \to \infty} p_n$ finite or infinite? (We will calculate the area of this snowflake in another section.)

10.3   Arithmetic Sequences

In the first section of this chapter we studied several different sequences as well as the sums of finite sequences. In this section we will develop many details related to one special sequence, the **arithmetic sequence**.

Definition | An **arithmetic sequence** is a sequence in which the difference between any term and the previous term is the same real number d, the **common difference**. Thus

$$a_n - a_{n-1} = d$$

With this common difference d we can write an arithmetic sequence with a first term of a_1 and add d to a term to create the next term as follows:

$$a_1 = a_1$$
$$a_2 = a_1 + d$$
$$a_3 = a_2 + d = (a_1 + d) + d = a_1 + 2d$$
$$a_4 = a_3 + d = (a_1 + 2d) + d = a_1 + 3d$$
$$a_5 = a_4 + d = (a_1 + 3d) + d = a_1 + 4d$$

and in general

$$a_n = a_1 + (n - 1)d$$

We can also prove this result by mathematical induction.

EXAMPLE 1   Show that $a_n = a_1 + (n - 1)d$ is the nth or general term of the arithmetic sequence with a first term of a_1 and a common d.

Solution   We will use mathematical induction.
(a) *"Firstness"*: If $n = 1$, then $a_1 = a_1 + (1 - 1)d = a_1$ the first term. Thus we have a valid first case.

(b) *"Nextness"*: We assume that this is true for $n = k$ and show that it must also be true for $n = k + 1$. For $n = k$ we have

$$a_k = a_1 + (k - 1)d$$

In an arithmetic sequence we add the common difference to both sides to obtain the next term. Thus we add d to both sides, and

$$a_k + d = a_1 + (k - 1)d + d$$

$$a_{k+1} = a_1 + (k - 1 + 1)d$$

$$a_{k+1} = a_1 + kd$$

Thus if one case is true, the next case is always true, and our proof is complete.

■

This formula $a_n = a_1 + (n - 1)d$ for the nth or general term will be most helpful in solving many problems with arithmetic sequences. The following examples use the definition of arithmetic sequences as well as the formula for the nth term.

EXAMPLE 2   Find the first four terms, the eleventh term, and the kth term for the arithmetic sequences in which (a) $a_1 = 3$ and $d = 2$ and (b) $a_1 = 15$ and $d = -3$.

Solution
(a) With $a_1 = 3$ and $d = 2$, then

$$a_1 = 3$$

$$a_2 = a_1 + d = 3 + 2 = 5$$

$$a_3 = a_2 + d = 5 + 2 = 7$$

$$a_4 = a_3 + d = 7 + 2 = 9$$

To find a_{11} and a_k we must use the formula $a_n = a_1 + (n - 1)d$. Thus

$$a_{11} = a_1 + (11 - 1)d \qquad \text{and} \qquad a_k = a_1 + (k - 1)d$$

$$= 3 + (10)2 \qquad\qquad\qquad\qquad = 3 + (k - 1)2$$

$$= 23 \qquad\qquad\qquad\qquad\qquad = 3 + 2k - 2$$

$$\qquad\qquad\qquad\qquad\qquad\qquad = 1 - 2k$$

Therefore $a_{11} = 23$ and $a_k = 1 - 2k$.

(b) Similarly, with $a_1 = 15$ and $d = -3$,

$$a_1 = 15$$

$$a_2 = a_1 + d = 15 - 3 = 12$$

$$a_3 = a_2 + d = 12 - 3 = 9$$

$$a_4 = a_3 + d = 9 - 3 = 6$$

Also

$$a_{11} = a_1 + (11 - 1)d \qquad \text{and} \qquad a_k = a_1 + (k - 1)d$$
$$= 15 + (10)(-3) \qquad\qquad = 15 + (k - 1)(-3)$$
$$= 15 - 30 \qquad\qquad = 15 - 3k + 3$$
$$= -15 \qquad\qquad = 18 - 3k$$

Therefore $a_{11} = -15$ and $a_k = 18 - 3k$.   ■

We can also use the formula for the nth term when solving sequence problems where any two terms are given, as in the next example.

EXAMPLE 3   Find a_7 in an arithmetic sequence when $a_3 = 11$ and $a_{11} = 43$.

Solution   Using the formula $a_n = a_1 + (n - 1)d$, we have

$$a_3 = a_1 + (3 - 1)d$$

or $11 = a_1 + 2d$. Similarly,

$$a_{11} = a_1 + (11 - 1)d$$

and $43 = a_1 + 10d$. Now we will solve the system

$$11 = a_1 + 2d$$
$$43 = a_1 + 10d$$

We multiply the first equation by (-1) and add:

$$-11 = -a_1 - 2d$$
$$\underline{43 = \quad a_1 + 10d}$$
$$32 = \qquad\quad 8d$$
$$4 = d$$

Then substituting $d = 4$ into $11 = a_1 + 2d$, we have

$$11 = a_1 + 2(4)$$
$$3 = a_1$$

Therefore

$$a_7 = a_1 + (7 - 1)d$$
$$= 3 + (6)4$$
$$= 27$$

Thus

$$a_7 = 27$$ ■

The **arithmetic mean** is a number between two given numbers that allows the three numbers to be consecutive terms in an arithmetic sequence. It is possible to have one, two, three, or more arithmetic means between two numbers. When there is only one arithmetic mean, it is also known as the *average* of the two numbers.

EXAMPLE 4

Find the two arithmetic means between 3 and 7.

Solution We let $a_1 = 3$ and $a_4 = 7$, so the two arithmetic means between a_1 and a_4 are a_2 and a_3. Since this is an arithmetic sequence, we have

$$a_1 = 3$$
$$a_2 = 3 + d$$
$$a_3 = 3 + 2d$$
$$a_4 = 3 + 3d = 7$$

Thus $3d = 4$ and $d = \frac{4}{3} = 1\frac{1}{3}$. Then

$$a_2 = 3 + 1\frac{1}{3} = 4\frac{1}{3}$$
$$a_3 = 4\frac{1}{3} + 1\frac{1}{3} = 5\frac{2}{3}$$

Therefore the two arithmetic means between 3 and 7 are

$$4\frac{1}{3} \quad \text{and} \quad 5\frac{2}{3}$$

$\blacksquare$

We will also consider the sums of terms in an arithmetic sequence.

Theorem

Arithmetic Partial Sums. For the arithmetic sequence $a_1, a_2, a_3 \cdots, a_n, \cdots$ with common difference d, the sum of the first n terms, **the partial sum S_n**, is

$$S_n = \sum_{i=1}^{n} a_i = \frac{n}{2}[2a_1 + (n-1)d] \quad \text{and} \quad S_n = \sum_{i=1}^{n} a_i = \frac{n}{2}(a_1 + a_n)$$

We will prove the first equation

(a) $$S_n = \sum_{i=1}^{n} a_i = \frac{n}{2}[2a_1 + (n-1)d]$$

using mathematical induction and then show that this is equivalent to the second equation

(b) $$S_n = \sum_{i=1}^{n} a_i = \frac{n}{2}(a_1 + a_n)$$

PROOF:

(a) We apply the two steps of mathematical induction to

$$S_n = \sum_{i=1}^{n} a_i = \frac{n}{2}[2a_1 + (n-1)d]$$

(i) *"Firstness"*: When $n = 1$, we have

$$S_1 = \sum_{i=1}^{1} a_i = a_1$$

and

$$\frac{n}{2}[2a_1 + (n-1)d] = \frac{1}{2}(2a_1 + 0) = a_1$$

Thus

$$S_n = \sum_{i=1}^{n} a_i = \frac{n}{2}[2a_1 + (n-1)d] \qquad \text{for } n = 1$$

(ii) *"Nextness"*: We assume that

$$S_k = \sum_{i=1}^{k} a_i = \frac{k}{2}[2a_1 + (k-1)d]$$

is true and add the next term a_{k+1} to both sides to show that

$$S_{k+1} = \sum_{i=1}^{k+1} a_i = \frac{k+1}{2}(2a_1 + kd)$$

Starting with the assumption, we have

$$\sum_{i=1}^{k} a_i = \frac{k}{2}[2a_1 + (k-1)d]$$

$$\sum_{i=1}^{k} a_i + a_{k+1} = \frac{k}{2}[2a_1 + (k-1)d] + a_{k+1}$$

$$\sum_{i=1}^{k+1} a_i = \frac{k[2a_1 + (k-1)d]}{2} + \frac{2a_{k+1}}{2} \qquad \textit{(Substitute } a_{k+1} = a_1 + kd\textit{)}$$

$$= \frac{2ka_1 + k(k-1)d + 2(a_1 + kd)}{2}$$

$$= \frac{2ka_1 + k^2d - kd + 2a_1 + 2kd}{2}$$

$$= \frac{2ka_1 + 2a_1 + k^2d - kd + 2kd}{2}$$

$$= \frac{(2k+2)a_1 + k^2d + kd}{2}$$

$$= \frac{(k+1)2a_1 + (k+1)kd}{2}$$

$$= \frac{(k + 1)(2a_1 + kd)}{2}$$

$$= \frac{k + 1}{2}(2a_1 + kd)$$

Therefore

$$\sum_{i=1}^{k+1} a_i = \frac{k + 1}{2}(2a_1 + kd)$$

which is the desired result. Thus by mathematical induction,

$$S_n = \sum_{i=1}^{n} a_i = \frac{n}{2}[2a_1 + (n - 1)d]$$

(b) To prove that

$$S_n = \sum_{i=1}^{n} a_i = \frac{n}{2}(a_1 + a_n)$$

we remember that $a_1 + (n - 1)d = a_n$ and substitute this into the result of part **(a)**. Thus we have

$$\sum_{i=1}^{n} a_i = \frac{n}{2}[2a_1 + (n - 1)d]$$

$$= \frac{n}{2}[a_1 + a_1 + (n - 1)d]$$

$$= \frac{n}{2}(a_1 + a_n)$$

Therefore

$$S_n = \sum_{i=1}^{n} a_i = \frac{n}{2}(a_1 + a_n) \qquad \Box$$

We can use these results to find the partial sums of arithmetic sequences. Notice that when we use the first formula,

$$S_n = \sum_{i=1}^{n} a_i = \frac{n}{2}[2a_1 + (n - 1)d]$$

we must know the first term a_1, the common difference d, and the number of terms n. When we use the second formula,

$$S_n = \sum_{i=1}^{n} a_i = \frac{n}{2}(a_1 + a_n)$$

we must know the first term a_1, the last term a_n, and the number of terms n. Each of these formulas is helpful when the required information is given in calculating the sum of the terms in an arithmetic sequence as in the next example.

EXAMPLE 5   Find the following sums:

(a) $\sum_{i=5}^{30} \frac{(1 + 3i)}{2}$ and   (b) $5 + 9 + 13 + 17 + \cdots + 121$.

Solution

(a) We first must determine whether

$$\sum_{i=5}^{30} \frac{(1 + 3i)}{2}$$

is the sum of an arithmetic sequence. If so, $d = a_k - a_{k-1}$ is a real number. We check this with

$$d = a_k - a_{k-1}$$
$$= \frac{1 + 3k}{2} - \frac{1 + 3(k - 1)}{2}$$
$$= \frac{1 + 3k}{2} - \frac{3k - 2}{2}$$
$$= \frac{1 + 3k - 3k + 2}{2}$$
$$= \frac{3}{2} \qquad \text{A real number for all values of } k$$

Thus $d = \frac{3}{2}$. To find this sum, we will find the first term, a_5, and the number of terms, n. To find n, we notice that i changes from 5 to 30. If i went from 1 to 30, we would have 30 terms. But since we start at 5, we do not have the first four terms. Thus there are $n = 30 - 4 = 26$ terms. When $i = 5$, the first term is

$$a_5 = \frac{1 + 3(5)}{2} = \frac{16}{2} = 8$$

Then we have

$$S_{26} = \sum_{i=5}^{30} a_i = \frac{26}{2}[2a_5 + (26 - 1)d]$$

$$\sum_{i=5}^{30} \frac{(1 + 3i)}{2} = \frac{26}{2}\left[2(8) + (25)\frac{3}{2}\right]$$

$$= \frac{26}{2}(16 + 37.5)$$

$$= 13(53.5)$$

$$= 695.5$$

(b) In the sum $5 + 9 + 13 + 17 + \cdots + 121$ we note that $a_1 = 5$ and $a_n = 121$. This is an arithmetic sequence, since the common difference d between terms is $d = 4$. To find n, we will evaluate $a_n = 121$ and solve for n:

$$a_n = a_1 + (n - 1)d$$
$$121 = 5 + (n - 1)4$$
$$121 = 5 + 4n - 4$$
$$120 = 4n$$
$$30 = n$$

Therefore

$$5 + 9 + 13 + 17 + \cdots + 121 = \frac{n}{2}(a_1 + a_n)$$
$$= \frac{30}{2}(5 + 121)$$
$$= 15(126)$$
$$= 1890 \qquad \blacksquare$$

In conclusion, remember that in an **arithmetic sequence** $a_1, a_2, a_3, \cdots, a_n, \cdots,$

$$a_k - a_{k-1} = d \qquad \textit{(The common difference)}$$
$$a_n = a_1 + (n - 1)d \qquad \textit{(The nth term)}$$
$$S_n = \sum_{i=1}^{n} a_i = \frac{n}{2}[2a_1 + (n - 1)d] \quad \text{and}$$
$$S_n = \sum_{i=1}^{n} a_i = \frac{n}{2}(a_1 + a_n) \qquad \textit{(The nth partial sum)}$$

Exercises 10.3

For each of the following arithmetic sequences, find the first five terms, the eleventh term, and the kth term.

1. $a_1 = 7$ and $d = 2$

2. $a_1 = 2$ and $d = 3$

3. $a_1 = 17$ and $d = -3$

4. $a_1 = 25$ and $d = -2$

5. $a_5 = 12$ and $d = -2$

6. $a_5 = -1$ and $d = 2$

7. $a_{15} = -1$ and $d = 2$

8. $a_{15} = 12$ and $d = -2$

Find a_6 and a_n for the following arithmetic sequences.

9. $7, 6.7, 6.4, 6.1, \cdots$

10. $10, 9.1, 8.2, 7.3, \cdots$

11. $x - 1, x + 2, x + 5, \cdots$

12. $x + 11, x + 5, x - 1, \cdots$

13. $\log 3, \log 6, \log 12, \cdots$

14. $\log 2, \log 6, \log 18, \cdots$

15. Find a_9 in the arithmetic sequence in which $a_4 = 6$ and $a_5 = 3$.

16. Find a_9 in the arithmetic sequence in which $a_4 = 9$ and $a_5 = 7$.

17. Find a_1 and d in the arithmetic sequence in which $a_4 = 15$ and $a_{11} = 43$.

18. Find a_1 and d in the arithmetic sequence in which $a_5 = 11$ and $a_{10} = 27$.

19. Find a_6 in the arithmetic sequence in which $a_3 = 5$ and $a_8 = 23$.

20. Find a_7 in the arithmetic sequence in which $a_3 = 11$ and $a_9 = 35$.

Evaluate the following sums.

21. $\displaystyle\sum_{i=1}^{11} (2i + 3)$

22. $\displaystyle\sum_{i=1}^{15} (3i + 2)$

23. $\displaystyle\sum_{i=5}^{50} (15 - 3i)$

24. $\displaystyle\sum_{i=6}^{31} (35 - 2i)$

25. Find S_{15} for the arithmetic sequence with $a_1 = 75$ and $d = -3$.

26. Find S_{13} for the arithmetic sequence with $a_1 = 13$ and $d = -1$.

27. Find S_{20} for the arithmetic sequence with $a_1 = 2$ and $d = 0.2$.

28. Find S_{30} for the arithmetic sequence with $a_1 = 2$ and $d = -0.1$.

29. Find S_{21} for the sequence in Exercise 9.

30. Find S_{17} for the sequence in Exercise 10.

31. Find S_{22} for the sequence in Exercise 11.

32. Find S_{22} for the sequence in Exercise 12.

33. Find S_{21} for the sequence in Exercise 13.

34. Find S_{21} for the sequence in Exercise 14.

35. How many integers between 33 and 295 are divisible by 3?

36. How many integers between 75 and 350 are divisible by 2? What is their sum?

37. Find the arithmetic mean between 5 and 10.

38. Find the arithmetic mean between 7 and 13.

39. Find two arithmetic means between 5 and 10.

40. Find two arithmetic means between 7 and 13.

41. Find three arithmetic means between 7 and 13.

42. Find three arithmetic means between 5 and 10.

43. How many terms are there in the arithmetic partial sum when the partial sum is 420, $a_1 = 7$, and $d = 3$?

44. How many terms are there in the arithmetic partial sum when the partial sum is 357, $a_1 = -3$, and $d = 2$?

45. A section of seats in the corner of a stadium contains 35 seats in row 1. Each row has two fewer seats until the last row contains 17 seats. How many seats are in this section?

46. A grocery store makes a pyramid of cans with 15 cans in the bottom row and one fewer can in each succeeding row until one can is at the top. How many rows are needed? How many cans are needed for the pyramid?

47. You are investigating two different employment opportunities. BioTronics will start you with a salary of $24,000 and an increase of $1000 a year for the first 7 years, while Health Care will start you at $20,000 and an increase of $1100 every 6 months for 7 years. Which company would provide the higher salary in 7 years? Which company would pay the most money over 7 years?

48. A contractor who does not meet his deadline is fined $500 per day for the first 7 days, and for each day after the seventh the fine is increased by $120 a day. What is the fine for a 20-day delay?

49. If the contractor in Exercise 48 must pay a $23,100 fine, by how many days did he exceed his deadline?

10.4   Geometric Sequences

We will now study another important sequence, called a **geometric sequence**. In many ways it is different from the arithmetic sequence, yet the geometric sequence does have an nth term formula, a mean, and a partial sum. Starting with the following definition, we will immediately see the basic difference between the arithmetic and geometric sequences.

Definition

A **geometric sequence** is a sequence in which the quotient (or ratio) of any term after the first term and the previous term, is the same real number r, the **common ratio**. Thus

$$\frac{a_n}{a_{n-1}} = r.$$

We can use the common ratio r to construct a geometric sequence with a first term a_1 by multiplying a term by r to obtain the next term as follows:

$$a_1 = a_1$$
$$a_2 = a_1 r$$
$$a_3 = (a_2)r = (a_1 r)r = a_1 r^2$$
$$a_4 = (a_3)r = (a_1 r^2)r = a_1 r^3$$
$$a_5 = (a_4)r = (a_1 r^3)r = a_1 r^4$$

and in general

$$a_n = a_1 r^{n-1}$$

This result for a_n, the nth or general term, can be proved by mathematical induction.

EXAMPLE 1

Show that $a_n = a_1 r^{n-1}$ is the nth or general term for the geometric sequence with a first term of a_1 and a common ratio of r.

Solution   We will prove this using mathematical induction.
(a) "*Firstness*": For $n = 1$ we have $a_1 = a_1 r^{1-1} = a_1 r^0 = a_1 \cdot 1 = a_1$. Thus we have a first case.
(b) "*Nextness*": We assume the result true when $n = k$ and show that it must also be true for $n = k + 1$. For $n = k$ we have

$$a_k = a_1 r^{k-1}$$

To get the next term in a geometric sequence, we multiply both sides by r:

$$a_k(r) = a_1 r^{k-1}(r)$$
$$a_{k+1} = a_1 r^{k-1+1}$$
$$a_{k+1} = a_1 r^k$$

Thus $a_n = a_1 r^{n-1}$ is true for a first case and every next case.   ■

The formula for the nth or general term, $a_n = a_1 r^{n-1}$, will help us to solve many problems with geometric sequences. The following examples show how to use the definition of a geometric sequence as well as the formula for the nth term.

EXAMPLE 2　Find the first four terms, the twenty-first term, and the kth term for the geometric sequence where

(a) $a_1 = 4$ and $r = \frac{1}{2}$　and　(b) $a_1 = \frac{1}{3}$ and $r = -3$.

Solution　We first use the definition to find the first four terms and then use the nth term formula to find the 21st and kth terms.

(a) With $a_1 = 4$ and $r = \frac{1}{2}$,

$$a_1 = 4$$

$$a_2 = (4)\frac{1}{2} = 2$$

$$a_3 = (2)\frac{1}{2} = 1$$

$$a_4 = (1)\frac{1}{2} = \frac{1}{2}$$

Now we use the nth or general term, $a_n = a_1 r^{n-1}$, to find a_{21} and a_k:

$$
\begin{aligned}
a_{21} &= 4\left(\frac{1}{2}\right)^{21-1} & \text{and} \qquad a_k &= 4\left(\frac{1}{2}\right)^{k-1} \\
&= 2^2\left(\frac{1}{2}\right)^{20} & &= 2^2(2^{-1})^{k-1} \\
&= 2^2(2)^{-20} & &= 2^2(2^{-k+1}) \\
&= 2^{-18} & &= 2^{-k+3} \\
& & &= 2^{3-k}
\end{aligned}
$$

Therefore $a_{21} = 2^{-18}$ and $a_k = 2^{3-k}$.

(b) Similarly, with $a_1 = \frac{1}{3}$ and $r = -3$ we have

$$a_1 = \frac{1}{3}$$

$$a_2 = \frac{1}{3}(-3) = -1$$

$$a_3 = (-1)(-3) = 3$$

$$a_4 = (3)(-3) = -9$$

Also

$$
\begin{aligned}
a_{21} &= a_1 r^{21-1} & \text{and} \qquad a_k &= a_1 r^{k-1} \\
&= \frac{1}{3}(-3)^{20} & &= \frac{1}{3}(-3)^{k-1} \\
&= \frac{1}{3}(3)^{20} & &= 3^{-1}(-1)^{k-1}(3)^{k-1} \\
&= 3^{19} & &= (-1)^{k-1}3^{k-2}
\end{aligned}
$$

Therefore $a_{21} = 3^{19}$ and $a_k = (-1)^{k-1}3^{k-2}$.　∎

We will also find the nth term helpful in solving sequence problems in which any two terms are given as in this next example.

EXAMPLE 3   Find a_7 in a geometric sequence when $a_3 = 8$ and $a_{10} = -\frac{1}{16}$.

Solution   Using the nth term, $a_n = a_1 r^{n-1}$, we have

$$a_3 = a_1 r^{3-1}$$

$$8 = a_1 r^2 \quad \text{or} \quad \frac{8}{r^2} = a_1$$

Similarly,

$$a_{10} = a_1 r^{10-1}$$

$$-\frac{1}{16} = a_1 r^9$$

Substituting $a_1 = \frac{8}{r^2}$ into $-\frac{1}{16} = a_1 r^9$, we have

$$-\frac{1}{16} = \frac{8}{r^2} r^9$$

$$-\frac{1}{2^4} = 2^3 r^7$$

$$-\frac{1}{2^4 \cdot 2^3} = r^7$$

$$-2^{-7} = r^7$$

Thus

$$r = -2^{-1} = -\frac{1}{2}$$

Substituting into

$$\frac{8}{r^2} = a_1, \text{ we have}$$

$$a_1 = \frac{8}{(-2^{-1})^2} = 8 \cdot 4 = 32$$

Therefore

$$a_7 = 32(-2^{-1})^6$$
$$= 2^5(-1)^6 2^{-6}$$
$$= 2^{-1}$$
$$= \frac{1}{2}$$

The **geometric mean** is a number between two given numbers that allows all three numbers to be in a geometric sequence. As with the arithmetic means, we can have one, two, three, or more geometric means between two numbers. We will find geometric means in the following example.

EXAMPLE 4

Find the two geometric means between 3 and 7.

Solution   We let $a_1 = 3$ and $a_4 = 7$ so that the two geometric means are a_2 and a_3. Since this is a geometric progression, we have

$$a_1 = 3$$
$$a_2 = 3r$$
$$a_3 = (3r)r = 3r^2$$
$$a_4 = (3r^2)r = 3r^3$$

Thus $3r^3 = 7$. Solving for r, we have

$$r^3 = \frac{7}{3}$$

$$r = \sqrt[3]{\frac{7}{3}} \quad \text{or} \quad r = \frac{\sqrt[3]{63}}{3}$$

Then

$$a_2 = 3\frac{\sqrt[3]{63}}{3} = \sqrt[3]{63}$$

and

$$a_3 = 3\left(\sqrt[3]{\frac{7}{3}}\right)^2 = 3\,\sqrt[3]{\frac{49}{9}}$$
$$= 3\frac{\sqrt[3]{49 \cdot 3}}{3}$$
$$= \sqrt[3]{147}$$

Therefore the two geometric means between 3 and 7 are $\sqrt[3]{63}$ and $\sqrt[3]{147}$. ■

It will be important for us to find the partial sum of terms in a geometric sequence.

Theorem

Geometric Partial Sums. The sum of the first n terms of the geometric sequence

$$a_1, a_2, a_3, \cdots, a_n, \cdots$$

with common ratio r, is called **the nth partial sum S_n** and is calculated as

$$S_n = \sum_{i=1}^{n} a_i = \frac{a_1(1 - r^n)}{1 - r}$$

PROOF: We will prove this concisely using mathematical induction.

(a) *"Firstness"*: When $n = 1$,

$$\sum_{i=1}^{1} a_i = a_1 \qquad \text{and} \qquad \frac{a_1(1 - r^1)}{1 - r} = a_1$$

Therefore

$$S_1 = \sum_{i=1}^{n} a_i = \frac{a_1(1 - r^n)}{1 - r} \qquad \text{for } n = 1$$

(b) *"Nextness"*: We assume that

$$\sum_{i=1}^{k} a_i = \frac{a_1(1 - r^k)}{1 - r}$$

is true and show that it implies that

$$\sum_{i=1}^{k+1} a_i = \frac{a_1(1 - r^{k+1})}{1 - r}$$

is true. We let $n = k$ and add the next term, a_{k+1}, to both sides:

$$\sum_{i=1}^{k} a_i = \frac{a_1(1 - r^k)}{1 - r}$$

$$\sum_{i=1}^{k} a_i + a_{k+1} = \frac{a_1(1 - r^k)}{1 - r} + a_{k+1} \qquad \textit{(Substitute } a_{k+1} = ar^k\textit{)}$$

$$\sum_{i=1}^{k+1} a_i = \frac{a_1(1 - r^k)}{1 - r} + a_1 r^k$$

$$= \frac{a_1(1 - r^k)}{1 - r} + \frac{a_1 r^k(1 - r)}{1 - r}$$

$$= \frac{a_1 - a_1 r^k + a_1 r^k - a_1 r^{k+1}}{1 - r}$$

$$= \frac{a_1 - a_1 r^{k+1}}{1 - r}$$

$$= \frac{a_1(1 - r^{k+1})}{1 - r}$$

Therefore

$$\sum_{i=1}^{n} a_i = \frac{a_1(1 - r^{k+1})}{1 - r} \qquad \qquad \square$$

We can now use this formula to find the partial sums of geometric sequences.

EXAMPLE 5   Find the sum $\displaystyle\sum_{i=1}^{10} 3^{i-4}$.

Solution   We first determine whether $\displaystyle\sum_{i=1}^{10} 3^{i-4}$ is the partial sum of a geometric sequence. We calculate r to see whether it is a real number:

$$
\begin{aligned}
r &= \frac{a_k}{a_{k-1}} \\
&= \frac{3^{k-4}}{3^{k-1-4}} \\
&= \frac{3^{k-4}}{3^{k-5}} \\
&= 3^{k-4-(k-5)} \\
&= 3^1 \qquad \text{for all real numbers } k
\end{aligned}
$$

Thus this is a geometric sequence with $r = 3$. Then the partial sum is

$$
\begin{aligned}
\sum_{i=1}^{10} 3^{i-4} &= \frac{a_1(1 - r^{10})}{1 - r} \\
&= \frac{3^{-3}(1 - 3^{10})}{1 - 3} \\
&= \frac{1 - 3^{10}}{3^3(-2)} \\
&= \frac{1 - 3^{10}}{-54} \\
&= \frac{3^{10} - 1}{54} \quad \text{or} \quad \frac{1}{54}(3^{10} - 1) \\
&\approx 1093.481
\end{aligned}
$$

■

EXAMPLE 6   Find the sum $\frac{1}{2} - \frac{1}{4} + \frac{1}{8} - \frac{1}{16} + \cdots - \frac{1}{256}$.

Solution   The sum $\frac{1}{2} - \frac{1}{4} + \frac{1}{8} - \frac{1}{16} + \cdots - \frac{1}{256}$ is a partial sum of a geometric sequence with $r = -\frac{1}{2}$. Using the nth term, we see that the last term in the sum is

$$
a_n = \frac{1}{2}\left(-\frac{1}{2}\right)^{n-1}
$$

$$
-\frac{1}{256} = (-1)^{n-1}\left(\frac{1}{2}\right)^n
$$

Thus n is even, since $(-1)^{n-1} = -1$. For $2^n = 256$ and $256 = 2^8$ we have $n = 8$. Therefore

$$\frac{1}{2} - \frac{1}{4} + \frac{1}{8} - \frac{1}{16} + \cdots - \frac{1}{256} = \frac{a_1(1 - r^8)}{1 - r}$$

$$= \frac{\frac{1}{2}\left[1 - \left(\frac{1}{2}\right)^8\right]}{1 - \left(-\frac{1}{2}\right)}$$

$$= \frac{\frac{1}{2}(1 - 2^{-8})}{\frac{3}{2}}$$

$$= \frac{1}{3}(1 - 2^{-8})$$

$$\approx 0.332 \qquad \blacksquare$$

When working with geometric sequences, we are not limited to finding partial sums but may also find the sum of the whole sequence. Doing this, we will find the sum of an infinite number of terms, S_∞. Finding S_∞ is possible only when $|r| < 1$.

Theorem

Infinite Geometric Sums. When $|r| < 1$, the infinite geometric sequence a_1, a_1r, a_1r^2, $\cdots$, a_1r^n, $\cdots$ has an **infinite sum** of

$$S_\infty = \sum_{i=1}^{\infty} a_1 r^{i-1} = \frac{a_1}{1 - r}$$

PROOF: Since a_1, a_1r, a_1r^2, $\cdots$, a_1r^n, $\cdots$ is a geometric sequence, the partial sum is

$$S_n = \frac{a_1(1 - r^n)}{1 - r}$$

As n gets larger and larger, we need to evaluate the following limit:

$$\lim_{n \to \infty} S_n = \lim_{n \to \infty} \frac{a_1(1 - r^n)}{1 - r}$$

This limit is dependent upon the limit of r^n, since this is the only part of the formula containing n. Remembering that $|r| < 1$, a fraction in the interval of $(-1, 1)$, we have the value of r^n getting closer to zero as n gets larger and

$$\lim_{n \to \infty} r^n = 0$$

Thus the limit of the sum is

$$\lim_{n \to \infty} S_n = \lim_{n \to \infty} \frac{a_1(1 - r^n)}{1 - r}$$

$$= \frac{a_1(1 - 0)}{1 - r}$$

$$= \frac{a_1}{1 - r}$$

Therefore we can conclude that the geometric sequence with $|r| < 1$ has a sum of

$$S_\infty = \sum_{i=1}^{\infty} a_1 r^{i-1} = \frac{a_1}{1 - r} \qquad \square$$

We can now use this formula to find the sum of all the terms in an infinite geometric sequence when $|r| < 1$, where r is the common ratio.

EXAMPLE 7   Find the sum of $S_\infty = \frac{1}{2} - \frac{1}{4} + \frac{1}{8} - \frac{1}{16} + \cdots$.

Solution   We first notice that $\frac{1}{2}, -\frac{1}{4}, \frac{1}{8}, -\frac{1}{16}, \cdots$ is an infinite geometric sequence with $r = -\frac{1}{2}$ and is the same geometric sequence in Example 6. Since $|r| < 1$, we have

$$S_\infty = \frac{a_1}{1 - r}$$

$$= \frac{\frac{1}{2}}{1 - \left(-\frac{1}{2}\right)}$$

$$= \frac{\frac{1}{2}}{1 + \frac{1}{2}}$$

$$= \frac{\frac{1}{2}}{\frac{3}{2}}$$

$$= \frac{1}{3}$$

It is interesting to note how close the partial sum S_8 of Example 6 is to the infinite sum S_∞ of Example 7.

In conclusion, remember that in a geometric sequence $a_1, a_2, a_3, \cdots, a_n, \cdots,$

$$\frac{a_k}{a_{k-1}} = r \qquad \text{(The common ratio)}$$

$$a_n = a_1 r^{n-1} \qquad \text{(The nth term)}$$

$$S_n = \sum_{i=1}^{n} a_i = \frac{a_1(1 - r^n)}{1 - r} \qquad \text{(The nth partial sum)}$$

$$S_\infty = \sum_{i=1}^{\infty} a_i = \frac{a_1}{1 - r} \qquad \text{when } |r| < 1 \quad \text{(The infinite sum)}$$

Exercises 10.4

For each of the following geometric sequences, find the first five terms, the eleventh term, and the kth term.

1. $a_1 = 3$ and $r = \frac{1}{3}$
2. $a_1 = 2$ and $r = 3$
3. $a_1 = 2$ and $r = -2$
4. $a_1 = 3$ and $r = -\frac{1}{3}$
5. $a_1 = 9$ and $r = \frac{1}{3}$
6. $a_1 = 4$ and $r = \frac{1}{2}$
7. $a_1 = 8$ and $r = -\frac{1}{2}$
8. $a_1 = \frac{1}{27}$ and $r = -3$

Find a_6 and a_n for the following geometric sequences.

9. $\frac{1}{3}, \frac{1}{9}, \frac{1}{27}, \cdots$
10. $\frac{1}{4}, \frac{1}{16}, \frac{1}{64}, \cdots$
11. $3, -0.9, 0.27, -0.081, \cdots$
12. $6, -0.18, 0.054, -0.0162, \cdots$
13. $-1, \sqrt{2}, -2, \sqrt{8}, \cdots$
14. $-\sqrt{3}, 3, -\sqrt{27}, 9, \cdots$
15. $5, 5^{1-x}, 5^{1-2x}, 5^{1-3x}, \cdots$
16. $3^{x+1}, 3^{2x+1}, 3^{3x+1}, 3^{4x+1}, \cdots$
17. $-1, x^2, -x^4, x^6, \cdots$
18. $1, -\frac{x}{2}, \frac{x^2}{4}, -\frac{x^3}{8}, \cdots$

19. Find a_8 in the geometric sequence in which $a_3 = \frac{2}{9}$ and $a_4 = -\frac{2}{3}$.

20. Find a_8 in the geometric sequence in which $a_3 = \frac{9}{4}$ and $a_4 = -\frac{3}{4}$.

21. Find a_{10} for the geometric sequence with positive terms $a_3 = 1$ and $a_7 = 4$.

22. Find a_3 for the geometric sequence with positive terms $a_5 = \frac{1}{3}$ and $a_9 = \frac{16}{243}$.

Find the following sums.

23. $\sum_{i=1}^{20} 8\left(\frac{1}{2}\right)^i$
24. $\sum_{i=1}^{30} 27\left(\frac{1}{3}\right)^i$
25. $\sum_{i=4}^{14} 81\left(-\frac{1}{3}\right)^{i-2}$
26. $\sum_{i=3}^{20} 64\left(-\frac{1}{2}\right)^{i-1}$
27. $\sum_{i=0}^{10} (-\sqrt{2})^i$
28. $\sum_{i=0}^{11} (\sqrt[3]{-3})^i$

Find the following infinite sums whenever they exist.

29. $5 - \frac{5}{2} + \frac{5}{4} - \frac{5}{8} + \cdots$
30. $2 - \frac{2}{3} + \frac{2}{9} - \frac{2}{27} + \cdots$
31. $\sqrt{2} + 1 + \frac{\sqrt{2}}{2} + \frac{1}{2} + \cdots$
32. $2 + \frac{2\sqrt{3}}{3} + \frac{2}{3} + \frac{2\sqrt{3}}{9} + \cdots$
33. $\frac{1}{2} + \frac{1}{8} + \frac{1}{32} + \frac{1}{128} + \cdots$
34. $\frac{4}{5} + \frac{4}{15} + \frac{4}{45} + \frac{4}{135} + \cdots$
35. $\frac{7}{8} + 1 + \frac{8}{7} + \frac{64}{49} + \cdots$
36. $\frac{\sqrt{2}}{3} + \frac{2}{3} + \frac{2\sqrt{2}}{3} + \frac{4}{3} + \cdots$
37. $7 + \frac{1}{4} + \frac{1}{8} + \frac{1}{16} + \cdots$
38. $-3 + 5 - 1 + \frac{1}{5} - \frac{1}{25} + \cdots$

39. Find the geometric mean between 8 and 16.

40. Find the geometric mean between 5 and 45.

41. Find the two geometric means between 5 and 135.

42. Find the two geometric means between 10 and 80.

43. A vacuum pump removes one-third of the air from a container with each stroke. After six strokes, what percentage of the original air has been removed?

44. If ocean kelp increases 10% every hour and there are 2000 pounds of kelp at time zero, how much kelp is there after 24 hours and after n hours?

45. Upon graduation you have the choice of working for a company that grants raises of 10% a year or a company that grants raises of 4% every six months. If the starting salary in both programs is $20,000, what would be your salary in each company after 10 years and how much would you have earned in each company after 10 years?

46. A golf ball is dropped from a 4-foot table and rebounds $\frac{1}{3}$ of the distance it falls on each successive bounce. Using a geometric sequence, find the vertical distance the ball travels before coming to rest.

47. The end of a pendulum swings out an arc of 18 inches. Each successive swing takes the same amount of time, but each swing is $\frac{3}{4}$ the distance of the previous arc. Using a geometric sequence, find the total distance traveled by the pendulum. How many swings does it take for the arc of the pendulum to be 1 inch or less?

48. A car hits a speed bump and bounces up 18 inches. The shock absorbers reduce the displacement of each successive vertical bounce, up and down, by 42%. How many bounces does it take before the displacement is 1 inch?

The bar in Exercises 49–54 indicates a repeating pattern of digits. Use the sum of an infinite geometric sequence to find a rational number for each of the following repeating decimals.

49. $7.35\overline{35}...$

50. $3.71\overline{71}...$

51. $5.125\overline{125}...$

52. $7.215\overline{215}...$

53. $7.9\overline{99}...$

54. $0.9\overline{99}...$

55. Construct von Koch's snowflakes as described in the Exercises at the end of Section 10.2, but let the area of the triangle in case 1 equal 1 square unit. Notice that the new triangles in case 2 have an area equal to $\frac{1}{9}$ of the area of the large triangle in case 1 (Fig. 10.7). Find the sequence $A_1, A_2, A_3, A_4, A_5, \cdots$, where A_n is the area enclosed by the snowflake in case n. Find a formula for A_n and the value of A_∞.

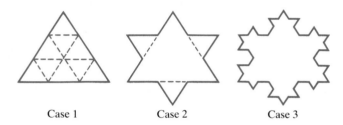

Case 1   Case 2   Case 3

Figure 10.7

10.5 The Binomial Theorem

We will now study another important sum that is the result of a binomial expression with a whole number exponent. The following whole number powers of binomials show some of these sums:

$$(a + b)^0 = 1$$
$$(a + b)^1 = a + b$$
$$(a + b)^2 = a^2 + 2ab + b^2$$
$$(a + b)^3 = a^3 + 3a^2b + 3ab^2 + b^3$$

In general, we would like to write the sum generated by $(a + b)^n$, where $n \in W$, in summation notation. We notice in the examples above that the exponent of a in the sum starts with the exponent of the binomial and then is decreased by 1 in each successive term. We also notice that the sum of the exponents of a and b in each term of the sum is always the same as the power of the binomial. This can be seen in $(a + b)^3$ as follows:

Descending powers of a: 3, $3 - 1 = 2,$ $2 - 1 = 1,$ $1 - 1 = 0$ $(a^0 = 1)$

$$(a + b)^3 = a^3 \quad + \quad 3a^2b \quad + \quad 3ab^2 \quad + \quad b^3$$

Sums of powers a and b: 3 = 3 $= 2 + 1$ $= 1 + 2$ $=$ 3

With the patterns above we can write the exponents for the sum of $(a + b)^3$ in summation notation. With C_i as the coefficient of the ith term we can write

$$(a + b)^3 = \sum_{i=0}^{3} C_i a^{3-i} b^i$$

To confirm this summation, we will expand the summation notation:

$$(a + b)^3 = \sum_{i=0}^{3} C_i a^{3-i} b^i$$

$$= C_0 a^3 b^0 + C_1 a^2 b^1 + C_2 ab^2 + C_3 a^0 b^3$$

$$= C_0 a^3 + C_1 a^2 b^1 + C_2 ab^2 + C_3 b^3$$

This summation provides the correct sum for $(a + b)^3$ when $C_0 = 1$, $C_1 = 3$, $C_2 = 3$, and $C_3 = 1$. Thus in general,

$$(a + b)^n = \sum_{i=0}^{n} C_i a^{n-i} b^i$$

Now that we understand the exponents, we will focus our attention on the coefficients C_i. Considering these coefficients, Blaise Pascal (1623–1662) noticed the following triangular pattern, called Pascal's triangle:

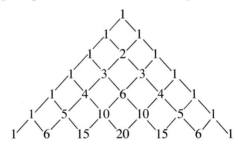

In this triangle he noticed that

(a) each row begins and ends with 1 and
(b) any other entry is the sum of the two entries above it.

These patterns are useful for finding some coefficients but are not very helpful if we need the third entry in the fifteenth row. Pascal found many uses for these coefficients, including probability and gambling theory. He also found a more general method for writing each term. Pascal's method for writing coefficients requires the new notation of **factorials**, which we will now define.

Definition

For $n \in N$, **n factorial**, written $n!$, is defined as

$$n! = n(n - 1)(n - 2) \cdots 3 \cdot 2 \cdot 1$$

with $0! = 1$ and $1! = 1$.

Before we proceed any further, we will first work the following example to better understand factorial evaluation.

EXAMPLE 1   Evaluate **(a)** $3!$ **(b)** $\dfrac{6!}{3!}$ and   **(c)** $\dfrac{6!}{2!3!}$.

Solution

(a) By the definition,

$$3! = 3 \cdot 2 \cdot 1$$
$$= 6$$

(b) Expanding by the definition and reducing, we have

$$\frac{6!}{3!} = \frac{6 \cdot 5 \cdot 4 \cdot 3 \cdot 2 \cdot 1}{3 \cdot 2 \cdot 1}$$
$$= \frac{6 \cdot 5 \cdot 4}{1}$$
$$= 120$$

We notice that $3 \cdot 2 \cdot 1 = 3!$ is a common factor in the numerator and denominator. Also note that $6!$ can be written in two ways as

$$6! = 6 \cdot 5 \cdot 4 \cdot 3 \cdot 2 \cdot 1 \quad \text{or}$$
$$6! = 6 \cdot 5 \cdot 4 \cdot 3!$$

This second expansion will provide a quicker way to complete this problem:

$$\frac{6!}{3!} = \frac{6 \cdot 5 \cdot 4 \cdot 3!}{3!}$$
$$= 6 \cdot 5 \cdot 4$$
$$= 120$$

(c) We will use the second method of part (b) above to evaluate $\dfrac{6!}{2!3!}$:

$$\frac{6!}{2!3!} = \frac{6 \cdot 5 \cdot 4 \cdot 3!}{2 \cdot 1 \cdot 3!}$$
$$= \frac{6 \cdot 5 \cdot 4}{2}$$
$$= 3 \cdot 5 \cdot 4$$
$$= 60$$

Using the factorial, we will write the coefficients for the expansion of $(a + b)^n$ as

$$C_i = \frac{n!}{(n - i)!i!}$$

as we will see in the next theorem.

The Binomial Theorem

For $a \in R$, $b \in R$, and $n \in N$,

$$(a + b)^n = \sum_{i=0}^{n} \frac{n!}{(n - i)!i!} a^{n-i} b^i$$

We will wait until the next section to prove the binomial theorem. It is interesting to note that the numerator of the coefficients

$$C_i = \frac{n!}{(n - i)!i!}$$

is the factorial of the exponent of the binomial and that the denominator is the product of the factorials of the exponents for that term.

Since we have not yet proven the binomial theorem, it would be good to confirm its validity by evaluating $(a + b)^3$ and $(a + b)^4$ in the next example.

EXAMPLE 2   Evaluate **(a)** $(a + b)^3$ and   **(b)** $(a + b)^4$.

Solution   Using the binomial theorem, we have

(a) $(a + b)^3 = \sum_{i=0}^{3} \frac{3!}{(3 - i)!i!} a^{3-i} b^i$

$$= \frac{3!}{3!0!} a^3 b^0 + \frac{3!}{2!1!} a^2 b^1 + \frac{3!}{1!2!} ab^2 + \frac{3!}{0!3!} a^0 b^3$$

$$= 1 \cdot a^3 \cdot 1 + \frac{3 \cdot 2!}{2!1!} a^2 b + \frac{3 \cdot 2!}{1!2!} ab^2 + 1 \cdot 1 \cdot b^3$$

$$= a^3 + 3a^2 b + 3ab^2 + b^3$$

(b) $(a + b)^4 = \sum_{i=0}^{4} \frac{4!}{(4 - i)!i!} a^{4-i} b^i$

$$= \frac{4!}{4!0!} a^4 b^0 + \frac{4!}{3!1!} a^3 b^1 + \frac{4!}{2!2!} a^2 b^2 + \frac{4!}{1!3!} a^1 b^3 + \frac{4!}{0!4!} a^0 b^4$$

$$= 1 \cdot a^4 \cdot 1 + \frac{4 \cdot 3!}{3!1!} a^3 b + \frac{4 \cdot 3 \cdot 2!}{2!2 \cdot 1} a^2 b^2 + \frac{4 \cdot 3!}{1!3!} ab^3 + 1 \cdot 1 \cdot b^4$$

$$= a^4 + 4a^3 b + 6a^2 b^2 + 4ab^3 + b^4$$

∎

Both parts of Example 2 follow the pattern established for the exponents and the coefficient values in Pascal's triangle.

We will use the pattern from the binomial theorem to evaluate other binomials as seen in the following example.

EXAMPLE 3   Write $(x - 2)^4$ as a sum.

Solution   We let $a = x$ and $b = -2$ and then use the binomial theorem. (The expansions for the coefficients C_i were evaluated in Example 2(b).)

$$(x - 2)^4 = \sum_{i=0}^{4} \frac{4!}{(4 - i)!i!} x^{4-i}(-2)^i$$

$$= \frac{4!}{4!0!} x^4(-2)^0 + \frac{4!}{3!1!} x^3(-2)^1 + \frac{4!}{2!2!} x^2(-2)^2 + \frac{4!}{1!3!} x^2(-2)^3 + \frac{4!}{0!4!} x^0(-2)^4$$

$$= 1x^4(1) + 4x^3(-2) + 6x^2(4) + 4x(-8) + 1 \cdot 1(16)$$

$$= x^4 - 8x^3 + 24x^2 - 32x + 16 \qquad \blacksquare$$

Thus we can use the binomial theorem to write a binomial to any natural number exponent.

To simplify the binomial theorem, we will use a new notation for the binomial coefficient as seen in this definition.

Definition   If n and r are nonnegative integers with $0 \le r \le n$, then the **binomial coefficient** is defined as

$$\frac{n!}{(n - r)!r!} = \binom{n}{r}$$

The number $\binom{n}{r}$ is called the binomial coefficient of n over r. Notice that the binomial coefficient $\binom{n}{r}$ does not contain a fraction bar between the n and r. This allows us to distinguish between the fraction $\left(\dfrac{n}{r}\right)$ and the binomial coefficient $\binom{n}{r}$. In general, $\left(\dfrac{n}{r}\right) \ne \binom{n}{r}$. This new notation now provides a new and quicker way to write the binomial theorem.

The Binomial Theorem   For $a \in R$, $b \in R$, and $n \in N$,

$$(a + b)^n = \sum_{i=0}^{n} \binom{n}{i} a^{n-i} b^i$$

Either representation of the binomial theorem is adequate, although we will find the latter more compact and easier to write.

Exercises 10.5

Evaluate the following expressions.

1. $5!$

2. $7!$

3. $\dfrac{6!}{2!4!}$

4. $\dfrac{8!}{6!2!}$

5. $\dfrac{10!}{8!2!}$

6. $\dfrac{12!}{10!2!}$

7. $\dfrac{n!}{(n-2)!2!}$

8. $\dfrac{(n+1)!}{n!1!}$

9. $\dbinom{5}{3}$

10. $\dbinom{7}{5}$

11. $\dbinom{4}{1}$

12. $\dbinom{7}{6}$

13. $\dbinom{7}{3}$

14. $\dbinom{8}{4}$

15. $\dbinom{n}{n-1}$

16. $\dbinom{n+1}{n-1}$

Write the following as a sum of terms. Simplify each term in each sum.

17. $(x-y)^4$

18. $(a-b)^4$

19. $(x+2)^3$

20. $(y-3)^3$

21. $\left(\frac{1}{3}x-3\right)^5$

22. $\left(\frac{1}{2}y-2\right)^5$

23. $(x^{1/2}+y^{1/2})^6$

24. $(x^{-2}+y)^6$

25. Find the first four terms of $(2x^{1/4}+x^{1/3})^{12}$.

26. Find the first four terms of $(x^{1/5}-2x^{1/3})^{15}$.

27. Find the third term of $\left(4x^{-2}-\frac{1}{2}x^2\right)^{12}$.

28. Find the fifth term of $\left(3x^2-\frac{1}{3}x^{-1}\right)^{10}$.

29. Find the term of

$$\left(5x-\frac{1}{3x}\right)^8$$

that does not contain x.

30. Find the term of

$$\left(3y-\frac{2}{3y}\right)^{10}$$

that does not contain y.

The following problems are necessary to prove the binomial theorem in the next section.

31. Verify that

$$\binom{k}{0}=\binom{k+1}{0} \text{ and } \binom{k}{k}=\binom{k+1}{k+1}$$

32. Verify that

$$\binom{n}{r}=\binom{n}{n-r}$$

33. Verify that

$$\binom{n}{i}+\binom{n}{i+1}=\binom{n+1}{i+1}$$

34. Verify that

$$\sum_{i=0}^{k}\binom{k}{i}a^{k-i}b^{i+1}=\sum_{i=1}^{k+1}\binom{k}{i-1}a^{k-i+1}b^{i}$$

10.6   The Proof of the Binomial Theorem

We will now prove the binomial theorem of Section 10.5 by using mathematical induction. To complete this proof, we will use the following facts and properties. For real numbers a and b and integers n and k:

(a) $\dbinom{k}{0}=\dbinom{k+1}{0}$

(b) $\dbinom{k}{k} = \dbinom{k+1}{k+1}$

(c) $\dbinom{n}{i} = \dbinom{n}{n-i}$

(d) $\dbinom{n}{i} + \dbinom{n}{i-1} = \dbinom{n+1}{i}$

(e) $\displaystyle\sum_{i=0}^{k} \dbinom{k}{i} a^{k-i}b^{i+1} = \sum_{i=1}^{k+1} \dbinom{k}{i-1} a^{k-i+1}b^{i}$

All of these properties were proven in the exercises at the end of Section 10.5.

The Binomial Theorem

For $a \in R$, $b \in R$, and $n \in N$,

$$(a + b)^n = \sum_{i=0}^{n} \dbinom{n}{i} a^{n-i}b^{i}$$

PROOF: We will prove this theorem using mathematical induction.

(a) *"Firstness"*: For $n = 1$ we have

$$(a + b)^n = (a + b)^1$$

$$\sum_{i=0}^{n} \dbinom{n}{i} a^{n-i}b^{i} = \sum_{i=0}^{1} \dbinom{1}{i} a^{1-i}b^{i}$$

$$= \dbinom{1}{0} a^1 b^0 + \dbinom{1}{1} a^0 b^1$$

$$= \frac{1!}{1!0!}a + \frac{1!}{0!1!}b$$

$$= 1 \cdot a + 1 \cdot b$$

$$= a + b$$

Therefore

$$(a + b)^n = \sum_{i=0}^{n} \dbinom{n}{i} a^{n-i}b^{i} \qquad \text{for } n = 1$$

(b) *"Nextness"*: Assume that it is true that the binomial theorem is valid when $n = k$ and prove that

$$(a + b)^{k+1} = \sum_{i=0}^{k+1} \dbinom{k+1}{i} a^{k+1-i}b^{i}$$

Starting with the binomial theorem for $n = k$ we multiply both sides by $(a + b)$ and simplify:

$$(a + b)^k = \sum_{i=0}^{k} \dbinom{k}{i} a^{k-i}b^{i}$$

$$(a + b)(a + b)^k = (a + b) \sum_{i=0}^{k} \dbinom{k}{i} a^{k-i}b^{i}$$

$$(a + b)^{k+1} = a \sum_{i=0}^{k} \binom{k}{i} a^{k-i} b^i + b \sum_{i=0}^{k} \binom{k}{i} a^{k-i} b^i$$

$$= \sum_{i=0}^{k} \binom{k}{i} a^{k-i+1} b^i + \sum_{i=0}^{k} \binom{k}{i} a^{k-i} b^{i+1}$$

We now use property (e) on the second sum, and

$$(a + b)^{k+1} = \sum_{i=0}^{k} \binom{k}{i} a^{k-i+1} b^i + \sum_{i=1}^{k+1} \binom{k}{i-1} a^{k-i+1} b^i$$

Separating the first term from the first summation and the last term from the second summation, we have

$$(a + b)^{k+1} =$$
$$\binom{k}{0} a^{k+1} b^0 + \sum_{i=1}^{k} \binom{k}{i} a^{k-i+1} b^i + \sum_{i=1}^{k} \binom{k}{i-1} a^{k-i+1} b^i + \binom{k}{k} a^0 b^{k+1}$$

We now use properties (a) and (b) and the properties for adding in summation notation to obtain

$$(a + b)^{k+1} =$$
$$\binom{k+1}{0} a^{k+1} b^0 + \sum_{i=1}^{k} \left[\binom{k}{i} + \binom{k}{i-1} \right] a^{k-i+1} b^i + \binom{k+1}{k+1} a^0 b^{k+1}$$

We now use property (d) to simplify the binomial coefficients in the sum:

$$(a + b)^{k+1} = \binom{k+1}{0} a^{k+1} b^0 + \sum_{i=1}^{k} \binom{k+1}{i} a^{k-i+1} b^i + \binom{k+1}{k+1} a^0 b^{k+1}$$

Next we include the first and last terms in the summation notation, and

$$(a + b)^{k+1} = \sum_{i=0}^{k+1} \binom{k+1}{i} a^{k+1-i} b^i$$

Therefore when the binomial theorem is true for $n = k$, it is also true for $n = k + 1$.   □

We did not take time and space to complete this proof in the last section, but we have now proven the binomial theorem. Now use the binomial theorem to complete the problems in the following exercises.

Exercises 10.6

Write each of the following as a sum of terms. Simplify each term in each sum.

1. $\left(1 - \dfrac{1}{x} \right)^5$

2. $\left(\dfrac{3}{x} + 3x \right)^6$

3. $\left(\dfrac{x}{y} + \dfrac{2y}{x} \right)^6$

4. $\left(8x - \dfrac{1}{4x} \right)^3$

Use the binomial theorem to evaluate each of the following with accuracy to two decimal places.

5. $(1.1)^8 = (1 + 0.1)^8$

6. $(2.01)^5 = (2 + 0.01)^5$

7. $(0.99)^7$

8. $(2.98)^5$

9. $(3.01)^5$

10. $(3.99)^7$

11. Prove that

$$\sum_{i=0}^{n} \binom{n}{i} = 2^n$$

12. Prove that

$$\sum_{i=0}^{n} (-1)^i \binom{n}{i} = 0$$

13. Prove that

$$\binom{n+1}{i} = \binom{n}{i-1} + \binom{n}{i}$$

which is the law of Pascal's triangle.

14. Prove that $(1 + x)^n \geq 1 + nx$ for $x > -1$, Bernoulli's inequality.

15. Prove that

$$\frac{n-r+1}{r}\binom{n}{r-1} = \binom{n}{r}$$

16. Prove by mathematical induction that

$$\sum_{i=1}^{n} \binom{i+1}{2} = \binom{n+2}{3}$$

17. Prove that

$$\binom{n}{r} = \binom{n-2}{r} + 2\binom{n-2}{r-1} + \binom{n-2}{r-2}$$

CHAPTER SUMMARY AND REVIEW

KEY TERMS

Sequence

Summation notation

Mathematical induction

Arithmetic sequence

Geometric sequence

Factorials

The binomial theorem

The binomial coefficient

CHAPTER EXERCISES

Find the first five terms for the following sequences.

1. $a_n = 5(1 - n)$

2. $a_n = (-1)^n 3^n$

3. $a_1 = 1$ and $a_n = 17 - 2a_{n-1}$

4. $a_1 = 3$ and $a_n = (a_{n-1})^{-1}$

5. The arithmetic sequence in which $a_1 = 13$ and $d = -3$.

6. The arithmetic sequence in which $a_1 = 2$ and $d = -\frac{1}{2}$.

7. The geometric sequence in which $a_1 = 2$ and $r = -\frac{1}{2}$.

8. The geometric sequence in which $a_1 = 81$ and $r = \frac{1}{3}$.

Evaluate each of the following.

9. $\sum_{i=1}^{4} (i^2 - 3)$

10. $\sum_{i=0}^{4} 7$

11. $\sum_{i=6}^{10} \left(5 - \frac{i}{2}\right)$

12. $\sum_{i=1}^{22} (2i - 3)$

13. $\sum_{i=3}^{32} (15 - 2i)$

14. $\sum_{i=1}^{72} \left(\frac{1}{3}\right)^i$

15. $\sum_{i=5}^{50} \left(-\frac{1}{4}\right)^{i-5}$

16. $\sum_{i=1}^{\infty} \frac{5}{2^i}$

Evaluate each of the following as a sum. Simplify each term in the sum.

17. $\left(x + \frac{2}{3}\right)^5$

18. $\left(\frac{1}{2}x - 2\right)^4$

19. Bales of hay are stacked in a pyramid with 25 bales in the bottom row and one fewer bale in each succeeding row until the top row has ten bales. If each bale is 2 feet high, how tall is this pyramid and how many bales does it contain?

20. A ball falls from 10 feet. The ball always rebounds $\frac{1}{4}$ of the distance it falls. Find the vertical distance this ball travels before it comes to rest.

Evaluate each of the following to the nearest three decimal places using the binomial theorem and your calculator.

21. $(7.01)^7$

22. $(5.98)^5$

23. Prove that
$$\sum_{i=1}^{n} 3^{i-1} \ge \frac{3}{2}(3^n - 1)$$

24. Prove that
$$\sum_{i=1}^{n} 8i < (2n + 1)^2$$

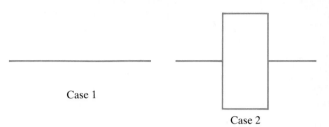

Case 1

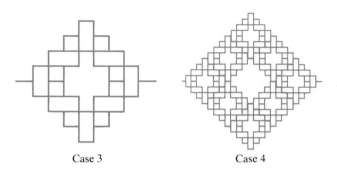

Case 2

Case 3

Case 4

Sierpinski's carpet is formed by dividing a segment into three equal segments and replacing the middle segment with a 1×2 rectangle as seen in cases 1 and 2 in Fig. 10.8. This process is then repeated on each segment to construct the next case as seen in cases 3, 4, and 5 in Fig. 10.8.

25. If the original length of the segment is 1 unit, find the sequence $p_0, p_1, p_2, p_3, \ldots, p_n$, where p_n is the sum of the lengths of all segments in the Sierpinski's carpet after n cases (divisions). Also find $\lim_{n \to \infty} p_n$ and the maximum area the carpet can cover.

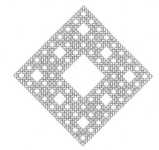

Figure 10.8

Case 5

Tables

TABLE 1 Exponential Functions

x	e^x	e^{-x}	x	e^x	e^{-x}
0.00	1.0000	1.0000	2.5	12.182	0.0821
0.05	1.0513	0.9512	2.6	13.464	0.0743
0.10	1.1052	0.9048	2.7	14.880	0.0672
0.15	1.1618	0.8607	2.8	16.445	0.0608
0.20	1.2214	0.8187	2.9	18.174	0.0550
0.25	1.2840	0.7788	3.0	20.086	0.0498
0.30	1.3499	0.7408	3.1	22.198	0.0450
0.35	1.4191	0.7047	3.2	24.533	0.0408
0.40	1.4918	0.6703	3.3	27.113	0.0369
0.45	1.5683	0.6376	3.4	29.964	0.0334
0.50	1.6487	0.6065	3.5	33.115	0.0302
0.55	1.7333	0.5769	3.6	36.598	0.0273
0.60	1.8221	0.5488	3.7	40.447	0.0247
0.65	1.9155	0.5220	3.8	44.701	0.0224
0.70	2.0138	0.4966	3.9	49.402	0.0202
0.75	2.1170	0.4724	4.0	54.598	0.0183
0.80	2.2255	0.4493	4.1	60.340	0.0166
0.85	2.3396	0.4274	4.2	66.686	0.0150
0.90	2.4596	0.4066	4.3	73.700	0.0136
0.95	2.5857	0.3867	4.4	81.451	0.0123
1.0	2.7183	0.3679	4.5	90.017	0.0111
1.1	3.0042	0.3329	4.6	99.484	0.0101
1.2	3.3201	0.3012	4.7	109.95	0.0091
1.3	3.6693	0.2725	4.8	121.51	0.0082
1.4	4.0552	0.2466	4.9	134.29	0.0074
1.5	4.4817	0.2231	5	148.41	0.0067
1.6	4.9530	0.2019	6	403.43	0.0025
1.7	5.4739	0.1827	7	1096.6	0.0009
1.8	6.0496	0.1653	8	2981.0	0.0003
1.9	6.6859	0.1496	9	8103.1	0.0001
2.0	7.3891	0.1353	10	22026	0.00005
2.1	8.1662	0.1225			
2.2	9.0250	0.1108			
2.3	9.9742	0.1003			
2.4	11.023	0.0907			

TABLE 2 Natural Logarithms

x	$\log_e x$	x	$\log_e x$	x	$\log_e x$
0.0	*	4.5	1.5041	9.0	2.1972
0.1	7.6974	4.6	1.5261	9.1	2.2083
0.2	8.3906	4.7	1.5476	9.2	2.2192
0.3	8.7960	4.8	1.5686	9.3	2.2300
0.4	9.0837	4.9	1.5892	9.4	2.2407
0.5	9.3069	5.0	1.6094	9.5	2.2513
0.6	9.4892	5.1	1.6292	9.6	2.2618
0.7	9.6433	5.2	1.6487	9.7	2.2721
0.8	9.7769	5.3	1.6677	9.8	2.2824
0.9	9.8946	5.4	1.6864	9.9	2.2925
1.0	0.0000	5.5	1.7047	10	2.3026
1.1	0.0953	5.6	1.7228	11	2.3979
1.2	0.1823	5.7	1.7405	12	2.4849
1.3	0.2624	5.8	1.7579	13	2.5649
1.4	0.3365	5.9	1.7750	14	2.6391
1.5	0.4055	6.0	1.7918	15	2.7081
1.6	0.4700	6.1	1.8083	16	2.7726
1.7	0.5306	6.2	1.8245	17	2.8332
1.8	0.5878	6.3	1.8405	18	2.8904
1.9	0.6419	6.4	1.8563	19	2.9444
2.0	0.6931	6.5	1.8718	20	2.9957
2.1	0.7419	6.6	1.8871	25	3.2189
2.2	0.7885	6.7	1.9021	30	3.4012
2.3	0.8329	6.8	1.9169	35	3.5553
2.4	0.8755	6.9	1.9315	40	3.6889
2.5	0.9163	7.0	1.9459	45	3.8067
2.6	0.9555	7.1	1.9601	50	3.9120
2.7	0.9933	7.2	1.9741	55	4.0073
2.8	1.0296	7.3	1.9879	60	4.0943
2.9	1.0647	7.4	2.0015	65	4.1744
3.0	1.0986	7.5	2.0149	70	4.2485
3.1	1.1314	7.6	2.0281	75	4.3175
3.2	1.1632	7.7	2.0412	80	4.3820
3.3	1.1939	7.8	2.0541	85	4.4427
3.4	1.2238	7.9	2.0669	90	4.4998
3.5	1.2528	8.0	2.0794	95	4.5539
3.6	1.2809	8.1	2.0919	100	4.6052
3.7	1.3083	8.2	2.1041		
3.8	1.3350	8.3	2.1163		
3.9	1.3610	8.4	2.1282		
4.0	1.3863	8.5	2.1401		
4.1	1.4110	8.6	2.1518		
4.2	1.4351	8.7	2.1633		
4.3	1.4586	8.8	2.1748		
4.4	1.4816	8.9	2.1861		

*Subtract 10 from $\log_e x$ entries for $x < 1.0$.

TABLE 3 Trigonometric Functions—Degrees and Minutes or Radians

Angle θ Degrees	Radians	sin θ	csc θ	tan θ	cot θ	sec θ	cos θ		
0° 00′	.0000	.0000	No value	.0000	No value	1.000	1.0000	1.5708	90° 00′
10	.0029	.0029	343.8	.0029	343.8	1.000	1.0000	1.5679	50
20	.0058	.0058	171.9	.0058	171.9	1.000	1.0000	1.5650	40
30	.0087	.0087	114.6	.0087	114.6	1.000	1.0000	1.5621	30
40	.0116	.0116	85.95	.0116	85.94	1.000	.9999	1.5592	20
50	.0145	.0145	68.76	.0145	68.75	1.000	.9999	1.5563	10
1° 00′	.0175	.0175	57.30	.0175	57.29	1.000	.9998	1.5533	89° 00′
10	.0204	.0204	49.11	.0204	49.10	1.000	.9998	1.5504	50
20	.0233	.0233	42.98	.0233	42.96	1.000	.9997	1.5475	40
30	.0262	.0262	38.20	.0262	38.19	1.000	.9997	1.5446	30
40	.0291	.0291	34.38	.0291	34.37	1.000	.9996	1.5417	20
50	.0320	.0320	31.26	.0320	31.24	1.001	.9995	1.5388	10
2° 00′	.0349	.0349	28.65	.0349	28.64	1.001	.9994	1.5359	88° 00′
10	.0378	.0378	26.45	.0378	26.43	1.001	.9993	1.5330	50
20	.0407	.0407	24.56	.0407	24.54	1.001	.9992	1.5301	40
30	.0436	.0436	22.93	.0437	22.90	1.001	.9990	1.5272	30
40	.0465	.0465	21.49	.0466	21.47	1.001	.9989	1.5243	20
50	.0495	.0494	20.23	.0495	20.21	1.001	.9988	1.5213	10
3° 00′	.0524	.0523	19.11	.0524	19.08	1.001	.9986	1.5184	87° 00′
10	.0553	.0552	18.10	.0553	18.07	1.002	.9985	1.5155	50
20	.0582	.0581	17.20	.0582	17.17	1.002	.9983	1.5126	40
30	.0611	.0610	16.38	.0612	16.35	1.002	.9981	1.5097	30
40	.0640	.0640	15.64	.0641	15.60	1.002	.9980	1.5068	20
50	.0669	.0669	14.96	.0670	14.92	1.002	.9978	1.5039	10
4° 00′	.0698	.0698	14.34	.0699	14.30	1.002	.9976	1.5010	86° 00′
10	.0727	.0727	13.76	.0729	13.73	1.003	.9974	1.5981	50
20	.0756	.0756	13.23	.0758	13.20	1.003	.9971	1.5952	40
30	.0785	.0785	12.75	.0787	12.71	1.003	.9969	1.5923	30
40	.0814	.0814	12.29	.0816	12.25	1.003	.9967	1.5893	20
50	.0844	.0843	11.87	.0846	11.83	1.004	.9964	1.5864	10
5° 00′	.0873	.0872	11.47	.0875	11.43	1.004	.9962	1.4835	85° 00′
10	.0902	.0901	11.10	.0904	11.06	1.004	.9959	1.4806	50
20	.0931	.0929	10.76	.0934	10.71	1.004	.9957	1.4777	40
30	.0960	.0958	10.43	.0963	10.39	1.005	.9954	1.4748	30
40	.0989	.0987	10.13	.0992	10.08	1.005	.9951	1.4719	20
50	.1018	.1016	9.839	.1022	9.788	1.005	.9948	1.4690	10
6° 00′	.1047	.1045	9.567	.1051	9.514	1.006	.9945	1.4661	84° 00′
10	.1076	.1074	9.309	.1080	9.255	1.006	.9942	1.4632	50
20	.1105	.1103	9.065	.1110	9.010	1.006	.9939	1.4603	40
30	.1134	.1132	8.834	.1139	8.777	1.006	.9936	1.4573	30
40	.1164	.1161	8.614	.1169	8.556	1.007	.9932	1.4544	20
50	.1193	.1190	8.405	.1198	8.345	1.007	.9929	1.4515	10
7° 00′	.1222	.1219	8.206	.1228	8.144	1.008	.9925	1.4486	83° 00′
10	.1251	.1248	8.016	.1257	7.953	1.008	.9922	1.4457	50
20	.1280	.1276	7.834	.1287	7.770	1.008	.9918	1.4428	40
30	.1309	.1305	7.661	.1317	7.596	1.009	.9914	1.4399	30
40	.1338	.1334	7.496	.1346	7.429	1.009	.9911	1.4370	20
50	.1367	.1363	7.337	.1376	7.269	1.009	.9907	1.4341	10
8° 00′	.1396	.1392	7.185	.1405	7.115	1.010	.9903	1.4312	82° 00′
10	.1425	.1421	7.040	.1435	6.968	1.010	.9899	1.4283	50
20	.1454	.1449	6.900	.1465	6.827	1.011	.9894	1.4254	40
30	.1484	.1478	6.765	.1495	6.691	1.011	.9890	1.4224	30
40	.1513	.1507	6.636	.1524	6.561	1.012	.9886	1.4195	20
50	.1542	.1536	6.512	.1554	6.435	1.012	.9881	1.4166	10
9° 00′	.1571	.1564	6.392	.1584	6.314	1.012	.9877	1.4137	81° 00′
		cos θ	sec θ	cot θ	tan θ	csc θ	sin θ	Radians	Degrees
								Angle θ	

TABLE 3　Trigonometric Functions—Degrees and Minutes or Radians
(continued)

Degrees	Radians	sin θ	csc θ	tan θ	cot θ	sec θ	cos θ		
9° 00′	.1571	.1564	6.392	.1584	6.314	1.012	.9877	1.4137	81° 00′
10	.1600	.1593	6.277	.1614	6.197	1.013	.9872	1.4108	50
20	.1629	.1622	6.166	.1644	6.084	1.013	.9868	1.4079	40
30	.1658	.1650	6.059	.1673	5.976	1.014	.9863	1.4050	30
40	.1687	.1679	5.955	.1703	5.871	1.014	.9858	1.4021	20
50	.1716	.1708	5.855	.1733	5.769	1.015	.9853	1.3992	10
10° 00′	.1745	.1736	5.759	.1763	5.671	1.015	.9848	1.3963	80° 00′
10	.1774	.1765	5.665	.1793	5.576	1.016	.9843	1.3934	50
20	.1804	.1794	5.575	.1823	5.485	1.016	.9838	1.3904	40
30	.1833	.1822	5.487	.1853	5.396	1.017	.9833	1.3875	30
40	.1862	.1851	5.403	.1883	5.309	1.018	.9827	1.3846	20
50	.1891	.1880	5.320	.1914	5.226	1.018	.9822	1.3817	10
11° 00′	.1920	.1908	5.241	.1944	5.145	1.019	.9816	1.3788	79° 00′
10	.1949	.1937	5.164	.1974	5.066	1.019	.9811	1.3759	50
20	.1978	.1965	5.089	.2004	4.989	1.020	.9805	1.3730	40
30	.2007	.1994	5.016	.2035	4.915	1.020	.9799	1.3701	30
40	.2036	.2022	4.945	.2065	4.843	1.021	.9793	1.3672	20
50	.2065	.2051	4.876	.2095	4.773	1.022	.9787	1.3643	10
12° 00′	.2094	.2079	4.810	.2126	4.705	1.022	.9781	1.3614	78° 00′
10	.2123	.2108	4.745	.2156	4.638	1.023	.9775	1.3584	50
20	.2153	.2136	4.682	.2186	4.574	1.024	.9769	1.3555	40
30	.2182	.2164	4.620	.2217	4.511	1.024	.9763	1.3526	30
40	.2211	.2193	4.560	.2247	4.449	1.025	.9757	1.3497	20
50	.2240	.2221	4.502	.2278	4.390	1.026	.9750	1.3468	10
13° 00′	.2269	.2250	4.445	.2309	4.331	1.026	.9744	1.3439	77° 00′
10	.2298	.2278	4.390	.2339	4.275	1.027	.9737	1.3410	50
20	.2327	.2306	4.336	.2370	4.219	1.028	.9730	1.3381	40
30	.2356	.2334	4.284	.2401	4.165	1.028	.9724	1.3352	30
40	.2385	.2363	4.232	.2432	4.113	1.029	.9717	1.3323	20
50	.2414	.2391	4.182	.2462	4.061	1.030	.9710	1.3294	10
14° 00′	.2443	.2419	4.134	.2493	4.011	1.031	.9703	1.3265	76° 00′
10	.2473	.2447	4.086	.2524	3.962	1.031	.9696	1.3235	50
20	.2502	.2476	4.039	.2555	3.914	1.032	.9689	1.3206	40
30	.2531	.2504	3.994	.2586	3.867	1.033	.9681	1.3177	30
40	.2560	.2532	3.950	.2617	3.821	1.034	.9674	1.3148	20
50	.2589	.2560	3.906	.2648	3.776	1.034	.9667	1.3119	10
15° 00′	.2618	.2588	3.864	.2679	3.732	1.035	.9659	1.3090	75° 00′
10	.2647	.2616	3.822	.2711	3.689	1.036	.9652	1.3061	50
20	.2676	.2644	3.782	.2742	3.647	1.037	.9644	1.3032	40
30	.2705	.2672	3.742	.2773	3.606	1.038	.9636	1.3003	30
40	.2734	.2700	3.703	.2805	3.566	1.039	.9628	1.2974	20
50	.2763	.2728	3.665	.2836	3.526	1.039	.9621	1.2945	10
16° 00′	.2793	.2756	3.628	.2867	3.487	1.040	.9613	1.2915	74° 00′
10	.2822	.2784	3.592	.2899	3.450	1.041	.9605	1.2886	50
20	.2851	.2812	3.556	.2931	3.412	1.042	.9596	1.2857	40
30	.2880	.2840	3.521	.2962	3.376	1.043	.9588	1.2828	30
40	.2909	.2868	3.487	.2944	3.340	1.044	.9580	1.2799	20
50	.2938	.2896	3.453	.3026	3.305	1.045	.9572	1.2770	10
17° 00′	.2967	.2924	3.420	.3057	3.271	1.046	.9563	1.2741	73° 00′
10	.2996	.2952	3.388	.3089	3.237	1.047	.9555	1.2712	50
20	.3025	.2979	3.357	.3121	3.204	1.048	.9546	1.2683	40
30	.3054	.3007	3.326	.3153	3.172	1.048	.9537	1.2654	30
40	.3083	.3035	3.295	.3185	3.140	1.049	.9528	1.2625	20
50	.3113	.3062	3.265	.3217	3.108	1.050	.9520	1.2595	10
18° 00′	.3142	.3090	3.236	.3249	3.078	1.051	.9511	1.2566	72° 00′
		cos θ	sec θ	cot θ	tan θ	csc θ	sin θ	Radians	Degrees
								Angle θ	

Angle θ

TABLE 3 Trigonometric Functions—Degrees and Minutes or Radians (continued)

Angle θ Degrees	Radians	sin θ	csc θ	tan θ	cot θ	sec θ	cos θ		
18° 00′	.3142	.3090	3.236	.3249	3.078	1.051	.9511	1.2566	72° 00′
10	.3171	.3118	3.207	.3281	3.047	1.052	.9502	1.2537	50
20	.3200	.3145	3.179	.3314	3.018	1.053	.9492	1.2508	40
30	.3229	.3173	3.152	.3346	2.989	1.054	.9483	1.2479	30
40	.3258	.3201	3.124	.3378	2.960	1.056	.9474	1.2450	20
50	.3287	.3228	3.098	.3411	2.932	1.057	.9465	1.2421	10
19° 00′	.3316	.3256	3.072	.3443	2.904	1.058	.9455	1.2392	71° 00′
10	.3345	.3283	3.046	.3476	2.877	1.059	.9446	1.2363	50
20	.3374	.3311	3.021	.3508	2.850	1.060	.9436	1.2334	40
30	.3403	.3338	2.996	.3541	2.824	1.061	.9426	1.2305	30
40	.3432	.3365	2.971	.3574	2.798	1.062	.9417	1.2275	20
50	.3462	.3393	2.947	.3607	2.773	1.063	.9407	1.2246	10
20° 00′	.3491	.3420	2.924	.3640	2.747	1.064	.9397	1.2217	70° 00′
10	.3520	.3448	2.901	.3673	2.723	1.065	.9387	1.2188	50
20	.3549	.3475	2.878	.3706	2.699	1.066	.9377	1.2159	40
30	.3578	.3502	2.855	.3739	2.675	1.068	.9367	1.2130	30
40	.3607	.3529	2.833	.3772	2.651	1.069	.9356	1.2101	20
50	.3636	.3557	2.812	.3805	2.628	1.070	.9346	1.2072	10
21° 00′	.3665	.3584	2.790	.3839	2.605	1.071	.9336	1.2043	69° 00′
10	.3694	.3611	2.769	.3872	2.583	1.072	.9325	1.2014	50
20	.3723	.3638	2.749	.3906	2.560	1.074	.9315	1.1985	40
30	.3752	.3665	2.729	.3939	2.539	1.075	.9304	1.1956	30
40	.3782	.3692	2.709	.3973	2.517	1.076	.9293	1.1926	20
50	.3811	.3719	2.689	.4006	2.496	1.077	.9283	1.1897	10
22° 00′	.3840	.3746	2.669	.4040	2.475	1.079	.9272	1.1868	68° 00′
10	.3869	.3773	2.650	.4074	2.455	1.080	.9261	1.1839	50
20	.3898	.3800	2.632	.4108	2.434	1.081	.9250	1.1810	40
30	.3927	.3827	2.613	.4142	2.414	1.082	.9239	1.1781	30
40	.3956	.3854	2.595	.4176	2.394	1.084	.9228	1.1752	20
50	.3985	.3881	2.577	.4210	2.375	1.085	.9216	1.1723	10
23° 00′	.4014	.3907	2.559	.4245	2.356	1.086	.9205	1.1694	67° 00′
10	.4043	.3934	2.542	.4279	2.337	1.088	.9194	1.1665	50
20	.4072	.3961	2.525	.4314	2.318	1.089	.9182	1.1636	40
30	.4102	.3987	2.508	.4348	2.300	1.090	.9171	1.1606	30
40	.4131	.4014	2.491	.4383	2.282	1.092	.9159	1.1577	20
50	.4160	.4041	2.475	.4417	2.264	1.093	.9147	1.1548	10
24° 00′	.4189	.4067	2.459	.4452	2.246	1.095	.9135	1.1519	66° 00′
10	.4218	.4094	2.443	.4487	2.229	1.096	.9124	1.1490	50
20	.4247	.4120	2.427	.4522	2.211	1.097	.9112	1.1461	40
30	.4276	.4147	2.411	.4557	2.194	1.099	.9100	1.1432	30
40	.4305	.4173	2.396	.4592	2.177	1.100	.9088	1.1403	20
50	.4334	.4200	2.381	.4628	2.161	1.102	.9075	1.1374	10
25° 00′	.4363	.4226	2.366	.4663	2.145	1.103	.9063	1.1345	65° 00′
10	.4392	.4253	2.352	.4699	2.128	1.105	.9051	1.1316	50
20	.4422	.4279	2.337	.4734	2.112	1.106	.9038	1.1286	40
30	.4451	.4305	2.323	.4770	2.097	1.108	.9026	1.1257	30
40	.4480	.4331	2.309	.4806	2.081	1.109	.9013	1.1228	20
50	.4509	.4358	2.295	.4841	2.066	1.111	.9001	1.1199	10
26° 00′	.4538	.4384	2.281	.4877	2.050	1.113	.8988	1.1170	64° 00′
10	.4567	.4410	2.268	.4913	2.035	1.114	.8975	1.1141	50
20	.4596	.4436	2.254	.4950	2.020	1.116	.8962	1.1112	40
30	.4625	.4462	2.241	.4986	2.006	1.117	.8949	1.1083	30
40	.4654	.4488	2.228	.5022	1.991	1.119	.8936	1.1054	20
50	.4683	.4514	2.215	.5059	1.977	1.121	.8923	1.1025	10
27° 00′	.4712	.4540	2.203	.5095	1.963	1.122	.8910	1.0996	63° 00′
		cos θ	sec θ	cot θ	tan θ	csc θ	sin θ	Radians	Degrees

Angle θ

TABLE 3 Trigonometric Functions—Degrees and Minutes or Radians (continued)

Degrees	Radians	$\sin \theta$	$\csc \theta$	$\tan \theta$	$\cot \theta$	$\sec \theta$	$\cos \theta$		
27° 00′	.4712	.4540	2.203	.5095	1.963	1.122	.8910	1.0996	63° 00′
10	.4741	.4566	2.190	.5132	1.949	1.124	.8897	1.0966	50
20	.4771	.4592	2.178	.5169	1.935	1.126	.8884	1.0937	40
30	.4800	.4617	2.166	.5206	1.921	1.127	.8870	1.0908	30
40	.4829	.4643	2.154	.5243	1.907	1.129	.8857	1.0879	20
50	.4858	.4669	2.142	.5280	1.894	1.131	.8843	1.0850	10
28° 00′	.4887	.4695	2.130	.5317	1.881	1.133	.8829	1.0821	62° 00′
10	.4916	.4720	2.118	.5354	1.868	1.134	.8816	1.0792	50
20	.4945	.4746	2.107	.5392	1.855	1.136	.8802	1.0763	40
30	.4974	.4772	2.096	.5430	1.842	1.138	.8788	1.0734	30
40	.5003	.4797	2.085	.5467	1.829	1.140	.8774	1.0705	20
50	.5032	.4823	2.074	.5505	1.816	1.142	.8760	1.0676	10
29° 00′	.5061	.4848	2.063	.5543	1.804	1.143	.8746	1.0647	61° 00′
10	.5091	.4874	2.052	.5581	1.792	1.145	.8732	1.0617	50
20	.5120	.4899	2.041	.5619	1.780	1.147	.8718	1.0588	40
30	.5149	.4924	2.031	.5658	1.767	1.149	.8704	1.0559	30
40	.5178	.4950	2.020	.5696	1.756	1.151	.8689	1.0530	20
50	.5207	.4975	2.010	.5735	1.744	1.153	.8675	1.0501	10
30° 00′	.5236	.5000	2.000	.5774	1.732	1.155	.8660	1.0472	60° 00′
10	.5265	.5025	1.990	.5812	1.720	1.157	.8646	1.0443	50
20	.5294	.5050	1.980	.5851	1.709	1.159	.8631	1.0414	40
30	.5323	.5075	1.970	.5890	1.698	1.161	.8616	1.0385	30
40	.5352	.5100	1.961	.5930	1.686	1.163	.8601	1.0356	20
50	.5381	.5125	1.951	.5969	1.675	1.165	.8587	1.0327	10
31° 00′	.5411	.5150	1.942	.6009	1.664	1.167	.8572	1.0297	59° 00′
10	.5440	.5175	1.932	.6048	1.653	1.169	.8557	1.0268	50
20	.5469	.5200	1.923	.6088	1.643	1.171	.8542	1.0239	40
30	.5498	.5225	1.914	.6128	1.632	1.173	.8526	1.0210	30
40	.5527	.5250	1.905	.6168	1.621	1.175	.8511	1.0181	20
50	.5556	.5275	1.896	.6208	1.611	1.177	.8496	1.0152	10
32° 00′	.5585	.5299	1.887	.6249	1.600	1.179	.8480	1.0123	58° 00′
10	.5614	.5324	1.878	.6289	1.590	1.181	.8465	1.0094	50
20	.5643	.5348	1.870	.6330	1.580	1.184	.8450	1.0065	40
30	.5672	.5373	1.861	.6371	1.570	1.186	.8434	1.0036	30
40	.5701	.5398	1.853	.6412	1.560	1.188	.8418	1.0007	20
50	.5730	.5422	1.844	.6453	1.550	1.190	.8403	.9977	10
33° 00′	.5760	.5446	1.836	.6494	1.540	1.192	.8387	.9948	57° 00′
10	.5789	.5471	1.828	.6536	1.530	1.195	.8371	.9919	50
20	.5818	.5495	1.820	.6577	1.520	1.197	.8355	.9890	40
30	.5847	.5519	1.812	.6619	1.511	1.199	.8339	.9861	30
40	.5876	.5544	1.804	.6661	1.501	1.202	.8323	.9832	20
50	.5905	.5568	1.796	.6703	1.492	1.204	.8307	.9803	10
34° 00′	.5934	.5592	1.788	.6745	1.483	1.206	.8290	.9774	56° 00′
10	.5963	.5616	1.781	.6787	1.473	1.209	.8274	.9745	50
20	.5992	.5640	1.773	.6830	1.464	1.211	.8258	.9716	40
30	.6021	.5664	1.766	.6873	1.455	1.213	.8241	.9687	30
40	.6050	.5688	1.758	.6916	1.446	1.216	.8225	.9657	20
50	.6080	.5712	1.751	.6959	1.437	1.218	.8208	.9628	10
35° 00′	.6109	.5736	1.743	.7002	1.428	1.221	.8192	.9599	55° 00′
10	.6138	.5760	1.736	.7046	1.419	1.223	.8175	.9570	50
20	.6167	.5783	1.729	.7089	1.411	1.226	.8158	.9541	40
30	.6196	.5807	1.722	.7133	1.402	1.228	.8141	.9512	30
40	.6225	.5831	1.715	.7177	1.393	1.231	.8124	.9483	20
50	.6254	.5854	1.708	.7221	1.385	1.233	.8107	.9454	10
36° 00′	.6283	.5878	1.701	.7265	1.376	1.236	.8090	.9425	54° 00′
		$\cos \theta$	$\sec \theta$	$\cot \theta$	$\tan \theta$	$\csc \theta$	$\sin \theta$	Radians	Degrees

Angle θ

TABLE 3 Trigonometric Functions—Degrees and Minutes or Radians (continued)

Angle θ Degrees	Radians	sin θ	csc θ	tan θ	cot θ	sec θ	cos θ		
36° 00′	.6283	.5878	1.701	.7265	1.376	1.236	.8090	.9425	54° 00′
10	.6312	.5901	1.695	.7310	1.368	1.239	.8073	.9396	50
20	.6341	.5925	1.688	.7355	1.360	1.241	.8056	.9367	40
30	.6370	.5948	1.681	.7400	1.351	1.244	.8039	.9338	30
40	.6400	.5972	1.675	.7445	1.343	1.247	.8021	.9308	20
50	.6429	.5995	1.668	.7490	1.335	1.249	.8004	.9279	10
37° 00′	.6458	.6018	1.662	.7536	1.327	1.252	.7986	.9250	53° 00′
10	.6487	.6041	1.655	.7581	1.319	1.255	.7969	.9221	50
20	.6516	.6065	1.649	.7627	1.311	1.258	.7951	.9192	40
30	.6545	.6088	1.643	.7673	1.303	1.260	.7934	.9163	30
40	.6574	.6111	1.636	.7720	1.295	1.263	.7916	.9134	20
50	.6603	.6134	1.630	.7766	1.288	1.266	.7898	.9105	10
38° 00′	.6632	.6157	1.624	.7813	1.280	1.269	.7880	.9076	52° 00′
10	.6661	.6180	1.618	.7860	1.272	1.272	.7862	.9047	50
20	.6690	.6202	1.612	.7907	1.265	1.275	.7844	.9018	40
30	.6720	.6225	1.606	.7954	1.257	1.278	.7826	.8988	30
40	.6749	.6248	1.601	.8002	1.250	1.281	.7808	.8959	20
50	.6778	.6271	1.595	.8050	1.242	1.284	.7790	.8930	10
39° 00′	.6807	.6293	1.589	.8098	1.235	1.287	.7771	.8901	51° 00′
10	.6836	.6316	1.583	.8146	1.228	1.290	.7753	.8872	50
20	.6865	.6338	1.578	.8195	1.220	1.293	.7735	.8843	40
30	.6894	.6361	1.572	.8243	1.213	1.296	.7716	.8814	30
40	.6923	.6383	1.567	.8292	1.206	1.299	.7698	.8785	20
50	.6952	.6406	1.561	.8342	1.199	1.302	.7679	.8756	10
40° 00′	.6981	.6428	1.556	.8391	1.192	1.305	.7660	.8727	50° 00′
10	.7010	.6450	1.550	.8441	1.185	1.309	.7642	.8698	50
20	.7039	.6472	1.545	.8491	1.178	1.312	.7623	.8668	40
30	.7069	.6494	1.540	.8541	1.171	1.315	.7604	.8639	30
40	.7098	.6517	1.535	.8591	1.164	1.318	.7585	.8610	20
50	.7127	.6539	1.529	.8642	1.157	1.322	.7566	.8581	10
41° 00′	.7156	.6561	1.524	.8693	1.150	1.325	.7547	.8552	49° 00′
10	.7185	.6583	1.519	.8744	1.144	1.328	.7528	.8523	50
20	.7214	.6604	1.514	.8796	1.137	1.332	.7509	.8494	40
30	.7243	.6626	1.509	.8847	1.130	1.335	.7490	.8465	30
40	.7272	.6648	1.504	.8899	1.124	1.339	.7470	.8436	20
50	.7301	.6670	1.499	.8952	1.117	1.342	.7451	.8407	10
42° 00′	.7330	.6691	1.494	.9004	1.111	1.346	.7431	.8378	48° 00′
10	.7359	.6713	1.490	.9057	1.104	1.349	.7412	.8348	50
20	.7389	.6734	1.485	.9110	1.098	1.353	.7392	.8319	40
30	.7418	.6756	1.480	.9163	1.091	1.356	.7373	.8290	30
40	.7447	.6777	1.476	.9217	1.085	1.360	.7353	.8261	20
50	.7476	.6799	1.471	.9271	1.079	1.364	.7333	.8232	10
43° 00′	.7505	.6820	1.466	.9325	1.072	1.367	.7314	.8203	47° 00′
10	.7534	.6841	1.462	.9380	1.066	1.371	.7294	.8174	50
20	.7563	.6862	1.457	.9435	1.060	1.375	.7274	.8145	40
30	.7592	.6884	1.453	.9490	1.054	1.379	.7254	.8116	30
40	.7621	.6905	1.448	.9545	1.048	1.382	.7234	.8087	20
50	.7650	.6926	1.444	.9601	1.042	1.386	.7214	.8058	10
44° 00′	.7679	.6947	1.440	.9657	1.036	1.390	.7193	.8029	46° 00′
10	.7709	.6967	1.435	.9713	1.030	1.394	.7173	.7999	50
20	.7738	.6988	1.431	.9770	1.024	1.398	.7153	.7970	40
30	.7767	.7009	1.427	.9827	1.018	1.402	.7133	.7941	30
40	.7796	.7030	1.423	.9884	1.012	1.406	.7112	.7912	20
50	.7825	.7050	1.418	.9942	1.006	1.410	.7092	.7883	10
45° 00′	.7854	.7071	1.414	1.000	1.000	1.414	.7071	.7854	45° 00′
		cos θ	sec θ	cot θ	tan θ	csc θ	sin θ	Radians	Degrees
								Angle θ	

TABLE 4   Common Logarithms

N	0	1	2	3	4	5	6	7	8	9
1.0	.0000	.0043	.0086	.0128	.0170	.0212	.0253	.0294	.0334	.0374
1.1	.0414	.0453	.0492	.0531	.0569	.0607	.0645	.0682	.0719	.0755
1.2	.0792	.0828	.0864	.0899	.0934	.0969	.1004	.1038	.1072	.1106
1.3	.1139	.1173	.1206	.1239	.1271	.1303	.1335	.1367	.1399	.1430
1.4	.1461	.1492	.1523	.1553	.1584	.1614	.1644	.1673	.1703	.1732
1.5	.1761	.1790	.1818	.1847	.1875	.1903	.1931	.1959	.1987	.2014
1.6	.2041	.2068	.2095	.2122	.2148	.2175	.2201	.2227	.2253	.2279
1.7	.2304	.2330	.2355	.2380	.2405	.2430	.2455	.2480	.2504	.2529
1.8	.2553	.2577	.2601	.2625	.2648	.2672	.2695	.2718	.2742	.2765
1.9	.2788	.2810	.2833	.2856	.2878	.2900	.2923	.2945	.2967	.2989
2.0	.3010	.3032	.3054	.3075	.3096	.3118	.3139	.3160	.3181	.3201
2.1	.3222	.3243	.3263	.3284	.3304	.3324	.3345	.3365	.3385	.3404
2.2	.3424	.3444	.3464	.3483	.3502	.3522	.3541	.3560	.3579	.3598
2.3	.3617	.3636	.3655	.3674	.3692	.3711	.3729	.3747	.3766	.3784
2.4	.3802	.3820	.3838	.3856	.3874	.3892	.3909	.3927	.3945	.3962
2.5	.3979	.3997	.4014	.4031	.4048	.4065	.4082	.4099	.4116	.4133
2.6	.4150	.4166	.4183	.4200	.4216	.4232	.4249	.4265	.4281	.4298
2.7	.4314	.4330	.4346	.4362	.4378	.4393	.4409	.4425	.4440	.4456
2.8	.4472	.4487	.4502	.4518	.4533	.4548	.4564	.4579	.4594	.4609
2.9	.4624	.4639	.4654	.4669	.4683	.4698	.4713	.4728	.4742	.4757
3.0	.4771	.4786	.4800	.4814	.4829	.4843	.4857	.4871	.4886	.4900
3.1	.4914	.4928	.4942	.4955	.4969	.4983	.4997	.5011	.5024	.5038
3.2	.5051	.5065	.5079	.5092	.5105	.5119	.5132	.5145	.5159	.5172
3.3	.5185	.5198	.5211	.5224	.5237	.5250	.5263	.5276	.5289	.5302
3.4	.5315	.5328	.5340	.5353	.5366	.5378	.5391	.5403	.5416	.5428
3.5	.5441	.5453	.5465	.5478	.5490	.5502	.5514	.5527	.5539	.5551
3.6	.5563	.5575	.5587	.5599	.5611	.5623	.5635	.5647	.5658	.5670
3.7	.5682	.5694	.5705	.5717	.5729	.5740	.5752	.5763	.5775	.5786
3.8	.5798	.5809	.5821	.5832	.5843	.5855	.5866	.5877	.5888	.5899
3.9	.5911	.5922	.5933	.5944	.5955	.5966	.5977	.5988	.5999	.6010
4.0	.6021	.6031	.6042	.6053	.6064	.6075	.6085	.6096	.6107	.6117
4.1	.6128	.6138	.6149	.6160	.6170	.6180	.6191	.6201	.6212	.6222
4.2	.6232	.6243	.6253	.6263	.6274	.6284	.6294	.6304	.6314	.6325
4.3	.6335	.6345	.6355	.6365	.6375	.6385	.6395	.6405	.6415	.6425
4.4	.6435	.6444	.6454	.6464	.6474	.6484	.6493	.6503	.6513	.6522
4.5	.6532	.6542	.6551	.6561	.6571	.6580	.6590	.6599	.6609	.6618
4.6	.6628	.6637	.6646	.6656	.6665	.6675	.6684	.6693	.6702	.6712
4.7	.6721	.6730	.6739	.6749	.6758	.6767	.6776	.6785	.6794	.6803
4.8	.6812	.6821	.6830	.6839	.6848	.6857	.6866	.6875	.6884	.6893
4.9	.6902	.6911	.6920	.6928	.6937	.6946	.6955	.6964	.6972	.6981
5.0	.6990	.6998	.7007	.7016	.7024	.7033	.7042	.7050	.7059	.7067
5.1	.7076	.7084	.7093	.7101	.7110	.7118	.7126	.7135	.7143	.7152
5.2	.7160	.7168	.7177	.7185	.7193	.7202	.7210	.7218	.7226	.7235
5.3	.7243	.7251	.7259	.7267	.7275	.7284	.7292	.7300	.7308	.7316
5.4	.7324	.7332	.7340	.7348	.7356	.7364	.7372	.7380	.7388	.7396
N		1	2	3	4	5	6	7	8	9

TABLE 4   Common Logarithms (continued)

N	0	1	2	3	4	5	6	7	8	9
5.5	.7404	.7412	.7419	.7427	.7435	.7443	.7451	.7459	.7466	.7474
5.6	.7482	.7490	.7497	.7505	.7513	.7520	.7528	.7536	.7543	.7551
5.7	.7559	.7566	.7574	.7582	.7589	.7597	.7604	.7612	.7619	.7627
5.8	.7634	.7642	.7649	.7657	.7664	.7672	.7679	.7686	.7694	.7701
5.9	.7709	.7716	.7723	.7731	.7738	.7745	.7752	.7760	.7767	.7774
6.0	.7782	.7789	.7796	.7803	.7810	.7818	.7825	.7832	.7839	.7846
6.1	.7853	.7860	.7868	.7875	.7882	.7889	.7896	.7903	.7910	.7917
6.2	.7924	.7931	.7938	.7945	.7952	.7959	.7966	.7973	.7980	.7987
6.3	.7993	.8000	.8007	.8014	.8021	.8028	.8035	.8041	.8048	.8055
6.4	.8062	.8069	.8075	.8082	.8089	.8096	.8102	.8109	.8116	.8122
6.5	.8129	.8136	.8142	.8149	.8156	.8162	.8169	.8176	.8182	.8189
6.6	.8195	.8202	.8209	.8215	.8222	.8228	.8235	.8241	.8248	.8254
6.7	.8261	.8267	.8274	.8280	.8287	.8293	.8299	.8306	.8312	.8319
6.8	.8325	.8331	.8338	.8344	.8351	.8357	.8363	.8370	.8376	.8382
6.9	.8388	.8395	.8401	.8407	.8414	.8420	.8426	.8432	.8439	.8445
7.0	.8451	.8457	.8463	.8470	.8476	.8482	.8488	.8494	.8500	.8506
7.1	.8513	.8519	.8525	.8531	.8537	.8543	.8549	.8555	.8561	.8567
7.2	.8573	.8579	.8585	.8591	.8597	.8603	.8609	.8615	.8621	.8627
7.3	.8633	.8639	.8645	.8651	.8657	.8663	.8669	.8675	.8681	.8686
7.4	.8692	.8698	.8704	.8710	.8716	.8722	.8727	.8733	.8739	.8745
7.5	.8751	.8756	.8762	.8768	.8774	.8779	.8785	.8791	.8797	.8802
7.6	.8808	.8814	.8820	.8825	.8831	.8837	.8842	.8848	.8854	.8859
7.7	.8865	.8871	.8876	.8882	.8887	.8893	.8899	.8904	.8910	.8915
7.8	.8921	.8927	.8932	.8938	.8943	.8949	.8954	.8960	.8965	.8971
7.9	.8976	.8982	.8987	.8993	.8998	.9004	.9009	.9015	.9020	.9025
8.0	.9031	.9036	.9042	.9047	.9053	.9058	.9063	.9069	.9074	.9079
8.1	.9085	.9090	.9096	.9101	.9106	.9112	.9117	.9122	.9128	.9133
8.2	.9138	.9143	.9149	.9154	.9159	.9165	.9170	.9175	.9180	.9186
8.3	.9191	.9196	.9201	.9206	.9212	.9217	.9222	.9227	.9232	.9238
8.4	.9243	.9248	.9253	.9258	.9263	.9269	.9274	.9279	.9284	.9289
8.5	.9294	.9299	.9304	.9309	.9315	.9320	.9325	.9330	.9335	.9340
8.6	.9345	.9350	.9355	.9360	.9365	.9370	.9375	.9380	.9385	.9390
8.7	.9395	.9400	.9405	.9410	.9415	.9420	.9425	.9430	.9435	.9440
8.8	.9445	.9450	.9455	.9460	.9465	.9469	.9474	.9479	.9484	.9489
8.9	.9494	.9499	.9504	.9509	.9513	.9518	.9523	.9528	.9533	.9538
9.0	.9542	.9547	.9552	.9557	.9562	.9566	.9571	.9576	.9581	.9586
9.1	.9590	.9595	.9600	.9605	.9609	.9614	.9619	.9624	.9628	.9633
9.2	.9638	.9643	.9647	.9652	.9657	.9661	.9666	.9671	.9675	.9680
9.3	.9685	.9689	.9694	.9699	.9703	.9708	.9713	.9717	.9722	.9727
9.4	.9731	.9736	.9741	.9745	.9750	.9754	.9759	.9763	.9768	.9773
9.5	.9777	.9782	.9786	.9791	.9795	.9800	.9805	.9809	.9814	.9818
9.6	.9823	.9827	.9832	.9836	.9841	.9845	.9850	.9854	.9859	.9863
9.7	.9868	.9872	.9877	.9881	.9886	.9890	.9894	.9899	.9903	.9908
9.8	.9912	.9917	.9921	.9926	.9930	.9934	.9939	.9943	.9948	.9952
9.9	.9956	.9961	.9965	.9969	.9974	.9978	.9983	.9987	.9991	.9996
N	0	1	2	3	4	5	6	7	8	9

Answers to
Odd-Numbered Exercises

EXERCISES 1.1

1. $\{1, 2, 3, 4, 5, 6, 7, 8\}$ **3.** $\{\ \} = \emptyset$
5. $\{1, 2, 3, 4, 5, 6, 8\}$ **7.** $\{\ \} = \emptyset$
9. True. Every $x \in A \cap B = \emptyset$ is also an element of
$A \cap C = \{2, 4\}$.   **11.** $\{2, 4, 6, 8, 10, \ldots\}$
13. $\{7, 9, 11, 13, 15, \ldots\}$
15. rational, $0.78888\ldots$, with 8 repeating
17. irrational, $1.57079632\ldots$, no repeating pattern
19. irrational, $1.32287656\ldots$, no repeating pattern
21. rational, $0.36363636\ldots$, with 36 repeating
23. irrational, $3.87298334\ldots$, no repeating pattern
25. $0.142857\overline{142857}\ldots$
27. $0.876543209\overline{876543209}\ldots$
29. $\frac{1}{2}$ **31.** $\frac{7}{8}$ **33.** $\frac{2}{5}$ **35.** -39 **37.** 11
39. $\frac{1}{4}$ **41.** $1\frac{2}{3}$ **43.** $x = 4$ **45.** $x = \frac{1}{2}$
47. $y = 5$ **49.** $x = -3$ **51.** $x = \frac{2}{3}$ **53.** $y = 0$
55. $t = 2a/g$ **57.** $m = E/c^2$ **59.** $l = (P - 2w)/2$
61. $r = (s - a)/s$ **63.** -1 **65.** -1 **67.** 9
69. 7 **71.** -25 **73.** $\pi - \sqrt{3}$ **75.** $\pi - 3$
77. $r = 24/\pi$ **79.** $B = 10$
81. Start with $a \cdot 0 = a[1 + (-1)]$.
87. $3.14 < \pi < \frac{22}{7}$

EXERCISES 1.2

1. x^3y^3 **3.** $-x^5y^3$ **5.** $72x^5y^2$ **7.** $\left(\frac{8}{27}\right)y^3 = 8y^3/27$
9. $8a^6b^3$ **11.** -5 **13.** $8/(xy)$ **15.** $27/(ab^4)$
17. $-y^9/(27x^6)$ **19.** $y^{10}/(32x^{15})$ **21.** $-128x^{13}/y^{12}$
23. $-\left(\frac{1}{3}\right)a^2b^6$ **25.** x^3y^2 **27.** x^4/y^3 **29.** $3x^3y$
31. $2x^2/y$ **33.** $x^3y^{1/4}$ **35.** $x^3y^{3/4}$ **37.** $4x^2y^4$
39. $xy^{5/6}$ **41.** $5\sqrt{2}$ **43.** $3\sqrt[3]{2}$ **45.** $-2\sqrt[3]{3}$
47. $6\sqrt{5}$ **49.** $4\sqrt[3]{2}$ **51.** $6\sqrt{2}$ **53.** $343xy^2$
55. $x\sqrt[3]{4y}$ **57.** x **59.** $\left(\frac{5}{7}\right)\sqrt{5} = (5\sqrt{5})/7$

61. $\left(\frac{1}{3}\right)\sqrt{6} = \sqrt{6}/3$ **63.** $\left(\frac{1}{4}\right)\sqrt{6} = \sqrt{6}/4$
65. $\left(\frac{1}{2}\right)\sqrt{2} = \sqrt{2}/2$ **67.** $\sqrt[3]{3}$ **69.** $\left(\frac{5}{2}\right)\sqrt[5]{2} = (5\sqrt[5]{2})/2$
71. $\left(\frac{7}{3}\right)\sqrt[4]{3} = (7\sqrt[4]{3})/3$ **73.** $\sqrt{2xy}/|y|$
75. 7.61×10^7 **77.** 3.23×10^{-5} **79.** 3×10^9
81. 9.46×10^{15} meters   **83.** 2.1×10^8 hairs
85. 2.5×10^{-7}, 5×10^{-12} meters
87. 1.6×10^{11} meters, 1.7×10^{-5} lt.-yr.
89. $r = 304.5$ km

EXERCISES 1.3

1. $x^4 + 3x^3 - 5x^2 + 4x - 1$
3. $-x^4 + 3x^3 + x^2 - 2x - 9$
5. $2y^5 + y^4 - 6y^3 - 4y^2 + 3$ **7.** $x^6 - y^6$
9. $x^4 - 1$ **11.** $9y^2 - 12y + 4$
13. $27t^3 + 54t^2 + 36t + 8$
15. $x^6 - 3x^4y^2 + 3x^2y^4 - y^6$ **17.** $x^2 + 2xy^2 + y^4$
19. $x^2 - y^4$
21. $x^3 + 2x^2y + 2xy^2 + y^3$ **23.** $5xy - y^2 + 3x$
25. $8xy\left(x - \frac{3}{4}y + \frac{1}{8}\right)$ **27.** $3x^{-2}\left(x^6 - \frac{5}{3}x^4 - 3x^2 + 4\right)$
29. $7y^{-3}\left(y^6 - \frac{1}{7}y^4 + 2y^2 - 3\right)$
31. $(x + 1)^{-1}[3(x + 1)^4 - 2(x + 1)^2 + 5]$
33. $(2x - y)(4x^2 + 2xy + y^2)$ **35.** $(x^3 - y)(x^3 + y)$
37. $(x + 3y)(x^2 - 3xy + 9y^2)$ **39.** $(x + 3)^2$
41. $5(x - 2)^2$ **43.** $(x + 3)(x^2 + 2)$
45. $(x - 1)(x + 1)(x - y)(x + y)$ **47.** $(x + 3)/(x - 2)$
49. 1 **51.** $5(x + 2)$ **53.** $(y - 1)/(y + 2)$
55. $1/(x - 1)$ **57.** $-\frac{33}{4} = -8\frac{1}{4}$ **59.** $(x + y)^2/(xy)$
61. $(x + y)/(x - y)$ **63.** $xy/(y - x)$
65. $-(2x + h)/[x^2(x + h)]$ **67.** $-1/[2x(2x + h)]$
69. $15\frac{5}{13}$ hours   **71.** $10:45$ **73.** 3 mph

EXERCISES 1.4

1. $2 - \sqrt{3}$　**3.** $\sqrt{11} - \pi$　**5.** $x - 3$　**7.** $x = \pm 5$
9. $x \in \{10, -4\}$　**11.** $x \in \left\{-\frac{3}{2}, \frac{9}{2}\right\}$　**13.** $x \in \emptyset$
15. $x \in \left\{\frac{1}{5}, 3\right\}$

19. $x > 18$, ⟶ (18, ∞)

21. $x < \frac{5}{3}$, ⟵ $\left(-\infty, \frac{5}{3}\right)$

23. $\frac{2}{3} \le x \le 5$, ⟶ $\left[\frac{2}{3}, 5\right]$

25. $-3 < x < 1$, ⟶ $(-3, 1)$

27. $-6 < x < 2$, ⟶ $(-6, 2)$

29. $-4 \le x \le 4$, ⟶ $[-4, 4]$

31. $x \le -17$ or $x \ge 17$, ⟶ ,
$(-\infty, -17] \cup [17, \infty)$
33. $x \in \emptyset$　**35.** $x \in R$
37. $-10 \le x \le 14$, ⟶ , $[-10, 14]$

39. $-1 \le x \le \frac{1}{3}$, ⟶ , $\left[-1, \frac{1}{3}\right]$

41. $x < -\frac{5}{2}$ or $x > \frac{5}{2}$, $\left(-\infty, -\frac{5}{2}\right) \cup \left(\frac{5}{2}, \infty\right)$ ⟶

43. $x < -2$ or $x > 4$, $(-\infty, 2) \cup (4, \infty)$ ⟶

45. $x < -4$ or $x > 2$, ⟶ , $(-\infty, -4) \cup (2, \infty)$

47. $(1/5) - (\varepsilon/10) < x < (1/5) + (\varepsilon/10)$,

⟶ , $(1/5 - \varepsilon/10, 1/5 + \varepsilon/10)$

49. $x > -1$, ⟶ , $(-1, \infty)$

51. $x < -3$, ⟵ , $(-\infty, -3)$

53. $x > 5$, ⟶ , $(5, \infty)$

55. $x < -1$, ⟵ , $(-\infty, -1)$

57. $x > -7$, ⟶ , $(-7, \infty)$

59. $|x - 5| \le 4$　**61.** $|x - 1| < 3$　**63.** $\left|x - \left(\frac{3}{2}\right)\right| < \frac{7}{2}$

65. $|x + 2| < \varepsilon$　**67.** $|x - 5| < \delta$
69. less than or equal to \$10,000　**71.** greater than 30
pounds　**73.** greater than 10 and less than 15
75. $\pi(4.7)^2$ m² $< A < \pi(5.1)^2$ m²
77. $\left(\frac{4}{3}\right)\pi(2.8)^3$ m³ $< V < \left(\frac{4}{3}\right)\pi(3.2)^3$ m³
79. $4.8\pi(2.4 + h)$ in.² $< S < 5.2\pi(2.6 - h)$ in.²

EXERCISES 1.5

1. $\{-5, 1\}$　**3.** $\{-7, 2\}$　**5.** $\{(-1 \pm \sqrt{29})/2\}$
7. $\{4, 5\}$　**9.** $\{-2, -1, 2\}$　**11.** $\{-1, 2\}$
13. $\{-1, 1, -3\}$　**15.** $\{1, -\sqrt{5}, \sqrt{5}\}$　**17.** $[-7, 2]$
19. $(-\infty, 4] \cup [5, \infty)$　**21.** $(-\infty, -1)$　**23.** $[-1, 3]$
25. $(-1, 5)$　**27.** $[-1, 1] \cup (3, \infty)$　**29.** $\left(1, \frac{7}{2}\right)$
31. $(-\infty, -2) \cup \left[-\frac{7}{5}, \infty\right)$　**33.** $(-\infty, -5) \cup \left[-\frac{24}{5}, \infty\right)$
35. $(-1, 0)$　**37.** $(-\infty, -2) \cup (-1, \infty)$
39. $\left[1\frac{4}{5}, 2\right) \cup \left(2, 2\frac{1}{5}\right)$　**41.** $[0, 5) \cup (5, 10]$
43. $(8, \infty)$　**45.** $[7, \infty)$　**47.** $(40, 260)$
49. Complete the square for $ax^2 + bx + c = 0$.

EXERCISES 1.6

1. $-6\sqrt{5}$　**3.** $\sqrt{5}$　**5.** $-2 - 2\sqrt{2}$
7. $4\sqrt[3]{x} - x\sqrt[3]{2}$　**9.** $i\sqrt{7}$　**11.** $3i$　**13.** $9i$
15. $3i\sqrt{2}$　**17.** $4i\sqrt{2}$　**19.** $3i\sqrt{5}$
21. i　**23.** 1
25. i　**27.** $10 - 5\sqrt{5} + 2\sqrt{2} - \sqrt{10}$
29. $37 - 12\sqrt{2}$　**31.** $9 - 2\sqrt{14}$　**33.** -2
35. $2 - 9i$　**37.** 2　**39.** $9 + 9i$　**41.** $15 + 35i$
43. $15 + 6i$　**45.** $8 - i$　**47.** 13　**49.** $(3 \pm i\sqrt{7})/2$
51. $1 \pm i\sqrt{3}$　**53.** $(1 \pm i\sqrt{15})/4$　**55.** $(1 - i)/2$
57. $(3 + i)/10$　**59.** $(2 - 3i)/13$　**61.** $(3 \pm 4i)/5$
63. $(-1 - \sqrt{3})/2$　**65.** $-\sqrt{2} - \sqrt{3}$
67. $(\sqrt{5} + 3 - \sqrt{3} - \sqrt{15})/2$

EXERCISES 1.7

1. $m = -\frac{1}{3}$　　　**3.** $m = -1.818\ldots$

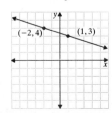

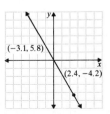

5.

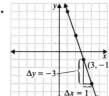

7.

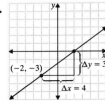

1.

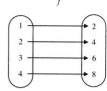

3.

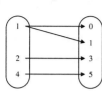

9. $y = \left(\frac{1}{2}\right)x + 1$ **11.** $y = \left(\frac{3}{2}\right)x - 3$ **13.** $y = -4$

15. $y = \left(-\frac{1}{3}\right)x + 2$ **17.** $y = -2x - 1$, $y = \left(\frac{1}{2}\right)x - \left(\frac{7}{2}\right)$

19. Use the slopes of the diagonals.   **23.** $x = 5$, $y = 2$

25. $y = 4$, $x = 16$ **27.** $w = 8$, $y = \pm 2$, $x = 3$

29. x varies inversely with y, and 5 is the constant of proportionality.   **31.** x is directly proportional to y^2, and 3 is the constant of proportionality.   **33.** x varies jointly with y^2 and w^3, and $\frac{1}{7}$ is the constant of proportionality.

35. x varies jointly with w and y^2, and the constant of proportionality is $\frac{1}{3}$.   **37.** $I = (kV)/R$

39. $G = 1/R$, $G = 0.01333\dots$ siemen

41. The new pitch is $\sqrt{3}$ times the original pitch.

43. The new intensity is $\frac{1}{4}$ the original intensity.

45. $I = nE/(R_0 + nb)$

CHAPTER REVIEW EXERCISES

1. $\{3, 4, 5, 6, 7\}$ **3.** no, since $6 \in A$ and $6 \notin B$

5. $\{3, 4, 5, 6, 7\}$ **7.** $W = (P - 2L)/2$

9. $-8/(27x^6y^3)$ **11.** x^9/y^6 **13.** xy^2

15. $27xy^3$ **17.** $\left(\frac{1}{2}\right)\sqrt[3]{4} = \sqrt[3]{4}/2$ **19.** 7.5×10^{-8}

21. $y^4 + y^3 + 2y^2 - y + 1$ **23.** $25x^2 - 49$

25. $(x - 2y)(x^2 + 2xy + 4y^2)$ **27.** $2x - 1$

29. $-5 < x < 1$, $\qquad$, $(-5, 1)$

31. $-2 < x < 8$, $\qquad$, $(-2, 8)$

33. $x > -3$, $\qquad$, $(-3, \infty)$

35. $(5 \pm \sqrt{33})/2$ **37.** $\{-6, -3, 3\}$

39. $(-\infty, -5) \cup \left(-3\frac{1}{3}, \infty\right)$ **41.** $3i$ **43.** $2 \pm i$

45. $11 + 3i$ **47.** $(5 + \sqrt{2})/23$ **49.** 7.5 hours

51. $x > 13$ inches   **53.** $y = 2x - 1$ **55.** $y = \frac{1}{2}x + 4$

EXERCISES 2.1

1. (a) $\{1, 2, 3, 4\}$ (b) $\{2, 4, 6, 8\}$ (c) yes (d) yes

3. (a) $\{1, 2, 4\}$ (b) $\{0, 1, 3, 5\}$ (c) no (d) no

5. (a) $\{1, 2, 3, 4\}$ (b) $\{1, 2\}$ (c) yes (d) no

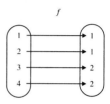

7. $f = \{(2, 5), (4, 5)\}$

9. $h = \{(2, 1), (4, 2), (6, 3), (8, 4), (10, 5), (12, 6), (14, 7), \dots\}$ **11.** $r = \{(0, 0), (\pm 1, 1), (\pm 2, 2), (\pm 3, 3), (\pm 4, 4), (\pm 5, 5), \dots\}$ **13.** (a) f, h, r (b) h

15. (a) x^2 (b) 1 (c) 4 (d) $\frac{1}{4}$ (e) $(x + 3)^2$ (f) $2x + h$

17. (a) 3 (b) -17 (c) $\frac{1}{2}$ (d) $5h - 2$ (e) $-5h + 2$ (f) $-5h - 2$ (g) $(5/x) - 2$ (h) $5x + 5h - 2$ (i) 5

19. (a) 1 (b) $-\frac{1}{3}$ (c) 2 (d) $1/h$ (e) $-1/h$ (f) $-1/h$ (g) x (h) $1/(x + h)$ (i) $-1/[x(x + h)]$

21. (a) 2 (b) 18 (c) $\frac{1}{2}$ (d) $2h^2$ (e) $-2h^2$ (f) $2h^2$ (g) $2/x^2$ (h) $2x^2 + 4xh + 2h^2$ (i) $4x + 2h$ **23.** (a) $C/2\pi$ (b) $\sqrt{A\pi}/\pi$ **25.** (a) $\sqrt{3\pi hV}/3\pi h$ (b) $h = 3V/\pi r^2$

27. domain $= \{V : V > 0\}$, range $= \{P : P > 0\}$

29. $c_1c_2/(c_1 + c_2)$ **31.** $n(E - bI)/I$

33. $H_S = AH_T - H_M$ **35.** (a) $\left(-\infty, \frac{4}{5}\right)$ (b) $[0, \infty)$

37. (a) $(-\infty, -1) \cup (-1, \infty)$ (b) $(-\infty, -1) \cup (-1, \infty)$

39. (a) $(-\infty, 2) \cup (2, \infty)$ (b) $\left(-\infty, \frac{1}{2}\right) \cup \left(\frac{1}{2}, \infty\right)$

41. (a) $(-\infty, 1) \cup (1, \infty)$ (b) $(0, 1]$

43. (a) $(-\infty, \infty)$ (b) $(-\infty, 0) \cup (0, \infty)$

45. (a) $(-\infty, -3) \cup (3, \infty)$ (b) $(0, \infty)$

47. $A = y\sqrt{144 - y^2}$ **49.** $A = \left[2 + \left(\frac{\pi}{8}\right)\right]x^2$

51. $A = 200r - \left(2 + \frac{\pi}{2}\right)r^2$

EXERCISES 2.2

1. domain $= \{-1, 0, 1, 2, 3, 4\}$, range $= \{-3, -1, 1, 2, 3\}$

3. domain $= \{-2, -1, 0, 1, 2, 3\}$, range $= \{-1, 0, 1, 2, 3, 4\}$ **5.** (a) $[-2, \infty)$ (b) $(-\infty, \infty)$

7. (a) $(-\infty, \infty)$ (b) $[-3, \infty)$ **9.** (a) $(-\infty, \infty)$ (b) $(-\infty, \infty)$

11. (a) $(-\infty, \infty)$ **(b)** $(-\infty, 4]$

13. (a) $[-1, \infty)$ **(b)** $(-\infty, \infty)$

15. (a) $(-\infty, \infty)$ **(b)** $(-\infty, 2)$

17. (a) F from 1 and 9, H from 3, 7, 11, and 15 **(b)** none

19. H from 7 **(a)** increasing on $[1, \infty)$ **(b)** decreasing on $(-\infty, 1]$; F from 9 **(a)** increasing on $(-\infty, -3)$ and on $[1, \infty)$ **(b)** decreasing on $[-3, 1]$; H from 15 **(a)** increasing on $(-\infty, -1]$ **((b))** decreasing on $[-1, \infty)$ **21.** $f(x) = 2x + 1$

23. $h(x) = -7x + 19$

25. $f(x) \approx 0.628x + 0.196$

27. $f(x) = -3x - 1$

29. $g(x) = x - 4$

31. (a) $f(x) = -2x - 1$ **(b)** $g(x) = \left(\frac{1}{2}\right)x - \left(\frac{7}{2}\right)$

33. (a) $f(x) = \left(\frac{2}{3}\right)x + 4$ **(b)** $g(x) = \left(-\frac{3}{2}\right)x + \left(\frac{13}{2}\right)$

35. $d(t) = 300 - 121t$

37. $y = \left(\frac{4}{7}\right)x$, an increasing linear function

39. $A(r) = 2\pi(r^2 - 10r + 50)$

41. $V(x) = 20x - 40x^2$, not a linear function

43. $a = (-1 + \sqrt{5})b/2$, $a/b = (-1 + \sqrt{5})/2$

45. $P = 0.88/C$, decreasing

EXERCISES 2.3

1. (a) $(-\infty, \infty)$ **(b)** $[0, \infty)$ **3. (a)** $(-\infty, \infty)$ **(b)** $(-\infty, \infty)$

5. (a) $(-\infty, \infty)$ **(b)** $\{0, \pm 2, \pm 4, \pm 6, \ldots\}$, set notation is more appropriate

7.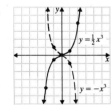

9.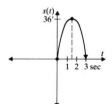

11.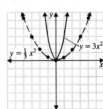

13.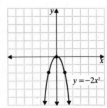

15.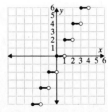

17.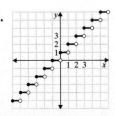

19.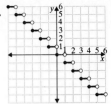

21.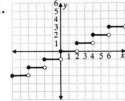

23.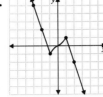

25.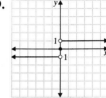

27.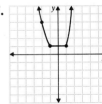

29.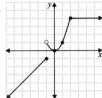

31. $y = |x|$ **33.** $y = x^3$

35. $y = \begin{cases} x^2 & \text{for } x \le 0 \\ 0 & \text{for } x > 0 \end{cases}$

37. domain: $[0, 3]$, range: $[0, 36]$

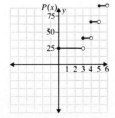

39. $p(x) = \begin{cases} 25\cent & \text{for } 0 \le x < 3 \\ (25 + 20\,[\![x - 2]\!])\cent & \text{for } x \ge 3 \end{cases}$

41. $t(x) = 2 + \left[\!\left[\left(\frac{1}{2}\right)x\right]\!\right]$

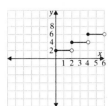

43. $V(x) = 20x - 40x^2$, a quadratic function, $\left(0, \frac{1}{2}\right)$

45. $i(x) = \begin{cases} 0.01825x & 0 < x \le 5000 \\ \$91.25 + 0.014x & \text{for } x > 5000 \end{cases}$

47. $d(t) = \sqrt{400 - t^2}$ where t is time, not a quadratic function

EXERCISES 2.4

1. domain: $\{x : x \le 1\}$,
range: $\{y : y > 1\}$

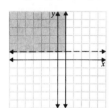

3. domain: $\{x : -2 < x < 3\}$,
range: $\{y : y \le 2\}$

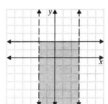

5. domain: $\{x : -3 \le x \le 3\}$,
range: $\{y : y < -2 \text{ or } y > 2\}$

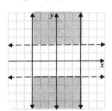

7. domain: $\{x : x \ge 3 \text{ or } x \le -3\}$,
range: $\{y : y > 1 \text{ or } y < -1\}$

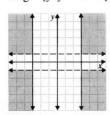

9. domain: $\{x : -2 < x < 2\}$,
range: $\{y : -2 < y < 2\}$

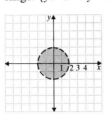

11. domain: $\{x : x \in R\}$,
range: $\{y : y \in R\}$

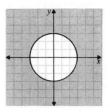

13. domain: $[-4, 4]$,
range: $[-4, 4]$

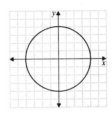

15. domain: $[-1, 5]$,
range: $[-2, 4]$

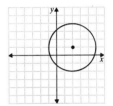

17. domain: $[-7, -1]$,
range: $[-4, 4]$

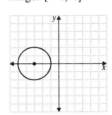

19. domain: $[-2, 4]$,
range: $[-2, 4]$

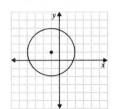

21. domain: $[-4, 2]$,
range: $[-1, 5]$

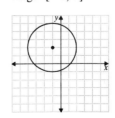

23. domain: $[-4, 2]$,
range: $[-2, 4]$

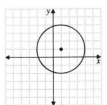

25. $(x + 1)^2 + (y - 4)^2 = 10$
27. $(x - 2)^2 + (y + 3)^2 = 9$ **29.** $(0, -1 \pm 2\sqrt{10})$
31. x-axis, y-axis, origin   **33.** y-axis   **35.** origin
37. x-axis, y-axis, origin, $x = y$ **39.** y-axis   **41.** none
43. For $-15 \le x \le 0$, $F(x) = \left(-\frac{9}{40}\right)(x^2 - 10x - 375)$ and
$G(x) = \left(\frac{1}{45}\right)(x^2 - 225)$
45. For $0 \le x \le 64$, $G(x) = -\left(\frac{1}{512}\right)x^2 + 8$

49. $y = -\sqrt{a^2 - x^2}$, domain: $\{x: -a \le x \le a\}$, range: $\{y: -a \le y \le 0\}$ **55.** even **57.** even **59.** odd **61.** even **63.** neither

EXERCISES 2.5

1. $h = 4$, $k = -4$, $(x - 4)^2 + (y + 4)^2 = r^2$
3. $h = -1$, $k = -5$, $y + 5 = \left(\frac{1}{4}\right)(x + 1)^2$
5. $h = -1$, $k = 5$, $y - 5 = \left(\frac{1}{2}\right)(x + 1)^3$
7. $h = 3$, $k = 1$, $y - 1 = |x - 3|$
9. $h = -2$, $k = -1$, $y + 1 = f(x + 2)$
11. $h = 2$, $k = -2$, $y + 2 = 1/(x - 2)$

13.

15.

17.

19.

21.

23.

25.

27.

29.

31.

33.

35.

37. $f(x) = |x| - 2$ **39.** $w_1(x) = 8 + 240x - 16x^2$
41. $f_2(x) = -x^2 + 12x - 24$ for $4 \le x \le 8$
$f_3(x) = -x^2 + 20x - 96$ for $8 \le x \le 12$
$f_4(x) = -x^2 + 28x - 192$ for $12 \le x \le 16$
43. $y = \left(\frac{1}{100}\right)(x - 1)^2 + \left(\frac{1}{2}\right)$

45.

47.

49.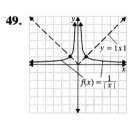

51.

53.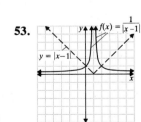

55.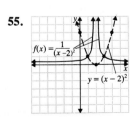

57.

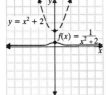

EXERCISES 2.6

1. {(1, 4), (2, 7), (3, 6), (5, 8)}
3. {(0, 5), (1, 0), (2, 1), (3, 4)}
5. (a) $5x - 1$ **(b)** $-x + 3$ **(c)** $6x^2 - x - 2$
(d) $(2x + 1)/(3x - 2)$ **(e)** $6x - 3$ **(f)** $6x + 1$
7. (a) $x^2 + 5x + 5$ **(b)** $-x^2 + 5x + 7$
(c) $5x^3 + 6x^2 - 5x - 6$ **(d)** $(5x + 6)/(x^2 - 1)$
(e) $5x^2 + 1$ **(f)** $25x^2 + 60x + 35$
9. (a) $2x^2 + 2x$ **(b)** $-2x - 2$ **(c)** $x^4 + 2x^3 - 2x - 1$
(d) $(x - 1)/(x + 1)$ **(e)** $(x + 1)^4 - 1$ **(f)** x^4
11. (a) $|3x|$ **(b)** $3|x + 2| - 2$
13. (a) x for $x \geq 1$ **(b)** x
15. (a) $1/(x + 2)$ for $x \neq -2$ **(b)** $(1/x) + 2$ for $x \neq 0$
17. (a) $x/(1 - x)$ for $x \neq 0$ and $x \neq 1$ **(b)** $x - 1$ for $x \neq 1$
19. $f(x)=x^3, g(x)=x+1$ **21.** $f(x)=|x|, g(x)=x+5$
23. $f(x) = [\![x]\!]$, $g(x) = 3x$ **25.** $f(x) = x - 5$, $g(x) = x^2$
27. $f(x) = x^2$, $g(x) = x - 5$
29. $f(x) = x^2$, $g(x) = 1/(x + 1)$ for $x \neq -1$
31. $f(x) = x^{5/3}$, $g(x) = x - 5$
33. $h(x) = \left(-\frac{89}{180}\right)x^2 - \left(\frac{40}{9}\right)x + \left(\frac{845}{9}\right)$ for $0 \leq x \leq 10$
35. $P(x) = 0.1x^3 - 21x^2 + 1100x$
37. $P(x) = 1000 + [(5000 - 3000x)/(x^2 + 5x)]$
39. $L(x) = \left(\frac{5}{3}\right)x$, $P(x) = \left(\frac{5}{4}\right)x$, $P \circ L(x) = \left(\frac{25}{12}\right)x$
41. (a) $p \circ c(x) = 2.50 - (100/x)$

(b)

x	10	50	100	200
$c(x)$	\$17.50	\$9.50	\$8.50	\$8.00
$p \circ c(x)$	\$-7.50	\$0.50	\$1.50	\$2.00

(c) $x = 40$, production should be greater than 40
43. $A(t) = \pi \sqrt[3]{t^2}$
45. one of many examples: $f(x) = x$, $g(x) = x - 2$
47. one of many examples: $f(x) = x^3$, $g(x) = \sqrt[3]{x}$
49. $V(h) = \begin{cases} 50h^2 & \text{for } 0 \leq h \leq 3 \\ 300h - 450 & \text{for } 3 \leq h \leq 4 \end{cases}$

$(c \circ V)(h) = \begin{cases} 0.014(50h^2) & \text{for } 0 \leq h \leq 3 \\ 0.014(300h - 450) & \text{for } 3 \leq h \leq 4 \end{cases}$

EXERCISES 2.7

1. {(3, 2), (1, 3), (5, 1), (2, 5)}
3. {(5, 2), (4, 3), (2, 4), (3, 5)}
5. {(7, 1), (5, 3), (3, 7), (1, 5)} **7.** $f^{-1}(x) = 2x - 5$
9. $g^{-1}(x) = \left(\frac{1}{2}\right)(x + 1)$ **11.** $h^{-1}(x) = \left(\frac{1}{4}\right)(3 - x)$
13. $f^{-1}(x) = x^{-1/3}$ for $x \neq 0$
15. $g^{-1}(x) = (1/x) - 1$ for $x \neq 0$ **17.** $h^{-1}(x) = \sqrt[3]{x + 1}$
19. $f^{-1}(x) = x^2 - 3$ for $x \geq 0$ **21.** $g^{-1}(x) = x^3 + 3$
23. $h^{-1}(x) = -x$

25.

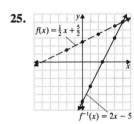

27.

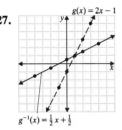

29.

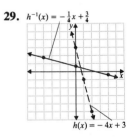

31.

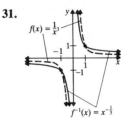

33.

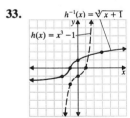

35. $f^{-1}(x) = (1/a)x - (b/a)$ **37.** yes
39. $p(s) = 4s, s(p) = \left(\frac{1}{4}\right)p$
41. $A(r) = \pi r^2$ for $r > 0$, $r(A) = (\sqrt{A\pi})/\pi$ for $r(A) > 0$
43. $C \circ r(A) = 2\sqrt{A\pi}$
45. $S^{-1}(x) = 110 + (\sqrt{16(110^2) - x}/4)$
47. $A \circ r(t) = \pi \sqrt[3]{t^2}$

CHAPTER REVIEW EXERCISES

1. domain = {1, 2, 3, 4}, range = {0, 1, 3}, a function
3. (a) -9 **(b)** $-\frac{3}{2}$ **(c)** $(5 - 4x)/x$ **(d)** $-5a - 4$
(e) $-5a + 4$ **(f)** $5x + 5a - 4$ **(g)** $5a$ **(h)** 5

5. (a) 3 **(b)** $\frac{3}{4}$ **(c)** $3/x^2$ **(d)** $3a^2$ **(e)** $-3a^2$
(f) $3x^2 + 6ax + 3a^2$ **(g)** $(6x + 3a)a$ **(h)** $6x + 3a$
7. $(-\infty, 2) \cup (2, 3) \cup (3, \infty)$ **9.** $(-\infty, 2) \cup (3, \infty)$
11. (a) $(-\infty, \infty)$ **(b)** $[1, \infty)$ **13.** $y = \left(\frac{3}{4}\right)x + 3$, yes
15. $f(x) = |x - 2| + 1$

17.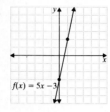

19.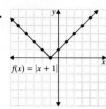

21. (a) a function **(b)** yes, one-to-one
23. (a) a function **((b))** not, one-to-one
25. $y = \left(-\frac{1}{4}\right)x + \left(\frac{9}{2}\right)$
27. even

29.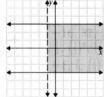

31.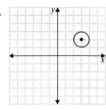

33. x-axis, y-axis, origin **35.** none **37.** y-axis
39. $g(x - 3) = y - 5$ **41.** $y - 3 = (x - 5)^3$

43.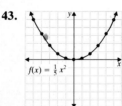

45.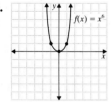

47.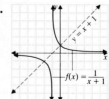

49. (a) $25x^2 - 1$ **(b)** $5x^2 + 10x - 1$
51. $f(x) = x^3$, $g(x) = x - 1$
53. $A(t) = 9\pi/t^2$, $t(A) = (3\sqrt{A\pi})/A$
55. $f^{-1}(x) = \sqrt{x + 5}$ for $x \geq -5$ **57.** $G = 1/R$, strictly decreasing
59. $s(t) = -16t^2 + 4800\,t - 2{,}000$

61. $V(h) = \begin{cases} (100h^2)/3 & \text{for } 0 \leq h \leq 3 \\ 200h - 300 & \text{for } 3 \leq h \leq 4 \end{cases}$

$(c \circ V)(h) = \begin{cases} 1.25h^2/3 & \text{for } 0 \leq h \leq 3 \\ 2.50h - 3.75 & \text{for } 3 \leq h \leq 4 \end{cases}$

EXERCISES 3.1

1.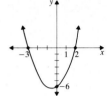

3.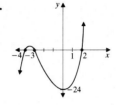

5.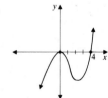

7.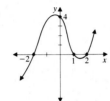

9.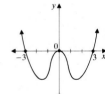

11.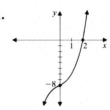

13.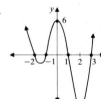

15.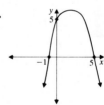

17.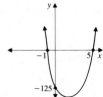

19.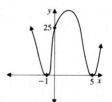

21. $\left\{-17, \frac{2}{3}\right\}$ **23.** $\{(-1 \pm \sqrt{13})/2\}$
25. $\{-3, 1\}$ **27.** $k = \pm 12$ **29.** $k = \pm 2\sqrt{6}$
31. $k = 2$ **33.** $k = \frac{25}{8}$
35. $k \in \{0, 2\}$, $g(x) = x^2 - 4$ and $g(x) = x^2 + 2x$

37.

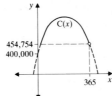

39.

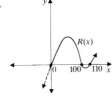

41. zeros $\{0, 4010\}$, maximum $R(x)$ when $x = 2005$

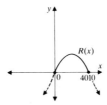

43. $[7, \infty)$

EXERCISES 3.2

1. $f(x)/g(x) = x + 1 + [3/(x + 2)]$,
$f(x) = (x + 2)(x + 1) + 3$
3. $f(x)/g(x) = 2x + 5 + [4x/(x^2 - 4)]$,
$f(x) = (x^2 - 4)(2x + 5) + 4x$
5. $f(x)/g(x) = 2x^2 + 9x + 14 + [8/(x - 2)]$,
$f(x) = g(x)(2x^2 + 9x + 14) + 8$
7. $f(x)/g(x) = \left(\frac{1}{3}\right)x^2 + 2x + \left(\frac{2}{3}\right) - [6/(x - 1)]$,
$f(x) = g(x)\left[\left(\frac{1}{3}\right)x^2 + 2x + \left(\frac{2}{3}\right)\right] - 18$
9. $f(x)/g(x) = x^2 + \left(\frac{9}{2}\right)x + \left(\frac{13}{4}\right) - [-35/(8x - 4]$,
$f(x) = g(x)\left[x^2 + \left(\frac{9}{2}\right)x + \left(\frac{3}{2}\right)\right] - \left(\frac{21}{4}\right)$
11. $f(x)/g(x) = x^3 + x + [(2x - 1)/(x^2 - 1)]$,
$f(x) = g(x)(x^3 + x) + 2x - 1$ **13.** $f(x)/g(x) =$
$x^6 - x^5 + x^4 - x^3 + x^2 - x + 1 - [2/(x + 1)]$,
$f(x) = g(x)(x^6 - x^5 + x^4 - x^3 + x^2 - x + 1) - 2$
15. $f(x)/g(x) = x^7 + x^6 + x^5 + x^4 + x^3 + x^2 + x + 1$,
$f(x) = g(x)(x^7 + x^6 + x^5 + x^4 + x^3 + x^2 + x + 1)$
17. $a(x^3 - 3x^2 - x + 3)$ for $a \neq 0$
19. $a(x^3 - 3x^2 - 7x + 21)$ for $a \neq 0$
21. $a(x^3 - x^2 + 3x + 5)$ for $a \neq 0$
23. $a(x^3 - x^2 - x^2 i - x + 1 + i)$ for $a \neq 0$
25. $k \in \{-3, 1\}, f(x) = ax^2 + 3ax - 10a$,
$f(x) = ax^2 - ax - 2a, a \neq 0$
27. Find $g(1)$.   **29.** Find $h(5)$.
31. $r = 1$ **33.** $r = -5$ **35.** Find $f(a)$.
37. Find $f(-a)$.   **39.** Find $f(-a)$.
41. Solve $f(d) = 0$.   **43. (a)** Find $f(1)$ **(b)** Find $f(-1)$
45. $x^3 + 4x^2 + 6x + 4$
47. $P(x) = 0.1x^3 - 21x^2 + 1100x$

EXERCISES 3.3

1. (a) $x^2 + 2x + 11$ **(b)** 50   **3. (a)** $x^2 + 4x - 5$ **(b)** 8
5. (a) $x^2 - 1$ **(b)** 0   **7. (a)** $x^2 + (1 + \sqrt{2})x + \sqrt{2}$
(b) 0   **9. (a)** $x^2 + (1 + i)x + i$ **(b)** 0
11. (a) $2x^3 - x^2 - 2x$ **(b)** -1
13. (a) $2x^3 - 7x^2 + 13x - 24$ **(b)** 47
15. (a) $2x^3 + 2x^2 - 8x - 8$ **(b)** -17
17. $\frac{1}{8}$ **19.** $\frac{19}{32}$ **21.** -4 **23.** 3
25. $P(35) = \$680,625, P(110) = \$810,000$,
$C(35) = \$3,754,750, C(110) = \$8,551,000$,
$R(35) = \$4,435,375, R(110) = \$9,361,000$
27. $P(25) = \$15,937.50, P(150) = \$30,000$,
$C(25) = \$399.38, C(150) = \$377.50, R(25) =$
$\$16,336.88, R(150) = \$30,377.50$
35. 3.4   **37.** $4x^2 + 2x + \left(\frac{1}{2}\right) + [11/(4x - 2)]$
39. $2x^3 - x + [a/(ax + b)]$
41. $f(x) = ax^4 - 8ax^3 + 16x^2$ for $a < 0$
43. $f(x) = \left(-\frac{2}{27}\right)x^3 - \left(\frac{2}{9}\right)x^2 + \left(\frac{2}{3}\right)x + 2$
45. $f(x) = \left(\frac{1}{6}\right)x^3 - \left(\frac{1}{2}\right)x^2 - \left(\frac{2}{3}\right)x + 2$
47. $f(x) = ax^4 - 6ax^3 + 8ax^2$ for $a > 0$

EXERCISES 3.4

1. $[-7, 1]$ **3.** $[-3, 4]$ **5.** $[-4, 5]$ **7.** $[-2, 4]$
9. $[-3, 3]$ **11.** $[-2, 2]$ **13.** $-5, -3, 1$
15. $2, 2, -3$ **17.** $-4, 1, 4$ **19.** $-1, -1, 4$
21. $-1, 2$ **23.** $-1, 2$

25.

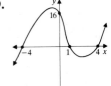

27.

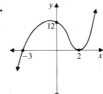

29.

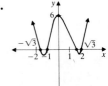

31.

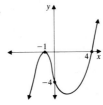

33.

35.

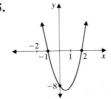

37. $a(x^4 - 6x^2 - 8x - 3)$ with $a \neq 0$
39. $a(x^4 - 2x^3 - 11x^2 + 12x + 36)$ with $a \neq 0$
41. $[-3, -1] \cup [7, \infty]$
43. $[-\infty, -3 - \sqrt{10}] \cup [-3 + \sqrt{10}, 2]$
45. $[0.6, 4]$ **47.** $(0, 3)$

EXERCISES 3.5

1. $\{\pm 1, \pm 2, \pm 3, \pm 6\}$ **3.** $\{\pm 1, \pm 3, \pm 5, \pm 15\}$
5. $\{\pm \frac{1}{2}, \pm 1, \pm 2, \pm \frac{5}{2}, \pm 5, \pm 10\}$
7. $\{\pm \frac{1}{3}, \pm 1, \pm \frac{7}{3}, \pm 3, \pm 7, \pm 21\}$
9. $\{\pm \frac{1}{4}, \pm \frac{1}{2}, \pm \frac{3}{4}, \pm 1, \pm \frac{3}{2}, \pm 2, \pm 3, \pm 4, \pm 6, \pm 12\}$
11. $\{\pm \frac{1}{6}, \pm \frac{1}{3}, \pm \frac{1}{2}, \pm \frac{2}{3}, \pm 1, \pm \frac{3}{2}, \pm 2, \pm 3, \pm \frac{9}{2},$
$\pm 6, \pm 9, \pm 18\}$ **13.** $a(x^4 - 7x^3 + 9x^2 + 7x - 10)$
15. $a(4x^4 + 8x^3 - 13x^2 - 2x + 3)$
17. $a(x^4 + 2x^3 - 2x - 1)$
19. $a(x^4 - 3x^3 + 2x^2 + 2x - 4)$
21. $a(x^4 - 3x^3 - 5x^2 + 9x - 2)$
23. $a(x^4 - 4x^3 + 8x^2 - 8x - 5)$
25. $-2, 2, 4$ **27.** $\frac{1}{2}, -\frac{1}{2}, -2$ **29.** $-1, -\frac{1}{2}, \frac{3}{2}$
31. $\frac{1}{3}, -2, \frac{1}{2}, -\frac{1}{2}$ **33.** $\frac{1}{2}, \frac{1}{3}, 3, -1$ **35.** $-2, 3, \pm \sqrt{2}$
37. $-2, 4, \pm i$ **39.** no rational zeros
41. no rational zeros   **43.** no rational zeros
45. $f(x) = (x - i)^2(2x - 1)(x + i)^2$

EXERCISES 3.6

1. $x = -3, x = 2, y = 0$ **3.** $y = 3, x = 2, x = 4$
5. (a) hole when $x = -3$ **7. (a)** hole when $x = 1$
(b) 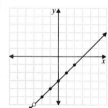   **(b)**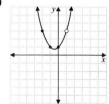

9. (a) vertical asymptotes of $x = 2$ and $x = -2$
(b)
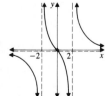

11. (a) -6 **(b)** -6 **13. (a)** 3 **(b)** 3
15. (a) $-\infty$ **(b)** ∞

17. (a) none
(b) none
(c)
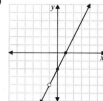

19. (a) $x = 3, x = -3$
(b) $y = 0$
(c)
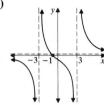

21. (a) $x = -3, x = 2$
(b) $y = 1$
(c)
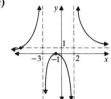

23. (a) none
(b) $y = 0$
(c)
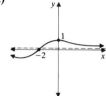

25.
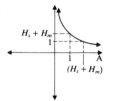

27.
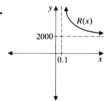

EXERCISES 3.7

1. $x = -1, x = 3, y = 2$
3. $x = -3, x = 2, y = -\frac{2}{3}x + 2$
5. $x = 1, x = -1, x = 3, y = 0$
7. $x = -3, x = 3, y = 2$
9. $x = -2, x = 2, y = 0$ with holes at $x = -1$
and $x = 1$
11. $x = -1, x = 1, y = 2$ **13.** $x = 1, y = x + 1$
15. $x = -2, y = x - 4$ **17.** $x = -1, x = 2, y = \frac{1}{3}x$
19. $x = 0, x = -2, y = x - 4$

5.
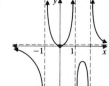

7.
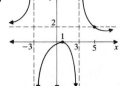

9.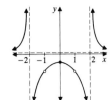

11.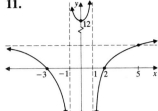

29. **(a)** 5¢ **(b)** 2¢

(c)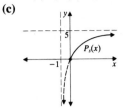

13.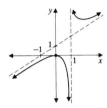

15.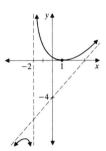

31. $y = x^2 - 2x + 4$ **33.** $x = 1, x = -1, y = x^3$

CHAPTER REVIEW EXERCISES

1.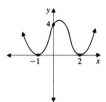

3.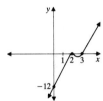

17.

19.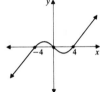

5.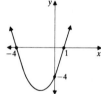

7.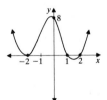

21. $x = 0, y = \left(\frac{1}{10}\right)x^2$

23. $x = 0, y = \left(\frac{1}{20}\right)x^3$

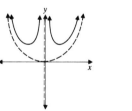

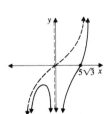

9.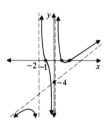

25. $x = \sqrt{3}, x = -\sqrt{3}, y = \left(\frac{1}{3}\right)x^2 + 1$

27. The current decreases as the resistance increases.

25.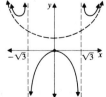

27.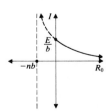

11. $a(x^3 - 7x + 6), a \neq 0$
13. $x^2 - x - 1 - [7/(x - 2)]$
15. 24   **17.** U.B.: 2, L.B.: 0
19. $\left\{\pm\frac{1}{5}, \pm\frac{2}{5}, \pm1, \pm2, \pm5, \pm10\right\}$
21. $y = 2, x = -1, x = -4$
23. $y = -x + 1, x = 3, x = -2$
25. $x = 1, x = -1, x = 3, y = 0$
27. $x = 2, y = x + 5$ **29.** $x = 1, y = x^2 + 2x + 2$
31. $y = 0$

33. 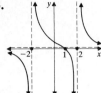   **35.** 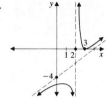   **13.** 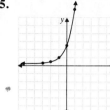   **15.**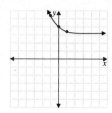

37. $a(x^5 + 3x^4 + 6x^3 + 18x^2 - 27x + 81)$, with $a \neq 0$

39. $P(100) = \$3,800$, $C(100) = \$110$, $R(100) = \$3,910$, $C(3500) = \$450$, $P(3500) = \$17,400$, $R(3500) = \$17,850$

17. 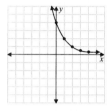   **19.**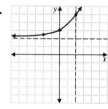

41.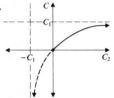

21.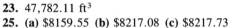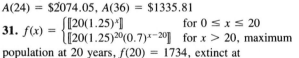

EXERCISES 4.1

1. 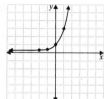   **3.**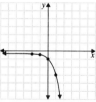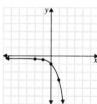

23. $47,782.11 \text{ ft}^3$

25. (a) $\$8159.55$ (b) $\$8217.08$ (c) $\$8217.73$

27. $b(n) = 2000\,(2)^n$ for n days

29. $A(n) = 5000\,(0.964)^n$ for n months, $A(12) = \$3220.29$, $A(24) = \$2074.05$, $A(36) = \$1335.81$

31. $f(x) = \begin{cases} [\![20(1.25)^x]\!] & \text{for } 0 \leq x \leq 20 \\ [\![20(1.25)^{20}(0.7)^{x-20}]\!] & \text{for } x > 20, \text{ maximum} \end{cases}$

population at 20 years, $f(20) = 1734$, extinct at approximately 41 years, $f(41) \approx [\![.9689]\!] = 0$

33. (a) e^{x^2-2x+3} (b) $e^{2x} - 2e^x + 3$ **35.** (a) $|3e^x|$ (b) $e^{|3x|}$

37. (a) $e^x(e^x - 3)^2$ (b) $e^{x(x-3)^2}$

39. $f(x) = 5x^2 - x + 1$, $g(x) = e^x$

41. $f(x) = x^{-1} + x$, $g(x) = e^x$

43. $g(x) = x^2$, $f(x) = e^{2x} - 1$

5. 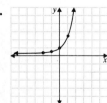   **7.**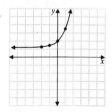

9. 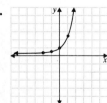   **11.**

45. 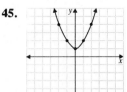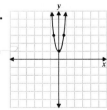   **47.**

49.

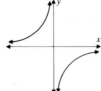

51. $i(t) = 10e^{-2t}$, $v(t) = 100(1 - e^{-2t})$

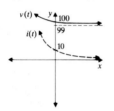

55. What happens when $a < 0$?

EXERCISES 4.2

1. $\log_3 243 = 5$ **3.** $\log_{10}(0.01) = -2$ **5.** $\log_5 3 = x$
7. $2^5 = 32$ **9.** $5^3 = x$ **11.** $3^x = 8$ **13.** $\text{pH} = 5.2$
15. $6.3 < \text{pH} < 6.6$ **17.** 1.5849×10^{-4}
19. $1.58 \times 10^{-4} < [\text{H}^+] < 1.58 \times 10^{-3}$ **21.** 6
23. 5   **25.** 0   **27.** $y - 1$ **29.** 7   **31.** 16
33. 125   **35.** 10   **37.** $\pm\sqrt{10}$ **39.** 3   **41.** 2
43. 5   **45.** 1   **47.** Ø   **49.** 2, 5   **51.** 2

EXERCISES 4.3

1. $\{x : x > 0\}$ **3.** $\{x : x > -2\}$ **5.** $\{x : x > 3\}$
7. $\{x : x \neq 0\}$ **9.** $\{x : x \neq -2\}$ **11.** $\{x : x > 3\}$

13.

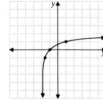

15.

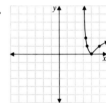

17.

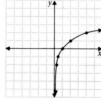

19.

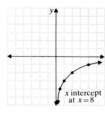

x intercept
at $x = 8$

21.

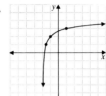

23.

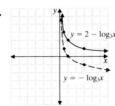

$y = 2 - \log_3 x$

$y = -\log_3 x$

25.

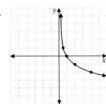

27.

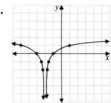

29.

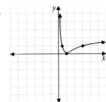

31.

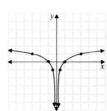

33.

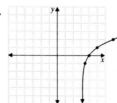

35. $f^{-1}(x) = 10^{x/3}$ **37.** $g^{-1}(x) = e^x - 1$
39. $h^{-1}(x) = e^{3x}$ **41.** $y = 9^x$ **43.** $y = 2^{-x}$
45. $y = \log_a x$, $a = 2^{4/9}$ **47.** $y = \log_a(x - 3)$, $a = 2^{4/9}$

EXERCISES 4.4

1. $2\log_5 x + 3\log_5 y - 3\log_5 3$
3. $2 + 3\log_4 x - \left(\frac{1}{2}\right)\log_4 y$
5. $\left(\frac{1}{2}\right)\log x + \left(\frac{1}{2}\right)\log y - \log 14$
7. $5\ln x + 3\ln y - 5\ln w$ **9.** $x + \left(\frac{1}{2}\right)\ln x$
11. $2\log(x + 2) - \log x - \log(x - 1)$
13. $\ln x + \ln(x + 2) - \left(\frac{1}{2}\right)\ln(x - 1)$ **15.** $\ln_5(y^{1/2}/x^2)$
17. $\log(x^{1/2}y^2/5)$ **19.** $\log[x/(y^{1/2}w^3)]$
21. $\ln[(5^{1/2}x^{3/4})/y^{1/2}]$ **23.** $\frac{3}{2}$ **25.** -3 **27.** $\frac{1}{4}$
29. 2   **31.** 1   **33.** 0   **35.** Ø   **37.** 5   **39.** Ø

41. $A = a10^R$ **43.** $A = 10^{6.3} = 1,995,262$
45. 100/1   **47.** 2500   **49.**

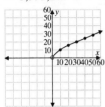

EXERCISES 4.5

1. 4.585   **3.** 3.5805   **5.** −5.129
7. 21.857   **9.** 0   **11.** $\ln 5 \approx 1.609$
13. $\left(\frac{1}{2}\right) \ln 7 \approx 0.973$ **15.** 1, 1000   **17.** 10^8
19. $\log \left[(y + \sqrt{y^2 - 4})/2\right]$
21. $\left(\frac{1}{2}\right) \ln (y + \sqrt{y^2 + 1})$ **23.** $\left(\frac{1}{2}\right) \ln x$ **25.** $2x - 6$
27. $1 + 3x \log x$ **29.** $\left(\frac{1}{4}\right) + \left(\frac{3}{4}\right) \log x$
31. 10.5 years   **33.** 23 years
35. $t = (1/k) \ln 2$ **37.** 8.6×10^{-6} for t in years
39. 16,064 years   **41.** $4115.27
43. 122.6 months = 10.22 years   **45.** 58.6 years
47. (a) $T(t) \approx 20 + 40e^{-4.7t \times 10^{-2}}$ (b) 39.76° (c) 25.6 min.
49. (a) $T(t) = 70 + 280e^{-9.88t \times 10^{-2}}$ (b) 78.8° (c) 7.8 min.

CHAPTER 4 REVIEW EXERCISES

1. 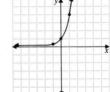   **3.**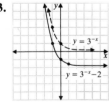

5. $\log_4 3 = x$ **7.** $x^3 = 7$ **9.** $x^2 - 1$

11.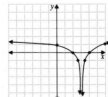

13. $\log x + \left(\frac{1}{2}\right) \log y - 2 \log w$
15. $\left(\frac{1}{2}\right) \ln(x + 1) - \left(\frac{1}{2}\right) \ln (x - 3)$ **17.** $\ln \left[(x^{1/2} w^5)/y^3\right]$

19. $\log \left[(x + 1)^3 (x - 1)^{-1/2}\right]$ **21.** $\ln \left[x^{1/2}(x - 1)^{-3/4}\right]$
23. 81   **25.** −0.3423   **27.** 3   **29.** 100   **31.** 7
33. $I = I_0 10^{8.2}$

35. (a)   (b)

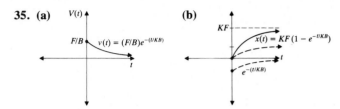

37. 2.688 years   **39.** $10,000   **41.** $3157.12
43. 32.5 months   **45.** $3753.08   **47.** 1.58×10^{-8}
49. 5.1   **51.** $0.1, 10^{-7}$ **53.** 0.2878   **55.** 0.2878
57. 25% + 75% = 100%, and the times are the same.

EXERCISES 5.1

1. $\pi/9 \approx 0.349$ **3.** $5\pi/12 \approx 1.309$ **5.** 0.657
7. 1.790   **9.** 75°0′0″   **11.** 67°30′   **13.** 76°46′35″
15. 327°9′32″   **17.** 1468.69 miles
19. 3006.5 miles
21. 1000π rad. per min = 500 rev. per min
23. 10.08″   **25.** 33,924.4 miles
27. 0.009 radian ≈ 0.5°
29. $\omega_p \approx 93772.8$ rad. per hour,
$v_p = 8.88$ miles per hour
31. $v_p = 1\frac{2}{3}$ miles per hour   **33.** 73.22 in.

EXERCISES 5.2

1. $\sqrt{3}/2$ **3.** 1   **5.** $\frac{1}{2}$ **7.** $\sqrt{2}/2$ **9.** 0
11. 0   **13.** 2   **15.** $\sqrt{3}/3$ **17.** $\frac{1}{2}$ **19.** 1
21. $2\sqrt{3}/3$ **23.** $\sqrt{2}$ **25.** 0   **27.** undefined
29. 2   **31.** 2   **33.** $c = 15\sqrt{2}, b = 15$
35. $c = 30, b = 15\sqrt{3}$ **37.** $a = 12.5, b = 12.5\sqrt{3}$
39. $a = 12.5\sqrt{3}, b = 12.5$ **41.** $c = 12\sqrt{3}, b = 6\sqrt{3}$
43. (a) 80.36 ft (b) 98.25 ft (c) 22.6%
45. 1345.74 ft, 184.31 ft   **47.** (a) 711.54 ft (b) 45.96%

EXERCISES 5.3

$\sin \alpha$	$\csc \alpha$
$\cos \alpha$	$\sec \alpha$
$\tan \alpha$	$\cot \alpha$

1.

$\frac{4}{5}$	$\frac{5}{4}$
$-\frac{3}{5}$	$-\frac{5}{3}$
$-\frac{4}{3}$	$-\frac{3}{4}$

3.

$-\frac{3}{5}$	$-\frac{5}{3}$
$\frac{4}{5}$	$\frac{5}{4}$
$-\frac{3}{4}$	$-\frac{4}{3}$

5.

$-\frac{3}{5}$	$-\frac{5}{3}$
$-\frac{4}{5}$	$-\frac{5}{4}$
$\frac{3}{4}$	$\frac{4}{3}$

7.

$-\frac{12}{13}$	$-\frac{13}{12}$
$\frac{5}{13}$	$\frac{13}{5}$
$-\frac{12}{5}$	$-\frac{5}{12}$

9.

$-2\sqrt{29}/29$	$-\sqrt{29}/2$
$-5\sqrt{29}/29$	$-\sqrt{29}/5$
$\frac{2}{5}$	$\frac{5}{2}$

11.

$\frac{1}{2}$	2
$-\sqrt{3}/2$	$-2\sqrt{3}/3$
$-\sqrt{3}/3$	$-\sqrt{3}$

13. 187.9 ft   **15.** $r \approx 718.24$ lb, $s \approx 1973.35$ lb
17. $-\sqrt{2}/2$, $\sqrt{2}$, -1 **19.** $-\frac{1}{2}$, $2\sqrt{3}/3$, $-\sqrt{3}/3$
21. $\sqrt{2}/2$, $-\sqrt{2}$, -1 **23.** $\sqrt{3}/2$, -2, $-\sqrt{3}$
25. $-\frac{1}{2}$, $-2\sqrt{3}/2$, $\sqrt{3}/3$ **27.** $\sqrt{2}/2$, $-\sqrt{2}$, -1
29. $-\frac{1}{2}$, $-2\sqrt{3}/3$, $\sqrt{3}/3$ **31.** $\frac{1}{2}$, $-2\sqrt{3}/3$, $-\sqrt{3}/3$
33. $\sqrt{2}/2$, $\sqrt{2}$, 1 **35.** $-\sqrt{3}/2$, 2, $-\sqrt{3}$
37. $h \approx 0.47$ miles, $x \approx 2.65$ miles, $d \approx 2.69$ miles
39. **(a)** $\sin 2x$ **(b)** $2 \sin x$
41. **(a)** $\cos(x^2 + 1)$ **(b)** $1 + \cos^2 x$
43. **(a)** $\tan x$ **(b)** $\sqrt{\tan(x^2)}$
45. **(a)** $|\cos x|$ **(b)** $\cos^2 \sqrt{x}$
51. $-3\sqrt{10}/10$, $-\sqrt{10}$, 3 **53.** $-2\sqrt{5}/5$, $\sqrt{5}$

EXERCISES 5.4

1. $17.46°$ **3.** $101.54°$ **5.** $56.31°$ **7.** $11.54°$
9. $-30° = -\pi/6$ **11.** $5\pi/6 = 150°$ **13.** $\pi/3 = 60°$
15. $\pi/3 = 60°$ **17.** **(b)** $y = \cos \theta$ **19.** **(d)** $y = \csc \theta$

21.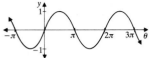

23.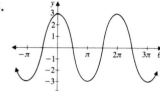

25.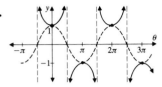

27.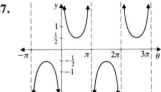

29.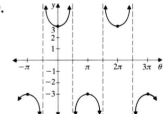

31.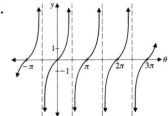

33.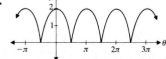

35.

$\frac{4}{5}$	$\frac{5}{4}$
$-\frac{3}{5}$	$-\frac{5}{3}$
$-\frac{4}{3}$	$-\frac{3}{4}$

37.

$-\sqrt{3}/2$	$-2\sqrt{3}/3$
$-\frac{1}{2}$	-2
$\sqrt{3}$	$\sqrt{3}/3$

39.

$-\sqrt{2}/2$	$-\sqrt{2}$
$\sqrt{2}/2$	$\sqrt{2}$
-1	-1

45. $88.9°$ **47.** (a) $23.58°$ (b) 0.438 miles   **49.** $36.87°$
51. (a) $15.32°$ (b) 45.7% (c) 3 overlapping   **53.** $16.86°$
55. odd

EXERCISES 5.5

1.

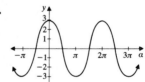

3.

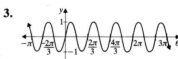

5.

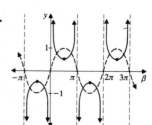

7.

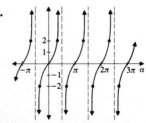

9.

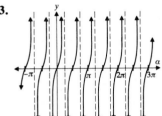

11.

13.

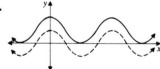

15.

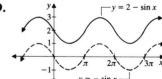

17.

19.

21.

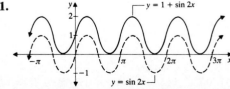

23. 4, 2π, $y = 4 \cos \theta$ **25.** 2, 4π, $y = 2 \sin \left[\left(\frac{1}{2}\right) \theta\right]$
27. 3, 2π, $y = 3 \sin [\theta - (\pi/4)]$ or $y = 3 \cos [\theta - (3\pi/4)]$

29.

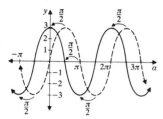

31.

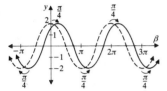

33.

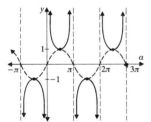

35.

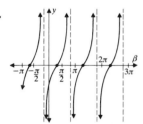

37.

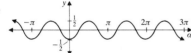

39.

41.

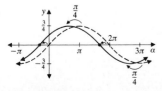

43.

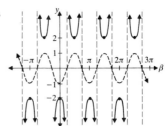

45. **(a)** $\theta = \omega t$ in radians
(b) $y = R \sin (\omega t)$,   **(c)** $y = R \cos (\omega t)$ or
$y = R \sin [\omega t + (\pi/2)]$

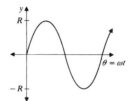

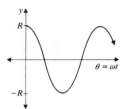

47. $y = \cos x$ or $y = \sin [x + (\pi/2)]$

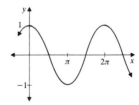

49. **(a)** $y = 32° \sin (\pi/45)[x - 8]$
(b) $y = -28.5° + 32° \sin (\pi/45)[x - 8]$

51.

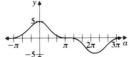

53.

55.

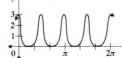

EXERCISES 5.6

1.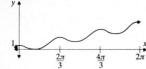

3.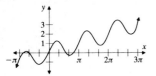

5.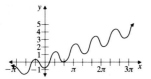

7.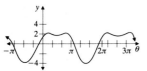

9.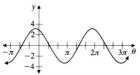

11.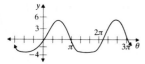

13.

15.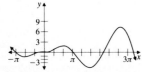

17.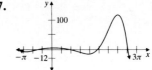

19.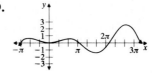

21.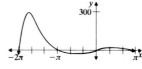

23.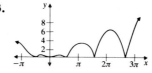

25.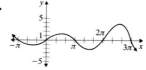

27.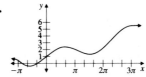

29.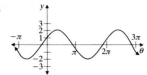

31.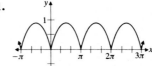

33.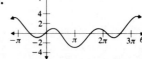

35.

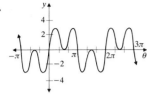

37. (a) $\theta = \omega t$

(b) $y = D + R + R \sin \omega t$

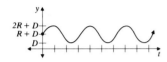

(c) $y = R + D + R \cos \omega t$

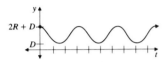

41. $f(x) = g(x) + 3 \sin[(2\pi/365)x]$

EXERCISES 5.7

1. $a \approx 135.5$, $c \approx 198.6$, $\gamma = 76.2°$
3. $a \approx 15.29$, $b \approx 5.04$, $\gamma = 109.1°$
5. $a \approx 19.2$, $\beta \approx 76.9°$, $\gamma \approx 46.1°$
7. $\alpha \approx 28.9°$, $\beta \approx 44.1°$, $\gamma \approx 107°$
9. $c_1 \approx 7.0$, $\alpha_1 \approx 54.9°$, $\gamma_1 \approx 88°$, $c_2 \approx 2.1$,
$\alpha_2 \approx 125.1°$, $\gamma_2 \approx 17.8°$ **11.** Ø
13. $a_1 \approx 122.2$, $\alpha_1 \approx 139.6°$, $\beta_1 \approx 30.1°$,
$a_2 \approx 71.4$, $\alpha_2 \approx 19.8°$, $\beta_2 \approx 149.9°$
15. $a \approx 13.8$, $\alpha \approx 45.8°$, $\beta \approx 9.2°$
17. $c_1 \approx 19.2$, $\alpha_1 \approx 45.4°$, $\gamma_1 \approx 94.5°$,
$c_2 \approx 1.8$, $\alpha_2 \approx 134.6°$, $\gamma_2 \approx 5.3$
19. $a \approx 7.7$, $\beta \approx 32.7°$, $\gamma \approx 130.3°$
21. $AB \approx 6.62$ miles, error ≈ 158.4 ft
23. $AB \approx 1.01$ miles, error ≈ 103.2 ft
25. 472.69 ft, 465.32 ft   **27.** 44,331.1 ft^2
29. $\alpha \approx 35.99°$, $\beta \approx 87.45°$, $\gamma \approx 56.56°$
31. 34,627.41 ft^2 **33.** 33.65 ft
35. 669.47 miles, N22.3° E   **37.** 51.53 ft.
39. $a \approx 10.94$ miles, $b \approx 10.33$ miles
41. 9.61 miles

CHAPTER 5 REVIEW EXERCISES

1. $\left(\frac{5}{12}\right)\pi$ **3.** $47\pi/36$ **5.** 135°   **7.** 240°

9. $\sqrt{2}/2$ **11.** $\sqrt{3}$ **13.** $2\sqrt{3}/3$
15. $b = 5\sqrt{3}$, $c = 10$
17. $a = 5\sqrt{2}$, $b = 5\sqrt{2}$ **19.** 14.2

21.

$-\frac{4}{5}$	$-\frac{5}{4}$
$-\frac{3}{5}$	$-\frac{5}{3}$
$\frac{4}{3}$	$\frac{3}{4}$

23.

$\sqrt{2}/2$	$\sqrt{2}$
$-\sqrt{2}/2$	$-\sqrt{2}$
-1	-1

25.

$-\sqrt{3}/2$	$-2\sqrt{3}/3$
$-\frac{1}{2}$	-2
$\sqrt{3}$	$\sqrt{3}/3$

27.

$-\frac{1}{2}$	-2
$-\sqrt{3}/2$	$-2\sqrt{3}/3$
$\sqrt{3}/3$	$\sqrt{3}$

29. $(f \circ g)(x) = 1 + \sin \sqrt{x}$, $(g \circ f)(x) = \sqrt{1 + \sin x}$
31. 45°   **33.** −45°   **35.** 51.3°

37.

$-\frac{4}{5}$	$-\frac{5}{4}$
$-\frac{3}{5}$	$-\frac{5}{3}$
$\frac{4}{3}$	$\frac{3}{4}$

39.

$\frac{3}{5}$	$\frac{5}{3}$
$-\frac{4}{5}$	$-\frac{5}{4}$
$-\frac{3}{4}$	$-\frac{4}{3}$

41. 20.8°   **43.** 86.7°

45.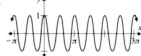

47.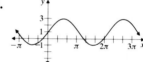

49.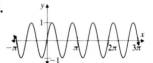

51.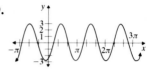

53.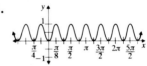

55.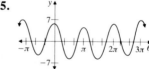

57.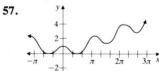

59.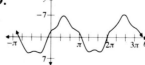

61.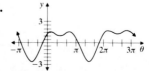

63.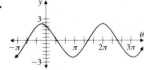

65. $A = 151.3$ square units   **67.** 663.5 miles
69. $x_n = \sin^{n+1} \theta$

EXERCISES 6.1

37. $\tan \theta$ **39.** $\cos \theta$ **41.** $\sin \theta$ **43.** false
45. false   **47.** false   **49.** false   **51.** false

EXERCISES 6.2

1. $\{90°, 120°, 240°\}$ **3.** $\{0, 30°, 180°, 330°\}$ **5.** $\{90°\}$
7. $\{0, 180°\}$ **9.** $\{0\}$ **11.** $\{\pi/4, 3\pi/4, 5\pi/4, 7\pi/4\}$
13. $\{\pi/4, \pi/3, 5\pi/4, 5\pi/3\}$ **15.** $\{\pi/4, 3\pi/4, 5\pi/4,$
$7\pi/4\}$ **17.** $\{0, \pi/2, \pi, 3\pi/2\}$ **19.** $\{\alpha : \alpha = k\pi$ for
$k \in Z\}$ **21.** $\{\beta :$for $k \in Z, \beta = \pi/3 + 2k\pi, \beta =$
$2\pi/3 + 2k\pi, \beta = 5\pi/6 + 2k\pi,$ or $\beta = 7\pi/6 + 2k\pi\}$
23. $\{\alpha :$for $k \in Z, \alpha = \pi/2 + k\pi$ or $\alpha = \pi/4 + k\pi/2\}$
25. $\{\beta :$for $k \in Z, \beta = \pi/2 + k\pi\}$
27. For $k \in Z, \alpha \in \{\pi/6 + 2k\pi, 2\pi/3 + 2k\pi,$
$5\pi/6 + 2k\pi, 4\pi/3 + 2k\pi\}$
29. $\{\beta : \beta = k\pi$ for $k \in Z\}$ **31.** $\{\pi\}$
33. $\{\pi/4, 5\pi/4, 7\pi/4\}$ **35.** $\{\pi/2\}$ **37.** $\{\pi/3,$
$5\pi/3, \pi\}$
39. $\{32.8°, 147.2°\}$ **41.** $\emptyset$ **43.** $x \approx 13.44°$

EXERCISES 6.3

1. $(\sqrt{2} + \sqrt{6})/4$ **3.** $(\sqrt{2} - \sqrt{6})/4$
5. $(\sqrt{6} - \sqrt{2})/4$ **7.** $2 + \sqrt{3}$ **9.** $(-\sqrt{6} - \sqrt{2})/4$
11. $\frac{1}{2}$ **13.** $\cos 69°$ **15.** $\cos 10°$
17. $\sin \pi/4$ or $\sin (\pi/4)$ **19.** $\sin 5$
21. $\cos (\alpha + \beta) = (6\sqrt{2} - 4)/15,$
$\sin (\alpha + \beta) = (8\sqrt{2} + 3)/15,$ QI
23. $\cos (\alpha + \beta) = 2\sqrt{5}/25,$
$\sin (\alpha + \beta) = 11\sqrt{5}/25,$ QI
41. $\{\pi/12, 5\pi/12, 13\pi/12, 17\pi/12\}$
43. $\{\pi/5, 3\pi/5, \pi, 7\pi/5, 9\pi/5\}$
47. $\tan^{-1} [(F/(S_2 \cos \beta) - \tan \beta]$

EXERCISES 6.4

1. $\frac{1}{2}\sqrt{2 - \sqrt{2}}$ **3.** $\frac{1}{2}\sqrt{2 + \sqrt{3}}$ **5.** $\sqrt{2} - 1$
7. $-\frac{1}{2}\sqrt{2 - \sqrt{3}}$ **9.** $-\frac{1}{2}\sqrt{2 + \sqrt{2}}$
11. $\sin 2\alpha = 5\sqrt{39}/32$, $\cos 2\alpha = -\frac{7}{32}$, QII
13. $\sin 2\alpha = \frac{24}{25}$, $\cos 2\alpha = -\frac{7}{25}$, QII
15. $\sin 2\alpha = 17\sqrt{35}/162$, $\cos 2\alpha = \frac{127}{162}$, QI
17. $\{\pi/6, \pi/2, 5\pi/6, 3\pi/2\}$ **19.** $\{0, 1.32, \pi, 4.97\}$
21. $\{\pi/24, 5\pi/24, 13\pi/24, 17\pi/24, 25\pi/24, 29\pi/24,$
$37\pi/24, 41\pi/24\}$ **23.** $\{0, \pi\}$ **25.** $\{1.2, 2.5, 3.8, 5.1\}$
27. $\varnothing$ **29.** $\{\pi/6, 5\pi/6, 3\pi/2\}$
39. $y = \sin 2x$ **41.** $y = \cos 2x$

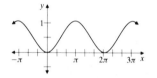   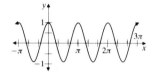

43. $y = \sin^2 (x/2)$

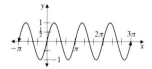

45. (a) $\frac{1}{2}\sqrt{2 + \sqrt{3}}$ **(b)** $(\sqrt{6} + \sqrt{2})/4$
(c) $2\sqrt{2 + \sqrt{3}} = \sqrt{6} + \sqrt{2}$

EXERCISES 6.5

1. $\sin (6\alpha) + \sin (4\alpha)$ **3.** $\frac{1}{2}[\sin (6\beta) + \sin (2\beta)]$
5. $\frac{1}{2}[\cos (8\alpha) - \cos (2\alpha)]$ **7.** $\frac{1}{2}[\cos (4\beta) + \cos (10\beta)]$
9. $2 \sin (7\alpha/2) \cos (3\alpha/2)$ **11.** $-2 \sin (7\beta/2) \sin (3\beta/2)$
13. $2 \cos (2\alpha) \cos \alpha$ **15.** $2 \cos (6\beta) \sin \beta$
33. $\{0, 90°\}$ **35.** $\{90°, 202.6°\}$ **37.** $\{90°, 343.8°\}$
39. $y = 2 \sin (x - 60°)$ **41.** $y = 5 \sin (x + 53.1°)$

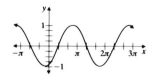   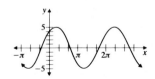

43. $y = 3 \sin (x + 41.8°)$ **45.** $y = 2 \sin x \cos^2 x$

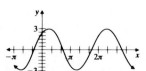   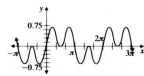

EXERCISES 6.6

1. $60°$ **3.** $30°$ **5.** $-45°$ **7.** $-45°$ **9.** $135°$
11. $90°$ **13.** $\sqrt{3}/2$ **15.** -1 **17.** $-\sqrt{2}/2$
19. $(\sqrt{6} + \sqrt{2})/4$ **21.** $(-\sqrt{2} - \sqrt{6})/4$

$\sin \theta$	$\csc \theta$
$\cos \theta$	$\sec \theta$
$\tan \theta$	$\cot \theta$

23.

$\sqrt{1 - x^2}$	$1/\sqrt{1 - x^2}$
x	$1/x$
$\sqrt{1 - x^2}/x$	$x/\sqrt{1 - x^2}$

25.

$\pm\sqrt{x^2 - 1}/x$	$\pm x/\sqrt{x^2 - 1}$
$1/x$	x
$\pm\sqrt{x^2 - 1}$	$\pm 1/\sqrt{x^2 - 1}$

27.

$\sqrt{x^2 - 1}$	$1/\sqrt{x^2 - 1}$
$\sqrt{2 - x^2}$	$1/\sqrt{2 - x^2}$
$\sqrt{x^2 - 1}/\sqrt{2 - x^2}$	$\sqrt{2 - x^2}/\sqrt{x^2 - 1}$

29.

$1/\sqrt{x^2 + 1}$	$\sqrt{x^2 + 1}$	
$x/\sqrt{x^2 + 1}$	$\sqrt{x^2 + 1}/x$	x must equal 0
$1/x$	x	

31.

0	undefined
1	1
0	undefined

33.

$\sqrt{1 - 9u^2}$	$1/\sqrt{1 - 9u^2}$
$3u$	$1/(3u)$
$\sqrt{1 - 9u^2}/3u$	$3u/\sqrt{1 - 9u^2}$

35.

$\pm\sqrt{x^2 - 1}/\sqrt{2x^2 - 1}$	$\pm\sqrt{2x^2 - 1}/\sqrt{x^2 - 1}$
$\pm x/\sqrt{2x^2 - 1}$	$\pm\sqrt{2x^2 - 1}/x$
$\sqrt{x^2 - 1}/x$	$x/\sqrt{x^2 - 1}$

37. $\pm 2x\sqrt{1 - x^2}$ **39.** $2x^2 - 1$ **41.** 1
43. $1 - 2x$ **45.** $\pm\frac{1}{2}\sqrt{2 - 2\sqrt{1 - x^2}}$
47. $2x\sqrt{1 - x^2}/(2x^2 - 1)$ **49.** $2x/(1 + x^2)$
57. (a) domain: $\{x: -\frac{1}{2} \le x \le \frac{1}{2}\}$; range: $\{y: 0 \le y \le 180°\}$
(b) domain: $\{x: -1 \le x \le 1\}$; range: $\{y: 0 \le y \le 360°\}$
59. (a) $y = \sqrt{1 - x^2}$ for x in $[-1, 1]$

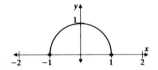

(b) $y = x$ for x in $(-1, 1)$

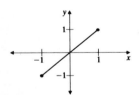

CHAPTER 6 REVIEW EXERCISES

1. $\frac{1}{2}\sqrt{2 - \sqrt{3}}$ **3.** $\frac{1}{2}\sqrt{2 + \sqrt{2}}$ **5.** $\frac{56}{65}$ **7.** $\frac{63}{65}$
9. $\frac{24}{25}$ **11.** $\frac{24}{7}$ **13.** $\frac{119}{169}$ **15.** $\sqrt{10}/10$
17. $5\sqrt{26}/26$ **19.** $\frac{3}{5}$ **21.** $\frac{12}{13}$ **23.** $\frac{4}{3}$ **25.** $\frac{56}{33}$
27. $\frac{24}{7}$ **41.** $\{\pi/6, 5\pi/6, 7\pi/6, 11\pi/6\}$ **43.** $\{\pi\}$
45. $\{1.1437, 5.1395\}$ **47.** $\{\pi/2, 7\pi/6, 3\pi/2, 11\pi/6\}$
49. $y = 3 \sin (4\theta)$ **51.** $y = \frac{1}{2} \cos^2 (\beta/2)$
53.

$\sqrt{1 - a^2}$	$1/\sqrt{1 - a^2}$
a	$1/a$
$\sqrt{1 - a^2}/a$	$a/\sqrt{1 - a^2}$

55.

$1/(2a)$	$2a$
$\sqrt{1 - 4a^2}$	$1/\sqrt{1 - 4a^2}$
$2a/\sqrt{1 - 4a^2}$	$\sqrt{1 - 4a^2}/2a$

57. $2a^2 - 1$ **59.** $2a/(1 + a^2)$
61. $AO = x = 1 + \sqrt{2}$, $BC = x - 1 = \sqrt{2}$

EXERCISES 7.1

1. $(2, 420°)$, $(-2, 240°)$ **3.** $(3, 510°)$, $(-3, 330°)$

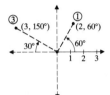

5. $(-2, 405°)$, $(2, 225°)$ **7.** $(3, 7\pi/6)$, $(-3, \pi/6)$

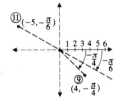

9. $(4, 7\pi/4)$, $(-4, 3\pi/4)$ **11.** $(-5, 11\pi/6)$, $(5, 5\pi/6)$

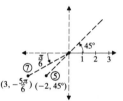

13. $(1, \sqrt{3})$ **15.** $\left(-3\sqrt{3}/2, \frac{3}{2}\right)$ **17.** $(-\sqrt{2}, -\sqrt{2})$
19. $\left(-3\sqrt{3}/2, -\frac{3}{2}\right)$
21. $(3, 45°)$, $(-3, 45°)$ **23.** $(0, 0)$

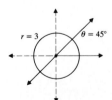

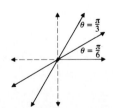

25. $(5, \pi/2), (5, 3\pi/2)$ **27.** Ø

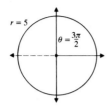

29. $r = \sqrt{7}$ **31.** $r^2 = 4 \sec 2\theta$ **33.** $\tan \theta = \frac{1}{3}$

35. $r^2 = 6 \csc 2\theta$ **37.** $r = 5 \sec \theta$ **39.** $r = 2 \cos \theta$

41. $r = -2 \sin \theta$ **43.** $x = 3$ **45.** $\left(x - \frac{3}{2}\right)^2 + y^2 = \frac{9}{4}$

47. $x^2 + \left(y + \frac{1}{2}\right)^2 = \frac{5}{4}$ **49.** $y^2 = 3(2x + 3)$

51. $y = x \tan [(x^2 + y^2)/2]$ **53.** $x^2 = -4(y - 1)$

55. $4x^2 + 3(y - 1)^2 = 12$ **57.** $(x^2 + y^2)^{3/2} = 2xy$

59. $\left(x - \frac{1}{2}\right)^2 + y^2 = \frac{1}{4}$ **61.** $y = \pm 2$

EXERCISES 7.2

1. (a) $\theta = \pi/2$ (b)   (c) cardioid

3. (a) $\theta = \pi/2$ (b)   (c) cardioid

5. (a) polar axis (b)   (c) circle

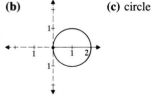

7. (a) polar axis (b)   (c) four-leaf rose

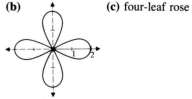

9. (a) $= \pi/2$ (b)   (c) three-leaf rose

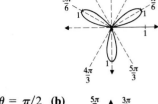

11. (a) $\theta = \pi/2$ (b)   (c) eight-leaf rose

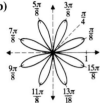

13. (a) the pole and the polar axis

(b)   (c) lemniscate

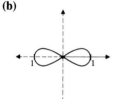

15. (a) the pole and $\theta = \pi/2$

(b)   (d) none of these

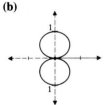

17. (a) $\theta = \pi/2$ (b)   (d) cardioid

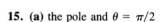

19. (a) polar axis (b)   (d) cardioid

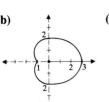

21.
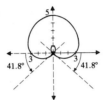

23.
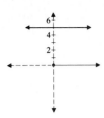

25.
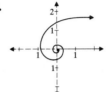

27.
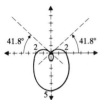

29.
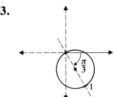

31.
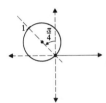

33.
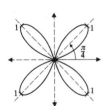

35.
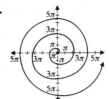

EXERCISES 7.3

1. $-i - 8j = \langle -1, -8 \rangle$ **3.** $5i + 21j = \langle 5, 21 \rangle$
5. $-i - \left(\frac{31}{4}\right)k = \langle -1, -\frac{31}{4} \rangle$
7. $(3\sqrt{3}/2)i + \left(\frac{3}{2}\right)j = \langle 3\sqrt{3}/2, \frac{3}{2} \rangle$
9. $(\sqrt{2}/4)i - (\sqrt{2}/4)j = \langle \sqrt{2}/4, -\sqrt{2}/4 \rangle$
11. $\left(-\frac{7}{4}\right)i + (7\sqrt{3}/4)j = \langle -7/4, 7\sqrt{3}/4 \rangle$
13. $i + 5j, 5i - 9j, -12i + 8j, -20$
15. $j, i - \left(\frac{1}{3}\right)j, 3i + 2j, -\frac{1}{36}$
17. $\langle -\frac{10}{3}, \frac{25}{3} \rangle, \langle -\frac{8}{3}, \frac{29}{3} \rangle, \langle -\frac{1}{3}, 1 \rangle, -5$
19. 69.4°, 42.72 lb   **21.** 39.15°, 375.01 mph
23. 119.7°, 169.4 nautical miles
25. 13.75°, 22.6 ft   **27.** 318.8°, 1164.25 mph
29. $v = 6.93i + 9.19j, v_{mg} = 9.19j$
33. $41\sqrt{58}/58$ **41.** $u = -2i - 7j, w = 2i + 7j$
45. $(\sqrt{153}/153)(3i + 12j)$

EXERCISES 7.4

1. $5\sqrt{2}(\cos 3\pi/4 + i \sin 3\pi/4)$
3. $2[\cos (11\pi/6) + i \sin (11\pi/6)]$
5. $4[\cos (5\pi/3) + i \sin (5\pi/3)]$
7. $\sqrt{2}[\cos (7\pi/4) + i \sin (7\pi/4)]$
9. $\sqrt{34}(\cos 59° + i \sin 59°)$
11. $5[\cos (\pi/2) + i \sin (\pi/2)]$ **13.** $7(\cos 0° + i \sin 0°)$
15. $2[\cos (3\pi/2) + i \sin (3\pi/2)]$
17. $10\sqrt{2}[\cos (7\pi/12) + i \sin (7\pi/12)]$,
$(5\sqrt{2}/2)[\cos (11\pi/12) + i \sin (11\pi/12)], -2500, 2^{12}$
19. $\sqrt{2}[\cos (7\pi/4) + i \sin (7\pi/4)]$,
$2\sqrt{2} [\cos (\pi/12) - i \sin (\pi/12)]$,
$4^4[\cos (2\pi/3) + i \sin (2\pi/3)], -2^6$
21. 34, $\cos 118° + i \sin 118°$, $34^2(\cos 236° + i \sin 236°)$,
$34^6(\cos 348° + i \sin 348°)$ **23.** -2^{164}
25. $2^{93}i$ **27.** 2^{16}
33. $\sin 4\theta = 4 \sin \theta \cos \theta, \cos 4\theta = \cos^4 \theta - 6 \sin^2 \theta \cos^2 \theta + \sin^4 \theta$

EXERCISES 7.5

1. $1, \frac{1}{2}(\cos \pi/6 - i \sin \pi/6), \frac{1}{4}(\cos \pi/3 - i \sin \pi/3)$,
$\sqrt{2} [\cos (\pi/12 + k\pi) + i \sin (\pi/12 + k\pi)]$ for $k \in \{0, 1\}$
3. $1, \frac{1}{8} + i\frac{1}{8}, i/32$,
$2^{5/4}[\cos ((-\pi/8) + k\pi) + i \sin ((-\pi/8) + k\pi)]$
for $k \in \{0, 1\}$
5. $1, \frac{1}{4}[\cos (5\pi/3) - i \sin (5\pi/3)]$,
$\frac{1}{16}[\cos (4\pi/3) - i \sin (4\pi/3)]$,
$2[\cos ((5\pi/6) + k\pi) + i \sin ((5\pi/6) + k\pi)]$ for $k \in \{0, 1\}$
7. $1, \frac{1}{5}(\cos 53° - i \sin 53°), \frac{1}{25}(\cos 106° - i \sin 106°)$,
$\sqrt{5}\left[\cos \left(26\frac{1}{2}° + 180°k\right) + i \sin \left(26\frac{1}{2}° + 180°k\right)\right]$
9. $1, \frac{1}{13}(\cos 112.6° - i \sin 112.6°)$,
$\frac{1}{169}(\cos 225.2° - i \sin 225.2°), \sqrt{13}[\cos (56.3° + 180°k) + i \sin (56.3° + 180°k)]$ for $k \in \{0, 1\}$
11. $1, (\sqrt{10}/10)(\cos 341.6° - i \sin 341.6°)$,
$\left(\frac{1}{10}\right)(\cos 323.2° - i \sin 323.2°), 10^{1/4}[\cos (170.8° + 180°k) + i \sin (170.8° + 180°k)]$ for $k \in \{0, 1\}$
13. $2[\cos (2\pi k/3) + i \sin (2\pi k/3)]$ for $k \in \{0, 1, 2\}$
15. $\cos [(\pi/6) + (2k\pi/3)] + i \sin [(\pi/6) + (2k\pi/3)]$
for $k \in \{0, 1, 2\}$
17. $2[\cos (\pi/4 + k\pi) + i \sin (\pi/4 + k\pi)]$ for $k \in \{0, 1\}$
19. $\cos [(3\pi/4) + k\pi] + i \sin [(3\pi/4) + k\pi]$ for $k \in \{0, 1\}$
21. $\cos [(\pi/10) + (2k\pi/5)] + i \sin [(\pi/10) + (2k\pi/5)]$
for $k \in \{0, 1, 2, 3, 4\}$
23. $2^{3/2}[\cos ((2k + 1)\pi/4) + i \sin ((2k + 1)\pi/4)]$
for $k \in \{0, 1, 2, 3\}$
25. $\cos [(\pi/4) + (k\pi/3)] + i \sin [(\pi/4) + (k\pi/3)]$
for $k \in \{0, 1, 2, 3, 4, 5\}$

27. $\sqrt{2}[\cos (7\pi/8 + \pi k) + i \sin (7\pi/8 + \pi k)]$
for $k \in \{0, 1\}$
29. $\sqrt[3]{13}[\cos (22.5° + 120°k) + i \sin (22.5° + 120°k)]$
for $k \in \{0, 1, 2\}$
31. $\cos [(\pi/6) + (2\pi k/3)] + i \sin [(\pi/6) + (2\pi k/3)]$
for $k \in \{0, 1, 2\}$
33. $\cos (k\pi/2) + i \sin (k\pi/2)$ for $k \in \{0, 1, 2, 3\}$
35. $\cos [(\pi/4) + k\pi] + i \sin [(\pi/4) + k\pi]$ for $k \in \{0, 1\}$
37. $\cos (2k\pi/5) + i \sin (2k\pi/5)$ for $k \in \{0, 1, 2, 3, 4\}$
39. $2[\cos ((\pi/6) + 2k\pi/3) + i \sin ((\pi/6) + 2k\pi/3)]$
for $k \in \{0, 1, 2\}$
41. $2\sqrt[3]{4}[\cos ((55\pi/18) + (2k\pi/3)) +$
$i \sin ((55\pi/18) + (2k\pi/3))]$ for $k \in \{0, 1, 2\}$
43. $\sqrt[3]{34}[\cos (201° + 120°k) + i \sin (201° + 120°k)]$
for $k \in \{0, 1, 2\}$
45. $2197\sqrt{13}[\cos (305.5° + 180°k) + i \sin (305.5° + 180°k)]$
for $k \in \{0, 1\}$ **47.** $z_2 = 2(\cos 135° + i \sin 135°)$,
$z_3 = 2 (\cos 255° + i \sin 255°)$
49. $z_2 = [\cos (5\pi/6) + i \sin (5\pi/6)]$,
$z_3 = [\cos (4\pi/3) + i \sin (4\pi/3)]$,
$z_4 = \cos (11\pi/6) + i \sin (11\pi/6)$
51. $z_2 = 3(\cos 200° + i \sin 200°)$
53. $z_2 = 4[\cos (11\pi/12) + i \sin (11\pi/12)]$,
$z_3 = 4[\cos (19\pi/12) + i \sin (19\pi/12)]$
55. for 47, $z = 8(\cos 45° + i \cos 45°)$;
for 49, $z = \cos (4\pi/3) + i \sin (4\pi/3)$
57. for 51, $z = 9(\cos 40° + i \sin 40°)$;
for 53, $z = 64[\cos (3\pi/4) + i \sin (3\pi/4)]$

CHAPTER 7 REVIEW EXERCISES

1. $(2, 7\pi/4)$ **3.** $(\sqrt{41}, 38.7°)$
5. $(3\sqrt{2}/2, -3\sqrt{2}/2)$ **7.** $(1, -\sqrt{3})$
9. $r = 2\sqrt{2}$ **11.** $\tan \theta = \frac{1}{3}$ **13.** $y = 4$
15. $\left(x - \frac{1}{2}\right)^2 + y^2 = \frac{9}{4}$

17.

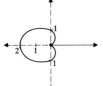

19.

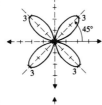

21.

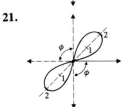

23.

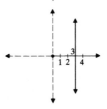

25.

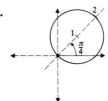

27. $5i - j, i - 7j, -6i + 8j, -6$ **29.** $\langle -8, 6 \rangle, \langle 6, 8 \rangle$,
$\langle -3, 21 \rangle, 0$ **31. (a)** $109.4°$ **(b)** $90°$ **33.** 392.37
miles, $203.07°$ **35.** $4\sqrt{2}[\cos (19\pi/12) + i \sin$
$(19\pi/12)], \sqrt{2}[\cos (\pi/12) - i \sin (\pi/12)]$,
$16\sqrt{2}[\cos (5\pi/4) + i \sin (5\pi/4)]$,
$\sqrt[3]{2}[\cos ((11\pi/18) + 2\pi/3k) + i \sin ((11\pi/18) + 2\pi/3k)]$
for $k \in \{0, 1, 2\}$
37. $128\sqrt{2}[\cos (7\pi/4) + i \sin (7\pi/4)]$ **39.** $-\left(\frac{1}{4}\right)i$
41. $2^{1/8}[\cos (7\pi/16 + k\pi/2) + i \sin (7\pi/16 + k\pi/2)]$
for $k \in \{0, 1, 2\}$
43. $\sqrt[3]{25}[\cos (84.6° + 120°k) + i \sin (84.6° + 120°k)]$
for $k \in \{0, 1, 2\}$
45. $1.06[\cos (0.916° + 60°k) + i \sin (0.916° + 60°k)]$
for $k \in \{0, 1, 2, 3, 4, 5\}$

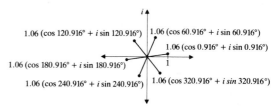

47. $\cos [(\pi/8) + (\pi k/2)] + i \sin [(\pi/8) + (\pi k/2)]$
for $k \in \{0, 1, 2, 3\}$

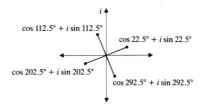

49. $5[\cos ((2k + 1)\pi/3) + i \sin ((2k + 1)\pi/3)]$
for $k \in \{0, 1, 2\}$

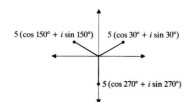

EXERCISES 8.1

1. (4, 3), consistent   **3.** $\{(x, y): y = \frac{1}{2}x + 3\}$, dependent
5. (10, 1), consistent   **7.** Ø, inconsistent   **9.** (16, 1), consistent   **11.** $(2, \sqrt{3})$, $(2, -\sqrt{3})$, $(-2, \sqrt{3})$, $(-2, \sqrt{3})$, consistent
13. (2, 0), (−2, 0), consistent
15. Ø, inconsistent   **17.** (1, −1), consistent
19. $\left(\frac{32}{17}, -\frac{20}{17}\right)$, consistent   **21.** (1, 1), (−1, 1), consistent
23. $(10^{1+\sqrt{2}}, 1 + \sqrt{2})$, $(10^{1-\sqrt{2}}, 1 - \sqrt{2})$, consistent
25. $\left(5\sqrt{3}/2, \pm\frac{1}{2}\right)$, $\left(-5\sqrt{3}/2, \pm\frac{1}{2}\right)$, consistent
27. (0, −2), $(\pm\sqrt{3}, 1)$, consistent   **29.** (±2, 0), consistent   **31.** (0, 0), (−1, 1), consistent
33. (1, 1), (2, 0), consistent   **35.** (2, 0), (6, 64), consistent   **37.** (2, 3, −1), consistent
39. 31 mph, 321 mph   **41.** 5 liters of 10% acid, 20 liters of 20% acid   **43.** 166 wide-angle lenses, 333 telephoto lenses

EXERCISES 8.2

1. (1, 2, 3)   **3.** (2, 1, −5)   **5.** (−1, 3, −2)
7. (4, −3, 2)   **9.** Ø, inconsistent   **11.** $\left(\frac{1}{2}, 1, -1\right)$
13. Ø, inconsistent   **15.** $\left(-\frac{1}{4}, 1, \frac{1}{3}\right)$
17. $x^2 + y^2 - 6x - 4y = 0$
19. $x^2 + y^2 - 8x - 4y + 15 = 0$
21. $x^2 + y^2 + 2x - 2y - 8 = 0$

EXERCISES 8.3

1. $((c + 14)/5, (4c + 1)/5, c)$ **3.** $((c + 10)/5, 7c/5, c)$
5. (3 − c, c, 0)   **7.** $((1 - 5c)/5, (5c + 11)/5, c)$
9. $((3 - 4c)/5, (19 - 7c)/5, c)$ **11.** (13c/4, 11c/4, c)
13. (c/3, 5c/3, c)   **15.** (0, 0, 0)
17. (−5c/3, 2c/3, c)   **19.** (0, 0, 0)
21. (−3c/11, 7c/11, c)   **23.** $\left(1 - \left(\frac{4}{11}\right)c, \left(\frac{23}{13}\right)c - 1, c\right)$
25. (−5c/3, −8c/3, c)   **27.** (−5c, −2c, c)
29. $\left(\left(\frac{13}{7}\right) - c, \frac{4}{7}, c\right)$ **31.** (5 − 13c, 5c, c)
33. Ø, inconsistent   **35.** (−5c − 75, 4c + 60, c) and for c=1, (−80, 64, 1)   **37.** x = number of pounds at $1, y = number of pounds at $2, z = number of pounds at $3, (1, 3, 11) and (2, 1, 12)

EXERCISES 8.4

1. −2   **3.** −17   **5.** 17   **7.** 0   **9.** −24
11. −60   **13.** {2, −1}   **15.** $1 \pm \sqrt{2}$ **17.** ±2
19. {0, 6}   **21.** −40   **23.** −32   **25.** 9

EXERCISES 8.5

1. $\left(\frac{59}{53}, -\frac{27}{53}\right)$ **3.** (0, 0)   **5.** $\left(-\frac{10}{23}, \frac{35}{23}, \frac{15}{23}\right)$
7. $\left(-\frac{4}{71}, \frac{28}{71}, \frac{68}{71}\right)$ **9.** $\left(\left(\frac{27}{11}\right) - c, c - \left(\frac{13}{11}\right), c\right)$
11. (3, −1, 2)   **13.** Ø, inconsistent
15. (−2, 5, 2)   **17.** (0, 0, 0)   **19.** $\left(-\frac{129}{26}, -\frac{49}{26}, \frac{7}{26}\right)$
21. rate of canoeist = 4 mph, rate of current = 2 mph
23. 5 pennies, 10 nickels, 5 dimes   **25.** 23 lb of $1 chocolate, 4 lb of $2 chocolate, 3 lb of $3 chocolate

EXERCISES 8.6

1. $A/(x + 1) + B/(x - 3)$ **3.** $A/(x - 2) + B/(x - 3)$
5. $A/(x + 1) + B/(x + 1)^2 + C/(x - 2)$
7. $A/(x + 2) + B/(x + 2)^2 + C/(x + 2)^3 + D/(x - 1) + E/(x - 1)^2$ **9.** $A/(x + 2) + (Bx + C)/(x^2 - 5x + 1)$
11. $A/(x + 2) + (Bx + C)/(x^2 + x + 6) + (Dx + E)/(x^2 + x + 6)^2 + F/(x + 2)^2$
13. $1/(x + 2) + 3/(x - 3)$
15. $5/(x - 2) + 2/(x + 2)$
17. $2/(x - 1) + 3/(x - 1)^2 + 5/(x - 1)^3$
19. $1/(x^2 + x + 1) + 3/(x - 2)$
21. $[-1/(x + 2)] + [7/(x + 2)^2]$
23. $5 + [(x + 4)/(x^2 - x + 9)]$
25. $[2/(x + 1)] - [2/(x + 2)] + [3/(x - 3)]$
27. $x + 3 + [3/(x - 1)] + [1/(x + 1)]$

EXERCISES 8.7

1. 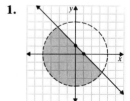   **3.**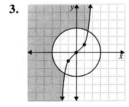

5. 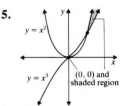   **7.**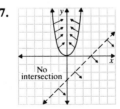

9. The maximum of P is 220 at (1, 4).
11. The minimum of C is 20 at (1, 4).
13. The maximum of P is 85 at (3, 2).

15. The maximum of P is 2,000 at (400, 400).
17. The maximum profit P is \$54,000 at (600, 100).
19. The maximum profit is \$355,000 with 7 sailboats and 5 yachts.

CHAPTER 8 REVIEW EXERCISES

1. (2, 625), (−1, 5)   **3.** (1, 1)   **5.** (5, 2, −1)
7. (−3c/2, c/2, c)   **9.** 16   **11.** $\left(\frac{5}{2}, 0, -\frac{5}{2}\right)$
13. (0, 0, 0)   **15.** $[3/(x − 1)] + [5/(x^2 + x + 3)]$
17. $−2 + /[3(x − 1)] + [8/(x + 2)] − [2/(x + 1)]$
19. The minimum of H is 2 at (1, 4).
The maximum of H is 27 at (4, 0).

EXERCISES 9.1

1. center (3, −2);
radius $r = 2$

3. center (−1, 1);
radius $r = 3$

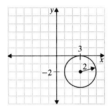

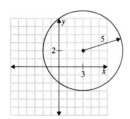

5. center (3, 2);
radius $r = 5$

7. center (2, 1);
radius $r = 5$

9. $\left(\frac{3}{2}, \frac{3}{2}\right)$, $r = 1/(2\sqrt{58})$
11. $(y − 2)^2 = 4(x − 1)$ **13.** $(x − 1)^2 = −4(y − 4)$
15. $(y − 2)^2 = 16(x + 1)$
17. $(x + 1)^2 = −12(y − 5)$
19. $(x − 2)^2 = 12(y + 1)$
21. $(y − 2)^2 = 8(x − 1)$ **23.** $\left(2, \frac{5}{4}\right)$

25. V: (1, −2); f: (1, 0)   **27.** V: (2, −3); f: (0, −3)

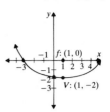

29. V: (2, −3); f: $\left(2, -\frac{5}{2}\right)$ **31.** V: (2, 3); f: (5, 3)

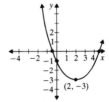   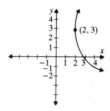

33. $p = 2\frac{1}{4}$ ft   **35.** $p = \frac{4}{3}$ in.   **37.** $p = 9$ in.
39. 492.8 ft

EXERCISES 9.2

1. $x^2/9 + y^2/4 = 1$ **3.** $x^2/16 + y^2/25 = 1$
5. $(x − 6)^2/16 + (y − 3)^2/4 = 1$
7. $(x − 6)^2 + (y − 3)^2/9 = 1$ **9.** $x^2/4 + y^2/16 = 1$
11. $x^2/4 + y^2/9 = 1$ **13.** $x^2/9 + y^2 = 1$

15. f: $(\pm\sqrt{7}, 0)$ **17.** f: $(0, \pm\sqrt{5})$

   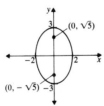

19. f: $(2, -2 \pm \sqrt{5})$ **21.** f: $(0, \pm 2\sqrt{2})$

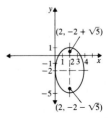   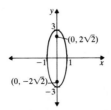

23. f: $(\pm\sqrt{21}, 0)$

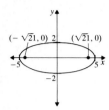

25. f: $(2 \pm \sqrt{5}, 1)$

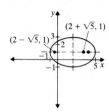

9. f: $(-2, 2 \pm \sqrt{10})$

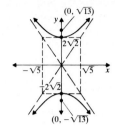

11. f: $(0, \pm\sqrt{13})$

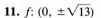

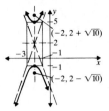

27. f: $(3, -2 \pm \sqrt{5})$

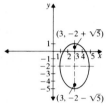

13. f_1: $(8, -2)$, f_2: $(-2, -2)$

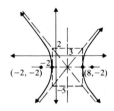

29. $x^2/16 + y^2/25 = 1$　　**31.** $x^2/25 + (y-2)^2/16 = 1$

33. $(x-2)^2/18 + (y-2)^2/9 = 1$

35. $y = \left(\frac{2}{5}\right)\sqrt{2500 - x^2}$　　**37.** $x^2/25 + y^2/16 = 1$

39. $x^2/4.18^2 + y^2/2.41^2 = 1$, aphelion $= 7.07 \times 10^8$ miles, perihelion $= 7.07 \times 10^7$ miles

41. 223.54 ft　　**45.** $x^2/24{,}000^2 + y^2/23{,}917^2 = 1$

15. $x^2/9 - y^2/7 = 1$　　**17.** $y^2/4 - x^2/5 = 1$

19. $x^2/4 - (y-3)^2/21 = 1$

21. $(y-1)^2 - (x-2)^2/3 = 1$

23. $y^2/9 - x^2/4 = 1$　　**25.** $y^2/16 - x^2 = 1$

27. $y^2/9 - x^2/9 = 1$

EXERCISES 9.3

1. f: $(\pm 2\sqrt{2}, 0)$

3. f: $(\pm\sqrt{13}, 0)$

29.

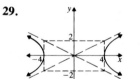

31.

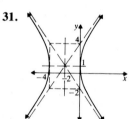

5. f: $(0, \pm 5)$

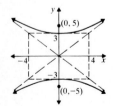

7. f: $(1 \pm \sqrt{13}, 2)$

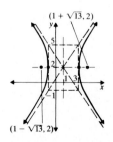

33.

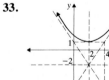

35.

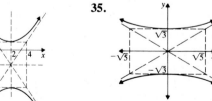

37.

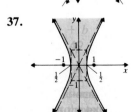

43.

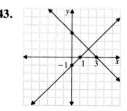

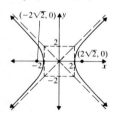
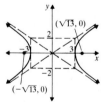
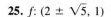

45.

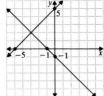

EXERCISES 9.4

1. (a) $\phi = \pi/4$ or $-\pi/4$ **(b)** for $\phi = \pi/4$,
$(x')^2 - (y')^2 = 16$; for $\phi = -\pi/4$, $(y')^2 - (x')^2 = 16$
(c)

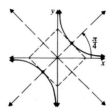

3. (a) $\phi = \pi/4$ or $-\pi/4$ **(b)** for $\phi = \pi/4$,
$(x')^2 - (y')^2 + 14x' + 2y' = 2$; for $\phi = -\pi/4$,
$(y')^2 - (x')^2 - 2x' + 14y' = 2$
(c)

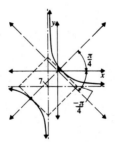

5. (a) $\phi = \pi/6$ **(b)** $6(x')^2 + 2(y')^2 = 18$
(c)

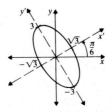

7. (a) $\phi = 26.565°$
(b) $(x')^2 - 9(y')^2 + 18x' - 9y' + 9 = 0$
(c)

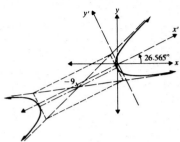

9. (a) $\phi = 18.435°$ **(b)** $(x')^2 + 11(y')^2 - 16x' + 22y' = 64$
(c)

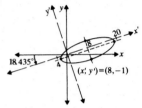

11. (a) $\phi = 63.435°$ or $\phi = -26.565°$ **(b)** for
$\phi = -26.565°$, $3(x')^2 - 2(y')^2 + 12x' - 4y' = 0$
(c)

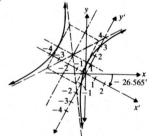

13. (a) $\phi = 26.565$
(b) $2(x')^2 + 7(y')^2 + 4(x') + 8y' + 3.6 = 0$
(c)

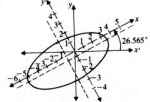

15. (a) $\phi = 73.565°$ or $\phi = -16.435°$

(b) $10(y')^2 + 0.632x' - 1.89y' = 0$

(c)

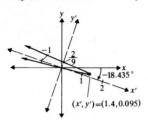

17.

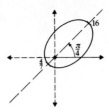

19.

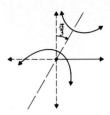

21.

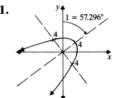

23.

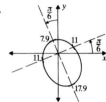

EXERCISES 9.5

1.

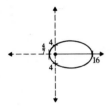

3.

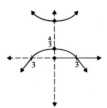

25.

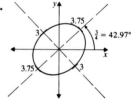

27.

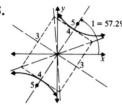

5.

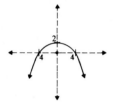

7.

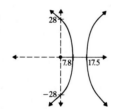

29. $(9.4, 40.15°), (9.4, 319.8°)$ **31.** $(8, 210°), (8, 330°)$

33. $r = 0.774/[1 \pm (0.9673) \cos \theta]$

35. $r = 9.56/[1 \pm (0.0056) \cos \theta]$ **37.** Saturn

9.

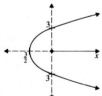

11.

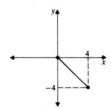

EXERCISES 9.6

1.

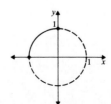

3.

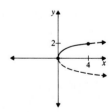

13.

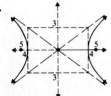

15.

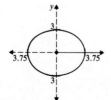

5.

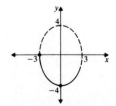

9.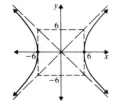

11.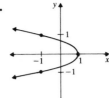

13.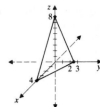

15.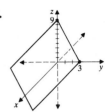

13.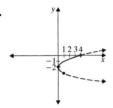

15.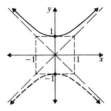

17.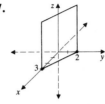

19.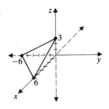

17.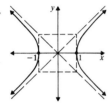

19.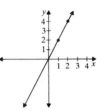

21.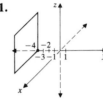

23.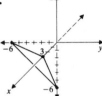

21. $x = \frac{1}{2}R(2t - \sin t)$, $y = \frac{1}{2}R(2 - \cos t)$

25.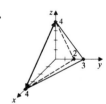

27.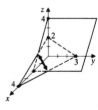

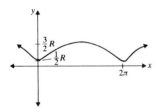

29.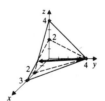

31.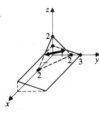

25. $x = (c + C) + a \sin (27t\pi/365) + A \sin (2t\pi/365)$
$y = b \cos (27t\pi/365) + B \cos (2t\pi/365)$

EXERCISES 9.7

1. $2\sqrt{6}$ **3.** $\sqrt{26}$ **5.** $\sqrt{22}$ **7.** $\sqrt{5}$

9.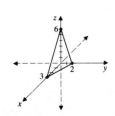

11.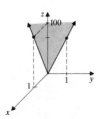

33. for 25, $(3c/2, 4 - 4c, c)$; for 29, $(2c - 1, 3 - 3c, c)$;
for 31, $((4 - c)/2, (12 - 9c)/8, c)$

35.

37.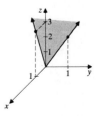

39.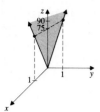

EXERCISES 9.8

1. 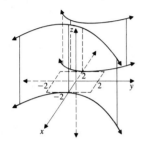   **3.**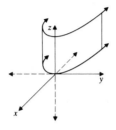

5. 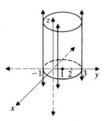   **7.**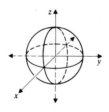

9. 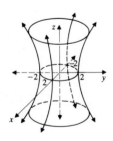   **11.**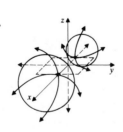

13. 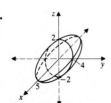   **15.**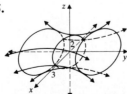

17. 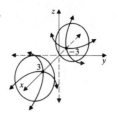   **19.**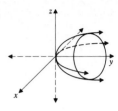

21. 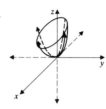   **23.**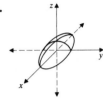

25.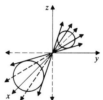

27. $(x - 2)^2 + (y - 3)^2 + (z - 4)^2 = 9$
29. $2x^2 + 2y^2 = 9z$ **31.** $x^2 + y^2 = z$
33. $16x^2 = 9y$ for $0 \le y \le 8$
35. $x^2/7.7^2 + y^2/5.8^2 + z^2/5.8^2 = 1$
37. $(x - 3)^2/a^2 - (y - 4)^2/b^2 + (z + 2)^2/c^2 = 1$

39. 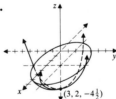   **41.**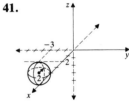

CHAPTER 9 REVIEW EXERCISES

1. $(x - 3)^2 + (y - 1)^2 = 9$
3. $(x - 3)^2 + (y + 3)^2 = 4$ **5.** $x^2/4 + y^2/16 = 1$
7. $(x + 1)^2/9 + (y + 1)^2 = 1$
9. $(x - 3)^2/4 + (y - 2)^2/16 = 1$
11. $x^2/4 - y^2/16 = 1$
13. $(x + 1)^2/9 - (y - 3)^2 = 1$

15. center: $(3, -1)$

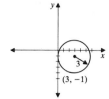

17. vertex: $(2, 2)$, focus: $(2, 5)$ **19.** f: $(0, \pm\sqrt{5})$

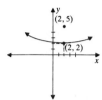 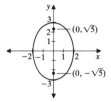

21. f: $(1, -2 \pm 2\sqrt{3})$ **23.** f_1: $(0, 0)$, f_2: $(4, 0)$

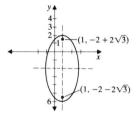 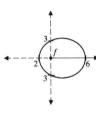

25. f in (x, y): $(0, 0)$, $(0, 8)$; f in (r, θ): $(0, 0)$, $(8, \pi/2)$

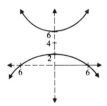

27. f: $(\pm 4, 0)$ **29.** f in (x, y): $(0, \pm\sqrt{5})$

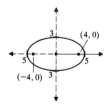 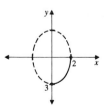

31. $2x^2 + 2y^2 = 25z$ for $0 \le z \le 2$, $p = 3$ ft 1.5 in
33. $x^2/20^2 + y^2/12^2 = 1\sqrt{63} \approx 7.9$ ft
35. $(x - 9.96)^2/(38.8)^2, + y^2/(37.5)^2 = 1$, $\varepsilon = 0.257$
37. $\phi = \pi/4$, an ellipse **39.** $\phi = \pi/4$, an ellipse

41.

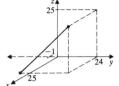

43.

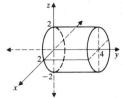

45.

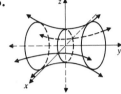

EXERCISES 10.1

1. 4, 3, 2, 1, 0, -5 **3.** 0, 2, 4, 6, 8, 18
5. -1, 3, -9, 27, -81, 3^9 **7.** 3, 3, 3, 3, 3, 3
9. $1, \frac{4}{3}, \frac{9}{7}, \frac{16}{15}, \frac{25}{31}, \frac{100}{1023}$ **11.** -2, 9, -13, 31, -57
13. $3, 1, \frac{1}{3}, \frac{1}{9}, \frac{1}{27}$ **15.** -1, 1, 1, 1, 1, 1
17. $2, 2, \sqrt{2}, \sqrt[6]{2}, \sqrt[24]{2}$ **19.** 5 **21.** -138
23. 270 **25.** 60 **27.** 1 **29.** 80

31. $1, \sqrt{2}, \sqrt{3}, 2, \sqrt{5}, \ldots, \sqrt{n}$ **33.** $2n; \sum_{i=1}^{50} 2i$

35. $n^3; \sum_{i=1}^{50} i^3$ **37.** $\left(-\frac{1}{2}\right)^{n-1}; \sum_{i=1}^{50} \left(-\frac{1}{2}\right)^{i-1}$

39. $n/(n + 1)^2; \sum_{i=1}^{50} i/(i + 1)^2$

41. $x^{n-1}/(n - 1); \sum_{i=1}^{50} x^{-1}/i - 1$

47. $a_n = \cos^{n+1} \theta; \lim_{n\to\infty} a_n = 0$

EXERCISES 10.2

11. $\left(\frac{1}{6}\right)(2n^3 + 3n^2 + n - 30)$
13. $\left(\frac{1}{6}\right)(2n^3 + 15n^2 + 37n - 150)$
15. $\left(\frac{1}{3}\right)(2n^3 + 12n^2 + 11n - 78)$
17. $\left(\frac{1}{4}\right)(4n^3 - n - 30)$
35. $S_n = \left(\frac{1}{3}\right)^{n-1}; \lim_{n\to\infty} S_n = 0$
37. $P_n = 3\left(\frac{4}{3}\right)^{n-1}; \lim_{n\to\infty} S_n$ approaches ∞

EXERCISES 10.3

1. 7, 9, 11, 13, 15, 27, . . . , $(2n + 5)$
3. 17, 14, 11, 8, 5, -13, . . . , $(20 - 3n)$
5. 20, 18, 16, 14, 12, 0, . . . , $(22 - 2n)$
7. $-29, -27, -25, -23, -21, -9$, . . . , $(2n - 31)$
9. $a_6 = 5.5$; $a_n = 7.3 - 0.3n$
11. $a_6 = x + 14$; $a_n = x + 3n - 4$
13. $a_6 = \log 96$; $a_n = \log [3(2^{n-1})]$ **15.** -9
17. $a_1 = 3$; $d = 4$ **19.** $a_1 = -\frac{11}{5}$; $d = \frac{18}{5}$ **21.** 165
23. -3105 **25.** 810   **27.** 78   **29.** 84
31. $22x + 671$ **33.** $21 \log (3 \cdot 2^{10})$ **35.** 88 terms
37. $7\frac{1}{2}$ **39.** $6\frac{2}{3}, 8\frac{1}{3}$ **41.** $8\frac{1}{2}, 10, 11\frac{1}{2}$ **43.** 15 terms
45. 260 seats   **47.** Health care pays a higher salary but BioTronics pays more over 7 years.
49. 21 days

EXERCISES 10.4

1. $3, 1, \frac{1}{3}, \frac{1}{9}, \frac{1}{27}, \left(\frac{1}{3}\right)^9, \left(\frac{1}{3}\right)^{n-2}$
3. $2, -4, 8, -16, 32, 2^{11}, (-1)^{n-1}2^n$
5. $9, 3, 1, \frac{1}{3}, \frac{1}{9}, \left(\frac{1}{3}\right)^8, \left(\frac{1}{3}\right)^{n-3}$
7. $8, -4, 2, -1, \frac{1}{2}, \left(\frac{1}{2}\right)^7, (-1)^{n-1}\left(\frac{1}{2}\right)^{n-4}$ **9.** $\left(\frac{1}{3}\right)^6, \left(\frac{1}{3}\right)^n$
11. $-3^6/10^5, (-1)^{n-1}3^n/10^{n-1}$ **13.** $4\sqrt{2}, (-1)^n 2^{(n-1)/2}$
15. $5^{1-5x}, 5^{1-xn+x}$ **17.** $x^{10}, (-1)x^{2n-2}$
19. -54 **21.** $\pm 8\sqrt{2}$ or $\pm 8i\sqrt{2}$ **23.** $8 - (1/2^{17})$
25. $(3^{10} + 1)/(4 \cdot 3^5)$ **27.** $63 - 31\sqrt{2}$ **29.** $\frac{10}{3}$
31. $2 + 2\sqrt{2}$ **33.** $\frac{2}{3}$ **35.** ∞ **37.** $\frac{15}{2}$
39. $\pm 8\sqrt{2}$ **41.** 15, 45   **43.** 50%
45. (a) \$47,158.95; \$318,748.49 **(b)** \$42,136.98; \$595,561.57. The 10% annual raise provides a higher 10-year salary. The 4% semiannual raise provides a higher 10-year

sum.   **47.** 72 ft, 1-in. arc after 11 swings   **49.** $\frac{728}{99}$
51. $\frac{5120}{999}$ **53.** 8   **55.** 1.6

EXERCISES 10.5

1. 120   **3.** 15   **5.** 45   **7.** $n(n - 1)/2$ **9.** 10
11. 4   **13.** 35   **15.** n
17. $x^4 - 4x^3y + 6x^2y^2 - 4xy^3 + y^4$
19. $x^3 + 6x^2 + 12x + 8$
21. $\left(\frac{1}{243}\right)x^5 + \left(\frac{5}{27}\right)x^4 + \left(\frac{10}{3}\right)x^3 - 30x^2 + 135x - 243$
23. $x^3 + 6x^2\sqrt{xy} + 15x^2y + 20xy\sqrt{xy} + 15xy^2 + 6y^2\sqrt{xy} + y^3$ **25.** $4{,}096x^3 + 24{,}576x^3\sqrt[12]{x} + 67{,}584x^3\sqrt[6]{x} + 112{,}640x^3\sqrt[4]{x}$
27. $33(2^{19})/x^{16}$ **29.** $\frac{43{,}750}{81}$

EXERCISES 10.6

1. $\displaystyle\sum_{i=0}^{5} \binom{5}{i}(-1/x)^i$ **3.** $\displaystyle\sum_{i=0}^{6} \binom{6}{i} 2^i(y/x)^{2i-6}$
5. 2.14   **7.** 0.93   **9.** 247.08

CHAPTER 10 REVIEW EXERCISES

1. $0, -5, -10, -15, -20$ **3.** $1, 15, -13, 43, -69$
5. 13, 10, 7, 4, 1   **7.** $2, -1, \frac{1}{2}, -\frac{1}{4}, \frac{1}{8}$
9. 18   **11.** 5   **13.** -600 **15.** $\left(\frac{4}{5}\right)[1 - (1/4^{46})]$
17. $x^5 + \left(\frac{10}{3}\right)x^4 + \left(\frac{40}{9}\right)x^3 + \left(\frac{80}{27}\right)x^2 + \left(\frac{80}{81}\right)x + \left(\frac{32}{243}\right)$
19. 32 ft high, 280 bales   **21.** 831,813.809
25. $p_n = \left(\frac{8}{3}\right)^n$; $\displaystyle\lim_{n\to\infty} p_n$ approaches ∞; maximum area $= \frac{1}{2}$

Index

TRIGONOMETRIC FUNCTIONS AND IDENTITIES

Definition | *Unit Circle Trigonometric Functions.* For any point $P = (x, y)$ on the unit circle with center at the origin.

$$\sin \theta = \frac{y}{1} = y \quad \text{and} \quad \cos \theta = \frac{x}{1} = x$$

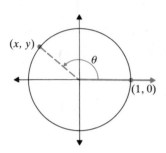

BASIC TRIGONOMETRIC IDENTITIES	
$\tan \theta = \dfrac{\sin \theta}{\cos \theta}$	$\csc \theta = \dfrac{1}{\sin \theta}$
$\cot \theta = \dfrac{1}{\tan \theta} = \dfrac{\cos \theta}{\sin \theta}$	$\sec \theta = \dfrac{1}{\cos \theta}$
$\sin^2 \theta + \cos^2 \theta = 1 \qquad \tan^2 \theta + 1 = \sec^2 \theta \qquad 1 + \cot^2 \theta = \csc^2 \theta$	
$\sin(-\theta) = -\sin \theta \qquad \cos(-\theta) = \cos \theta \qquad \tan(-\theta) = -\tan \theta$	

SPECIAL ANGLES

$\theta =$	0	$\dfrac{\pi}{6}$	$\dfrac{\pi}{4}$	$\dfrac{\pi}{3}$	$\dfrac{\pi}{2}$
$\sin \theta$	0	$\dfrac{1}{2}$	$\dfrac{\sqrt{2}}{2}$	$\dfrac{\sqrt{3}}{2}$	1
$\cos \theta$	1	$\dfrac{\sqrt{3}}{2}$	$\dfrac{\sqrt{2}}{2}$	$\dfrac{1}{2}$	0

COFUNCTION IDENTITIES

$$\cos\left(\frac{\pi}{2} - \theta\right) = \sin \theta \quad \text{or} \quad \cos(90° - \theta) = \sin \theta$$

$$\sin\left(\frac{\pi}{2} - \theta\right) = \cos \theta \quad \text{or} \quad \sin(90° - \theta) = \cos \theta$$

$$\tan\left(\frac{\pi}{2} - \theta\right) = \cot \theta \quad \text{or} \quad \tan(90° - \theta) = \cot \theta$$

FUNCTIONS OF SUMS AND DIFFERENCES

$$\sin(\alpha + \beta) = \sin \alpha \cos \beta + \cos \alpha \sin \beta$$

$$\sin(\alpha - \beta) = \sin \alpha \cos \beta - \cos \alpha \sin \beta$$

$$\cos(\alpha + \beta) = \cos \alpha \cos \beta - \sin \alpha \sin \beta$$

$$\cos(\alpha - \beta) = \cos \alpha \cos \beta + \sin \alpha \sin \beta$$

$$\tan(\alpha + \beta) = \frac{\tan \alpha + \tan \beta}{1 - \tan \alpha \tan \beta} \qquad \tan(\alpha - \beta) = \frac{\tan \alpha - \tan \beta}{1 + \tan \alpha \tan \beta}$$